AN INTRODUCTION TO RESERVOIR SIMULATION USING MATLAB/GNU OCTAVE

This book provides a self-contained introduction to the simulation of flow and transport in porous media, written by a developer of numerical methods. The reader will learn how to implement reservoir simulation models and computational algorithms in a robust and efficient manner. The book contains a large number of numerical examples, all fully equipped with online code and data, allowing the reader to reproduce results and use them as a starting point for their own work. All of the examples in the book are based on the MATLAB Reservoir Simulation Toolbox (MRST), an open-source toolbox popular in both academic institutions and the petroleum industry. The book can also be seen as a user guide to the MRST software. It will prove invaluable for researchers, professionals, and advanced students using reservoir simulation methods. This title is also available as Open Access on Cambridge Core at https://doi.org/9781108591416.

KNUT-ANDREAS LIE is chief scientist at SINTEF in Oslo, Norway, where he heads a research group in computational methods for geoscience applications. Over the last 20 years he has developed commercial and in-house software solutions for the international petroleum industry. In particular, he is a founding father of two open-source community software tools: MRST and OPM. Lie holds a PhD in mathematics from the Norwegian University of Science and Technology in Trondheim, where he also has a position as adjunct professor. He has authored 140 scientific papers and supervised 65 MSc and PhD students. He currently serves as executive editor for the *SPE Journal* of the Society of Petroleum Engineers.

AN INTRODUCTION TO RESERVOIR SIMULATION USING MATLAB/GNU OCTAVE

User Guide for the MATLAB Reservoir Simulation Toolbox (MRST)

KNUT-ANDREAS LIE
SINTEF

CAMBRIDGE
UNIVERSITY PRESS

University Printing House, Cambridge CB2 8BS, United Kingdom

One Liberty Plaza, 20th Floor, New York, NY 10006, USA

477 Williamstown Road, Port Melbourne, VIC 3207, Australia

314–321, 3rd Floor, Plot 3, Splendor Forum, Jasola District Centre, New Delhi – 110025, India

79 Anson Road, #06–04/06, Singapore 079906

Cambridge University Press is part of the University of Cambridge.

It furthers the University's mission by disseminating knowledge in the pursuit of education, learning, and research at the highest international levels of excellence.

www.cambridge.org
Information on this title: www.cambridge.org/9781108492430
DOI: 10.1017/9781108591416

© Knut-Andreas Lie 2019

This work is in copyright. It is subject to statutory exceptions and to the provisions of relevant licensing agreements; with the exception of the Creative Commons version the link, which is provided below, no reproduction of any part of this work may take place without the written permission of Cambridge University Press.

An online version of this work is published at doi.org/9781108591416 under a Creative Commons Open Access license CC-BY-NC-ND 4.0, which permits reuse, distribution, and reproduction in any medium for noncommercial purposes providing appropriate credit to the original work is given. You may not distribute derivative works without permission. To view a copy of this license, visit
https://creativecommons.org/licenses/by-nc-nd/4.0

All versions of this work may contain content reproduced under license from third parties.

Permission to reproduce this third-party content must be obtained from these third-parties directly. When citing this work, please include a reference to the DOI 10.1017/9781108591416

First published 2019

Printed in the United Kingdom by TJ International Ltd., Padstow, Cornwall

A catalogue record for this publication is available from the British Library.

Library of Congress Cataloging-in-Publication Data
Names: Lie, Knut-Andreas, author.
Title: An introduction to reservoir simulation using MATLAB/GNU Octave : user guide for the MATLAB reservoir simulation toolbox (MRST) / Knut-Andreas Lie, SINTEF.
Description: Cambridge, United Kingdom ; New York, NY, USA : Cambridge University Press, 2019. | Includes bibliographical references and index.
Identifiers: LCCN 2018049101 | ISBN 9781108492430 (hardback : alk. paper)
Subjects: LCSH: Hydrocarbon reservoirs–Computer simulation. | Oil reservoir engineering–Data processing. | MATLAB. | GNU Octave.
Classification: LCC TN870.53 .L54 2019 | DDC 622/.338–dc23
LC record available at https://lccn.loc.gov/2018049101

ISBN 978-1-108-49243-0 Hardback

Cambridge University Press has no responsibility for the persistence or accuracy of URLs for external or third-party internet websites referred to in this publication and does not guarantee that any content on such websites is, or will remain, accurate or appropriate.

Contents

Preface			*page* xiii
1	Introduction		1
	1.1	Petroleum Recovery	3
	1.2	Reservoir Simulation	6
	1.3	Outline of the Book	9
	1.4	The First Encounter with MRST	14
	Part I	**Geological Models and Grids**	**19**
2	Modeling Reservoir Rocks		21
	2.1	Formation of Sedimentary Rocks	21
	2.2	Creation of Crude Oil and Natural Gas	26
	2.3	Multiscale Modeling of Permeable Rocks	28
		2.3.1 Geological Characterization	29
		2.3.2 Representative Elementary Volumes	31
		2.3.3 Microscopic Models: The Pore Scale	33
		2.3.4 Mesoscopic Models	34
	2.4	Modeling Rock Properties	34
		2.4.1 Porosity	35
		2.4.2 Permeability	36
		2.4.3 Other Parameters	38
	2.5	Property Modeling in MRST	39
		2.5.1 Homogeneous Models	40
		2.5.2 Random and Lognormal Models	40
		2.5.3 The 10th SPE Comparative Solution Project: Model 2	41
		2.5.4 The Johansen Formation	44
		2.5.5 SAIGUP: Shallow-Marine Reservoirs	46
3	Grids in Subsurface Modeling		55
	3.1	Structured Grids	57
	3.2	Unstructured Grids	62

	3.2.1	Delaunay Tessellation	63
	3.2.2	Voronoi Diagrams	66
	3.2.3	General Tessellations	68
	3.2.4	Using an External Mesh Generator	69
3.3	Stratigraphic Grids	72
	3.3.1	Corner-Point Grids	73
	3.3.2	2.5D Unstructured Grids	85
3.4	Grid Structure in MRST	88
3.5	Examples of More Complex Grids	96
	3.5.1	SAIGUP: Model of a Shallow-Marine Reservoir	97
	3.5.2	Composite Grids	101
	3.5.3	Control-Point and Boundary Conformal Grids	103
	3.5.4	Multiblock Grids	104

Part II Single-Phase Flow   111

4   Mathematical Models for Single-Phase Flow   113
- 4.1 Fundamental Concept: Darcy's Law   113
- 4.2 General Flow Equations for Single-Phase Flow   115
- 4.3 Auxiliary Conditions and Equations   120
  - 4.3.1 Boundary and Initial Conditions   120
  - 4.3.2 Injection and Production Wells   121
  - 4.3.3 Field Lines and Time-of-Flight   128
  - 4.3.4 Tracers and Volume Partitions   129
- 4.4 Basic Finite-Volume Discretizations   131
  - 4.4.1 Two-Point Flux-Approximation   131
  - 4.4.2 Discrete `div` and `grad` Operators   135
  - 4.4.3 Time-of-Flight and Tracer   141

5   Incompressible Solvers for Single-Phase Flow   143
- 5.1 Basic Data Structures in a Simulation Model   144
  - 5.1.1 Fluid Properties   144
  - 5.1.2 Reservoir States   145
  - 5.1.3 Fluid Sources   145
  - 5.1.4 Boundary Conditions   146
  - 5.1.5 Wells   147
- 5.2 Incompressible Two-Point Pressure Solver   149
- 5.3 Upwind Solver for Time-of-Flight and Tracer   153
- 5.4 Simulation Examples   156
  - 5.4.1 Quarter Five-Spot   157
  - 5.4.2 Boundary Conditions   160
  - 5.4.3 Structured versus Unstructured Stencils   165
  - 5.4.4 Using Peaceman Well Models   169

6	Consistent Discretizations on Polyhedral Grids		174
	6.1	The TPFA Method Is Not Consistent	174
	6.2	The Mixed Finite-Element Method	177
		6.2.1 Continuous Formulation	178
		6.2.2 Discrete Formulation	180
		6.2.3 Hybrid Formulation	182
	6.3	Finite-Volume Methods on Mixed Hybrid Form	185
	6.4	The Mimetic Method	188
	6.5	Monotonicity	197
	6.6	Discussion	199
7	Compressible Flow and Rapid Prototyping		202
	7.1	Implicit Discretization	202
	7.2	A Simulator Based on Automatic Differentiation	204
		7.2.1 Model Setup and Initial State	205
		7.2.2 Discrete Operators and Equations	206
		7.2.3 Well Model	208
		7.2.4 The Simulation Loop	209
	7.3	Pressure-Dependent Viscosity	213
	7.4	Non-Newtonian Fluid	215
	7.5	Thermal Effects	220

Part III Multiphase Flow — 229

8	Mathematical Models for Multiphase Flow		231
	8.1	New Physical Properties and Phenomena	232
		8.1.1 Saturation	232
		8.1.2 Wettability	234
		8.1.3 Capillary Pressure	235
		8.1.4 Relative Permeability	239
	8.2	Flow Equations for Multiphase Flow	243
		8.2.1 Single-Component Phases	244
		8.2.2 Multicomponent Phases	245
		8.2.3 Black-Oil Models	246
	8.3	Model Reformulations for Immiscible Two-Phase Flow	248
		8.3.1 Pressure Formulation	248
		8.3.2 Fractional-Flow Formulation in Phase Pressure	249
		8.3.3 Fractional-Flow Formulation in Global Pressure	254
		8.3.4 Fractional-Flow Formulation in Phase Potential	255
		8.3.5 Richards' Equation	256
	8.4	The Buckley–Leverett Theory of 1D Displacements	258
		8.4.1 Horizontal Displacement	258
		8.4.2 Gravity Segregation	264
		8.4.3 Front Tracking: Semi-Analytical Solutions	266

9	Discretizing Hyperbolic Transport Equations		272
	9.1 A New Solution Concept: Entropy-Weak Solutions		272
	9.2 Conservative Finite-Volume Methods		274
	9.3 Centered versus Upwind Schemes		275
		9.3.1 Centered Schemes	275
		9.3.2 Upwind or Godunov Schemes	277
		9.3.3 Comparison of Centered and Upwind Schemes	279
		9.3.4 Implicit Schemes	283
	9.4 Discretization on Unstructured Polyhedral Grids		286
10	Solvers for Incompressible Immiscible Flow		289
	10.1 Fluid Objects for Multiphase Flow		290
	10.2 Sequential Solution Procedures		292
		10.2.1 Pressure Solvers	293
		10.2.2 Saturation Solvers	294
	10.3 Simulation Examples		297
		10.3.1 Buckley–Leverett Displacement	298
		10.3.2 Inverted Gravity Column	301
		10.3.3 Homogeneous Quarter Five-Spot	303
		10.3.4 Heterogeneous Quarter Five-Spot: Viscous Fingering	307
		10.3.5 Buoyant Migration of CO_2 in a Sloping Sandbox	311
		10.3.6 Water Coning and Gravity Override	313
		10.3.7 The Effect of Capillary Forces – Capillary Fringe	319
		10.3.8 Norne: Simplified Simulation of a Real-Field Model	323
	10.4 Numerical Errors		326
		10.4.1 Splitting Errors	326
		10.4.2 Grid Orientation Errors	330
11	Compressible Multiphase Flow		337
	11.1 Industry-Standard Simulation		337
	11.2 Two-Phase Flow without Mass Transfer		342
	11.3 Three-Phase Relative Permeabilities		348
		11.3.1 Relative Permeability Models from ECLIPSE 100	349
		11.3.2 Evaluating Relative Permeabilities in MRST	353
		11.3.3 The SPE 1, SPE 3, and SPE 9 Benchmark Cases	355
		11.3.4 A Simple Three-Phase Simulator	358
	11.4 PVT Behavior of Petroleum Fluids		359
		11.4.1 Phase Diagrams	360
		11.4.2 Reservoir Types and Their Phase Behavior during Recovery	364
		11.4.3 PVT and Fluid Properties in Black-Oil Models	370
	11.5 Phase Behavior in ECLIPSE Input Decks		379
	11.6 The Black-Oil Equations		388

		11.6.1	The Water Component	389
		11.6.2	The Oil Component	390
		11.6.3	The Gas Component	392
		11.6.4	Appearance and Disappearance of Phases	392
	11.7	Well Models		394
		11.7.1	Inflow-Performance Relationships	394
		11.7.2	Multisegment Wells	395
	11.8	Black-Oil Simulation with MRST		399
		11.8.1	Simulating the SPE 1 Benchmark Case	399
		11.8.2	Comparison against a Commercial Simulator	405
		11.8.3	Limitations and Potential Pitfalls	406
12	The AD-OO Framework for Reservoir Simulation			413
	12.1	Overview of the Simulator Framework		414
	12.2	Model Hierarchy		420
		12.2.1	PhysicalModel – Generic Physical Models	422
		12.2.2	ReservoirModel – Basic Reservoir Models	426
		12.2.3	Black-Oil Models	430
		12.2.4	Models of Wells and Production Facilities	433
	12.3	Solving the Discrete Model Equations		434
		12.3.1	Assembly of Linearized Systems	434
		12.3.2	Nonlinear Solvers	436
		12.3.3	Selection of Time-Steps	439
		12.3.4	Linear Solvers	440
	12.4	Simulation Examples		449
		12.4.1	Depletion of a Closed/Open Compartment	449
		12.4.2	An Undersaturated Sector Model	451
		12.4.3	SPE 1 Instrumented with Inflow Valves	455
		12.4.4	The SPE 9 Benchmark Case	460
	12.5	Improving Convergence and Reducing Runtime		470
	Part IV	**Reservoir Engineering Workflows**		**475**
13	Flow Diagnostics			477
	13.1	Flow Patterns and Volumetric Connections		478
		13.1.1	Volumetric Partitions	479
		13.1.2	Time-of-Flight Per Partition Region: Improved Accuracy	482
		13.1.3	Well Allocation Factors	482
	13.2	Measures of Dynamic Heterogeneity		483
		13.2.1	Flow and Storage Capacity	483
		13.2.2	Lorenz Coefficient and Sweep Efficiency	486
	13.3	Residence-Time Distributions		489
	13.4	Case Studies		495

		13.4.1	Tarbert Formation: Volumetric Connections	495
		13.4.2	Heterogeneity and Optimized Well Placement	501
	13.5	Interactive Flow Diagnostics Tools		505
		13.5.1	Synthetic 2D Example: Improving Areal Sweep	510
		13.5.2	SAIGUP: Flow Patterns and Volumetric Connections	514
14	Grid Coarsening			518
	14.1	Grid Partitions		518
		14.1.1	Uniform Partitions	519
		14.1.2	Connected Partitions	520
		14.1.3	Composite Partitions	522
	14.2	Coarse Grid Representation in MRST		524
		14.2.1	Subdivision of Coarse Faces	526
	14.3	Partitioning Stratigraphic Grids		529
		14.3.1	The Johansen Aquifer	529
		14.3.2	The SAIGUP Model	532
		14.3.3	Near Well Refinement for CaseB4	536
	14.4	More Advanced Coarsening Methods		540
	14.5	A General Framework for Agglomerating Cells		541
		14.5.1	Creating Initial Partitions	541
		14.5.2	Connectivity Checks and Repair Algorithms	542
		14.5.3	Indicator Functions	544
		14.5.4	Merge Blocks	545
		14.5.5	Refine Blocks	547
		14.5.6	Examples	550
	14.6	Multilevel Hierarchical Coarsening		553
	14.7	General Advice and Simple Guidelines		556
15	Upscaling Petrophysical Properties			558
	15.1	Upscaling for Reservoir Simulation		559
	15.2	Upscaling Additive Properties		561
	15.3	Upscaling Absolute Permeability		563
		15.3.1	Averaging Methods	564
		15.3.2	Flow-Based Upscaling	569
	15.4	Upscaling Transmissibility		575
	15.5	Global and Local–Global Upscaling		578
	15.6	Upscaling Examples		580
		15.6.1	Flow Diagnostics Quality Measure	581
		15.6.2	A Model with Two Facies	581
		15.6.3	SPE 10 with Six Wells	585
		15.6.4	Complete Workflow Example	589
		15.6.5	General Advice and Simple Guidelines	595

Appendix		The MATLAB Reservoir Simulation Toolbox	597
	A.1	Getting Started with the Software	598
		A.1.1 Core Functionality and Add-on Modules	598
		A.1.2 Downloading and Installing	601
		A.1.3 Exploring Functionality and Getting Help	602
		A.1.4 Release Policy and Version Numbers	606
		A.1.5 Software Requirements and Backward Compatibility	606
		A.1.6 Terms of Usage	608
	A.2	Public Data Sets and Test Cases	609
	A.3	More About Modules and Advanced Functionality	610
		A.3.1 Operating the Module System	611
		A.3.2 What Characterizes a Module?	612
		A.3.3 List of Modules	613
	A.4	Rapid Prototyping Using MATLAB and MRST	620
	A.5	Automatic Differentiation in MRST	623
References			631
Index			650
Usage of MRST Functions			658

Preface

There are many books that describe mathematical models for flow in porous media and present numerical methods used to discretize and solve the corresponding systems of partial differential equations; a comprehensive list can be found in the References. However, neither of these books fully describes how you should implement the models and numerical methods to form a robust and efficient simulator. Some books may present algorithms and data structures, but most leave it up to you to figure out all the nitty-gritty details you need in order to get your implementation up and running. Likewise, you may read papers presenting models or computational methods that may be exactly what you need for your work. After the initial enthusiasm, however, you very often end up quite disappointed, or at least I do when I realize that the authors have not presented all the details of their methods, or that it will probably take me weeks or months to get my own implementation working.

In this book, I try to be a bit different and give a reasonably self-contained introduction to the simulation of flow and transport in porous media that also discusses how to implement the models and algorithms in a robust and efficient manner. In the presentation, I have tried to let the discussion of models and numerical methods go hand in hand with numerical examples that come fully equipped with codes and data, so that you can rerun and reproduce the results yourself and use them as a starting point for your own research and experiments. You will get most out of the book if you continuously switch between reading and experimenting with the many code snippets and tutorial examples on your own computer. All examples in the book are based on the MATLAB Reservoir Simulation Toolbox (MRST), which has been developed by my group and published online as free open-source code under the GNU General Public License since 2009.

The book can alternatively be seen as a comprehensive user-guide to MRST. Over the years, MRST has become surprisingly popular. At the time of writing (July 2018), the software has more than 17,000 unique downloads, 120 students have used MRST in their master or PhD theses, and more than 190 papers written by authors outside of SINTEF include numerical experiments run in MRST. This book tries to give an in-depth introduction to MRST and explain the two different programming paradigms you can find in the software. The book is up to date with respect to the latest developments in data structures and syntax, both for the original procedural approaches that primarily focus on incompressible flow, as well as for the more recent object-oriented, automatic-differentiation (AD-OO) framework we developed to simulate compressible, multiphase flow.

The book has grown much longer than I anticipated when I started writing. Initially, my ambition was to provide introductory material on single-phase and two-phase incompressible flow, and discuss how such models could be implemented in a flexible and efficient manner by use of MATLAB. While I was writing, the software expanded rapidly and I had a hard time keeping pace. Inspired in part by the many people who have downloaded (and cited) the preliminary editions I have published online, but also as a result of numerous requests, I decided to expand the book and include a detailed discussion of the physics underlying industry-standard black-oil simulators, and give a thorough introduction to how such models have been implemented in MRST. Now that this is done, it is time to stop. However, MRST has more to offer, including solvers for various enhanced oil recovery (EOR) models, compositional flow, fractured media, geomechanics, geochemistry, as well as a comprehensive set of tools for modeling CO_2 storage in large-scale aquifer systems, but documentation of these will have to be another book, or perhaps also another author. At the moment, I am more than happy that I finally managed to finish this book.

First of all, I am very grateful to Equinor for the generous grant that enabled Gold Open Access publication of this book. I would also like to thank my current and former colleagues at SINTEF with whom I have collaborated over many year to develop MRST; primarily Olav Møyner, Halvor Møll Nilsen, Stein Krogstad, Jostein R. Natvig, Odd Andersen, Bård Skaflestad, and Xavier Raynaud. I have also had a great number of inspiring discussions with Alf Birger Rustad and Vegard Kippe at Equinor over the past decade. The chapter on flow diagnostics is partially the result of many discussions with Brad Mallison from Chevron. I am grateful to the University of Bergen and the Norwegian University of Science and Technology (NTNU) for funding through my Professor II positions. Victor Calo and Yalchin Efendiev invited me to KAUST, where important parts of the chapters on grids and petrophysics were written. Likewise, Margot Gerritsen invited me to Stanford and gave me the opportunity to develop Jolts videos that complement the first chapters of the book. Significant parts of the chapter on black-oil models were written during the three weeks I participated in the Long Program on Computational Issues in Oil Field Applications at UCLA in 2017. Let me also thank all colleagues and students who have given suggestions, pointed out errors and misprints, asked questions, and given me inspiration to continue working. Even though your name is not mentioned explicitly, I have not forgotten your important contributions. Last, but not least, I thank my wife Anne: the many evenings, late nights, and early mornings it took to write this book would never have been possible without your support and understanding.

Finally to the reader: As you may understand from the book, I strongly believe in reproducible computational science and sharing your work. I hope the book manages to convince you that the combination of a high-level scripting language like MATLAB/GNU Octave and development of reusable open-source software is an excellent approach to both being productive and having a significant impact on others; I elaborate more on this in an essay I wrote to my former supervisor, Helge Holden, for his 60th birthday [189]. Whereas the book is static, the MRST software will continue to develop in new directions as a result of continued research and contributions from external users. Your help is invaluable to making this happen; please do not hesitate to contact me if you have suggestions for improvements or new functionality, or if you have developed your independent add-on functionality and want advice on how to best share it with the general community.

Knut-Andreas Lie
Knut-Andreas.Lie@sintef.no

1
Introduction

Subsurface flow phenomena cover some of the most important technological challenges of our time. The road toward sustainable use and management of the earth's freshwater reserves necessarily involves modeling of hydrological systems in order to understand fluid movement in groundwater basins, quantify limits of sustainable use, monitor transport of pollutants in the subsurface, and appraise schemes for groundwater remediation. An equally important problem is the reduction of the emission of greenhouse gases like CO_2 into the atmosphere. Carbon sequestration in subsurface rock formations has been suggested as a possible means to that end. The primary concern is how fast the injected CO_2 will escape back to the atmosphere. Repositories do not necessarily need to store CO_2 forever, just long enough to allow the natural carbon cycle to reduce the atmospheric CO_2 to near pre-industrial levels. Nevertheless, making qualified estimates of leakage rates from potential storage facilities is a nontrivial task, and demands interdisciplinary research and software based on state-of-the-art numerical methods for modeling subsurface flow. Other obvious concerns are whether the injected CO_2 will leak into freshwater aquifers, or migrate to habitated or different legislative areas.

A third challenge is petroleum recovery. The civilized world will likely continue to depend on utilization of petroleum resources both as an energy carrier and as raw material for a wide variety of petrochemical products (fertilizers, detergents, dyes, synthetic rubber, plastics, etc.) in the foreseeable future. In the North Sea region, the recovery of existing and discovery of new conventional petroleum resources has declined significantly in recent years. Optimal utilization of known resources is therefore of utter importance to meet the demands for petroleum and lessen the pressure on exploration in vulnerable areas like in the arctic regions. Meanwhile, there has been a dramatic increase in the utilization of unconventional petroleum resources (shale gas/oil), and there is a strong need to understand how these can be produced in an economic way that minimizes harm to the environment.

Reliable computer modeling of subsurface flow is much needed to overcome these three challenges, but is also needed to exploit deep geothermal energy or to inject compressed gas in the subsurface to store energy, ensure safe storage of nuclear waste, improve remediation technologies to remove contaminants from the subsurface, etc. Indeed, the need for tools that help us understand flow processes in the subsurface is probably greater than ever, and increasing. More than fifty years of prior research in this area has led to some degree of agreement in terms of how subsurface flow processes can be modeled adequately with numerical simulation technology.

This book introduces and discusses mathematical models describing flow processes in porous rocks on a macroscopic scale. The presentation focuses primarily on physical processes that take place during hydrocarbon recovery. Simulation of such processes is often referred to as reservoir simulation. This means that even though the mathematical models, numerical methods, and software implementations presented can be applied to any of the applications just mentioned, the specific examples use vocabulary, physical scales, and balances of driving forces specific to petroleum recovery. As an example of this vocabulary, we can consider the ability of a porous medium to transmit fluids. In petroleum engineering this is typically given in terms of the "permeability," which is a pure rock property, whereas one in water resource engineering is more concerned with the "hydraulic conductivity" that also takes the viscosity and density of the fluid into account. In CO_2 sequestration you can see both quantities used.

As an example of physical scales, let us compare oil recovery by water flooding and the long-term geological storage of CO_2. Petroleum resources can only accumulate when the natural upward movement of hydrocarbons relative to water is prevented by confinements in the overlying rocks. An oil reservoir is therefore usually a relatively closed system with a spatial extent of hundreds to thousands of meters, from which oil can be recovered over tens of years. The most promising candidates for geological CO_2 storage are huge aquifer systems that stretch out for hundreds of kilometers. The injection of CO_2 is mainly driven by pressure differences as in oil recovery. However, as CO_2 moves into the aquifer and the effects of the injection pressure ceases, the fluid movement is gradually dominated by buoyant forces that cause the lighter CO_2 phase to migrate upward in the open aquifer system. This process can potentially continue over thousands of years. The basic flow physics and governing equations are the same in both cases, but the balances between physical forces are different, and this should be accounted for when formulating the overall mathematical models and appropriate numerical methods.

Models and numerical methods developed to study subsurface flow are also applicable to other natural and man-made porous media such as soils, biological tissues and plants, electrochemical devices like batteries and fuel cells, concrete and other porous constructional materials, food and sanitary products, textiles, industrial filtering processes, polymer composites, water desalination, etc. Porous media models are also used to describe human physiology, e.g., waterways in the brain or flow in the capillary part of the vascular system. An interesting example is in-tissue drug delivery, where the challenge is to minimize the volume swept by the injected fluid, unlike in petroleum recovery, where one seeks to maximize the volumetric sweep to push out as much petroleum as possible.

1.1 Petroleum Recovery

We start by a conceptual sketch of the basic mechanisms by which hydrocarbon can be recovered from a subsurface reservoir. In good reservoir rock, the void spaces between mineral grains form networks of connected pores that can store and transmit large amounts of fluids. Petroleum discoveries vary in size from small pockets of hydrocarbon that may be buried just a few meters beneath the surface of the earth and can easily be produced, to huge reservoirs[1] stretching out several square kilometers beneath remote and stormy seas. However, as a simple mental picture, you can think of a hydrocarbon reservoir as a bent, rigid sponge that is confined inside an insulating material and has all its pores filled with hydrocarbons, which may appear in the form of a liquid oleic and a gaseous phase as illustrated in Figure 1.1. Natural gas will be dissolved in oil under high volumetric pressure like carbon dioxide inside a soda can. If the pressure inside the pristine reservoir is above the bubble point, the oil is undersaturated and still able to dissolve more gas. If the pressure is below the bubble point, the oil will be fully saturated with gas and any excess gas will form a gas cap on top of the oil since it is lighter.

To extract oil from the reservoir, we start by drilling a well into the oil zone. If the pristine pressure inside the reservoir is sufficiently high, it will push oil up to the surface and start what we will refer to as *primary production*. Alternatively, one may use a pump to lower the pressure in the wellbore below the point where oil starts flowing. How large

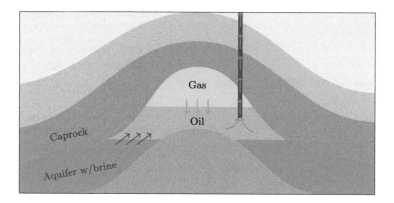

Figure 1.1 Conceptual illustration of a petroleum reservoir during primary production. Over millions of years, hydrocarbons have accumulated under a caprock that has a low ability to transmit fluids and thus prevents their upward movement. Inside the trap, the fluids will distribute according to density, with light hydrocarbons in a gaseous phase on top, heavier hydrocarbons in an oleic phase in the middle, and brine at the bottom. If the pressure difference between the oil zone and the wellbore is sufficiently high, the oleic phase will flow naturally out of the reservoir. As the oleic phase is produced, the pressure inside the reservoir will decline, which in turn may introduce other mechanisms that help to maintain pressure and push more oil out of the well.

---

[1] The largest reservoir in the world is found in Ghawar in the Saudi Arabian desert and is approximately 230 km long, 30 km wide, and 90 m thick.

pressure differential one needs for hydrocarbons to flow, depends on the permeability of the rock; the higher the permeability, the easier the hydrocarbons will flow toward the well.

As hydrocarbons are extracted, the pressure inside the reservoir will decay and the recovery will gradually decline. However, declining pressure will often induce other physical processes that contribute to maintain recovery:

- In a *water drive*, the pore space below the hydrocarbons is filled with salt water, which is slightly compressible and hence will expand a little as the reservoir pressure is lowered. If the total water volume is large compared with the oil zone, even a small expansion will create significant water volumes that will push oil toward the well and hence contribute to maintain pressure. Sometimes the underlying water is part of a large aquifer system that has a natural influx that replenishes the extracted oil by water and maintains pressure.
- *Solution gas drive* works like when you shake and open a soda can. Initially, the pristine oil will be in a pure liquid state and contain no free gas. Extraction of fluids will gradually lower the reservoir pressure below the bubble point, which causes free gas to develop and form expanding gas bubbles that force oil into the well. Inside the well, the gas bubbles rise with the oil and make the combined fluid lighter and hence easier to push upward to the surface. At a certain point, however, the bubbles may reach a critical volume fraction and start to flow as a single gas phase, which has lower viscosity than the oil and hence moves faster. This rapidly depletes the energy stored inside the reservoir and causes the production to falter. Gas coming out of solution can also migrate to the top of the structure and form a gas cap above the oil; the gas cap pushes down on the liquid oil and hence helps to maintain pressure.
- In a *gas cap drive*, the reservoir contains more gas than what can be dissolved in the oil. When pressure is lowered, the gas cap expands and pushes oil into the well. Over time, the gas cap will gradually infiltrate the oil and cause the well to produce increasing amounts of gas.
- If a reservoir is highly permeable, gravity will force oil to move downward relative to gas and upward relative to water. This is called *gravity drive*.
- A *combination drive* has water below the oil zone and a gas cap above that both will push oil to the well at the same time as reservoir pressure is reduced.

These natural (or primary) drive mechanisms can only maintain the pressure for a limited period and the production will gradually falter as the reservoir pressure declines. How fast the pressure declines and how much hydrocarbons one can extract before the production ceases, varies with the drive mechanism. Solution gas drives can have a relatively rapid decline, whereas water and gas cap drives are able to maintain production for longer periods. Normally only 30% of the oil can be extracted using primary drive mechanisms.

To keep up the production and increase the recovery factor, many reservoirs employ some kind of engineered drive mechanisms for *secondary production*. Figure 1.2 illustrates two examples of voidage replacement in which water and/or gas is injected to support pressure in the reservoir. Water can also be injected to sweep the reservoir, displace the oil, and push it toward the wells. In some cases, one may choose to inject produced formation

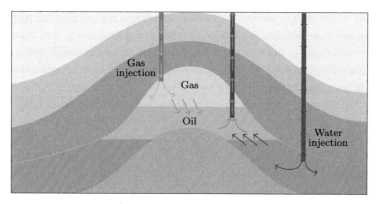

Figure 1.2 Conceptual illustration of voidage replacement, which is an example of a secondary production strategy in which gas and/or water is injected to maintain the reservoir pressure.

water, which is contaminated with hydrocarbons and solid particles and hence must be disposed of in some manner. Alternatively, one can extract formation water from a nearby aquifer. In offshore production it is also common to inject seawater. A common problem for all waterflooding methods is to maximize the sweep efficiency so that water does not move rapidly through high-flow zones in the reservoir and leave behind large volumes of unswept, mobile oil. Maintaining good sweep efficiency is particularly challenging for reservoirs with high-viscosity oil. If the injected water has lower viscosity than the resident oil, it will tend to form viscous fingers that rapidly expand through the oil and cause early water breakthrough in producers. (Think of water being poured into a cup of honey.) To improve the sweep efficiency, one can add polymers to the water to increase its viscosity and improve the mobility ratio between the injected and displaced fluids. Polymers have also been used to create flow diversions by plugging high-flow zones so that the injected fluid contacts and displaces more oil. For heavy oils, adverse mobility ratios can be improved by using steam injection or some other thermal method to heat the oil to reduce its viscosity.

Polymer injection and steam injection are examples of methods for so-called *enhanced oil recovery* (EOR), also called *tertiary production*. Another example is miscible and chemical injection, in which one injects solvents or surfactants that mix with the oleic phase in the reservoir to make it flow more readily. The solvent may be a gas such as carbon dioxide or nitrogen. However, the most common approach is to inject natural gas produced from the reservoir when there is no market that will accept the gas. Surfactants are similar to detergents used for laundry. Alkaline or caustic solutions, for instance, can react with organic acids occuring naturally in the reservoir to produce soap. The effect of all these substances is that they reduce the interfacial tension between water and oil, which enables small droplets of oil that were previously immobile to flow (more) freely. This is the same type of process that takes place when you use detergent to remove waxy and greasy stains from textiles. A limiting factor of these methods is that the chemicals are quickly adsorbed and lost into the reservoir rock.

Often, one will want to combine methods that improve the sweep efficiency of mobile oil with methods that mobilize immobile oil. Miscible gas injection, for instance, can be used after a waterflood to flush out residually trapped oil and establish new pathways to the production wells. *Water-alternating-gas* (WAG) is the most successful and widely used EOR method. Injecting large volumes of gas is expensive, and by injecting alternating slugs of water, one reduces the injected volume of gas required to maintain pressure. Similarly, the presence of mobile water reduces the tendency of the injected gas to finger through the less mobile oil. In polymer flooding, it is common to add surfactants to mobilize immobile oil by reducing or removing the interface tension between oil and water, and likewise, add alkaline solutions to reduce the adsorption of chemicals onto the rock faces.

Whereas the mechanisms of all the previously described methods for EOR are reasonably well studied and understood, there are other methods whose mechanisms are much debated. This includes injection of low-salinity water, which is not well understood, even though it has proved to be highly effective in certain cases. Another example is microbial EOR, which relies on microbes that digest long hydrocarbon molecules to form biosurfactants or emit carbon dioxide that will reduce interfacial tension and mobilize immobile oil. Microbial activity can either by achieved by injecting bacterial cultures mixed with a food source, or by injecting nutrients that will activate microbes that already reside in the reservoir.

Use of secondary recovery mechanisms has been highly successful. On the Norwegian continental shelf, for instance, water flooding and miscible gas injection have helped to increase the average recovery factor to 46%, which is significantly higher than the worldwide average of 22%. In other parts of the world, chemical methods have proved to be very efficient for onshore reservoirs having relatively short distances between wells. For offshore fields, however, the potential benefits of using chemical methods are much debated. First of all, it is not obvious that such methods will be effective for reservoirs characterized by large inter-well distances, as rapid adsorption onto the pore walls generally makes it difficult to transport the active ingredients long distances into a reservoir. Chemicals are also costly, need to be transported in large quantities, and consume platform space.

Even small improvements in recovery rates can lead to huge economic benefits for the owners of a petroleum asset and for this reason much research and engineering work is devoted to improving the understanding of mobilization and displacement mechanisms, and to design improved methods for primary and enhanced oil recovery. Mathematical modeling and numerical reservoir simulation play key roles in this endeavor.

1.2 Reservoir Simulation

Reservoir simulation is the means by which we use a numerical model of the geological and petrophysical characteristics of a subsurface reservoir, the (multiphase) fluid system, and the production equipment (wells and surface facilities) to analyze and predict how fluids flow through the reservoir rock to the stock tank or transport pipeline over time. It is

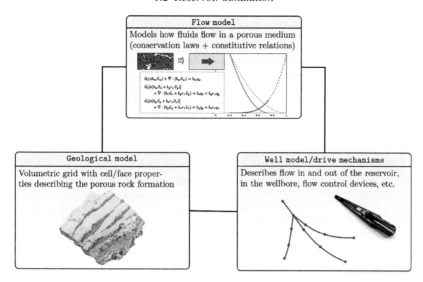

Figure 1.3 The three main constituents of a reservoir simulation model.

generally very difficult to observe and understand dynamic fluid behavior inside a reservoir, describe the physical processes, and measure all the parameters that influence the flow behavior. Predicting how a reservoir will produce over time and respond to different drive and displacement mechanisms therefore has a large degree of uncertainty attached. Simulation of petroleum reservoirs started in the mid 1950s as a means to quantify and reduce this uncertainty, and has become an important tool for qualitative and quantitative prediction of the flow of fluid phases. Reservoir simulation is a complement to field observations, pilot and laboratory tests, well testing, and analytical models, and is used by reservoir engineers to investigate displacement processes, compare and contrast the characteristics of different production scenarios, or as part of inverse modeling to calibrate reservoir parameters by integrating static and dynamic (production) data. In the big picture, reservoir simulation is mostly used to guide two different types of decisions: (i) to optimize development plans for new fields, and (ii) to assist with operational and investment decisions.

To describe the subsurface flow processes mathematically, our simulation model will be made up of three main constituents. First, we need a mathematical *flow model* that describes how fluids flow in a porous medium. These models are typically given as a set of partial differential equations describing the mass conservation of fluid phases, accompanied by a suitable set of constitutive relations that describe the relationship among different physical quantities. Second, we need a *geological model* that describes the given porous rock formation (the reservoir). The geological model is realized as a grid populated with petrophysical properties that are used as input to the flow model, and together they make up the reservoir simulation model. Last, but not least, we need a model for the *wells and production facilities* that provide pressure and fluid communication between the reservoir and the surface.

Accurate prediction of reservoir flow scenarios is a difficult task. One reason is that we can never get a complete and accurate characterization of the rock parameters that influence the flow pattern. Even if we did, we would not be able to run simulations that exploit all available information, since this would require a tremendous amount of computer resources that far exceed the capabilities of modern multiprocessor computers. On the other hand, we neither need nor do we seek a simultaneous description of the flow scenario on all scales down to the pore scale. For reservoir management it is usually sufficient to describe the general trends in the reservoir flow pattern.

In the early days of the computer, reservoir simulation models were built from two-dimensional slices with 10^2–10^3 Cartesian grid cells representing the whole reservoir. In contrast, contemporary reservoir characterization methods can model the porous rock formations by the means of grid-blocks down to the meter scale. This gives three-dimensional models consisting of millions of cells. Stratigraphic grid models, based on extrusion of 2D areal grids to form volumetric descriptions, have been popular for many years and are the current industry standard. However, more complex methods based on unstructured grids are gaining in popularity.

Despite an astonishing increase in computer power and intensive research on computation techniques, commercial reservoir simulators are rarely used to run simulations directly on geological grid models. Instead, coarse grid models with grid-blocks that are typically ten to hundred times larger are built using some kind of upscaling of the geophysical parameters. How one should perform this upscaling is not trivial. In fact, upscaling has been, and probably still is, one of the most active research areas in the oil industry. This effort reflects the general opinion that given the ever increasing size and complexity of the geological reservoir models, one cannot generally expect to run simulations directly on geological models in the foreseeable future.

Along with the development of better computers, new and more robust upscaling techniques, and more detailed reservoir and fluid characterization, there has also been an equally significant development in the area of numerical methods. State-of-the-art simulators employ numerical methods that can take advantage of multiple processors, distributed memory computer architectures, adaptive grid refinement strategies, and iterative techniques with linear complexity. For the simulation, there exists a wide variety of different numerical schemes that all have their pros and cons. With all these techniques available, we see a trend where methods from the research forefront are being tuned to a special set of applications and mathematical models, as opposed to traditional methods that were developed for a large class of differential equations.

Altogether, these developments have enabled geologists and reservoir engineers to build increasingly complex geological and reservoir models. Continuing to improve such models to gain a better understanding of the reservoir and reduce uncertainty is obviously important. Nevertheless, you should not forget that the purpose of a simulation study usually is to help your company make better decisions, and thus the value of the study generally depends on to what extent it influences decisions and leads to higher profit, e.g., by increasing recovery and/or reducing capital expenditures (CAPEX) or operational expenses (OPEX).

Developing a very advanced model that has all the physical effects imaginable included may therefore not be necessary to answer a specific question. Indeed, to paraphrase the famous Occam's principle: *a simulation model should be as simple as possible, but not simpler*. In the coming years, we will probably see a larger degree of hybrid modeling that combines models based on physical principles with data-driven approaches.

1.3 Outline of the Book

The book is intended to serve several purposes. First of all, you can use it as a self-contained introduction to the physics of flow in porous media, its basic mathematical theory, and the numerical methods used to solve the underlying differential equations. Hopefully, the book will also give you a hands-on introduction to practical modeling of flow in porous media, focusing in particular on models and problems that are relevant to the petroleum industry. The discussion of mathematical models and numerical methods is accompanied by a large number of illustrative examples, ranging from idealized and highly simplified setups to cases involving models of real-life reservoirs. Last, but not least, the book will introduce you to a widely used open-source software, and teach you some of the principles that have been used in developing it.

The Software Aspect: User Guide, Examples, and Exercises

All examples in the book have been created using the MATLAB Reservoir Simulation Toolbox (MRST). This open-source software consists on one hand of a set of reservoir simulators and workflow tools that you can modify to suit your own purposes, and on the other hand it can be seen as an enhancement of MATLAB/GNU Octave toward reservoir modeling in the form of a large a collection of flexible and efficient software libraries and data structures, which you can use to design your own simulators or computational workflows. The use of MRST permeates more traditional textbook material, and the book can therefore be seen as a user guide to MRST, in which you get introduced to the software gradually, example by example. Appendix A gives a more focused introduction to MRST, which starts by explaining how the software is organized, how you can install it, explore its functionality, and find help much in the same way as in MATLAB/GNU Octave.

The book can alternatively be viewed as a discussion by example of how a numerical scripting language like MATLAB/GNU Octave can for be used for rapid prototyping, testing, and verification on realistic reservoir problems with a high degree of complexity. Through the many examples, I also try to gradually teach you some of the techniques and programming concepts that have been used to create MRST, so that you later can use similar ideas to ensure flexibility and high efficiency in your own programs.

In the introductory part of the book that covers grids, petrophysical parameters and basic discretizations and solvers for single-phase flow, I have tried to make all code examples as self-contained as possible. To this end, all code lines necessary to produce the numerical results and figures are presented and discuss in detail. However, occasionally I omit minor details that either have been discussed elsewhere or should be part of your basic MATLAB repertoire. As we move to more complex examples, in particular for multiphase flow and reservoir engineering workflows, it is no longer expedient to discuss scripts in full detail. In most cases, however, complete scripts containing all code lines necessary to run the examples can be found in a dedicated book module in MRST. I strongly encourage that you use your own computer to run as many as possible of the examples in the book, as well as other examples and tutorials that are distributed with the software. Your understanding will be further enhanced if you also modify the examples, e.g., by changing the input parameters, or extend them to solve problems that are related, but (slightly) different. To help you in this direction, I have included a number of computer exercises that modify and extend some of the examples, combine ideas from two or more examples, or investigate in more detail aspects that are not covered by any of the worked examples. For some of the exercises you can find solution proposals in the book module.

Finally, I point out that MRST is an open-source software, and if reading this book gives you ideas about new functionality, or you discover things that are not working as they should or could, you are free to improve the toolbox and expand it in new directions. If you do so, I strongly encourage you to pay us back by releasing your code publicly for the benefit of the reservoir simulation community.

Part I: Geological Models and Grids

The first part of the book discusses how to represent a geological medium as a discrete computer model that can be used to study the flow of one or more fluid phases. Chapter 2 gives you a crash course in petroleum geology and geological modeling, written by a non-geologist. In this chapter, I try to explain key processes that lead to the formation of a hydrocarbon reservoir, discuss modeling of permeable rocks across multiple spatial scales, and introduce you to the basic physical properties used to describe porous media in general. For many purposes, reservoir geology can be represented as a collection of maps and surfaces. However, if the geological model is to be used as input to a macroscale fluid simulation, we must assume a continuum hypothesis and represent the reservoir in terms of a volumetric grid, in which each cell is equipped with a set of petrophysical properties. On top of this grid, one can then impose mathematical models that describe the macroscopic continuum physics of one or more fluid phases flowing through the microscopic network of pores and throats between mineral grains that are present on the subscale inside the porous rock of each grid block.

Chapter 3 describes in more detail how to represent and generate grids, with special emphasis on the types of grids commonly used in reservoir simulation. The chapter presents a wide variety of examples to illustrate and explain different grid formats, from simple structured grids, via unstructured grids based on Delaunay tessellations and Voronoi diagrams, to stratigraphic grids represented on the industry-standard corner-point format or as 2.5D and 3D unstructured grids. The examples also demonstrate various methods for generating grids that represent plausible reservoir models, and they discuss some of the realistic data sets that can be downloaded along with the MRST software.

1.3 Outline of the Book

Through Chapters 2 and 3, you will be introduced to the data structures used to represent grids and petrophysical data in MRST. Understanding these basic data structures, and the various methods used to create and manipulate them, is fundamental if you want to understand the inner workings of a majority of the routines implemented in MRST or use the software as a platform to implement your own computational methods. Through the many examples, you will also be introduced to various functionality in MRST for plotting data associated with cells and faces (interface between two neighboring cells) as well as various strategies for traversing the grid and picking subsets of data; these techniques will prove very useful later in the book.

Part II: Single-Phase Flow

The second part of the book is devoted entirely to single-phase flow and will introduce you to many of the key concepts for modeling flow in porous media, including basic flow equations, closure relationships, auxiliary conditions and models, spatial and temporal discretizations, and linear and nonlinear solvers.

Chapter 4 starts by introducing the two fundamental principles necessary to describe flow in porous media: conservation of mass and Darcy's law. Mass conservation is a fundamental physical property, whereas Darcy's law is a phenomenological description of the conservation of momentum that introduces permeability, a rock property, to relate volumetric flow rate to pressure differentials. The chapter outlines different forms these two equations can take in various special cases. To form a full model, the basic flow equations must be extended with various constitutive laws and extra equations describing external forces that drive fluid flow; these can either be boundary conditions and/or wells that inject or produce fluids. The last section of Chapter 4 introduces the classical two-point finite-volume method, which is the current industry standard for discretizing flow equations. In particular, we demonstrate how to write this discretization in a very compact way by introducing discrete analogues of the divergence and gradient operators. These operators will be used extensively later when developing solvers for compressible single and multiphase flow with the AD-OO framework.

Chapter 5 focuses on the special case of an incompressible fluid flowing in a completely rigid medium, for which the flow model can be written as a Poisson-type *partial differential equation* (PDE) with a varying coefficient. We start by introducing the various data structures that are necessary to make a full simulator, including fluid properties, reservoir states describing the primary unknowns, fluid sources, boundary conditions, and models of injection and production wells. We then discuss in detail the implementation of a two-point pressure solver, as well as upwind solvers for linear advection equations, which, e.g., can be used to compute time lines (time-of-flight) in the reservoir and steady-state tracer distributions that delineate the reservoir into various influence regions. We end the chapter with four simulation examples: the first introduces the classical quarter five-spot pattern, which is a standard test case in reservoir simulation, while the next three explain how to incorporate boundary conditions and Peaceman well models and discuss the difference between using structured and unstructured grids.

Grids describing real reservoirs typically have unstructured topology and irregular cell geometries with high aspect ratios. Two-point discretizations are unfortunately only consistent if a certain relationship between the grid and the permeability tensor is satisfied, which is quite restrictive and difficult to fulfill when modeling complex geology. Chapter 6 discusses methods that are consistent for general polyhedral cell geometries. This includes mixed finite-element methods, multipoint flux approximation methods, and mimetic finite-difference methods that are still being researched by academia.

Chapter 7, the last in Part II, is devoted to compressible flow, which in the general case is modeled by a nonlinear, time-dependent, parabolic PDE. Using this relatively simple model, we introduce many of the concepts that will later be used to develop multiphase simulators of full industry-standard complexity. To discretize the transient flow equation, we combine the two-point method introduced for incompressible flow with an implicit temporal discretization. The standard approach to solving the nonlinear system of discrete equations arising from complex multiphase models is to compute the Jacobian matrix of first derivatives for the nonlinear system, and use Newton's method to successively find better approximations to the solution. Deriving and implementing analytic expressions for Jacobian matrices is both error-prone and time-consuming, in particular if the flow equations contain complex fluid models, well descriptions, thermodynamical behavior, etc. In MRST, we have therefore chosen to construct Jacobian matrices using automatic differentiation, which is a technique to numerically evaluate the derivatives of functions specified by a computer program to working precision accuracy. Combining this technique with discrete averaging and differential operators enables you to write very compact simulator codes in which flow models are implemented almost in the same form as they are written in the underlying mathematical model. This greatly simplifies the task of writing new simulators: all you have to do is to implement the new model equations in residual form and specify which variables should be considered as primary unknowns. When the software then evaluates all the elementary operations necessary to compute the residual, it simultaneously uses elementary differentiation rules to compute the analytical derivative of each elementary operation at the specific function value. These values are then gathered using the standard chain rule and assembled into a block matrix containing the correct partial derivatives with respect to all primary variables. To demonstrate the utility and power of the resulting framework, we discuss how one can quickly change functional dependencies in the single-phase pressure solver and extend it to include new functional dependencies like pressure-dependent viscosity, thermal effects, or non-Newtonian fluid rheology.

Part III: Multiphase Flow

The third part of the book outlines how to extend the ideas from Part II to multiphase flow. Chapter 8 starts by introducing new physical phenomena and properties that appear for multiphase flows, including fluid saturations, wettability and capillary pressure, relative permeability, etc. With this introduced, we move on to outline the general flow equations for multiphase flow, before we discuss various model reformulations and (semi-)analytical solutions for the special case of immiscible, two-phase flow. The main difference between single and multiphase flow is that we now, in addition to an equation for fluid pressure or density, get additional equations for the transport of fluid phases and/or chemical species. These equations are generally parabolic, but will often simplify to or behave like hyperbolic equations. Chapter 9 therefore introduces basic concept from the theory of hyperbolic conservation laws and introduces a few classical schemes. We also introduce the basic discretization that will be used for transport equations later in the book.

Chapter 10 follows along the same lines as Chapter 5 and explains how incompressible solvers developed for single-phase flow can be extended to account for multiphase flow effects using the so-called fractional-flow formulation introduced in Chapter 8 and simulated using sequential methods in which pressure effects and transport of fluid saturations and/or component concentrations are computed in separate steps. The chapter includes a number of test cases highlighting various effect of multiphase flow and illustrating error mechanisms inherent in sequential simulation methods.

Once the basics of incompressible, multiphase flow has been discussed, we move on to discuss more advanced multiphase flow models, focusing primarily on the black-oil formulation, which can be found in contemporary commercial simulators. The black-oil equations lump all hydrocarbons into two pseudo-components: light components that exist in a gas phase at surface conditions and heavier components that exist in a liquid oil phase. At reservoir conditions, the two components form a gaseous and an oleic phase, but can also dissolve fully or partially in the other phase. The dissolution depends on pressure (and temperature), and the simplified pressure, volume, and temperature (PVT) behavior of the resulting fluids follows analytical or tabulated relationships. Together with water, the hydrocarbon system forms a three-phase, three-component system, whose behavior can be surprisingly intricate. Chapter 11 gives an in-depth discussion of the underlying physical principles, whereas Chapter 12 discusses how one can use variants of the automatic-differentiation methods introduced in Chapter 7, combined with modern principles of object orientation, to develop robust simulators. The two chapters also include a number of illustrative simulation cases.

Part IV: Reservoir Engineering Workflows

The fourth and last part of the book is devoted to discussing additional computational methods and tools that can be used to address common tasks within reservoir engineering workflows.

Chapter 13 is devoted to flow diagnostics, which are simple numerical experiments that can be used to probe a reservoir to understand flow paths and communication patterns. Herein, all types of flow diagnostics will be based on the computation of time-of-flight, which defines natural time lines in the porous medium, and steady-state distribution of numerical tracers, which delineate the reservoir into subregions that can be uniquely associated with distinct sources of the inflow/outflow. Both quantities will be computed using finite-volume methods introduced in Chapters 4 and 5, and you do not have to read the chapters in between to understand the essential ideas of Chapter 13. The concept of flow diagnostics also includes several measures of dynamic heterogeneity, which can be used as simple proxies for more comprehensive multiphase simulations in various reservoir engineering workflows including placement of wells, rate optimization, etc.

As you will see in Chapter 2, porous rocks are heterogeneous at a large variety of length scales. There is therefore a general trend toward building complex, high-resolution models for geological characterization to represent small-scale geological structures. Likewise, large ensembles of equiprobable models are routinely generated to systematically quantify model uncertainty. In addition, many companies develop hierarchies of models that cover a wide range of physical scales to systematically propagate the effects of small-scale geological variations up to the reservoir scale. In either case, one quickly ends up with geological models that contain more details than what can or should be used in subsequent flow simulation studies. Hence, there is a strong need for mathematical and numerical techniques for formulating reduced models, or communicating effective parameters and properties between models of different spatial resolution. Such methods are discussed in Chapters 14 and 15. Chapter 14 introduces data structures and various methods for partitioning a fine-scale grid into a coarse-scale grid. This chapter hardly assumes any familiarity with flow solvers and can be read directly after Chapters 2 and 3. However, before you continue to read about upscaling in Chapter 15, which refers to the process in which petrophysical properties in all cells that make up a coarse block are averaged into a single effective value for each coarse block, I suggest that you read the chapters about incompressible, single-phase flow.

After this quick tour of the book, you are probably eager to start digging into the material. Before continuing to the next chapter, we present a simple example that will give you a first taste of simulating flow in porous media.

1.4 The First Encounter with MRST

The purpose of this first example is to show the ten code lines needed to set up and solve simple flow problem. We also show how to visualize the geological model and the computed flow solution. To this end, we consider a very simple problem: compute the pressure variation $p(z)$ inside a $[0,1] \times [0,1] \times [0,30]$ m^3 rock column for a single-phase fluid of constant density ρ assuming a datum pressure $p(z_0) = 100$ bar at the top of the column. The solution is trivial and follows by integration,

$$p(z) = p(z_0) + \int_{z_0}^{z} g\rho \, dz = p(z_0) + g\rho(z - z_0). \tag{1.1}$$

Here, g is the gravity constant along the z axis, which by convention in reservoir simulation is assumed to point downwards so that z increases toward greater depth.

We will now compute the same solution using a simple incompressible flow solver. To this end, we need two of the three constituents discussed in Section 1.2: (i) a flow model describing the fluid behavior and (ii) a model describing the reservoir rock. As explained already, the flow model consists of an equation describing conservation of mass and Darcy's law, which essentially represents conservation of momentum:

1.4 The First Encounter with MRST

$$\nabla \cdot \vec{v} = 0, \qquad \vec{v} = -\frac{K}{\mu}\bigl[\nabla p - \rho g \nabla z\bigr], \tag{1.2}$$

Darcy's law relates the flow rate $\vec{v}$ to the gradient of the flow potential $p - \rho g z$. The constant of proportionality is given as the ratio between the macroscopic rock permeability K and the fluid viscosity μ. By eliminating $\vec{v}$, we can reduce (1.2) to an elliptic Poisson equation, whose solution is given by (1.1) for constant μ and K.

Most examples in MRST feature some geological model, which by convention is created first. The basic part of a geological model is a grid describing the geometry of the reservoir rock, here chosen as a regular $1 \times 1 \times 30$ Cartesian grid. The grid is generated in two steps using standard routines from MRST:

```
G = cartGrid([1, 1, 30], [1, 1, 30]*meter^3);
G = computeGeometry(G);
```

The first function constructs the grid topology and cell geometries. The second function computes derived quantities such as cell volumes and centroids, areas, and normals of all cell faces, which you need if you want to solve flow equations on the grid. More details about grids are given in Chapter 3. To plot the grid, we can use the command

```
plotGrid(G); view(3);
```

A geological model should also describe petrophysical and other geological properties. In our case, the only parameter is the permeability, which is set to 100 millidarcy (md or mD). MRST also requires porosity (i.e., the void fraction of the bulk volume) to be given for all models. Since it is not used here, we set it to a typical value of 0.2:

```
rock = makeRock(G, 0.1*darcy(), 0.2);
```

MRST works in SI units and we must therefore be careful to specify the correct units for all physical quantities. The result of this construction is a MATLAB structure that contains two vectors of 30 elements each, giving the permeability and porosity in each cell. Petrophysical properties are discussed in more detail in Chapter 2. The grid colored by permeability values is plotted as follows:

```
cla, plotCellData(G, rock.perm)
```

To start representing the flow part of the model, we first need a structure to hold the reservoir state, which in the basic form will consist of the pressure and the flux across cell faces. We can construct an empty container with all values set to zero as follows

```
sol = initResSol(G, 0.0);
```

Since we are solving (1.2) on a finite domain, we must also describe conditions on all boundaries. The default assumption is that there will be no flow over the reservoir boundary ($\vec{v} \cdot \vec{n} = 0$). In our case, we also need to prescribe $p = 100$ bar at the top of the column

```
bc = pside([], G, 'TOP', 100.*barsa());
```

To discretize Darcy's law in (1.2), we use a standard two-point flux-approximation scheme, which relates the flux between two neighboring cells i and j to their pressure difference, $v_{ij} = -T_{ij}(p_i - p_j)$. The constant of proportionality T_{ij}, is called the transmissibility and is computed as the harmonic average of two other constants $T_{i,j}$ and $T_{j,i}$ associated with cell i and j, respectively. Each of these quantities on depend on geometrical grid properties and the permeability tensor of a single cell and can be computed once and for all independent of the particular flow model once we have defined the grid and petrophysical parameters:

```
hT = computeTrans(G, rock);
```

This scheme will be discussed in more detail in Section 4.4. More advanced discretizations can be found in other add-on modules and are discussed further in Chapter 6.

The next thing we need to define is the fluid properties. Unlike grids, petrophysical data, and boundary conditions, data structures for representing fluid properties are not part of the basic functionality of MRST. The reason is that the way fluid properties are specified is tightly coupled with the mathematical and numerical formulation of the flow equations, and may differ a lot between different types of simulators. Here, we have assumed incompressible flow and can therefore use fluid models from the incomp add-on module,

```
mrstModule add incomp
gravity reset on
fluid = initSingleFluid('mu', 1*centi*poise, 'rho', 1014*kilogram/meter^3);
```

The second line ensures that gravity is set to its default value, whereas the third line defines the fluid to have a density of 1014 kg/m^3, which is representative for water. The constant dynamic viscosity does not affect the pressure distribution and is arbitrarily set to 1 cP. More details about the fluid object will be given in Section 5.1.1. All we need to know here is that we can query it for viscosity and density by the following call:

```
[mu,rho] = fluid.properties();
```

We now have all we need to set up and solve a discrete version of (1.2). The first sub-equation says that the fluxes across all faces of each cell should sum to zero, i.e., $\sum_j T_{ij}(p_i - p_j) = 0$, for all i. For Cartesian grids and homogeneous permeability, this scheme coincides with the classical seven-point finite-difference scheme for Poisson's problem and is the only discretization in the incomp module. As a final step, we use the transmissibilities, the fluid object, and the boundary conditions to assemble and solve the discrete system:

```
sol = incompTPFA(sol, G, hT, fluid,'bc', bc);
```

Having computed the solution, we plot the pressure given in units "bar," which equals 0.1 MPa and is referred to as "barsa" in MRST since "bar" is a built-in command in MATLAB:

1.4 The First Encounter with MRST

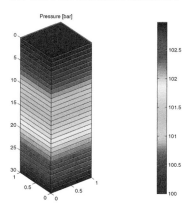

Figure 1.4 Hydrostatic pressure distribution in a gravity column computed by MRST. This example is taken from the MRST tutorial `flowSolverTutorial1.m` (`gravityColumn.m` in older versions of MRST).

```
plotFaces(G, 1:G.faces.num, convertTo(sol.facePressure, barsa()));
set(gca, 'ZDir', 'reverse'), title('Pressure [bar]')
view(3), colorbar, set(gca,'DataAspect',[1 1 10])
```

From the plot shown in Figure 1.4, we see that our solution correctly reproduces the linear pressure increase given in (1.1).

Before you continue reading, I encourage you to consult Appendix A, which describes MRST in more detail and explains how to install and get started with the software. Having the software operational is not a prerequisite for reading this book, but will contribute significantly to increase your understanding. And in case you should worry, you do not need to type in all the individual MATLAB/MRST commands; complete source codes for almost all the cases discussed later in the book can be found in a special add-on module, which you activate by calling `mrstModule add book`.

Part I
Geological Models and Grids

2
Modeling Reservoir Rocks

Aquifers and natural petroleum reservoirs consist of a subvolume of sedimentary rocks that have sufficient porosity and permeability to store and transmit fluids. This chapter gives an overview of how such rocks are modeled, so as to become part of a simulation model. We start by describing briefly how sedimentary rocks and hydrocarbons are formed. In doing so, we introduce some geological concepts that you may encounter while working with subsurface modeling. We then move on to describe how rocks that contain hydrocarbons or aquifer systems are modeled in terms of a volumetric grid and a set of accompanying discrete petrophysical properties. Finally, we discuss how rock properties are represented in MRST, and show several examples of rock models with varying complexity, ranging from an idealized shoebox rock body with homogeneous properties, via the widely used SPE 10 model, to two realistic models, one synthetic and one representing a large-scale aquifer from the North Sea.

2.1 Formation of Sedimentary Rocks

Sedimentary rocks are created by mineral or organic particles that are deposited and accumulated on the Earth's surface or within bodies of water to create layer upon layer of sediments. The sedimentary rocks found in reservoirs come from sedimentary basins, inside which large-scale sedimentation processes have taken place. Sedimentary basins are formed as the result of stretching and breaking of the continental crust. As the crust is stretched, hot rocks deeper in the earth come closer to the surface. When the stretching stops, the hot rocks start to cool, which causes the crustal rock to gradually subside and move downward to create a basin. Such processes are taking place today. The Great Rift Valley of Africa is a good example of a so-called rift basin, where a rift splits the continental plate so that opposite sides of the valley are moving a millimeter apart each year. This gradually creates a basin inside which a new ocean may appear over the next hundred million years.

On the part of the Earth's surface that lies above sea level, wind and flowing water will remove soil and rock from the crust and transport it to another place where it is deposited when the forces transporting the mineral particles can no longer overcome gravity and friction. Mineral particles can also be transported and deposited by other geophysical

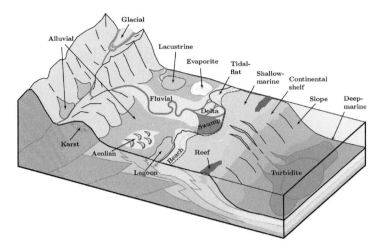

Figure 2.1 Illustration of various forms of depositional environments. Aeolian sediments are created by wind, evaporite minerals are left behind as water evaporates, fluvial sediments are deposited by rivers, and lacustrine sediments are deposited in lakes. Alluvial refers to the case where the geological process is not well described.

The illustration is based on an image by PePeEfe on Wikimedia Commons. The modified image is published under the Creative Commons Attribution-ShareAlike 3.0 Unported license.

processes, like mass movement and glaciers. Over millions of years, layers of sediments will build up in deep waters, in shallow-marine waters along the coastline, or on land in lakes, rivers, sand deltas, and lagoons, see Figure 2.1. The sediments accumulate a depth of a few centimeters every one hundred years to form sedimentary beds (or strata) that may extend many kilometers in the lateral directions.

Over time, the weight of the upper layers of accumulating sediment will push the lower layers downward. A combination of heat from the Earth's center and pressure from the overburden will cause the sediments to undergo various chemical, physical, and biological processes (commonly referred to as diagenesis). These processes cause the sediments to consolidate so that the loose materials form a compact, solid substance that may come in varying forms. This process is called lithification. Sedimentary rocks have a layered structure with different mixtures of rock types with varying grain size, mineral types, and clay content. The composition of a rock depends strongly upon a combination of geological processes, the type of sediments that are transported, and the environment in which it is deposited. *Sandstone* are formed by mineral particles that are broken off from a solid material by weathering and erosion in a source area and transported by water to a place where they settle and accumulate. *Limestone* is formed by the skeletons of marine organisms such as corals, which have a high calcium content. Sandstone and limestone will typically contain relative large void spaces between mineral particles in which fluids can move easily. *Mudrocks*, on the other hand, are formed by compression of clay, mud, and silt, and will consist of relatively fine-grained particles. These rocks, which are sometimes also referred to as *shale*, will have small pores and therefore have limited ability to transmit

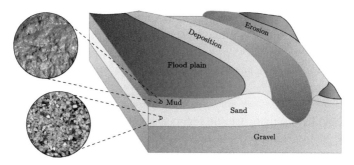

Figure 2.2 Illustration of a meandering river. Because water flows faster along the outer than along the inner curve of the bend, the river will erode along the outer bank and deposit at the inner bank. This causes the river to move sideways over time. Occasionally, the river will overflow its banks and cover the lower-lying flood plain with water and mud deposits that makes the soil fertile. Flood plains are therefore very important for agriculture.

fluids. Chemical rocks are formed by minerals that precipitate from a solution, e.g., salts that are left behind when oceans evaporate. Like mudrocks, salts are impermeable to fluids. Sediments will also generally contain organic particles that originate from the remains of plants, living creatures, and small organisms living in water.

To understand a particular rock formation, one must understand the prehistoric sedimentary environment from which it originates. It is common to distinguish between continental and marine environments. The primary source of sediments in a continental environment is rivers, inside which the moving water has high energy so that sediments will mainly be composed of fragments of preexisting minerals and rock, called clasts. Resulting rocks are therefore often referred to as *clastic rocks*. A flood plain is the area of the land adjacent to the river bank that will be covered by water when a river breaks its bank and floods during periods of high water discharge. Flood plains are created by bends in the river (meanders) that erode sideways; see Figure 2.2. A flood plain is the natural place for a river to diminish its kinetic energy. As the water moves, particles of rock are carried along by frictional drag of water on the particle, and water must flow at a certain velocity to suspend and transport a particle. Large particles are therefore deposited where the water flows rapidly, whereas finer particles will settle where the current is weak. River systems also create alluvial fans, which are fan-shaped sediment deposits built up by streams that carry debris from a single source. Other examples of continental environments are lagoons, lakes, and swamps that contain quiet water in which fine-grained sediments mingled with organic material are deposited.

For *marine rocks* formed in a sea or ocean, one usually makes the distinction between deep and shallow environments. In deep waters (200 m or more), the water moves relatively slowly over the bottom, and deposits will mainly consist of fine clay and skeletons of small microorganisms. However, if the sea bottom is slightly inclined, the sediments can become unstable and induce sudden currents of sediment-laden water to move rapidly downslope. These so-called turbidity currents are caused by density differences between the suspended sediments and the surrounding water and will generally be highly turbulent. Because of the high density, a turbidity current can transport larger particles than what pure water would be able to at the same velocity. These currents will therefore cause what can be considered as instantaneous deposits of large amounts of clastic sediments into deep oceans. The resulting rocks are called *turbidite*.

Shallow-marine environments are found near the coastline and contain greater kinetic energy caused by wave activity. Clastic sediments will therefore generally be coarser than in deep-marine environments and will consists of small-grained sand, clay, and silt that has been washed out and transported from areas of higher water energy on the continent. Far from the continent, the transport of clastic sediments is small, and deposits are dominated by biological activity. In warm waters, there are multitudes of small organisms that build carbonate skeletons, and, when deposited together with mud, these skeletons will turn into limestone, which is an example of a *carbonate rock*. Carbonate rocks can also be made from the skeletons of larger organisms living on coral reefs. Carbonates are soluble in slightly acidic water, and this may create *karst rocks* that contain large regions of void space (caves, caverns).

To model sedimentary rocks in a simulation model, one must be able to describe the geometry and the physical properties of different rock layers. A *bed* denotes the smallest unit of rock that is distinguishable from an adjacent rock layer unit above or below it, and can be seen as bands of different colors or textures in hillsides, cliffs, river banks, road cuts, etc. Each band represents a specific sedimentary environment, or mode of deposition, and can range from a few centimeters to several meters thick, often varying in the lateral direction. A sedimentary rock is characterized by its bedding, i.e., sequence of beds and lamina (less-pronounced layers). The bedding process is typically horizontal, but beds may also be deposited at a small angle, and parts of the beds may be weathered down or completely eroded away during deposition, allowing newer beds to form at an angle with older ones. Figure 2.3 shows two photos of sedimentary rock outcrops from Svalbard, which is one of the few places in Northern Europe where one can observe large-scale outcrops of sedimentary rocks. Figure 2.4 shows two more pictures of layered structures on meter to kilometer scale and centimeter scale, respectively.

 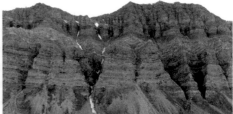

Figure 2.3 Outcrops of sedimentary rocks from Svalbard, Norway. The length scale is a few hundred meters.

Figure 2.4 Layered geological structures occur on both large and small scales in sedimentary rocks. The left picture is from Grand Canyon, the right picture is courtesy of Silje Støren Berg, University of Bergen.

Each sedimentary environment has its own characteristic deposits and forms what is called a *depositional facies*, i.e., a body of rock with distinct characteristics. Different sedimentary environments usually exist alongside each other in a natural succession. Small stones, gravel, and large sand particles are heavy and are deposited at the river bottom, whereas small sand particles are easily transported and are found at river banks and on the surrounding plains along with mud and clay. Following a rock layer of a given age, one will therefore see changes in the facies (rock type). Similarly, depositional environments change with time: shorelines move with changes in the sea level and land level or because of formation of river deltas, rivers change their course because of erosion or flooding, etc. Dramatic events like floods create more abrupt changes. At a given position, the accumulated sequence of beds will therefore contain different facies.

As time passes by, more and more sediments accumulate and the stack of beds piles up. Simultaneously, severe geological activity takes place: Cracking of continental plates and volcanic activity will change our eventual reservoir from a relatively smooth, layered sedimentary basin into a complex structure. Previously, continuous layers of sediments are compressed and pushed against each other to form arches, which are referred to as *anticlines*, and depressions, which are referred to as *synclines*. If the deposits contain large accumulations of salts, these will tend to flow upward through the surrounding layers to form large domes, since salts are lighter than other mineral particles. Likewise, as the sediments may be stretched, cut, shifted, or twisted in various directions, their movement may introduce fractures and faults. *Fractures* are cracks or breakage in the rock, across which there has been no movement. *Faults* are fractures or discontinuities in a volume of rock, across which movement in the crust has caused rock volumes on opposite sides to be displaced relative to each other. Some faults are small and localized, whereas others are part of the vast system of boundaries between tectonic plates that crisscrosses the crust of the Earth. Faults are described in terms of their *strike*, which is the compass direction in which the fault intersects the horizontal plane, and their *dip*, which is the angle that the fault plane makes with the horizontal, measured perpendicular to the strike. Faults are further classified

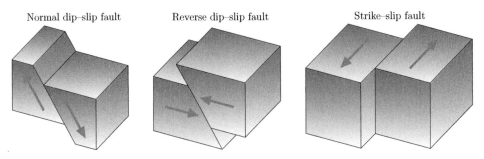

Figure 2.5 Fault types: In strike–slip faults, shear stress causes the rock volumes to slide past each other to the left and right in the lateral direction with very little vertical motion. In dip–slip faults, the rock volumes move predominantly in the vertical direction. A normal dip–slip fault occurs when the crust is extended and is sometimes called a divergent fault. Reverse dip–slip faults occur as a result of compressive shortening of the crust, and are sometimes called convergent faults, since rock volumes on opposite sides of the fault plane move horizontally toward one another.

by their *slip*, which is the displacement vector that describes the relative movement of rock volumes on opposite sides of the fault plane. The dip is usually separated into its vertical component, called *throw*, and its horizontal component, called *heave*. Figure 2.5 illustrates these different types of faults.

2.2 Creation of Crude Oil and Natural Gas

Deposits not only consist of sand grains, mud, and small rock particles, but will also contain remains of water-based plankton, algae, and other organisms that die and fall to the bottom where they are covered in mud. During their life, these organisms have absorbed heat from the sunlight and when they die and mix with the sediments, they take energy with them. As the sediments get buried deeper and deeper, the increasing heat from the Earth's core and pressure from the overburden will compress and "cook" the organic material that consists of cellulose, fatty oils, proteins, starches, sugar, waxes, and so on into an intermediate waxy product called kerogen. Whereas organic material contains carbon, hydrogen, and oxygen, the kerogen contains less oxygen and has a higher ratio of hydrogen to carbon. The maturation process that eventually turns kerogen into crude oil and natural gas depends highly upon temperature, and may take from a million years at $170°C$ to a hundred million of years at $100°C$. Most of the oil and natural gas we extract today has been formed from the remains of prehistoric algae and zooplankton living in the ocean. Coal, on the other hand, is formed from the remains of dead plants. The rock in which this "cooking" process takes place is commonly referred to as the source rock. The chances of forming a source rock containing a significant amount of oil and gas increases if there has been an event that caused mass death of microorganisms in an ocean basin.

Pressure from sediments lying above the source rock will force the hydrocarbons to migrate upward through the void space in these newer sediments. The lightest hydrocarbons (methane and ethane) that form natural gas usually escape quickly, while the liquid oils move more slowly toward the surface. In this process, the natural gas will separate from the oil, and the oil will separate from the resident brine (salty water). At certain sites, the migrating hydrocarbons will be collected and trapped in structural or stratigraphic traps. Structural traps are created by tectonic activity that forms layers of sediments into anticlines, domes, and folds. Stratigraphic traps form because of changes in facies (e.g., in clay content) within the bed itself or when the bed is sealed by an impermeable layer such as mudstone, which consists of small and densely packed particles, and thus has strong capillary forces that cannot easily be overcome by the buoyancy forces that drive migrating hydrocarbons upward. If a layer of mudstone is folded into an anticline above a more permeable sandstone or limestone, it will therefore act as a caprock that prevents the upward migration of hydrocarbons. Stratigraphic traps are especially important in the search for hydrocarbons; approximately 4/5 of the world's oil and gas reserves are found in anticlinal traps.

Figure 2.6 illustrates the various forms in which migrating oil and natural gas can be trapped and accumulate to form petroleum reservoirs. *Anticlines* are typically produced when lateral pressure causes strata to fold, but can also result from draping and subsequent compaction of sediments accumulating over local elevations in the topography. *Stratigraphic traps* accumulate hydrocarbons because of changes in the rock type. For example, sand banks deposited by a meandering river is later covered by mud from the flood plain. Similarly, stratigraphic traps can form when rock layers are tilted and eroded away and subsequently covered by other low-permeable strata. *Fault traps* are formed when strata are moved in opposite directions along a fault line so that permeable rocks come in contact with impermeable rocks. The fault itself can also be an effective trap if it contains clays that

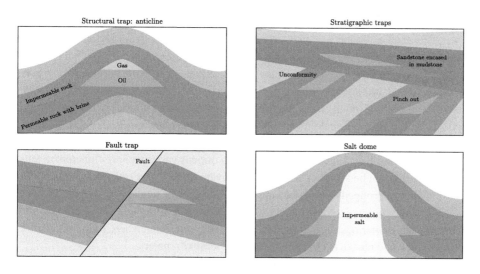

Figure 2.6 Various forms of traps in which migrating hydrocarbons can accumulate.

are smeared as the layers of rock move past each other. *Salt domes* are created by buried salt deposits that rise unevenly through the surrounding strata. Oil can either rest against the salt itself, or the salt induces chemical reactions in the surrounding rock that makes it impermeable.

The first evidence of hydrocarbons beneath the earth's surface were found in so-called seeps, which are found in many areas throughout the world. Seeps are formed when the seal above a trap is breached so that oil and gas can migrate all the way through the geological layers and escape to the surface. As the hydrocarbons break through the surface, the lighter components will continue to escape to the atmosphere and leave behind heavier products like bitumen, pitch, asphalt, and tar have been exploited by mankind since paleolithic times. In 1859, Edwin L. Drake, also known as Colonel Drake, drilled the world's first exploration well into an anticline in Titusville, Pennsylvania to find oil 69.5 ft below the surface. Since then, an enormous amount of wells have been drilled to extract oil from onshore reservoirs in the Middle East, North America, etc. Drilling of submerged oil wells started just before the turn of the nineteenth century: from platforms built on piles in the fresh waters of Grand Lake St. Marys (Ohio) in 1891, and from piers extending into the salt water of the Santa Barbara Channel (California) in 1896. Today, hydrocarbons are recovered from offshore reservoirs that are located thousands of meters below the sea bed and in waters with a depth up to 2,600 m in the Gulf of Mexico, the North Sea, and offshore from Brazil.

2.3 Multiscale Modeling of Permeable Rocks

All sedimentary rocks consist of a solid matrix with an interconnected void. The void pore space allows the rocks to store and transmit fluids. The ability to store fluids is determined by the volume fraction of pores (*rock porosity*), and the ability to transmit fluids (*rock permeability*) is given by the interconnection of the pores.

Rock formations found in natural petroleum reservoirs are typically heterogeneous at all length scales, from the micrometer scale of pore channels between the solid particles making up the rock to the kilometer scale of a full reservoir formation. On the scale of individual grains, there can be large variation in grain sizes, giving a broad distribution of void volumes and interconnections. Moving up the scale, laminae may exhibit large contrasts on the mm–cm scale in the ability to store and transmit fluids because of alternating layers of coarse and fine-grained material. Laminae are stacked to form beds, which are the smallest stratigraphic units. The thickness of beds varies from millimeters to tens of meters, and different beds are separated by thin layers with significantly lower permeability. Beds are, in turn, grouped and stacked into parasequences or sequences (parallel layers that have undergone similar geologic history). Parasequences represent the deposition of marine sediments, during periods of high sea level, and tend to be somewhere in the range from 1 to 100 m thick and have a horizontal extent of several kilometers.

The trends and heterogeneity of parasequences depend on the depositional environment. For instance, whereas shallow-marine deposits may lead to rather smoothly varying

2.3 Multiscale Modeling of Permeable Rocks

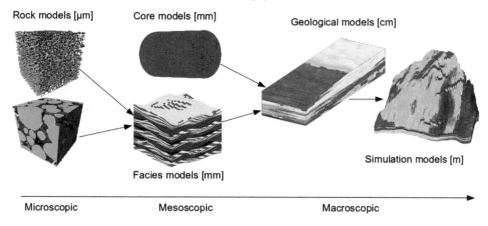

Figure 2.7 Illustration of the hierarchy of flow models used in subsurface modeling. The length scales are the vertical sizes of typical elements in the models.

permeability distributions with correlation lengths in the order of 10–100 m, fluvial reservoirs may contain intertwined patterns of sand bodies on a background with high clay content; see Figure 2.12. The reservoir geology can also consist of other structures such as shale layers (impermeable clays), which are the most abundant sedimentary rocks. Fractures and faults, on the other hand, are created by stresses in the rock and may extend from a few centimeters to tens or hundreds of meters. Faults may have a significantly higher or lower ability to transmit fluids than the surrounding rocks, depending upon whether the void space has been filled with clay material.

All these different length scales can have a profound impact on fluid flow. However, it is generally not possible to account for all pertinent scales that impact the flow in a single model. Instead, one has to create a hierarchy of models for studying phenomena occurring at reduced spans of scales. This is illustrated in Figure 2.7. Microscopic models represent the void spaces between individual grains and are used to provide porosity, permeability, electrical, and elastic properties of rocks from core samples and drill cuttings. Mesoscopic models are used to upscale these basic rock properties from the mm–cm scale of internal laminations, through the lithofacies scale ($\sim$50 cm), to the macroscopic facies association scale ($\sim$100 m) of geological models. In this book, we will primarily focus on another scale, simulation models, which represent the last scale in the model hierarchy. Simulation models are obtained by upscaling geological models and are either introduced out of necessity because geological models contain more details than a flow simulator can cope with, or out of convenience to provide faster calculation of flow responses.

2.3.1 Geological Characterization

To model a reservoir on a macroscopic scale, we basically need to represent its geology at a level of detail that is sufficient for the purpose the model is built to serve: to visualize how different experts perceive the reservoir, to provide estimates of hydrocarbon volumes, to assist well planning and geosteering, or as input to geophysical analysis (seismic modeling, rock mechanics) or flow simulations. For flow simulation, which is our primary concern in this book, we need a volumetric description that decomposes the reservoir into a set of grid cells (small 3D polygonal volumes) that are petrophysically and/or geometrically distinct from each other. With a slight abuse of terminology, we will refer to this as the *geological model*, which we distinguish from models describing the reservoir fluids and the forces that cause their movement.

Geological models are generally built in a sequence of steps, using a combination of stratigraphy (the study of rock layers and layering), sedimentology (study of sedimentary rocks), structural geology (the study of how rock layers are deformed over time by geological activity), diagenesis (the study of chemical, physical, and biological processes that transform sediments to rock), and interpretation of measured data. The starting point is usually a seismic interpretation, from which one obtains a representation of faults and geological horizons that bound different geological units. The seismic interpretation is used alongside a conceptual model in which geologists express how they believe the reservoir looks like based on studies of geological history and geological outcrops. The result can be expressed as a geometric model that consists of vertical or inclined surfaces representing faults, and horizontal or slightly sloping surfaces representing horizons that subdivide the reservoir volume into different geological units (zones). This zonation is obtained by combining seismic interpretation that describes the gross geometry of the reservoir with stratigraphic modeling and thickness information (isochores) obtained from well logs that define the internal layering. Once a model of the structural and stratigraphic architecture of the reservoir is established, the surface description can be turned into a 3D grid that is populated with cell and face properties that reflect the geological framework.

Unfortunately, building a geological model for a reservoir is like finishing a puzzle with most of the pieces missing. The amount of data available is limited due to the costs of acquiring it, and the data that is obtained is measured on scales that may be quite disparate from the geological features one needs to model. Seismic surveys give a sort of X-ray image of the reservoir, but are both expensive and time-consuming and can only give limited resolution; you cannot expect to see structures thinner than ten meters from seismic data. Information on finer scales is available from various measuring tools lowered into the wells to gather information of the rock in near-well region, e.g., by radiating the reservoir and measuring the response. Well logs have quite limited resolution, rarely down to centimeter scale. The most detailed information is available from rock samples (cores) extracted from the well. The industry uses X-ray, CT scan, as well as electron microscopes to gather high-resolution information from the cores, and the data resolution is only limited by the apparatus at hand. However, information from cores and well logs can only tell you how the rock looks like near the well, and extrapolating this information to the rest of the reservoir is subject to great uncertainty. Moreover, due to high costs, one cannot expect well logs and cores to be taken from every well. All these techniques give separately small contributions that can help build a geological model. However, in the end we still have very limited information available, considering that a petroleum reservoir can have complex geological features that span across all types of length scales from a few millimeters to several kilometers.

In summary, the process of making a geological model is generally strongly underdetermined. It is therefore customary to use a combination of deterministic and probabilistic modeling to estimate the subsurface characteristics between the wells. Deterministic modeling is used to specify large-scale structures such as faults, correlation, trends, and layering, which are used as input and controls to *geostatistical* techniques [258, 57, 82] that build detailed grid models satisfying statistical properties assumed for the petrophysical heterogeneity. In recent years, process models that seek to mimic deposition and subsequent structural changes have also gained more popularity since they tend to produce less artifacts in the geocellular models. Trends and heterogeneity in petrophysical properties depend strongly on the structure of sedimentary deposits, and high-resolution petrophysical realizations are thus rarely not built directly. Instead, one starts by building a *rock model* that is based on the structural model and consists of a set of discrete rock bodies (facies) that are specified to improve petrophysical classification and spatial shape. For carbonates, the modeled facies are highly related to diagenesis, while the facies modeled for sandstone reservoirs are typically derived from the depositional facies. By supplying knowledge of the depositional environment (fluvial, shallow-marine, deep-marine, etc.) and conditioning this to observed data, one can determine the geometry of the facies and how they are mixed.

To populate the modeled facies with petrophysical properties, it is common to replace the facies with a volumetric grid, and use stochastic simulation to generate multiple discrete realizations of the petrophysical properties defined on this grid. Each grid model has a plausible heterogeneity and can contain millions of cells. The collection of all realizations gives a measure of the uncertainty involved in the modeling. Hence, if the sample of realizations (and the upscaling procedure that converts the geological models into simulation models) is unbiased, then it is possible to supply predicted production characteristics, such as the cumulative oil production, obtained from simulation studies with a measure of uncertainty.

You can find a lot more details about the complex process of characterizing a reservoir and making geological models in the excellent textbook by Ringrose and Bentley [265]. The rest of this section will first introduce the concept of representative elementary volumes, which underlies the continuum models used to describe subsurface flow and transport, and then briefly discuss microscopic and mesoscopic modeling. This discussion is not essential to understand what follows in the next few chapters, and if you are not interested, you can therefore safely skip to Section 2.4, which discusses macroscopic modeling of reservoir rocks.

2.3.2 Representative Elementary Volumes

Choosing appropriate modeling scales is often done by intuition and experience, and it is hard to give general guidelines. An important concept in choosing model scales is the

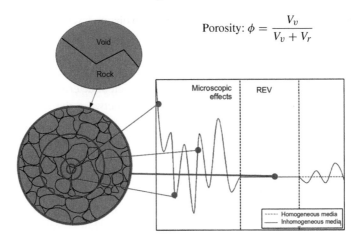

Figure 2.8 The concept of a representative elementary volume (REV), here illustrated for porosity, which measures the fraction of void space to bulk volume.

notion of representative elementary volumes (REVs), which is the smallest volume over which a measurement can be made and be representative of the whole. This concept is based on the idea that petrophysical flow properties are constant on some intervals of scale; see Figure 2.8. Representative elementary volumes, if they exist, mark transitions between scales of heterogeneity, and present natural length scales for modeling.

To identify a range of length scales where REVs exist, e.g., for porosity, we move along the length-scale axis from the micrometer-scale of pores toward the kilometer-scale of the reservoir. At the pore scale, the porosity is a rapidly oscillating function equal to zero (in solid rock) or to one (in the pores). Hence, obviously no REVs can exist at this scale. At the next characteristic length scale, the core scale level, we find laminae deposits. Because the laminae consist of alternating layers of coarse and fine grained material, we cannot expect to find a common porosity value for the different rock structures. Moving further along the length-scale axis, we may find long thin layers, perhaps extending throughout the entire horizontal length of the reservoirs. Each of these individual layers may be nearly homogeneous because they are created by the same geological process, and probably contain approximately the same rock types. Hence, at this scale it sounds reasonable to speak of an REV. If we move to the high end of the length-scale axis, we start to group more and more layers into families with different sedimentary structures, and REVs for porosity will probably not exist.

The discussion gives grounds to claim that reservoir rock structures contain scales where REVs may exist. From a general point of view, however, the existence of REVs in porous media is highly disputable. A faulted reservoir, for instance, can have faults distributed continuously both in length and aperture throughout the reservoir, and will typically have no REVs. Moreover, no two reservoirs are identical, so it is difficult to capitalize from previous experience. Indeed, porous formations in reservoirs may vary greatly, also in terms of scales. Nevertheless, the concept of REVs can serve as a guideline when deciding what scales to model.

2.3.3 Microscopic Models: The Pore Scale

Pore-scale models, as illustrated on the left-hand side of Figure 2.7, may be about the size of a sugar cube and are based on measurements from core plugs obtained from well trajectories during drilling. These rock samples are necessarily confined (in dimension) by the radius of the well, although they lengthwise are only confined by the length of the well. Three such rock samples are shown in Figure 2.9. The main methods for obtaining pore-scale models from a rock sample are by studying thin slices using an electron microscope with micrometer resolution or by CT scans. In the following, we will give a simplified overview of flow modeling on this scale.

At the pore scale, the porous medium is either represented by a volumetric grid or by a graph (see e.g., [242]). A graph is a pair (V, E), where V is a set whose elements are called vertices (or nodes), and E is a subset of $V \times V$ whose elements are called edges. The vertices are taken to represent pores, and the edges represent pore throats (i.e., connections between pores). The flow process, in which one fluid invades the void space filled by another fluid, is generally described as an invasion–percolation process dominated by capillary forces, in which a fluid phase can invade a pore only if a neighboring pore is already invaded. Another approach to multiphase modeling is through the use of the lattice Boltzmann method, which represents the fluids as a set of particles that propagate and collide according to a set of rules defined for interactions between particles of the same fluid phase, between particles of different fluid phases, and between the fluids and the walls of the void space. A further presentation of pore-scale modeling is beyond our scope here, but the interested reader is encouraged to consult, e.g., [242, 44] and references therein.

From an analytical point of view, pore-scale modeling is very important as it represents flow at the fundamental scale (or more loosely, where the flow really takes place), and

Figure 2.9 Three core plugs with a diameter of 3.81 cm and a height of 5 cm.

hence provides the proper framework for understanding the fundamentals of porous media flow. From a practical point of view, pore-scale modeling has a huge potential. Modeling flow at all other scales can be seen as averaging of flow at the pore scale, and properties describing the flow at larger scales are usually a mixture of pore-scale properties. At larger scales, the complexity of flow modeling is often overwhelming, with large uncertainties in determining flow parameters. Hence being able to single out and estimate the various factors determining flow parameters is invaluable, and pore-scale models can be instrumental in this respect. However, to extrapolate properties from the pore scale to an entire reservoir is very challenging, even if the entire pore space of the reservoir was known. In real life you will of course not be anywhere close to knowing the entire pore space of a reservoir.

2.3.4 Mesoscopic Models

Models based on flow experiments on core plugs are by far the most common mesoscopic models. The main equations describing flow are continuity of fluid phases and Darcy's law, which states that flow rate is proportional to pressure drop. The purpose of core-plug experiments is to determine capillary pressure curves and the proportionality constant in Darcy's law that measures the ability to transmit fluids; see (1.2) in Section 1.4. To this end, the sides of the core are insulated and flow is driven through the core. By measuring flow rate versus pressure drop, one can estimate the proportionality constant for both single-phase and multiphase flows.

In conventional reservoir modeling, the effective properties from core-scale flow experiments are extrapolated to the macroscopic geological model, or directly to the simulation model. Cores should therefore ideally be representative for the heterogeneous structures in a typical grid block of the geological model, but flow experiments are usually performed on relatively homogeneous cores that rarely exceed one meter in length. Flow at the core scale is also more influenced by capillary forces than flow on a reservoir scale.

As a supplement to core-flooding experiments, it has in recent years become popular to build 3D grid models to represent small-scale geological details like the bedding structure and lithology (composition and texture). One example of such a model is shown in Figure 2.7. Effective flow properties for the 3D model can now be estimated in the same way as for core plugs by replacing the flow experiment by flow simulations using rock properties that are, e.g., based on the input from microscopic models. This way, one can incorporate fine-scale geological details from lamina into the macroscopic reservoir models. However, the process of extrapolating information from cores to build a geological model is largely underdetermined and must generally be supplied with geological data from other sources.

2.4 Modeling Rock Properties

Describing the flow through a porous rock structure is largely a question of the scale of interest. The size of the rock bodies forming a typical petroleum reservoir will be from 10 to 100 meters in the vertical direction and several hundred meters or a few kilometers in the lateral direction. On this scale, it is clearly impossible to describe the storage and transport in individual pores and pore channels, as discussed in Section 2.3.3, or through individual laminae, as in Section 2.3.4. To obtain a description of the reservoir geology, one builds models that attempt to reproduce the true geological heterogeneity in the reservoir rock at the macroscopic scale by introducing macroscopic petrophysical properties based on a continuum hypothesis and volume averaging over a sufficiently large REV. These petrophysical properties are engineering quantities that are used as input to flow simulators, and are not true geological or geophysical properties of the underlying media.

A geological model is a conceptual, three-dimensional representation of a reservoir, whose main purpose is to give a spatial representation of the ability to store and transmit fluids, besides giving location and geometry of the reservoir. The rock body itself is modeled in terms of a volumetric grid, in which the layered structure of sedimentary beds and the geometry of faults and large-scale fractures in the reservoir are represented by the geometry and topology of the grid cells. Cell sizes in geocellular models can vary a lot, but typical sizes are in the range of 0.1–1 m in the vertical direction and 10–50 m in the horizontal direction. The petrophysical properties of the rock are represented as constant values inside each grid cell (porosity and permeability) or as values attached to cell faces (fault multipliers, fracture apertures, etc.). In the following sections, we will describe the main rock properties in more detail. More details about the grid modeling will follow in Chapter 3.

2.4.1 Porosity

The porosity ϕ of a porous medium is defined as the fraction of the bulk volume that represents void space, which means that $0 \leq \phi < 1$. Likewise, $1 - \phi$ is the fraction occupied by solid material (rock matrix). The void space generally consists of two parts: interconnected pore space available to fluid flow, and disconnected pores (dead ends) that are unavailable to flow. Only the first part is interesting for flow simulation, and porosity is therefore usually a shorthand for the "effective porosity" that measures the fraction of connected void space to bulk volume.

For a completely rigid medium, porosity is a static, dimensionless quantity that can be measured in the absence of flow. Porosity is mainly determined by the pore and grain-size distribution. Rocks with nonuniform grain size usually have smaller porosity than rocks with a uniform grain size, because smaller grains tend to fill pores formed by larger grains. For a bed of solid spheres of uniform diameter, the porosity depends on the packing, varying between 0.2595 for a rhomboidal packing to 0.4764 for cubic packing. When sediments are first deposited in water, they usually have a porosity of approximately 0.5, but as they are buried, the water is gradually squeezed out and the void space between the mineral particles decreases as the sediments are consolidated into rocks. For sandstone and limestone, ϕ is in the range 0.05–0.5, although values outside this range may be infrequently observed. Sandstone porosity is usually determined by the sedimentological process by which the rock was deposited, whereas carbonate porosity is mainly a result of changes taking place after deposition. Increase compaction (and cementation) causes porosity to decrease with depth in sedimentary rocks. The porosity can also be reduced by minerals that are deposited as water moves through pore spaces. For sandstone, the loss in porosity is small, whereas shales loose their porosity very quickly. Shales are therefore unlikely to be good reservoir rocks in the conventional sense, and will instead act like caprocks having porosities that are orders of magnitude lower than those found in good sandstone and carbonates.

For non rigid rocks, the porosity is usually modeled as a pressure-dependent parameter. That is, one says that the rock is *compressible*, having a rock compressibility defined by

$$c_r = \frac{1}{\phi}\frac{d\phi}{dp} = \frac{d\ln(\phi)}{dp}, \qquad (2.1)$$

where p is the overall reservoir pressure. Compressibility can be significant in some cases, e.g., as evidenced by the subsidence observed in the Ekofisk area in the North Sea. For a rock with constant compressibility, (2.1) can be integrated to give

$$\phi(p) = \phi_0 e^{c_r(p-p_0)}, \qquad (2.2)$$

and for simplified models, it is common to use a linearization so that:

$$\phi = \phi_0\bigl[1 + c_r(p - p_0)\bigr]. \qquad (2.3)$$

Because the dimension of the pores is very small compared to any interesting scale for reservoir simulation, one normally assumes that porosity is a piecewise continuous spatial function. However, ongoing research aims to understand better the relation between flow models on pore scale and on reservoir scale.

2.4.2 Permeability

Permeability is a basic flow property of a porous medium and measures its ability to transmit a single fluid when the void space is completely filled with this fluid. This means that permeability, unlike porosity, is a parameter that cannot be defined apart from fluid flow. The precise definition of the (absolute, specific, or intrinsic) permeability K is as the proportionality factor between the flow rate and an applied pressure or potential gradient $\nabla\Phi$,

$$\vec{v} = -\frac{K}{\mu}\nabla\Phi. \qquad (2.4)$$

This relationship is called Darcy's law after the french hydrologist Henry Darcy, who first observed it in 1856 while studying flow of water through beds of sand [77]. In (2.4), μ is the fluid viscosity and $\vec{v}$ is the *superficial velocity*, i.e., the flow rate divided by the cross-sectional area perpendicular to the flow. This should not be confused with the interstitial velocity $\vec{u} = \phi^{-1}\vec{v}$, i.e., the rate at which an actual fluid particle moves through the medium. We will come back to a more detailed discussion of Darcy's law in Section 4.1.

2.4 Modeling Rock Properties

The SI-unit for permeability is m², which reflects the fact that permeability is determined entirely by the geometric characteristics of the pores. However, it is more common to use the unit "darcy" (D). The precise definition of 1 D ≈ $0.987 \cdot 10^{-12}$ m² involves transmission of a 1 cP fluid through a homogeneous rock at a speed of 1 cm/s due to a pressure gradient of 1 atm/cm. Translated to reservoir conditions, 1 D is a relatively high permeability and it is therefore customary to specify permeabilities in millidarcys (md or mD). Rock formations like sandstones tend to have many large or well-connected pores and therefore transmit fluids readily. They are therefore described as permeable. Other formations, like shales, may have smaller, fewer, or less interconnected pores and are hence described as impermeable. Conventional reservoirs have permeabilities ranging from 0.1 mD to 20 D for liquid flow, and down to 10 mD for gases. In recent years, however, there has been an increasing interest in unconventional resources, that is, gas and oil locked in extraordinarily impermeable and hard rocks, with permeability values ranging from 0.1 mD and down to 1 μD or lower. *"Tight"* reservoirs are defined as those having permeability less than 0.1 mD. Compared with conventional resources, the potential volumes of tight gas, shale gas, and shale oil are enormous, but cannot be easily produced at economic rates unless stimulated, e.g., using a pressurized fluid to break up the rock to create new flow paths (hydraulic fracturing). This book focuses exclusively on simulation of conventional resources.

In (2.4), we tacitly assumed that K is a scalar quantity. However, permeability will generally be a full tensor,

$$\mathbf{K} = \begin{bmatrix} K_{xx} & K_{xy} & K_{xz} \\ K_{yx} & K_{yy} & K_{yz} \\ K_{zx} & K_{zy} & K_{zz} \end{bmatrix}. \tag{2.5}$$

Here, the diagonal terms $\{K_{xx}, K_{yy}, K_{zz}\}$ represent how the flow rate in one axial direction depends on the pressure drop in the same direction. The off-diagonal terms $\{K_{xy}, K_{xz}, K_{yx}, K_{yz}, K_{zx}, K_{zy}\}$ account for dependence between flow rate in one axial direction and the pressure drop in perpendicular directions. A full tensor is needed to model local flow in directions at an angle to the coordinate axes. Let us for instance consider a layered system, for which the dominant direction of flow will generally be along the layers. However, if the layers form an angle to the coordinate axes, a pressure drop in one coordinate direction will produce flow at an angle to this direction. This type of flow can only be modeled correctly with a permeability tensor with nonzero off-diagonal terms. If the permeability can be represented by a scalar function $K(\vec{x})$, we say that the permeability is *isotropic* as opposed to the *anisotropic* case where we need a full tensor $\mathbf{K}(\vec{x})$. To model a physical system, the anisotropic permeability tensor must be symmetric because of the Onsager principle of reciprocal relations and positive definite because the flow component parallel to the pressure drop should be in the same direction as the pressure drop. As a result, a full-tensor permeability $\mathbf{K}$ may be diagonalized by a change of basis.

Since the porous medium is formed by deposition of sediments over thousands of years, there is often a significant difference between permeability in the vertical and lateral directions, but no difference between the permeabilities in the two lateral directions. The permeability is obviously also a function of porosity. Assuming a laminar flow (low Reynolds numbers) in a set of capillary tubes, one can derive the Carman–Kozeny relation,

$$K = \frac{1}{8\tau A_v^2} \frac{\phi^3}{(1-\phi)^2}, \qquad (2.6)$$

which relates permeability to porosity ϕ, but also shows that the permeability depends on local rock texture described by tortuosity τ and specific surface area A_v. The *tortuosity* is defined as the squared ratio of the mean arc–chord length of flow paths, i.e., the ratio between the length of a flow path and the distance between its ends. The *specific surface area* is an intrinsic and characteristic property of any porous medium that measures the internal surface of the medium per unit volume. Clay minerals, for instance, have large specific surface areas and hence low permeability. The quantities τ and A_v can be calculated for simple geometries, e.g., for engineered beds of particles and fibers, but are rarely measured for reservoir rocks. Moreover, the relationship in (2.6) is highly idealized and only gives satisfactory results for media consisting of grains that are approximately spherical and have a narrow size distribution. For consolidated media and cases in which rock particles are far from spherical and have a broad size distribution, the simple Carman–Kozeny equation does not apply. Instead, permeability is typically obtained through macroscopic flow measurements.

Permeability is generally heterogeneous in space because of different sorting of particles, degree of cementation (filling of clay), and transitions between different rock formations. Indeed, the permeability may vary rapidly over several orders of magnitude; local variations in the range 1 mD to 10 D are not unusual in field and sector models. The heterogeneous structure of a porous rock formation is a result of the deposition and geological history and will therefore vary strongly from one formation to another, as we will see in a few of the examples in Section 2.5.

Production of fluids may also change the permeability. When temperature and pressure are changed, microfractures may open and significantly change the permeability. Furthermore, since the definition of permeability involves a certain fluid, different fluids will experience different permeability in the same rock sample. Such rock–fluid interactions are discussed in Chapter 8.

2.4.3 Other Parameters

Not all the bulk volume in a reservoir zone consist of reservoir rocks. To account for the fact that some portion of a cell may contain impermeable shale, it is common to introduce the so-called *net-to-gross* (N/G) property, which is a number in the range 0–1 that represents the fraction of reservoir rock in a cell. To get the effective porosity of a given cell, one must multiply the porosity and N/G value of the cell. (The N/G values also act as multipliers for lateral transmissibilities, which we will come back to later in the book.) A zero value means that the corresponding cell only contains shale (either because the porosity, the N/G value, or both are zero), and such cells are by convention typically not included in the active model.

Faults can either act as conduits for fluid flow in subsurface reservoirs or create flow barriers and introduce compartmentalization that severely affects fluid distribution and/or reduces recovery. On a reservoir scale, faults are generally volumetric objects that can be described in terms of displacement and petrophysical alteration of the surrounding host rock. However, lack of geological resolution in simulation models means that fault zones are commonly modeled as surfaces that explicitly approximate the faults' geometrical properties. To model the hydraulic properties of faults, one introduces so-called *multipliers* that alter the ability to transmit fluid between two neighboring cells. Multipliers can also model other types of subscale features that affect the communication between grid blocks, e.g., thin mud layers resulting from a flooding event, which may partially cover the sand bodies and reduce vertical communication. It is also common to (ab)use multipliers to increase or decrease the flow in certain parts of the model to calibrate simulated reservoir responses to historic data (production curves from wells, etc.). More details about multipliers will be given later in the book.

2.5 Property Modeling in MRST

All flow and transport solvers in MRST assume that rock parameters are represented as fields in a structure array (struct), which by convention is called `rock`. You do not need to follow this convention, but `rock` must contain two subfields called `poro` and `perm`. The porosity field `rock.poro` is a vector with one value for each active cell in the associated grid model. The permeability field `rock.perm` can either contain a single column for an isotropic permeability, two or three columns for a diagonal permeability (in 2D and 3D, respectively), or six columns for a symmetric, full-tensor permeability. In the latter case, cell number i has the permeability tensor

$$\mathbf{K}_i = \begin{bmatrix} K_1(i) & K_2(i) \\ K_2(i) & K_3(i) \end{bmatrix}, \qquad \mathbf{K}_i = \begin{bmatrix} K_1(i) & K_2(i) & K_3(i) \\ K_2(i) & K_4(i) & K_5(i) \\ K_3(i) & K_5(i) & K_6(i) \end{bmatrix},$$

where $K_j(i)$ is the entry in column j and row i of `rock.perm`. Full-tensor, nonsymmetric permeabilities are currently not supported. In addition to porosity and permeability, MRST supports a field called `ntg` representing the net-to-gross ratio, which either consists of a scalar or a single column with one value per active cell.

The rest of this section demonstrate how to generate and specify rock parameters in MRST and briefly outlines a few realistic models. The discussion will also expose you to a standard visualization capabilities in MRST. Scripts containing the code lines necessary to reproduce all figures can be found in the `rock` subdirectory of the `book` module.

2.5.1 Homogeneous Models

Homogeneous models are very simple to specify. To exemplify, consider a square 10×10 grid model with a uniform porosity of 0.2 and isotropic permeability equal 200 mD:

```
G = cartGrid([10 10]);
rock = makeRock(G, 200*milli*darcy, 0.2);
```

Because MRST works in SI units, it is important to convert from the field unit "darcy" to the SI unit "meters^2." Here, we did this by multiplying with `milli` and `darcy`, which are two functions that return the corresponding conversion factors. Alternatively, we could have used the conversion function `convertFrom(200, milli*darcy)`. Homogeneous, anisotropic permeability can be specified in the same way:

```
rock = makeRock(G, [100 100 10].*milli*darcy, 0.2);
```

2.5.2 Random and Lognormal Models

Given the difficulty of measuring rock properties, it is common to use geostatistical methods to make realizations of porosity and permeability. MRST contains two *very* simplified methods for generating geostatistical realizations. For more realistic geostatistics, you should use GSLIB [82] or a commercial geomodeling software.

In our first example, we will generate the porosity ϕ as a Gaussian field. To get a crude approximation to the permeability–porosity relationship, we assume that our medium is made up of uniform spherical grains of diameter $d_p = 10$ μm, for which the specific surface area is $A_v = 6/d_p$. Using the Carman–Kozeny relation (2.6), we can then calculate the isotropic permeability K from

$$K = \frac{1}{72\tau} \frac{\phi^3 d_p^2}{(1-\phi)^2},$$

where we further assume that $\tau = 0.81$. As a simple approximation to a Gaussian porosity field, we generate a field of independent normally distributed variables using MATLAB's built-in `rand` function, convolve it with a Gaussian kernel, and scale the result to the interval $[0.2, 0.4]$:

```
G = cartGrid([50 20]);
p = gaussianField(G.cartDims, [0.2 0.4], [11 3], 2.5);
K = p.^3.*(1e-5)^2./(0.81*72*(1-p).^2);
rock = makeRock(G, K(:), p(:));
```

The left plot of Figure 2.10 shows the resulting porosity field, whereas the right plot shows the permeability obtained for a 3D realization with $50 \times 20 \times 10$ cells generated in the same way.

2.5 Property Modeling in MRST

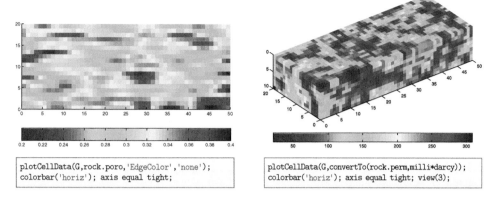

Figure 2.10 The left plot shows a 50 × 20 porosity field generated as a Gaussian field with a larger filter size in x-direction than in the y-direction. The right plot shows the permeability field computed from the Carman–Kozeny relation for a similar 50 × 20 × 10 porosity realization computed with filter size [3, 3, 3].

In the second example, we use the same methodology to generate layered realizations, defined so that the permeability is lognormally distributed in each geological layer and independent of the other layers. Each layer can be represented by several grid cells in the vertical direction. Rather than using a simple Cartesian grid, we generate a stratigraphic grid with wavy geological faces and a single fault. Chapter 3 describes this type of industry-standard grids in more detail.

```
G = processGRDECL(simpleGrdecl([50 30 10], 0.12));
K = logNormLayers(G.cartDims, [100 400 50 350], 'indices', [1 2 5 7 11]);
```

Here, we have specified four geological layers of different thickness. From top to bottom, which is the default numbering in stratigraphic grids, the first layer is one cell thick and has a mean permeability value of 100 mD, the second layer is three cells thick and has mean permeability of 400 mD, the third layer is two cells thick and has mean value 50 mD, and the fourth layer is four cells thick and has mean value 350 mD. To specify this, we use an indirection map; that is, if Km is the n-vector of mean permeabilities and L is the $(n+1)$-vector of indices, the value Km(i) is assigned to vertical layers number L(i) to L(i+1)-1. Figure 2.11 shows the resulting permeability field.

2.5.3 The 10th SPE Comparative Solution Project: Model 2

The Society of Petroleum Engineers (SPE) has developed a series of benchmarks for comparing computational methods and simulators. The first nine benchmarks focus on black-oil, compositional, dual-porosity, thermal, and miscible simulations, as well as horizontal wells and gridding techniques. The 10th SPE Comparative Solution Project [71]

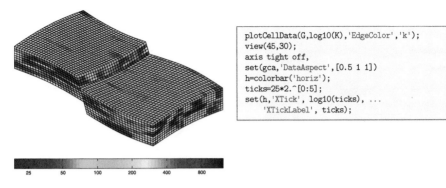

Figure 2.11 A stratigraphic grid with a single fault and four geological layers, each with a lognormal permeability distribution.

was posed as a benchmark for upscaling methods, but the second data set of this benchmark has later become very popular as a general test case within the academic community. The data set is a 3D geostatistical realization from the Jurassic Upper Brent formations from the North Sea, which, e.g., contains the giant fields of Statfjord, Gullfaks, Oseberg, and Snorre. The main features of the model are the permeability and porosity fields given on a $60 \times 220 \times 85$ Cartesian grid, in which each cells is of size 20 ft $\times$ 10 ft $\times$ 2 ft. In this specific model, the top 35 cell layers having a total height of 70 ft represent the shallow-marine Tarbert formation, and the lower 50 layers having a height of 100 ft represent the fluvial Ness formation. The original geostatistical model was developed in the PUNQ project [111], and later the horizontal dimensions were scaled by a factor 1/3 to make it more heterogeneous. The model is structurally simple but is highly heterogeneous, and, for this reason, some describe it as a "simulator-killer." On the other hand, the fact that the flow is dictated by the strong heterogeneity means that streamline methods will be particularly efficient for this model [4].

The SPE 10 data set is used in a large number of publications and is publicly available from the SPE website (http://www.spe.org/web/csp/). MRST supplies a module called spe10 that downloads, reorganizes, and stores the data set in *.mat files for later use. The module also contains routines that extract (subsets of) the petrophysical data and set up simulation models and appropriate data structures representing grids, petrophysics, and wells. Alternatively, the data set can be downloaded using mrstDatasetGUI. We start by loading the petrophysical data

```
% load SPE 10 data set
mrstModule add spe10;
rock = getSPE10rock(); p=rock.poro; K=rock.perm;
```

Because of the simple Cartesian grid topology, we can use standard MATLAB functionality to visualize the heterogeneity in the permeability and porosity (full details can be found in the script rocks/showSPE10.m):

2.5 Property Modeling in MRST

```
slice( reshape(p,60,220,85), [1 220], 60, [1 85]);
shading flat, axis equal off, set(gca,'zdir','reverse'), box on;
colorbar('horiz');

% show Kx
slice( reshape(log10(K(:,1)),60,220,85), [1 220], 60, [1 85]);
shading flat, axis equal off, set(gca,'zdir','reverse'), box on;
h=colorbar('horiz');
set(h,'XTickLabel',10.^[get(h,'XTick')]);
set(h,'YTick',mean(get(h,'YLim')),'YTickLabel','mD');
```

Figure 2.12 shows porosity and permeability; the permeability tensor is diagonal with equal permeability in the two horizontal coordinate directions. Both formations are characterized by large permeability variations, 8–12 orders of magnitude, but are qualitatively different. The Tarbert consists of sandstone, siltstone, and shales, and comes from a tidally influenced, transgressive, shallow-marine deposit; in other words, a deposit that has taken place close to the coastline; see Figure 2.1. The formation has good communication in the vertical and horizontal directions. The fluvial Ness formation has been deposited by rivers or running water in a delta-plain, continental environment (see Figures 2.1 and 2.2), leading to a spaghetti of well-sorted, high-permeable sandstone channels with good communication

Figure 2.12 Rock properties for the SPE 10 model. The upper plot shows the porosity, the lower left the horizontal permeability, and the lower right the vertical permeability. (The permeabilities are shown using a logarithmic color scale.)

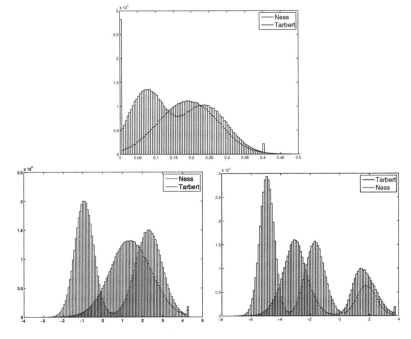

Figure 2.13 Histogram of rock properties for the SPE 10 model: ϕ (upper plot), $\log \mathbf{K}_x$ (lower left), and $\log \mathbf{K}_z$ (lower right) The Tarbert formation is shown in blue and the Ness formation in red.

(long correlation lengths) imposed on a low-permeable background of shales and coal, which gives low communication between different sand bodies. The porosity field has a large span of values, and approximately 2.5% of the cells have zero porosity and should be considered as being inactive.

Figure 2.13 shows histograms of the porosity and the logarithm of the horizontal and vertical permeabilities. The nonzero porosity values and the horizontal permeability of the Tarbert formation appear to follow a normal and lognormal distribution, respectively. The vertical permeability follows a bimodal distribution. For the Ness formation, the nonzero porosities and the horizontal permeability follow bi-modal normal and lognormal distributions, respectively, as is to be expected for a fluvial formation. The vertical permeability is trimodal. To five significant digits, the ratio of vertical to horizontal permeability has two unique values 10^{-4} and 0.3, which together with the K index can be used to distinguish the three different rock types (facies) in the model: sheet sand and mudstone in the Tarbert formation, and channel sand and mudstone in Upper Ness.

2.5.4 The Johansen Formation

The Johansen formation is located in the deeper part of the Sognefjord delta, 40–90 km offshore Mongstad on the west coast of Norway. Some years ago, a gas power plant with carbon capture and storage was planned at Mongstad, and the water-bearing Johansen formation was a possible candidate for storing the captured CO_2. The Johansen formation is part of the Dunlin group, and is interpreted as a large sandstone delta 2,200–3,100 m below sea level, which is limited above by the Dunlin shale and below by the Amundsen shale. The average thickness of the formation is roughly 100 m, and the aquifer extends laterally up to 100 km in the north–south direction and 60 km in the east–west direction. The aquifer has good sand quality and lies at a depth where CO_2 would undoubtedly be in supercritical phase, and would thus be ideal for carbon storage. With average porosities of approximately 25%, this implies that the theoretical storage capacity of the Johansen formation is more than one gigatonne of CO_2 [97]. The Troll field, one of the largest gas fields in the North Sea, is located some 500 m above the northwestern parts of the Johansen formation.

A previous research project generated several models of the Johansen formation using so-called corner-point grids. We will discuss this format in more detail in Section 3.3.1. For now, it is sufficient to note that corner-point grids consist of (distorted) hexahedral cells organized according to an underlying Cartesian topology. Altogether, there are five models: one full-field model ($149 \times 189 \times 16$ grid), three homogeneous sector models ($100 \times 100 \times n$ for $n = 11, 16, 21$), and one heterogeneous sector model ($100 \times 100 \times 11$). All models are publicly available on the MRST website and can, for instance, be downloaded to MRST using the `mrstDatasetGUI` function. Herein, we consider the heterogeneous sector model. You can find all statements used in the following analysis in the script `rocks/showJohansenNPD.m`.

We start by reading the grid geometry from an industry-standard input format, which we discuss in more detail in the next section, and then construct a corresponding MRST grid structure. The rock properties are given as plain ASCII files that each contains 110,000 values. Some of these values correspond to logical cell indices that are not part of the actual simulation model and can be discarded if we use the indirection map `G.cells.indexMap` to extract the correct subset:

```
G = processGRDECL(readGRDECL('NPD5.grdecl'));
p = load('NPD5_Porosity.txt')'; p = p(G.cells.indexMap);
K = load('NPD5_Permeability.txt')'; K=K(G.cells.indexMap);
```

In the model, the Johansen formation is represented by five layers of cells, whereas the low-permeable Dunlin shale above and the Amundsen shale below are represented by five and one cell layers. The Johansen formation consists of approximately 80% sandstone and 20% claystone, whereas the Amundsen formation consists of siltstones and shales; see [97] for more details. In the left plot of Figure 2.14, the Johansen sand is clearly distinguished as a wedge shape that is pinched out in the front part of the model and splits the shales laterally in two at the back. The right plot shows the good reservoir rocks distinguished as cells with a porosity value larger than 0.1. The permeability tensor is assumed to be diagonal, with the vertical permeability equaling one-tenth of the horizontal permeability. Hence, only the x-component $\mathbf{K}_x$ is given in the data file. Figure 2.15 shows three different plots of the

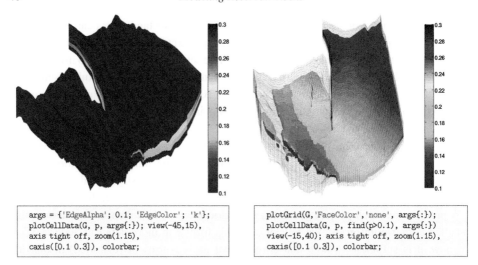

Figure 2.14 Porosity for the Johansen data set "NPD5." The left plot shows porosity for the whole model, whereas in the right plot we have masked the low-porosity cells in the Amundsen and Dunlin formations.

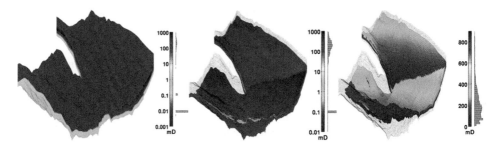

Figure 2.15 Permeability for the Johansen data set "NPD5." The left plot shows permeability for the whole model, the middle plot shows the Johansen sand and Amundsen shale, whereas the right plot only shows permeability of the Johansen sand.

permeability. The first plot presents the whole permeability field with a logarithmic color scale. In the second plot, we have filtered out the Dunlin shale above Johansen but not the Amundsen shale below. The third plot depicts the permeability in the Johansen formation using a linear color scale, which clearly shows the depth trend that was used to model the heterogeneity.

2.5.5 SAIGUP: Shallow-Marine Reservoirs

Most commercial simulators use a combination of an *input language* and a set of data files to describe and set up a simulation model of a reservoir. In this book, we only discuss the

ECLIPSE input format, which has emerged as an industry standard for describing static and dynamic properties of a reservoir system, from the reservoir rock, via production and injection wells and up to connected top-side facilities. ECLIPSE input decks use keywords to signify and separate the different data elements that comprise a full model. These keywords define a detailed language you can use to specify how the data elements should be put together and modify each other to form a full spatio-temporal model of a reservoir. In the most general form, an ECLIPSE input deck consists of eight sections of keywords that must come in a prescribed order. However, some of the sections are optional and may not always be present. The order of the keywords within each section is arbitrary, except in the section that defines wells and gives operating schedule, etc. Altogether, the ECLIPSE format consists of thousands of keywords, and describing them all is far beyond the scope of this book. Other simulators use a somewhat different syntax, but the principles of the overall input description is often similar.

In the following, we briefly outline some of the most common keywords from the GRID section that describes the reservoir geometry and petrophysical properties. To provide you with a basic understanding of the required input for simulating real-life reservoir models, we focus on the model ingredients and not on the specific syntax. For brevity, we also do not go through all MATLAB and MRST statements used to visualize the different data elements. All the necessary details can be found in the script rocks/show SAIGUP.m.

As our example of a realistic petroleum reservoir, we consider a model from the SAIGUP study [203], whose purpose was to conduct a sensitivity analysis of how geological uncertainties impacts production forecasting in clastic reservoirs. To this end, the study generated a large suite of geostatistical realizations and structural models to represent a wide span of shallow-marine sedimentological reservoirs. The SAIGUP study focused on shoreface reservoirs in which the deposition of sediments is caused by variation in sea level, so that facies form belts in a systematic pattern (river deposits create curved facies belts, wave deposits create parallel belts, etc.). Sediments are in general deposited when the sea level is increasing. No sediments are deposited during decreasing sea levels; instead, the receding sea may affect the appearing shoreline and cause the creation of a barrier. All models are synthetic, but contain representative examples of the complexities seen in real-life reservoirs.

One of the many SAIGUP realizations is publicly available from the MRST website and comes in the form of a GZip-compressed TAR file (SAIGUP.tar.gz) that contains the structural model as well as petrophysical parameters. You can download the data set using the mrstDatasetGUI function. Here, however, we unpack the data set manually for completeness of presentation. Assuming that we have already downloaded the archive file SAIGUP.tar.gz with the model realization, we extract the data files and place them in a standardized path relative to the root directory of MRST:

```
untar('SAIGUP.tar.gz', fullfile(ROOTDIR, 'examples', 'data', 'SAIGUP'))
```

This will create a new directory containing seventeen data files that comprise the structural model, various petrophysical parameters, etc.:

```
028_A11.EDITNNC       028.MULTX  028.PERMX  028.SATNUM     SAIGUP.GRDECL
028_A11.EDITNNC.001   028.MULTY  028.PERMY  SAIGUP_A1.ZCORN
028_A11.TRANX         028.MULTZ  028.PERMZ  SAIGUP.ACTNUM
028_A11.TRANY         028.NTG    028.PORO   SAIGUP.COORD
```

The main file is SAIGUP.GRDECL, which lists the sequence of keywords that specifies how the data elements found in the other files should be put together to make a complete model of the reservoir rock. The remaining files represent different keywords: the grid geometry is given in files SAIGUP_A1.ZCORN and SAIGUP.COORD, the porosity in 028.PORO, the permeability tensor in the three 028.PERM* files, net-to-gross properties in 028.NTG, the list of active cells in SAIGUP.ACTNUM, transmissibility multipliers that modify the flow connections between different cells are given in 028.MULT*, etc. For now, we rely entirely on MRST's routines for reading ECLIPSE input files; more details about corner-point grids and the ECLIPSE input format will follow later in the book, starting in Chapter 3.

The SAIGUP.GRDECL file contains seven of the eight possible sections of a full input deck. The deckformat module in MRST contains a comprehensive set of input routines that enable you to read the most important keywords and options supported in these sections. For the SAIGUP model, it is mainly the sections describing static reservoir properties that contain complete and useful information, and we will therefore use the much simpler function readGRDECL from MRST core to read and interpret the GRID section of the input deck:

```
grdecl = readGRDECL(fullfile(ROOTDIR, 'examples', ...
    'data', 'SAIGUP','SAIGUP.GRDECL'));
```

This statement parses the input file and stores the content of all keywords it recognizes in the structure grdecl:

```
grdecl =
    cartDims: [40 120 20]
       COORD: [29766x1 double]
       ZCORN: [768000x1 double]
      ACTNUM: [96000x1 int32]
       PERMX: [96000x1 double]
       PERMY: [96000x1 double]
       PERMZ: [96000x1 double]
       MULTX: [96000x1 double]
       MULTY: [96000x1 double]
       MULTZ: [96000x1 double]
        PORO: [96000x1 double]
         NTG: [96000x1 double]
      SATNUM: [96000x1 double]
```

The first four data fields describe the grid, and we will come back to these in Section 3.3.1. In the following, we discuss the next eight data fields, which contain the petrophysical parameters. We also look briefly at the last data field, which delineates the reservoir into different (user-defined) rock types that can be used to associate different rock-fluid properties.

2.5 Property Modeling in MRST

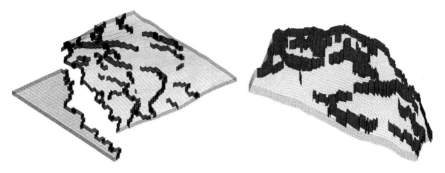

Figure 2.16 The structural SAIGUP model. The left plot shows the full model with faults marked in red and inactive cells marked in yellow, whereas the right plot shows only the active parts of the model seen from the opposite direction.

Recall that MRST uses strict SI conventions in all of its internal calculations. The SAIGUP model follows the ECLIPSE "METRIC" conventions (permeabilities in mD and so on). We thus use functions getUnitSystem and convertInputUnits to convert the input data to MRST's internal unit conventions.

```
usys   = getUnitSystem('METRIC');
grdecl = convertInputUnits(grdecl, usys);
```

Having converted the units properly, we generate a space-filling grid and extract the petrophysical properties:

```
G = processGRDECL(grdecl);
G = computeGeometry(G);
rock = grdecl2Rock(grdecl, G.cells.indexMap);
```

The first statement takes the description of the grid geometry and constructs an unstructured MRST grid represented with the data structure outlined in Section 3.4. The second statement computes a few geometric primitives like cell volumes, centroids, etc., as discussed on page 93. The third statement constructs a rock object containing porosity, permeability, and net-to-gross.

For completeness, we first show a bit more details of the structural model in Figure 2.16. The left plot shows the whole $40 \times 120 \times 20$ grid model,[1] where we in particular should note the disconnected cells marked in yellow that are not part of the active model. The large fault throw that disconnects the two parts is most likely a modeling artifact introduced to clearly distinguish the active and inactive parts of the model. A shoreface reservoir is bounded by faults and geological horizons, but faults also appear inside the reservoir, as the right plot in Figure 2.16 shows. Faults and barriers will typically have a pronounced effect on the

---

[1] To not confuse the reader, I emphasize that only the active part of the model is read with the processGRDECL routine. How to also include the inactive part, will be explained in more detail in Chapter 3.

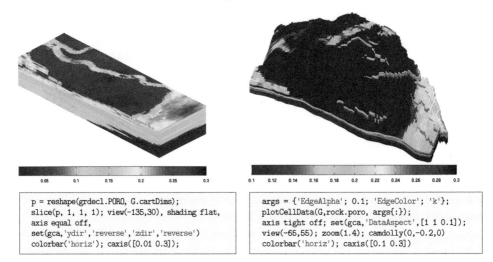

Figure 2.17 Porosity for the SAIGUP model. The left plot shows porosity as generated by geostatistics in logical ijk space. The right plot shows the porosity mapped to the structural model shown in Figure 2.16.

flow pattern, and having an accurate representation is important to produce reliable flow predictions.

The petrophysical parameters for the model were generated by a geostatistical algorithm on a regular $40 \times 120 \times 20$ Cartesian grid, as illustrated in the left plot of Figure 2.17, and then mapped onto the structural model, as shown in the plot to the right. A bit simplified, one can view the Cartesian grid model as representing the rock body at geological "time zero," when the sediments have been deposited and have formed a stack of horizontal grid layers. From geological time zero and up to now, geological activity has introduced faults and deformed the layers, resulting in the structural model seen in the left plot of Figure 2.17.

Having seen the structural model, we continue to study the petrophysical parameters. The grid cells are thought to be larger than the laminae of our imaginary reservoir, hence each grid block will generally contain both reservoir rock (with sufficient permeability) and impermeable shale. This is modeled using the net-to-gross ratio, rock.ntg, shown in Figure 2.18 along with the horizontal and vertical permeability. The plotting routines are exactly the same as for the porosity in Figure 2.17, but with different data and slightly different specification of the colorbar. From the figure, we clearly see that the model has a large content of shale and thus low permeability along the top, but we also see high-permeable sand bodies that cut through the low-permeable top. In general, the permeabilities seem to correlate well with the sand content given by the net-to-gross parameter.

Some parts of the sand bodies are partially covered by mud that strongly reduces the vertical communication, most likely because of flooding events. These mud-draped surfaces occur on a sub-grid scale and are modeled through a multiplier value (MULTZ) associated

2.5 Property Modeling in MRST

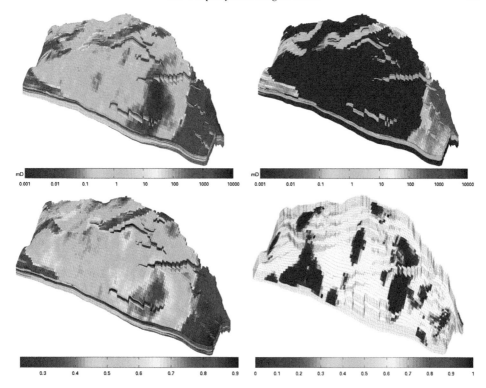

Figure 2.18 The upper plots show the permeability for the SAIGUP model, using a logarithmic color scale, with horizontal permeability to the left and vertical permeability to the right. The lower-left plot shows net-to-gross, i.e., the fraction of reservoir rock per bulk volume. The lower-right plot shows regions of the reservoir where reduced vertical communication is modeled by vertical multiplier values less than unity.

with each cell, which takes values between zero and one and can be used to manipulate the effective communication (the transmissibility) between a given cell (i, j, k) and the cell immediately above $(i, j, k + 1)$. For completeness, we remark that the horizontal multiplier values (MULTX and MULTY) play a similar role for vertical faces, but are equal one in (almost) all cells for this particular realization.

To further investigate the heterogeneity of the model, we next look at histograms of the porosity and the permeabilities, as we did for the SPE 10 example (the MATLAB statements are omitted since they are almost identical). In Figure 2.19, we clearly see that the distributions of porosity and horizontal permeability are multi-modal in the sense that five different modes can be distinguished, corresponding to the five different facies used in the petrophysical modeling.

It is common modeling practice that different rock types are assigned different rock-fluid properties (relative permeability and capillary functions), more details about such properties will be given later in the book. Different rock types are represented using the

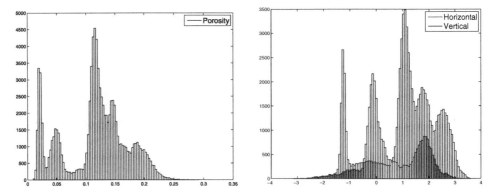

Figure 2.19 Histogram of the porosity (left) and the logarithm of the horizontal and vertical permeability (right) for the shallow-marine SAIGUP model. Since the reservoir contains five different facies, the histograms are multimodal. See also Figure 2.20.

SATNUM keyword, and by inspection of the SATNUM field in the input data, we see that the model contains six different rock types as depicted in Figure 2.20. For completeness, the figure also shows the permeability distribution inside each rock type. Interestingly, the permeability distribution is multimodal for at least two of the rock types.

Finally, to demonstrate the large difference in heterogeneity resulting from different depositional environments, we compare the realization studied so far in this example with another realization; Figure 2.21 shows porosities and rock-type distribution. Whereas the original realization corresponds to a depositional environment with a flat shoreline, the other realization has a two-lobed shoreline, giving distinctively different facies belts. The figure also clearly demonstrates how the porosity (which depends on the grain-size distribution and packing) varies with the rock types. This can be confirmed by a quick analysis:

```
for i=1:6,  pavg(i) = mean(rock.poro(SN==i));
            navg(i) = mean(rock.ntg(SN==i)); end
```

```
pavg =  0.0615    0.1883    0.1462    0.1145    0.0237    0.1924
navg =  0.5555    0.8421    0.7554    0.6179    0.3888    0.7793
```

In other words, rock types two and six are good sands with high porosity, three and four have intermediate porosity, whereas one and five correspond to less quality sand with a high clay content and hence low porosity.

COMPUTER EXERCISES

2.5.1 Look at the correlation between the porosity and the permeability for the SPE 10 data set. Do you see any artifacts, and if so, how would you explain them? (Hint: plot ϕ versus $\log K$.)

2.5 Property Modeling in MRST

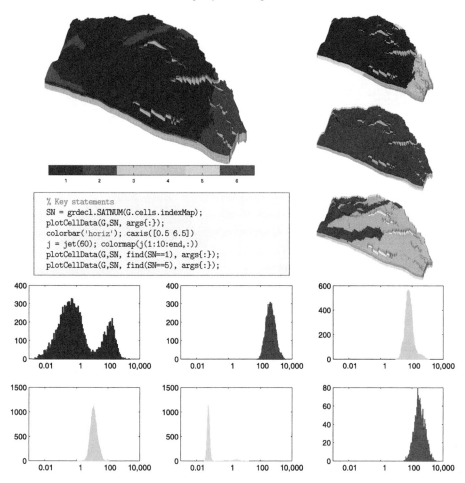

Figure 2.20 The upper-left plot shows the rock type distribution for SAIGUP. The right column shows the six rock types grouped in pairs; from top to bottom, rock types number 1 and 5, 2 and 4, and 3 and 6. The bottom part of the figure shows histograms of the lateral permeability in units [mD] for each of the six rock types found in the model.

2.5.2 Download the `CaseB4` models that represent a sector model with intersecting faults. Pick at least one model realization and set homogeneous and random petrophysical data as discussed in Sections 2.5.1 and 2.5.2.

2.5.3 The $58 \times 48 \times 10$ permeability field given in `rock1.mat` in the book module contains an unusual geological structure. Can you find what it is? (Hint: if you want to explore the model interactively, you can try to use `plotToolbar` from the `mrst-gui` module).

2.5.4 Download the `BedModels1` and `BedModel2` data sets that represent sedimentary beds similar to the facies model shown in Figure 2.7. Use the techniques introduced in Sections 2.5.3–2.5.5 to familiarize yourself with these models:

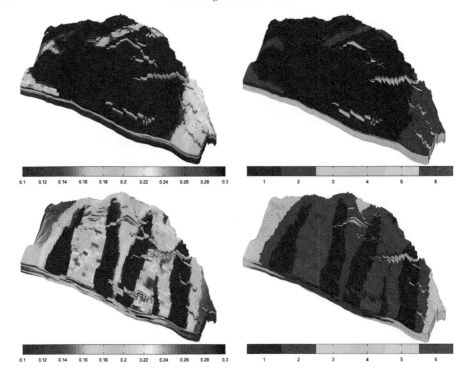

Figure 2.21 Comparison of porosity (left) and the distribution of rock types (right) for two different SAIGUP realizations.

- look at porosities and permeabilities in physical space
- compare with the same quantities in ijk space
- find models that have facies information and look at the distribution of petrophysical properties inside each facies

2.5.5 Modify the `simpleGravityColumn` example from Section 1.4 so that it uses the geometry and petrophysical data in the `mortarTestModel` or `periodic Tilted` models from the `BedModels1` data set instead. Can you explain what you observe?

3
Grids in Subsurface Modeling

The basic geological description of a petroleum reservoir or an aquifer system will typically consist of two sets of surfaces. Geological horizons are lateral surfaces describing the bedding planes that delimit the rock strata, whereas faults are vertical or inclined surfaces along which the strata may have been displaced by geological processes. In this chapter, we discuss how to turn the basic geological description into a discrete model that can be used to formulate various computational methods, e.g., for solving the equations that describe fluid flow.

A *grid* is a tessellation of a planar or volumetric object by a set of contiguous simple shapes referred to as *cells*. Grids can be described and distinguished by their *geometry*, which gives the shape of the cells that form the grid, and their *topology* that tells how the individual cells are connected. In 2D, a cell is in general a closed polygon for which the geometry is defined by a set of *vertices* and a set of *edges* that connect pairs of vertices and define the interface between two neighboring cells. In 3D, a cell is a closed polyhedron for which the geometry is defined by a set of vertices, a set of edges that connect pairs of vertices, and a set of *faces* (surfaces delimited by a subset of the edges) that define the interface between two different cells; see Figure 3.1. Herein, we assume that all cells are nonoverlapping, so that each point in the planar/volumetric object represented by the grid is either inside a single cell, lies on an interface or edge, or is a vertex. Two cells that share a common face are said to be connected. Likewise, one can also define connections based on edges and vertices. The topology of a grid is defined by the total set of connections, which is sometimes also called the *connectivity* of the grid.

When implementing grids in modeling software, one always has the choice between generality and efficiency. To represent an arbitrary grid, it is necessary to explicitly store the geometry of each cell in terms of vertices, edges, and faces, as well as storing the connectivity among cells, faces, edges, and vertices. However, as we will see later, huge simplifications can be made for particular classes of grids by exploiting regularity in the geometry and structures in the topology. Consider, for instance, a planar grid consisting of rectangular cells of equal size. Here, the topology can be represented by two indices, and one only needs to specify a reference point and the two side lengths of the rectangle to describe the geometry. This way, one ensures minimal memory usage and optimal efficiency when accessing the grid. On the other hand, exploiting the simplified description

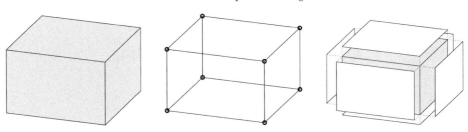

Figure 3.1 Illustration of a single cell, vertices and edges, and cell faces.

explicitly in your flow or transport solver inevitably means that the solver must be reimplemented if you later decide to use another grid format.

The most important goal for our development of MRST is to provide a toolbox that both allows *and* enables the use of various grid types. To avoid having a large number of different, and potentially incompatible, grid representations, we have therefore chosen to store *all grid types* using a general unstructured format in which cells, faces, vertices, and connections between cells and faces are explicitly represented. This means that we, for the sake of generality, have sacrificed some of the efficiency one can obtain by exploiting special structures in a particular grid type, and instead have focused on obtaining a flexible grid description that is not overly inefficient. Moreover, our grid structure can be extended by other properties that are required by various discretization schemes for flow and transport simulations. A particular discretization may need the *volume* or the *centroid* (grid-point, midpoint, or generating point) of each cell. Likewise, for cell faces one may need to know the face areas, the face normals, and the face centroids. Although these properties can be computed from the geometry (and topology) of the grid, it is often useful to precompute and include them explicitly in the grid representation.

The first third of this chapter is devoted to standard grid formats that are available in MRST. We introduce examples of structured grids, including regular Cartesian, rectilinear, and curvilinear grids, and briefly discuss unstructured grids, including Delaunay triangulations and Voronoi grids. The purpose of our discussion is to demonstrate the basic grid functionality in MRST and show some key principles you can use to implement new structured and unstructured grid formats. In the second part of the chapter, we discuss industry-standard formats for stratigraphic grids that are based on extrusion of 2D shapes (corner-point, prismatic, and 2.5D PEBI grids). Although these grids have an inherent logical structure, representation of faults, erosion, pinch-outs, and so on lead to cells that can have quite irregular shapes and an (almost) arbitrary number of faces. In the last part of the chapter, we discuss how the grids introduced in the first two parts of the chapter can be partitioned to form flexible coarse descriptions that preserve the geometry of the underlying fine grids. The ability to represent a wide range of grids, structured or unstructured on the fine and/or coarse scale, is a strength of MRST compared to the majority of research codes arising from academic institutions.

A number of videos that complement the material presented in this chapter can be found in the second MRST Jolt [189]. This Jolt introduces different types of grids, discusses how such grids can be represented, and outlines functionality you can use to generate your own grids.

3.1 Structured Grids

As we saw in this chapter's introduction, a grid is a tessellation of a planar or volumetric object by a set of simple shapes. In a *structured* grid, only one basic shape is allowed and this basic shape is laid out in a regular repeating pattern so that the topology of the grid is constant in space. The most typical structured grids are based on quadrilaterals in 2D and hexahedrons in 3D, but in principle it is also possible to construct grids with a fixed topology using certain other shapes. Structured grids can be generalized to so-called multiblock grids (or hybrid grids), in which each block consists of basic shapes that are laid out in a regular repeating pattern.

Regular Cartesian Grids

The simplest form of a structured grid consists of unit squares in 2D and unit cubes in 3D, so that all vertices in the grid are integer points. More generally, a regular Cartesian grid can be defined as consisting of congruent rectangles in 2D and rectilinear parallelepipeds in 3D, etc. Hence, the vertices have coordinates $(i_1 \Delta x_1, i_2 \Delta x_2, \dots)$ and the cells can be referenced using the multi-index $(i_1, i_2, \dots)$. Herein, we only consider finite Cartesian grids that consist of a finite number $n_2 \times n_2 \times \cdots \times n_k$ of cells that cover a bounded domain $[0, L_1] \times [0, L_2] \times \cdots \times [0, L_k]$.

Regular Cartesian grids can be represented very compactly by storing n_i and L_i for each dimension. In MRST, however, Cartesian grids are represented as if they were fully unstructured using a general grid structure that will be described in more detail in Section 3.4. Cartesian grids therefore have special constructors,

```
G = cartGrid([nx, ny], [Lx Ly]);
G = cartGrid([nx, ny, nz], [Lx Ly Lz]);
```

that set up the data structures representing the basic geometry and topology of the grid. The second argument is optional.

Rectilinear Grids

A rectilinear grid (also called a tensor grid) consists of rectilinear shapes (rectangles or parallelepipeds) that are not necessarily congruent to each other. In other words, whereas a regular Cartesian grid has a uniform spacing between its vertices, the grid spacing can vary along the coordinate directions in a rectilinear grid. The cells can still be referenced using a multi-index $(i_1, i_2, \dots)$, but the mapping from indices to vertex coordinates is nonuniform.

In MRST, one can construct a rectilinear grid by specifying vectors with the grid vertices along the coordinate directions:

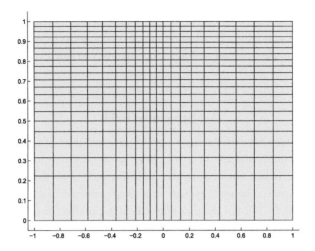

Figure 3.2 Example of a rectilinear grid.

```
G = tensorGrid(x, y);
G = tensorGrid(x, y, z);
```

This syntax is the same as for the MATLAB functions meshgrid and ndgrid.

As an example of a rectilinear grid, we construct a 2D grid that covers the domain $[-1,1] \times [0,1]$ and is graded toward $x = 0$ and $y = 1$ as shown in Figure 3.2.

```
dx = 1-0.5*cos((-1:0.1:1)*pi);
x = -1.15+0.1*cumsum(dx);
y = 0:0.05:1;
G = tensorGrid(x, sqrt(y));
plotGrid(G); axis([-1.05 1.05 -0.05 1.05]);
```

Curvilinear Grids

A curvilinear grid is a grid with the same topological structure as a regular Cartesian grid, but in which the cells are quadrilaterals rather than rectangles in 2D and cuboids rather than parallelepipeds in 3D. The grid is given by the coordinates of the vertices, but there exists a mapping that will transform the curvilinear grid to a uniform Cartesian grid so that each cell can still be referenced using a multi-index $(i_1, i_2, \dots)$.

For the time being, MRST has no constructor for curvilinear grids. Instead, you can create such grids by first instantiating a regular Cartesian grid or a rectilinear grid and then manipulating the vertices. This method is quite simple as long as there is a one-to-one mapping between the curvilinear grid in physical space and the logically Cartesian grid in reference space. The method will *not* work if the mapping is not one-to-one so that vertices with different indices coincide in physical space. In this case, the user should create an

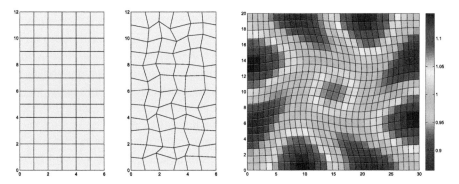

Figure 3.3 The middle plot shows a rough grid created by perturbing all internal nodes of the regular 6 × 12 Cartesian grid in the left plot. The right plot shows a curvilinear grid created using the function `twister` that uses a combination of sin functions to perturb a rectilinear grid. The color is determined by the cell volumes.

ECLIPSE input file [270, 33] with keywords COORD[XYZ] (see Section 3.3.1) and use the function `buildCoordGrid` to create the grid.

To illustrate the discussion, we show two examples of how to create curvilinear grids. In the first example, we create a rough grid by perturbing all internal nodes of a regular Cartesian grid (see Figure 3.3):

```
nx = 6; ny=12;
G = cartGrid([nx, ny]);
subplot(1,2,1); plotGrid(G);
c = G.nodes.coords;
I = any(c==0,2) | any(c(:,1)==nx,2) | any(c(:,2)==ny,2);
G.nodes.coords(~I,:) = c(~I,:) + 0.6*rand(sum(~I),2)-0.3;
subplot(1,2,2); plotGrid(G);
```

In the second example, we use the routine `twister` to perturb the internal vertices. The function maps the grid back to the unit square, perturbs the vertices according to the mapping

$$(x_i, y_i) \mapsto \left(x_i + f(x_i, y_i), y_i - f(x_i, y_i)\right), \quad f(x,y) = 0.03 \sin(\pi x) \sin\left(3\pi(y - \tfrac{1}{2})\right),$$

and then maps the grid back to its original domain. The right plot of Figure 3.3 shows the resulting grid. To illuminate the effect of the mapping, we have colored the cells according to their volume, which has been computed using the function `computeGeometry`, which we will come back to later in the chapter.

```
G = cartGrid([30, 20]);
G.nodes.coords = twister(G.nodes.coords);
G = computeGeometry(G);
plotCellData(G, G.cells.volumes, 'EdgeColor', 'k'), colorbar
```

Fictitious Domains

One obvious drawback with Cartesian and rectilinear grids, as just defined, is that they can only represent rectangular domains in 2D and cubic domains in 3D. Curvilinear grids, on the other hand, can represent more general shapes by introducing an appropriate mapping, and can be used in combination with rectangular/cubic grids in multiblock grids for efficient representation of realistic reservoir geometries. However, finding a mapping that conforms to a given boundary is often difficult, in particular for complex geologies, and using a mapping in the interior of the domain will inadvertently lead to cells with rough geometries that deviate far from being rectilinear. Such cells may in turn introduce problems if the grid is to be used in a subsequent numerical discretization, as we will see later.

As an alternative, complex geometries can be easily modeled using structured grids by a so-called fictitious domain method. In this method, the complex domain is embedded into a larger "fictitious" domain of simple shape (a rectangle or cube) using, e.g., a Boolean indicator value in each cell to tell whether the cell is part of the domain or not. The observant reader will notice that we already have encountered the use of this technique for the SAIGUP dataset (Figure 2.16) and the Johansen dataset in Chapter 2. In some cases, one can also adapt the structured grid by moving the nearest vertices to the domain boundary.

MRST has support for fictitious domain methods through the function removeCells, which we will demonstrate in the next example, where we create a regular Cartesian grid that fills the volume of an ellipsoid:

```
x = linspace(-2,2,21);
G = tensorGrid(x,x,x);
subplot(1,2,1); plotGrid(G);view(3); axis equal

subplot(1,2,2); plotGrid(G,'FaceColor','none');
G = computeGeometry(G);
c = G.cells.centroids;
r = c(:,1).^2 + 0.25*c(:,2).^2+0.25*c(:,3).^2;
G = removeCells(G, r>1);
plotGrid(G); view(-70,70); axis equal;
```

Worth observing here is the use of computeGeometry to compute cell centroids that are not part of the basic geometry representation in MRST. Figure 3.4 shows plots of the grid before and after removing the inactive parts. Because of the fully unstructured representation used in MRST, calling computeGeometry actually removes the inactive cells from the grid structure, but from the outside, the structure behaves as if we had used a fictitious domain method.

You can find more examples of how you can make structured grids and populate them with petrophysical properties in the fourth video of the second MRST Jolt [189].

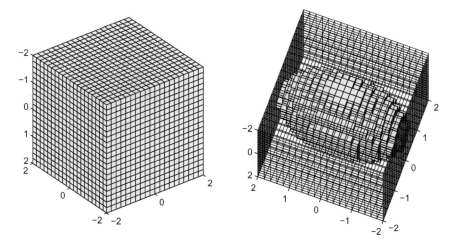

Figure 3.4 Example of a regular Cartesian grid representing a domain in the form of an ellipsoid. The underlying logical Cartesian grid is shown in the left plot and as a wireframe in the right plot. The active part of the model is shown in yellow color in the right plot.

COMPUTER EXERCISES

3.1.1 Make the grid shown:

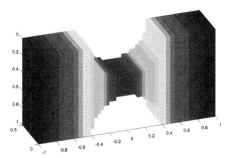

Hint: the grid spacing in the x-direction is given by $\Delta x(1 - \frac{1}{2}\cos(\pi x))$ and the colors signify cell volumes.

3.1.2 Metaballs are commonly used in computer graphics to generate organic-looking objects. Each metaball is defined as a smooth function that has finite support. One example is

$$m(\vec{x}, r) = \left[1 - \min\left(\frac{|\vec{x}|^2}{r^2}, 1\right)\right]^4.$$

Metaballs can be used to define objects implicitly, e.g., as all the points $\vec{x}$ that satisfy

$$\sum_i m(\vec{x} - \vec{x}_i, r_i) \leq C, \qquad C \in \mathbb{R}^+$$

Use this approach and try to make grids similar to the ones shown here:

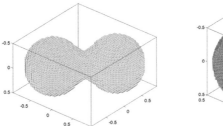
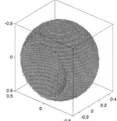

3.1.3 A simple way to make test models with funny geometries is to use the method of fictitious domains and let an image define the domain of interest. In the figures, the image was taken from penny, which is one of the standard data sets that are distributed with MATLAB, and then used to define the geometry of the grid and assign permeability values

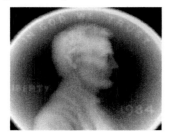

Pick your own favorite image or make one in a drawing program and use imread to load the image into MATLAB as a 3D array, which you can use to define your geometry and petrophysical values. If you do not have an image at hand, you can use penny or spine. For penny, in particular, you may have to experiment a bit with the threshold used to define your domain to ensure that all cells are connected, i.e., that the grid you obtain consists of only one piece.

3.2 Unstructured Grids

An unstructured grid consists of a set of simple shapes that are laid out in an irregular pattern so that any number of cells can meet at a single vertex. The topology of the grid will therefore change throughout space. An unstructured grid can generally consist of a combination of polyhedral cells with varying numbers of faces, as we will see in this section. However, the most common forms of unstructured grids are based on triangles in 2D and tetrahedrons in 3D. These grids are very flexible and are relatively easy to adapt to complex domains and structures, or refine to provide increased local resolution.

Unlike structured grids, unstructured grids cannot generally be efficiently referenced using a structured multi-index. Instead, one must describe a list of connectivities that specifies the way a given set of vertices make up individual cells and cell faces, and how these cells are connected to each other via faces, edges, and vertices.

3.2 Unstructured Grids

To understand the properties and construction of unstructured grids, we start by a brief discussion of two concepts from computational geometry: Delaunay tessellation and Voronoi diagrams. Both these concepts are supported by standard functionality in MATLAB.

3.2.1 Delaunay Tessellation

A tessellation of a set of *generating points* $\mathcal{P} = \{x_i\}_{i=1}^n$ is defined as a set of simplices that completely fills the convex hull of $\mathcal{P}$. The *convex hull* H of $\mathcal{P}$ is the convex minimal set that contains $\mathcal{P}$ and can be described constructively as the set of convex combinations of a finite subset of points from $\mathcal{P}$,

$$H(\mathcal{P}) = \left\{ \sum_{i=1}^{\ell} \lambda_i x_i \;\Big|\; x_i \in \mathcal{P}, \lambda_i \in \mathbb{R}, \lambda_i \geq 0, \sum_{i=1}^{\ell} \lambda_i = 1, 1 \leq \ell \leq n \right\}.$$

Delaunay tessellation is by far the most common method of generating a tessellation based on a set of generating points. In 2D, the Delaunay tessellation consists of a set of triangles defined so that three points form the corners of a Delaunay triangle only when the *circumcircle* that passes through them contains no other generating points; see Figure 3.5. The definition using circumcircles can readily be generalized to higher dimensions using simplices and hyperspheres.

The center of the circumcircle is called the *circumcenter* of the triangle. We will come back to this quantity when discussing Voronoi diagrams in the next subsection. When four (or more) points lie on the same circle, the Delaunay triangulation is not unique. As an example, consider four points defining a rectangle. Using either of the two diagonals will give two triangles satisfying the Delaunay condition.

The Delaunay triangulation can alternatively be defined using the so-called *max-min angle criterion*, which states that the Delaunay triangulation is the one that maximizes the minimum angle of all angles in a triangulation; see Figure 3.6. Likewise, the Delaunay triangulation minimizes the largest circumcircle and minimizes the largest min-containment circle, which is the smallest circle that contains a given triangle. Additionally, the closest two generating points are connected by an edge of a Delaunay triangulation. This is called

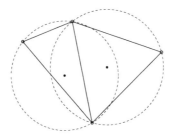

Figure 3.5 Two triangles and their circumcircles.

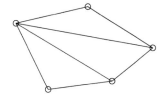 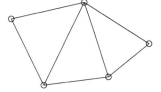

Figure 3.6 Example of two triangulations of the same five points; the triangulation to the right satisfies the min-max criterion.

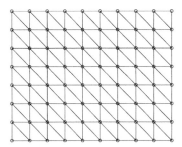 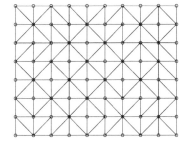

Figure 3.7 Two different Delaunay tessellations of a rectangular point mesh.

the *closest-pair property*, and the two neighboring points are referred to as natural neighbors. This way, the Delaunay triangulation can be seen as the natural tessellation of a set of generating points.

Delaunay tessellation is a popular research topic and there exists a large body of literature on theoretical aspects and computer algorithms. Likewise, there are a large number of software implementations available on the Internet. For this reason, MRST does not have any routines for generating tessellations based on simplexes. Instead, we have provided simple routines for mapping a set of points and edges, as generated by MATLAB's Delaunay triangulation routines, to the internal data structure used to represent grids in MRST. How these routines work will be illustrated in terms of a few simple examples.

In the first example, we triangulate a rectangular mesh and then convert the result by use of the MRST routine `triangleGrid`:

```
[x,y] = meshgrid(1:10,1:8);
t = delaunay(x(:),y(:));
G = triangleGrid([x(:) y(:)],t);
plot(x(:),y(:),'o','MarkerSize',8);
plotGrid(G,'FaceColor','none');
```

Depending on what version you have of MATLAB, the 2D Delaunay routine `delaunay` will produce one of the triangulations shown in Figure 3.7. In older versions of MATLAB, the implementation of `delaunay` was based on Qhull (see www.qhull.org), which produces the unstructured triangulation shown in the right plot. MATLAB 7.9 and newer has

3.2 Unstructured Grids

improved routines for 2D and 3D computational geometry, and here delaunay will produce the structured triangulation shown in the left plot. However, the n-D tessellation routine delaunayn([x(:)y(:)]) is still based on Qhull and will generally produce an unstructured tessellation, as shown in the right plot.

If the set of generating points is structured, e.g., as one would obtain by calling either meshgrid or ndgrid, it is straightforward to make a structured triangulation. The following skeleton of a function makes a 2D triangulation and can easily be extended by the interested reader to 3D:

```
function t = mesh2tri(n,m)
[I,J]=ndgrid(1:n-1, 1:m-1); p1=sub2ind([n,m],I(:),J(:));
[I,J]=ndgrid(2:n  , 1:m-1); p2=sub2ind([n,m],I(:),J(:));
[I,J]=ndgrid(1:n-1, 2:m  ); p3=sub2ind([n,m],I(:),J(:));
[I,J]=ndgrid(2:n  , 1:m-1); p4=sub2ind([n,m],I(:),J(:));
[I,J]=ndgrid(2:n  , 2:m  ); p5=sub2ind([n,m],I(:),J(:));
[I,J]=ndgrid(1:n-1, 2:m  ); p6=sub2ind([n,m],I(:),J(:));
t = [p1 p2 p3; p4 p5 p6];
```

In Figure 3.8, we have used the demo case seamount supplied with MATLAB as an example of a more complex unstructured grid

```
load seamount;
plot(x(:),y(:),'o');
G = triangleGrid([x(:) y(:)]);
plotGrid(G,'FaceColor',[.8 .8 .8]); axis off;
```

The observant reader will notice that we did not explicitly generate a triangulation before calling triangleGrid; if the second argument is omitted, the routine uses MATLAB's built-in delaunay triangulation as default.

For 3D grids, MRST supplies a conversion routine tetrahedralGrid(P, T) that constructs a valid grid definition from a set of points P ($m \times 3$ array of node coordinates)

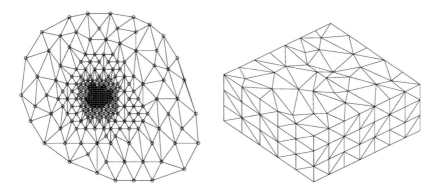

Figure 3.8 The left plot shows the triangular grid from the seamount demo case. The right plot shows a tetrahedral tessellation of a 3D point mesh.

and a tetrahedron list T (*n* array of node indices). The tetrahedral tessellation shown to the right in Figure 3.8 was constructed from a set of generating points defined by perturbing a regular hexahedral mesh:

```
N=7; M=5; K=3;
[x,y,z] = ndgrid(0:N,0:M,0:K);
x(2:N,2:M,:) = x(2:N,2:M,:) + 0.3*randn(N-1,M-1,K+1);
y(2:N,2:M,:) = y(2:N,2:M,:) + 0.3*randn(N-1,M-1,K+1);
G = tetrahedralGrid([x(:) y(:) z(:)]);
plotGrid(G, 'FaceColor',[.8 .8 .8]); view(-40,60); axis tight off
```

3.2.2 *Voronoi Diagrams*

The Voronoi diagram of a set of points $\mathcal{P} = \{x_i\}_{i=1}^n$ is the partitioning of Euclidean space into n (possibly unbounded) convex polytopes[1] such that each polytope contains exactly one generating point x_i and every point inside the given polytope is closer to its generating point than any other point in $\mathcal{P}$. The convex polytopes are called Voronoi cells or Voronoi regions. Mathematically, the *Voronoi region* $V(x_i)$ of a generating point x_i in $\mathcal{P}$ can be defined as

$$V(x_i) = \left\{ x \mid \|x - x_i\| < \|x - x_j\| \; \forall j \neq i \right\}. \tag{3.1}$$

A Voronoi region is not closed in the sense that a point that is equally close to two or more generating points does not belong to the region defined by (3.1). Instead, these points are said to lie on the Voronoi segments and can be included in the Voronoi cells by defining the closure of $V(x_i)$, using "$\leq$" rather than "$<$" in (3.1).

The Voronoi regions for all generating points lying at the convex hull of $\mathcal{P}$ are unbounded, all other Voronoi regions are bounded. For each pair of two points x_i and x_j, one can define a hyperplane with co-dimension one consisting of all points that lie equally close to x_i and x_j. This hyperplane is the perpendicular bisector to the line segment between x_i and x_j and passes through the midpoint of the line segment. The Voronoi diagram of a set of points can be derived directly as the *dual* of the Delaunay triangulation of the same points. To understand this, we consider the planar case; see Figure 3.9. For every triangle, there is a polyhedron in which the vertices occupy complementary locations:

- The circumcenter of a Delaunay triangle corresponds to a vertex of a Voronoi cell.
- Each vertex in the Delaunay triangulation corresponds to, and is the center of, a Voronoi cell.

Moreover, for locally orthogonal Voronoi diagrams, an edge in the Delaunay triangulation corresponds to a segment in the Voronoi diagram and the two intersect each other orthogonally. However, as we can see in Figure 3.9, this is not always the case. If the

---

[1] A polytope is a generic term that refers to a polygon in 2D, a polyhedron in 3D, and so on.

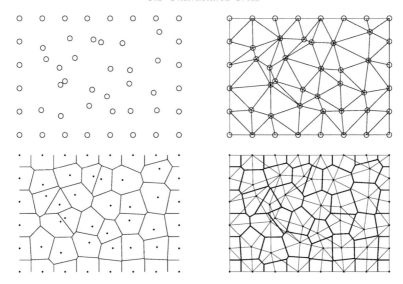

Figure 3.9 Duality between Voronoi diagrams and Delaunay triangulation. From top left across to bottom right: generating points, Delaunay triangulation, Voronoi diagram, and Voronoi diagram (thick lines) and Delaunay triangulation (thin lines).

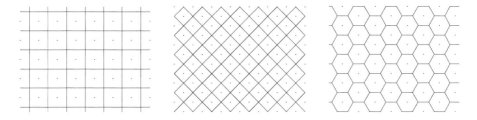

Figure 3.10 Three examples of Voronoi diagrams generated from 2D point lattices. From left to right: square lattice, square lattice rotated 45 degrees, and lattice forming equilateral triangles.

circumcenter of a triangle lies outside the triangle itself, the Voronoi segment does not intersect the corresponding Delaunay edge. To avoid this situation, one can perform a *constrained Delaunay triangulation* and insert additional points where the constraint is not met (i.e., the circumcenter is outside its triangle).

Figure 3.10 shows three examples of planar Voronoi diagrams generated from 2D point lattices using the MATLAB function `voronoi`. MRST does not yet have a similar function that generates a Voronoi grid from a point set, but offers `V=pebi(T)` that generates a locally orthogonal, Voronoi grid `V` as a dual to a triangular grid `T`. The grids are constructed by connecting the perpendicular bisectors of the edges of the Delaunay triangulation, hence the name perpendicular bisector (PEBI) grids. To demonstrate the functionality, we first generate a honeycombed grid similar to the one shown in the right plot in Figure 3.10:

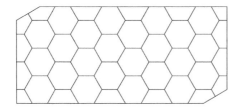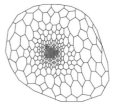

Figure 3.11 Two examples of Voronoi grids. The left plot shows a honeycombed PEBI grid and the right plot shows the PEBI grid derived from the `seamount` demo case.

```
[x,y] = meshgrid([0:4]*2*cos(pi/6),0:3);
x = [x(:); x(:)+cos(pi/6)];
y = [y(:); y(:)+sin(pi/6)];
G = triangleGrid([x,y]);
plotGrid(pebi(G), 'FaceColor','none'); axis equal off
```

Figure 3.11 shows the resulting grid. As a second example, we reiterate the `seamount` examples shown in Figure 3.8:

```
load seamount
V = pebi(triangleGrid([x y]));
plotGrid(V,'FaceColor',[.8 .8 .8]); axis off;
```

Several of the examples discussed in this section can also be found in the fifth video of the second MRST Jolt [189]. Since `pebi` is a 2D code, we cannot apply it directly to the 3D tetrahedral grid shown in Figure 3.8 to generate a dual 3D Voronoi grid. The `upr` module [42, 169] offers several more advanced tools for generating 2D and 3D Voronoi grids; we will return to this in Section 3.5.3. Likewise, Section 3.3.2 discusses how 2D Voronoi grids can be extruded to 3D giving so-called 2.5D grid; this is a standard approach in geological modeling to preserve geological layers.

3.2.3 General Tessellations

Tessellations come in many other forms than the Delaunay and Voronoi types discussed in the two previous sections. Tessellations are more commonly referred to as *tilings* and are patterns made up of geometric forms (tiles) that are repeated over and over without overlapping or leaving any gaps. Such tilings can be found in many patterns of nature, like in honeycombs, giraffe skin, pineapples, snake skin, tortoise shells, to name a few. Tessellations have also been extensively used for artistic purposes since ancient times, from the decorative tiles of Ancient Rome and Islamic art to the amazing artwork of M. C. Esher. For completeness (and fun), MRST offers the function `tessellationGrid` that can take a tessellation consisting of symmetric n-polygonals and turn it into a correct grid structure. While this may not be very useful in modeling petroleum reservoirs, it can easily be used to generate irregular grids that can be used to stress test various discretization methods. To exemplify, the following commands will make a standard $n \times m$ Cartesian mesh:

```
[x,y] = meshgrid(linspace(0,1,n+1),linspace(0,1,m+1));
I = reshape(1:(n+1)*(m+1),m+1,n+1);
T = [reshape(I(1:end-1,1:end-1),[],1)'; reshape(I(1:end-1,2:end  ),[],1)';
     reshape(I(2:end,  2:end  ),[],1)'; reshape(I(2:end,  1:end-1),[],1)'];
G = tessellationGrid([x(:) y(:)], T);
```

Here, the vertices and cells are numbered first in the y direction and then in the x-direction, so that the first two lines in T read [1 m+2 m+3 2; 2 m+3 m+4 3], and so on. There are obviously many ways to make more general tilings. The script `showTessellation` in the book module shows some simple but amazing examples.

3.2.4 Using an External Mesh Generator

Using a Delaunay triangulation as discussed earlier, we can generate grids that fit a given set of vertices. However, in most applications, the vertex points are not given a priori and all one wants is a reasonable grid that fits an exterior outline and possibly also a set of interior boundaries and/or features, which in the case of subsurface media could be faults, fractures, or geological horizons. To this end, you will typically have to use a mesh generator. There are a large number of mesh generators available online, and in principle any of these can be used in combination with MRST's grid factory routines, as long as they produce triangulations on the form outlined earlier. The `triangle` module implements a simple MATLAB interface to the `Triangle` generator (www.cs.cmu.edu/~quake/triangle.html).

My personal favorite, however, is `DistMesh` by Persson and Strang [248]. While most mesh generators tend to be complex and quite inaccessible codes, `DistMesh` is a relatively short and simple MATLAB code written in the same spirit as MRST. The performance of the code may not be optimal, but the user can go in and inspect all algorithms and modify them to his or her purpose. A slightly modified version of `DistMesh` is, for instance, used on the inside of the upr module for generating high-quality Voronoi grids, which will be briefly outlined in Section 3.5.3. In the following, we use `DistMesh` to generate a few examples of more complex triangular and Voronoi grids in 2D.

`DistMesh` is distributed under the GNU GPL license (which is the same license that MRST uses) and can be downloaded from the software's website. The simplest way to integrate `DistMesh` with MRST is to install it as a third-party module. Assuming that you are connected to Internet, this is done as follows:

```
pth = fullfile(ROOTDIR,'utils','3rdparty','distmesh');
mkdir(pth)
unzip('http://persson.berkeley.edu/distmesh/distmesh.zip', pth);
mrstPth('reregister','distmesh', pth);
```

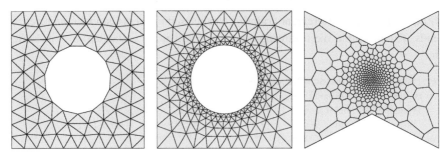

Figure 3.12 Grids generated by the `DistMesh` grid generator.

and you are ready to use `DistMesh`. If you intend to use it many times, you should copy the last line to the `startup_user.m` file in MRST's root directory.

In `DistMesh`, the perimeter of the domain is represented using a signed distance function $d(x, y)$, which by definition is set to be negative inside the region. The software offers a number of utility functions that makes it simple to describe relatively complex geometries, as we shall see in the following. Let us start with a simple example taken from the `DistMesh` website: Consider a square domain $[-1, 1] \times [-1, 1]$ with a circular cutout of radius 0.5 centered at the origin. We start by making a grid that has a uniform target size $h = 0.2$

```
mrstModule add distmesh;
fd=@(p) ddiff(drectangle(p,-1,1,-1,1), dcircle(p,0,0,0.5));
[p,t]=distmesh2d(fd, @huniform, 0.2, [-1,-1;1,1], [-1,-1;-1,1;1,-1;1,1]);
G = triangleGrid(p, t);
```

Here, we have used utility functions `ddif`, `drectangle` and `dcircle` to compute the signed distance from the outer and inner perimeter. Likewise, `huniform` is set to enforce uniform cell size and equal distance between the initial points, here this distance is 0.2. The fourth argument is the bounding box of our domain, and the fifth argument consists of fixed points that the algorithm is not allowed to move. After the triangulation has been computed by `distmesh2d`, we pass it to `triangleGrid` to make an MRST grid structure. The resulting grid is shown to the left in Figure 3.12.

As a second test, let us make a graded grid that has a mesh size of approximately 0.05 at the inner circle and 0.2–0.35 at the outer perimeter. To enforce this, we replace the `huniform` function by another function that gives the correct mesh size distribution:

```
fh=@(p) 0.05+0.3*dcircle(p,0,0,0.5);
[p,t]=distmesh2d(fd, fh, 0.05,[-1,-1;1,1], [-1,-1;-1,1;1,-1;1,1]);
```

The middle plot of Figure 3.12 shows that the resulting grid has the expected grading from the inner boundary and outwards to the perimeter.

3.2 Unstructured Grids

In our last example, we create a graded triangulation of a polygonal domain and then use `pebi` to compute its Voronoi diagram

```
pv = [-1 -1; 0 -.5; 1 -1; 1 1; 0 .5; -1 1; -1 -1];
fh = @(p,x) 0.025 + 0.375*sum(p.^2,2);
[p,t] = distmesh2d(@dpoly, fh, 0.025, [-1 -1; 1 1], pv, pv);
G = pebi(triangleGrid(p, t));
```

Here, we use the utility function `dpoly(p,pv)` to compute the signed distance of any point set p to the polygon with vertices in pv. Notice that pv must form a closed path. Likewise, since the signed distance function and the grid density function are assumed to take the same number of arguments, `fh` is created with a dummy argument x. The argument sent to these functions is passed as the sixth argument to `distmesh2d`. The resulting grid is shown to the right in Figure 3.12.

DistMesh also has routines for creating nD triangulations and triangulations of surfaces, but these are beyond the scope of the current presentation.

COMPUTER EXERCISES

3.2.1 Create MRST grids from the standard data sets `trimesh2d` and `tetmesh`. How would you assign lognormal petrophysical parameters to these grids so that the spatial correlation is preserved?

3.2.2 All grids generated by `triangleGrid` are assumed to be planar so that each vertex can be given by a 2D coordinate. However, triangular grids are commonly used to represent nonplanar surfaces in 3D. Can you extend the function so it can construct both 2D and 3D grids? You can use the data set `trimesh3d` as an example of a triangulated 3D surface.

3.2.3 Load the `triangle` module and use it to mesh the polygonal outline to the right in Figure 3.12. Try to also make other convex and concave outlines.

3.2.4 Download and install `distmesh` and try to make MRST grids from all the triangulations shown in [248, Fig. 5.1].

3.2.5 Running the tutorial `showTessellation` from the `book` module will produce figures like these:

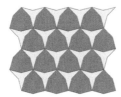 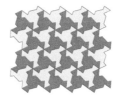

What would you do to fit the tessellations so that they fill a rectangular box without leaving any gaps along the border?

3.2.6 MRST does not yet have a grid factory routine to generate structured grids with local, nested refinement, as shown in the figure to the left:

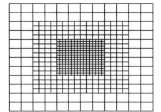
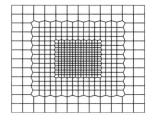

Try to use a combination of `triangleGrid` and `pebi` to make a good approximation to such a grid as shown to the right in the figure. (Hint: to get rid of artifacts, one layer of cells were removed along the outer boundary.)

3.2.7 The figure shows an unstructured hexagonal grid that has been adapted to two faults in the interior of the domain and padded with rectangular cells near the boundary. Try to implement a routine that generates a similar grid.

Hint: in this case, the strike direction of the faults are $\pm 30°$ and the start and endpoints of the faults have been adjusted so that they coincide with the generating points of hexagonal cells.

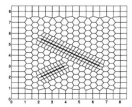

3.3 Stratigraphic Grids

When discussing how to model reservoir rocks, we saw that the grid is closely connected to the description of the petrophysical characteristics of a reservoir. Not only does the grid describe the outer geometry of the reservoir, but it is also used to represent internal structures such as fractures and faults, as well as surfaces representing marked changes the rock's physical characteristics (*lithography*). This means that the grid is closely related to the parameter description of the flow model and, unlike in many other disciplines, cannot be chosen independently to provide a certain numerical accuracy. Indeed, the grid is typically chosen by a geologist who tries to describe the rock body by as few volumetric cells as possible and who does not necessarily care too much about potential numerical difficulties the choice of grid may cause in subsequent flow simulations. Although grossly simplified, this observation is important to bear in mind throughout the rest of this chapter.

The industry standard to represent reservoir geology in a flow simulator is to use a *stratigraphic grid* built based on geological horizons and fault surfaces. The volumetric grid is typically built by extruding 2D tessellations of the geological horizons in the vertical direction or in a direction following major fault surfaces. For this reason, some stratigraphic grids, like the PEBI grids you will meet in Section 3.3.2, are often called 2.5D rather than 3D grids. These grids may be unstructured in the lateral direction, but have a clear structure in the vertical direction to reflect the layering of the reservoir.

Because of the role grid models play in representing geological formations, real-life stratigraphic grids tend to be highly complex and have unstructured connections induced by the displacements that have occured over faults. Another characteristic feature is high aspect ratios. Typical reservoirs extend several hundred or thousand meters in the lateral direction, but the zones carrying hydrocarbon may be just a few tens of meters in the vertical direction and consist of several layers with (largely) different rock properties. Getting the stratigraphy correct is crucial, and high-resolution geological modeling will typically result in a high number of (very) thin grid layers in the vertical direction, resulting in two or three orders of magnitude aspect ratios.

A full exposition of stratigraphic grids is far beyond the scope of this book. In the next two subsections, we discuss the basics of the two most commonly used forms of stratigraphic grids. A complementary discussion is given in videos 2, 6, and 7 of the second MRST Jolt [189].

3.3.1 Corner-Point Grids

To model the geological structures of petroleum reservoirs, the industry-standard approach is to introduce what is called a corner-point grid [254], which we already encountered in Section 2.5. A corner-point grid consists of a set of hexahedral cells that are topologically aligned in a Cartesian fashion so that the cells can be numbered using a logical ijk index. In its simplest form, a corner-point grid is specified in terms of a set of vertical or inclined *pillars* defined over an areal Cartesian 2D mesh in the lateral direction. Each cell in the volumetric grid is defined by eight *logical* corner points specified as two *depth coordinates* on each of the four *coordinate lines* that define a pillar; see Figure 3.13. The grid consists of $n_x \times n_y \times n_z$ grid cells and the cells are ordered with the i-index (x-axis) cycling fastest, then the j-index (y-axis), and finally the k-index (negative z-direction). All cellwise property data are assumed to follow the same numbering scheme.

As discussed previously, a fictitious domain approach is used to embed the reservoir in a logically Cartesian shoebox. This means that inactive cells that are not part of the physical model, e.g., as shown in Figure 2.16, are present in the topological ijk-numbering but are

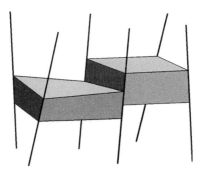

Figure 3.13 Each cell in the corner-point grid is restricted by two points on each of the four grid lines that make up a pillar cells. The plot shows cells and grid lines from two neighboring pillars.

indicated by a zero porosity or net-to-gross value, as discussed in Section 2.4, or marked by a special boolean indicator (called ACTNUM in the input files).

So far, the topology and geometry of a corner-point grid have not deviated from that of the mapped Cartesian grids studied in the previous section. Somewhat simplified, one may view the logical ijk numbering as a reflection of the sedimentary rock bodies as they may have been deposited in geological time, so that cells with the same k index have been deposited at approximately the same time. To model geological features like erosion and pinch-outs of geological layers, the corner-point format allows point-pairs to collapse along coordinate lines. This creates degenerate hexahedral cells that may have less than six faces, as illustrated in Figure 3.14. The corner points can even collapse along all four lines of a pillar, so that a cell completely disappears. This implicitly introduces a new local topology, which is sometimes referred to as *non-neighboring connections*, in which cells that are not logical ijk neighbors can be neighbors in physical space and share a common face. Figure 3.15 shows an example of a model that contains both eroded geological layers and fully collapsed cells. In a similar manner, (simple) vertical and inclined faults can be easily modeled by aligning the pillars with fault surfaces and displacing the corner points defining neighboring cells on one or both sides of the fault. This way, one creates nonmatching geometries and non-neighboring connections in the underlying ijk topology.

To illustrate the concepts introduced so far, we consider a low-resolution version of the model from Figure 2.11 on page 42 created by the simpleGrdecl routine, which generates an input stream containing the basic keywords that describe a corner-point grid in the ECLIPSE input deck [270, 33]

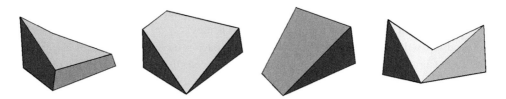

Figure 3.14 Examples of deformed and degenerate hexahedral cells arising in corner-point grid models.

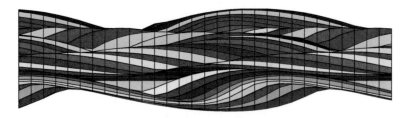

Figure 3.15 Side view in the xz-plane of corner-point grid with vertical pillars modeling a stack of sedimentary beds (each layer indicated by a different color).

3.3 Stratigraphic Grids

```
grdecl = simpleGrdecl([4, 2, 3], .12, 'flat', true);
```

```
grdecl =
    cartDims: [4 2 3]
      COORD: [90x1 double]
      ZCORN: [192x1 double]
     ACTNUM: [24x1 int32]
```

The COORD field contains the pairs of 3D coordinates for the 15 coordinate lines that define the 4×2 mesh of pillars, whereas the ZCORN field gives the z-values that determine vertical positions uniquely along each coordinate line for the eight corner-points of the 24 cells. To extract these data, we use two MRST routines

```
[X,Y,Z] = buildCornerPtPillars(grdecl,'Scale',true);
[x,y,z] = buildCornerPtNodes(grdecl);
```

We now plot the coordinate lines and the corner-points and mark those lines on which the corner-points of logical ij neighbors do not coincide,

```
% Plot pillars
plot3(X',Y',Z','k');
set(gca,'zdir','reverse'), view(35,35), axis off, zoom(1.2);

% Plot points on coordinate lines, mark those with faults red
hold on; I=[3 8 13];
hpr = plot3(X(I,:)',Y(I,:)',Z(I,:)','r','LineWidth',2);
hpt = plot3(x(:),y(:),z(:),'o'); hold off;
```

From the plots in the upper row of Figure 3.16 we clearly see how the coordinate lines change slope from the east and west side toward the fault in the middle, and how the grid points sit like beads on a string along each coordinate line.

Cells along a pillar are now defined by connecting pairs of points from four neighboring coordinate lines that make up a cell in the areal mesh. To see this, we plot two pillars as well as the whole grid with the fault surface marked in blue:

```
% Create grid and plot two stacks of cells
G = processGRDECL(grdecl);
args = {'FaceColor'; 'r'; 'EdgeColor'; 'k'};
hcst = plotGrid(G,[1:8:24 7:8:24],'FaceAlpha', .1, args{:});

% Plot cells and fault surface
delete([hpt; hpr; hcst]);
plotGrid(G,'FaceAlpha', .15, args{:});
plotFaces(G, G.faces.tag>0,'FaceColor','b','FaceAlpha',.4);
```

The upper-left plot in Figure 3.17 shows the same model sampled with even fewer cells. We have used different coloring of the cell faces on each side of the fault to highlight nonmatching faces along the fault plane. MRST represents corner-point grids as *matching*

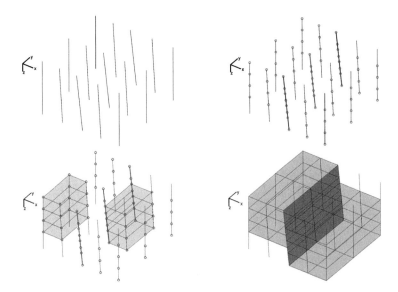

Figure 3.16 Specification of a corner-point grid. Starting from the coordinate lines defining pillars (upper left), we add corner-points and identify lines containing nonmatching corner marked in red (upper right). A stack of cells is created for each set of four lines defining a pillar (lower left), and then the full grid is obtained (lower right). In the last plot, the fault faces have been marked in blue.

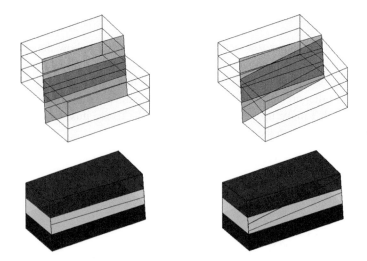

Figure 3.17 Subdivision of fault face in two three-dimensional models. In the left column, the subfaces are all rectangular. In the right columns they are not. In both the upper plots, the faces marked in red belong only to the cells behind the fault surface, the blue faces belong only to the cells in front of the fault surface, and the magenta ones belong to cells on both sides. The lower plot shows the cells behind the surface, where each cell has been given its own color.

3.3 Stratigraphic Grids

unstructured grids obtained by subdividing all nonmatching cell faces instead of using the more compact nonmatching hexahedral form. This means that the four cells with non-neighboring connections across the fault plane will have seven and not six faces. For each such cell, two of the seven faces lie along the fault plane. Here, the subdivision resulted in new faces that all have four corners and rectangular geometry. This is generally not the case; the right column of Figure 3.17 shows a grid where we can see cells with six, seven, or eight faces, and faces with three, four, and five corners. Indeed, for real-life models, subdivision of nonmatching fault faces can lead to cells having much more than six faces.

Using the inherent flexibility of the corner-point format, it is possible to construct very complex geological models that come a long way in matching the geologist's perception of the underlying rock formations. Because of their many appealing features, corner-point grids have been an industry standard for years, and the format is supported in most commercial software for reservoir modeling and simulation.

A Simple Grid Generator

The internal and external geology of a real reservoir is usually describe by a large set of horizons and fault surfaces. As explained earlier, horizons are lateral surfaces that deliminate the reservoir in the upward and downward direction and represent the main stratigraphic layers. These surfaces can either be triangulated/tessellated or represented using splines. Major fault surfaces are typically vertical or inclined, but need to be planar. Generating corner-point grids that adapt to these surfaces is a challenging task and requires specialized geomodeling software and integration of a lot of different data types. MRST does not offer any such capabilities, and complex geomodels must therefore be created using external software. However, we do offer a relatively simple grid generator that takes a set of horizons as input and generates a corner-point grid by interpolating vertically between pairs of consecutive horizons. The function converHorizonsToGrid cannot model faults, and assumes that each horizon is represented over a rectilinear areal mesh. The horizons need not use the same areal mesh points, but the areal meshes must have a certain minimal overlap inside which the corner-point will be generated. The following example illustrates the basic use of the grid generator:

```
[y,x,z]  = peaks(15); z = z+8;
horizons = {struct('x',x,'y',y,'z',z),struct('x',x,'y',y,'z',z+8)};
grdecl   = convertHorizonsToGrid(horizons,'layers', 4);
G        = processGRDECL(grdecl);
```

Here, we use the peaks function from MATLAB to generate a surface and create two horizons that deliminate the reservoir upward and downward. We then create a grid with four layers of cells in the vertical direction. When no extra parameters are specified, the areal mesh defining the corner-point grid has the same number of points as the horizons; see Figure 3.18. When the areal domain of definition for the horizons do not coincide, the interpolation region is set to be the minimum rectangle that contains the areal bounding boxes of all the horizons. Inside this rectangle, the routine creates a Cartesian mesh and

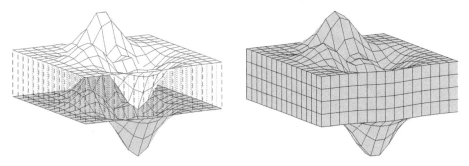

Figure 3.18 Example of a corner-point grid generated by interpolating vertically between two geological horizons.

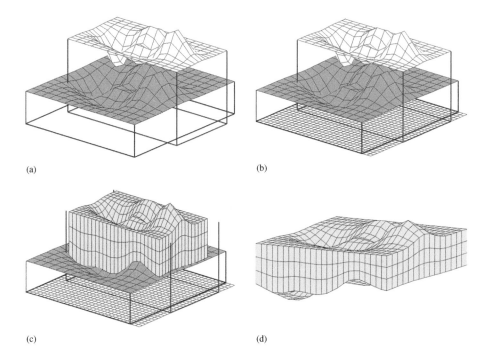

Figure 3.19 Example of a corner-point grid generated by interpolating vertically between two partially overlapping horizons. (a) Horizons and projected bounding boxes. (b) Interpolation region with 31 × 31 areal mesh covers the two bounding boxes. (c) Corner-point grid generated inside the maximum overlap of the two horizons. (d) The full 3D grid generated by the routine.

uses this to interpolate vertically. In the output grid, the routine discards all cells whose lateral projection is not contained inside the maximum rectangle that fits inside the areal bounding boxes of all horizons. Figure 3.19 shows illustrates how the grid resulting from the following commands is created:

```
[n,m]    = deal(30);
horizons = {struct('x',x,'y',y,'z',z),struct('x',x+.5,'y',2*y+1,'z',z+10)};
grdecl   = convertHorizonsToGrid(horizons,'dims',[n m], 'layers', 3);
G        = processGRDECL(grdecl);
```

Here, we have only used two horizons to simplify the illustration. Section 15.6.4 presents a more complex case with multiple horizons.

A Synthetic Faulted Reservoir

In our second example we consider a synthetic model of two intersecting faults that make up the letter Y in the lateral direction. The two fault surfaces are highly deviated, making an angle far from 90 degrees with the horizontal direction. To model this scenario using corner-point grids, we basically have two different choices. The first choice, which is quite common, is to let the coordinate lines (and hence the extrusion direction) follow the main fault surfaces. For highly deviated faults, like in the current case, this will lead to extruded cells that are far from K-orthogonal and hence susceptible to grid-orientation errors in a subsequent simulation, as will be discussed in more detail in Chapter 6. Alternatively, we can choose a vertical extrusion direction and replace deviated fault surfaces by *stair-stepped* approximations so that the faults zigzag in a direction not aligned with the grid. This will create cells that are mostly K-orthogonal and less prone to grid-orientation errors, but will only represent the fault approximately.

Figure 3.20 shows two different grid models, taken from the CaseB4 data set. In the stair-stepped model, the use of cells with orthogonal faces causes the faults to be represented as zigzag patterns. The pillar grid correctly represents the faults as inclined planes, but has cells with degenerate geometries and cells that deviate strongly from being orthogonal in the lateral direction. Likewise, some coordinate lines have close to 45 degrees inclination, which will likely give significant grid-orientation effects in a standard two-point scheme.

A Simulation Model of the Norne Field

Norne is an oil and gas field lies located in the Norwegian Sea. The reservoir is found in Jurassic sandstone at a depth of 2,500 meters below sea level, and was originally estimated to contain 90.8 million m^3 oil, mainly in the Ile and Tofte formations, and 12 billion m^3 gas in the Garn formation. The field is operated by Statoil and production started in November 1997, using a floating production, storage, and offloading (FPSO) ship connected to seven subsea templates at a water depth of 380 meters. The oil is produced with water injection as the main drive mechanisms, and the expected ultimate oil recovery is more than 60%, which is very high for a subsea oil reservoir. During thirteen years of production, five 4D seismic surveys of high quality have been recorded. In cooperation with NTNU, operator Statoil and partners have released large amounts of subsurface data from the Norne field for

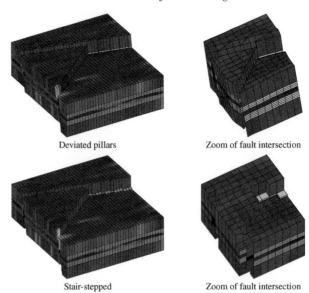

Figure 3.20 Modeling the intersection of two deviated faults using deviated pillars (top) and stair-stepped approximation (bottom). CaseB4 grids courtesy of Statoil.

research and education purposes.[2] More recently, the Open Porous Media (OPM) initiative (opm-project.org) released the full simulation model as an open data set on Github (github.com/OPM/opm-data). For several years, Norne was the only full simulation model of a real oil reservoir that was freely available to the general public. Recently, Equinor (former Statoil) released a much more comprehensive data set of the Volve field. The Norne model can either be downloaded and installed using `mrstDatasetGUI` or directly from the command line:

```
makeNorneSubsetAvailable() && makeNorneGRDECL()
```

Once in place, we can load the grid and the petrophysical data from the simulation model as follows:

```
grdecl = readGRDECL(fullfile(getDatasetPath('norne'), 'NORNE.GRDECL'));
grdecl = convertInputUnits(grdecl, getUnitSystem('METRIC'));
```

The model consists of $46 \times 112 \times 22$ corner-point cells. We start by plotting the whole model, including inactive cells. To this end, we need to override[3] the ACTNUM field before

---

[2] The Norne Benchmark data sets are hosted and supported by the Center for Integrated Operations in the Petroleum Industry (IO Center) at NTNU (www.ipt.ntnu.no/~norne/). The data set used herein was first released as part of "Package 2: Full field model" (2013)

[3] At this point I hasten to warn you that inactive cells often contain junk data and may generally not be inspected in this manner. Here, however, most inactive cells are defined in a reasonable way. By not performing basic sanity checks on the resulting grid (option `'checkgrid'=false`), we manage to process the grid and produce reasonable graphical output. In general, however, I advise that `'checkgrid'` remains set in its default state of `true`.

3.3 Stratigraphic Grids

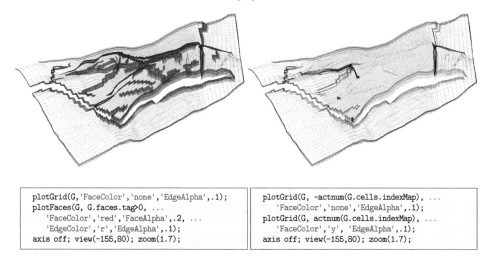

Figure 3.21 The Norne field from the Norwegian Sea. The plots show the whole grid with fault faces marked in red (left) and active cells marked in yellow (right).

we start processing the input, because if the ACTNUM flag is set, all inactive cells will be ignored when the unstructured grid is built

```
actnum = grdecl.ACTNUM;
grdecl.ACTNUM = ones(prod(grdecl.cartDims),1);
G = processGRDECL(grdecl, 'checkgrid', false);
```

Having processed and converted the grid to the correct unstructured format, we first plot the outline of the whole model and highlight all faults and the active part of the model; see Figure 3.21. During the processing, all fault faces are tagged with a positive number. This can be utilized to highlight the faults: we simply find all faces with a positive tag, and color them with a specific color as shown in the left box in the figure. To continue with the active model only, we reset the ACTNUM field to its original values so that inactive cells are ignored when we process the ECLIPSE input stream [270, 33]. This gives a grid with two components: the main reservoir and a detached stack of twelve cells that should be ignored.

To examine some parts of the model in more detail, we can use the function cutGrdecl to extract a rectangular box in index space from the ECLIPSE input stream, e.g., as follows:

```
cut_grdecl = cutGrdecl(grdecl, [6 15; 80 100; 1 22]);
g = processGRDECL(cut_grdecl);
```

Figure 3.22 shows a zoom on four different regions. The first region (red color), is sampled near a laterally stair-stepped fault, i.e., a curved fault surface that has been approximated by a surface that zigzags in the lateral direction. We also notice how the fault

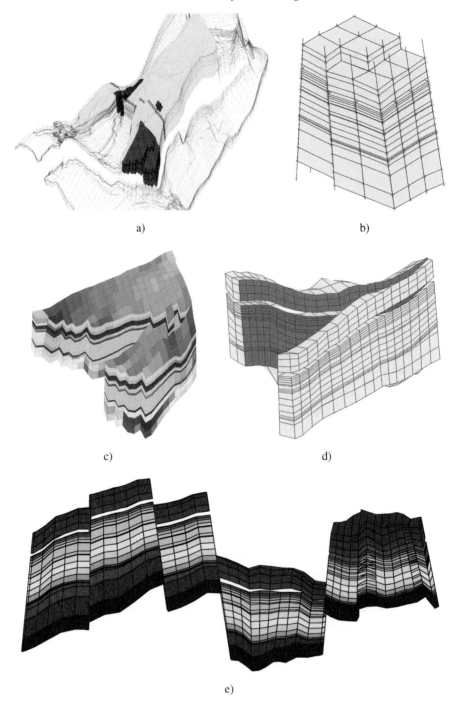

Figure 3.22 Detailed view of subsets from the Norne simulation model. a) The whole model with active and inactive cells and four regions of interest marked in different colors. b) Zoom of the red region with pillars and corner-points shown as red circles. c) The magenta region with coloring according to cell volumes, which vary by a factor 700. d) The blue region in which fault faces have been colored gray and the corresponding coordinate lines have been colored blue. e) The green cross-section with coloring according to layer number from top to bottom.

displacement leads to cells that are nonmatching across the fault surface, and the presence of some very thin layers (the thinnest layers may actually appear as thick lines in the plot). The thin layers are also clearly seen in the second region (magenta color), which represents a somewhat larger sample from an area near the tip of one of the "fingers" in the model. Here, we clearly see how similar layers have been strongly displaced across the fault zone. In the third (blue) region, we have colored the fault faces to clearly show the displacement and the hole through the model in the vertical direction, which likely corresponds to a shale layer that has been eliminated from the active model. Gaps and holes, and displacement along fault faces, are even more evident for the vertical cross section (green region) for which the layers have been given different colors, as in Figure 3.15. Altogether, the four views of the model demonstrate typical patterns that you can see in realistic models. We return to this simulation model in Section 10.3.8.

Extensions, Difficulties, and Challenges

The original corner-point format has been extended in several ways, e.g., to enable vertical intersection of two coordinate lines in the shape of the letter Y. Coordinate lines may also be defined by piecewise polynomial curves, resulting in what is sometimes called S-faulted grids. Likewise, two neighboring coordinate lines can collapse so that the basic grid shape becomes a prism rather than a hexahedron. However, there are several features you cannot easily model, like multiple fault intersections (e.g., as in the letter F), and the industry is thus constantly in search for improved griding methods. One example will be discussed in the next subsection. First, however, we will discuss some difficulties and challenges, seen from the side of a computational scientist seeking to use corner-point grids for his or her computations.

The flexible cell geometry of the corner-point format poses several challenges for numerical implementations. Indeed, a geocellular grid is typically chosen by a geologist who tries to describe the rock body by as few volumetric cells as possible and who may not be aware of potential numerical difficulties his or her choice of geometries and topologies may cause in subsequent flow simulations.

Writing a robust grid-processing algorithm to compute geometry and topology or to determine an equivalent matching, polyhedral grid proved to be much more challenging than we anticipated when we started developing MRST. Displacements across faults will lead to geometrically complex, nonconforming grids, e.g., as illustrated in Figure 3.22. Since each face of a grid cell is specified by four (arbitrary) points, the cell interfaces in the grid will generally be bilinear and possibly strongly curved surfaces. Geometrically, this can lead to several complications. Cell faces on different sides of a fault may intersect each other so that cells overlap volumetrically, or they may leave void spaces in between them. There may be tiny overlap areas between cell faces on different sides a fault, and so on. All these factors contribute to make fault geometries hard to interpret in a geometrically consistent way: a subdivision into triangles is, for instance, not unique. Likewise, top and bottom surfaces may intersect for highly curved cells with high aspect ratios, cell centroids may be outside the cell volume, etc.

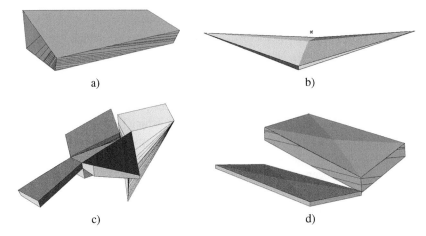

Figure 3.23 Illustration of difficulties associated with real-life corner-point geometries. a) Many faces resulting from subdivision to give matching grid at faults. b) A curved $800 \times 800 \times 0.25$ m cell, whose centroid lies outside the cell. c) Difficult geometries. d) Small interface between two cells.

The presence of collapsed corner-points implies that the grid cells will generally be polyhedral and possibly contain both triangular and bilinear faces (see Figure 3.14). Likewise, individual cells can have many neighbors; the Norne model has cells with up to 20 neighbors. Altogether, this calls for flexible discretizations that work on unstructured topologies and are not sensitive to the geometry of each cell or the number of faces and corner points. Simple discretization methods that only account for pressure differences between cells sharing a common face (see Section 4.4.1 are fairly easy to implement for general unstructured grids, but more advanced discretization methods that rely on the use of dual grids or reference elements can be quite challenging to implement on highly irregular grids. Figure 3.23 illustrates some geometrical and topological challenges seen in standard grid models.

Cell geometries will often deviate significantly from being orthogonal to adapt to complex horizons and fault surfaces. This may introduce unphysical preferential flow directions in the numerical methods (as we will see later). Stratigraphic grids will also typically have aspect ratios that are two or three orders of magnitude, which can introduce severe numerical difficulties because the majority of the flow in and out of a cell occurs across the faces with the smallest area. Similarly, the possible presence of strong heterogeneities and anisotropies in the permeability fields, e.g., as seen in the SPE 10 example in Chapter 2, typically introduces large condition numbers in the discretized flow equations, which make them hard to solve.

Corner-point grids generated by geological modeling typically contain too many cells. Once created by the geologist, the grid is handed to a reservoir engineer, whose first job is to reduce the number of cells if he or she is to have any hope of getting the model through a simulator. The generation of good coarse grids for use in upscaling, and the upscaling procedure itself, is generally work-intensive, error-prone, and not always sufficiently robust, as we will come back to in Chapter 15.

3.3.2 2.5D Unstructured Grids

So-called 2.5D grids have been designed to combine the advantages of two different griding methods: the (areal) flexibility of unstructured grids and the simple topology of Cartesian grids in the vertical direction. These grids are constructed in much the same way as corner-point grids, except that coordinate lines now are defined over an unstructured lateral grid so that pillars will have polygonal base area. One starts by defining an areal tessellation on a surface that either aligns with the lateral direction or one of the major geological horizons. Then a coordinate line is introduced through each vertex in the areal grid. The lines can either be vertical, inclined to gradually align with major fault planes, as shown for the corner-point grid in Figure 3.20, or be defined so that they connect pairs of vertices in areal tessellations of two different geological horizons.

Figure 3.24 shows the key steps in the construction of a simple 2.5D PEBI grid. Starting from a Delaunay triangulation of set of generating points, we form a perpendicular bisector grid. Through each vertex in this areal tessellation, we define a coordinate line, whose angle of inclination will change from 90 degrees for vertices on the far left to 45 degrees for vertices on the far right. coordinate lines associated with individual areas cells form pillars, and are next used to extrude the areal tessellation to a volumetric grid. The resulting volumetric grid is unstructured in the lateral direction, but has a layered structure in the vertical direction and can thus be indexed using an ik index pair. This format has quite large freedom in choosing the size and shape of the grid cells to adapt to complex features such as curved faults or to improve the areal resolution in near-well zones.

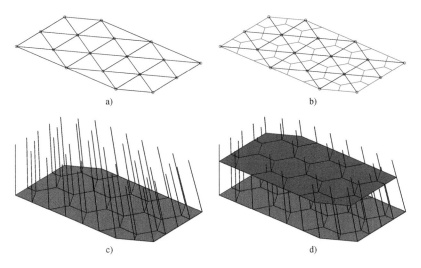

Figure 3.24 Illustration of a typical process for generating 2.5D PEBI grids. a) Triangulated point set. b) Perpendicular bisector grid. c) Coordinate lines aligned with faults. d) Volumetric extrusion.

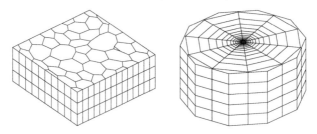

Figure 3.25 The left plot shows a 2.5D Voronoi grid derived from a perturbed 2D point mesh extruded in the *z*-direction, whereas the right plot shows a radial grid.

Example 1: We first construct an areal Voronoi grid from a set of generating points obtained by perturbing the vertices of a regular Cartesian grid and then use the function makeLayeredGrid to extrude this Voronoi grid to 3D along vertical pillars in the *z*-direction.

```
N=7; M=5; [x,y] = ndgrid(0:N,0:M);
x(2:N,2:M) = x(2:N,2:M) + 0.3*randn(N-1,M-1);
y(2:N,2:M) = y(2:N,2:M) + 0.3*randn(N-1,M-1);
aG = pebi(triangleGrid([x(:) y(:)]));
G = makeLayeredGrid(aG, 3);
plotGrid(G, 'FaceColor',[.8 .8 .8]); view(-40,60); axis tight off
```

The resulting grid is shown in the left plot of Figure 3.25 and should be contrasted to the 3D tetrahedral tessellation shown to the right in Figure 3.8

Example 2: Radial symmetric grids graded towards the origin are commonly used to increase resolution near wells. Figure 3.25 shows one example that can be generated as follows:

```
P = [];
for r = exp(-3.5:.25:0),
    [x,y,z] = cylinder(r,16); P = [P [x(1,:); y(1,:)]];
end
P = unique([P'; 0 0],'rows');
G = makeLayeredGrid(pebi(triangleGrid(P)), 5);
plotGrid(G,'FaceColor',[.8 .8 .8]); view(30,50), axis tight off
```

Example 3: The model shown in Figure 3.26 is a realistic representation of a reservoir in which an unstructured areal grid has been extruded vertically to model different geological layers. Some of the layers are very thin and appear as if they were thick lines in plot **a**. Near the perimeter of the model (plot **b**), we notice that the lower layers (yellow to red colors) have been eroded away in most of the grid columns, and although the vertical dimension is strongly exaggerated, we see that the layers contain steep slopes. To a non-geologist looking at subplot **e**, it may appear as if the reservoir was formed by sediments being deposited along a sloping valley that ends in a flat plain. Subplots **c** and **d** show more details of the

3.3 Stratigraphic Grids

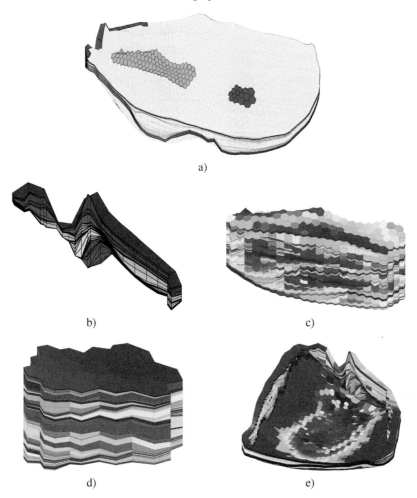

Figure 3.26 A real petroleum reservoir modeled by a 2.5D PEBI grid having 1,174 cells in the lateral direction and 150 cells along each pillar. Only 90,644 out of the 176,100 cells are active. The plots show the whole model as well as selected details. a) The whole model with three areas of interest marked in different colors. b) Layers 41 to 99 of the red region with colors representing the k-index. c) Horizontal permeability in the green region. d) Horizontal permeability in the magenta region. e) Horizontal permeability along the perimeter and bottom of the model.

permeability field inside the model. The layering is particularly distinct in plot **d**, which is sampled from the flatter part of the model. The cells in plot **c**, on the other hand, show examples of pinch-outs. The layering provides a certain ik structure in the model, similar to the logical ijk index for corner-point grids, where i refers to the areal numbering and k to the different layers. Some also prefer to associate a virtual logically Cartesian grid as an overlay to the 2.5D grid that can be used, e.g., to simplify lookup of cells in visualization. In this setup, more than one grid cell may be associated with a cell in the virtual grid.

Figure 3.27 Example of a virtual 37 × 20 × 150 grid used for fast lookup in a 2.5D PEBI grid with dimensions 1174 × 150.

3.4 Grid Structure in MRST

The two previous sections introduced various structured and unstructured grid types that can be created using the many grid-factory routines in MRST. In this section, we go into more detail about the internal data structure used to represent all these different grid formats. This data structure is in many ways the most fundamental part of MRST since almost all solvers, workflow tools, and visualization routines require an instance of a grid as input argument. By convention, instances of the grid structure are denoted G. If you are mainly interested in *using* solvers and visualization routines already available in MRST, you need no further knowledge of the grid structure beyond what has been encountered in the examples presented so far, and you can safely skip the remains of this section. On the other hand, if you wish to use the software to prototype new computational methods, knowledge of the inner workings of the grid structure is essential.

To provide a uniform way of accessing different grid types, all grids are stored using the same unstructured grid format that explicitly represents cells, faces, vertices, and connections between cells and faces. The main grid structure G contains three fields – `cells`, `faces`, and `nodes` – that specify individual properties for each individual cell/face/vertex in the grid. Grids in MRST can either be volumetric or lie on a 2D or 3D surface. The field `griddim` distinguishes volumetric and surface grids; all cells in a grid are polygonal surface patches if `griddim=2` and polyhedral volumetric entities otherwise. In addition, the grid contains a field `type` consisting of a cell array of strings describing the history of grid constructors and modifier functions that have been applied to create and modify the grid, e.g., `'tensorGrid'`. For grids having an underlying logical Cartesian structure, we also include the field `cartDims`.

As with the rest of the basic functionality in MRST, the grid structure is thoroughly documented in the code. To read the documentation, type

```
help grid_structure
```

3.4 Grid Structure in MRST

Before we continue to discuss the individual data members in the structure, we first have to introduce you to two essential implementation tricks in MATLAB.

Indirection maps and run-length encoding. In an unstructured grid, individual cells may have a different number of faces, and faces may have different number of edges/vertices. A compact way to represent such data is to use a data container and an indirection map. To illustrated, let us consider a simple example of a data object b, whose first members consist of five, three, and four numbers. This can be stored using an indirection map m and a data array $a = [a_1, \ldots, a_{12}]$, as illustrated:

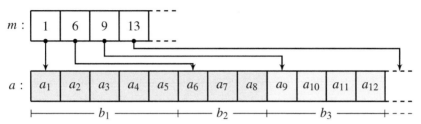

Using this data structure, we can quickly find all the a values that belong to a particular b_i. In some cases, we may wish to go the other way around and determine the b_i element a particular value a_j corresponds to. A simple solution would be to introduce an extra row in a that contains the values $[1,1,1,1,1,2,2,2,3,3,3,3]$. However, this contains a lot of redundant information and can be stored much more compactly using run-length encoding, which is a simple lossless data compression method that removes repeated values. In this particular example, the run-length encoding would give a compressed array $b = [1,2,3]$ and a repetition count $n = [5,3,4]$. This data compression and decompression technique is used abundantly at a low level in MRST to avoid storing redundant information, as you will see shortly.

Representing grid cells. The cell structure G.cells consists of the mandatory fields:

- **num:** the number n_c of cells in the global grid.
- **facePos:** an indirection map of size [num+1,1] into the faces array. Specifically, the face information of cell i is found in the submatrix

    faces(facePos(i): facePos(i+1)-1, :)

    The number of faces per cell may be computed as diff(facePos), whereas the total number of faces is $n_f = $ facePos(end)-1.
- **faces:** an $n_f \times 2$ array specifying the global faces connected to a given cell. Specifically, faces(i,1)==j if face with global number faces(i,2) is connected to cell number j. The array may optionally have a third component, faces(i,3), that for certain types of grids contains a name tag used to distinguish face directions: West, East, South, North, Bottom, Top.

The first column of faces is redundant: it consists of each cell index j repeated facePos(j+1)-facePos(j) times and can therefore be reconstructed by decompressing a run-length encoding with the cell indices 1:num as the encoded vector and the number of faces per cell as repetition count. Hence, to conserve memory, only the last two columns of faces are stored, while the first column can be reconstructed using the statement:

```
rldecode(1:G.cells.num, diff(G.cells.facePos), 2) .'
```

This construction is used a lot throughout MRST and has therefore been implemented as a utility function inside mrst-core/utils/gridtools

```
f2cn = gridCellNo(G);
```

- **indexMap:** an optional $n_c \times 1$ array that maps internal cell indices to external cell indices. For models with no inactive cells, indexMap equals $1 : n_c$. For cases with inactive cells, indexMap contains the indices of the active cells sorted in ascending order. For grids with an underlying Cartesian organization, a map of cell numbers to logical indices can be constructed using the following statements in 2D and 3D:

```
[ij{1:2}] = ind2sub(dims, G.cells.indexMap(:));    % In 2D
ij        = [ij{:}];
[ijk{1:3}] = ind2sub(dims, G.cells.indexMap(:));   % In 3D
ijk       = [ijk{:}];
```

In the latter case, ijk(i:) is the global (I, J, K) index of cell i.

In addition, the cell structure can contain the following optional fields that typically will be added by a call to computeGeometry:

- **volumes:** an $n_c \times 1$ array of cell volumes
- **centroids:** an $n_c \times d$ array of cell centroids in $\mathbb{R}^d$

Representing cell faces. The mandatory fields of the face structure, G.faces, are quite similar to those of the cells in the sense that they provide mappings to the lower-dimensional objects (nodes) that delimit the faces:

- **num:** the number n_f of global faces in the grid.
- **nodePos:** an indirection map of size [num+1,1] into the nodes array. Specifically, the node information of face i is found in the submatrix
  nodes(nodePos(i): nodePos(i+1)-1, :)
  The number of nodes of each face may be computed using the statement diff (nodePos). Likewise, the total number of nodes is given as $n_n = $ nodePos(end)-1.
- **nodes:** an $n_n \times 2$ array of vertices in the grid. If nodes(i,1)==j, the local vertex i is part of global face number j and corresponds to global vertex nodes(i,2). For each face the nodes are assumed to be oriented such that a right-hand rule

determines the direction of the face normal. As for cells.faces, the first column of nodes is redundant and can be easily reconstructed. Hence, to conserve memory, only the last column is stored, while the first column can be constructed using the statement:

```
rldecode(1:G.faces.num, diff(G.faces.nodePos), 2) .'
```

- **neighbors:** an $n_f \times 2$ array of neighboring information. Global face i is shared by global cells neighbors(i,1) and neighbors(i,2). One of the entries in neighbors(i,:), but not both, can be zero, to indicate that face i is an external face that belongs to only one cell (the nonzero entry).

In addition to the mandatory fields, G.faces has optional fields containing geometry information that typically are added by a call to computeGeometry:

- **areas:** an $n_f \times 1$ array of face areas.
- **normals:** an $n_f \times d$ array of **area weighted**, directed face normals in $\mathbb{R}^d$. The normal on face i points from cell neighbors(i,1) to cell neighbors(i,2).
- **centroids:** an $n_f \times d$ array of face centroids in $\mathbb{R}^d$.

Moreover, G.faces can sometimes contain an $n_f \times 1$ (int8) array, G.faces.tag, with user-defined face indicators used, e.g., to specify that the face belongs to a fault.

Representing vertices. The vertex structure, G.nodes, consists of two fields:

- **num:** number n_n of global nodes (vertices) in the grid,
- **coords:** an $n_n \times d$ array of physical nodal coordinates in $\mathbb{R}^d$. Global node i is at physical coordinate coords(i,:).

The grid is constructed according to a right-handed coordinate system where the z coordinate is interpreted as depth. Consequently, plotting routines such as plotGrid display the grid with a reversed z axis. To illustrate how the grid structure works, we consider two examples.

Example 1: We start by considering a regular 3×2 grid, in which we take away the second cell in the logical numbering,

```
G = removeCells( cartGrid([3,2]), 2)
```

This produces the output

```
G = 
       cells: [1x1 struct]
       faces: [1x1 struct]
       nodes: [1x1 struct]
    cartDims: [3 2]
        type: {'tensorGrid'  'cartGrid'  'removeCells'}
     griddim: 2
```

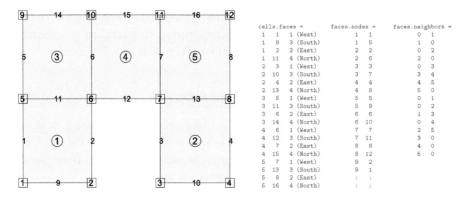

Figure 3.28 Illustration of the `cell` and `faces` fields of the grid structure: cell numbers are marked by circles, node numbers by squares, and face numbers have no marker.

Examining the output from the call, we notice that the field `G.type` contains three values, `'cartGrid'` indicates the creator of the grid, which again relies on `'tensorGrid'`, whereas the field `'removeCells'` indicates that cells have been removed from the Cartesian topology. The resulting 2D geometry consists of five cells, twelve nodes, and sixteen faces. All cells have four faces and hence `G.cells.facePos = [1 5 9 13 17 21]`. Figure 3.28 shows[4] the geometry and topology of the grid, including the content of the fields `cells.faces`, `faces.nodes`, and `faces.neighbors`. We notice, in particular, that all interior faces (6, 7, 11, and 13) are represented twice in `cells.faces` as they belong to two different cells. Likewise, for all exterior faces, the corresponding row in `faces.neighbors` has one zero entry. Finally, being logically Cartesian, the grid structure contains a few optional fields:

- `G.cartDims` equals [3 2],
- `G.cells.indexMap` equals [1 3 4 5 6], since the second cell in the logical numbering has been removed from the model, and
- `G.cells.faces` contains a third column with tags that distinguish global directions for the individual faces.

Example 2: We consider the Delaunay triangulation of seven points in 2D:

```
p = [ 0.0, 1.0, 0.9, 0.1, 0.6, 0.3, 0.75; ...
      0.0, 0.0, 0.8, 0.9, 0.2, 0.6, 0.45]'; p = sortrows(p);
G = triangleGrid(p)
```

---

[4] To create the plot in Figure 3.28, we first called `plotGrid` to plot the grid, then called `computeGeometry` to compute cell and face centroids, which were used to place a marker and a text label with the cell/face number in the correct position.

3.4 Grid Structure in MRST

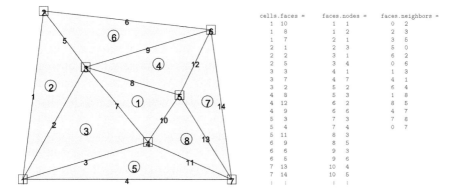

Figure 3.29 Illustration of the `cell` and `faces` fields of the grid structure: cell numbers are marked by circles, node numbers by squares, and face numbers have no marker.

which produces the output

```
G =
    faces: [1x1 struct]
    cells: [1x1 struct]
    nodes: [1x1 struct]
     type: {'triangleGrid'}
   griddim: 2
```

The grid contains no structured parts, and thus G consists of the three mandatory fields, `cells`, `faces`, and `nodes`, that are sufficient to determine the geometry and topology of the grid; the `type` tag that names its creator, and `griddim` that gives that it is a surface grid. Altogether, the grid consists of eight cells, fourteen faces, and seven nodes, which are shown in Figure 3.29, along with the contents of the fields `cells.faces`, `faces.nodes`, and `faces.neighbors`. Notice, in particular, the absence of the third column in `cells.faces`, since face tags associated with the axial directions generally do not make sense for a (fully) unstructured grid. Likewise, the `cells` structure does not contain any `indexMap`, as all cells in the model are active.

Computing geometry information. All grid-factory routines in MRST generate the basic geometry and topology of a grid, that is, how nodes are connected to make up faces, how faces are connected to form cells, and how cells are connected over common faces. Whereas this information is sufficient for many purposes, more geometric information may be required in many cases. As explained earlier, such information is provided by the routine `computeGeometry`, which computes cell centroids and volumes as well as face areas, centroids, and normals. Computing this information is straightforward for simplexes and Cartesian grids, but is not so for general polyhedral grids that may contain curved polygonal faces. In the following we will therefore go through how it is done in MRST. Our discussion follows the original implementation, which in recent releases has been superseded by a more optimized routine. You can find the complete original code in `computeGeometryOrig`.

For each cell, the basic grid structure provides us with a list of vertices, a list of cell faces, etc., as shown in plots **a** and **e** of Figure 3.30. The routine starts by computing face quantities (areas, centroids, and normals). To utilize MATLAB efficiently, the computations are programmed using vectorization so that each derived quantity is computed for all points, all faces, and all cells in one go. To keep the current presentation as simple as possible, we will herein only give formulas for a single face and a single cell. Let us consider a single face given by the points $\vec{p}(i_1), \ldots, \vec{p}(i_m)$ and let $\alpha = (\alpha_1, \ldots, \alpha_m)$ denote a multi-index describing how these points are connected to form the outline of the faces. For the face with global number j, the multi-index is given by the vector

```
G.faces.nodes(G.faces.nodePos(j):G.faces.nodePos(j+1)-1)
```

Let us consider two faces. Global face number two in Figure 3.30a is planar and consists of points $\vec{p}(2), \vec{p}(4), \vec{p}(6), \vec{p}(8)$ with the ordering $\alpha = (2, 4, 8, 6)$. Likewise, we consider global face number one in Figure 3.30e, which is curved and consists of points $\vec{p}(1), \ldots, \vec{p}(5)$ with the ordering $\alpha = (4, 3, 2, 1, 5)$. For curved faces, we need to make a choice of how to interpret the surface spanned by the node points. In MRST (and some commercial simulators) this is done as follows: we start by defining a so-called *hinge point* $\vec{p}_h$, which is often given as part of the input specification of the grid. If not, the hinge point can be computed as the center point of the m points that make up the face, $\vec{p}_h = \sum_{k=1}^{m} \vec{p}(\alpha_k)/m$. The hinge point can now be used to tessellate the face into m triangles, as shown in Figures 3.30b and 3.30f. The triangles are defined by the points $\vec{p}(\alpha_k), \vec{p}(\alpha_{\mathrm{mod}(k,m)+1})$, and $\vec{p}_h$ for $k = 1, \ldots, m$. Each triangle has a center point $\vec{p}_c^k$ defined in the usual way as the average of its three vertexes and a normal vector and area given by

$$\vec{n}^k = \left(\vec{p}(\alpha_{\mathrm{mod}(k,m)+1}) - \vec{p}(\alpha_k)\right) \times \left(\vec{p}_h - \vec{p}(\alpha_k)\right) = \vec{v}_1^k \times \vec{v}_2^k$$

$$A^k = \sqrt{\vec{n}^k \cdot \vec{n}^k}.$$

The face area, centroid, and normal are now computed as follows

$$A_f = \sum_{k=1}^{m} A^k, \qquad \vec{c}_f = (A_f)^{-1} \sum_{k=1}^{m} \vec{p}_c^k A^k, \qquad \vec{n}_f = \sum_{k=1}^{m} \vec{n}^k. \qquad (3.2)$$

The result is shown in plot **c** of Figure 3.30, where the observant reader will see that the centroid $\vec{c}_f$ does not coincide with the hinge point $\vec{p}_h$ unless the planar face is a square. This effect is more pronounced for the curved faces of the PEBI cell in Figure 3.30g.

The computation of centroids in (3.2) is only valid if the grid does not have faces with zero area, because otherwise the second formula would involve a division by zero and hence incur centroids with NaN values. The reader interested in creating his/her own grid-factory routines for grids that may contain degenerate (pinched) cells should be aware of this and make sure that all faces with zero area are removed in a preprocessing step.

3.4 Grid Structure in MRST

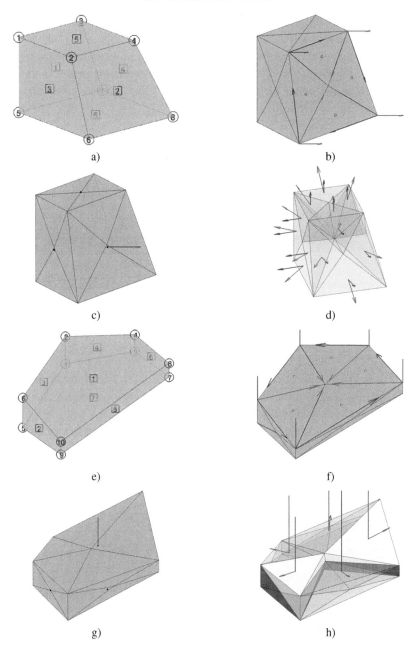

Figure 3.30 Steps in the computation of geometry information for a single corner-point cell and a single PEBI cell using `computeGeometry`. a) A single corner-point cell with face numbers (squares) and node numbers (circles). b) Tessellation of faces with vectors $\vec{v}_1^k$ (blue), $\vec{v}_2^k$ (green), and $\vec{n}^k$ (red). c) Face centroids and normal vectors computed from tessellation. d) Triangulation of cell volume with vectors $\vec{n}^k$ (blue) and $\vec{c}_r^k$ (green). e) A single PEBI cell with face numbers (squares) and node numbers (circles). f) Tessellation of faces with vectors $\vec{v}_1^k$ (blue), $\vec{v}_2^k$ (green), and $\vec{n}^k$ (red). g) Face centroids and normal vectors computed from tessellation. h) Triangulation of cell volume with vectors $\vec{n}^k$ (blue) and $\vec{c}_r^k$ (green).

To compute the cell centroid and volume, we start by computing the center point $\vec{c}_c$ of the cell, which we define as the average of the face centroids, $\vec{c}_c = \sum_{k=1}^{m_f} \vec{c}_f / m_f$, where m_f is the number of faces of the cell. By connecting this center point to the m_t face triangles, we define a unique triangulation of the cell volume, as shown in Figures 3.30d and 3.30h. For each tetrahedron, we define the vector $\vec{c}_r^k = \vec{p}_c^k - \vec{c}_c$ and compute the volume (which may be negative if the center point $\vec{c}_c$ lies outside the cell)

$$V^k = \tfrac{1}{3}\vec{c}_r^k \cdot \vec{n}^k.$$

The triangle normals $\vec{n}^k$ will point outward or inward depending upon the orientation of the points used to calculate them, and to get a correct computation we therefore must modify the triangle normals so that they point outward. Finally, we can define the volume and the centroid of the cell as follows

$$V = \sum_{k=1}^{m_t} V^k, \qquad \vec{c} = \vec{c}_c + \frac{3}{4V} \sum_{k=1}^{m_t} V^k \vec{c}_r^k. \qquad (3.3)$$

In the original implementation, all cell quantities were computed inside a loop, which may not be as efficient as the computation of the face quantities. This has been improved in recent releases.

COMPUTER EXERCISES

3.4.1 Go back to Exercise 3.1.1. What would you do to randomly perturb all nodes in the grid except for those that lie on an outer face whose normal vector has no component in the y-direction?

3.4.2 Exercise 3.2.2 extended the function `triangleGrid` to triangulated surfaces in 3D. Verify that `computeGeometry` computes cell areas, cell centroids, face centroids, and face lengths correctly for 3D surfaces.

3.4.3 How would you write a function that purges all cells that have an invalid vertex (with value `NaN`) from a grid?

3.5 Examples of More Complex Grids

To help users generate test cases, MRST supplies a number of routines for generating example grids. We have previously encountered `twister`, which perturbs the x and y coordinates in a grid. Likewise, in Section 2.5 we used `simpleGrdecl` to generate a simple ECLIPSE input stream for a stratigraphic grid describing a wavy structure with a single deviated fault. The routine has several options that enable you to specify the magnitude of the fault displacement, flat rather than a wavy top and bottom surfaces, and vertical rather than inclined coordinate lines; see Figure 3.31.

The routine with the somewhat cryptic name `makeModel3` generates a corner-point input stream that models parts of an anticline structure that is cut through by two faults; see Figure 3.32. Similarly, `extrudedTriangleGrid` generates a 2.5D prismatic grid with a laterally

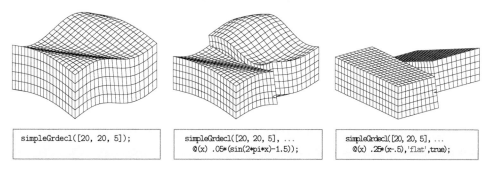

Figure 3.31 The `simpleGrdecl` routine can be used to produce faulted, two-block grids of different shapes.

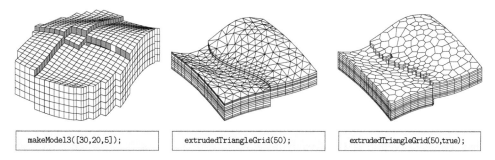

Figure 3.32 Three different example grids created by the grid example functions `makeModel3` and `extrudedTriangleGrid`.

curved fault in the middle. Alternatively, the routine can generate a 2.5D PEBI grid in which the curved fault is laterally stair-stepped; see Figure 3.32.

3.5.1 SAIGUP: Model of a Shallow-Marine Reservoir

Having discussed the corner-point format in some detail, it is now time to return to the SAIGUP model. In the following discussion, we will look at the grid representation in more detail and show some examples of how to interact and visualize different features of the grid (see also the last video of the second MRST Jolt on grids and petrophysical data [189]). You can find the complete source code `showGridSAIGUP`. In Section 2.5, we saw that parsing the input file creates the following structure

```
grdecl =
    cartDims: [40 120 20]
      COORD: [29766x1 double]
      ZCORN: [768000x1 double]
     ACTNUM: [96000x1 int32]
      PERMX: [96000x1 double]
         :    :    :
```

The first four fields describe the grid and the remaining the petrophysical propertes. Here, we will (mostly) consider the first four fields, which represent the following information:

1. The dimension of the underlying logical Cartesian grid: ECLIPSE keyword SPECGRID, equal $40 \times 120 \times 20$.
2. The coordinate lines: ECLIPSE keyword COORD, top and bottom coordinate per vertex in the logical 40×120 areal grid, i.e., $6 \times 41 \times 121$ values.
3. The coordinates along the individual coordinate lines: ECLIPSE keyword ZCORN, eight values per cell, i.e., $8 \times 40 \times 120 \times 20$ values.
4. The Boolean indicator for active cells: ECLIPSE keyword ACTNUM, one value per cell, i.e., $40 \times 120 \times 20$ values.

To process the ECLIPSE input stream and turn the corner-point grid into MRST's unstructured description, we use the processGRDECL routine. The interested reader may ask the processing routine to display diagnostic output

```
G = processGRDECL(grdecl, 'Verbose', true);
G = computeGeometry(G)
```

and consult the SAIGUP tutorial (showSAIGUP) or the technical documentation of the processing routine for a more detailed explanation of the resulting output.

The model has been created with vertical pillars having lateral resolution of 75 meters and vertical resolution of 4 meters, giving a typical aspect ratio of 18.75. You can see this, e.g., by extracting the pillars and corner points and analyzing the results as follows:

```
[X,Y,Z] = buildCornerPtPillars(grdecl,'Scale',true);
dx = unique(diff(X)).'
[x,y,z] = buildCornerPtNodes(grdecl);
dz = unique(reshape(diff(z,1,3),1,[]))
```

The resulting grid has 78,720 cells that are almost equal in size (consult the histogram hist(G.cells.volumes)), with cell volumes varying between 22,500 m^3 and 24,915 m^3. Altogether, the model has 264,305 faces: 181,649 vertical faces on the outer boundary and between lateral neighbors, and 82,656 lateral faces on the outer boundary and between vertical neighbors. Most of the vertical faces are not part of a fault and are therefore parallelograms with area equal 300 m^2. However, the remaining 26,000–27,000 faces are a result of the subdivision introduced to create a matching grid along the (stair-stepped) faults. Figure 3.33 shows where these faces appear in the model and a histogram of their areas: the smallest face has an area of $5.77 \cdot 10^{-4}$ m^2 and there are 43, 202, and 868 faces with areas smaller than 0.01, 0.1, and 1 m^2, respectively. The processGRDECL routine has an optional parameter 'Tolerance' that sets the minimum distance for distinguishing points along coordinate lines (default value is zero). By setting this to parameter to 5, 10, 25, or 50 cm, the area of the smallest face is increased to 0.032, 0.027, 0.097, or 0.604 m^2, respectively. In general, we advice against aggressive use of this tolerance parameter; one should

3.5 Examples of More Complex Grids 99

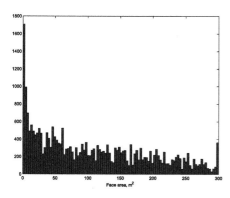

Figure 3.33 Faces that have been subdivided for the SAIGUP mode. The left plot shows a histogram of the faces areas. The right plot shows all fault faces (yellow) and fault faces having area less than 290 m^2 (red).

instead develop robust discretization schemes and, if necessary, suitable postprocessing methods that eliminate or ignore faces with small areas.

Visualizing subsets of the grid. Next, we show a few examples of visualizations highlighting various mechanisms for interacting with the grid and accessing parts of it. As a first example, we start by plotting the layered structure of the model. To this end, we use a simple trick: create a matrix with ones in all cells of the logical Cartesian grid and then do a cumulative summation in the vertical direction to get increasing values,

```
val = cumsum(ones(G.cartDims),3);
```

which we then plot using a standard call to plotCellData; see the left plot in Figure 3.34. To reveal the layering in the interior of the model, we extract and visualize only the cells that are adjacent to a fault:

```
cellList = G.faces.neighbors(G.faces.tag>0, :);
cells    = unique(cellList(cellList>0));
```

The first statement uses logical indexing to loop through all faces and extract the neighboring cells of all faces that are marked with a tag (i.e., lie at a fault face). The list may have repeated entries if a cell is attached to more than one fault face and contain zeros if a fault face is part of the outer boundary. We get rid of these in the second statement, and can then plot the result list of cells, giving the plot shown to the right in Figure 3.34

```
plotCellData(G,val(G.cells.indexMap),cells)
```

Next, let us inspect the fault structure in the lower-right corner of the plot. If we disregard using cutGrdecl as discussed on page 81, there are basically two ways we can extract parts

Figure 3.34 Visualizing the layered structure of the SAIGUP model.

of the model, which both rely on the construction of a map of cell numbers of logical indices. In the first method, we first construct a logical set for the cells in a logically Cartesian bounding box and then use the built-in function `ismember` to extract the members of `cells` that lie within this bounding box:

```
[ijk{1:3}] = ind2sub(G.cartDims, G.cells.indexMap); IJK = [ijk{:}];
[I,J,K] = meshgrid(1:9,1:30,1:20);
bndBox = find(ismember(IJK,[I(:), J(:), K(:)],'rows'));
inspect = cells(ismember(cells,bndBox));
```

The `ismember` function has an operational count of $\mathcal{O}(n \log n)$. A faster alternative is to use logical operations having an operational count of $\mathcal{O}(n)$. That is, we construct a vector of boolean numbers that are `true` for the entries we want to extract and `false` for the remaining entries

```
I = false(G.cartDims(1),1); I(1:9)=true;
J = false(G.cartDims(2),1); J(1:30)=true;
K = false(G.cartDims(3),1); K(1:20)=true;
pick  = I(ijk{1}) & J(ijk{2}) & K(ijk{3});
pick2 = false(G.cells.num,1); pick2(cells) = true;
inspect = find(pick & pick2);
```

Both approaches produce the same index set; the resulting plot is shown in Figure 3.35. To mark the fault faces in this subset of the model, we do the following steps

```
cellno  = gridCellNo(G);
faces   = unique(G.cells.faces(pick(cellno), 1));
inspect = faces(G.faces.tag(faces)>0);
plotFaces(G, inspect, [.7 .7 .7], 'EdgeColor','r');
```

The first statement constructs a list of cells attached to all faces in the model, the second extracts a unique list of face numbers associated with the cells in the logical vector `pick` (which represents the bounding box in logical index space), and the third statement extracts the faces within this bounding box that are marked as fault faces.

3.5 Examples of More Complex Grids   101

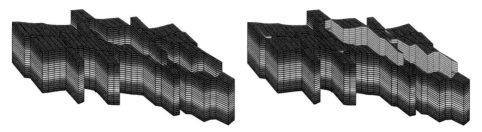

Figure 3.35 Details from the SAIGUP model showing a zoom of the fault structure in the lower-right corner of the right plot in Figure 3.34. The left plot shows the cells attached to the fault faces, and in the right plot the fault faces have been marked with gray color and red edges.

Logical operations are also useful in other circumstances. As an example, we will extract a subset of cells forming a sieve that can be used to visualize the petrophysical quantities in the interior of the model:

```
% Every fifth cell in the x-direction
I = false(G.cartDims(1),1); I(1:5:end)=true;
J = true(G.cartDims(2),1);
K = true(G.cartDims(3),1);
pickX = I(ijk{1}) & J(ijk{2}) & K(ijk{3});

% Every tenth cell in the y-direction
I = true(G.cartDims(1),1);
J = false(G.cartDims(2),1); J(1:10:end) = true;
pickY = I(ijk{1}) & J(ijk{2}) & K(ijk{3});

% Combine the two picks
plotCellData(G,rock.poro, pickX | pickY, 'EdgeColor','k','EdgeAlpha',.1);
```

3.5.2 Composite Grids

Using an unstructured grid description enables you to define composite grids whose geometries and topologies vary throughout the model. That is, different types of cells or different grid resolution may be used locally to adapt to well trajectories and flow and geological constraints; see e.g., [123, 209, 113, 202, 51, 83, 292] and references therein. You may already have encountered a composite grid if you did Exercise 3.2.7 on page 72, where we sought an unstructured grid that adapted to two skew faults and was padded with rectangular cells near the boundary. As another example, we will generate a Cartesian grid that has a radial refinement around two wells in the interior of the domain. This composite grid will be constructed from a set of control points using the pebi routine. To this end, we first construct the generating points for a unit refinement, as discussed in Figure 3.25 on page 86

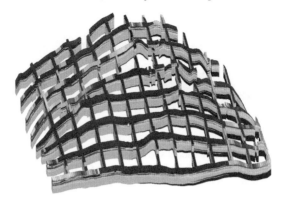

Figure 3.36 A "sieve" plot of the porosity in the SAIGUP model. Using this technique, one can more easily see the structure in the interior of the model.

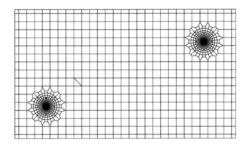

Figure 3.37 A composite grid consisting of a regular Cartesian mesh with radial refinement around two well positions.

```
Pw = [];
for r = exp(-3.5:.2:0),
   [x,y,z] = cylinder(r,28); Pw = [Pw [x(1,:); y(1,:)]];
end
Pw = [Pw [0; 0]];
```

Then this point set is translated to the positions of the wells and glued into a standard regular point lattice (generated using `meshgrid`):

```
Pw1 = bsxfun(@plus, Pw, [2; 2]);
Pw2 = bsxfun(@plus, Pw, [12; 6]);
[x,y] = meshgrid(0:.5:14, 0:.5:8);
P = unique([Pw1'; Pw2'; x(:) y(:)], 'rows');
G = pebi(triangleGrid(P));
```

Figure 3.37 shows the resulting grid. To get a good grid, it is important that the number of points around the cylinder has a reasonable match with the density of the points in the regular lattice. If not, the transition cells between the radial and the regular grid may exhibit

3.5 Examples of More Complex Grids

Figure 3.38 Examples of composite grids. The left plot shows an areal grid consisting of Cartesian, hexagonal, and radial parts. The right plot shows the same grid extruded to 3D with two faults added. Both faults are stair-stepped in the areal direction. The fault in the north–south direction has no displacement and thus the adjacent cells have no extra non-neighboring connections.

quite unfeasible geometries. The observant reader will also notice the layer of small cells at the boundary, which is an effect of the particular distribution of the generating points (see the left plot in Figure 3.10 on page 67) and can, if necessary be avoided by a more meticulous choice of points.

In the left plot of Figure 3.38, we have combined these two approaches to generate an areal grid consisting of three characteristic components: Cartesian grid cells at the outer boundary, hexagonal cells in the interior, and a radial grid with exponential radial refinement around two wells. The right plot shows a 2.5D grid in which the areal Voronoi grid has been extruded to 3D along vertical pillars. In addition, structural displacement has been modeled along two areally stair-stepped faults that intersect near the west boundary. Petrophysical parameters have been sampled from layers 40–44 of the SPE 10 data set [71].

3.5.3 Control-Point and Boundary Conformal Grids

To correctly split the reservoir volume into sub-volumes, build flow units with similar or correlated petrophysical properties, and resolve flow patterns, it is important that a volumetric simulation grid conforms as closely as possible to fault surfaces and horizons representing the structural architecture of the reservoir. This can be achieved by introducing unstructured grids whose cell interfaces adapt to curved lines in 2D and triangulated surfaces in 3D. Likewise, it is of interest to have grids that can adapt the centroids of certain cells to 2D lines or 3D lines or planes. This is particularly relevant if one were to provide accurate description of well paths and the near-well region. Modern wells are typically long and horizontal and may have complex geometry that consists of multiple branches. Such

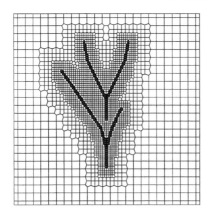 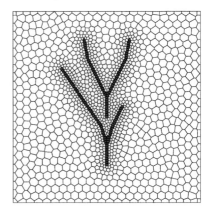

Figure 3.39 Two different 2D grids adapting to curved well paths. The hybrid grid shown to the left has Cartesian cells away from the well path, with two levels of refinement in the near-well zone. Polygonal cells are used on the transition between refinement levels and to track the well path. The Voronoi grid shown to the right has gradual refinement towards the well and is computed as the dual to a triangulation generated by `DistMesh`.

wells are increasingly becoming the main determining factor for reservoir flow patterns because of their long reach, but also as a result of various techniques for modifying the near-well region to increase injectivity or use of (intelligent) inflow devices to control fluid production.

MRST has a third-party module called `upr`, which offers such adaptivity in the form of 2D Voronoi grids and/or composite grids as well as fully 3D Voronoi grids. The module was initially developed by one of my previous students as part of his master thesis [42]. The most recent version offers two types of conformity to lower-dimensional objects: (i) control-point alignment of cell centroids to accurately represent horizontal and multilateral wells or fracture networks and (ii) boundary alignment of cell faces to accurately preserve layers, fractures, faults, pinchouts, etc.

Each constraint is represented as a lower-dimensional grid, and intersections of two objects is associated with a grid with dimension one lower than the objects. The grids are hierarchically consistent in the sense that cell faces of a higher-dimensional grid conform to the cells of all lower-dimensional grids. I will not go into more details about the underlying methods; you can find more details, as well as references to other related work, in [42, 169]. Instead, I have included Figures 3.39–3.42 to illustrate the flexibility in geometric modeling inherent in such grids.

3.5.4 Multiblock Grids

A somewhat different approach to get grids whose geometry and topology vary throughout the physical domain is to use multiblock grids in which different types of structured or

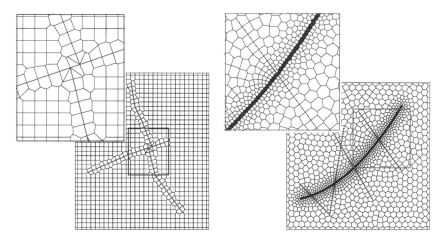

Figure 3.40 The left grid adapts to three intersecting faults, whereas the grid to the right adapts to both cell and face constraints.

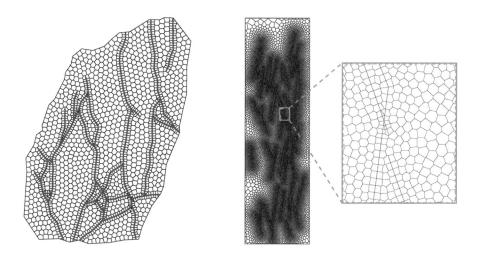

Figure 3.41 The left grid is an areal description of a complex fault network taken from Branets et al. [46] and is generated as a 2D centroidal Voronoi diagram (CVD) found by minimizing an energy functional. In the right grid, there are 51 randomly distributed fractures, and the resolution is graded toward these.

unstructured subgrids are glued together. The resulting grid can be nonmatching across block interfaces (see e.g., [313, 25, 312]) or have grid lines that are continuous (see e.g., [148, 180]). MRST does not have any grid-factory routine for generating advanced multi-block grids, but offers the function `glue2DGrid` for gluing together rectangular blocks in

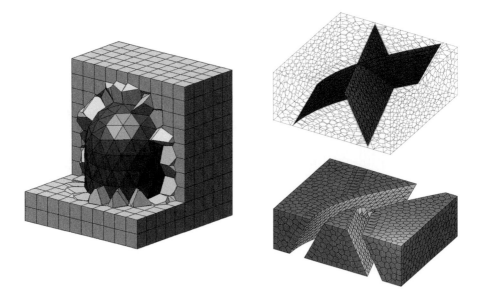

Figure 3.42 The left grid shows a conceptual model of a salt dome, with triangulated surface extracted from a 3D triangulation generated by `DistMesh`. The plots to the right shows a 3D Voronoi grid adapted to a curved and a planar fault. At the intersection of these, only the curved fault is represented exactly.

2D. These grids can, if necessary, be extruded to 3D by the use of `makeLayeredGrid`. In the following, we will show a few examples of such grids.

As our first example, let us generate a curvilinear grid that has a local refinement at its center as shown in Figure 3.43 (see also Exercise 3.2.6 on page 71). To this end, we start by generating three different block types shown in red, green, and blue colors to the left in the figure:

```
G1 = cartGrid([ 5  5],[1 1]);
G2 = cartGrid([20 20],[1 1]);
G3 = cartGrid([15  5],[3 1]);
```

Once these are in place, we translate the blocks and glue them together and then apply the `twister` function to make a curvilinear transformation of each grid line:

```
G = glue2DGrid(G1, translateGrid(G2,[1 0]));
G = glue2DGrid(G,  translateGrid(G1,[2 0]));
G = glue2DGrid(G3, translateGrid(G, [0 1]));
G = glue2DGrid(G,  translateGrid(G3,[0 2]));
G = twister(G);
```

Let us now replace the central block by a patch consisting of triangular cells. To this end, we start by generating a new grid G2

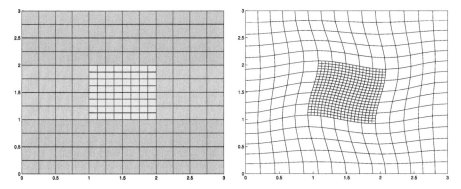

Figure 3.43 Using a multiblock approach to construct a rectilinear grid with refinement.

```
[N,M]=deal(10,15);
[x,y] = ndgrid( linspace(0,1,N+1), linspace(0,1,M+1));
x(2:N,2:M) = x(2:N,2:M) + 0.3*randn(N-1,M-1)*max(diff(xv));
y(2:N,2:M) = y(2:N,2:M) + 0.3*randn(N-1,M-1)*max(diff(yv));

G2 = computeGeometry(triangleGrid([x(:) y(:)]));
```

The glue2DGrid routine relies on face tags as explained in Figure 3.4 on page 89 that can be used to identify the external faces facing east, west, north, and south. Generally, such tags do not make much sense for triangular grids and are therefore not supplied. However, to be able to find the correct interface to glue together, we need to supply tags on the perimeter of the triangular patch, where the normal vectors follow the axial directions and tags therefore make sense. To this end, we start by computing the true normal vectors:

```
hf   = G2.cells.faces(:,1);
hf2cn = gridCellNo(G2);
sgn  = 2*(hf2cn == G2.faces.neighbors(hf, 1)) - 1;
N    = bsxfun(@times, sgn, G2.faces.normals(hf,:));
N    = bsxfun(@rdivide, N, G2.faces.areas(hf,:));
n    = zeros(numel(hf),2); n(:,1)=1;
```

Then, all interfaces that face east are those whose dot-product with the vector $(1,0)$ is identical to -1:

```
G2.cells.faces(:,2) = zeros(size(hf));
i = sum(N.*n,2)==-1; G2.cells.faces(i,2) = 1;
```

Similarly, we can identify all the interfaces facing west, north, and south. The left plot in Figure 3.44 shows the resulting multiblock grid. Likewise, the middle plot shows another multiblock grid where the central refinement is the dual to the triangular patch and G3 has

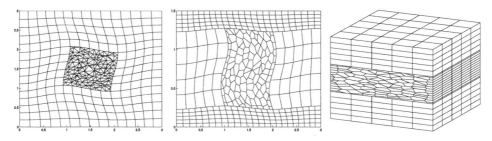

Figure 3.44 Using a multiblock approach to construct rectilinear grid with triangular (left) and polygonal refinements (middle). The right plot shows a 3D multiblock grid.

been scaled in the y-direction and refined in the x-direction so that the grid lines are no longer matching with the grid lines of G1.

As a last example, let us use this technique to generate a 3D multiblock grid that consists of three blocks in the vertical direction

```
G = glue2DGrid(G1, translateGrid(G2,[0 1]));
G = glue2DGrid(G,  translateGrid(G1,[0 2]));
G = makeLayeredGrid(G, 5);
G.nodes.coords = G.nodes.coords(:,[3 1 2]);
```

That is, we first generate an areal grid in the xy-plane, extrude it to 3D along the z direction, and then permute the axis so that their relative orientation is correctly preserved (notice that simply flipping [1 2 3] → [1 3 2], for instance, will *not* create a functional grid).

COMPUTER EXERCISES

3.5.1 How would you populate the grids shown in Figures 3.31 and 3.32 with petrophysical properties so that spatial correlation and displacement across the fault(s) is correctly accounted for? As an illustrative example, you can try to sample petrophysical properties from the SPE 10 data set.

3.5.2 Select at least one of the models in the data sets BedModels1 or BedModel2 and try to find all inactive cells and then all cells that do not have six faces. Hint: it may be instructive to visualize these models both in physical space and in index space.

3.5.3 Extend your function from Exercise 3.2.7 on page 72 to also include radial refinement in near-well regions as shown in Figure 3.37.

3.5.4 Make a grid similar to the one shown to the right in Figure 3.38. Hint: although it is not easy to see, the grid is matching across the fault, which means that you can use the method of fictitious domain to make the fault structure.

3.5.5 As pointed out in Exercise 3.2.6, MRST does not yet have a grid factory routine to generate structured grids with local nested refinement as shown in the figure to the right. While it is not very difficult to generate the necessary vertices if each refinement patch is rectangular and matches the grid cells on the coarser level, building the grid structure may prove to be a challenge. Try to develop an efficient algorithm and implement it in MRST.

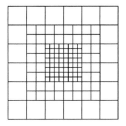

Part II

Single-Phase Flow

4
Mathematical Models for Single-Phase Flow

If you have read the chapters of the book in chronological order, you have already encountered the equations modeling flow of a single, incompressible fluid through a porous media twice: first in Section 1.4, which showed how vertical equilibrium inside can be computed by use of MRST, and then in Section 2.4.2, in which we discussed the concept of rock permeability. In this chapter, we review mathematical models for single-phase flow in more detail, including auxiliary boundary and initial conditions, subscale models for injection and production wells, and steady-state advection equations. We then introduce basic spatial discretizations for the resulting equations and show how these discretizations can be formulated as abstract discrete differentiation operators. A more thorough discussion of how these discretizations can be turned into efficient solvers is deferred to Chapters 5 and 6, which focus on the incompressible case, and Chapter 7, which discusses compressible flow and how the discrete differentiation operators can be combined with automatic differentiation for rapid prototyping of new solvers.

4.1 Fundamental Concept: Darcy's Law

Macroscale modeling of single-phase flow in porous media started with the work of Henry Darcy, a French hydraulic engineer, who in the middle of the nineteenth century was engaged to enlarge and modernize the waterworks of the city of Dijon. To understand the physics of flow through the sand filters that were used to clean the water supply, Darcy designed a vertical experimental tank filled with sand, in which water was injected at the top and allowed to flow out at the bottom; Figure 4.1 shows a conceptual illustration. Once the sand pack is filled with water, and the inflow and outflow rates are equal, the hydraulic head at the inlet and at the outlet can be measured using mercury-filled manometers. The hydraulic head is given as $h = E/mg = z - p/\rho g$, relative to a fixed datum. As water flows through the porous medium, it will experience a loss of energy. In a series of experiments, Darcy measured the water volumetric flow rate out of the tank and compared this rate with the loss of hydrostatic head from top to bottom of the column. From the experiments, he established that for the same sand pack, the discharge (flow rate) Q [m^3/s] is proportional

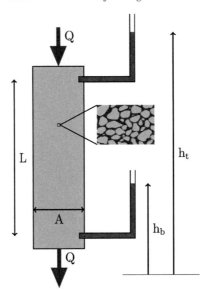

Figure 4.1 Conceptual illustration of Darcy's experiment.

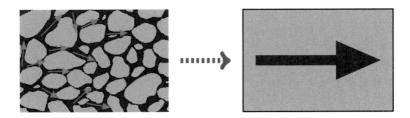

Figure 4.2 The macroscopic Darcy velocity represents an average of microscopic fluid fluxes.

to the cross-sectional area A [m^2] and the difference in hydraulic head (height of the water) $h_t - h_b$ [m], and inversely proportional to the flow length of the tank L [m]. Altogether, this can be summarized as

$$\frac{Q}{A} = \kappa \frac{h_t - h_b}{L}. \tag{4.1}$$

The result was presented in 1856 as an appendix to [77] entitled "Determination of the laws of flow of water through sand" and is the direct predecessor to what we today call Darcy's law. In (4.1), κ [m/s] denotes the hydraulic conductivity, which is a function both of the medium and the fluid flowing through it. It follows from a dimensional analysis that $\kappa = \rho g K/\mu$, where g [m/s^2] is the gravitational acceleration, μ [kg/ms] is the dynamic viscosity, and K [m^2] is the intrinsic permeability of a given sand pack.

The specific discharge $v = Q/A$, or Darcy flux, through the sand pack represents the volume of fluid flowing per total area of the porous medium and has dimensions [m/s].

Somewhat misleading, v is often referred to as a velocity. However, since only a fraction of the cross-sectional area is available for flow (the majority of the area is blocked by sand grains), v is not a velocity in the microscopic sense. Instead, v is the apparent macroscopic velocity obtained by averaging the microscopic fluxes inside representative elementary volumes (REVs), which were discussed in Section 2.3.2. The macroscopic fluid velocity, defined as volume of fluid flowing per area occupied by fluid, is therefore given by v/ϕ, where ϕ is the porosity associated with the REV.

Henceforth, we will, with a slight abuse of notation, refer to the specific discharge as the *Darcy velocity*. In modern differential notation, first proposed by Nutting [239] and Wyckoff et al. [320], Darcy's law for a single-phase fluid reads,

$$\vec{v} = -\frac{\mathbf{K}}{\mu}(\nabla p - g\rho \nabla z), \qquad (4.2)$$

where p is the fluid pressure and z is the vertical coordinate. The equation expresses conservation of momentum and was derived from the Navier–Stokes equations by averaging and neglecting inertial and viscous effects by Hubbert [137] and later from Stokes flow by Whitaker [315]. The observant reader will notice that Darcy's law (4.2) is analogous to Fourier's law (1822) for heat conduction, Ohm's law (1827) in the field of electrical networks, or Fick's law (1855) for fluid concentrations in diffusion theory, except that for Darcy there are two driving forces, pressure and gravity. Notice also that Darcy's law assumes a reversible fluid process, which is a special case of the more general physical laws of irreversible processes that were first described by Onsager (1931).

4.2 General Flow Equations for Single-Phase Flow

To derive a mathematical model for single-phase flow on the macroscopic scale, we first make a continuum assumption based on the existence of REVs as discussed in the previous section and then look at a control volume as shown in Figure 4.3. From the fundamental

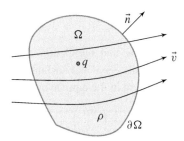

Figure 4.3 Illustration of a control volume Ω on which one can apply the principle of conservation to derive macroscopic continuity equations.

law of mass conservation, we know that the accumulation of mass inside this volume must equal the net flux over the boundaries,

$$\frac{\partial}{\partial t}\int_\Omega \phi\rho\, d\vec{x} + \int_{\partial\Omega} \rho\vec{v}\cdot\vec{n}\, ds = \int_\Omega \rho q\, d\vec{x}, \qquad (4.3)$$

where ρ is the density of the fluid, ϕ is the rock porosity, $\vec{v}$ is the macroscopic Darcy velocity, $\vec{n}$ denotes the normal at the boundary $\partial\Omega$ of the computational domain Ω, and q denotes fluid sources and sinks, i.e., outflow and inflow of fluids per volume at certain locations. Applying Gauss's theorem, this conservation law can be written in the alternative integral form

$$\int_\Omega \left[\frac{\partial}{\partial t}\phi\rho + \nabla\cdot(\rho\vec{v})\right] d\vec{x} = \int_\Omega \rho q\, d\vec{x}. \qquad (4.4)$$

This equation is valid for any volume Ω, and in particular volumes that are infinitesimally small. Hence, it follows that the macroscopic behavior of the single-phase fluid must satisfy the continuity equation

$$\frac{\partial(\phi\rho)}{\partial t} + \nabla\cdot(\rho\vec{v}) = \rho q. \qquad (4.5)$$

Equation (4.5) contains more unknowns than equations, and to derive a closed mathematical model, we need to introduce what are commonly referred to as *constitutive equations* that give the relationship between different states of the system (pressure, volume, temperature, etc.) at given physical conditions. Darcy's law, discussed in the previous section, is an example of a constitutive relation that has been derived to provide a phenomenological relationship between the macroscale $\vec{v}$ and the fluid pressure p. In Section 2.4.1 we introduced the rock compressibility $c_r = d\ln(\phi)/dp$, which describes the relationship between the porosity ϕ and the pressure p. In a similar way, we can introduce the fluid compressibility to relate the density ρ to the fluid pressure p.

A change in density will generally cause a change in both pressure p and temperature T. The usual way of describing these changes in thermodynamics is to consider the change of volume V for a fixed number of particles,

$$\frac{dV}{V} = \frac{1}{V}\left(\frac{\partial V}{\partial p}\right)_T dp + \frac{1}{V}\left(\frac{\partial V}{\partial T}\right)_p dT, \qquad (4.6)$$

where the subscripts T and p indicate that the change takes place under constant temperature and pressure, respectively. Since ρV is constant for a fixed number of particles, $V d\rho = -\rho dV$, and (4.6) can be written in the equivalent form

$$\frac{d\rho}{\rho} = \frac{1}{\rho}\left(\frac{\partial\rho}{\partial p}\right)_T dp + \frac{1}{\rho}\left(\frac{\partial\rho}{\partial T}\right)_p dT = c_f dp + \alpha_f dT. \qquad (4.7)$$

Here, c_f denotes the *isothermal compressibility* and α_f the *thermal expansion coefficient*. In many subsurface systems, the density changes slowly so that heat conduction keeps the temperature constant, in which case (4.7) simplifies to

4.2 General Flow Equations for Single-Phase Flow

$$c_f = \frac{1}{\rho}\frac{d\rho}{dp} = \frac{d\ln(\rho)}{dp}. \tag{4.8}$$

The factor c_f, which we henceforth will refer to as the *fluid compressibility*, is nonnegative and will generally depend on both pressure and temperature, i.e., $c_f = c_f(p,T)$.

Introducing Darcy's law and fluid and rock compressibilities in (4.5), we obtain the parabolic equation for the fluid pressure

$$c_t \phi \rho \frac{\partial p}{\partial t} - \nabla \cdot \left[\frac{\rho \mathbf{K}}{\mu}(\nabla p - g\rho \nabla z)\right] = \rho q, \tag{4.9}$$

where $c_t = c_r + c_f$ denotes the *total compressibility*. Notice that this equation is generally nonlinear since both ρ and c_t may depend on p. In the following, we will look briefly at several special cases in which the governing single-phase equation becomes a linear equation for the primary unknown; more extensive discussions can be found in standard textbooks like [246, chap. 1], [69, chap. 2]. For completeness, we also briefly review the concept of an equation of state.

Incompressible flow. In the special case of an incompressible rock and fluid – that is, ρ and ϕ are independent of p so that $c_t = 0$ – (4.9) simplifies to an elliptic equation with variable coefficients,

$$-\nabla \cdot \left[\frac{\mathbf{K}}{\mu}\nabla(p - g\rho z)\right] = q. \tag{4.10}$$

If we introduce the fluid potential, $\Phi = p - g\rho z$, (4.10) can be recognized as a variable-coefficient Poisson's equation $-\nabla \cdot \mathbf{K}\nabla\Phi = q$ or as the Laplace equation $\nabla \cdot \mathbf{K}\nabla\Phi = 0$ if there are no volumetric fluid sources or sinks. In the next section, we will discuss in detail how to discretize the second-order spatial Laplace operator $\mathcal{L} = \nabla \cdot \mathbf{K}\nabla$, which is a key technological component that will enter almost any software for simulating flow in porous rock formations.

Constant compressibility. If the fluid compressibility is constant and independent of pressure, (4.8) can be integrated from a known density ρ_0 at a pressure datum p_0 to give the equation

$$\rho(p) = \rho_0 e^{c_f(p-p_0)}, \tag{4.11}$$

which applies well to most liquids that do not contain large quantities of dissolved gas. To develop the differential equation, we first assume that the porosity and the fluid viscosity do not depend on pressure. Going back to the definition of fluid compressibility (4.8), it also follows from this equation that $\nabla p = (c_f \rho)^{-1}\nabla \rho$, which we can use to eliminate ∇p from Darcy's law (4.2). Inserting the result into (4.5) gives us the continuity equation

$$\frac{\partial \rho}{\partial t} - \frac{1}{\mu \phi c_f}\nabla \cdot \left(\mathbf{K}\nabla \rho - c_f g \rho^2 \mathbf{K}\nabla z\right) = \rho q. \tag{4.12}$$

In the absence of gravity forces and source terms, this is a linear equation for the fluid density that is similar to the classical heat equation with variable coefficients,

$$\frac{\partial \rho}{\partial t} = \frac{1}{\mu \phi c_f} \nabla \cdot (\mathbf{K} \nabla \rho). \tag{4.13}$$

Slightly compressible flow. If fluid compressibility is small, it is sufficient to use a linear relationship to evaluate the exponential form as

$$\rho = \rho_0 \left[1 + c_f (p - p_0) \right]. \tag{4.14}$$

We further assume that ϕ has a similar functional dependence and that μ is constant. For simplicity, we also assume that g and q are both zero. Then, we can simplify (4.9) as follows:

$$\left[(c_r + c_f) \phi \rho \right] \frac{\partial p}{\partial t} = \frac{c_f \rho}{\mu} \nabla p \cdot \mathbf{K} \nabla p + \frac{\rho}{\mu} \nabla \cdot (\mathbf{K} \nabla p).$$

If c_f is sufficiently small, in the sense that $c_f \nabla p \cdot \mathbf{K} \nabla p \ll \nabla \cdot (\mathbf{K} \nabla p)$, we can neglect the first term on the right-hand side to derive a linear equation similar to (4.13) for the fluid pressure (introducing total compressibility $c = c_r + c_f$)

$$\frac{\partial p}{\partial t} = \frac{1}{\mu \phi c} \nabla \cdot (\mathbf{K} \nabla p). \tag{4.15}$$

Ideal gas. If the fluid is a gas, compressibility can be derived from the gas law, which for an ideal gas can be written in two alternative forms,

$$pV = nRT, \qquad \rho = p(\gamma - 1)e. \tag{4.16}$$

In the first form, T is temperature, V is volume, $R = 8.314 \text{ J K}^{-1} \text{mol}^{-1}$ is the gas constant, and $n = m/M$ is the amount of substance of the gas in moles, where m is the mass and M is the molecular weight. In the second form, γ is the adiabatic constant, i.e., the ratio of specific heat at constant pressure and constant volume, and e is the specific internal energy (internal energy per unit mass). In either case, it follows from (4.8) that $c_f = 1/p$.

Gravity effects are negligible for gases and for brevity we once again we assume that ϕ is a function of $\vec{x}$ only. Inserting (4.16) into (4.9) then gives

$$\frac{\partial (\rho \phi)}{\partial t} = \phi (\gamma - 1) e \frac{\partial p}{\partial t} = \frac{1}{\mu} \nabla \cdot (\rho \mathbf{K} \nabla p) = \frac{(\gamma - 1) e}{\mu} \nabla \cdot (p \mathbf{K} \nabla p),$$

from which it follows that

$$\phi \mu \frac{\partial p}{\partial t} = \nabla \cdot (p \mathbf{K} \nabla p) \qquad \Leftrightarrow \qquad \frac{\phi \mu}{p} \frac{\partial p^2}{\partial t} = \nabla \cdot \left(\mathbf{K} \nabla p^2 \right). \tag{4.17}$$

Equation of state. Equation (4.11), (4.14), and (4.16) are all examples of what is commonly referred to as equations of state, which provide constitutive relationships between mass, pressures, temperature, and volumes at thermodynamic equilibrium. Another popular form of these equations are the so-called cubic equations of state, which can be written as cubic functions of the molar volume $V_m = V/n = M/\rho$ involving constants that depend on pressure p_c, temperature T_c, and the molar volume V_c at the critical point, i.e., the point at which $\left(\frac{\partial p}{\partial V}\right)_T = \left(\frac{\partial^2 p}{\partial V^2}\right)_T \equiv 0$. A few particular examples include the Redlich–Kwong equation of state

$$p = \frac{RT}{V_m - b} - \frac{a}{\sqrt{T}\,V_m(V_m + b)},$$
$$a = \frac{0.42748 R^2 T_c^{5/2}}{p_c}, \quad b = \frac{0.08662 R T_c}{p_c}, \tag{4.18}$$

the modified version called Redlich–Kwong–Soave

$$p = \frac{RT}{V_m - b} - \frac{a\alpha}{\sqrt{T}\,V_m(V_m + b)},$$
$$a = \frac{0.427 R^2 T_c^2}{p_c}, \quad b = \frac{0.08664 R T_c}{p_c}, \tag{4.19}$$
$$\alpha = \left[1 + \left(0.48508 + 1.55171\omega - 0.15613\omega^2\right)\left(1 - \sqrt{T/T_c}\right)\right]^2,$$

as well as the Peng–Robinson equation of state,

$$p = \frac{RT}{V_m - b} - \frac{a\alpha}{V_m^2 + 2bV_m - b^2)},$$
$$a = \frac{0.4527235 R^2 T_c^2}{p_c}, \quad b = \frac{0.077796 R T_c}{p_c}, \tag{4.20}$$
$$\alpha = \left[1 + \left(0.37464 + 1.54226\omega - 0.26992\omega^2\right)\left(1 - \sqrt{T/T_c}\right)\right]^2.$$

Here, ω denotes the acentric factor of the species, which is a measure of the centricity (deviation from spherical form) of the molecules in the fluid. The Peng–Robinson model is much better at predicting liquid densities than the Redlich–Kwong–Soave model, which was developed to fit pressure data of hydrocarbon vapor phases. If we introduce an attractive factor A, a repulsive factor B, and the gas deviation factor Z (to be discussed in Section 11.4.3 on page 373)

$$A = \frac{a\alpha p}{(RT)^2}, \quad B = \frac{bp}{RT}, \quad Z = \frac{pV}{RT},$$

the Redlich–Kwong–Soave equation (4.19) and the Peng–Robinson equation (4.20) can be written in alternative polynomial forms,

$$0 = Z^3 - Z^2 + Z(A - B - B^2) - AB, \tag{4.21}$$
$$0 = Z^3 - (1 - B)Z^2 + (A - 2B - 3B^2)Z - (AB - B^2 - B^3), \tag{4.22}$$

which illustrates why they are called cubic equations of state.

4.3 Auxiliary Conditions and Equations

The governing equations for single-phase flow discussed in the previous section are all parabolic, except for the incompressible case in which the governing equation is elliptic. For the solution to be well-posed[1] inside a finite domain for any of the equations, one needs to supply conditions that determine the behavior on the external boundary. For the parabolic equations describing unsteady flow, we also need to impose an initial condition that determines the initial state of the fluid system. This section discusses these conditions in more detail. We also discuss models for representing flow in and out of the reservoir rock through wellbores. Flow near wellbores typically takes place on a length scale much smaller than the length scales of the global flow inside the reservoir, and it is customary to model it using special analytical models. Finally, we also discuss a set of auxiliary equations for describing the movement of fluid elements and/or neutral particles that follow the single-phase flow without affecting it.

4.3.1 Boundary and Initial Conditions

In reservoir simulation one is often interested in describing closed flow systems that have no fluid flow across its external boundaries. This is a natural assumption when studying full reservoirs that have trapped and contained petroleum fluids for million of years. Mathematically, no-flow conditions across external boundaries are modeled by specifying *homogeneous Neumann conditions*,

$$\vec{v} \cdot \vec{n} = 0 \quad \text{for } \vec{x} \in \partial\Omega. \tag{4.23}$$

With no-flow boundary conditions, any pressure solution of (4.10) is immaterial and only defined up to an additive constant, unless a datum value is prescribed at some internal point or along the boundary.

It is also common that parts of the reservoir may be in communication with a larger aquifer system that provides external pressure support, which can be modeled in terms of a *Dirichlet condition* of the form

$$p(\vec{x}) = p_a(\vec{x}, t) \quad \text{for } \vec{x} \in \Gamma_a \subset \partial\Omega. \tag{4.24}$$

The function p_a can, for instance, be given as a hydrostatic condition. Alternatively, parts of the boundary may have a certain prescribed influx, which can be modeled in terms of an *inhomogeneous Neumann condition*,

$$\vec{v} \cdot \vec{n} = u_a(\vec{x}, t) \quad \text{for } \vec{x} \in \Gamma_a \subset \partial\Omega. \tag{4.25}$$

Combinations of these conditions are used when studying parts of a reservoir (e.g., sector models). There are also cases, for instance when describing groundwater systems or CO_2

---

[1] A solution is well-posed if it exists, is unique, and depends continuously on the initial and boundary conditions.

sequestration in saline aquifers, where (parts of) the boundaries are open, or the system contains a background flow. More information of how to set boundary conditions will be given in Section 5.1.4. In the compressible case in (4.9), we also need to specify an initial pressure distribution. Typically, this pressure distribution will be hydrostatic, as in the gravity column we discussed briefly in Section 1.4, and hence be given by the ordinary differential equation,

$$\frac{dp}{dz} = \rho g, \qquad p(z_0) = p_0. \qquad (4.26)$$

4.3.2 Injection and Production Wells

In a typical reservoir simulation, the inflow and outflow in wells occur on a subgrid scale. In most discretized flow models, the pressure is modeled using a single pressure value inside each grid cell. The size of each grid cell must therefore be chosen so small that the pressure variation inside the cell can be approximated accurately in terms of its volumetric average. Far away from wells, spatial variations in pressure tend to be relatively slow, at least in certain directions, and one can therefore choose cell sizes in the order of tens or hundreds of meters, which is a reasonable size compared with the extent of the reservoir. Near the well, however, the pressure will have large variations over short distances, and to compute a good approximation of these pressure variations, one would need grid cells than are generally smaller than what is computationally tractable. As a result, one ends up with a setup similar to the illustration in Figure 4.4, where the radius of the well is typically between 1/100 and 1/1000 of the horizontal dimensions of the grid cell. The largest percentage of the pressure drop associated with a well occurs near the well and the pressure at the well radius will thus deviate significantly from the volumetric pressure average inside the cell. Special analytical models are therefore developed to represent the subgrid variations in the particular flow patterns near wells.

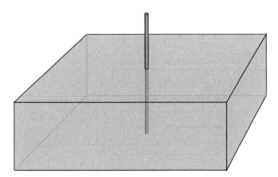

Figure 4.4 Illustration of a well inside a grid cell. The proportions are not fully to scale: whereas the diameter of a well varies from 5 to 40 inches, a grid block may extend from tens to hundreds of meters in the lateral direction and from a few decimeters to ten meters in the vertical direction.

Inflow-Performance Relation

Normally, fluids are injected from a well at either constant *surface rate* or at constant *bottom-hole pressure*, which is also called wellbore flowing pressure and refers to the pressure at a certain datum point inside the wellbore. Similarly, fluids are produced at constant bottom-hole pressure or at constant surface liquid rate. The main purpose of a well model is then to accurately compute the pressure at the well radius when the injection or production rate is known, or to accurately compute the flow rate in or out of the reservoir when the pressure at well radius is known. The resulting relation between the bottom-hole pressure and surface flow rate is often called the *inflow-performance relation* or IPR.

The simplest and most widely used inflow–performance relation is the linear law

$$q_o = J(p_R - p_w), \qquad (4.27)$$

which states that the flow rate is directly proportional to the pressure drawdown near the well; that is, flow rate is proportional to the difference between the average reservoir pressure p_R in the grid cell and the pressure p_w at the wellbore. The constant of proportionality J is called the *productivity index* (PI) for production wells, or the *well injectivity index* (WI) for injectors, and accounts for all rock and fluid properties, as well as geometric factors that affect the flow. In MRST, we do not distinguish between productivity and injectivity indices, and henceforth we will only use the shorthand WI.

Steady-State Flow in a Radial System

To derive the basic linear relation (4.27), we consider a vertical well that drains a rock with uniform permeability. The well penetrates the rock completely over a height h and is open in the radial direction. Gravity is neglected and the single-phase fluid is assumed to flow radially inward from an infinite domain; see Figure 4.5. From mass conservation written in cylinder coordinates, we have

$$\frac{1}{r}\frac{\partial(rvB^{-1})}{\partial r} = 0 \quad \longrightarrow \quad v = BC/r.$$

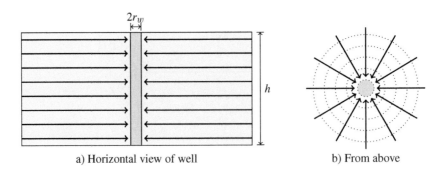

Figure 4.5 Radial flow used to derive inflow–performance relationship.

Here, the formation-volume factor B is the ratio between the volume of the fluid at reservoir conditions and the volume of the fluid at surface conditions.

The integration constant C is determined by integrating around a small cylinder surrounding the fluid sink,

$$q = \oint B^{-1} \vec{v} \cdot \vec{n} \, ds = -2\pi h C.$$

Inserting this into Darcy's law in cylinder coordinates, we have

$$v = -\frac{qB}{2\pi r h} = -\frac{K}{\mu}\frac{dp}{dr}.$$

Even if several different flow patterns can be expected when fluids flow toward a vertical wellbore, two-dimensional radial flow is considered to be the most representative for vertical oil and gas wells. We now integrate this equation from the wellbore r_w and to an arbitrary radius,

$$2\pi K h \int_{p_w}^{p} \frac{1}{q\mu B} \, dp = \int_{r_w}^{r} \frac{1}{r} \, dr.$$

Here, B decreases with pressure and μ increases. The composite effect is that $(\mu B)^{-1}$ decreases almost linearly with pressure. We can therefore approximate μB by $(\mu B)_{avg}$ evaluated at the average pressure $(p_w + p)/2$, to obtain an explicit expression for the pressure (dropping the subscript on μB for brevity)

$$p(r) = p_w + \frac{q\mu B}{2\pi K h} \ln(r/r_w). \tag{4.28}$$

Figure 4.6 shows a plot of this solution on a scale that is somewhat larger than the typical size of grid cells in field-scale models. Notice, in particular, that most of the pressure

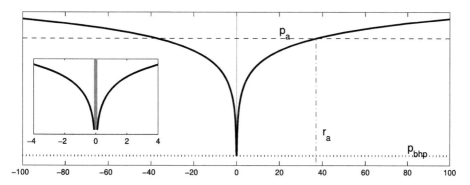

Figure 4.6 Illustration of the pressure distribution inside a circular region with a radius of 100 meters computed from (4.28), assuming the well is producing fluids from an infinite domain. Here, p_a is the volumetric pressure average inside the domain and r_a is the radius at which this value is found. The inset shows a zoom of the near-well zone.

Pseudo-Steady State in a Radial System

We can also derive a linear inflow relationship of the form (4.27) under somewhat more realistic conditions by considering *pseudo-steady* flow conditions. Assuming constant compressibility and fluid viscosity, the single-phase flow equation written in cylinder coordinates reads

$$\frac{1}{r}\frac{\partial}{\partial r}\left(r\frac{\partial p}{\partial r}\right) = \frac{\mu\phi c}{K}\frac{\partial p}{\partial t}. \qquad (4.29)$$

At pseudo-steady state, $\partial p/\partial t$ is constant. To determine the pressure differential, we consider the global mass-balance, which states that the amount of produced fluid must equal the change of mass inside the reservoir. Writing this on integral form and differentiating gives us

$$\int_0^t q(t)\,dt = -\int_{p_0}^{\bar{p}} \frac{Ah\phi c}{B}\,d\bar{p} \quad\longrightarrow\quad \frac{\partial p}{\partial t} = \frac{d\bar{p}}{dt} = -\frac{Bq}{\pi\left(r_d^2 - r_w^2\right)h\phi c}.$$

Here, q is the volumetric rate at surface conditions, p_0 is the initial reservoir pressure, and $\bar{p}$ is the average pressure inside the cylindrical drainage region at time t. In the expression above, we have introduced the formation-volume factor B, defined as the ratio between the volume of the fluid at reservoir conditions and the volume of the fluid at surface conditions. Inserting the constant pressure change into (4.29) gives

$$\frac{1}{r}\frac{d}{dr}\left(r\frac{dp}{dr}\right) = -\frac{q\mu B}{hK\pi\left(r_d^2 - r_w^2\right)} = -C.$$

To solve this equation, we first integrate with respect to r and use the fact that $dp/dr = 0$ at the outer boundary,

$$\int d\left(r\frac{dp}{dr}\right) = -\int Cr\,dr \quad\xrightarrow{\frac{dp}{dr}|_{r_d}=0}\quad r\frac{dp}{dr} = \tfrac{1}{2}C\left(r_d^2 - r^2\right).$$

Then, we integrate outward from r_w to determine $p(r)$,

$$p(r) - p_w = \tfrac{1}{2}C\int_{r_w}^r \left(\frac{r_d^2}{r} - r\right) dr = \frac{q\mu B}{2\pi Kh}\left[\frac{r_d^2}{r_d^2 - r_w^2}\ln\left(\frac{r}{r_w}\right) - \frac{1}{2}\frac{r^2 - r_w^2}{r_d^2 - r_w^2}\right].$$

4.3 Auxiliary Conditions and Equations

This equation can be used to develop the desired relationship between the average reservoir pressure and the wellbore flowing pressure p_w,

$$p_a = \frac{1}{V}\int_{r_w}^{r_d} p(r)\,dV = \frac{2\pi h \phi}{\pi(r_d^2 - r_w^2)h\phi}\int_{r_w}^{r_d} p(r)r\,dr$$

$$= \frac{2}{r_d^2 - r_w^2}\int_{r_w}^{r_d}\left[p_w r + \frac{C}{2}\left(r_d^2 r \ln\left(\frac{r}{r_w}\right) - \tfrac{1}{2}r\left(r^2 - r_w^2\right)\right)\right]dr$$

$$= p_w + \frac{2}{(r_d^2 - r_w^2)}\frac{q\mu B}{2\pi h K\left(r_d^2 - r_w^2\right)}[I_1 + I_2].$$

The two integrals can be computed as follows,

$$I_1 = \int_{r_w}^{r_d} r\ln(r/r_w)\,dr = \tfrac{1}{2}r_d^2\ln(r_d/r_w) - \tfrac{1}{4}\left(r_d^2 - r_w\right)^2,$$

$$I_2 = \int_{r_w}^{r_d}\tfrac{1}{2}r\left(r^2 - r_w^2\right)dr = \tfrac{1}{8}\left(r_d^4 - r_w^4\right) - \tfrac{1}{4}r_w^2\left(r_d^2 - r_w^2\right).$$

We can now insert this back into the equation and use the fact that $r_d \gg r_w$ to derive our desired expression,

$$p_a - p_w = \frac{q\mu B}{2\pi h K}\left[\underbrace{\frac{r_d^4}{(r_d^2 - r_w^2)^2}}_{\approx 1}\ln\left(\frac{r_d}{r_w}\right) - \frac{1}{2}\underbrace{\frac{r_d^2}{r_d^2 - r_w^2}}_{\approx 1} - \frac{1}{4}\underbrace{\frac{r_d^2 + r_w^2}{r_d^2 - r_w^2}}_{\approx 1} + \frac{1}{2}\underbrace{\frac{r_w^2}{r_d^2 - r_w^2}}_{\approx 0}\right]$$

$$\approx \frac{q\mu B}{2\pi h K}\left[\ln\left(\frac{r_d}{r_w}\right) - \frac{3}{4}\right].$$

Rearranging terms, we obtain an expression for the fluid rate of the producer

$$q = \frac{2\pi K h}{\mu B \left(\ln(r_d/r_w) - 0.75\right)}(p_a - p_w). \tag{4.30}$$

For an injector, we likewise obtain

$$q = \frac{2\pi K h}{\mu B \left(\ln(r_d/r_w) - 0.75\right)}(p_w - p_a).$$

The relation (4.30) was developed for an ideal well under several simplifying assumptions: homogeneous and isotropic formation of constant thickness, clean wellbore, etc. A well will rarely experience these ideal conditions in practice. Typically, the permeability is altered close to the wellbore under drilling and completion, the well will only be partially completed, and so on. The actual pressure performance will therefore deviate from (4.30). To model this, it is customary to include a *skin factor S* to account for extra pressure loss due to alterations in the inflow zone. The resulting equation is

$$q = \frac{2\pi K h}{\mu B \left(\ln(r_d/r_w) - 0.75 + S\right)}(p_a - p_w). \tag{4.31}$$

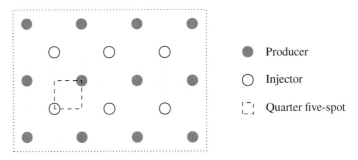

Figure 4.7 Excerpts of a repeated five-spot pattern.

Often the constant -0.75 is included in the skin factor S, which could be negative for stimulated wells. Sometimes h is modified to ch, where c is the completion factor, i.e., a dimensionless number between zero and one that describes the fraction of the wellbore open to flow.

Peaceman-Type Well Models

To use the radial model in conjunction with a reservoir model, the volumetric average pressure in the radial model must be related to the computed cell pressure p. Exact solutions are generally not known, since real reservoirs have complicated geometries and irregular boundaries. Well models are therefore developed using highly idealized reservoir geometries. One such example is the so-called *repeated five-spot pattern*, which consists of a thin, infinitely large, horizontal reservoir with a staggered pattern of injection and production wells as (see Figure 4.7) that repeats itself to infinity in all directions. The name comes from the fact that each injector is surrounded by four producers, and vice versa, hence creating tiles of five-spot patterns. If all wells operate at equal rates, the flow pattern has certain symmetries and it is common to only consider one quarter of the five spot, as shown in Figure 4.7, subject to no-flow boundary conditions. Muskat and Wyckoff [219] derived the following explicit solution for the pressure drop between the injection and production wells,

$$\Delta p = \frac{q\mu B}{\pi K h}\Big(\ln(d/r_w) - C\Big), \qquad (4.32)$$

where d is the distance between the wells, and C is given by an infinite series. The authors originally used $C = 0.6190$, but a more accurate value, $C = 0.61738575$, was later derived by Peaceman [245], who used (4.32) to determine an equivalent radius r_e at which the cell pressure is equal to average of the exact pressure. Assuming isotropic permeabilities, square grid blocks, single-phase flow, and a well at a center of an interior block, Peaceman [247] showed that the equivalent radius is

$$r_e \approx 0.20788\sqrt{\Delta x\, \Delta y}$$

for the two-point discretization that will be discussed in more detail in Section 4.4.1. To give you an idea, let us present a somewhat simplified derivation in the incompressible case

4.3 Auxiliary Conditions and Equations

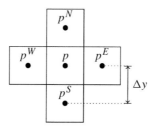

Figure 4.8 Five-point stencil on a grid with square grid blocks ($\Delta x = \Delta y$).

without gravity and assume that the viscosity and formation-volume factor are constant. For a single-layered square grid (Figure 4.8), the two-point scheme then simplifies to a standard five-point difference scheme

$$-\frac{Kh}{\mu B}\left[4p - p^W - p^N - p^E - p^S\right] = q.$$

The solution is symmetric for the repeated five-spot pattern, so that $p^E = p^N = p^W = p^S$ (which explains why μB could be factored out). Hence, we can eliminate the last three pressure values and write the five-point stencil as

$$p = p^E - \frac{q\mu B}{4Kh}.$$

Next, we assume that the radial well solution is an accurate approximation to the pressure values in the neighboring cells,

$$p^E = p_w + \frac{q\mu B}{2\pi Kh}\ln(\Delta x/r_w).$$

Combining the two equations, we have

$$\begin{aligned}p &= p_w + \frac{q\mu B}{2\pi Kh}\ln(\Delta x/r_w) - \frac{q\mu B}{4Kh}\\ &= p_w + \frac{q\mu B}{2\pi Kh}\left[\ln(\Delta x/r_w) - \frac{\pi}{2}\right] = p_w + \frac{q\mu B}{2\pi Kh}\ln\left(e^{-\pi/2}\Delta x/r_w\right).\end{aligned}$$

Now, $\exp(-\pi/2) \approx 0.20788$, and hence we have derived Peaceman's classical result.

This basic model has later been extended to cover a lot of other cases, e.g., off-center wells, multiple wells, non-square grids, anisotropic permeability, horizontal wells; see for instance [18, 102, 13]. For anisotropic permeabilities $\mathbf{K} = \text{diag}(K_x, K_y, K_z)$, the permeability is replaced by an effective permeability

$$K_e = \sqrt{K_x K_y}, \tag{4.33}$$

which follows by a coordinate transformation $x \to \hat{x}\sqrt{K_x}$, etc. For a grid with rectangular grid blocks ($\Delta x \neq \Delta y$), the equivalent radius reads [245]

$$r_e = 0.28 \frac{\left(\sqrt{K_y/K_x}\Delta x^2 + \sqrt{K_x/K_y}\Delta y^2\right)^{1/2}}{\left(K_y/K_x\right)^{1/4} + \left(K_x/K_y\right)^{1/4}}. \tag{4.34}$$

The same formulas apply to horizontal wells, with the necessary permutations; that is, for a well in the x-direction $\Delta x \to \Delta z$ and $K_x \to K_z$.

If we include gravity forces in the well and assume hydrostatic equilibrium, the well model reads

$$q_i = \frac{2\pi h K_e}{\mu_i B_i [\ln(r_e/r_w) + S]} \left(p_R - p_w - \rho_i(z - z_{bh})g\right), \tag{4.35}$$

where K_e is given by (4.33) and r_e is given by (4.34). For deviated wells, h denotes the length of the grid block in the major direction of the wellbore and *not* the length of the wellbore.

At this point we should add a word of caution. The equivalent radius of a numerical method generally depends on how the method approximates pressure inside the grid cell containing the well perforation. The formulas given in this subsection are, strictly speaking, only valid if you use the specific two-point discretization they were developed for. When using another discretization method, you may have to compute other values for the equivalent radius, e.g., as discussed in [192, 197].

4.3.3 Field Lines and Time-of-Flight

Equation (4.10) together with a set of suitable and compatible boundary conditions constitute all you need to determine the pressure distribution and flow velocity of an incompressible fluid inside an incompressible rock. In the remainder of this section, we will discuss a few simple concepts and auxiliary equations you can use to visualize, analyze, and improve your understanding of the computed flow fields.

A simple way to visualize a flow field is to use field lines resulting from the vector field: streamlines, streaklines, and pathlines. In steady flow, the three are identical. However, if the flow is not steady, i.e., when $\vec{v}$ changes with time, they differ. *Streamlines* are associated with an instant snapshot of the flow field and consist of a family of curves that are everywhere tangential to $\vec{v}$ and show the direction a fluid element will travel at this specific point in time. That is, if $\vec{x}(r)$ is a parametric representation of a single streamline at this instance $\hat{t}$ in time, then

$$\frac{d\vec{x}}{dr} \times \vec{v}(\vec{x}, \hat{t}) = 0, \quad \text{or equivalently,} \quad \frac{d\vec{x}}{dr} = \frac{\vec{v}(\hat{t})}{|\vec{v}(\hat{t})|}. \tag{4.36}$$

In other words, streamlines are calculated throughout the fluid from an *instantaneous* snapshot of the flow field. Because two streamlines from the same instance in time cannot cross, there cannot be flow across a streamline, and if we align a coordinate along a bundle of streamlines, the flow through them will be one-dimensional.

4.3 Auxiliary Conditions and Equations

Pathlines are the trajectories individual fluid elements will follow over a certain period. In each moment of time, the path a fluid particle takes will be determined by the streamlines associated with the velocity field at this instance in time. If $\vec{y}(t)$ represents a single path line starting in $\vec{y}_0$ at time t_0, then

$$\frac{d\vec{y}}{dt} = \vec{v}(\vec{y},t), \qquad \vec{y}(t_0) = \vec{y}_0. \tag{4.37}$$

A *streakline* is the line traced out by all fluid particles that have passed through a prescribed point throughout a certain period of time. (Think of dye injected into the fluid at a specific point.) If we let $\vec{z}(t,s)$ denote a parametrization of a streakline and $\vec{z}_0$ the specific point through which all fluid particles have passed, then

$$\frac{d\vec{z}}{dt} = \vec{v}(\vec{z},t), \qquad \vec{z}(s) = \vec{z}_0. \tag{4.38}$$

Like streamlines, two streaklines cannot intersect each other. In summary: *streamline patterns change over time, but are easy to generate mathematically. Pathlines and streaklines are recordings of the passage of time obtained through experiments.*

Streamlines are far more used within reservoir simulation than pathlines and streaklines. Instead of using the arc length r to parametrize streamlines, it is common to introduce an alternative parametrization called *time-of-flight*, which takes into account the reduced volume available for flow, i.e., the porosity ϕ. Time-of-flight is defined by the following integral

$$\tau(r) = \int_0^r \frac{\phi(\vec{x}(s))}{|\vec{v}(\vec{x}(s))|} \, ds, \tag{4.39}$$

where τ expresses the time it takes a fluid particle to travel a distance r along a streamline (in the interstitial velocity field $\vec{v}/\phi$). Alternatively, computing the directional derivative of τ along a streamline and applying the fundamental theorem of calculus to (4.39),

$$\left(\frac{\vec{v}}{|\vec{v}|} \cdot \nabla\right) \tau = \frac{d}{dr} \tau = \frac{\phi}{|\vec{v}|}$$

from which it follows that τ can be expressed by a differential equation [78, 79]

$$\vec{v} \cdot \nabla \tau = \phi. \tag{4.40}$$

For lack of a better name, we will refer to this as the *time-of-flight equation*.

4.3.4 Tracers and Volume Partitions

A tracer can somewhat simplified be seen as a neutral particle that passively flow with the fluid without altering its flow properties. The concentration of a tracer is given by a continuity equation on the same form as (4.5),

$$\frac{\partial(\phi C)}{\partial t} + \nabla \cdot (\vec{v} C) = q_C. \tag{4.41}$$

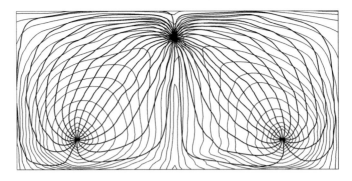

Figure 4.9 Illustration of time-of-flight, shown as gray isocontour lines, and streamlines shown as thick black lines.

Volumetric connections within a reservoir can be determined by simulating the evolution of artificial, nondispersive and nondiffusive tracers whose concentration does not change upon fluid compression or expansion. A simple flow diagnostic is to set the tracer concentration equal to one in a particular fluid source or at a certain part of the inflow boundary, and compute the solution approached at steady-state conditions from the nonconservative equation,

$$\vec{v} \cdot \nabla C = q_C, \qquad C|_{\text{inflow}} = 1. \qquad (4.42)$$

The resulting *tracer distribution* gives the portion of the total fluid volume passing through a point in the reservoir that can be attributed to a given fluid source or parts of the inflow boundary. Likewise, by reversing the sign of the flow field and assigning unit tracers to a particular fluid sink or parts of the outflow, one can compute the portion of the fluid content at a certain point in the reservoir that eventually will arrive at the given fluid sink or outflow boundary. By repeating this process for all parts of the inflow, we obtain a partition of the instantaneous flow field: Let C_i denote the tracer associated with fluid source Q_i. This fluid source will influence the flow through all points $\vec{x}$ for which $C_i(\vec{x}) > 0$. If $C(\vec{x}) = 1$, then Q_i is the only fluid source affecting the flow through $\vec{x}$. Otherwise, the point must be affected by at least one other fluid source Q_j. The tracer distributions $\{C_i(\vec{x})\}_{i=1}^m$ can thus be seen as measures of how strongly the flow through a point is influenced by various fluid sources. We will return with detailed discussion in Chapter 13 on flow diagnostics.

You can also use streamlines and time-of-flight to define an alternative curvilinear and flow-based coordinate system in three dimensions. To this end, we introduce the bi-streamfunctions ψ and χ [34], for which $\vec{v} = \nabla\psi \times \nabla\chi$. In the streamline coordinates (τ, ψ, χ), the gradient operator is expressed as

$$\nabla_{(\tau,\psi,\chi)} = (\nabla\tau)\frac{\partial}{\partial\tau} + (\nabla\psi)\frac{\partial}{\partial\psi} + (\nabla\chi)\frac{\partial}{\partial\chi}. \qquad (4.43)$$

Moreover, a streamline Ψ is defined by the intersection of a constant value for ψ and a constant value for χ. Because $\vec{v}$ is orthogonal to $\nabla\psi$ and $\nabla\chi$, it follows from (4.40) that

$$\vec{v} \cdot \nabla_{(\tau,\psi,\chi)} = (\vec{v} \cdot \nabla\tau)\frac{\partial}{\partial \tau} = \phi\frac{\partial}{\partial \tau}. \tag{4.44}$$

Therefore, the coordinate transformation $(x, y, z) \to (\tau, \psi, \chi)$ will reduce the three-dimensional transport equation (4.41) to a family of one-dimensional transport equations along each streamline [78, 164]. For incompressible flow this reads

$$\frac{\partial C}{\partial t} + \frac{\partial C}{\partial \tau} = 0. \tag{4.45}$$

There is no exchange of the quantity C among streamlines, and we can thus view each streamline as an isolated flow system. Assuming a prescribed concentration history $C_0(t)$ at the inflow gives a time-dependent boundary-value problem for the concentration at the outflow. Here, the response is given as (see [78]),

$$C(t) = C_0(t - \tau), \tag{4.46}$$

which is easily verified by inserting the expression into (4.45) and using the fact that the solution is unique [134]. For the special case of continuous and constant injection, the solution is particularly simple

$$C(t) = \begin{cases} 0, & t < \tau, \\ C_0, & t > \tau. \end{cases}$$

4.4 Basic Finite-Volume Discretizations

Extensive research on numerical methods for the Laplace/Poisson equation has led to a large number of different finite-difference and finite-volume methods, as well as finite-element methods based on standard, mixed, or discontinuous Galerkin formulations. In the rest of this section we present the simplest example of a finite-volume discretization, the two-point flux-approximation scheme, which is used extensively throughout industry and also is the default discretization method in MRST. We give a detailed derivation of the two-point method and briefly outline how to discretize the time-of-flight and the stationary tracer equations.

The two-point method is simple to implement, gives sparse linear systems that are not too costly to invert, and is robust in the sense that it produces monotone pressure approximations. Unfortunately, the method is only conditionally consistent, and will introduce numerical artifacts and not provide convergent solutions for increased grid resolution if the grid does not satisfy a certain orthogonality condition with respect to the permeability tensor. We discuss this in more detail in Chapter 6, which also introduces more advanced and consistent discretizations that are guaranteed to be convergent on general polyhedral grids and for anisotropic permeabilities.

4.4.1 Two-Point Flux-Approximation

To keep technical details at a minimum, we will in the following without loss of generality consider the simplified single-phase flow equation

132   *Mathematical Models for Single-Phase Flow*

$$\nabla \cdot \vec{v} = q, \qquad \vec{v} = -\mathbf{K}\nabla p, \qquad \text{in } \Omega \subset \mathbb{R}^d. \tag{4.47}$$

In classical finite-difference methods, partial differential equations are approximated by replacing the derivatives with appropriate divided differences between point-values on a discrete set of points in the domain. Finite-volume methods, on the other hand, have a more physical motivation and are derived from conservation of (physical) quantities over cell volumes. Thus, in a finite-volume method the unknown quantities are represented in terms of average values defined over a set of finite volumes. The PDE model is integrated and required to hold in an averaged sense rather than in a pointwise sense. In this sense, finite-difference and finite-volume methods have fundamentally different interpretation and derivation, but the names are often used interchangeably in the scientific literature. The main reason for this is probably that the discrete equations derived for the cell-centered values in a low-order, mass-conservative, finite-difference method are identical to the discrete equations for the cell averages in corresponding finite-volume methods. Herein, we will stick to this convention and not make a strict distinction between the two types of methods.

To develop a finite-volume discretization for (4.47), we start by rewriting the equation in integral form using a single cell Ω_i in the discrete grid as control volume

$$\int_{\partial \Omega_i} \vec{v} \cdot \vec{n}\, ds = \int_{\Omega_i} q\, d\vec{x}. \tag{4.48}$$

This is a simpler form of (4.3), where the accumulation term has disappeared because ϕ and ρ are independent of time and the constant ρ has been eliminated. Equation (4.48) ensures that mass is conserved for each grid cell. The next step is to use Darcy's law to compute the flux across each face of the cell,

$$v_{i,k} = \int_{\Gamma_{ik}} \vec{v} \cdot \vec{n}\, ds, \qquad \Gamma_{i,k} = \partial \Omega_i \cap \partial \Omega_k. \tag{4.49}$$

We refer to the faces $\Gamma_{i,k}$ as *half-faces*, since they are associated with a particular grid cell Ω_i and a certain normal vector $\vec{n}_{i,k}$. However, since the grid is assumed to be matching, each interior half face will have a twin half-face $\Gamma_{k,i}$ that has identical area $A_{k,i} = A_{i,k}$ but opposite normal vector $\vec{n}_{k,i} = -\vec{n}_{i,k}$. If we now approximate the integral over the cell face in (4.49) by the midpoint rule, we can use Darcy's law to write the flux as

$$v_{i,k} \approx A_{i,k} \vec{v}(\vec{x}_{i,k}) \cdot \vec{n}_{i,k} = -A_{i,k} (\mathbf{K}\nabla p)(\vec{x}_{i,k}) \cdot \vec{n}_{i,k}, \tag{4.50}$$

where $\vec{x}_{i,k}$ denotes the centroid on $\Gamma_{i,k}$. The idea is now to use a one-sided finite difference to express the pressure gradient as the difference between the pressure $\pi_{i,k}$ at the face centroid and the pressure at some point inside the cell. However, in a finite-volume method, we only know the cell averaged value of the pressure inside the cell. We must therefore make some additional assumption to be able to reconstruct the point values we need to estimate the pressure gradient in Darcy's law. If we assume that the pressure is linear (or constant) inside each cell, the reconstructed pressure value at the cell center is identical to the average pressure p_i inside the cell, and hence it follows that (see Figure 4.10)

$$v_{i,k} \approx A_{i,k} \mathbf{K}_i \frac{(p_i - \pi_{i,k})\vec{c}_{i,k}}{|\vec{c}_{i,k}|^2} \cdot \vec{n}_{i,k} = T_{i,k}(p_i - \pi_{i,k}). \tag{4.51}$$

4.4 Basic Finite-Volume Discretizations

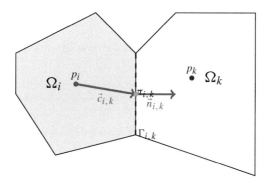

Figure 4.10 Two cells used to define the two-point finite-volume discretization of the Laplace operator.

Here, we have introduced one-sided *transmissibilities* $T_{i,k}$ that are associated with a single cell and give a two-point relation between the flux across a cell face and the difference between the pressure at the cell and face centroids. We refer to these one-sided transmissibilities as *half-transmissibilities*, since they are associated with a half face.

To derive the final discretization, we impose continuity of fluxes across all faces, $v_{i,k} = -v_{k,i} = v_{ik}$ and continuity of face pressures $\pi_{i,k} = \pi_{k,i} = \pi_{ik}$. This gives us two equations,

$$T_{i,k}^{-1} v_{ik} = p_i - \pi_{ik}, \qquad -T_{k,i}^{-1} v_{ik} = p_k - \pi_{ik}.$$

By eliminating the interface pressure π_{ik}, we end up with the following two-point flux-approximation (TPFA) scheme,

$$v_{ik} = \left[T_{i,k}^{-1} + T_{k,i}^{-1}\right]^{-1} (p_i - p_k) = T_{ik}(p_i - p_k), \qquad (4.52)$$

where T_{ik} is the transmissibility associated with the connection between the two cells. As the name suggests, the TPFA scheme uses two "points," the cell averages p_i and p_k, to approximate the flux across the interface Γ_{ik} between cells Ω_i and Ω_k. In our derivation so far, we have parametrized the cell fluxes in terms of the index of the neighboring cell. Extending the derivation to also include fluxes on exterior faces is trivial since we either know the flux explicitly for Neumann boundary conditions (4.23) or (4.25), or know the interface pressure for Dirichlet conditions (4.24).

By inserting the expression for v_{ik} into (4.48), we see that the TPFA scheme for (4.47), in compact form, seeks a set of cell averages that satisfy the following system of equations

$$\sum_k T_{ik}(p_i - p_k) = q_i, \qquad \forall \Omega_i \subset \Omega. \qquad (4.53)$$

This system is clearly symmetric, and a solution is only defined up to an arbitrary constant, as is also the case for the corresponding Poisson problem. We can make the system positive-definite and preserve symmetry by specifying the pressure in a single point. In MRST, we

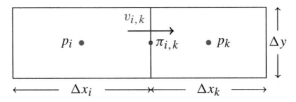

Figure 4.11 Two cells used to derive the TPFA discretization for a 2D Cartesian grid.

have chosen to set $p_1 = 0$ by adding a positive constant to the first diagonal of the matrix $\mathbf{A} = [a_{ij}]$, where:

$$a_{ij} = \begin{cases} \sum_k T_{ik} & \text{if } j = i, \\ -T_{ij} & \text{if } j \neq i. \end{cases}$$

The matrix $\mathbf{A}$ is sparse and will have a banded structure for structured grids: tridiagonal for 1D grids and penta- and heptadiagonal for logically Cartesian grids in 2D and 3D, respectively. The TPFA scheme is monotone, robust, and relatively simple to implement, and is currently the industry standard with reservoir simulation.

Example 4.4.1 *To tie the links with standard finite-difference methods on Cartesian grids, we derive the two-point discretization for a 2D Cartesian grid with isotropic permeability. Consider the flux in the x-direction between two cells* i *and* k *as illustrated in Figure 4.11. As above, we impose mass conservation inside each cell. For cell* i *this reads:*

$$v_{i,k} = \Delta y \frac{(p_i - \pi_{i,k})}{\left(\frac{1}{2}\Delta x_i\right)^2} \left(\tfrac{1}{2}\Delta x_i, 0\right) K_i \left(1, 0\right)^T = \Delta y \frac{2K_i}{\Delta x_i} \left(p_i - \pi_{i,k}\right)$$

and likewise for cell k:

$$v_{k,i} = \Delta y \frac{(p_k - \pi_{k,i})}{\left(\frac{1}{2}\Delta x_k\right)^2} \left(-\tfrac{1}{2}\Delta x_k, 0\right) K_k \left(-1, 0\right)^T = \Delta y \frac{2K_k}{\Delta x_k} \left(p_k - \pi_{k,i}\right)$$

Next, we impose continuity of fluxes and face pressures,

$$v_{i,k} = -v_{k,i} = v_{ik}, \qquad \pi_{i,k} = \pi_{k,i} = \pi_{ik},$$

which gives us two equations

$$\frac{\Delta x_i}{2K_i \Delta y} v_{ik} = p_i - \pi_{ik}, \qquad -\frac{\Delta x_k}{2K_k \Delta y} v_{ik} = p_k - \pi_{ik}.$$

Finally, we eliminate π_{ik} *to obtain*

$$v_{ik} = 2\Delta y \left(\frac{\Delta x_i}{K_i} + \frac{\Delta x_k}{K_k}\right)^{-1} (p_i - p_k),$$

which shows that the transmissibility is given by the harmonic average of the permeability values in the two adjacent cells, as one would expect.

In [2], two former colleagues and I showed how you can implement an efficient and self-contained MATLAB program that in approximately 30 compact lines solves the incompressible flow equation (4.47) using the two-point method outlined in this section. The program was designed for Cartesian grids with no-flow boundary conditions only, and relied strongly on the logical ijk numbering of grid cells. For this reason, the program has limited applicability beyond highly idealized cases such as the SPE 10 model. However, in its simplicity, it presents an interesting contrast to the general-purpose implementation in MRST which is designed to handle unstructured grids, wells, and more general boundary conditions. I encourage you to read the paper and try the accompanying program and example scripts, which you can download from my personal website.

4.4.2 Discrete div and grad Operators

The double-index notation $v_{i,k}$ and v_{ik} used in the previous section is simple and easy to comprehend when working with a single interface between two neighboring cells. It becomes more involved when we want to introduce the same type of discretizations for more complex models than the Poisson equation for incompressible flow. To prepare for discussions later in the book, we will now introduce a more abstract way of writing the two-point finite-volume discretization. The idea is to introduce discrete operators for the divergence and gradient operators that mimic their continuous counterparts. This will enable us to write the discretized version of Poisson's equation (4.47) in the same form as its continuous counterpart. These operators will necessarily have to incorporate information about the grid topology in the same way as the transmissibility incorporates information about grid geometry. However, once defined, we can apply them to any quantities defined over the cells and faces of a grid without knowing anything about the underlying grid topology. This gives us a powerful tool to write solvers in a way that effectively is grid agnostic.

We start by a quick recap of the definition of unstructured grids. As discussed in detail in Section 3.4, the grid structure in MRST consists of three objects: *cells*, *faces*, and *nodes*. Each cell is delimited by a set of faces, and each face by a set of *edges*, which again are determined by a pair of nodes. Each object has given geometrical properties (volume, areas, centroids). As before, let n_c and n_f denote the number of cells and faces respectively. To define the topology of the grid, we can use two mappings. The first is given by $F : \{1, \ldots, n_c\} \to \{0, 1\}^{n_f}$ and maps a cell to the set of faces that delimit this cell. In the grid structure G, this is represented as the G.cells.faces array, where the first column specifying the cell numbers is not stored, since it is redundant and instead must be computed by a call f2cn = gridCellNo(G). The second mapping consists of two components that for a given face give the corresponding neighboring cells, $C_1, C_2 : \{1, \ldots, n_f\} \to \{1, \ldots, n_c\}$. In the grid structure G, C_1 is given by G.faces.neighbors(:,1) and C_2 by G.faces.neighbors(:,2). This is illustrated in Figure 4.12.

Let us now construct the discrete analogues of the divergence and gradient operators, which we denote div and grad. The divergence operator div is a linear mapping from

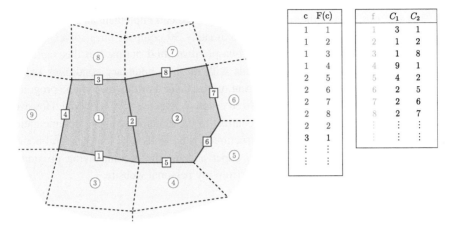

Figure 4.12 Illustration of the mappings from cells to faces and from faces to cells used to define the discrete divergence and gradient operators.

faces to cells, which typically will be applied to a discrete flux $v \in \mathbb{R}^{n_f}$. We write $v[f]$ to denote the flux evaluated on a face f and assume that the orientation of this flux is from $C_1(f)$ to $C_2(f)$. Hence, the total amount of matter leaving a cell c is given by the sum of outfluxes minus the sum of influxes

$$\mathtt{div}(v)[c] = \sum_{f \in F(c)} v[f] \mathbf{1}_{\{c=C_1(f)\}} - \sum_{f \in F(c)} v[f] \mathbf{1}_{\{c=C_2(f)\}}. \tag{4.54}$$

Here, element c of the vector $\mathbf{1}_{\{c=C_1(f)\}}$ equals one if cell c is neighbor to face f and zero otherwise. The $\mathtt{grad}$ operator maps $\mathbb{R}^{n_c}$ to $\mathbb{R}^{n_f}$ and is defined as

$$\mathtt{grad}(p)[f] = p[C_2(f)] - p[C_1(f)], \tag{4.55}$$

for any $p \in \mathbb{R}^{n_c}$. In the continuous case, the gradient operator is the adjoint of the divergence operator (up to a sign), as we have

$$\int_\Omega p \nabla \cdot \vec{v} \, d\vec{x} + \int_\Omega \vec{v} \cdot \nabla p \, d\vec{x} = 0, \tag{4.56}$$

for vanishing boundary conditions. Let us prove that this property also holds in the discrete case. To simplify notation, we set $S_c = \{1, \ldots, n_c\}$ and $S_f = \{1, \ldots, n_f\}$. For any $v \in \mathbb{R}^{n_f}$ and $p \in \mathbb{R}^{n_c}$, we have

$$\sum_{c \in S_c} \mathtt{div}(v)[c] \, p[c] = \sum_{c \in S_c} p[c] \left(\sum_{f \in F(c)} v[f] \mathbf{1}_{\{c=C_1(f)\}} - \sum_{f \in F(c)} v[f] \mathbf{1}_{\{c=C_2(f)\}} \right)$$

$$= \sum_{c \in S_c} \sum_{f \in S_f} v[f] \, p[c] \, \mathbf{1}_{\{c=C_1(f)\}} \mathbf{1}_{\{f \in F(c)\}}$$

$$- \sum_{c \in S_c} \sum_{f \in S_f} v[f] \, p[c] \, \mathbf{1}_{\{c=C_2(f)\}} \mathbf{1}_{\{f \in F(c)\}}. \tag{4.57}$$

4.4 Basic Finite-Volume Discretizations

We can switch the order in the sums above and obtain

$$\sum_{c\in S_c}\sum_{f\in S_f} v[f]\,p[c]\mathbf{1}_{\{c=C_1(f)\}}\mathbf{1}_{\{f\in F(c)\}} = $$
$$\sum_{f\in S_f}\sum_{c\in S_c} v[f]\,p[c]\,\mathbf{1}_{\{c=C_1(f)\}}\mathbf{1}_{\{f\in F(c)\}}.$$

For a given face f, we have that $\mathbf{1}_{\{c=C_1(f)\}}\mathbf{1}_{\{f\in F(c)\}}$ is nonzero if and only if $c = C_1(f)$, and therefore

$$\sum_{f\in S_f}\sum_{c\in S_c}\mathbf{1}_{\{c=C_1(f)\}}\mathbf{1}_{\{f\in F(c)\}}v[f]p[c] = \sum_{f\in S_f} v[f]p[C_1(f)].$$

In the same way, we have

$$\sum_{c\in S_c}\sum_{f\in S_f} v[f]\,p[c]\,\mathbf{1}_{\{c=C_2(f)\}}\mathbf{1}_{\{f\in F(c)\}} = \sum_{f\in S_f} v[f]\,p[C_2(f)],$$

so that (4.57) yields

$$\sum_{c\in S_c}\text{div}(v)[c]\,p[c] + \sum_{f\in S_f}\text{grad}(p)[f]\,v[f] = 0. \tag{4.58}$$

Until now, we have ignored boundary conditions. We can include them by introducing one more cell number $c = 0$ to denote the exterior. Then, we can consider external faces and extend the mappings C_1 and C_2 to $S_c \cup \{0\}$ so that a given face f is external if it satisfies $C_1(f) = 0$ or $C_2(f) = 0$. Note that the grad operator only defines values on internal faces. Now taking external faces into account, we obtain

$$\sum_{c\in S_c}\text{div}(v)[c]\,p[c] + \sum_{f\in S_f}\text{grad}(p)[f]\,v[f]$$
$$= \sum_{f\in \overline{S}_f\setminus S_f}\Big(p[C_1(f)]\,\mathbf{1}_{\{C_2(f)=0\}} - p[C_2(f)]\,\mathbf{1}_{\{C_1(f)=0\}}\Big)v[f], \tag{4.59}$$

where $\overline{S}_f$ denotes the extended set of faces, consisting of both internal and external faces. Identity (4.59) is the discrete counterpart to

$$\int_\Omega p\nabla\cdot\vec{v}\,d\vec{x} + \int_\Omega \vec{v}\cdot\nabla p\,d\vec{x} = \int_{\partial\Omega} p\vec{v}\cdot\vec{n}\,ds. \tag{4.60}$$

Let us now use these operators to discretize the Poisson equation. Going back to (4.47), let the vector $v \in \mathbb{R}^{n_f}$ be a discrete approximation of the flux on faces; that is, for $f \in S_f$, we have

$$v[f] \approx \int_{\Gamma_f} \vec{v}(x)\cdot\vec{n}_f\,ds,$$

where $\vec{n}_f$ is the normal to face f with orientation given by the grid. The relation between the discrete pressure $\boldsymbol{p} \in \mathbb{R}^{n_c}$ and the discrete flux is given by the two-point flux approximation discussed in the previous section,

$$v[f] = -T[f]\,\mathrm{grad}(p)[f] \approx -\int_{\Gamma_f} \mathbf{K}(x)\nabla p \cdot \vec{n}_f \, ds. \tag{4.61}$$

Here, $T[f]$ denotes the transmissibility associated with face f, as defined in (4.52). Hence, the discretization of (4.47) can be written as

$$\mathrm{div}(v) = q \tag{4.62a}$$
$$v = -T\,\mathrm{grad}(p), \tag{4.62b}$$

where the multiplication in (4.62b) holds element-wise.

We end our discussion by two examples that demonstrate how these discrete operators can be used to write very compact solvers that easily can be extended to solve problems of increasing complexity with minimal modifications to the code.

Example 4.4.2 *To illustrate the use of the discrete operators, let us set up and solve the classical Poisson equation on a simple box geometry,*

$$-\mathrm{div}(T\,\mathrm{grad}(\boldsymbol{p})) = \boldsymbol{q}, \qquad \Omega = [0,1] \times [0,1], \tag{4.63}$$

subject to no-flow boundary conditions with q consisting of a point source at (0,0) and a point sink at (1,1). First, we construct a small Cartesian grid

```
G = computeGeometry(cartGrid([5 5],[1 1]));
```

for which T equals a scalar multiple of the identity matrix and is therefore dropped for simplicity. The natural way to represent the `div` *and* `grad` *operators is to use sparse matrices. To this end, we use (4.58), which implies that the sparse matrix representing* `div` *is the negative transpose of the matrix that defines* `grad` *using (4.55). Moreover, since we assume no-flow boundary conditions, we only need to let C_1 and C_2 account for internal connections:*

```
C = G.faces.neighbors;
C = C(all(C ~= 0, 2), :);
nf = size(C,1);
nc = G.cells.num;
D = sparse([(1:nf)'; (1:nf)'], C, ...
           ones(nf,1)*[-1 1], nf, nc);
grad = @(x) D*x;
div  = @(x) -D'*x;
```

$$D = \begin{bmatrix} \ddots \\ & \ddots \\ & & \ddots \end{bmatrix} \genfrac{}{}{0pt}{}{\frac{\partial}{\partial x}}{\frac{\partial}{\partial y}}$$

Once we have the discrete operators, we can write (4.63) in residual form, $\boldsymbol{f}(\boldsymbol{p}) = \boldsymbol{Ap} + \boldsymbol{q} = 0$ and use automatic differentiation as discussed in Example A.5.3 to obtain $\boldsymbol{A}$ by computing $\partial \boldsymbol{f}/\partial \boldsymbol{p}$

4.4 Basic Finite-Volume Discretizations

```
p = initVariablesADI(zeros(nc,1));
q = zeros(nc, 1);           % source term
q(1) = 1; q(nc) = -1;       % -> quarter five-spot

eq     = div(grad(p))+q;    % equation
eq(1)  = eq(1) + p(1);      % make solution unique
p      = -eq.jac{1}\eq.val; % solve equation
```

Next, we try to solve the same type of flow problem on a non-rectangular domain. That is, we still consider the unit square, but remove two half-circles of radius 0.4 centered at (0.5,0) and (0.5,1) respectively. To construct the corresponding grid, we use the fictitious grid approach from Section 3.1 (see Exercise 3.1.1):

```
G = cartGrid([20 20],[1 1]);
G = computeGeometry(G);
r1 = sum(bsxfun(@minus,G.cells.centroids,[0.5 1]).^2,2);
r2 = sum(bsxfun(@minus,G.cells.centroids,[0.5 0]).^2,2);
G = extractSubgrid(G, (r1>0.16) & (r2>0.16));
```

The construction of the discrete operators is agnostic to the exact layout of the grid, and because the transmissibility matrix T is still a multiple of the identity matrix, since the grid cells are equidistant squares, we can simply reuse the exact same set-up as in the previous code box:

```
% Grid information
C = G.faces.neighbors;
:
% Operators
D = sparse([(1:nf)'; (1:nf)'], C, ...
           ones(nf,1)*[-1 1], nf, nc);
:
% Assemble and solve equations
p = initVariablesADI(zeros(nc,1));
q = zeros(nc, 1);
q(1) = 1; q(nc) = -1;
eq     = div(grad(p))+q;
eq(1)  = eq(1) + p(1);
p      = -eq.jac{1}\eq.val;
plotCellData(G,p);
```

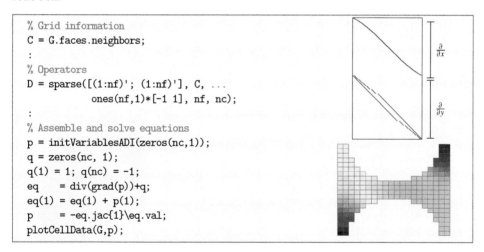

Notice that the D matrix has almost the same sparsity structure as in our first problem, except that the nonzero bands now are curved because the number of cells in each column/row of the grid changes throughout the domain.

Example 4.4.3 *To illustrate the power of the combination of an unstructured grid format and discrete differential operators, we also go through how you can use this technique to* solve the Poisson equation on an unstructured grid. As an example, we use the Voronoi grid generated from the `seamount` data set shown in Figure 3.11 on page 68. Now comes *the important point:* Because the discrete differential operators are defined in terms of the two general matrices C_1 and C_2 that describe the internal connections in the grid, their construction remains *exactly* the same as for the Cartesian grid:

```
load seamount
G = pebi(triangleGrid([x(:) y(:)], delaunay(x,y)));
G = computeGeometry(G);

C  = G.faces.neighbors;
C  = C(all(C ~= 0, 2), :);
nf = size(C,1);
nc = G.cells.num;
D  = sparse([(1:nf)'; (1:nf)'], C, ...
     ones(nf,1)*[-1 1], nf, nc);
grad = @(x) D*x;
```

Here, the directional derivatives do not follow the axial directions and hence D will have a general sparse structure and not the banded structure we saw for the Cartesian grids in the previous example. Moreover, because the cell centers are no longer equidistant points on a uniform mesh, the diagonal entries in the transmissibility matrix will not be the same constant for all cells and hence cannot be scaled out of the discrete system. For historical reasons, MRST only supplies a routine for computing half-transmissibilities defined in (4.51) on page 132. These are defined for *all* faces in the grid. Since we have assumed no-flow boundary conditions, we only need to find the half-transmissibilities associated with the interior *faces*, and compute their harmonic average to get the transmissibilities defined in (4.52):

```
hT = computeTrans(G, struct('perm', ones(nc,1)));
cf = G.cells.faces(:,1);
T  = 1 ./ accumarray(cf, 1 ./ hT, [G.faces.num, 1]);
T  = T(all(C~=0,2),:);

p  = initVariablesADI(zeros(nc,1));
q  = zeros(nc, 1);   q([135 282 17]) = [-1 .5 .5];
eq = div(T.*grad(p))+q;
eq(1) = eq(1) + p(1);
p  = -eq.jac{1}\eq.val;
```

You may also notice that we have changed our source terms slightly so that there is now a fluid sink at the center and fluid sources to the northwest and southeast. We will return to a more detailed discussion of the computation of transmissibilities and assembly of discrete linear systems in Section 5.2.

4.4.3 Time-of-Flight and Tracer

The transport equation (4.40) and (4.42) are examples of a more general class of steady-state transport problems,

$$\nabla \cdot (u\vec{v}) = h(\vec{x}, u), \qquad (4.64)$$

where $u = \tau$ and $h = \phi + \tau \nabla \cdot \vec{v}$ for time-of-flight and $u = C$ and $h = q_C + C \nabla \cdot \vec{v}$ for the artificial tracer. To discretize the steady transport equation (4.64), we integrate it over a single grid cell Ω_i and use Gauss's divergence theorem to obtain

$$\int_{\partial \Omega_i} u\vec{v} \cdot \vec{n}\, ds = \int_{\Omega_i} h(\vec{x}, u(\vec{x}))\, d\vec{x}.$$

In Section 4.4.1 we discussed how to discretize the flux over an interface Γ_{ik} between two cells Ω_i and Ω_k for the case that $u \equiv 1$. To be consistent with the notation used above, we will call this flux v_{ik}. If we can define an appropriate value u_{ik} at the interface Γ_{ik}, we can write the flux across the interface as

$$\int_{\Gamma_{ik}} u\vec{v} \cdot \vec{n}\, ds = u_{ik} v_{ik}. \qquad (4.65)$$

The obvious idea of setting $u_{ik} = \frac{1}{2}(u_i + u_k)$ gives a centered scheme that is unfortunately notoriously unstable. To get a better approximation, we can inspect the direction information propagates in the transport equation. If the flux v_{ik} is positive, the content of u in cell Ω_i is continuously streaming into cell Ω_k, and it is thus natural to choose u_i as the interface value, or vice versa. This gives us the so-called upwind value

$$u_{ik} = \begin{cases} u_i, & \text{if } v_{ik} \geq 0, \\ u_k, & \text{otherwise.} \end{cases} \qquad (4.66)$$

Using the upwind value can be thought of as adding extra numerical dispersion to stabilize the resulting scheme so that it does not introduce spurious oscillations; more details about such upwind schemes are given in Chapter 9.

For completeness, let us also write this discretization using the abstract notation defined in the previous section. If we discretize u by a vector $\boldsymbol{u} \in \mathbb{R}^{n_c}$ and h by a vector function $\boldsymbol{h}(\boldsymbol{u}) \in \mathbb{R}^{n_c}$, the transport equation (4.64) can be written in the discrete form

$$\text{div}(\boldsymbol{uv}) = \boldsymbol{h}(\boldsymbol{u}). \qquad (4.67)$$

We also substitute the expression for $\boldsymbol{v}$ from (4.62b) and use (4.66) to define u at each face f. Then, we define for each face $f \in S_f$,

$$(\boldsymbol{uv})[f] = \boldsymbol{u}^{uw}[f]\, \boldsymbol{T}[f]\, \text{grad}(\boldsymbol{p})[f], \qquad (4.68)$$

where

$$\boldsymbol{u}^{uw}[f] = \begin{cases} \boldsymbol{u}[C_1(f)]), & \text{if } \text{grad}(\boldsymbol{p})[f] > 0, \\ \boldsymbol{u}[C_2(f)]), & \text{otherwise.} \end{cases} \qquad (4.69)$$

Time-of-flight and tracer distributions can of course also be computed if we trace individual streamlines by solving the ordinary differential equation (4.36). The most commonly used method for tracing streamlines on hexahedral grids is a semi-analytical tracing algorithm introduced by Pollock [253], which uses analytical expressions of the streamline paths inside each cell, based on the assumption that the velocity field is piecewise linear locally. Pollock's method is only correct for regular grids but is still used for highly skewed and irregular grids. Other approaches for tracing on unstructured grids and the associated accuracy are discussed in [74, 256, 150, 205, 122, 204, 166]. On unstructured polygonal grids, tracing of streamlines becomes significantly more involved. Because the general philosophy of MRST is that solvers should work independent of grid type – so that you can seamlessly switch from structured to fully unstructured, polygonal grids – we prefer to use finite-volume methods rather than streamline tracing to compute time-of-flight and tracer distributions. The disadvantages of this choice are that finite-volume methods have lower pointwise accuracy and that averaged time-of-flight value have a less intuitive interpretation. We will return to a more in-depth discussion of how these quantities can be used to investigate timelines and volumetric connections in the reservoir in Chapter 13.

COMPUTER EXERCISES

4.4.1   Compare the discrete differentiation operators for selected grids from Chapter 3, e.g., Exercises 3.1.1–3.1.3, 3.2.6, and 3.2.7. Can you explain the differences?

4.4.2   Go back to Example 4.4.2 and extend the solver to also compute time-of-flight by using the discrete operators defined in the previous exercise in this chapter.

4.4.3   Extend the solver from Example 4.4.3 to compute one tracer associated with each source term. Together, the two tracers should define a partition of unity for the reservoir volume. How would you visualize it?

5
Incompressible Solvers for Single-Phase Flow

A simulation model can be considered to consist of three main parts; the first part describes the reservoir rock, the second part describes the mathematical laws that govern fluid behavior, and the last represents wells and other drive mechanisms. We have already discussed how to model the reservoir rock and its petrophysical properties in Chapters 2 and 3, and shown how the resulting models are represented in MRST using a grid object, usually called G, that describes the geometry of the reservoir, and a rock object, usually called rock, that describes petrophysical parameters. Likewise, Chapter 4 discussed the fundamental flow equations for single-phase flow and presented basic numerical discretizations for elliptic Poisson-type equations and the gradient/divergence operators that appear in flow models.

This chapter discusses the additional parts you need to make a full model and implement a simulator. We show how to represent fluid behavior in terms of fluid objects that contain basic properties such as density, viscosity, and compressibility. We will later extend these fluid objects to model more complex behavior by including properties like relative permeability and capillary pressure that describe interaction among multiple fluid phase and the porous rock. Likewise, we discuss necessary data structures to represent forcing terms such as boundary conditions, (volumetric) source terms, and models of injection and production wells. It is also convenient to introduce a state object holding the primary unknowns and derived quantities like pressure, fluxes, and face pressures.

There are two different ways the data objects just outlined can be combined to form a full simulator. In Section 4.4.2, we saw how to use discrete differential operators to write the flow equations in residual form and then employ automatic differentiation to linearize and form a linear system. Whereas this technique is elegant and will prove highly versatile for compressible flow models later in the book, it is an overkill for incompressible single-phase flow, since the flow equations already are linear. In this chapter, we therefore outline how to use a classic procedural approach to implement the discretized flow equations. We start by outlining the data structures and constructors needed to set up fluid properties and forcing terms, and once this is done, we move on to discuss in detail how to build two-point discretizations and assemble and solve corresponding linear systems. For pedagogical purposes, we present a somewhat simplified version of the basic flow solvers for incompressible flow that are implemented in the add-on modules incomp and diagnostics of

MRST. At the end of the chapter we go through several simulation cases and give all code lines necessary for full simulation setups with various drive mechanisms.

5.1 Basic Data Structures in a Simulation Model

The simple flow solvers we discussed in the previous chapter did not contain any fluid properties and assumed no-flow boundary conditions and point sources as the only forcing term. In this section we outline basic data structures you can use to set up more comprehensive single-phase simulation cases.

5.1.1 Fluid Properties

The only fluid properties we need in the basic single-phase flow equations are the viscosity and the fluid density for incompressible models and the fluid compressibility for compressible models. More complex single-phase and multiphase models require additional fluid and rock-fluid properties. To simplify the communication of fluid properties between flow and transport solvers, it is good practice to introduce a common API. To this end, MRST uses so-called *fluid objects* that contain a predefined set of basic fluid properties as well as function handles used to evaluate rock-fluid properties that are only relevant for multiphase flow. This basic structure can be expanded by optional parameters and functions to represent more advanced fluid models. The following shows how to initialize the most basic fluid object that only requires viscosity and density as input

```
fluid = initSingleFluid('mu' , 1*centi*poise, ...
                        'rho', 1014*kilogram/meter^3);
```

After initialization, the fluid object contains pointers to functions that can be used to evaluate petrophysical properties of the fluid:

```
fluid =
    properties: @(varargin)properties(opt,varargin{:})
    saturation: @(x,varargin)x.s
       relperm: @(s,varargin)relperm(s,opt,varargin{:})
```

Only the first function is relevant for single-phase flow, and returns the viscosity when called with a single output argument, and the viscosity and the density when called with two output arguments. The other two functions can be considered as dummy functions that ensure that the single-phase fluid object is compatible with solvers written for more advanced fluid models. The `saturation` function accepts a reservoir state as argument (see Section 5.1.2) and returns the corresponding saturation (volume fraction of the fluid phase), which will either be empty or set to unity, depending upon how the reservoir state has been initialized. The `relperm` function accepts a fluid saturation as argument and returns the relative permeability, i.e., the reduction in permeability due to the presence of other fluid phases. This function should always be identical to one for single-phase models.

5.1.2 Reservoir States

To hold the dynamic state of the reservoir, MRST uses a special data structure. We refer to realizations of this structure as the *state objects*. In its basic form, the structure contains three elements: a vector pressure with one pressure per cell in the grid, a vector flux with one value per face in the grid, and a vector s that should either be empty or be a vector with a unit entry for each cell, since we only have a single fluid. The state object is initialized by a call to the function

```
state = initResSol(G, p0, s0);
```

where p0 is the initial pressure and s0 is an optional parameter, giving the initial saturation (which should be identical to one for single-phase models). Contrary to what the name may imply, this function *does not* initialize the fluid pressure to be in hydrostatic equilibrium. If such a condition is needed, it must be enforced explicitly by the user. In the case of wells in the reservoir, you should use the alternative function:

```
state = initState(G, W, p0, s0);
```

This gives a state object with an additional field wellSol, which is a vector with one entry per well. Each element in the vector is a structure that contains two fields, wellSol.pressure and wellSol.flux. These two fields are vectors of length equal the number of completions in the well and contain the bottom-hole pressure and flux for each completion.

5.1.3 Fluid Sources

The simplest way to describe flow in or out from interior points in the reservoir is to use volumetric source terms. You can create source terms as follows

```
src = addSource([], cells, rates);
src = addSource(src, cells, rates, 'sat', sat);
```

Here, the input/output values are:

- **src:** array of MATLAB structures describing separate sources. If the first input argument is empty, the routine will output a single structure. Otherwise, it will append the new structure to the array of existing sources sent as input. Each source structure contains the following fields:
  - **cell:** cells containing explicit sources,
  - **rate:** rates for these explicit sources,
  - **value:** pressure or flux value for the given condition,
  - **sat:** fluid composition of injected fluids in cells with rate>0.
- **cells:** indices to the cells in the grid model in which this source term should be applied.

- **rates:** vector of volumetric flow rates, one scalar value for each cell in `cells`. Note that these values are interpreted as flux rates (typically in units of [m^3/day]) rather than as flux density rates (which must be integrated over the cell volumes to obtain flux rates).
- **sat:** optional parameter that specifies the composition of the fluid injected from this source. In this $n \times m$ array of fluid compositions, n is the number of elements in `cells` and m is the number of fluid phases. For $m = 3$, the columns are interpreted as: 1="aqua," 2="liquid," and 3="vapor." This field is for the benefit of multiphase transport solvers, and is ignored for all sinks (at which fluids flow *out* of the reservoir). The default value is `sat = []`, which corresponds to single-phase flow. If `size(sat,1)==1`, this saturation value will be repeated for all cells specified by `cells`.

For convenience, `rates` and `sat` *may* contain a single value; this value is then used for all faces specified in the call.

There can only be a single net source term per cell in the grid. Moreover, for incompressible flow with no-flow boundary conditions, the source terms *must* sum to zero if the model is to be well posed, or alternatively sum to the flux across the boundary. If not, we would either inject more fluids than we extract, or vice versa, and hence implicitly violate the assumption of incompressibility.

5.1.4 Boundary Conditions

As discussed in Section 4.3.1, all outer faces of a grid are assumed as no-flow boundaries unless other conditions are specified explicitly. The basic mechanism for specifying Dirichlet and Neumann boundary conditions is to use the function:

```
bc = addBC(bc, faces, type, values);
bc = addBC(bc, faces, type, values, 'sat', sat);
```

Here, the input values are:

- **bc:** array of MATLAB structures describing separate boundary conditions. If the first input argument is empty (`bc==[]`), the routine will output a single structure. Otherwise, it will append the new structure to the array of existing boundary conditions sent as input. Each structure contains the following fields:
  - **face:** external faces for which explicit conditions are set,
  - **type:** cell array of strings denoting type of condition,
  - **value:** pressure or flux value for the given condition,
  - **sat:** composition of fluids passing through inflow faces, not used for single-phase models.
- **faces:** array of external faces at which this boundary condition is applied.
- **type:** type of boundary condition. Supported values are `'pressure'` and `'flux'`, or a cell array of such strings.

5.1 Basic Data Structures in a Simulation Model

- `values`: vector of boundary conditions, one scalar value for each face in `faces`. Interpreted as a pressure value in units [Pa] when `type` equals `'pressure'` and as a flux value in units [m^3/s] when `type` is `'flux'`. In the latter case, positive values in `values` are interpreted as injection fluxes *into* the reservoir, while negative values signify extraction fluxes, i.e., fluxes *out of* the reservoir.
- `sat`: optional parameter that specifies the composition of the fluid injected across inflow faces. Similar setup as explained for source terms in Section 5.1.3.

There can only be a single boundary condition per face in the grid. Solvers assume that boundary conditions are given on the boundary; conditions in the interior of the domain yield unpredictable results. Moreover, for incompressible flow and only Neumann conditions, the boundary fluxes *must* sum to zero if the model is to be well posed. If not, we would either inject more fluids than we extract, or vice versa, and hence implicitly violate the assumption of incompressibility.

For convenience, MRST also offers two additional routines for setting Dirichlet and Neumann conditions at all outer faces in a certain direction for grids having a logical IJK numbering:

```
bc = pside(bc, G, side, p);
bc = fluxside(bc, G, side, flux)
```

The `side` argument is a string that must match one out of the following six alias groups:

```
1:  'West',   'XMin',  'Left'
2:  'East',   'XMax',  'Right'
3:  'South',  'YMin',  'Back'
4:  'North',  'YMax',  'Front'
5:  'Upper',  'ZMin',  'Top'
6:  'Lower',  'ZMax',  'Bottom'
```

These groups correspond to the cardinal directions mentioned as the first alternative in each group. You should also be aware of an important difference in how fluxes are specified in `addBC` and `fluxside`. Specifying a scalar value in `addBC` means that this value will be copied to all faces the boundary condition is applied to, whereas a scalar value in `fluxside` sets the cumulative flux for all faces that make up the global side to be equal the specified value.

5.1.5 Wells

Wells are similar to source terms in the sense that they describe injection or extraction of fluids from the reservoir, but differ in the sense that they not only provide a volumetric flow rate, but also contain a model that couples this flow rate to the difference between the average reservoir in the grid cell and the pressure inside the wellbore. As discussed in Section 4.3.2, this relation can be written for each perforation as

$$v_p = J(p_i - p_f), \tag{5.1}$$

where J is the well index, p_i is the pressure in the perforated grid cell, and p_f is the flowing pressure in the wellbore. The wellbore is assumed to be in hydrostatic equilibrium so that p_f in each completion can be found from the pressure at the top of the well and the density along the wellbore. For single-phase, incompressible flow, this hydrostatic balance reads $p_f = p_{wh} + \rho \Delta z_f$, where p_{wh} is the pressure at the well head and Δz_f is the vertical distance from this point and to the perforation. By convention, the structure used to represent wells in MRST is called W, and consists of the following fields:

- **cells:** an array index to cells perforated by this well.
- **type:** string describing which variable is controlled (i.e., assumed to be fixed), either 'bhp' or 'rate'.
- **val:** the target value of the well control; pressure value for type='bhp' or rate for type='rate'.
- **r:** the wellbore radius (double).
- **dir:** a char describing the direction of the perforation ('x', 'y' or 'z').
- **WI:** the well index: either the productivity index or the well injectivity index depending on whether the well is producing or injecting.
- **dZ:** the height differences from the well head, which is defined as the topmost contact (i.e., the contact with the minimum z-value counted amongst all cells perforated by this well).
- **name:** string giving the name of the well.
- **compi:** fluid composition, only used for injectors.
- **refDepth:** reference depth of control mode.
- **sign:** defines whether the well is intended to be producer or injector.

Well structures are created by a call to the function

```
W = addWell(W, G, rock, cellInx);
W = addWell(W, G, rock, cellInx, 'pn', pv, ..);
```

Here, cellInx is a vector of indices to the cells perforated by the well, and 'pn'/pv denote one or more keyword/value pairs that can be used to specify optional parameters in the well model:

- **type:** string specifying well control, 'bhp' (default) means that the well is controlled by bottom-hole pressure, whereas 'rate' means that the well is rate controlled.
- **val:** target for well control. Interpretation of this values depends upon type. For 'bhp' the value is assumed to be in unit Pascal, and for 'rate' the value is given in unit [m^3/sec]. Default value is 0.
- **radius:** wellbore radius in meters. Either a single, scalar value that applies to all perforations, or a vector of radii, with one value for each perforation. The default radius is 0.1 m.
- **dir:** well direction. A single CHAR applies to all perforations, while a CHAR array defines the direction of the corresponding perforation.

- `innerProduct`: used for consistent discretizations discussed in Chapter 6.
- `WI`: well index. Vector of length equal the number of perforations in the well. The default value is `-1` in all perforations, whence the well index will be computed from available data (cell geometry, petrophysical data, etc.) in grid cells containing well completions.
- `Kh`: permeability times thickness. Vector of length equal the number of perforations in the well. The default value is `-1` in all perforations, whence the thickness will be computed from the geometry of each perforated cell.
- `skin`: skin factor for computing effective well bore radius. Scalar value or vector with one value per perforation. Default value: `0.0` (no skin effect).
- `Comp_i`: fluid composition for injection wells. Vector of saturations. Default value: `Comp_i=[1,0,0]` (water injection).
- `Sign`: well type: production (`sign=-1`) or injection (`sign=1`). Default value: `[]` (no type specified).
- `name`: string giving the name of the well. Default value is `'Wn'`, where *n* is the number of this well, i.e., `n=numel(W)+1`.

For convenience, MRST also provides the function

```
W = verticalWell(W, G, rock, I, J, K)
W = verticalWell(W, G, rock, I,    K)
```

for specifying vertical wells in models described by Cartesian grids or grids that have some kind of extruded structure. Here,

- `I,J`: gives the horizontal location of the well heel. In the first mode, both `I` and `J` are given and then signify logically Cartesian indices so that `I` is the index along the first logical direction, whereas `J` is the index along the second logical direction. This mode is only supported for grids having an underlying Cartesian (logical) structure such as purely Cartesian grids or corner-point grids.

  In the second mode, only `I` is described and gives the *cell index* of the topmost cell in the column through which the vertical well is completed. This mode is supported for logically Cartesian grids containing a three-component field `G.cartDims` or for otherwise layered grids that contain the fields `G.numLayers` and `G.layerSize`.
- `K`: a vector of layers in which this well should be completed. If `isempty(K)` is true, then the well is assumed to be completed in all layers in this grid column and the vector is replaced by `1:num_layers`.

5.2 Incompressible Two-Point Pressure Solver

The two-point flux-approximation (TPFA) scheme introduced in Section 4.4.1 is implemented as two different routines. The first routine,

```
hT = computeTrans(G,rock)
```

computes the half-face transmissibilities and does not depend on the fluid model, the reservoir state, or the driving mechanisms and is hence part of is part of MRST's core functionality. The second routine

```
state = incompTPFA(state, G, hT, fluid, 'mech1', obj1, ..)
```

takes the complete model description as input and assembles and solves the two-point system. The routine is specific to incompressible flow and is thus placed in the incomp module. Here, 'mech' specifies the drive mechanisms ('src', 'bc', and/or 'wells') using correctly defined objects obj, as discussed in Sections 5.1.3–5.1.5.

Notice that computeTrans may fail to compute sensible transmissibilities if rock.perm is not given in SI units. Likewise, incompTPFA may produce strange results if the inflow and outflow specified by the boundary conditions, source terms, and wells do not sum to zero and hence violate the assumption of incompressibility. However, if fixed pressure is specified in wells or on parts of the outer boundary, there will be an outflow or inflow that will balance the net rate specified elsewhere.

The remainder of this section presents details of the inner workings of the incompressible solver. By going through the essential code lines needed to compute half-transmissibilities and solve and assemble the global system, we demonstrate how simple it is to implement the TPFA method on general polyhedral grid. If you are not interested in these details, you can jump directly to Section 5.4, which contains several examples demonstrating the use of the incompressible solver for single-phase flow.

To focus on the discretization and keep the discussion simple, we will not look at the full implementation of the two-point solver in incomp. Instead, we discuss excerpts from two simplified functions, simpleComputeTrans and simpleIncompTPFA, located in the 1phase directory of the mrst-book module. Together, these form a simplified single-phase solver, which has been created for pedagogical purposes.

Assume we have a standard grid G containing cell and face centroids, e.g., as computed by the computeGeometry function discussed in Section 3.4. Then, the essential code lines of simpleComputeTrans are as follows: First, we define the vectors $\vec{c}_{i,k}$ from cell centroids to face centroids; see Figure 5.1. To this end, we first need to determine the map

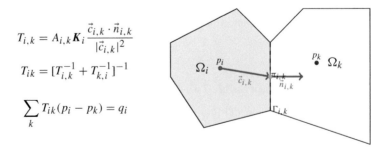

Figure 5.1 Two-point discretization on general polyhedral cells

5.2 Incompressible Two-Point Pressure Solver

from faces to cell number so that the correct cell centroid is subtracted from each face centroid.

```
hf = G.cells.faces(:,1);
hf2cn = gridCellNo(G);
C = G.faces.centroids(hf,:) - G.cells.centroids(hf2cn,:);
```

Face normals in MRST are assumed to have length equal the corresponding face areas, and hence correspond to $A_{i,k}\vec{n}_{i,k}$ in (4.51). To get the correct sign, we look at the neighboring information specifying which cells share the face: if the current cell number is in the first column, the face normal has positive sign. If not, it gets a negative sign:

```
sgn = 2*(hf2cn == G.faces.neighbors(hf, 1)) - 1;
N   = bsxfun(@times, sgn, G.faces.normals(hf,:));
```

The permeability tensor may be stored in different formats, as discussed in Section 2.5, and we therefore use an utility function to extract it:

```
[K, i, j] = permTensor(rock, G.griddim);
```

Finally, we compute the half transmissibilities, $C^T K N / C^T C$. To limit memory use, this is done in a for-loop (which is rarely used in MRST):

```
hT = zeros(size(hf2cn));
for k=1:size(i,2),
   hT = hT + C(:,i(k)) .* K(hf2cn, k) .* N(:,j(k));
end
hT = hT./ sum(C.*C,2);
```

The actual code has a few additional lines that perform various safeguards and consistency checks.

Once the half transmissibilities have been computed, they can be passed to the `simpleIncompTPFA` solver. The first thing this solver needs to do is adjust the half transmissibilities to account for fluid viscosity, since they were derived for a fluid with unit viscosity:

```
mob = 1./fluid.properties(state);
hT  = hT .* mob(hf2cn);
```

Then, we loop through all faces and compute the face transmissibility as the harmonic average of the half-transmissibilities

```
T = 1 ./ accumarray(hf, 1 ./ hT, [G.faces.num, 1]);
```

The MATLAB function `accumarray` constructs an array by accumulation (see Appendix A.4). A call to `a = accumarray(subs,val)` will use the subscripts in `subs` to create an array `a` based on the values `val`. Each element in `val` has a corresponding row in subs. The function collects all elements that correspond to identical subscripts in subs and stores their sum in the element of a corresponding to the subscript. In our case, G.cells.faces(:,1) gives the global face number for each half face, and hence the call to accumarray will sum the transmissibilities of the half-faces that correspond to a given global face and store the result in the correct place in a vector of G.faces.num elements. The function accumarray is very powerful and is used a lot in MRST in place of nested for-loops. In fact, we also employ this function to loop over all cells in the grid and collect and sum the transmissibilities of the faces of each cell to define the diagonal of the TPFA matrix:

```
nc = G.cells.num;
i  = all(G.faces.neighbors ~= 0, 2);
n1 = G.faces.neighbors(i,1);
n2 = G.faces.neighbors(i,2);
d  = accumarray([n1; n2], repmat(T(i),[2,1]),[nc, 1]);
```

Now that we have computed both the diagonal and the off-diagonal elements of A, the discretization matrix itself can be constructed by a straightforward call to MATLAB's sparse function:

```
I = [n1; n2; (1:nc)'];
J = [n2; n1; (1:nc)'];
V = [-T(i); -T(i); d]; clear d;
A = sparse(double(I), double(J), V, nc, nc);
```

Finally, we check if Dirichlet boundary conditions are imposed on the system, and if not, we modify the first element of the system matrix to fix the pressure in the first cell to zero, before solving the system:

```
A(1) = 2*A(1);
p = mldivide(A, rhs);
```

To solve the system, we rely on MATLAB's default solver mldivide, which uses a complex flow chart to check the structure of the matrix to see whether it is square, diagonal, (permuted) triangular, tridiagonal, banded, or Hermitian, and then chooses a specialized and efficient solver accordingly. By running the command spparms('spumoni',2) before you call mldivide, you can tell MATLAB to output information about which linear solver it chooses and the tests leading up the specific choice. For the type of sparse matrices we consider here, the end result is a call to a direct solver from UMFPACK implementing unsymmetric, sparse, multifrontal LU factorization [89, 88]. Such a direct solver is efficient for small-to-medium-sized systems, but for larger systems it is more efficient to use sparse iterative solvers such as a (preconditioned) multilevel method. The linear solver can be passed as a function-pointer argument to both incompTPFA and simpleIncompTPFA,

```
mrstModule add agmg
state = incompTPFA(state, G, hT, fluid, 'wells', W, 'LinSolve', @(A,b) agmg(A,b,1));
```

Here, we have used the aggregation-based algebraic multigrid solver AGMG [238, 15], which integrates well with MRST and is vailable at no cost for academic research and teaching. Section 12.3.4 discusses various specialized linear solver for compressible multiphase flow simulations.

Once the cell pressures have been computed, we can compute pressure values at the face centroids using the half-face transmissibilities

```
fp = accumarray(G.cells.faces(:,1), p(hf2cn).*hT, [G.faces.num,1])./ ...
     accumarray(G.cells.faces(:,1), hT, [G.faces.num,1]);
```

and then construct fluxes across the interior faces

```
ni   = G.faces.neighbors(i,:);
flux = -accumarray(find(i), T(i).*(p(ni(:,2))-p(ni(:,1))), [nf, 1]);
```

In the code excerpts given above, we did not account for gravity forces and general Dirichlet or Neumann boundary conditions, which both will complicate the code beyond the scope of the current presentation. The interested reader should consult the actual code to work out these details. The standard `computeTrans` function can also be used for different representations of petrophysical parameters, and includes functionality to modify the discretization by overriding the definition of cell and face centers and/or including multipliers that modify the values of the half-transmissibilities; see e.g., Sections 2.4.3 and 2.5.5. Likewise, the `incompTPFA` solver from the `incomp` module is implemented for a general, incompressible flow model with multiple fluid phases with flow driven by a general combination of boundary conditions, fluid sources, and well models.

We will shortly present several examples of how this solver can be used for flow problems on structured and unstructured grids. However, before doing so, we outline another flow solver from the `diagnostics` module, which will prove useful to visualize flow patterns.

5.3 Upwind Solver for Time-of-Flight and Tracer

The `diagnostics` module, discussed in more detail in Chapter 13, provides various functionality to probe a reservoir model to establish communication patterns between inflow and outflow regions, timelines for fluid movement, and various measures of reservoir heterogeneity. At the hart of this module, lies the function

```
tof = computeTimeOfFlight(state, G, rock, 'mech1', obj1, ..)
```

which implements the upwind, finite-volume discretization introduced in Section 4.4.3 for solving the time-of-flight equation $\vec{v} \cdot \nabla \tau = \phi$. As you probably recall, time-of-flight is the time it takes a neutral particle to travel from the nearest fluid source or inflow boundary to each point in the reservoir. Here, the `'mech'` arguments represent drive mechanisms (`'src'`, `'bc'`, and/or `'wells'`) specified in terms of specific objects `obj`, as discussed in Sections 5.1.3 to 5.1.5. You can also compute the backward time-of-flight – the time it takes to travel from any point in the reservoir to the nearest fluid sink or outflow boundary – with the same equation if we change sign of the flow field and modify the boundary conditions and/or source terms accordingly. In the following, we will go through the main parts of how this discretization is implemented.

We start by identifying all volumetric sources of inflow and outflow, which may be described as source/sink terms in `src` and/or as wells in `W`, and collect the results in a vector q of source terms having one value per cell

```
[qi,qs] = deal([]);
if ~isempty(W),
   qi = [qi; vertcat(W.cells)];
   qs = [qs; vertcat(state.wellSol.flux)];
end
if ~isempty(src),
   qi = [qi; src.cell];
   qs = [qs; src.rate];
end
q = sparse(qi, 1, qs, G.cells.num, 1);
```

We also need to compute the accumulated inflow and outflow from boundary fluxes for each cell. This will be done in three steps. First, we create an empty vector `ff` with one entry per global face, find all faces that have Neumann conditions, and insert the corresponding value in the correct row

```
ff    = zeros(G.faces.num, 1);
isNeu = strcmp('flux', bc.type);
ff(bc.face(isNeu)) = bc.value(isNeu);
```

The flux is not specified on faces with Dirichlet boundary conditions and must be extracted from the solution computed by the pressure solver, i.e., from the `state` object that holds the reservoir state. We also need to set the correct sign so that fluxes *into* a cell are positive and fluxes *out of* a cell are negative. The sign of the flux across an outer face is correct if `neighbors(i,1)==0`, but if `neighbors(i,2)==0` we need to reverse the sign (we should also check that i is not empty)

```
i = bc.face(strcmp('pressure', bc.type));
ff(i) = state.flux(i) .* (2*(G.faces.neighbors(i,1)==0) - 1);
```

The last step is to sum all the fluxes across outer faces and collect the result in a vector qb that has one value per cell

5.3 Upwind Solver for Time-of-Flight and Tracer

```
outer = ~all(double(G.faces.neighbors) > 0, 2);
qb = sparse(sum(G.faces.neighbors(outer,:), 2), 1, ff(is_outer), G.cells.num, 1);
```

Here, sum(G.faces.neighbors(outer,:), 2) gives an array containing the indices of each cell attached to an outer face.

Once the contributions to inflow and outflow are collected, we can start building the upwind flux discretization matrix A. The off-diagonal entries are defined such that $A_{ji} = \max(v_{ij}, 0)$ and $A_{ij} = -\min(v_{ij}, 0)$, where v_{ij} is the flux computed by the TPFA scheme discussed in the previous section.

```
i   = ~any(G.faces.neighbors==0, 2);
out = min(state.flux(i), 0);
in  = max(state.flux(i), 0);
```

The diagonal entry equals the outflux minus the divergence of the velocity, which can be obtained by summing the off-diagonal rows. This will give the correct equation in all cell except for those with a positive fluid source. Here, the net outflux equals the divergence of the velocity and we hence end up with an undetermined equation. In these cells, we can as a reasonable approximation[1] set the time-of-flight to be equal the time it takes to fill the cell, which means that the diagonal entry should be equal the fluid rate inside the cell.

```
n      = double(G.faces.neighbors(i,:));
inflow = accumarray([n(:, 2); n(:, 1)], [in; -out]);
d      = inflow + max(q+qb, 0);
```

Having obtained diagonal and all the nonzero off-diagonal elements, we can assemble the full matrix

```
nc = G.cells.num;
A  = sparse(n(:,2), n(:,1), in, nc, nc) + sparse(n(:,1), n(:,2), -out, nc, nc);
A  = -A + spdiags(d, 0, nc, nc);
```

We have now established the complete discretization matrix, and time-of-flight can be computed by a simple matrix inversion

```
tof  = A \ poreVolume(G,rock);
```

If there are no gravity forces and the flux has been computed by the two-point method (or some other monotone scheme), one can show that the discretization matrix A can be permuted to a lower-triangular form [221, 220]. In the general case, the permuted matrix will be block triangular with irreducible diagonal blocks. Such systems can be inverted very

---

[1] Notice, however, that to get the correct values for 1D cases, it is more natural to set time-of-flight equal *half* the time it takes to fill the cell.

efficiently using a permuted back-substitution algorithm as long as the irreducible diagonal blocks are small. MATLAB is quite good at detecting such structures, and using the simple backslash (\) operator is therefore efficient, even for quite large models. However, for models of real petroleum assets described on stratigraphic grids (see Section 3.3), it is often necessary to preprocess flux fields to get rid of numerical clutter that would otherwise introduce large irreducible blocks inside stagnant regions. By specifying optional parameters to computeTimeOfFlight, the function will get rid of such small cycles in the flux field and set the time-of-flight to a prescribed upper value in all cells that have sufficiently small influx. This tends to reduce the computational cost significantly for large models with complex geology and/or significant compressibility effects.

The same routine can also compute *stationary tracers*, as discussed in Section 4.3.4. This is done by passing an optional parameter,

```
tof = computeTimeOfFlight(state, G, rock, .., 'tracer',tr)
```

where tr is a cell-array of vectors that each gives the indices of cells that emit a unique tracer. For incompressible flow, the discretization matrix of the tracer equation is the same as that for time-of-flight, and all we need to do to extend the solver is to assemble the right-hand side

```
numTrRHS = numel(tr);
TrRHS = zeros(nc,numTrRHS);
for i=1:numTrRHS,
    TrRHS(tr{i},i) = 2*qp(tr{i});
end
```

Since we have doubled the rate in any cells with a positive source when constructing the matrix A, the rate on the right-hand side must also be doubled.

With the extra right-hand sides assembled, we can solve the combined time-of-flight/tracer problem as a linear system with multiple right-hand sides,

```
T = A \ [poreVolume(G,rock) TrRHS];
```

which means that we essentially get the tracer for free as long as the number of tracers does not exceed the number of right-hand columns MATLAB can handle in one solve. We will return to a more thorough discussion of the tracer partitions Chapter 13 and show how these can be used to delineate connectivities within the reservoir. In the rest of this chapter, we will consider time-of-flight and streamlines as a means to study flow patterns in reservoir models.

5.4 Simulation Examples

We have now introduced you to all the functionality from the incomp module necessary to solve a single-phase flow problem, as well as the time-of-flight solver from the diagnostics module, which can be used to compute time lines in the reservoir. In the following, we discuss several examples and demonstrate step by step how to set up a flow model, solve it, and visualize and analyze the resulting flow field. Complete codes can be found in the 1phase directory of the book module.

5.4.1 Quarter Five-Spot

As our first example, we show how to solve $-\nabla \cdot (\mathbf{K}\nabla p) = q$ with no-flow boundary conditions and two source terms at diagonally opposite corners of a 2D Cartesian grid covering a 500×500 m^2 area. This setup mimics a standard quarter five-spot well pattern, which we encountered in Figure 4.7 on page 126 when discussing well models. The full code is available in the script quarterFiveSpot.m. We use a rectangular grid with homogeneous petrophysical data ($K = 100$ mD and $\phi = 0.2$):

```
[nx,ny] = deal(32);
G = cartGrid([nx,ny],[500,500]);
G = computeGeometry(G);
rock = makeRock(G, 100*milli*darcy, .2);
```

As we saw above, all we need to develop the spatial discretization is the reservoir geometry and the petrophysical properties. This means that we can compute the half transmissibilities without knowing any details about the fluid properties and the boundary conditions and/or sources/sinks that will drive the global flow:

```
hT = simpleComputeTrans(G, rock);
```

The result of this computation is a vector with one value per local face of each cell in the grid, i.e., a vector with G.cells.faces entries.

The reservoir is horizontal and gravity forces are therefore not active. We create a fluid with properties that are typical for water:

```
gravity reset off
fluid = initSingleFluid('mu' , 1*centi*poise, ...
                       'rho', 1014*kilogram/meter^3);
```

To drive flow, we use a fluid source in the southwest corner and a fluid sink in the northeast corner. The time scale of the problem is defined by the strength of the source terms. Here, we set these terms so that a unit time corresponds to the injection of one pore volume of fluids. By convention, all flow solvers in MRST automatically assume no-flow conditions on all outer (and inner) boundaries if no other conditions are specified explicitly.

```
pv  = sum(poreVolume(G,rock));
src = addSource([], 1, pv);
src = addSource(src, G.cells.num, -pv);
```

The data structure used to represent the fluid sources contains three elements:

```
cell: [2x1 double]
rate: [2x1 double]
 sat: []
```

We recall that `src.cell` gives cell numbers where source terms are nonzero, and the vector `src.rate` specifies the fluid rates, which by convention are positive for inflow into the reservoir and negative for outflow from the reservoir. The last data element `src.sat` specifies fluid saturations, which only has meaning for multiphase flow models and hence is set to be empty here.

Strictly speaking, state structure need not be initialized for an incompressible model in which none of the fluid properties depend on the reservoir state. However, to avoid treatment of special cases, MRST requires that the structure is initialized and passed as argument to the pressure solver. We therefore initialize it with a dummy pressure value of zero and a unit fluid saturation (fraction of void volume filled by fluid), since we only have a single fluid

```
state = initResSol(G, 0.0, 1.0);
display(state)
```

```
state =
    pressure: [1024x1 double]
        flux: [2112x1 double]
           s: [1024x1 double]
```

This completes the setup of the model. To solve for pressure, we simply pass reservoir state, grid, half transmissibilities, fluid model, and driving forces to the flow solver, which assembles and solves the incompressible equation.

```
state = simpleIncompTPFA(state, G, hT, fluid, 'src', src);
display(state)
```

```
state =
        pressure: [1024x1 double]
            flux: [2112x1 double]
               s: [1024x1 double]
    facePressure: [2112x1 double]
```

As explained here, `simpleIncompTPFA` solves for pressure as the primary variable and then uses transmissibilities to reconstruct the face pressure and intercell fluxes. After a call to the pressure solver, the `state` object is therefore expanded by a new field `facePressure` that contains pressures reconstructed at the face centroids. Figure 5.2 shows the resulting pressure distribution. To improve the visualization of the flow field, we show streamlines. The `streamlines` add-on module to MRST implements Pollock's method [253] for semi-analytical tracing of streamlines. Here, we use this functionality to trace streamlines forward and backward, starting from the midpoint of all cells along the northwest–southeast diagonal in the grid

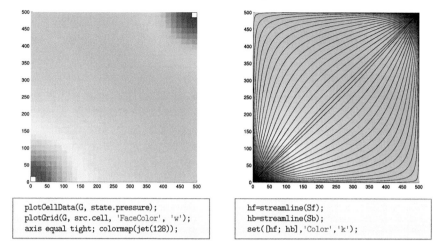

Figure 5.2 Solution of the quarter five-spot problem on a 32 × 32 uniform grid. The left plot shows the pressure distribution and in the right plot we have imposed streamlines passing through centers of the cells on the northwest–southeast diagonal.

```
mrstModule add streamlines;
seed = (nx:nx-1:nx*ny).';
Sf = pollock(G, state, seed, 'substeps', 1);
Sb = pollock(G, state, seed, 'substeps', 1, 'reverse', true);
```

The `pollock` routine produces a cell array of individual streamlines, which we pass onto MATLAB's `streamline` routine for plotting, as shown to the right in Figure 5.2.

To get a better picture of how fast the fluid flows through our domain, we solve the time-of-flight equation (4.40) subject to the condition that $\tau = 0$ at the inflow, i.e., at all points where $q > 0$. For this purpose, we use the `computeTimeOfFlight` solver discussed in Section 5.3, which can compute time-of-flight both forward from inflow points and into the reservoir,

```
toff = computeTimeOfFlight(state, G, rock, 'src', src);
```

and from outflow points and backwards into the reservoir

```
tofb = computeTimeOfFlight(state, G, rock, 'src', src, 'reverse', true);
```

Isocontours of time-of-flight define natural time lines in the reservoir. To emphasize this, the left plot in Figure 5.3 shows time-of-flight plotted using only a few colors to make a rough contouring effect. The sum of forward and backward time-of-flight gives the total time it takes a fluid particle to pass from an inflow point to an outflow point. We can thus use this total residence time to visualize high-flow and stagnant regions, as demonstrated in the right plot of Figure 5.3.

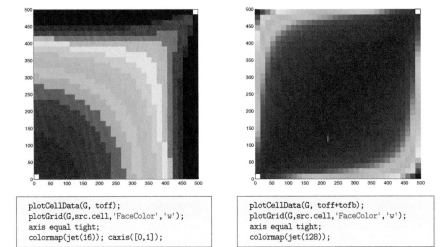

Figure 5.3 Quarter five-spot problem on a 32 × 32 grid. The left plot shows time-of-flight plotted with a few color levels to create a crude contouring effect. The right plots total travel time clearly distinguishing high-flow and stagnant regions.

COMPUTER EXERCISES

5.4.1 Run the quarter five-spot example with the following modifications:

a. Replace the Cartesian grid by a curvilinear grid, e.g., using `twister` or a random perturbation of internal nodes as shown in Figure 3.3.
b. Replace the grid by the locally refined grid from Exercise 3.2.6.
c. Replace the homogeneous permeability by a heterogeneous permeability derived from the Carman–Kozeny relation (2.6).
d. Set the domain to be a single layer of the SPE 10 model. Hint: use `getSPE10rock()` to sample the petrophysical parameters and remember to convert to SI units.

Notice that the `pollock` function may not work for non-Cartesian grids.

5.4.2 Construct a grid similar to the one in Exercise 3.1.1, except that the domain is given a 90-degree flip so that axis of the cylindrical cutouts align with the z-direction. Modify the code presented in this section so that you can compute a five-spot setup with one injector near each corner and a producer in the narrow middle section between the cylindrical cutouts.

5.4.2 Boundary Conditions

To demonstrate how to specify boundary conditions, we go through essential code lines of three different examples. In all three examples, the reservoir is 50 m thick, is located at a depth of approximately 500 m, and is restricted to a 1×1 km^2 area. The permeability is uniform and anisotropic, with a diagonal (1,000, 300, 10) mD tensor, and the porosity is uniform and equal 0.2. In the first two examples, the reservoir is represented as a $20 \times 20 \times 5$ rectangular grid, and in the third example the reservoir is given as a corner-point grid of the same Cartesian dimension, but with an uneven uplift and four intersecting faults (as shown in the left plot of Figure 3.32):

```
[nx,ny,nz] = deal(20, 20, 5);
[Lx,Ly,Lz] = deal(1000, 1000, 50);
switch setup
   case 1,
      G = cartGrid([nx ny nz], [Lx Ly Lz]);
   case 2,
      G = cartGrid([nx ny nz], [Lx Ly Lz]);
   case 3,
      G = processGRDECL(makeModel3([nx ny nz], [Lx Ly Lz/5]));
      G.nodes.coords(:,3) = 5*(G.nodes.coords(:,3)-min(G.nodes.coords(:,3)));
end
G.nodes.coords(:,3) = G.nodes.coords(:,3) + 500;
```

Setting rock and fluid parameters, computing transmissibilities, and initializing the reservoir state can be done as explained in the previous section, and details are not included for brevity; you find the complete scripts in boundaryConditions.

Linear Pressure Drop

In the first example (setup=1), we specify Neumann conditions with total inflow of 5,000 m^3/day on the east boundary and Dirichlet conditions with fixed pressure of 50 bar on the west boundary:

```
bc = fluxside(bc, G, 'EAST', 5e3*meter^3/day);
bc = pside   (bc, G, 'WEST', 50*barsa);
```

This completes the definition of the model, and we can pass the resulting objects to the simpleIncompTFPA solver to compute the pressure distribution shown to the right in Figure 5.4. In the absence of gravity, these boundary conditions will result in a linear pressure drop from east to west inside the reservoir.

Hydrostatic Boundary Conditions

In the next example, we use the same model, except that we now include the effects of gravity and assume hydrostatic equilibrium at the outer vertical boundaries of the model. First, we initialize the reservoir state according to hydrostatic equilibrium, which is straightforward to compute if we for simplicity assume that the overburden pressure is caused by a column of fluids with the exact same density as in the reservoir:

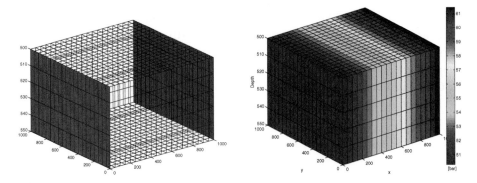

Figure 5.4 First example of a flow driven by boundary conditions. In the left plot, faces with Neumann conditions are marked in blue and faces with Dirichlet conditions are marked in red. The right plot shows the resulting pressure distribution.

```
state = initResSol(G, G.cells.centroids(:,3)*rho*norm(gravity), 1.0);
```

There are at least two different ways to specify hydrostatic boundary conditions. The simplest approach is to use the function psideh, i.e.,

```
bc = psideh([], G, 'EAST', fluid);   bc = psideh(bc, G, 'WEST', fluid);
bc = psideh(bc, G, 'SOUTH', fluid);  bc = psideh(bc, G, 'NORTH', fluid);
```

Alternatively, we can do it manually ourselves. To this end, we need to extract the reservoir perimeter defined as all exterior faces are vertical, i.e., whose normal vector has no z-component,

```
f = boundaryFaces(G);
f = f(abs(G.faces.normals(f,3))<eps);
```

To get the hydrostatic pressure at each face, we can either compute it directly by using the face centroids,

```
fp = G.faces.centroids(f,3)*rho*norm(gravity);
```

or we use the initial equilibrium that has already been established in the reservoir by sampling from the cells adjacent to the boundary

```
cif = sum(G.faces.neighbors(f,:),2);
fp  = state.pressure(cif);
```

The latter may be useful if the initial pressure distribution has been computed by a more elaborate procedure than what is currently implemented in psideh. In either case, the boundary conditions can now be set by the call

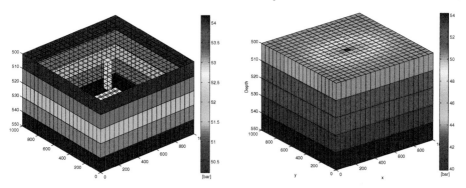

Figure 5.5 A reservoir with hydrostatic boundary condition and fluid extracted from a sink penetrating two cells in the upper two layers of the model. The left plot shows the boundary and the fluid sink, while the right plot shows the resulting pressure distribution.

```
bc = addBC(bc, f, 'pressure', fp);
```

To make the problem a bit more interesting, we also include a fluid sink at the midpoint of the upper two layers in the model,

```
ci = round(.5*(nx*ny-nx));
ci = [ci; ci+nx*ny];
src = addSource(src, ci, repmat(-1e3*meter^3/day,numel(ci),1));
```

Figure 5.5 shows the boundary conditions and source terms to the left and the resulting pressure distribution to the right. The fluid sink causes a pressure drawdown, which has ellipsoidal shape because of the anisotropic permeability field.

Conditions on Non-Rectangular Domain

For the reservoir shown in Figure 5.6, we cannot simply use tags for to select faces on the east and west perimeter. The problem is that `fluxside` and `pside` define cardinal sides to consist of all exterior faces whose normal vector point in the correct cardinal direction. This is illustrated in the upper-left plot of Figure 5.6, where we have tried to specify boundary conditions using the same procedure as in Figure 5.4. We can easily get rid of faces that really lie on the north and south boundary if we use the subrange feature of `fluxside` and `pside` to restrict the boundary conditions to a subset of the global side, as shown in the upper-right plot of Figure 5.6

```
bc = fluxside([], G, 'EAST', 5e3*meter^3/day, 4:15, 1:5);
bc = pside   (bc, G, 'WEST', 50*barsa,        7:17, []);
```

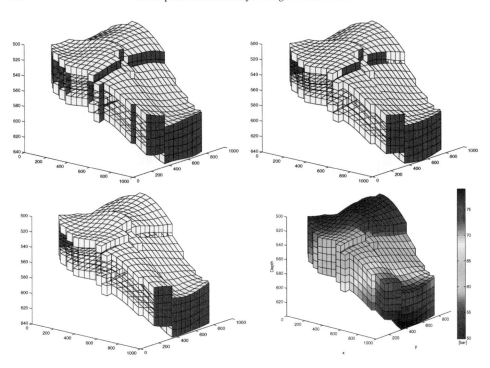

Figure 5.6 Specifying boundary conditions along the outer perimeter of a corner-point model. The upper-left plot shows the use of `fluxside/pside` (blue/red color) and to set boundary conditions on the east and west global boundaries. In the upper-right point, the same functions have been used with a specification of subranges in the global sides. In the lower-left plot, we have utilized user-supplied information to correctly set the conditions only along the perimeter. The lower-right plot shows the resulting pressure solution.

Unfortunately, `fluxside/pside` still picks exterior faces at the faults, classified as pointing east/west because of their normal. Similar problems may arise in other models because of pinched, eroded, or inactive cells.

To find the east- and west-most faces only, we need to use advanced tactics. In our case, the natural perimeter is defined as those faces that lie on the bounding box of the model, on which we distribute the total flux to individual faces according to the face area. For the Neumann condition we therefore get

```
x = G.faces.centroids(f,1);
[xm,xM] = deal(min(x), max(x));
ff = f(x>xM-1e-5);
bc = addBC(bc, ff, 'flux', (5e3*meter^3/day) ...
             * G.faces.areas(ff)/ sum(G.faces.areas(ff)));
```

We can specify the Dirichlet condition in a similar manner. The lower-right plot shows the correct linear pressure drop.

COMPUTER EXERCISES

5.4.3 Consider a 2D box with a sink at the midpoint and inflow across the perimeter specified either in terms of a constant pressure or a constant flux. Are there differences in the two solutions, and if so, can you explain why? Hint: use time-of-flight, total travel time, and/or streamlines to investigate.

5.4.4 Apply the production setup from Figure 5.5, with hydrostatic boundary conditions and fluids extracted from two cells at the midpoint of the model, to the model depicted in Figure 5.6.

5.4.5 Compute flow for all models in data sets BedModels1 and BedModel2 subject to linear pressure drop in the x and then in the y-direction. These models of small-scale heterogeneity are developed to compute representative properties in simulation models on a larger scale. A linear pressure drop is the most widespread computational setup used for flow-based upscaling. What happens if you try to specify flux conditions?

5.4.6 Consider models from the CaseB4 data set. Use appropriate boundary conditions to drive flow across the faults and compare flow patterns computed on the pillar and the stair-stepped grid for the two different model resolutions. Can you explain any differences you observe?

5.4.3 Structured versus Unstructured Stencils

We have so far only discussed hexahedral grids having structured cell numbering. The two-point schemes can also be applied to fully unstructured and polyhedral grids. To demonstrate this, we define a non-rectangular reservoir by scaling the grid from Figure 3.8 to cover a 1×1 km^2 area.

```
load seamount;
T = triangleGrid([x(:) y(:)], delaunay(x,y));
[Tmin,Tmax] = deal(min(T.nodes.coords), max(T.nodes.coords));
T.nodes.coords = bsxfun(@times, ...
    bsxfun(@minus, T.nodes.coords, Tmin), 1000./(Tmax - Tmin));
T = computeGeometry(T);
```

We assume a homogeneous and isotropic permeability of 100 mD and use the same fluid properties as in the previous examples. Constant pressure of 50 bar is set at the outer perimeter and fluid is drained at a constant rate of one pore volume over 50 years from a well located at (450,500). (Script: stencilComparison.m.)

For comparison, we generate two Cartesian grids covering the same domain, one with approximately the same number of cells as the triangular grid and a 10×10 refinement of this grid to provide a reference solution,

```
G = computeGeometry(cartGrid([25 25], [1000 1000]));
inside = isPointInsideGrid(T, G.cells.centroids);
G = removeCells(G, ~inside);
```

The function `isPointInsideGrid` implements a simple algorithm for finding whether points lie inside the circumference of a grid. First, all boundary faces are extracted and then the corresponding nodes are sorted so that they form a closed polygon. Then, we use MATLAB's built-in function `inpolygon` to check if the points are inside this polygon. To also construct a radial grid, we reuse the code from page 86 to set up points inside $[-1, 1] \times [-1, 1]$ graded radially towards the origin

```
P = [];
for r = exp([-3.5:.2:0, 0, .1]),
    [x,y] = cylinder(r,25); P = [P [x(1,:); y(1,:)]];
end
P = unique([P'; 0 0],'rows');
```

We scale the points and translate so that their origin coincides with the fluid sink

```
[Pmin,Pmax] = deal(min(P), max(P));
P = bsxfun(@minus, bsxfun(@times, ...
        bsxfun(@minus, P, Pmin), 1200./(Pmax-Pmin)), [150 100]);
```

We remove all points outside of the triangular grid before we use the point set to first generate a triangular and then a Voronoi grid:

```
inside = isPointInsideGrid(T, P);
V = computeGeometry( pebi( triangleGrid(P(inside,:)) ));
```

Once we have constructed the grids, the setup of the remaining part of the model is the same in all cases. To avoid unnecessary replication of code, we collect the grids in a cell array and use a simple for-loop to set up and simulate each model realization:

```
g = {G, T, V, Gr};
for i=1:4
    rock = makeRock(g{i}, 100*milli*darcy, 0.2);
    hT = simpleComputeTrans(g{i}, rock);
    pv = sum(poreVolume(g{i}, rock));

    tmp = (g{i}.cells.centroids - repmat([450, 500],g{i}.cells.num, [])).^2;
    [~,ind] = min(sum(tmp,2));
    src{i} = addSource(src{i}, ind, -.02*pv/year);

    bc{i} = addBC([], boundaryFaces(g{i}), 'pressure', 50*barsa);

    state{i} = incompTPFA(initResSol(g{i},0,1), ...
        g{i}, hT, fluid, 'src', src{i}, 'bc', bc{i}, 'MatrixOutput', true);

    [tof{i},A{i}] = computeTimeOfFlight(state{i}, g{i}, rock,...
        'src', src{i},'bc',bc{i}, 'reverse', true);
end
```

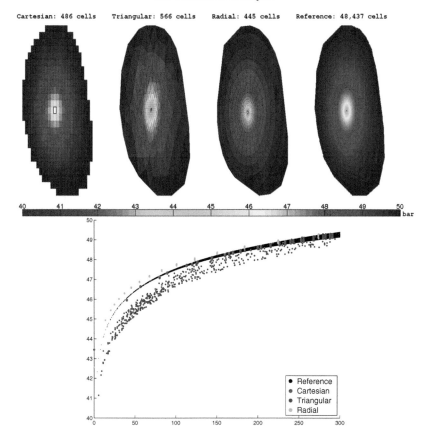

Figure 5.7 Comparison of the pressure solution for three different grid types: uniform Cartesian, triangular, and a graded radial grid. The scattered points used to generate the triangulated domain and limit the reservoir are sampled from the `seamount` data set and scaled to cover a 1×1 km^2 area. Fluids are drained from the center of the domain, assuming a constant pressure of 50 bar at the perimeter.

Figure 5.7 shows the pressure solutions computed on the four different grids, whereas Figure 5.8 reports the sparsity patterns of the corresponding linear systems for the three coarse grids. As expected, the Cartesian grid gives a banded matrix consisting of five diagonals corresponding to each cell and its four neighbors in the cardinal directions. Even though this discretization is not able to predict the complete drawdown at the center (the reference solution predicts a pressure slightly below 40 bar), it captures the shape of the drawdown region quite accurately; the region appears ellipsoidal because of the non-unit aspect ratio in the plot. In particular, we see that the points in the radial plot follow those of the fine-scale reference closely. The spread in the points as $r \to 300$ is not a grid orientation effect, but the result of variations in the radial distance to the fixed pressure at the outer boundary on all four grids.

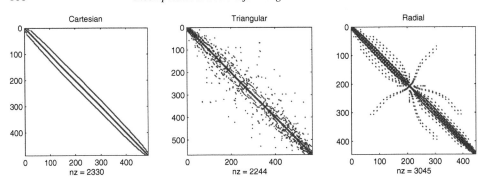

Figure 5.8 Sparsity patterns for the TPFA stencils on the three different grid types shown in Figure 5.7.

The unstructured triangular grid is more refined near the well and is hence able to predict the pressure drawdown in the near-well region more accurately. However, the overall structure of this grid is quite irregular, as you can see from the sparsity pattern of the linear system shown in Figure 5.8, and the irregularity gives significant grid orientation effects. You can see this from the irregular shape of the color contours in the upper part of Figure 5.7, as well as from the spread in the scatter plot. In summary, this grid is not well suited for resolving the radial symmetry of the pressure drawdown in the near-well region. But to be fair, the grid was not generated for this purpose either.

Except for close to the well and close to the exterior boundary, the topology of the radial grid is structured in the sense that each cell has four neighbors, two in the radial direction and two in the angular direction, and the cells are regular trapezoids. This should, in principle, give a banded sparsity pattern if the cells are ordered starting at the natural center point and moving outward, one ring at the time. To verify this claim, you can execute the following code:

```
[~,q] = sort(state{3}.pressure);
spy(state{3}.A(q,q));
```

However, as a result of how the grid was generated by first triangulating and then forming the dual, the cells are numbered from west to east, which explains why the sparsity pattern is so far from being a simple banded structure. This may potentially affect the efficiency of a linear solver, but has no impact on the accuracy of the numerical approximation, which is good because of the grading towards the well and the symmetry inherent in the grid. Slight differences in the radial profile compared with the Cartesian grid(s) are mainly the result of the fact that the source term and the fixed pressure conditions are not located at the exact same positions in the simulations.

In Figure 5.9, we also show the sparsity pattern of the linear system used to compute the reverse time-of-flight from the well and back into the reservoir. Using the default cell ordering, the sparsity pattern of each upwind matrix will appear as a less dense version

5.4 Simulation Examples

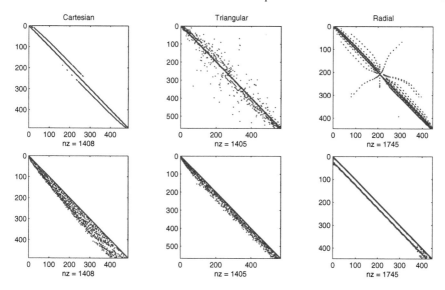

Figure 5.9 Sparsity patterns for the upwind stencils used to compute time-of-flight on the three different grid types shown in Figure 5.7. In the lower row, the matrices have been permuted to lower-triangular form by sorting the cell pressures in ascending order.

of the pattern for the corresponding TPFA matrix. However, whereas the TPFA matrices represent an elliptic equation in which information propagates in both directions across cell interfaces, the upwind matrices are based on one-way connections arising from fluxes between pairs of cells that are connected in the TPFA discretization. To reveal the true nature of the system, we can permute the system by either sorting the cell pressures in ascending order (potential ordering) or using the function dmperm to compute a Dulmage–Mendelsohn decomposition. As pointed out in Section 5.3, the result is a lower triangular matrix, from which it is simple to see the unidirectional propagation of information you should expect for a hyperbolic equation having only positive characteristics.

COMPUTER EXERCISES

5.4.7 Compare the sparsity patterns resulting from potential ordering and use of dmperm for both the upwind and the TPFA matrices.

5.4.8 Investigate the flow patterns in more details using forward time-of-flight, travel time, and streamlines.

5.4.9 Replace the boundary conditions by a constant influx, or set pressure values sampled from a radially symmetric pressure solution in an infinite domain.

5.4.4 Using Peaceman Well Models

Flow in and out of a wellbore takes place on a scale much smaller than a single grid cell in typical sector and field models and is therefore commonly modeled using a semi-analytical model of the form (4.35). This section presents two examples to demonstrate how such models can be included in the simulation setup using data objects and utility functions introduced in Section 5.1.5.

Box Reservoir

We consider a reservoir consisting of a homogeneous $500 \times 500 \times 25$ m^3 sand box with an isotropic permeability of 100 mD, represented on a regular $20 \times 20 \times 5$ Cartesian grid. The fluid is the same as in the examples as in the earlier examples in this section. You can find all code lines necessary to set up the model, solve the flow equations, and visualize the results in the script firstWellExample.m. Setting up the model is quickly done once you have gotten familiar with MRST:

```
[nx,ny,nz] = deal(20,20,5);
G    = computeGeometry( cartGrid([nx,ny,nz], [500 500 25]) );
rock = makeRock(G, 100*milli*darcy, .2);
fluid = initSingleFluid('mu',1*centi*poise,'rho',1014*kilogram/meter^3);
hT   = computeTrans(G, rock);
```

The reservoir is produced by a well pattern consisting of a vertical injector and a horizontal producer. The injector is located in the southwest corner of the model and operates at a constant rate of 3,000 m^3 per day. The producer is completed in all cells along the upper east rim and operates at a constant bottom-hole pressure of 1 bar (i.e., 10^5 Pascal in SI units):

```
W = verticalWell([], G, rock, 1, 1, 1:nz, 'Type', 'rate', 'Comp_i', 1,...
                 'Val', 3e3/day(), 'Radius', .12*meter, 'name', 'I');
W = addWell(W, G, rock, nx : ny : nx*ny, 'Type', 'bhp',  'Comp_i', 1, ...
            'Val', 1.0e5, 'Radius', .12*meter, 'Dir', 'y', 'name', 'P');
```

In addition to specifying the type of control on the well ('bhp' or 'rate'), we must specify wellbore radius and fluid composition, which is '1' for a single phase. After initialization, the array W contains two data objects, one for each well:

```
    Well #1:                    |     Well #2:
       cells: [5x1 double]      |        cells: [20x1 double]
        type: 'rate'            |         type: 'bhp'
         val: 0.0347            |          val: 100000
           r: 0.1000            |            r: 0.1000
         dir: [5x1 char]        |          dir: [20x1 char]
          WI: [5x1 double]      |           WI: [20x1 double]
          dZ: [5x1 double]      |           dZ: [20x1 double]
        name: 'I'               |         name: 'P'
       compi: 1                 |        compi: 1
    refDepth: 0                 |     refDepth: 0
        sign: 1                 |         sign: []
```

This concludes the model specification, and we now have all the information we need to initialize the reservoir state, and assemble and solve the system

5.4 Simulation Examples

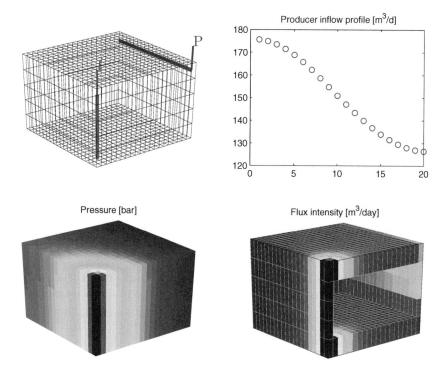

Figure 5.10 Solution of a single-phase, incompressible flow problem inside a box reservoir with a vertical injector and a horizontal producer.

```
gravity reset on;
resSol = initState(G, W, 0);
state  = incompTPFA(state, G, hT, fluid, 'wells', W);
```

Figure 5.10 shows how the inflow rate decays with the distance to the injector as expected. We can compute the flux intensity depicted in the lower-right plot using the following command, which first maps the vector of face fluxes to a vector with one flux per half face and then sums the absolute value of these fluxes to get a flux intensity per cell:

```
cf = accumarray(getCellNoFaces(G), abs(faceFlux2cellFlux(G, state.flux)));
```

Shallow-Marine Reservoir

In the final example, we return to the SAIGUP model from Section 3.5.1. This model does not represent a real reservoir, but is one out of a large number of models built to be plausible realizations that contain the types of structural and stratigraphic features one could encounter in models of real clastic reservoirs. Continuing from Section 3.5.1, we simply assume that the grid and the petrophysical model has been loaded and processed.

The script `saigupWithWells.m` reports all details of the setup and also explains how to speed up the grid processing by using two C-accelerated routines for constructing a grid from ECLIPSE input and computing areas, centroids, normals, volumes, etc.

The permeability input is an anisotropic tensor with zero vertical permeability in a number of cells. As a result, some parts of the reservoir may be completely sealed off from the wells. This will cause problems for the time-of-flight solver, which requires that all cells in the model must be flooded after some finite time that can be arbitrarily large. To avoid this potential problem, we assign a small constant times the minimum positive vertical permeability to the cells that have zero cross-layer permeability.

```
is_pos = rock.perm(:, 3) > 0;
rock.perm(~is_pos, 3) = 1e-6*min(rock.perm(is_pos, 3));
```

Similar safeguards are implemented in most commercial simulators.

We recover fluid from the reservoir using six producers spread throughout the middle of the reservoir; each producer operates at a fixed bottom-hole pressure of 200 bar. Eight injectors located around the perimeter provide pressure support, each operating at a prescribed and fixed rate. The wells are described by a Peaceman model as in the previous example. To make the code as compact as possible, all wells are vertical with location specified in the logical ij subindex available in the corner-point format. The following code specifies the injectors:

```
nz = G.cartDims(3);
I = [ 3, 20,  3, 25,  3, 30,  5, 29];
J = [ 4,  3, 35, 35, 70, 70,113,113];
R = [ 1,  3,  3,  3,  2,  4,  2,  3]*500*meter^3/day;
W = [];
for i = 1 : numel(I),
   W = verticalWell(W, G, rock, I(i), J(i), 1:nz, 'Type', 'rate', ...
                   'Val', R(i), 'Radius', .1*meter, 'Comp_i', 1, ...
                   'name', ['I$_{', int2str(i), '}$']);
end
```

The producers are specified in the same way. Figure 5.11 shows the well positions and the pressure distribution. We see a clear pressure buildup along the east, south, and west rims of the model. Similarly, there is a pressure drawdown in the middle of the model around producers P2, P3, and P4. The total injection rate is set so that one pore volume will be injected in a little less than 40 years.

This is a single-phase simulation, but let us for a while think of our setup in terms of injection and production of different fluids (since the fluids have identical properties, we can think of a blue fluid being injected into a black fluid). In an ideal situation, one would wish that the blue fluid would sweep the whole reservoir before it breaks through to the production wells, as this would maximize the displacement of the black fluid. Even in the simple quarter five-spot examples in Section 5.4.1 (see Figure 5.3), we saw that this was not the case, and you cannot expect that this will happen here, either. The lower plot in Figure 5.11 shows all cells in which the total travel time (sum of forward and backward

5.4 Simulation Examples

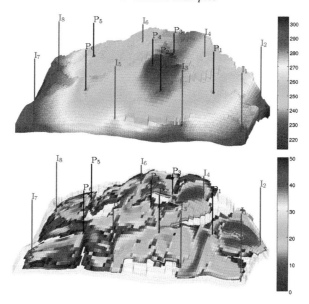

Figure 5.11 Incompressible, single-phase simulation of the SAIGUP model. The upper plot shows pressure distribution, and the lower plot shows cells with total travel time of less than 50 years.

time-of-flight) is less than 50 years. By looking at such a plot, you can get a quite a good idea of regions where there is very limited communication between the injectors and producers (i.e., areas without colors). If this was a multiphase flow problem, these areas would typically contain bypassed or unswept oil and be candidates for infill drilling or other mechanisms that would improve the volumetric sweep. We will come back to a more detailed discussion of flow patterns and volumetric connections in Section 13.5.2.

COMPUTER EXERCISES

5.4.10 Change the parameter 'Dir' from 'y' to 'z' in the box example and rerun the case. Can you explain why you get a different result?

5.4.11 Switch the injector in the box example to be controlled by a bottom-hole pressure of 200 bar. Where would you place the injector to maximize production rate if you can only perforate (complete) it in five cells?

5.4.12 Consider the SAIGUP model: can you improve the well placement and/or the distribution of fluid rates. Hint: is it possible to utilize time-of-flight information?

5.4.13 Use the function getSPE10setup to set up an incompressible, single-phase version of the full SPE 10 benchmark. Compute pressure, time-of-flight and tracer concentrations associated with each well. Hint: You may need to replace MATLAB's standard backslash-solver by a highly-efficient iterative solver like AGMG [238, 15] to get reasonable computational performance. Also, beware that you may run out of memory if your computer is not sufficiently powerful.

6
Consistent Discretizations on Polyhedral Grids

The two-point flux-approximation (TPFA) scheme is robust in the sense that it generally gives a linear system that has a solution regardless of the variations in **K** and the geometrical and topological complexity of the grid. The resulting solutions will also be monotone, but the scheme is only consistent for certain combinations of grids and permeability tensors **K**. This implies that a TPFA solution will not necessarily approach the true solution when we increase the grid resolution. It also means that the scheme may produce different solutions depending upon how the grid is oriented relative to the main flow directions; we will discuss this in more detail in Chapter 10. In this chapter, we first explain the lack of consistency for TPFA, before we introduce a few consistent schemes implemented in MRST. If you have only limited interest in discretizations, my advice is that you read Sections 6.1 and 6.3 before continuing to Chapter 7.

6.1 The TPFA Method Is Not Consistent

To provide some background for the following discussion, we start by giving a quick and informal recap of some important concepts from basic numerical analysis. When we approximate the Laplace operator $\mathcal{L} = \nabla \cdot \mathbf{K} \nabla$ by a numerical scheme, the approximation can be written as follows:

$$\mathcal{L} p = \mathcal{L}_h p + \mathcal{T}(h),$$

where h is the characteristic size of the cells in our grid, $\mathcal{L}_h$ is the approximation to the Laplace operator, and $\mathcal{T}$ is the *truncation error*, i.e., the difference between the true operator and our approximation. A numerical scheme is said to be *consistent* if the truncation error vanishes as h tends to zero, or in other words, if the difference between the true operator and our approximation diminishes as the grid is refined. A fundamental result in numerical analysis (Lax's equivalence theorem) states that *a consistent numerical method is convergent if and only if the method is stable*. By *convergent*, we mean that our approximations will approach the true solution as the size of the grid cells approaches zero. By *stable*, we mean that errors we make in approximating parameters determining the solution (for Poisson's equation: the source term q and the permeability **K**) do not give unbounded amplification of the error in the computed solution. Here, a word of caution is

6.1 The TPFA Method Is Not Consistent

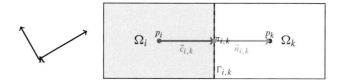

Figure 6.1 Two cells in a Cartesian grid and a full permeability tensor whose principal axes are not aligned with the coordinate system.

in order: *a scheme does not need to be consistent to be convergent*; showing consistency and stability is only a convenient approach to prove convergence.

Example 6.1.1 *To show that the TPFA scheme is generally not consistent, we start by a simple example. Recall from Section 4.4.1 that if p_i and p_k denote the average pressures in two neighboring cells Ω_i and Ω_k, then the flux across the interface Γ_{ik} between them is given as*

$$v_{ik} = T_{ik}(p_i - p_k), \tag{6.1}$$

where the transmissibility T_{ik} depends on the geometry of the two cells and the associated permeability tensors $\mathbf{K}_i$ and $\mathbf{K}_k$. To see that the method is not consistent, we assume that $\mathbf{K}$ is a homogeneous symmetric tensor

$$\mathbf{K} = \begin{bmatrix} \mathbf{K}_{xx} & \mathbf{K}_{xy} \\ \mathbf{K}_{xy} & \mathbf{K}_{yy} \end{bmatrix}.$$

Consider now the flux across an interface Γ_{ik} between two cells Ω_i and Ω_k in a 2D Cartesian grid, whose normal vector $\vec{n}_{i,k}$ points in the x-direction so that $\vec{n}_{i,k} = \vec{c}_{i,k} = (1,0)$; see Figure 6.1. If follows from Darcy's law that $\vec{v} \cdot \vec{n} = -(\mathbf{K}_{xx}\partial_x p + \mathbf{K}_{xy}\partial_y p)$, from which it is easy to see that the flux across the interface will have two components, one orthogonal and one transverse,

$$v_{ik} = -\int_{\Gamma_{ik}} \mathbf{K}\nabla p \cdot \vec{n}\, ds = -\int_{\Gamma_{ik}} \left(\mathbf{K}_{xx}\, \partial_x p + \mathbf{K}_{xy}\, \partial_y p\right) ds.$$

The two flux components correspond to the x-derivative and the y-derivative of the pressure, respectively. In the two-point approximation (6.1), the only points we use to approximate the flux are the pressures p_i and p_k. Because these two pressures are associated with the same y-value, their difference can only be used to estimate $\partial_x p$ and not $\partial_y p$. This means that the TPFA method cannot account for the transverse flux contribution $\mathbf{K}_{xy}\partial_y p$ and will hence generally not be consistent. The only exception is if $\mathbf{K}_{xy} \equiv 0$ in all cells in the grid, in which case there are no fluxes in the transverse direction.

To link the derivation in Example 6.1.1 to the geometry of the cells, we can use the one-sided definition of fluxes (4.51) of the two-point scheme to write

$$v_{i,k} \approx T_{i,k}(p_i - \pi_{i,k}) = \frac{A_{i,k}}{|\vec{c}_{i,k}|^2}\left(\mathbf{K}\vec{c}_{i,k}\right) \cdot \vec{n}_{i,k}(p_i - \pi_{i,k}).$$

More generally, we have that the TPFA scheme is only convergent for *K-orthogonal grids*. An example of a K-orthogonal grid is a grid in which all cells are parallelepipeds in 3D or parallelograms in 2D and satisfy the condition

$$\vec{n}_{i,j} \cdot \mathbf{K}\vec{n}_{i,k} = 0, \qquad \forall j \neq k, \tag{6.2}$$

where $\vec{n}_{i,j}$ and $\vec{n}_{i,k}$ denote normal vectors to faces of cell number i.

What about grids that are not parallelepipeds or parallelograms? For simplicity, we only consider the 2D case. Let $\vec{c} = (c_1, c_2)$ denote the vector from the center of cell Ω_i to the centroid of face Γ_{ik}; see Figure 4.10. If we let $\vec{c}_\perp = (c_2, -c_1)$ denote a vector that is orthogonal to $\vec{c}$, then any constant pressure gradient inside Ω_i can be written as $\nabla p = p_1 \vec{c} + p_2 \vec{c}_\perp$, where p_1 can be determined from p_i and $\pi_{i,k}$, and p_2 is some unknown constant. Inserted into the definition of the face flux, this gives

$$v_{i,k} = -\int_{\Gamma_{ik}} \left[p_1 (\mathbf{K}\vec{c}) \cdot \vec{n} + p_2 (\mathbf{K}\vec{c}_\perp) \cdot \vec{n} \right] ds.$$

For the two-point scheme to be correct, the second term in the integrand must be zero. Setting $\vec{n} = (n_1, n_2)$, we find that

$$0 = (\mathbf{K}\vec{c}_\perp) \cdot \vec{n} = (K_1 c_2, -K_2 c_1) \cdot (n_1, n_2) = (K_1 n_1 c_2 - K_2 n_2 c_1) = (\mathbf{K}\vec{n}) \times \vec{c}.$$

In other words, a sufficient condition for a polygonal grid to be K-orthogonal is that $(\mathbf{K}\vec{n}_{i,k}) \parallel \vec{c}_{i,k}$ for all cells in the grid, in which case the transverse flux contributions are all zero and hence need not be estimated. Consequently, the TPFA method will be consistent.

The lack of consistency in the TPFA scheme for grids that are not K-orthogonal may lead to significant *grid orientation effects* [8, 319], i.e., artifacts or errors in the solution that will appear in varying degree depending upon the angles the cell faces make with the principal directions of the permeability tensor. We have already seen an example of grid effects in Figure 5.7 in Section 5.4.3, where an irregular triangular grid caused large deviations from what should have been an almost symmetric pressure drawdown. To demonstrate grid orientation effects and lack of convergence more clearly, we look at a simple example that has been widely used to teach reservoir engineers the importance of aligning the grid with the principal permeability directions.

Example 6.1.2 *Consider a homogeneous reservoir in the form of a 2D rectangle* $[0,4] \times [0,1]$. *Flow is driven from the north to the south side of the reservoir by a fluid source with unit rate located at* $(2, 0.975)$ *along the north side, and two fluid sinks located at* $(0.5, 0.025)$ *and* $(3.5, 0.025)$ *along the south side. The two source terms each have rate equal one half and are symmetric to the line* $x = 2$ *and will therefore result in a symmetric pressure distribution.*

To challenge TPFA, we discretize the domain by a skewed grid that is uniform along the north side and graded towards east along the south side:

```
G = cartGrid([41,20],[2,1]);
makeSkew = @(c) c(:,1) + .4*(1-(c(:,1)-1).^2).*(1-c(:,2));
G.nodes.coords(:,1) = 2*makeSkew(G.nodes.coords);
```

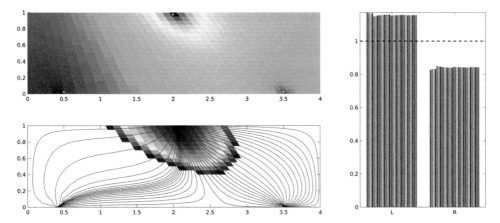

Figure 6.2 Solution of a symmetric flow problem in a homogeneous domain using the TPFA method on a skew grid that is not K-orthogonal. The upper-left plot shows the pressure distribution, whereas the lower-left plot shows streamlines and values of time-of-flight less than 0.25 PVI. The plot to the right shows time-of-flight in the left (L) and right (R) sink for a series of refined grids of dimension $(20n + 1) \times 10n$ for $n = 1, \ldots, 30$. The solutions do not converge toward the analytical solution shown as the dashed red line.

If we assume an isotropic and homogeneous permeability field, the grid is not K-orthogonal, and we therefore cannot expect that the TPFA method will converge to the correct physical solution. You can find a complete setup of the problem in the script `1phase/showInconsistentTPFA.m`.

Figure 6.2 reports the result of a convergence study. All the computed solutions exhibit the same lack of symmetry you see in the solution on the 41×20 grid. Moreover, the large difference in travel times from the injector to the two producers does not decay with increasing grid resolution, which confirms the expected lack of convergence for an inconsistent scheme.

Grid orientation errors are discussed in more detail in Section 10.4.2. It is a good advice to try to make grids so that cell faces align with the principle directions of the permeability tensor. A main motivation for using PEBI grids is to maintain flexibility in areal representation while keeping grid orientation errors at a minimum; see [128, 121, 210].

6.2 The Mixed Finite-Element Method

There are several ways to formulate consistent discretization methods. The mixed finite-element method [53] is based on a quite different formulation than the two-point method. The first new idea is that instead of discretizing the second-order Poisson equation, we form a system that consists of two discrete equations representing mass conservation and Darcy's law. Hence, instead of solving for pressure and computing discrete fluxes by postprocessing the pressure solution, we solve for pressure and fluxes simultaneously. The second new idea is to look for solutions that satisfy the flow equations in a weak sense; that is, look for solutions that fulfill the equation when it is multiplied with a suitable *test function* and integrated in space. The third new idea is to express the unknown solution as a linear combination of a set of *basis functions*, i.e., piecewise polynomials with localized support.

MRST does not implement the mixed finite-element method directly in the classical sense, but the key ideas of this method are instrumental when we develop a general class of finite-volume discretizations in Section 6.3 and discussed in more details in the remaining parts of the chapter. As a precursor to this discussion, we review the mixed finite-element method in some detail.

6.2.1 Continuous Formulation

We start by restating the continuous flow equation written as a first-order system:

$$\nabla \cdot \vec{v} = q, \qquad \vec{v} = -\mathbf{K}\nabla p, \qquad \vec{x} \in \Omega \subset \mathbb{R}^d \tag{6.3}$$

with boundary conditions $\vec{v} \cdot \vec{n} =$ for $\vec{x} \in \partial\Omega$. Gravity has been omitted for brevity, but can easily be included. For compatibility, we require that $\int_\Omega q\, d\vec{x} = 0$. Because this is a pure Neumann boundary-value problem, pressure p is only defined up to an arbitrary constant, and as an extra constraint, we require that $\int_\Omega p\, d\vec{x} = 0$.

In the mixed method, we look for solutions in an abstract function space. To this end, we need two spaces: $L^2(\Omega)$ is the space of square integrable functions, and H_0^{div} is a so-called Sobolev space defined as

$$H_0^{\text{div}}(\Omega) = \{\vec{v} \in L^2(\Omega)^d : \nabla \cdot \vec{v} \in L^2(\Omega) \text{ and } \vec{v} \cdot \vec{n} \text{ on } \partial\Omega\}, \tag{6.4}$$

i.e., as the set of square-integrable, vector-valued functions with compact support in Ω, whose divergence is also square integrable. The *mixed formulation* of (6.3) now reads: find a pair $(p, \vec{v})$ that lies in $L^2(\Omega) \times H_0^{\text{div}}(\Omega)$ and satisfies

$$\int_\Omega \vec{u} \cdot \mathbf{K}^{-1}\vec{v}\, d\vec{x} - \int_\Omega p\nabla \cdot \vec{u}\, d\vec{x} = 0, \qquad \forall \vec{u} \in H_0^{\text{div}}(\Omega),$$
$$\int_\Omega w\nabla \cdot \vec{v}\, d\vec{x} = \int_\Omega qw\, d\vec{x}, \qquad \forall w \in L^2(\Omega). \tag{6.5}$$

The first equation follows by multiplying Darcy's law by $\mathbf{K}^{-1}$ and applying the divergence theorem to the term involving the pressure gradient. (Recall that $\vec{u}$ is zero on $\partial\Omega$ by definition so that the term $\int_{\partial\Omega} p\vec{u} \cdot \vec{n}\, ds = 0$.)

We can write equation (6.5) on a more compact form if we introduce the following three inner products

$$b(\cdot, \cdot) : H_0^{\text{div}} \times H_0^{\text{div}} \to \mathbb{R}, \qquad b(\vec{u}, \vec{v}) = \int_\Omega \vec{u} \cdot \mathbf{K}^{-1}\vec{v}\, d\vec{x}$$

$$c(\cdot, \cdot) : H_0^{\text{div}} \times L^2 \to \mathbb{R}, \qquad c(\vec{u}, p) = \int_\Omega p\nabla \cdot \vec{u}\, d\vec{x}$$

6.2 The Mixed Finite-Element Method

$$(\cdot,\cdot) : L^2 \times L^2 \to \mathbb{R}, \qquad (q,w) = \int_\Omega qw\, d\vec{x}, \qquad (6.6)$$

and write

$$b(\vec{u},\vec{v}) - c(\vec{u},p) = 0, \qquad c(\vec{v},w) = (q,w). \qquad (6.7)$$

Alternatively, we can derive (6.7) by minimizing the energy functional

$$I(\vec{v}) = \frac{1}{2} \int_\Omega \vec{v} \cdot \mathbf{K}^{-1} \vec{v}\, d\vec{x},$$

over all mass-conservative functions $\vec{v} \in H_0^{\mathrm{div}}(\Omega)$, i.e., subject to the constraint

$$\nabla \cdot \vec{v} = q.$$

The common strategy for solving such minimization problems is to introduce a Lagrangian functional, which in our case reads

$$L(\vec{v},p) = \int_\Omega \left(\tfrac{1}{2} \vec{v} \cdot \mathbf{K}^{-1} \vec{v} - p(\nabla \cdot \vec{v} - q) \right) d\vec{x} = \frac{1}{2} b(\vec{v},\vec{v}) - c(\vec{v},p) + (p,q),$$

where p is a so-called Lagrangian multiplier. At a minimum of L, we must have $\partial L/\partial \vec{v} = 0$. Looking at an increment $\vec{u}$ of $\vec{v}$, we have the requirement that

$$0 = \frac{\partial L}{\partial \vec{v}} \vec{u} = \lim_{\varepsilon \to 0} \frac{1}{\varepsilon} \left[L(\vec{v} + \varepsilon \vec{u}, p) - L(\vec{v}, p) \right] = b(\vec{u}, \vec{v}) - c(\vec{u}, p)$$

for all $u \in H_0^{\mathrm{div}}(\Omega)$. Similarly, we can show that

$$0 = \frac{\partial L}{\partial p} w = \lim_{\varepsilon \to 0} \frac{1}{\varepsilon} \left[L(\vec{v}, p + \varepsilon w) - L(\vec{v}, p) \right] = -c(\vec{v}, w) + (q, w)$$

for all $w \in L^2(\Omega)$. To show that the solution $\vec{v}$ is a minimal point, we consider a perturbation $\vec{v} + \vec{u}$ that satisfies the constraints so that $\vec{u} \in H_0^{\mathrm{div}}$ and $\nabla \cdot \vec{u} = 0$. For the energy functional we have that

$$I(\vec{v} + \vec{u}) = \frac{1}{2} \int_\Omega (\vec{v} + \vec{u}) \cdot \mathbf{K}^{-1} (\vec{v} + \vec{u})\, d\vec{x} = I(\vec{v}) + I(\vec{u}) + b(\vec{u}, \vec{v})$$
$$= I(\vec{v}) + I(\vec{u}) + c(\vec{u}, p) = I(\vec{v}) + I(\vec{u}) > I(\vec{v}),$$

which proves that $\vec{v}$ is indeed a minimum of I. We can also prove that the solution $(p, \vec{v})$ is a saddle-point of the Lagrange functional, i.e., that $L(\vec{v}, w) \leq L(\vec{v}, p) \leq L(\vec{u}, p)$ for all $\vec{u} \neq \vec{v}$ and $w \neq p$. The right inequality can be shown as follows

$$L(\vec{v} + \vec{u}, p) = \tfrac{1}{2} b(\vec{v} + \vec{u}, \vec{v} + \vec{u}) - c(\vec{v} + \vec{u}, p) + (q, p)$$
$$= L(\vec{v}, p) + I(\vec{u}) + b(\vec{u}, \vec{v}) - c(\vec{u}, p) = L(\vec{v}, p) + I(\vec{u}) > L(\vec{v}, p).$$

The left inequality follows in a similar manner.

6.2.2 Discrete Formulation

To discretize the mixed formulation, (6.5), we introduce a grid $\Omega_h = \cup \Omega_i$, replace $L^2(\Omega)$ and $H_0^{\text{div}}(\Omega)$ by finite-dimensional subspaces U and V defined over Ω_h, and rewrite the inner products (6.6) as sums of integrals localized to cells

$$b(\cdot,\cdot)_h : V \times V \to \mathbb{R}, \qquad b(\vec{u},\vec{v})_h = \sum_i \int_{\Omega_i} \vec{u} \cdot \mathbf{K}^{-1} \vec{v} \, d\vec{x},$$

$$c(\cdot,\cdot)_h : V \times U \to \mathbb{R}, \qquad c(\vec{u},p)_h = \sum_i \int_{\Omega_i} p \nabla \cdot \vec{u} \, d\vec{x}, \qquad (6.8)$$

$$(\cdot,\cdot)_h : U \times U \to \mathbb{R}, \qquad (q,w)_h = \sum_i \int_{\Omega_i} qw \, d\vec{x}.$$

To obtain a practical numerical method, the spaces U and V are typically defined as piecewise polynomial functions that are nonzero on a small collection of grid cells. For instance, in the Raviart–Thomas method [262, 53] of lowest order for triangular, tetrahedral, or regular parallelepiped grids, $L^2(\Omega)$ is replaced by

$$U = \{p \in L^2(\cup \Omega_i) : p|_{\Omega_i} \text{ is constant } \forall \Omega_i \subset \Omega\} = \text{span}\{\chi_i\},$$

where χ_i is the characteristic function of grid cell Ω_i. Likewise, $H_0^{\text{div}}(\Omega)$ is replaced by a space V consisting of functions $\vec{v} \in H_0^{\text{div}}(\cup \Omega_i)$ that have linear components on each grid cell $\Omega_i \in \Omega$, have normal components $\vec{v} \cdot \vec{n}_{ik}$ that are constant on each cell interface Γ_{ik}, and are continuous across these interfaces. These requirements are satisfied by functions $\vec{v} \in \mathcal{P}_1(\Omega_i)^d$, i.e., first-order polynomials that on each grid cell take the form $\vec{v}(\vec{x}) = \vec{a} + B\vec{x}$, where $\vec{a}$ is a constant vector and B is a matrix. These functions can be parametrized in terms of the faces between two neighboring cells. This means that we can write $V = \text{span}\{\vec{\psi}_{ik}\}$, where each function $\vec{\psi}_{ik}$ is defined as

$$\vec{\psi}_{ik} \in \mathcal{P}_1(\Omega_i)^d \cup \mathcal{P}_1(\Omega_k)^d \quad \text{and} \quad (\vec{\psi}_{ik} \cdot \vec{n}_{jl})|_{\Gamma_{jl}} = \begin{cases} 1, & \text{if } \Gamma_{jl} = \Gamma_{ik}, \\ 0, & \text{otherwise.} \end{cases}$$

Figure 6.3 illustrates the four nonzero basis functions for the special case of a Cartesian grid in 2D.

To derive a fully discrete method, we use χ_i and $\vec{\psi}_{ik}$ as our *trial functions* and express the unknown $p(\vec{x})$ and $\vec{v}(\vec{x})$ as sums over these trial functions, $p = \sum_i p_i \chi_i$ and $\vec{v} = \sum_{ik} v_{ik} \vec{\psi}_{ik}$. We now have two different degrees of freedom, p_i associated with each cell and v_{ik} associated with each interface Γ_{ik}, which we collect in two vectors $\boldsymbol{p} = \{p_i\}$ and $\boldsymbol{v} = \{v_{ik}\}$. Using the same functions as test functions, we derive a linear system of the form

$$\begin{bmatrix} B & -C^T \\ C & 0 \end{bmatrix} \begin{bmatrix} v \\ p \end{bmatrix} = \begin{bmatrix} 0 \\ q \end{bmatrix}. \qquad (6.9)$$

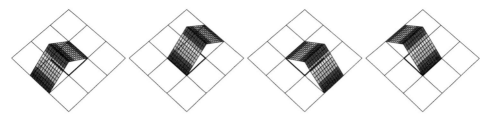

Figure 6.3 Illustration of the velocity basis for the lowest-order Raviart–Thomas method for a 2D Cartesian grid. There are four basis functions that have nonzero support in the interior of the center cell. The basis functions correspond to degrees of freedom associated with the west, east, south, and north cell faces.

Here, $\mathbf{B} = [b_{ik,jl}]$, $\mathbf{C} = [c_{i,kl}]$, and $\mathbf{q} = [q_i]$, where:

$$
\begin{aligned}
b_{ik,jl} &= b(\vec{\psi}_{ik}, \vec{\psi}_{jl})_h = \sum_{\ell} \int_{\Omega_\ell} \vec{\psi}_{ik} \cdot \mathbf{K}^{-1} \vec{\psi}_{jl} \, d\vec{x}, \\
c_{i,kl} &= c(\vec{\psi}_{kl}, \chi_i)_h = \int_{\Omega_i} \nabla \cdot \vec{\psi}_{kl} \, d\vec{x}, \\
q_i &= (q, \chi_i)_h = \int_{\Omega_i} q \, d\vec{x}.
\end{aligned}
\qquad (6.10)
$$

Note that for the first-order Raviart–Thomas finite elements, we have

$$
c_{i,kl} = \begin{cases} 1, & \text{if } i = k, \\ -1, & \text{if } i = l, \\ 0, & \text{otherwise.} \end{cases}
$$

The matrix entries $b_{ik,jl}$ depend on the geometry of the grid cells and whether $\mathbf{K}$ is isotropic or anisotropic. Unlike the two-point method, mixed methods are consistent and will therefore be convergent also on grids that are not K-orthogonal.

In [2], we presented a simple MATLAB code that in approximately seventy-five code lines implements the lowest-order Raviart–Thomas mixed finite-element method on regular hexahedral grids for flow problems with diagonal tensor permeability in two or three spatial dimensions. The code is divided into three parts: assembly of the $\mathbf{B}$ block, assembly of the $\mathbf{C}$ block, and a main routine that loads data (permeability and grid), assembles the whole matrix, and solves the system. Figure 6.4 illustrates the sparsity pattern of the mixed system for a $n_x \times n_y \times n_z$ Cartesian grid. The system matrix clearly has the block structure given in (6.9). The matrix blocks $\mathbf{B}$ and $\mathbf{C}$ each have three nonzero blocks that correspond to velocity basis functions oriented along the three axial directions. That is, degrees of freedom at interfaces (fluxes) have been numbered in the same way as the grid cells; i.e., first faces orthogonal to the x-direction, then faces orthogonal to the y-direction, and finally faces orthogonal to the z-direction, resulting in a heptadiagonal structure.

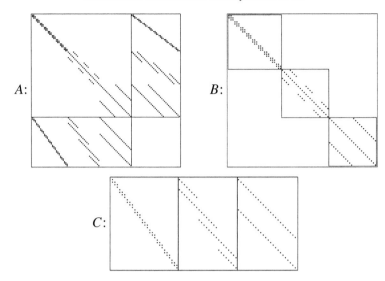

Figure 6.4 Sparsity patterns for the full mixed finite-element matrix A, and the matrix blocks B and C for a $4 \times 3 \times 3$ Cartesian grid.

The code presented in [2] exploits the simple geometry of the Cartesian grid to integrate the Raviart–Thomas basis functions exactly and perform a direct assembly of the matrix blocks, giving a MATLAB code that is both compact and efficient. For more general grids, you would typically perform an element-wise assembly by looping over all cells in the grid, map each cell them back to a reference element, and then integrate the individual inner production by use of a suitable numerical quadrature rule. In our experience, this procedure is cumbersome to implement for general stratigraphic and polyhedral grids and would typically require the use of many different reference elements or subdivision of cells.

MRST is designed to work on general polyhedral grids in 3D and does not supply any direct implementation of mixed finite-element methods. Instead, the closest we get to a mixed method is a finite-volume method in which the half-transmissibilities are equivalent to the discrete inner products for the lowest-order Raviart–Thomas (RT0) method on rectangular cuboids; Section 6.4 on page 188 explains more details. In the rest of the current section, we introduce an alternative formulation of the mixed method that gives a smaller linear system that is better conditioned.

6.2.3 Hybrid Formulation

In the same way as we proved that the solution of the mixed problem is a saddle point, we can show that the linear system (6.9) is indefinite. Indefinite systems are harder to solve and generally require special linear solvers. We therefore introduce an alternative formulation that, when discretized, gives a positive-definite discrete system and thereby simplifies the computation of a discrete solution. Later in the chapter, this so-called *hybrid formulation*

6.2 The Mixed Finite-Element Method

will form the basis for a general family of finite-volume discretizations on polygonal and polyhedral grids.

A hybrid formulation avoids solving a saddle-point problem by lifting the constraint that the normal velocity must be continuous across cell faces and instead integrate (6.3) to get a weak form that contains jump terms at the cell interfaces. Continuity of the normal velocity component is then reintroduced by adding an extra set of equations, in which the pressure π at the cell interfaces plays the role of Lagrange multipliers. (Recall how Lagrange multipliers were used to impose mass conservation as a constraint in the minimization procedure used to derive the mixed formulation in Section 6.2.1.) Introducing Lagrange multipliers does not change $\vec{v}$ or p, but enables the recovery of pressure values at cell faces, in addition to introducing a change in the structure of the weak equations. Mathematically, the mixed hybrid formulation reads: find $(\vec{v}, p, \pi) \in H_0^{\mathrm{div}}(\Omega_h) \times L^2(\Omega_h) \times H^{\frac{1}{2}}(\Gamma_h)$ such that

$$\sum_i \int_{\Omega_i} \left(\vec{u} \cdot \mathbf{K}^{-1} \vec{v} - p \nabla \cdot \vec{u} \right) d\vec{x} + \sum_{ik} \int_{\Gamma_{ik}} \pi \vec{u} \cdot \vec{n} \, ds = 0,$$

$$\sum_i \int_{\Omega_i} w \nabla \cdot \vec{v} \, d\vec{x} = \int_{\Omega_h} q w \, d\vec{x}, \qquad (6.11)$$

$$\sum_{ik} \int_{\Gamma_{ik}} \mu \vec{v} \cdot \vec{n} \, ds = 0$$

for all test functions $\vec{u} \in H_0^{\mathrm{div}}(\Omega_h)$, $w \in L^2(\Omega_h)$, and $\mu \in H^{\frac{1}{2}}(\Gamma_h)$. Here, Γ_h denotes all the interior faces of the grid and $H^{\frac{1}{2}}(\Gamma_h)$ is the space spanned by the traces[1] of functions in $H^1(\Omega_h)$, i.e., the space of square integrable functions whose derivatives are also square integrable. As for the mixed formulation, we can introduce inner products to write the weak equations (6.11) in a more compact form,

$$b(\vec{u}, \vec{v}) - c(\vec{u}, p) + d(\vec{u}, \pi) = 0$$
$$c(\vec{v}, w) = (q, w) \qquad (6.12)$$
$$d(\vec{v}, \mu) = 0,$$

where the inner products $b(\cdot, \cdot)$, $c(\cdot, \cdot)$, and $(\cdot, \cdot)$ are defined as in (6.6), and $d(\cdot, \cdot)$ is a new inner product defined over the interior faces,

$$d(\cdot, \cdot)_h : H_0^{\mathrm{div}}(\Omega_h) \times H^{\frac{1}{2}}(\Gamma_h) \to \mathbb{R}, \qquad d(\vec{v}, \pi) = \sum_{ik} \int_{\Gamma_{ik}} \pi \vec{v} \cdot \vec{n} \, ds. \qquad (6.13)$$

To derive a fully discrete problem, we proceed in the exact same way as for the mixed problem by first replacing the function spaces L^2, H_0^{div}, and $H^{\frac{1}{2}}$ by finite-dimensional subspaces V and U that are spanned by piecewise polynomial basis functions with local

---

[1] If you are not familiar with the notion of a trace operator, think of it as the values of a function along the boundary of the domain this function is defined on.

support, as discussed earlier. In the lowest-order approximation, the finite-dimensional space Π consists of functions that are constant on each face,

$$\Pi = \mathrm{span}\{\mu_{ik}\}, \qquad \mu_{ik}(\vec{x}) = \begin{cases} 1, & \text{if } \vec{x} \in \Gamma_{ik}, \\ 0, & \text{otherwise}. \end{cases} \qquad (6.14)$$

Using these basis functions as test and trial functions, one can derive a discrete linear system of the form,

$$\begin{bmatrix} B & C & D \\ C^\mathsf{T} & 0 & 0 \\ D^\mathsf{T} & 0 & 0 \end{bmatrix} \begin{bmatrix} v \\ -p \\ \pi \end{bmatrix} = \begin{bmatrix} 0 \\ q \\ 0 \end{bmatrix}, \qquad (6.15)$$

where the vectors v, p, and π collect the degrees of freedom associated with fluxes across the cell interfaces, face pressures, and cell pressures, respectively, the matrix blocks B and C are defined as in (6.10), and D has two nonzero entries per column (one entry for each side of the cell interfaces).

The linear system (6.15) is an example of a sparse, symmetric, indefinite system, i.e., a system A whose quadratic form $x^T A x$ takes both positive and negative values. Several methods for solving such systems can be found in the literature, but these are generally not as efficient as solving a symmetric, positive-definite system. We will therefore use a so-called Schur-complement method to reduce the mixed hybrid system to a positive-definite system. The Schur-complement method basically consists of using a block-wise Gaussian elimination of (6.15) to form a positive-definite system (the Schur complement) for the face pressures,

$$\left(D^\mathsf{T} B^{-1} D - F^\mathsf{T} L^{-1} F\right)\pi = F^\mathsf{T} L^{-1} q. \qquad (6.16)$$

Here, $F = C^\mathsf{T} B^{-1} D$ and $L = C^\mathsf{T} B^{-1} C$. Given the face pressures, the cell pressures and fluxes can be reconstructed by back-substitution, i.e., by solving

$$L p = q + F\pi, \qquad B u = C p - D\pi. \qquad (6.17)$$

Unlike the mixed method, the hybrid method has a so-called explicit flux representation, which means that the intercell fluxes can be expressed as a linear combination of neighboring values for the pressure. This property is particularly useful in fully implicit discretizations of time-dependent problems, as discussed in Chapter 7.

This is all we will discuss about mixed and mixed-hybrid methods herein. If you want to learn more about these methods, and how to solve the corresponding linear systems, you should consult some of the excellent books on the subject, for instance [50, 45, 53]. In the rest of the chapter, we discuss how ideas from mixed-hybrid methods can be used to formulate finite-volume methods that are consistent and hence convergent on non-K-orthogonal grids.

6.3 Finite-Volume Methods on Mixed Hybrid Form

We will now present a large class of consistent, finite-volume methods. Our formulation borrows ideas from the mixed methods, but is much simpler to formulate and implement for general polygonal and polyhedral grids. For simplicity, we assume that all methods can be written on the following local form

$$v_i = T_i(e_i p_i - \pi_i). \tag{6.18}$$

This is essentially the same formulation we used for the one-sided fluxes to develop the TPFA method in (4.51), except that now v_i is a vector of all one-sided fluxes associated with cell Ω_i, π_i is a vector of face pressures, T_i is a matrix of one-sided transmissibilities, and $e_i = (1, \ldots, 1)^\mathsf{T}$ has one unit value per face of the cell. If we define $M_i = T_i^{-1}$, the local discretization (6.18) can alternatively be written as

$$M_i v_i = e_i p_i - \pi_i. \tag{6.19}$$

Consistent with the discussion in Section 6.2, we refer to the matrix M_i as the *local inner product*. From the local discretizations (6.18) or (6.18) on each cell, we derive a linear system of discrete global equations written on mixed or hybridized mixed form. Both forms are supported in MRST, but herein we only discuss the hybrid form for brevity. In the two-point method, we assembled the linear system by combining mass conservation and Darcy's law into one second-order discrete equation for the pressure. In the mixed formulation, mass conservation and Darcy's law were kept as separate first-order equations that together formed a coupled system for pressure and face fluxes. In the hybrid formulation, the continuity of pressures across cell faces was introduced as a third equation that together with mass conservation and Darcy's law constitute a coupled system for pressure, face pressure, and fluxes. Not all consistent methods need to, or can be, formulated in this form. Nevertheless, using the mixed hybrid formulation will enable us to give a uniform presentation of a large class of schemes.

Going back to the TPFA method formulated in Section 4.4.1, it follows immediately from (4.52) that we can write the method as in (6.18) and that the resulting matrix T_i is diagonal with entries

$$(T_i)_{kk} = \vec{n}_{ik} \cdot \mathbf{K} \vec{c}_{ik} / |\vec{c}_{ik}|^2, \tag{6.20}$$

where the length of the normal vector $\vec{n}_{ik}$ is assumed to be equal the area of the corresponding face. The equivalent form (6.19) follows trivially by setting $M_i = T_i^{-1}$. Examples of consistent methods that can be written in the form (6.18) or (6.19) include the lowest-order RT0 method seen in the previous section, multipoint flux approximation (MPFA) schemes [9, 94, 7], and recently developed mimetic finite-difference methods [52], as well as the coarse-scale formulation of several multiscale methods [95]. For all these methods, the corresponding T_i and M_i will be full matrices. We will come back to more details about specific schemes later in the chapter.

To derive the global linear system on mixed hybrid form, we augment (6.19) with flux and pressure continuity across cell faces. Assembling contributions from all cells in the grid, gives the following linear system

$$\begin{bmatrix} B & C & D \\ C^\mathsf{T} & 0 & 0 \\ D^\mathsf{T} & 0 & 0 \end{bmatrix} \begin{bmatrix} v \\ -p \\ \pi \end{bmatrix} = \begin{bmatrix} 0 \\ q \\ 0 \end{bmatrix}, \qquad (6.21)$$

where the first row in the block-matrix equation corresponds to (6.19) for all grid cells. Here, vector v contains the outward fluxes associated with half-faces ordered cell-wise so that fluxes over interior interfaces in the grid appear twice, once for each half-face, with opposite signs. Likewise, the vector p contains the cell pressures, and π the face pressures. The matrices B and C are block diagonal with each block corresponding to a cell,

$$B = \begin{bmatrix} M_1 & 0 & \dots & 0 \\ 0 & M_2 & \dots & 0 \\ \vdots & \vdots & \dots & \vdots \\ 0 & 0 & \dots & M_n \end{bmatrix}, \qquad C = \begin{bmatrix} e_1 & 0 & \dots & 0 \\ 0 & e_2 & \dots & 0 \\ \vdots & \vdots & \dots & \vdots \\ 0 & 0 & \dots & e_n \end{bmatrix}, \qquad (6.22)$$

where $M_i = T_i^{-1}$. Similarly, each column of D corresponds to a unique interface in the grid and has two unit entries for interfaces between cells in the interior of the grid. The two nonzero entries appear at the indexes of the corresponding internal half-faces in the cell-wise ordering. Similarly, D has a single nonzero entry in each column that corresponds to an interface between a cell and the exterior.

The hybrid system (6.21) is obviously much larger than the linear system for the standard TPFA method, but can be reduced to a positive-definite system for the face pressures, as discussed in Section 6.2.3, and then solved using either MATLAB's standard linear solvers or a highly efficient, third-party solver like AGMG [238, 15]. From (6.16) and (6.17), we see that to compute the Schur complement to form the reduced linear system for face pressures, and to reconstruct cell pressures and face fluxes, we only need B^{-1}. Moreover, the matrix L is by construction diagonal, and computing fluxes is therefore an inexpensive operation. Many schemes – including the mimetic method, the MPFA-O method, and the standard TPFA scheme – yield algebraic approximations for the B^{-1} matrix. Thus, (6.21) encompasses a family of discretization schemes whose properties are determined by the choice of B, which we will discuss in more detail in Section 6.4.

Before digging into details about specific, consistent methods, we revisit the example discussed on page 177 to demonstrate how use of a consistent method significantly reduces the grid orientation effects seen in Figure 6.2. In MRST, the mimetic methods are implemented in a separate module. Replacing TPFA by a mimetic solver is easy; we first load the `mimetic` module by calling

```
mrstModule add mimetic
```

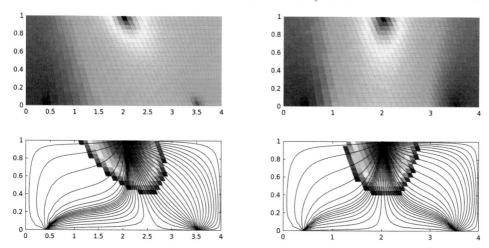

Figure 6.5 Solution of a symmetric flow problem in a homogeneous domain using the TPFA method (left) and the mimetic method (right) on a skew grid that is not K-orthogonal.

Then, we can replace computation of half-face transmissibilities by a call that constructs the local mimetic half-transmissibility T_i or its equivalent inner product $M_i = T_i^{-1}$ for each cell

```
% hT = computeTrans(G, rock);
  S  = computeMimeticIP(G, rock);
```

Likewise, we replace each call to the two-point solver by a call to a function that assembles and solves the global mixed hybrid system

```
% state = incompTPFA(state, G, hT, fluid);
  state = incompMimetic(state, G, S, fluid);
```

We can also use the same routine to assemble a mixed system, and in certain cases also a TPFA-type system. In Figure 6.5 we have applied the mimetic solver to the exact same setup as in Figure 6.2. The approximate solution computed by the mimetic method is almost symmetric and represents a significant improvement compared with the TPFA method. In particular, the difference in travel times between the injector and each producer is reduced from 17% for TPFA to less than 2% for the mimetic method. Moreover, repeating a similar grid refinement study as reported in Figure 6.2 verifies that the consistent mimetic method converges toward the correct solution.

The two calls just outlined constitute all you need to obtain a consistent discretization on general polygonal and polyhedral grids, and if you are not interested in getting to know the inner details of various consistent methods, you can safely jump to the next chapter.

COMPUTER EXERCISES

6.3.1 Run `mimeticExample1` to familiarize yourself with the mimetic solver, the mixed-hybrid formulation, the Schur-component reduction, etc.

6.3.2 Go back to Exercise 5.4.1a and verify that using the mimetic method cures the grid orientation effects you can observe on the non-K-orthogonal grid.

6.3.3 Can you cure the grid orientation problems observed in Figure 5.7 by using a consistent discretization?

6.3.4 Compare the solutions computed by TPFA and the mimetic method for the SAIGUP model as set up in Section 5.4.4. In particular, you should compare discrepancies in pressure and time-of-flight values at the well perforations. Which solution do you trust most?

Hints: (i) Remember to use the correct inner product when setting up well models. (ii) The mimetic linear system is much larger than the TPFA system and you may have to use an iterative solver like AGMG to reduce runtime and memory consumption.

6.4 The Mimetic Method

Mimetic finite-difference methods are examples of so-called *compatible spatial discretizations* that are constructed so that they not only provide accurate approximation of the mathematical models but also inherit or mimic fundamental properties of the differential operators and mathematical solutions they approximate. Examples of properties include conservation, symmetries, vector calculus identities, etc. Such methods have become very popular in recent years and are currently used in wide range of applications [40].

Mimetic methods can be seen as a finite-volume generalization of finite-differences or (low-order) mixed-finite element methods to general polyhedral grids. The methods are defined in a way that introduces a certain freedom of construction that naturally leads to a family of methods. By carefully picking the parameters needed to fully specify a method, one can construct mimetic methods that coincide with other known methods, or reduce to these methods (e.g., the two-point method, the RT0 mixed finite-element method, or the MPFA-O multipoint method) on certain types of grids.

The mimetic methods discussed herein can all be written on the equivalent forms (6.18) or (6.19) and are constructed so that they are exact for linear pressure fields and give a symmetric positive-definite matrix M. In addition, the methods use discrete pressures and fluxes associated with cell and face centroids, respectively, and consequently resemble finite-difference methods.

If we write a linear pressure field in the form $p = \vec{x} \cdot \vec{a} + b$ for a constant vector $\vec{a}$ and scalar b, the corresponding Darcy velocity is $\vec{v} = -\mathbf{K}\vec{a}$. Let $\vec{n}_{i,k}$ denote the area-weighted normal vector to Γ_{ik} and $\vec{c}_{i,k}$ be the vector pointing from the centroid of cell Ω_i to the centroid of face Γ_{ik}, as seen in Figure 6.6. Using Darcy's law and the notion of a one-sided transmissibility $T_{i,k}$ (which will generally not be the same as the two-point transmissibility), the flux and pressure drop can be related as follows,

$$v_{i,k} = -\vec{n}_{i,k}\mathbf{K}\vec{a} = T_{i,k}(p_i - \pi_{i,k}) = -T_{i,k}\vec{c}_{i,k} \cdot \vec{a}. \qquad (6.23)$$

To get these relations in one of the local forms (6.18) and (6.19), we collect all vectors $\vec{c}_{i,k}$ and $\vec{n}_{i,k}$ defined for cell Ω_i as rows in two matrices C_i and N_i. Because relation (6.23) is

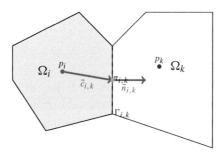

Figure 6.6 Two neighboring cells with quantities used to define mimetic methods.

required to hold for all linear pressure drops, i.e., for an arbitrary vector $\vec{a}$, we see that the matrices M_i and T_i must satisfy the following consistency conditions

$$MNK = C, \qquad NK = TC, \tag{6.24}$$

where we have dropped the subscript i that identifies cell number. These general equations give us (quite) large freedom in how to specify a specific discretization method: any method in which the local inner product M_i, or equivalently the local transmissibility matrix T_i, is positive definite and satisfies (6.24) will give a consistent, first-order accurate discretization. In the rest of the section, we discuss various choices of valid inner products M_i or reverse inner products T_i.

General Family of Inner Products

The original article [52] that first introduced the mimetic methods considered herein, discussed inner products for discrete velocities. However, as we have seen in previous chapters, it is more common in the simulation of flow in porous media to consider intercell fluxes as the primary unknowns. We therefore consider inner products of fluxes rather than velocities in the following. The relation between the two is trivial: an inner product of velocities becomes an inner product for fluxes by pre- and post-multiplying by the inverse of the area of the corresponding cell faces. In other words, if A is a diagonal matrix with element A_{jj} equal the area of the j-th face, the flux inner product M_{flux} is related to the velocity inner product M_{vel} through

$$M_{\text{flux}} = A^{-1} M_{\text{vel}} A^{-1}. \tag{6.25}$$

To simplify the derivation of valid solutions, we start by stating a geometrical property that relates C and N as follows (for a proof, see [52]):

$$C^\mathsf{T} N = V = \text{diag}(|\Omega_i|). \tag{6.26}$$

Next, we multiply the left matrix equation in (6.24) by $K^{-1} C^\mathsf{T} N$

$$C\,(K^{-1} C^\mathsf{T} N) = (MNK)\,(K^{-1} C^\mathsf{T} N) = MNC^\mathsf{T} N = MNV,$$

from which it follows that there exists a family of valid solution of the form

$$M = \frac{1}{|\Omega_i|} C K^{-1} C^\mathsf{T} + M_2, \tag{6.27}$$

where M_2 is a matrix defined such that $M_2 N = 0$, i.e., any matrix whose rows lie in the left nullspace of N^T. Moreover, to make a sensible method, we must require that M is symmetric positive definite. In other words, any symmetric and positive-definite inner product that fulfills these requirements can be represented in two alternative compact forms,

$$\begin{aligned} M &= \frac{1}{|\Omega_i|} C K^{-1} C^\mathsf{T} + Q_N^\perp S_M {Q_N^\perp}^\mathsf{T} \\ &= \frac{1}{|\Omega_i|} C K^{-1} C^\mathsf{T} + P_N^\perp S_M P_N^\perp. \end{aligned} \tag{6.28}$$

Here, S_M denotes any symmetric positive-definite matrix, $Q_N^\perp$ is an orthonormal basis for the left nullspace of N^T, and $P_N^\perp$ is the nullspace projection $I - Q_N Q_N^\mathsf{T}$, in which Q_N is a basis for spaces spanned by the columns of N. Similarly, we can derive a closed expression for the inverse inner product T by multiplying the right matrix equation in (6.24) by V^{-1} from the left and by V from the right and using the identity (6.26),

$$TC = V^{-1}(NK)V = V^{-1}(NK)(C^\mathsf{T} N)^\mathsf{T} = V^{-1}(NKN^\mathsf{T})C.$$

By the same argument as for M, mutatis mutandis, we obtain the following family of inverse inner products,

$$\begin{aligned} T &= \frac{1}{|\Omega_i|} N K N^\mathsf{T} + Q_C^\perp S_T {Q_C^\perp}^\mathsf{T} \\ &= \frac{1}{|\Omega_i|} N K N^\mathsf{T} + P_C^\perp S_T P_C^\perp, \end{aligned} \tag{6.29}$$

where $Q_C^\perp$ is an orthonormal basis for the left nullspace of C^T and $P_C^\perp = I - Q_C Q_C^\mathsf{T}$ is the corresponding nullspace projection.

The matrices M and T in (6.28) and (6.29) are evidently symmetric, so we only need to prove that they are positive definite. We start by writing the inner product as $M = M_1 + M_2$, and observe that each of these matrices are positive semi-definite. Hence, our result follows if we can prove that this implies that M is positive definite. Let z be an arbitrary nonzero vector, which we split uniquely as $z = Nx + y$, where x lies in the column space of N and y lies in the left nullspace of N^T. If y is zero, we have

$$\begin{aligned} z^\mathsf{T} M z &= x^\mathsf{T} N^\mathsf{T} M_1 N x \\ &= |\Omega_i|^{-1} x^\mathsf{T} (N^\mathsf{T} C) K^{-1} (C^\mathsf{T} N) x = |\Omega_i| x^\mathsf{T} K^{-1} x > 0 \end{aligned}$$

because K^{-1} is a positive definite matrix. If y is nonzero, we have

$$z^\mathsf{T} M z = z^\mathsf{T} M_1 z + y^\mathsf{T} M_2 y > 0$$

6.4 The Mimetic Method

because $z^T M_1 z \geq 0$ and $y^T M_2 y > 0$. An analogous argument holds for the matrix T, and hence we have proved that (6.28) and (6.29) give a well-defined family of consistent discretizations.

So far, we have not put any restrictions on the matrices S_M and S_T that will determine the specific methods, except for requiring that they should be positive definite. In addition, these matrices should mimic the scaling properties of the continuous equation, which is invariant under affine transformations of space and permeability,

$$\vec{x} \mapsto \sigma \vec{x} \quad \text{and} \quad \mathbf{K} \mapsto \sigma^T \mathbf{K} \sigma, \tag{6.30}$$

To motivate how we should choose the matrices S_M and S_T to mimic these scaling properties, we look at a simple 1D example:

Example 6.4.1 *Consider the grid cell $x \in [-1, 1]$, for which $N = C = [1, -1]^T$ and $Q_N^\perp = Q_C^\perp = \frac{1}{\sqrt{2}}[1, 1]^T$. Hence*

$$M = \frac{1}{2}\begin{bmatrix} 1 \\ -1 \end{bmatrix}\frac{1}{K}[1, -1] + \frac{1}{2}\begin{bmatrix} 1 \\ 1 \end{bmatrix} S_M [1, 1],$$

$$T = \frac{1}{2}\begin{bmatrix} 1 \\ -1 \end{bmatrix} K [1, -1] + \frac{1}{2}\begin{bmatrix} 1 \\ 1 \end{bmatrix} S [1, 1]. \tag{6.31}$$

The structure of the inner product should be invariant under scaling of K and thus we can write the inner product as a one-parameter family of the form

$$M = \frac{1}{2K}\left(\begin{bmatrix} 1 & -1 \\ -1 & 1 \end{bmatrix} + \frac{2}{t}\begin{bmatrix} 1 & 1 \\ 1 & 1 \end{bmatrix}\right),$$

$$T = \frac{K}{2}\left(\begin{bmatrix} 1 & -1 \\ -1 & 1 \end{bmatrix} + \frac{t}{2}\begin{bmatrix} 1 & 1 \\ 1 & 1 \end{bmatrix}\right). \tag{6.32}$$

Having established a plausible way of scaling the inner products, we are now in a position to go through various specific choices that are implemented in the `mimetic` module of MRST and look at correspondence between these methods and the standard two-point method, the lowest-order RT0 mixed method, and the MPFA-O method. Our discussion follows [192].

General Parametric Family

Motivated by Example 6.4.1, we propose to choose the matrix S_T as the diagonal of the first matrix term in (6.29) so that these two terms scale similarly under transformations of the type (6.30). Using this definition of S_T and allowing that M should be equal to T^{-1} suggests the following general family of inner products that only differ in the constant in front of the second (regularization) matrix term:

$$M = \frac{1}{|\Omega_i|} C K^{-1} C^T + \frac{|\Omega_i|}{t} P_N^\perp \operatorname{diag}(N K N^T)^{-1} P_N^\perp,$$

$$T = \frac{1}{|\Omega_i|}\left[N K N^T + t P_C^\perp \operatorname{diag}(N K N^T) P_C^\perp\right]. \tag{6.33}$$

In MRST, this family of inner products is called `'ip_qfamily'`, and the parameter t is supplied in a separate option:

```
S = computeMimeticIP(G, rock, 'InnerProduct', 'ip_qfamily', 'qparam', t);
```

As we will see shortly, mimetic inner products that reduce to the standard TPFA method or the RT0 mixed finite-element method on simple grids are members of this family.

Two-Point Type Methods

A requirement of the two-point method is that the transmissibility matrix T (and hence also the matrix M of the inner product) should be diagonal. Looking at (6.24), we see that this is only possible if the vectors $K\vec{N}_{ik}$ and $\vec{c}_{ik}$ are parallel, which is the exact same condition for K-orthogonality that we argued was sufficient to guarantee consistency of the two-point method on page 176. An explicit expression for the diagonal entries of the two-point method on K-orthogonal grids has already been given in (6.20). Strictly speaking, this relation does not always yield a positive value for the two-point transmissibility on any grid. For instance, for corner-point grids it is normal to define the face centroids as the arithmetic mean of the associated corner-point nodes and the cell centroids as the arithmetic mean of the centroids of the top and bottom cell faces, which for most grids will guarantee a positive transmissibility.

The extension of the two-point to non-orthogonal grids is not unique. Here, we present a mimetic method that coincides with the two-point methods in this method's region of validity and at the same time gives a valid mimetic inner product and hence a consistent discretization for all types of grids and permeability tensors. One advantage of this method, compared with multipoint flux-approximation methods, is that the implementation is simpler for general unstructured grids. The most instructive way to introduce the general method is to look at a simple example that will motivate the general construction.

Example 6.4.2 *We consider the grid cell $[-1, 1] \times [-1, 1]$ in 2D and calculate the components that make up the inverse inner product T for a diagonal and a full permeability tensor (K_1 and K_2, respectively) and compare with the corresponding two-point discretization. We start by setting up the permeability and the geometric properties of the cell*

```
K1 = eye(2);   K2 = [1 .5; .5 1];
C = [-1 0; 1 0; 0 -1; 0 1]; N = 2*C; vol = 4;
```

Using the definition (6.20), we see that the transmissibility matrix resulting from the standard two-point discretization can be computed as:

```
T = diag(diag(N*K*C')./sum(C.*C,2))
```

The result is `T=diag([2 2 2 2])` *for both permeability tensors, which clearly demonstrates that the scheme is not consistent for K_2. To construct the inverse inner product, we start by computing the nullspace projection*

```
Q = orth(C);
P = eye(size(C,1)) - Q*Q';
```

We saw earlier that as a simple means of providing a positive definite matrix S_T that scales similarly to the first term in the definition of T in (6.29), we could choose S_T as a multiple of the diagonal of this matrix:

```
W = (N * K * N') ./ vol;
St = diag(diag(W));
```

Collecting our terms, we see that the inverse inner product will be made up of the following two terms for the case with the diagonal permeability tensor:

```
W =   1  -1   0   0       P*St*P =  0.5  0.5    0    0
     -1   1   0   0                 0.5  0.5    0    0
      0   0   1  -1                   0    0  0.5  0.5
      0   0  -1   1                   0    0  0.5  0.5
```

If we now define the inverse inner product as

```
T = W + 2*P*St*P
```

the resulting method will coincide with the diagonal transmissibility matrix for diagonal permeability and give a full matrix for the tensor permeability,

```
T1 = 2.0    0    0    0     T2 =  2.0    0   0.5  -0.5
       0  2.0    0    0             0  2.0  -0.5   0.5
       0    0  2.0    0           0.5 -0.5   2.0     0
       0    0    0  2.0          -0.5  0.5     0   2.0
```

Altogether, we have derived a discrete inner product that generalizes the two-point method to full tensor permeabilities on 2D Cartesian grids and gives a consistent discretization also when the grid is not K-orthogonal.

Motivated by this example, we define a quasi-two-point inner product that simplifies to the standard TPFA method on Cartesian grids with diagonal permeability tensor:

$$M = \frac{1}{|\Omega_i|} C K^{-1} C^T + \frac{|\Omega_i|}{2} P_N^\perp \operatorname{diag}(N K N^T)^{-1} P_N^\perp,$$

$$T = \frac{1}{|\Omega_i|}\left[N K N^T + 2 P_C^\perp \operatorname{diag}(N K N^T) P_C^\perp \right]. \quad (6.34)$$

The observant reader will notice that this is a special case for $t = 2$ of the general family (6.33) of inner products. In MRST, this inner product is constructed by

```
S = computeMimeticIP(G, rock, 'InnerProduct', 'ip_quasitpf');
```

For completeness, the `mimetic` module also supplies a standard, diagonal two-point inner product that is not generally consistent. This is invoked by:

```
S = computeMimeticIP(G, rock, 'InnerProduct', 'ip_tpf');
```

Raviart–Thomas-Type Inner Product

To compute the mixed finite-element inner product on cells that are not simplexes or hexahedrons aligned with the coordinate axes, the standard approach is to map the cell back to a unit reference cell on which the mixed basis functions are defined and perform the integration there. For a general hexahedral cell in 3D, the resulting integral would conceptually look something like

$$\iiint_{\Omega_i} f(\vec{x})\, d\vec{x} = \iiint_{[0,1]^3} f(\vec{x}(\xi,\eta,\zeta))\, |J(\xi,\eta,\zeta)|\, d\xi\, d\eta\, d\zeta,$$

where $J = \partial(x,y,z)/\partial(\xi,\eta,\zeta)$ denotes the Jacobian matrix of the coordinate transformation. The determinant of J is generally nonlinear and it is therefore not possible to develop a mimetic inner product that equals the lowest-order, RT0 inner product on general polygonal or polyhedral cells, and at the same time is simpler to compute.

Instead, we develop an inner product that is equivalent to RT0 on grids that are orthogonal and aligned with the principal axes of the permeability tensor. To motivate the definition of this inner product, we first look at a simple example.

Example 6.4.3 *The RT0 basis functions defined on the reference element in two spatial dimensions read*

$$\vec{\psi}_1 = \begin{pmatrix} 1-x \\ 0 \end{pmatrix}, \quad \vec{\psi}_2 = \begin{pmatrix} x \\ 0 \end{pmatrix}, \quad \vec{\psi}_3 = \begin{pmatrix} 0 \\ 1-y \end{pmatrix}, \quad \vec{\psi}_4 = \begin{pmatrix} 0 \\ y \end{pmatrix}.$$

We compute the elements of the corresponding inner product matrix for a diagonal permeability with unit entries, $\mathbf{K} = \mathbf{I}$. This entails careful treatment of a sticky point: whereas the mixed inner product involves velocities, the mimetic inner product involves fluxes that are considered positive when going out of a cell and negative when going into the cell. To get comparative results, we thus reverse the sign of $\vec{\psi}_1$ and $\vec{\psi}_3$. Using symmetry and the fact that $\vec{\psi}_i \cdot \vec{\psi}_k = 0$ if $i = 1,2$ and $k = 3,4$, we only have to compute two integrals:

$$\int_0^1 \int_0^1 \vec{\psi}_1 \cdot \vec{\psi}_1\, dx dy = \int_0^1 (x-1)^2\, dx = \tfrac{1}{3}$$

$$\int_0^1 \int_0^1 \vec{\psi}_1 \cdot \vec{\psi}_2\, dx dy = \int_0^1 (x-1)x\, dx = -\tfrac{1}{6}.$$

This means that the RT0 inner product in 2D reads

$$\mathbf{M} = \begin{bmatrix} \tfrac{1}{3} & -\tfrac{1}{6} & 0 & 0 \\ -\tfrac{1}{6} & \tfrac{1}{3} & 0 & 0 \\ 0 & 0 & \tfrac{1}{3} & -\tfrac{1}{6} \\ 0 & 0 & -\tfrac{1}{6} & \tfrac{1}{3} \end{bmatrix}.$$

Similarly as in Example 6.4.2, we can compute the two matrices used to form the mimetic inner product

```
C  = .5*[-1 0; 1 0; 0 -1; 0 1]; N = 2*C; vol = 1; K = eye(2);
Q  = orth(N);
P  = eye(size(C,1)) - Q*Q';
Sm = diag(1 ./ diag(N*K*N'))*vol;
M1 = C*(K\C')./vol
M2 = P*Sm*P
```

The result of this computation is

```
M1 =  0.25  -0.25      0      0      M2 =  0.50   0.50      0      0
     -0.25   0.25      0      0            0.50   0.50      0      0
         0      0   0.25  -0.25               0      0   0.50   0.50
         0      0  -0.25   0.25               0      0   0.50   0.50
```

from which it follows that M_2 should be scaled by $\frac{1}{6}$ if the mimetic inner product is to coincide with the RT0 mixed inner product.

For a general grid, we define a quasi-RT0 inner product that simplifies to the standard RT0 inner product on orthogonal grids with diagonal permeability tensors, as well as on all other cases that can be transformed to such a grid by an affine transformation of the form (6.30). The quasi-RT0 inner product reads

$$M = \frac{1}{|\Omega_i|} C K^{-1} C^\mathsf{T} + \frac{|\Omega_i|}{6} P_N^\perp \operatorname{diag}(N K N^\mathsf{T})^{-1} P_N^\perp,$$

$$T = \frac{1}{|\Omega_i|}\Big[N K N^\mathsf{T} + 6\, P_C^\perp \operatorname{diag}(N K N^\mathsf{T}) P_C^\perp \Big]. \qquad (6.35)$$

and is a special case of the general family (6.33) of inner products for $t = 6$. In MRST, this inner product is constructed as follows

```
S = computeMimeticIP(G, rock, 'InnerProduct', 'ip_quasitpf');
```

For completeness, the `mimetic` module also supplies a standard RT0 inner product, `'ip_rt'`, that is only valid on Cartesian grids.

Default Inner Product in MRST

For historical reasons, the default discretization in the `mimetic` module of MRST corresponds to a mimetic method with half-transmissibilities defined by the following expression,

$$T = \frac{1}{|\Omega_i|}\Big[N K N^\mathsf{T} + \frac{6}{d} \operatorname{tr}(K) A\, (I - Q Q^\mathsf{T})\, A \Big]. \qquad (6.36)$$

Here, A is the diagonal matrix containing face areas and Q is an orthogonal basis for the range of AC. The inner product, referred to as `'ip_simple'`, was inspired by [52] and introduced in [6] to resemble the RT0 inner product; they are equal for scalar permeability on orthogonal grids, which can be verified by inspection. Because the inner product is based on velocities, it involves pre- and post-multiplication of (inverse) face areas and is in this sense different from the general class of inner products discussed in (6.33). It is nevertheless possible to show that the eigenspace corresponding to the nonzero eigenvalues of the second matrix term is equal to the nullspace for C.

Local-Flux Mimetic Method

Another large class of consistent discretization methods that have received a lot of attention in recent years is the multipoint flux-approximation (MPFA) schemes [9, 94, 7]. Discussing different variants of this method is beyond the scope of the current presentation, mainly because efficient implementation of a general class of MPFA schemes requires some additional mappings that are not yet part of the standard grid structure outlined in Section 3.4. These mappings are available and have been used in some of the add-on modules for geomechanics, and it should therefore not be too difficult for the competent and interested reader to also implement other MPFA type schemes.

Having said this, there *is* a module `mpfa` in MRST that gives a simple implementation of the MPFA-O method. In this implementation, we have utilized the fact that some variants of the method can be formulated as a mimetic method. This was first done by [281, 199] and is called the local-flux mimetic formulation of the MPFA method. In this approach, each face in the grid is subdivided into a set of subfaces, one subface for each of the nodes that make up the face. The inner product of the local-flux mimetic method gives exact results for linear flow and is block diagonal with respect to the faces corresponding to each node of the cell, but it is not symmetric. The block-diagonal property makes it possible to reduce the system into a cell-centered discretization for the cell pressures. This naturally leads to a method for calculating the MPFA transmissibilities. The crucial point is to have the corner geometry in the grid structure and handle the problems with corners that do not have three unique half-faces associated.

The local-flux mimetic formulation of the MPFA-O method can be used with a calling sequence that is similar to that of the TPFA and the mimetic methods:

```
mrstModule add mpfa;
hT = computeMultiPointTrans(G, rock);
state = incompMPFA(state, G, hT, fluid)
```

By default, `computeMultiPointTrans` uses MATLAB to invert a sequence of small matrices that arise in the discretization. This operation is relatively slow and the routine therefore takes an optional parameter `'invertBlocks'` that can be set equal to `'MEX'` to invoke C-accelerated routines for the inversion of the small matrix blocks. This option may not work if MRST has not been properly set up to automatically compile C-code on your computer.

6.5 Monotonicity

The main motivation for developing consistent discretizations is to get schemes that are convergent for rough grids and full-tensor permeabilities and less influenced by grid orientation errors. All consistent methods discussed in this chapter fulfill this requirement. On the other hand, these methods are all linear and not guaranteed to satisfy discrete counterparts of the maximum principle fulfilled by the continuous solution. For the Poisson equation $\nabla \cdot \mathbf{K} \nabla p = q$, this principle implies that if there is a single source within the domain, the pressure will decrease monotonically from the source toward the boundary. The discrete solution should also follow a similar principle, in which case we say that the underlying scheme is *monotone*.

If a scheme is not monotone, the discrete solutions may contain oscillations that are not present in the underlying physical system, particularly for cases with full-tensor permeabilities and large anisotropy ratios and/or high aspect ratio grids. While such oscillations are not very serious for single-phase, incompressible flow, they may have a significant impact on multiphase flows, which can be highly sensitive to small pressure variations; think of an oil–gas system operating slightly above the bubble point, for which artificial gas is liberated if the pressure falls below the bubble point; see Chapter 11. A classical result says that a discretization scheme is monotone if it generates an *M-matrix*, i.e., a matrix whose off-diagonal elements are less than or equal zero and whose eigenvalues have positive real parts. Two-point schemes are strictly monotone if all transmissibilities are nonnegative. Multipoint schemes, on the other hand, do not necessarily give an M-matrix, and extensive research has gone into investigate their monotonicity properties [236, 234, 235, 12, 157] and proving convergence [308, 26, 103, 104, 96, 314, 10, 167, 11]. In 2D, this has lead to the somewhat negative result [157]: *no linear nine-point control volume method can be constructed for quadrilateral grids in 2D that is exact for linear solutions while remaining monotone for general problems..*

For brevity, we will only present an example that illustrates lack of monotonicity for linear, consistent schemes. We refer the interested reader to [235] and references therein for a thorough discussion of the underlying mathematical properties that cause these unphysical oscillations.

Example 6.5.1 *(Example 2 from [192]) To illustrate the different behavior of TPFA and linear, consistent discretizations, we consider a 2D domain of size $n \times n$, where n is also the number of grid cells in each spatial direction. To generate a non-orthogonal grid, we use the* `twister` *routine, which you have encountered multiple times already. This routine normalizes all coordinates in a rectilinear grid to the interval $[0,1]$, then perturbs all interior points by adding $0.03 \sin(\pi x) \sin(3\pi(y - 1/2))$ to the x-coordinates and subtracting the same value from the y-coordinate before transforming back to the original domain. This creates a non-orthogonal, but smoothly varying, logically Cartesian grid. The domain has Dirichlet boundary conditions on the left and right-hand sides, with values one and zero, respectively, and zero Neumann conditions at the top and bottom.*

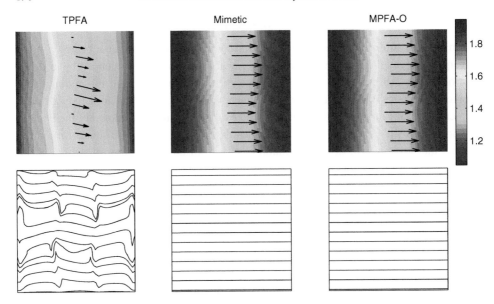

Figure 6.7 Grid orientation effects for a flow described by a linear pressure drop from the left to the right boundary. Approximate solutions computed by three different incompressible solvers from the `incomp`, `mimetic`, and `mpfa` modules, respectively. The permeability field is homogeneous with anisotropy ratio 1:1,000 aligned with the grid.

An important design principle for consistent discretizations is that they, unlike the TPFA method, should reproduce flows that result from a linear pressure drop exactly on grids that are not K-orthogonal. We therefore start by considering a diagonal permeability tensor with anisotropy ratio 1:1,000 aligned with the x-axis. Figure 6.7 compares the discrete solutions computed TPFA, the local-flux mimetic (MPFA-O) method, and the mimetic method with inner product `'ip_simple'` *on a 101 × 101 grid. Whereas the mimetic and MPFA-O schemes produce the expected result, the pressure and velocity solutions computed by TPFA show significant grid-orientation effects and are obviously wrong.*

To illustrate non-monotonic behavior, we repeat the experiment with the tensor rotated by $\pi/6$ on a grid with 21 × 21 cells; see Figure 6.8. The solution computed by TPFA is monotone, but fails to capture the correct flow pattern in which streamlines starting near the upper-left corner should exit near the lower-right corner. Judging from the color plots, the pressure values computed by the mimetic and MPFA-O methods appear to be correct and have no discernible oscillations. However, both the streamline distribution and the velocity vectors plotted for cells along the midsection of the grid clearly show that the resulting velocity fields are highly oscillatory, in particular for the mimetic method. If we change the inner product from its default value (`'ip_simple'` *) to, e.g., the quasi-two-point inner product (* `'ip_quasitpf'` *), the solution is qualitatively the same as for the MFPA-O method.*

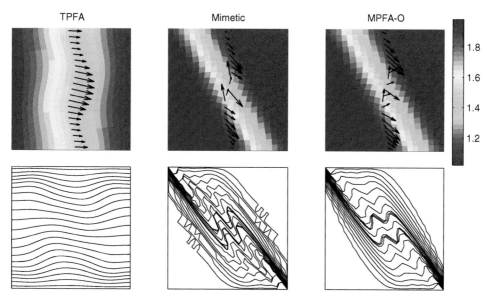

Figure 6.8 Monotonicity illustrated with the same setup as in Figure 6.8, but with the anisotropy ratio 1:1,000, making an angle $\pi/6$ with the x-direction.

6.6 Discussion

In this chapter, we have shown that TPFA is only consistent on K-orthogonal grids and hence may give significant grid orientation errors for tensor permeabilities and general cell geometries. On the other hand, the method is monotone and coercive (sum of fluxes is always positive), has a rather sparse stencil, and is hence computationally efficient. It is also possible to reduce grid orientation errors in the TPFA method by modifying the stencil. For tetrahedral grids satisfying the Delaunay criterion, one can design consistent two-point schemes by computing transmissibilities using the circumcenter rather than the cell centroid. You can also reduce numerical artifacts for general cell geometries by either optimizing the point used to compute intercell transmissibilities and/or use fluxes from a consistent method to adjust the transmissibilities. As a general precaution, I would recommend that you compute solutions using both a two-point and a consistent method to get an idea of the amount of grid orientation when working with stratigraphic or other unstructured representations of complex geology. This applies particularly to multiphase flow, for which you can get a good idea of the grid orientation errors by solving a representative single-phase problem.

Which consistent method should you choose amongst those implemented in MRST? The consistent MPFA method is cell-centered like the TPFA method and thus has the same number of unknowns, but the resulting stencil is denser and the method is therefore significantly less computationally efficient. Compared with mimetic methods, MPFA is quite involved to implement on general polyhedral meshes, and the method is not very robust on very skewed and deformed grids and for high aspect ratios or strong anisotropy [267]. In my experience, the mimetic method seems more robust and would therefore be my favorite choice, despite a higher number of unknowns.

There are also many other consistent methods than those discussed herein. Virtual element methods [38, 39] can be seen as a variational analog of mimetic methods and have proved to be versatile and robust for discretizing diffusion, dispersion, and mechanical effects. Unfortunately, the methods are not locally conservative when applied to Poisson-type flow equations, and thus require postprocessing to compute useful fluxes. MRST offers an implementation in the recent vem module [168], but this is currently primarily of academic interest when it comes to flow equations.

To develop schemes that are both consistent *and* monotone, several authors have started looking into *nonlinear two-point methods* [179, 198, 226, 272]. In these methods, the flux approximation is written

$$v_{i,k} = T_i(\vec{p})p_i - T_k(\vec{p})p_k,$$

where the transmissibilities depend on one or more pressure values so that the method becomes nonlinear. The disadvantage is that you have to solve a nonlinear system, but this additional cost may not present a significant drawback when the method is used for compressible and multiphase flow, where you need to use a nonlinear solver to account for other nonlinear couplings in the model equations, as we will see later in the book. At the time of writing, MRST only contains a preliminary implementation of nonlinear TPFA schemes. You can find more detailed comparisons of all the methods discussed previously in [169].

Another example of a recent and promising method is the vertex approximation gradient (VAG) method [105, 106, 267]. Unlike other nodal methods, this method has unknowns associated with both vertices and cell centers and relies on a local triangulation of the grid cells to enable simple extension of finite-elements to general cell geometries. The scheme can also be written on finite-volume form, introducing explicit cell-vertex fluxes that connect each cell to its vertex. This method is not yet implemented in MRST.

Computer exercises

6.6.1 The two tutorials `mimeticExample2` and `mimeticExample3` compare the TPFA method implemented in the `incomp` module with its mimetic counterpart. Do the methods produce the same results if you perturb the grid cells using `twister` or specify an anisotropic permeability field? What happens if you use a different inner product?

6.6.2 Redo the comparison of structured and unstructured grids in Section 5.4.3 using a mimetic solver and possibly the MPFA-O scheme. Try to also compare computational efficiency with the TPFA scheme in terms of number of unknowns, number of nonzero elements in the matrix, and condition numbers.

6.6.3 A standard method to construct analytical solutions to the Laplace equation in 2D is to write the unknown as the real or imaginary part of an complex analytic function

$f(z)$. We choose the function $f(z) = (z + \frac{1}{2}z^2)$ and set our analytical solution to be $p_a(x, y) = \text{Im}(f(x + iy)) = y + xy$. By prescribing $p_a(x, y)$ along the boundary, we obtain a Dirichlet problem with known solution. Use this technique to study the convergence of the TPFA method and at least two different consistent methods on a sequence of rough non-Cartesian grids of increasing resolution. Try to compare grids with cells that are always non-Cartesian at all resolutions with grids in which the cells are non-Cartesian at coarse resolutions but gradually approach a Cartesian shape as the grid is refined.

7
Compressible Flow and Rapid Prototyping

In previous chapters we have outlined and explained in detail how to discretize and solve incompressible flow problems. This chapter will teach you how to discretize the basic equations for single-phase, compressible flow by use of the discrete differential and averaging operators that were introduced in Section 4.4.2. As briefly shown in Examples 4.4.2 and 4.4.3 in the same section, these discrete operators enable you to implement discretized flow equations in a compact form similar to the continuous mathematical description. Use of automatic differentiation (see Appendix A.5 for more details) then ensures that no analytical derivatives have to be programmed explicitly as long as the discrete flow equations and constitutive relationships are implemented as a sequence of algebraic operations. MRST makes it possible to combine discrete operators and automatic differentiation with a flexible grid structure, a highly vectorized and interactive scripting language, and a powerful graphical environment. This is in my opinion the main reason why the software has proved to be an efficient tool for developing new computational methods and workflow tools. In this chapter, I try to substantiate this claim by showing several examples of rapid prototyping. We first develop a compact and transparent solver for compressible flow and then extend the basic single-phase model to include pressure-dependent viscosity, non-Newtonian fluid behavior, and temperature effects. As usual, you can find complete scripts for all examples in a subdirectory (ad-1ph) of the book module.

7.1 Implicit Discretization

As our basic model, we consider the single-phase continuity equation,

$$\frac{\partial}{\partial t}(\phi\rho) + \nabla \cdot (\rho\vec{v}) = q, \qquad \vec{v} = -\frac{\mathbf{K}}{\mu}(\nabla p - g\rho\nabla z). \tag{7.1}$$

The primary unknown is usually the fluid pressure p. Additional equations are supplied to provide relations between p and the other quantities in the equation, e.g., by using an equation of state to relate fluid density to pressure $\rho = \rho(p)$, specifying porosity as function of pressure $\phi(p)$ through a compressibility factor, and so on; see the discussion in Section 4.2. Notice also that q is defined slightly differently in (7.1) than in (4.5).

Using the discrete operators introduced in Section 4.4.2, the basic implicit discretization of (7.1) reads

$$\frac{(\boldsymbol{\phi}\rho)^{n+1} - (\boldsymbol{\phi}\rho)^n}{\Delta t^n} + \mathrm{div}(\rho \boldsymbol{v})^{n+1} = q^{n+1}, \tag{7.2a}$$

$$\boldsymbol{v}^{n+1} = -\frac{K}{\mu^{n+1}}\left[\mathrm{grad}(\boldsymbol{p}^{n+1}) - g\rho^{n+1}\mathrm{grad}(z)\right]. \tag{7.2b}$$

Here, $\boldsymbol{\phi} \in \mathbb{R}^{n_c}$ denotes the vector with one porosity value per cell, $\boldsymbol{v}$ is the vector of fluxes per face, and so on. The superscript refers to discrete times at which one wishes to compute the unknown reservoir states and Δt denotes the distance between two such consecutive points in time.

In many cases of practical interest it is possible to simplify (7.2). For instance, if the fluid is only slightly compressible, several terms can be neglected so that the nonlinear equation reduces to a linear equation in the unknown pressure p^{n+1}, which we can write on residual form as

$$\frac{p^{n+1} - p^n}{\Delta t^n} - \frac{1}{c_t \mu \phi}\mathrm{div}\bigl(K\,\mathrm{grad}(p^{n+1})\bigr) - q^n = 0. \tag{7.3}$$

The assumption of slight compressibility is not always applicable and for generality we assume that ϕ and ρ depend nonlinearly on p so that (7.2) gives rise to a nonlinear system of equations that needs to be solved in each time step. As we will see later in this chapter, viscosity may also depend on pressure, flow velocity, and/or temperature, which adds further nonlinearity to the system. If we now collect all the discrete equations, we can write the resulting system of nonlinear equations in short vector form as

$$\boldsymbol{F}(\boldsymbol{x}^{n+1}; \boldsymbol{x}^n) = \boldsymbol{0}. \tag{7.4}$$

Here, $\boldsymbol{x}^{n+1}$ is the vector of unknown state variables at the next time step and the vector of current states $\boldsymbol{x}^n$ can be seen as a parameter.

We will use the Newton–Raphson method to solve the nonlinear system (7.4): assume that we have a guess $\boldsymbol{x}_0$ and want to move this towards the correct solution, $\boldsymbol{F}(\boldsymbol{x}) = \boldsymbol{0}$. To this end, we write $\boldsymbol{x} = \boldsymbol{x}_0 + \Delta \boldsymbol{x}$, use a Taylor expansion for linearization, and solve for the approximate increment $\delta \boldsymbol{x}$

$$\boldsymbol{0} = \boldsymbol{F}(\boldsymbol{x}_0 + \Delta \boldsymbol{x}) \approx \boldsymbol{F}(\boldsymbol{x}_0) + \nabla \boldsymbol{F}(\boldsymbol{x}_0)\delta \boldsymbol{x}.$$

This gives rise to an iterative scheme in which the approximate solution $\boldsymbol{x}^{i+1}$ in the $(i+1)$-th iteration is obtained from

$$\frac{d\boldsymbol{F}}{d\boldsymbol{x}}(\boldsymbol{x}^i)\delta \boldsymbol{x}^{i+1} = -\boldsymbol{F}(\boldsymbol{x}^i), \qquad \boldsymbol{x}^{i+1} \leftarrow \boldsymbol{x}^i + \delta \boldsymbol{x}^{i+1}. \tag{7.5}$$

Here, $J = dF/dx$ is the Jacobian matrix, while δx^{i+1} is referred to as the *Newton update* at iteration number $i + 1$. Theoretically, the Newton process exhibits quadratic convergence under certain smoothness and differentiability requirements on F. Obtaining such convergence in practice, however, will crucially depend on having a sufficiently accurate Jacobian matrix. For complex flow models, the computation of residual equations typically requires evaluation of many constitutive laws that altogether make up complex nonlinear dependencies. Analytical derivation and subsequent coding of the Jacobian can therefore be very time-consuming and prone to human errors. Fortunately, the computation of the Jacobian matrix can in almost all cases be broken down to nested differentiation of elementary operations and functions and is therefore a good candidate for automation using automatic differentiation. This will add an extra computational overhead to your code, but in most cases the increased CPU time is completely offset by the shorter time it takes you to develop a proof-of-concept code. Likewise, unless your model problem is very small, the dominant computational cost of solving a nonlinear PDE comes from the linear solver called within each Newton iteration.

The idea of using automatic differentiation to develop reservoir simulators is not new. This technique was introduced in an early version of the commercial Intersect simulator [80], but has mainly been pioneered through a reimplementation of the GPRS research simulator [58]. The new simulator, called AD-GPRS, is primarily based on fully implicit formulations [303, 325, 302], in which independent variables and residual equations are AD structures implemented using ADETL, a library for forward-mode AD realized by expression templates in C++ [323, 322]. This way, the Jacobi matrices needed in the nonlinear Newton-type iterations can be constructed from implicitly computed derivatives when evaluating the residual equations. In [185], the authors discuss how to use the alternative backward-mode differentiation to improve computational efficiency. Automatic differentiation is also used in the open-source `Flow` simulator from the Open Porous Media (OPM) initiative. `OPM Flow` can be considered as a C++ sibling of MRST, which originally used a similar vector-oriented AD library. This has later been replaced by a localized, cell-based AD library for improved efficiency.

7.2 A Simulator Based on Automatic Differentiation

We will now present step-by-step how you can use the AD library in MRST to implement an implicit solver for the compressible, single-phase continuity equation (7.1). In particular, we revisit the discrete spatial differentiation operators from Section 4.4.2 and introduce additional discrete averaging operators that together enable us to write the discretized equations in an abstract residual form that resembles the semi-continuous form of the implicit discretization in (7.2). Starting from this residual form, it is relatively simple to obtain a linearization using automatic differentiation and set up a Newton iteration.

7.2.1 Model Setup and Initial State

For simplicity, we consider a homogeneous box model:

```
[nx,ny,nz] = deal( 10, 10, 10);
[Lx,Ly,Lz] = deal(200, 200, 50);
G = cartGrid([nx, ny, nz], [Lx, Ly, Lz]);
G = computeGeometry(G);

rock = makeRock(G, 30*milli*darcy, 0.3);
```

Beyond this point, our implementation is agnostic to details about the grid, except when we specify well positions on page 206, which would typically involve more code lines for a complex corner-point model like the Norne and SAIGUP models discussed in Sections 3.3.1 and 3.5.1.

We assume constant rock compressibility c_r. Accordingly, the pore volume pv obeys the differential equation[1] $c_r \text{pv} = d\text{pv}/dp$ or

$$\text{pv}(p) = \text{pv}_r \, e^{c_r(p-p_r)}, \qquad (7.6)$$

where pv_r is the pore volume at reference pressure p_r. To define the relation between pore volume and pressure, we use an anonymous function:

```
cr   = 1e-6/barsa;
p_r  = 200*barsa;
pv_r = poreVolume(G, rock);

pv   = @(p) pv_r .* exp( cr * (p - p_r) );
```

The fluid is assumed to have constant viscosity $\mu = 5$ cP, and as for the rock, we assume constant fluid compressibility c, resulting in the differential equation $c\rho = d\rho/dp$ for fluid density. Accordingly,

$$\rho(p) = \rho_r e^{c(p-p_r)}, \qquad (7.7)$$

where ρ_r is the density at reference pressure p_r. With this set, we can define the equation of state for the fluid:

```
mu    = 5*centi*poise;
c     = 1e-3/barsa;
rho_r = 850*kilogram/meter^3;
rhoS  = 750*kilogram/meter^3;
rho   = @(p) rho_r .* exp( c * (p - p_r) );
```

The assumption of constant compressibility will only hold for a limited range of temperatures. Moreover, surface conditions are not inside the validity range of the constant

---

[1] To highlight the close correspondence between the computer code and the mathematical equation, we here deliberately violate the advice to never use a compound symbol to denote a single mathematical quantity.

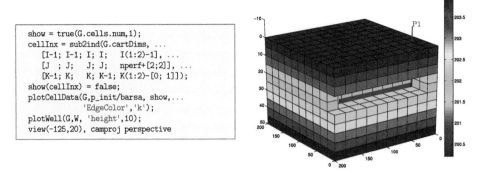

```
show = true(G.cells.num,1);
cellInx = sub2ind(G.cartDims, ...
    [I-1; I-1; I; I;   I(1:2)-1], ...
    [J  ; J;   J; J;   nperf+[2;2]], ...
    [K-1; K;   K; K-1; K(1:2)-[0; 1]]);
show(cellInx) = false;
plotCellData(G,p_init/barsa, show, ...
        'EdgeColor','k');
plotWell(G,W, 'height',10);
view(-125,20), camproj perspective
```

Figure 7.1 Model with initial pressure and single horizontal well.

compressibility assumption. We therefore set the fluid density ρ_S at surface conditions separately because we will need it later to evaluate surface volume rate in our model of the well, which consists of a horizontal wellbore perforated in eight cells:

```
nperf = 8;
I = repmat(2, [nperf, 1]);
J = (1:nperf).'+1;
K = repmat(5, [nperf, 1]);
cellInx = sub2ind(G.cartDims, I, J, K);
W = addWell([ ], G, rock, cellInx, 'Name', 'producer', 'Dir', 'x');
```

Assuming the reservoir is initially at equilibrium implies that we must have $dp/dz = g\rho(p)$. In our simple setup, this differential equation can be solved analytically, but for demonstration purposes, we use one of MATLAB's built-in ODE-solvers to compute the hydrostatic distribution numerically, relative to a fixed datum point $p(z_0) = p_r$. Without lack of generality, we set $z_0 = 0$ since the reservoir geometry is defined relative to this height:

```
gravity reset on, g = norm(gravity);
[z_0, z_max] = deal(0, max(G.cells.centroids(:,3)));
equil   = ode23(@(z,p) g .* rho(p), [z_0, z_max], p_r);
p_init = reshape(deval(equil, G.cells.centroids(:,3)), [], 1);
```

This finishes the model setup, and at this stage we plot the reservoir with well and initial pressure as shown in Figure 7.1.

7.2.2 Discrete Operators and Equations

We are now ready to discretize the model. As seen in Section 4.4.2, the discrete version of the gradient operator maps from the set of cells to the set of faces. For a pressure field, it computes the pressure difference between neighboring cells of each face. Likewise, the discrete divergence operator is a linear mapping from the set of faces to the set of cells. For a flux field, it sums the outward fluxes for each cell. The complete code needed to form the grad and div operators has already been presented in Examples 4.4.2 and 4.4.3, but here we repeat it in order to make the example more self-contained.

To define the discrete operators, we must first compute the map between interior faces and cells

```
C = double(G.faces.neighbors);
C = C(all(C ~= 0, 2), :);
```

Exterior faces need not be included since they have zero flow, given our assumption of no-flow boundary conditions. It now follows that $\text{grad}(x) = x(C(:,2)) - x(C(:,1)) = Dx$, where D is a sparse matrix with values ± 1 in columns $C(i,2)$ and $C(i,1)$ for row i. As a linear mapping, the discrete div-function is simply the negative transpose of grad; this follows from the discrete version of the Gauss–Green theorem, (4.58). In addition, we define an *averaging* operator that for each face computes the arithmetic average of the neighboring cells, which we will need to evaluate density values at grid faces:

```
n    = size(C,1);
D    = sparse([(1:n)'; (1:n)'], C, ...
              ones(n,1)*[-1 1], n, G.cells.num);
grad = @(x) D*x;
div  = @(x) -D'*x;
avg  = @(x) 0.5 * (x(C(:,1)) + x(C(:,2)));
```

This is all we need to define the spatial discretization for a homogeneous medium on a grid with cubic cells. To make a generic spatial discretization that also can account for more general cell geometries and heterogeneities, we must include transmissibilities. To this end, we first compute one-sided transmissibilities $T_{i,j}$ using the function computeTrans, which was discussed in detail in Section 5.2, and then use harmonic averaging to obtain face-transmissibilities. That is, for neighboring cells i and j, we compute $T_{ij} = (T_{i,j}^{-1} + T_{j,i}^{-1})^{-1}$ as in (4.52) on page 133.

```
hT = computeTrans(G, rock);            % Half-transmissibilities
cf = G.cells.faces(:,1);
nf = G.faces.num;
T  = 1 ./ accumarray(cf, 1 ./ hT, [nf, 1]); % Harmonic average
T  = T(intInx);                         % Restricted to interior
```

Having defined the necessary discrete operators, we are in a position to use the basic implicit discretization from (7.2). We start with Darcy's law (7.2b),

$$\vec{v}[f] = -\frac{T[f]}{\mu}\bigl(\text{grad}(p) - g\,\rho_a[f]\,\text{grad}(z)\bigr), \qquad (7.8)$$

where the density at the interface is evaluated using the arithmetic average

$$\rho_a[f] = \tfrac{1}{2}\bigl(\rho[C_1(f)] + \rho[C_2(f)]\bigr). \qquad (7.9)$$

Similarly, we can write the continuity equation for each cell c as

$$\frac{1}{\Delta t}\left[\left(\boldsymbol{\phi}(\boldsymbol{p})[c]\,\boldsymbol{\rho}(\boldsymbol{p})[c]\right)^{n+1} - \left(\boldsymbol{\phi}(\boldsymbol{p})[c]\,\boldsymbol{\rho}(\boldsymbol{p})[c]\right)^{n}\right] + \mathtt{div}(\boldsymbol{\rho}_a \boldsymbol{v})[c] = \mathbf{0}. \tag{7.10}$$

The two residual equations (7.8) and (7.10) are implemented as anonymous functions of pressure:

```
gradz = grad(G.cells.centroids(:,3));
v    = @(p)   -(T/mu).*( grad(p) - g*avg(rho(p)).*gradz );

presEq = @(p,p0,dt)  (1/dt)*(pv(p).*rho(p) - pv(p0).*rho(p0)) ...
                     + div( avg(rho(p)).*v(p) );
```

In the code above, p0 is the pressure field at the *previous* time step (i.e., $\boldsymbol{p}^n$), whereas p is the pressure at the *current* time step ($\boldsymbol{p}^{n+1}$). Having defined the discrete expression for Darcy fluxes, we can check that this is in agreement with our initial pressure field by computing the magnitude of the flux, norm(v(p_init))*day. The result is 1.5×10^{-6} m^3/day, which should convince us that the initial state of the reservoir is sufficiently close to equilibrium.

7.2.3 Well Model

The production well will appear as a source term in the pressure equation. We therefore need to define an expression for flow rate in all cells the well is connected to the reservoir (which we refer to as well connections). Inside the well, we assume instantaneous flow so that the pressure drop is always hydrostatic. For a horizontal well, the hydrostatic term is zero and could obviously be disregarded, but we include it for completeness and as a robust precaution in case we later want to reuse the code with a different well path. Approximating the fluid density in the well as constant, computed at bottom-hole pressure, the pressure $\boldsymbol{p}_c[w]$ in connection w of well $N_w(w)$ is given by

$$\boldsymbol{p}_c[w] = \boldsymbol{p}_{bh}[N_w(w)] + g\,\Delta z[w]\,\rho(\boldsymbol{p}_{bh}[N_w(w)]), \tag{7.11}$$

where $\Delta z[w]$ is the vertical distance from the bottom-hole to the connection. We use the standard Peaceman model introduced in Section 4.3.2 to relate the pressure at the well connection to the average pressure inside the grid cell. Using the well-indices from W, the mass flow-rate at connection c reads

$$\boldsymbol{q}_c[w] = \frac{\rho(\boldsymbol{p}[N_c(w)])}{\mu}\,\mathrm{WI}[w]\,\left(\boldsymbol{p}_c[w] - \boldsymbol{p}[N_c(w)]\right), \tag{7.12}$$

where $\boldsymbol{p}[N_c(w)]$ is the pressure in cell $N_c(w)$ containing connection w. In our code, this model is implemented as follows:

```
wc = W(1).cells;  % connection grid cells
WI = W(1).WI;     % well-indices
dz = W(1).dZ;     % depth relative to bottom-hole

p_conn = @(bhp)    bhp + g*dz.*rho(bhp); %connection pressures
q_conn = @(p, bhp) WI .* (rho(p(wc)) / mu) .* (p_conn(bhp) - p(wc));
```

We also include the total volumetric well-rate at surface conditions as a free variable. This is simply given by summing all mass well-rates and dividing by the surface density:

```
rateEq = @(p, bhp, qS)  qS-sum(q_conn(p, bhp))/rhoS;
```

With free variables p, bhp, and qS, we lack exactly one equation to close the system. This equation should account for *boundary conditions* in the form of a well control. Here, we choose to control the well by specifying a fixed bottom-hole pressure

```
ctrlEq = @(bhp)  bhp-100*barsa;
```

7.2.4 The Simulation Loop

What now remains is to set up a simulation loop that will evolve the transient pressure. We start by initializing the AD variables. For clarity, we append _ad to all variable names to distinguish them from doubles. The initial bottom-hole pressure is set to the corresponding grid-cell pressure.

```
[p_ad, bhp_ad, qS_ad] = initVariablesADI(p_init, p_init(wc(1)), 0);
```

This gives the following AD pairs that make up the unknowns in our system:

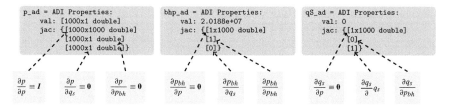

To solve the global flow problem, we must stack all the equations into one big system, compute the corresponding Jacobian, and perform a Newton update. We therefore set indices for easy access to individual variables

```
[p_ad, bhp_ad, qS_ad] = initVariablesADI(p_init, p_init(wc(1)), 0);
nc = G.cells.num;
[pIx, bhpIx, qSIx] = deal(1:nc, nc+1, nc+2);
```

Next, we set parameters to control the time steps in the simulation and the iterations in the Newton solver:

```
[numSteps, totTime] = deal(52, 365*day);   % time-steps/ total simulation time
[tol, maxits]       = deal(1e-5, 10)       % Newton tolerance / maximum Newton its
dt = totTime / numSteps;
```

Simulation results from all time steps are stored in a structure `sol`. For efficiency, this structure is preallocated and initialized so that the first entry is the initial state of the reservoir:

```
sol = repmat(struct('time',[], 'pressure',[], 'bhp',[], 'qS',[]), [numSteps+1, 1]);
sol(1) = struct('time', 0, 'pressure', double(p_ad), ...
                'bhp', double(bhp_ad), 'qS', double(qS_ad));
```

We now have all we need to set up the time-stepping algorithm, which consists of an outer and an inner loop. The outer loop updates the time step, advances the solution one step forward in time, and stores the result in the `sol` structure. This procedure is repeated until we reach the desired final time:

```
t = 0; step = 0;
while t < totTime,
   t = t + dt; step = step + 1;
   fprintf('\nTime step %d: Time %.2f -> %.2f days\n', ...
           step, convertTo(t - dt, day), convertTo(t, day));
   % Newton loop
   [resNorm, nit] = deal(1e99, 0);
   p0  = double(p_ad); % Previous step pressure
   while (resNorm > tol) && (nit <= maxits)
     : % Newton update
     :
     resNorm = norm(res);
     nit     = nit + 1;
     fprintf('  Iteration %3d:  Res = %.4e\n', nit, resNorm);
   end
   if nit > maxits, error('Newton solves did not converge')
   else % store solution
      sol(step+1) = struct('time', t, 'pressure', double(p_ad), ...
                           'bhp', double(bhp_ad), 'qS', double(qS_ad));
   end
end
```

The inner loop performs the Newton iteration by computing and assembling the Jacobian of the global system and solving the linearized residual equation to compute an iterative update. The first step to this end is to evaluate the residual for the flow pressure equation and add source terms from wells:

```
eq1     = presEq(p_ad, p0, dt);
eq1(wc) = eq1(wc) - q_conn(p_ad, bhp_ad);
```

Most of the lines we have implemented so far are fairly standard, except perhaps for the definition of the residual equations as anonymous functions. Equivalent statements can be found in almost any computer program solving this type of time-dependent equation by an implicit method. Now, however, comes what is normally the tricky part: linearization of the equations that make up the whole model and assembly of the resulting Jacobian matrices to generate the Jacobian for the full system. And here you have the magic of automatic differentiation: *you do not have to do this at all!* The computer code necessary to evaluate all Jacobians has been defined implicitly by the functions in the AD library in MRST, which overloads the elementary operators used to define the residual equations. An example of a complete calling sequence for a simple calculation is shown in Figure A.7 on page 625. The sequence of operations we use to compute the residual equations is obviously more complex than this example, but the operators used are in fact only the three elementary operators plus, minus, and multiply applied to scalars, vectors, and matrices, as well as element-wise division by a scalar and evaluation of exponential functions. When the residuals are evaluated by use of the anonymous functions defined in Sections 7.2.2 and 7.2.3, the AD library also evaluates the derivatives of each equation with respect to each independent variable and collects the corresponding sub-Jacobians in a list. To form the full system, we simply evaluate the residuals of the remaining equations (the rate equation and the equation for well control) and concatenate the three equations into a cell array:

```
eqs = {eq1, rateEq(p_ad, bhp_ad, qS_ad), ctrlEq(bhp_ad)};
eq  = cat(eqs{:});
```

In doing this, the AD library will correctly combine the various sub-Jacobians and set up the Jacobian for the full system. Then, we can extract this Jacobian, solve for the Newton increment, and update the three primary unknowns:

```
J   = eq.jac{1};   % Jacobian
res = eq.val;      % residual
upd = -(J \ res);  % Newton update

% Update variables
p_ad.val   = p_ad.val   + upd(pIx);
bhp_ad.val = bhp_ad.val + upd(bhpIx);
qS_ad.val  = qS_ad.val  + upd(qSIx);
```

The sparsity pattern of the Jacobian is shown in the plot to the left of the code for the Newton update. The use of a two-point scheme on a 3D Cartesian grid gives a Jacobi matrix that has a heptadiagonal structure, except for the off-diagonal entries in the two red rectangles. These arise from the well equation and correspond to derivatives of this equation with respect to cell pressures.

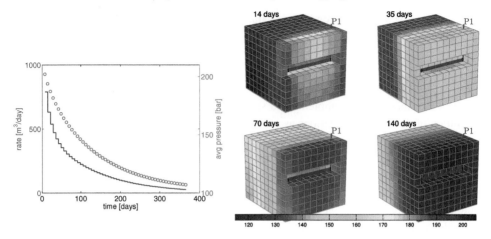

Figure 7.2 Time evolution of the pressure solution for the compressible single-phase problem. The plot to the left shows the well rate (blue line) and average reservoir pressure (green circles) as function of time, and the plots to the right show the pressure after 2, 5, 10, and 20 pressure steps.

Figure 7.2 plots how the dynamic pressure evolves with time. Initially, the pressure is in hydrostatic equilibrium as shown in Figure 7.1. When the well starts to drain the reservoir, the pressure drawdown near the well will start to gradually propagate outward from the well. As a result, the average pressure inside the reservoir is reduced, which again causes a decay in the production rate.

COMPUTER EXERCISES

7.2.1 Apply the compressible pressure solver to the quarter five-spot problem from Section 5.4.1.

7.2.2 Rerun compressible simulations for the three different grid models that were derived from the `seamount` data set Section 5.4.3. Replace the fixed boundary conditions by a no-flow condition.

7.2.3 Use the implementation from Section 7.2 as a template to develop a solver for slightly compressible flow (7.3). More details about this model can be found on page 118 in Section 4.2. How large can c_f be before the assumptions of slight compressibility become inaccurate? Use different heterogeneities, well placements, and model geometries to investigate this.

7.2.4 Extend the compressible solver developed in this section to incorporate other boundary conditions than no flow.

7.2.5 Try to compute time-of-flight by extending the equation set to also include the time-of-flight equation (4.40). Hint: the time-of-flight and the pressure equations need not be solved as a coupled system.

7.2.6 Same as the previous exercise, except that you should try to reuse the solver from in Section 5.3. Hint: you must first reconstruct fluxes from the computed pressure and then construct a state object to communicate with the TOF solver.

7.3 Pressure-Dependent Viscosity

One particular advantage of using automatic differentiation in combination with the discrete differential and averaging operators is that it simplifies the testing of new models and alternative computational approaches. In this section, we discuss two examples that hopefully demonstrate this aspect.

In the model discussed in the previous section, the viscosity was assumed to be constant. However, viscosity will generally increase with increasing pressures and this effect may be significant for the high pressures seen inside a reservoir, as we will see later in the book when discussing black-oil models in Chapter 11. To illustrate, we introduce a linear dependence, rather than the exponential pressure-dependence used for pore volume (7.6) and the fluid density (7.7). That is, we assume the viscosity is given by

$$\mu(p) = \mu_0 \big[1 + c_\mu (p - p_r) \big]. \tag{7.13}$$

Having a pressure dependence means that we have to change two parts of our discretization: the approximation of the Darcy flux across a cell face (7.8) and the flow rate through a well connection (7.12). Starting with the latter, we evaluate the viscosity using the same pressure as was used to evaluate the density, i.e.,

$$q_c[w] = \frac{\rho(p[N_c(w)])}{\mu(p[N_c(w)])} \, \text{WI}[w] \, \big(p_c[w] - p[N_c(w)] \big). \tag{7.14}$$

For the Darcy flux (7.8), we have two choices: either use a simple arithmetic average as in (7.9) to approximate the viscosity at each cell face,

$$v[f] = -\frac{T[f]}{\mu_a[f]} \big(\text{grad}(p) - g \, \rho_a[f] \, \text{grad}(z) \big), \tag{7.15}$$

or replace the quotient of the transmissibility and the face viscosity by the harmonic average of the mobility $\lambda = \mathbf{K}/\mu$ in the adjacent cells. Both choices introduce changes in the structure of the discrete nonlinear system, but because we are using automatic differentiation, all we have to do is code the corresponding formulas. Let us look at the details of the implementation, starting with the arithmetic approach.

Arithmetic Average

First, we introduce a new anonymous function to evaluate the relation between viscosity and pressure:

```
[mu0,c_mu] = deal(5*centi*poise, 2e-3/barsa);
mu         = @(p) mu0*(1+c_mu*(p-p_r));
```

Then, we can replace the definition of the Darcy flux (changes marked in red):

```
v = @(p) -(T./mu(avg(p))).*( grad(p) - g*avg(rho(p)).*gradz );
```

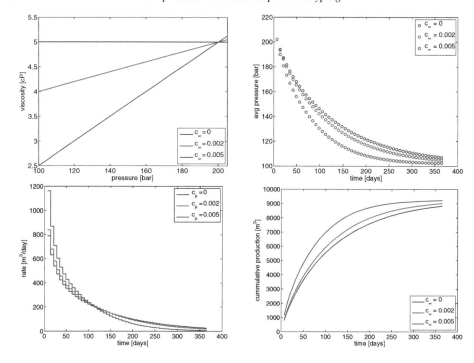

Figure 7.3 The effect of increasing the degree of pressure-dependence for the viscosity.

and similarly for flow rate through each well connection:

```
q_conn = @(p,bhp) WI.*(rho(p(wc))./ mu(p(wc))) .* (p_conn(bhp) - p(wc));
```

Figure 7.3 illustrates the effect of increasing the pressure dependence of the viscosity. Since the reference value is given at $p = 200$ bar, which is close to the initial pressure inside the reservoir, the more we increase c_μ, the lower μ will be in the pressure-drawdown zone near the well. Hence, we see a significantly higher initial production rate for $c_\mu = 0.005$ than for $c_\mu = 0$. On the other hand, the higher value of c_μ, the faster the drawdown effect of the well will propagate into the reservoir, inducing a reduction in reservoir pressure that eventually will cause production to cease. In terms of overall production, a stronger pressure dependence may be more advantageous as it leads to a higher total recovery and higher cumulative production early in the production period.

Face Mobility: Harmonic Average

A more correct approximation is to write Darcy's law based on mobility instead of using the quotient of the transmissibility and an averaged viscosity:

$$v[f] = -\Lambda[f]\big(\text{grad}(p) - g\,\rho_a[f]\,\text{grad}(z)\big). \tag{7.16}$$

7.4 Non-Newtonian Fluid

The face mobility $\Lambda[f]$ can be defined in the same way as the transmissibility is defined in terms of the half transmissibilities using harmonic averages. That is, if $T[f, c]$ denotes the half transmissibility associated with face f and cell c, the face mobility $\Lambda[f]$ for face f can be written as

$$\Lambda[f] = \left(\frac{\mu[C_1(f)]}{T[f, C_1(f)]} + \frac{\mu[C_2(f)]}{T[f, C_2(f)]} \right)^{-1}. \tag{7.17}$$

In MRST, the corresponding code reads:

```
hf2cn = getCellNoFaces(G);
nhf   = numel(hf2cn);
hf2f  = sparse(double(G.cells.faces(:,1)),(1:nhf)',1);
hf2if = hf2f(intInx,:);
fmob  = @(mu,p) 1./(hf2if*(mu(p(hf2cn))./hT));

v = @(p) -fmob(mu,p).*( grad(p) - g*avg(rho(p)).*gradz );
```

Here, hf2cn represents the maps C_1 and C_2, which enable us to sample the viscosity value in the correct cell for each half-face transmissibility, whereas hf2if represents a map from half-faces (i.e., faces seen from a single cell) to global faces (which are shared by two cells). The map has a unit value in row i and column j if half-face j belongs to global face i. Hence, premultiplying a vector of half-face quantities by hf2if amounts to summing the contributions from cells $C_1(f)$ and $C_2(f)$ for each global face f.

Using the harmonic average for a homogeneous model should produce simulation results that are identical (to machine precision) to those produced by the arithmetic average. With heterogeneous permeability, there will be small differences in well rates and averaged pressures for the specific parameters considered herein. For sub-samples of the SPE 10 data set, we typically observe maximum relative differences in well rates of the order 10^{-3}.

COMPUTER EXERCISES

7.3.1 Investigate the claim that the difference between using an arithmetic average of the viscosity and a harmonic average of the fluid mobility is typically small. To this end, you can for instance use the following sub-sample from the SPE 10 data set: rock = getSPE10rock(41:50,101:110,1:10).

7.4 Non-Newtonian Fluid

Viscosity is the material property that measures a fluid's resistance to flow, i.e., the resistance to a change in shape, or to the movement of neighboring portions of the fluid relative to each other. The more viscous a fluid is, the less easily it will flow. In Newtonian fluids, the shear stress or the force applied per area tangential to the force at any point is proportional to the strain rate (the symmetric part of the velocity gradient) at that point, and the viscosity is the constant of proportionality. For non-Newtonian fluids, the relationship is no longer linear. The most common nonlinear behavior is *shear thinning*, in which the viscosity of the system decreases as the shear rate increases. An example is paint, which should flow easily when leaving the brush, but stay on the surface and not drip once it has been applied. The second type of nonlinearity is *shear thickening*, in which the viscosity increases with increasing shear rate. A common example is the mixture of cornstarch and water. If you search YouTube for "cornstarch pool" you will find several spectacular videos of pools filled with this mixture. When stress is applied to the mixture, it exhibits properties like a solid and you may be able to run across its surface. However, if you go too slow, the fluid behaves more like a liquid and you fall in.

Solutions of large polymeric molecules are another example of shear-thinning liquids. In enhanced oil recovery, polymer solutions may be injected into reservoirs to improve unfavorable mobility ratios between oil and water and improve the sweep efficiency of the injected fluid. At low flow rates, the polymer molecule chains tumble around randomly and present large resistance to flow. When the flow velocity increases, the viscosity decreases as the molecules gradually align themselves in the direction of increasing shear rate. A model of the rheology is given by

$$\mu = \mu_\infty + (\mu_0 - \mu_\infty)\left(1 + \left(\frac{K_c}{\mu_0}\right)^{\frac{2}{n-1}} \dot{\gamma}^2\right)^{\frac{n-1}{2}}, \qquad (7.18)$$

where μ_0 represents the Newtonian viscosity at zero shear rate, μ_∞ represents the Newtonian viscosity at infinite shear rate, K_c represents the consistency index, and n represents the power-law exponent ($n < 1$). The shear rate $\dot{\gamma}$ in a porous medium can be approximated by

$$\dot{\gamma}_{\text{app}} = 6\left(\frac{3n+1}{4n}\right)^{\frac{n}{n-1}} \frac{|\vec{v}|}{\sqrt{K\phi}}. \qquad (7.19)$$

Combining (7.18) and (7.19), we can write our model for the viscosity as

$$\mu = \mu_0\left(1 + \bar{K}_c \frac{|\vec{v}|^2}{K\phi}\right)^{\frac{n-1}{2}}, \quad \bar{K}_c = 36\left(\frac{K_c}{\mu_0}\right)^{\frac{2}{n-1}}\left(\frac{3n+1}{4n}\right)^{\frac{2n}{n-1}}, \qquad (7.20)$$

where we for simplicity have assumed that $\mu_\infty = 0$.

Rapid Development of Proof-of-Concept Codes

We now demonstrate how easy it is to extend the simple simulator developed so far in this chapter to model non-Newtonian fluid behavior (see `nonNewtonianCell.m`). To simulate injection, we increase the bottom-hole pressure to 300 bar. Our rheology model has parameters:

```
mu0 = 100*centi*poise;
nmu = .3;
Kc  = .1);
Kbc = (Kc/mu0)^(2/(nmu-1))*36*((3*nmu+1)/(4*nmu))^(2*nmu/(nmu-1));
```

7.4 Non-Newtonian Fluid

In principle, we could continue to solve the system using the same primary unknowns as before. However, it has proved convenient to write (7.20) in the form $\mu = \eta\mu_0$, and introduce η as an additional unknown. In each Newton step, we start by solving the equation for the shear factor η exactly for the given pressure distribution. This is done by initializing an AD variable for η, but not for p in etaEq so that this residual now only has one unknown, η. This will take out the implicit nature of Darcy's law and hence reduce the nonlinearity and simplify the solution of the global system.

```
while (resNorm > tol) && (nit < maxits)
  % Newton loop for eta (shear multiplier)
  [resNorm2,nit2] = deal(1e99, 0);
  eta_ad2 = initVariablesADI(eta_ad.val);
  while (resNorm2 > tol) && (nit2 <= maxits)
    eeq = etaEq(p_ad.val, eta_ad2);
    res = eeq.val;
    eta_ad2.val = eta_ad2.val - (eeq.jac{1} \ res);
    resNorm2 = norm(res);
    nit2    = nit2+1;
  end
  eta_ad.val = eta_ad2.val;
```

Once the shear factor has been computed for the values in the previous iterate, we can use the same approach as earlier to compute a Newton update for the full system. (Here, etaEq is treated as a system with two unknowns, p and η.)

```
eq1    = presEq(p_ad, p0, eta_ad, dt);
eq1(wc) = eq1(wc) - q_conn(p_ad, eta_ad, bhp_ad);
eqs = {eq1, etaEq(p_ad, eta_ad), ...
       rateEq(p_ad, eta_ad, bhp_ad, qS_ad), ctrlEq(bhp_ad)};
eq  = cat(eqs{:});
upd = -(eq.jac{1} \ eq.val); % Newton update
```

To finish the solver, we need to define the flow equations and the extra equation for the shear multiplier. The main question now is how we should compute $|\vec{v}|$? One solution could be to define $|\vec{v}|$ on each face as the flux divided by the face area. In other words, use a code like

```
phiK   = avg(rock.perm.*rock.poro)./G.faces.areas(intInx).^2;
v      = @(p, eta)   -(T./(mu0*eta)).*( grad(p) - g*avg(rho(p)).*gradz );
etaEq  = @(p, eta)   eta - (1 + Kbc*v(p,eta).^2./phiK).^((nmu-1)/2);
```

Although simple, this approach has three potential issues: First, it does not tell us how to compute the shear factor for the well perforations. Second, it disregards contributions from any tangential components of the velocity field. Third, the number of unknowns in the linear system increases by almost a factor six since we now have one extra unknown per internal face. The first issue is easy to fix: To get a representative value in the well cells, we simply average the η values from the cells' faces. If we now recall how the discrete divergence operator was defined, we realize that this operation is almost implemented for us already: if `div(x)=-D'*x` computes the discrete divergence in each cell of the field x defined at the faces, then `wavg(x)=1/6*abs(D)'*x` computes the average of x for each cell. In other words, our well equation becomes:

```
wavg   = @(eta) 1/6*abs(D(:,W.cells))'*eta;
q_conn = @(p, eta, bhp) ...
   WI .* (rho(p(wc)) ./ (mu0*wavg(eta))) .* (p_conn(bhp) - p(wc));
```

The second issue would have to be investigated in more detail, and this is not within the scope of this book. The third issue is simply a disadvantage.

To get a method that consumes less memory, we can compute one η value per cell. Using the following formula, we can reconstruct an approximate velocity $\vec{v}_i$ at the center of cell i

$$\vec{v}_i = \sum_{j \in N(i)} \frac{v_{ij}}{V_i} (\vec{c}_{ij} - \vec{c}_i), \tag{7.21}$$

where $N(i)$ is the map from cell i to its neighboring cells, v_{ij} is the flux between cell i and cell j, $\vec{c}_{ij}$ is the centroid of the corresponding face, and $\vec{c}_i$ is the centroid of cell i. For a Cartesian grid, this formula simplifies so that an approximate velocity can be obtained as the sum of the absolute value of the flux divided by the face area over all faces that make up a cell. Using a similar trick as we used in the `wavg` operator to compute η in the well cells, our implementation follows trivially. We first define the averaging operator to compute cell velocity

```
aC = bsxfun(@rdivide, 0.5*abs(D), G.faces.areas(intInx))';
cavg = @(x) aC*x;
```

In doing so, we also rename our old averaging operator `avg` as `favg` to avoid confusion and make it more clear that this operator maps from cell values to face values. Then we can define the needed equations:

```
phiK   = rock.perm.*rock.poro;
gradz  = grad(G.cells.centroids(:,3));
v = @(p, eta)   -(T./(mu0*favg(eta))).*( grad(p) - g*favg(rho(p)).*gradz );
etaEq = @(p, eta)
    eta - ( 1 + Kbc* cavg(v(p,eta)).^2 ./phiK ).^((nmu-1)/2);
presEq= @(p, p0, eta, dt)  ...
    (1/dt)*(pv(p).*rho(p) - pv(p0).*rho(p0)) + div(favg(rho(p)).*v(p, eta));
```

With this approach, the well equation becomes particularly simple, since all we need to do is sample the η value from the correct cell:

```
q_conn = @(p, eta, bhp) ...
    WI .* (rho(p(wc)) ./ (mu0*eta(wc))) .* (p_conn(bhp) - p(wc));
```

7.4 Non-Newtonian Fluid

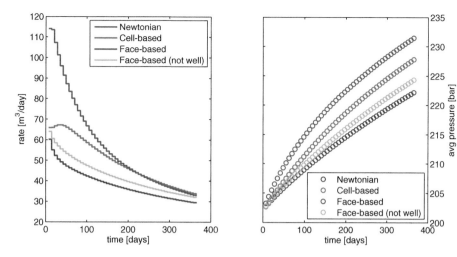

Figure 7.4 Single-phase injection of a highly viscous, shear-thinning fluid computed by four different simulation methods: (i) fluid assumed to be Newtonian, (ii) shear multiplier η computed in cells, (iii) shear multiplier computed at faces, and (iv) shear multiplier computed at faces, but $\eta \equiv 1$ used in well model.

A potential drawback of this second approach is that it may introduce numerical smearing, but this will, on the other hand, most likely increase the robustness of the resulting scheme.

Figure 7.4 compares the predicted flow rates and average reservoir pressure for two different fluid models: one that assumes a standard Newtonian fluid (i.e., $\eta \equiv 1$) and one that models shear thinning. With shear thinning, the higher pressure in the injection well causes a decrease in the viscosity, which leads to significantly higher injection rates than for the Newtonian fluid and hence a higher average reservoir pressure. Perhaps more interesting is the large discrepancy in rates and pressures predicted by the face-based and cell-based simulation algorithms. If we disregard the shear multiplier q_conn in the face-based method, the predicted rate and pressure buildup is smaller than what is predicted by the cell-based method, and closer to the Newtonian fluid case. We take this as evidence that the differences between the cell and the face-based methods to a large extent can be explained by differences in the discretized well models and their ability to capture the formation and propagation of the strong initial transient. To further back this up, we have included results from a simulation with ten times as many time steps in Figure 7.5, which also includes plots of the evolution of $\min(\eta)$ as function of time. Whereas the face-based method predicts a large, immediate drop in viscosity in the near-well region, the viscosity drop predicted by the cell-based method is much smaller during the first 20–30 days. This results in a delay in the peak of the injection rate and a much smaller injected volume.

We leave the discussion here. The parameters used in the example were chosen quite haphazardly to demonstrate a pronounced shear-thinning effect. Which method is the most correct for real computations is a question that goes beyond the current scope, and could

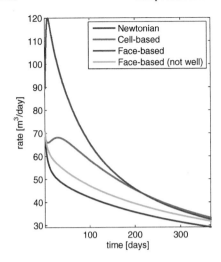 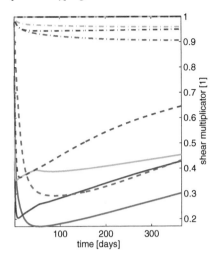

Figure 7.5 Single-phase injection of a highly viscous, shear-thinning fluid; simulation with $\Delta t = 1/520$ year. The right plot shows the evolution of η as a function of time: solid lines show min(η) over all cells, dashed lines min(η) over the perforated cells, and dash-dotted lines average η value.

probably best be answered by verifying against observed data for a real case. Our point here was mainly to demonstrate the capability for rapid implementation of proof-of-concept codes that comes with the use of MRST. However, as the example shows, this lunch is not completely free: you still have to understand features and limitations of the models and discretizations you choose to implement.

COMPUTER EXERCISES

7.4.1 Investigate whether the large differences observed in Figures 7.4 and 7.5 between the cell-based and face-based approaches to the non-Newtonian flow problem is a result of insufficient grid resolution.

7.4.2 The non-Newtonian fluid has a strong transient during the first 30–100 days. Try to implement adaptive time steps that utilize this fact. Can you come up with a strategy that automatically choose good time steps?

7.5 Thermal Effects

As another example of rapid prototyping, we extend the single-phase flow model (7.1) to account for thermal effects. That is, we assume that $\rho(p,T)$ is now a function of pressure and temperature T and extend our model to also include conservation on energy,

$$\frac{\partial}{\partial t}\left[\phi\rho\right] + \nabla\cdot\left[\rho\vec{v}\right] = q, \qquad \vec{v} = -\frac{\mathbf{K}}{\mu}\left[\nabla p - g\rho\nabla z\right], \qquad (7.22a)$$

$$\frac{\partial}{\partial t}\left[\phi\rho E_f(p,t) + (1-\phi)E_r\right] + \nabla\cdot\left[\rho H_f\vec{v}\right] - \nabla\cdot\left[\kappa\nabla T\right] = q_e. \qquad (7.22b)$$

7.5 Thermal Effects

Here, the rock and the fluid are assumed to be in local thermal equilibrium. In the energy equation (7.22b), E_f is energy density per mass of the fluid, $H_f = E_f + p/\rho$ is enthalpy density per mass, E_r is energy per volume of the rock, and κ is the heat conduction coefficient of the rock. Fluid pressure p and temperature T are used as primary variables.

As in the original isothermal simulator, we must first define constitutive relationships that express the new physical quantities in terms of the primary variables. The energy equation includes heating of the solid rock, and we therefore start by defining a quantity that keeps track of the solid volume, which also depends on pressure:

```
sv = @(p) G.cells.volumes - pv(p);
```

For the fluid model, we use

$$\rho(p,T) = \rho_r \big[1 + \beta_T(p - p_r)\big] e^{-\alpha(T-T_r)},$$
$$\mu(p,T) = \mu_0 \big[1 + c_\mu(p - p_r)\big] e^{-c_T(T-T_r)}. \quad (7.23)$$

Here, $\rho_r = 850$ kg/m^3 is the density and $\mu_0 = 5$ cP the viscosity of the fluid at reference conditions with pressure $p_r = 200$ bar and temperature $T_r = 300$ K. The constants are $\beta_T = 10^{-3}$ bar^{-1}, $\alpha = 5 \times 10^{-3}$ K^{-1}, $c_\mu = 2 \times 10^{-3}$ bar^{-1}, and $c_T = 10^{-3}$ K^{-1}. This translates to the following code:

```
[mu0,cmup] = deal( 5*centi*poise, 2e-3/barsa);
[cmut,T_r] = deal( 1e-3, 300);
mu  = @(p,T) mu0*(1+cmup*(p-p_r)).*exp(-cmut*(T-T_r));

[alpha, beta] = deal(5e-3, 1e-3/barsa);
rho_r = 850*kilogram/meter^3;
rho = @(p,T) rho_r .* (1+beta*(p-p_r)) .* exp(-alpha*(T-T_r));
```

We use a simple linear relation for the enthalpy, which is based on the thermodynamical relations that give

$$dH_f = c_p\, dT + \left(\frac{1 - \alpha T_r}{\rho}\right) dp, \qquad \alpha = -\frac{1}{\rho}\frac{\partial \rho}{\partial T}\bigg|_p, \quad (7.24)$$

where $c_p = 4 \times 10^3$ J/kg. The code for enthalpy/energy densities reads:

```
Cp = 4e3;
Hf = @(p,T) Cp*T+(1-T_r*alpha).*(p-p_r)./rho(p,T);
Ef = @(p,T) Hf(p,T) - p./rho(p,T);
Er = @(T)   Cp*T;
```

We defer discussing details of these new relationships and only note that it is important that the thermal potentials E_f and H_f are consistent with the equation of state $\rho(p,T)$ to get a physically meaningful model.

Having defined all constitutive relationships in terms of anonymous functions, we can set up the equation for mass conservation and Darcy's law (with transmissibility renamed to Tp to avoid name clash with temperature):

```
v    = @(p,T) -(Tp./mu(avg(p),avg(T))).*(grad(p) - avg(rho(p,T)).*gdz);
pEq  = @(p,T,p0,T0,dt) ...
       (1/dt)*(pv(p).*rho(p,T) - pv(p0).*rho(p0,T0)) ...
       + div( avg(rho(p,T)).*v(p,T) );
```

In the energy equation (7.22b), the accumulation and the heat-conduction terms are on the same form as the operators appearing in (7.22a) and can hence be discretized in the same way. We use an artificial "rock object" to compute transmissibilities for κ instead of **K**:

```
tmp = struct('perm',4*ones(G.cells.num,1));
hT  = computeTrans(G, tmp);
Th  = 1 ./ accumarray(cf, 1 ./ hT, [nf, 1]);
Th  = Th(intInx);
```

The remaining term in (7.22b), $\nabla \cdot [\rho H_f \vec{v}]$, represents advection of enthalpy and has a differential operator on the same form as the transport equations discussed in Section 4.4.3 and must hence be discretized by an upwind scheme. To this end, we introduce a new discrete operator that will compute the correct upwind value for the enthalpy density,

$$\text{upw}(\boldsymbol{H})[f] = \begin{cases} \boldsymbol{H}[C_1(f)], & \text{if } \boldsymbol{v}[f] > 0, \\ \boldsymbol{H}[C_2(f)], & \text{otherwise.} \end{cases} \tag{7.25}$$

With this, we can set up the energy equation in residual form in the same way as we previously have done for Darcy's law and mass conservation:

```
upw  = @(x,flag)  x(C(:,1)).*double(flag)+x(C(:,2)).*double(~flag);
hEq  = @(p, T, p0, T0, dt) ...
       (1/dt)*(pv(p ).*rho(p, T ).*Ef(p ,T ) + sv(p ).*Er(T ) ...
             - pv(p0).*rho(p0,T0).*Ef(p0,T0) - sv(p0).*Er(T0)) ...
       + div( upw(Hf(p,T),v(p,T)>0).*avg(rho(p,T)).*v(p,T) ) ...
       + div( -Th.*grad(T));
```

With this, we are almost done. As a last technical detail, we must also make sure that the energy transfer in injection and production wells is modeled correctly using appropriate upwind values:

```
qw        = q_conn(p_ad, T_ad, bhp_ad);
eq1       = pEq(p_ad, T_ad, p0, T0, dt);
eq1(wc)   = eq1(wc) - qw;
hq        = Hf(bhp_ad,bhT).*qw;
Hcells    = Hf(p_ad,T_ad);
hq(qw<0)  = Hcells(wc(qw<0)).*qw(qw<0);
eq2       = hEq(p_ad,T_ad, p0, T0,dt);
eq2(wc)   = eq2(wc) - hq;
```

Here, we evaluate the enthalpy using *cell values* for pressure and temperature for production wells (for which qw<0) and pressure and temperatures at the *bottom hole* for injection wells.

7.5 Thermal Effects

What remains are trivial changes to the iteration loop to declare the correct variables as AD structures, evaluate the discrete equations, collect their residuals, and update the state variables. These details can be found in the complete code given in `singlePhaseThermal.m` and have been left out for brevity.

Understanding Thermal Expansion

Except for the modifications discussed in the previous subsection, the setup is the exact same as in Section 7.2. That is, the reservoir is a $200 \times 200 \times 50$ m^3 rectangular box with homogeneous permeability of 30 mD, constant porosity 0.3, and a rock compressibility of 10^{-6} bar^{-1}, realized on a $10 \times 10 \times 10$ Cartesian grid. The reservoir is realized on a $10 \times 10 \times 10$ Cartesian grid. Fluid is drained from a horizontal well perforated in cells with indices $i = 2$, $j = 2, \ldots, 9$, and $k = 5$, and operating at a constant bottom-hole pressure of 100 bar. Initially, the reservoir has constant temperature of 300 K and is in hydrostatic equilibrium with a datum pressure of 200 bar specified in the uppermost cell centroids.

In the same way as in the isothermal case, the open well will create a pressure drawdown that propagates into the reservoir. As more fluid is produced from the reservoir, the pressure will gradually decay towards a steady state with pressure values between 101.2 and 104.7 bar. Figure 7.6 shows that the simulation predicts a faster pressure drawdown, and hence a faster decay in production rates, if thermal effects are taken into account.

The change in temperature of an expanding fluid will not only depend on the initial and final pressure, but also on the type of process in which the temperature is changed:

- In a *free expansion*, the internal energy is preserved and the fluid does no work. That is, the process can be described by the following differential:

$$\frac{dE_f}{dp} \Delta p + \frac{dE_f}{dT} \Delta T = 0. \tag{7.26}$$

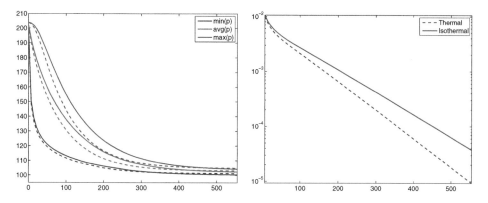

Figure 7.6 To the left, time evolution for pressure for an isothermal simulation (solid lines) and a thermal simulation with $\alpha = 5 \times 10^{-3}$ (dashed lines). To the right, decay in production rate at the surface.

When the fluid is an ideal gas, the temperature is constant, but otherwise the temperature will either increase or decrease during the process depending on the initial temperature and pressure.

- In a reversible process, the fluid is in thermodynamical equilibrium and does positive work while the temperature decreases. The linearized function associated with this *adiabatic expansion* reads

$$dE + \frac{p}{\rho V} dV = dE + p\, d\left(\frac{1}{\rho}\right) = 0. \tag{7.27}$$

- In a Joule–Thomson process, the enthalpy remains constant while the fluid flows from higher to lower pressure under steady-state conditions and without change in kinetic energy. That is,

$$\frac{dH_f}{dp}\Delta p + \frac{dH_f}{dT}\Delta T = 0. \tag{7.28}$$

Our case is a combination of these three processes and their interplay will vary with the initial temperature and pressure as well as with the constants in the fluid model for $\rho(p,T)$. To better understand a specific case, we can use (7.26–7.28) to compute the temperature change that would take place for an observed pressure drawdown if only one of the processes took place. Computing such linearized responses for thermodynamical functions is particularly simple using automatic differentiation. Assuming we know the reference state (p_r, T_r) at which the process starts and the pressure p_e after the process has taken place, we initialize the AD variables and compute the pressure difference:

```
[p,T] = initVariablesADI(p_r,T_r);
dp    = p_e - p_r;
```

Then, we can solve (7.26) or (7.28) for ΔT and use the result to compute the temperature change resulting from a free expansion or a Joule–Thomson expansion:

```
E   = Ef(p,T); dEdp = E.jac{1}; dEdT = E.jac{2};
Tfr = T_r - dEdp*dp/dEdT;

hf  = Hf(p,T); dHdp = hf.jac{1}; dHdT = hf.jac{2};
Tjt = T_r - dHdp*dp/dHdT;
```

The temperature change after a reversible (adiabatic) expansion is not described by a total differential. In this case we have to specify that p should be kept constant. This is done by replacing the AD variable p by an ordinary variable double(p) in the code at the specific places where p appears in front of a differential; see (7.27).

```
E    = Ef(p,T) + double(p)./rho(p,T);
dEdp = hf.jac{1};
dEdT = hf.jac{2};
Tab  = T_r - dEdp*dp/dEdT;
```

7.5 Thermal Effects

The same kind of manipulation can be used to study alternative linearizations of systems of nonlinear equations and the influence of neglecting some of the derivatives when forming Jacobians.

To illustrate how the interplay between the three processes can change significantly and lead to quite different temperature behavior, we compare the predicted evolution of the temperature field for $\alpha = 5 \times 10^{-n}$, $n = 3, 4$, as shown in Figures 7.7 and 7.8. The change in behavior between the two figures is associated with the change in sign of $\partial E/\partial p$,

$$dE = \left(c_p - \frac{\alpha T}{\rho}\right) dT + \left(\frac{\beta_T p - \alpha T}{\rho}\right) dp, \qquad \beta_T = \frac{1}{\rho}\frac{\partial \rho}{\partial p}\bigg|_T. \qquad (7.29)$$

In the isothermal case and for $\alpha = 5 \times 10^{-4}$, we have that $\alpha T < \beta_T p$ so that $\partial E/\partial p > 0$. The expansion and flow of fluid will cause an instant heating near the well-bore, which is what we see in the initial temperature increase for the maximum value in Figure 7.7. The Joule–Thomson coefficient $(\alpha T - 1)/(c_p \rho)$ is also negative, which means that the fluid gets heated if it flows from high pressure to low pressure in a steady-state flow. This is seen by observing the temperature in the well perforations. The fast pressure drop in these cells causes an almost instant cooling effect, but soon after we see a transition in which most of the cells containing a well perforation start having the highest temperature in the reservoir because of heating from the moving fluids. For $\alpha = 5 \times 10^{-3}$, we have that $\alpha T > \beta_T p$ so that $\partial E/\partial p < 0$ and likewise the Joule–Thomson coefficient is positive. The moving fluids will induce a cooling effect and hence the minimum temperature is observed at the well for a longer time. The weak kink in the minimum temperature curve is the result of the point of minimum temperature moving from being at the bottom front side to the far back of the reservoir. The cell with lowest temperature is where the fluid has done most work, neglecting heat conduction. In the beginning, this is the cell near the well because the pressure drop is largest there. Later, it will be the cell furthest from the well, since this is where the fluid can expand most. The discussion in this subsection is only meant to illustrate physical effect and does not necessarily represent realistic wells.

Computational Performance

If you are observant, you may have realized that the code presented in this chapter contains a number of redundant function evaluations that may potentially add significantly to the overall computational cost: in each nonlinear iteration we keep reevaluating quantities that depend on p0 and T0 even though these stay constant for each time step. We can easily avoid this by moving the definition of the anonymous functions evaluating the residual equations inside the outer time loop. The main contribution to potential computational overhead, however, comes from repeated evaluations of fluid viscosity and density. Because each residual equation is defined as an anonymous function, v(p,T) appears three times for each residual evaluation, once in pEq and twice in hEq. This, in turn, translates to three calls to mu(avg(p),avg(T)) and *seven* calls to rho(p,T), and so on. In practice, the number of actual function evaluations is smaller, since the MATLAB interpreter most likely has some kind of built-in intelligence to spot and reduce redundant function evaluations.

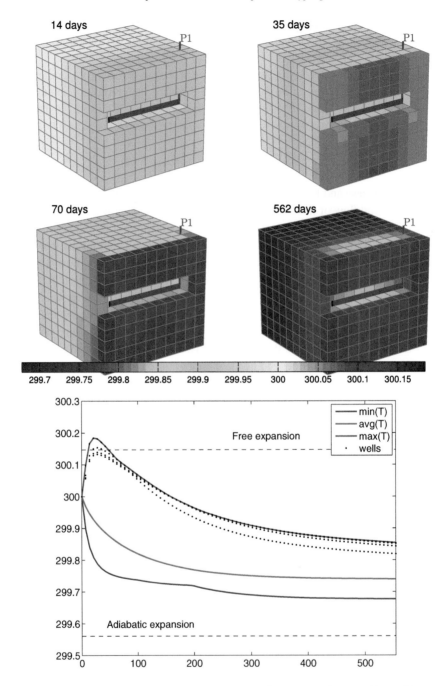

Figure 7.7 Time evolution of temperature for a compressible, single-phase problem with $\alpha = 5 \cdot 10^{-4}$. The upper plots show four snapshots of the temperature field. The lower plot shows minimum, average, maximum, and well-perforation values.

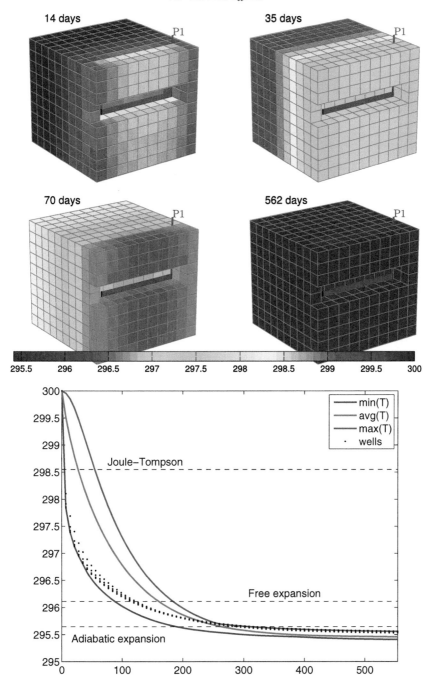

Figure 7.8 Time evolution of temperature for a compressible, single-phase problem with $\alpha = 5 \times 10^{-3}$. The upper plots show four snapshots of the temperature field. The lower plot shows minimum, average, maximum, and well-perforation values.

Nonetheless, to cure this problem, we can move the computations of residuals inside a function so that the constitutive relationships can be computed one by one and stored in temporary variables. The disadvantage is that we increase the complexity of the code and move one step away from the mathematical formulas describing the method. This type of optimization should therefore only be introduced after the code has been profiled and redundant function evaluations have proved to have a significant computational cost.

COMPUTER EXERCISES

7.5.1 Perform a more systematic investigation of how changes in α affect the temperature and pressure behavior. To this end, you should change α systematically, e.g., from 0 to 10^{-2}. What is the effect of changing β, the parameters c_μ and c_T for the viscosity, or c_p in the definition of enthalpy?

7.5.2 Use the MATLAB profiling tool to investigate to what extent the use of nested anonymous functions causes redundant function evaluations or introduces other types of computational overhead. Hint: to profile the CPU usage, you can use the following call sequence

```
profile on, singlePhaseThermal; profile off; profile report
```

Try to modify the code as suggested above to reduce the CPU time. How low can you get the ratio between the cost of constructing the linearized system and the cost of solving it?

Part III
Multiphase Flow

8
Mathematical Models for Multiphase Flow

Up to now, we have only considered flow of a single fluid phase. For most applications of reservoir simulation, however, one is interested in modeling how one fluid phase displaces one or more other fluid phases. In hydrocarbon recovery, typical examples are water or gas flooding, in which injected water or gas displaces the resident hydrocarbon phase(s). Likewise, in geological carbon sequestration, the injected CO_2 forms a supercritical fluid phase that displaces the resident brine. In both cases, more than one phase will flow simultaneously throughout the porous medium when viewed from the scale of a representative elementary volume, even if the fluids are immiscible and do not mix on the microscale.

To model such flows, we introduce three new physical properties of multiphase models (saturation, relative permeability, and capillary pressure) and discuss how one can use these to extend Darcy's law to multiphase flow and combine it with conservation of mass for each fluid phase to develop models describing multiphase displacements. The resulting system of partial differential equations is parabolic in the general case, but has a mixed elliptic-hyperbolic mathematical character. Fluid pressures tend to behave as following a near-elliptic equation, whereas the transport of fluid phases has a strong hyperbolic character. It is therefore common to use a so-called fractional-flow formulation to write the flow equations as a coupled system consisting of a pressure equation describing the evolution of one of the fluid pressures and (a system of) saturation equation(s) that describes the transport of fluid phases. The mixed elliptic-hyperbolic nature is particularly evident in the case of immiscible, incompressible flow. In this case, the pressure equation simplifies to a Poisson equation of the same form we have studied in previous chapters.

This chapter describes the new physical effects that appear in multiphase flow and discusses general multiphase models in some detail. We also derive the fractional-flow formulation in the special case of immiscible flow and analyze the mathematical character of the system in certain limiting cases. Chapter 9 introduces various methods for solving hyperbolic transport equation and reviews some of the supporting theory. Then, in Chapter 10, we focus entirely on the incompressible case and show how we can easily reuse the elliptic solvers developed in the previous chapters and combine them with a set of simple first-order transport solvers that are implemented in the `incomp` module of MRST. To simplify the discussion, this and the next two chapters will mainly focus on two-phase, immiscible systems, but the most crucial equations will be stated and developed for the general multiphase case. In Chapters 11 and 12, we return to the general case and discuss the compressible models and numerical methods that are used for three-phase flow in most contemporary commercial simulators.

8.1 New Physical Properties and Phenomena

As we have seen previously, a Darcy-type continuum description of a reservoir fluid system means that any physical quantity defined at a point $\bar{x}$ represents an average over a representative elementary volume (REV). Let us consider a system with two or more fluid phases that are *immiscible* so that no mass transfer takes place among the phases. This means that the fluid phases will not mix and form a solution on the microscale but rather stay as separate volumes or layers separated by a curved meniscus as illustrated in Figure 8.1. Nevertheless, when considering the flow averaged over a REV, the fluid phases will generally not be separated by a sharp interface, and two or more phases may occupy the same point in the continuum description. In this section, we introduce the new fundamental concepts necessary to understand multiphase flow and formulate continuum models that describe the simultaneous flow of two or more fluid phases taking place at the same point in a reservoir. Unless stated otherwise, the two fluids are assumed to be oil and water when we discuss two-phase systems.

8.1.1 Saturation

Saturation S_α is defined as the fraction of the pore volume occupied by phase α. In the single-phase models discussed in previous chapters, we assumed that the void space between the solid particles that make up the porous medium was completely filled by fluid. Similarly, for multiphase models we assume that the void space is completely filled with one or more fluid phases, so that

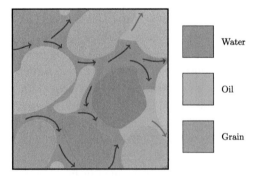

Figure 8.1 Averaging of a multiphase flow over a representative elementary volume (REV) to obtain a Darcy-scale continuum model.

8.1 New Physical Properties and Phenomena

$$\sum_\alpha S_\alpha = 1. \tag{8.1}$$

In reservoir simulation, it is most common to consider three phases: an aqueous phase (w), an oleic phase (o), and a gaseous (g) phase. Each saturation can vary from 0, which means that the phase is not present at all at this point in space, to 1, which means that the phase completely fills the local pore volume. In most practical cases, however, the range of variability is smaller.

Consider a rock originally deposited in an aqueous environment. During deposition, the pores between rock particles are completely filled with water. Later, as hydrocarbons start to migrate into what is to become our reservoir, the resident water will be displaced and the saturation gradually reduced to some small value, typically 5–40%, at which the water can no longer flow and exists as small drops trapped between mineral particles or encapsulated by the invading hydrocarbon phases. The saturation at which water goes from being mobile (funicular state) to being immobile (pendular state) is called the *irreducible water saturation* and is usually denoted S_{wir} or S_{wr}. The irreducible water saturation is determined by the topology of the pore space and the water's affinity to wet the mineral particles relative to that of the invading hydrocarbons; this affinity is determined by the chemical composition of the fluids and the mineral particles.

Petroleum literature also talks of the *connate water saturation*, usually denoted S_{wc}, which is the water saturation that exists upon discovery of the reservoir. The quantities S_{wir} and S_{wc} may or may not coincide, but should not be confused. Sometimes, one also sees the notation S_{wi}, which may refer to any of the two. If water is later injected to displace the oil, it is generally not possible to flush out all the oil and parts of the pore space will be occupied by isolated oil droplets, as illustrated in Figure 8.1. The corresponding *residual oil saturation* is denoted as S_{or}. In most systems, the water has a stronger affinity for the rock, which means that S_{or} is usually higher than S_{wr} and S_{wc}: typically they are in the range 10–50%.

In many models, each phase may also contain one or more *components*. These may be unique hydrocarbon species like methane, ethane, propane, etc., or other chemical species like polymers, salts, surfactants, tracers, etc. Since the number of hydrocarbon components can be quite large, it is common to group components into *pseudo-components*. Because of the varying and extreme conditions in a reservoir, the composition of the different phases can change throughout a simulation and may sometimes be difficult to determine uniquely. We therefore need to describe this composition. There are several ways to do this. Herein, we use the mass fraction of component ℓ in phase α, denoted by c_α^ℓ and defined as

$$c_\alpha^\ell = \frac{\rho_\alpha^\ell}{\rho_\alpha} \tag{8.2}$$

where ρ_α denotes the bulk density of phase α and ρ_α^ℓ the effective density of component ℓ in phase α. In each phase, the mass fractions should add up to unity, so that for M different components in a system consisting of an aqueous, a gaseous, and an oleic phase, we have:

$$\sum_{\ell=1}^{M} c_w^\ell = \sum_{\ell=1}^{M} c_g^\ell = \sum_{\ell=1}^{M} c_o^\ell = 1. \tag{8.3}$$

We will return to models having three phases and more than one component per phase in Chapter 11. For now, however, we assume that our system consists of two immiscible phases.

8.1.2 Wettability

At the microscale, which is significantly larger than the molecular scale, immiscible fluid phases are separated by well-defined, infinitely thin interfaces. Because cohesion forces between molecules are different on opposite sides, each interface has an associated *surface tension* (or surface energy), which measures the forces the interface must overcome to change its shape. In the absence of external forces, minimization of surface energy will cause the interface of a droplet of one phase contained within another phase to assume a spherical shape. The interface tension will keep the fluids apart, regardless of the size of the droplet.

The microscale flow of our oil–water system is strongly affected by how the phases attach to the interface of the solid rock. The ability of a liquid phase to maintain contact with a solid surface is called *wettability* and is determined by intermolecular interactions when the liquid and solid are brought together. Adhesive forces between different molecules in the liquid phase and the solid rock will cause liquid droplets to spread across the mineral surface. Likewise, cohesive forces between similar molecules within the liquid phases will cause the droplets to avoid contact with the surface and ball up. When two fluid phases are present in the same pore space, one phase will be more attracted to the mineral particles than the other phase. We refer to the preferential phase as the *wetting phase*, while the other is called the *non-wetting phase*. The balance of the adhesive and cohesive forces determines the *contact angle* θ shown in Figure 8.2, which is a measure of the wettability of a system that can be related to the interface energies by Young's equation:

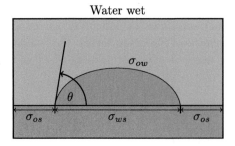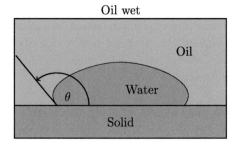

Figure 8.2 Contact angle θ and surface tension σ for two different oil–water systems. In water-wet systems, $0 \le \theta < 90°$, whereas $90° < 180°$ in oil-wet systems.

$$\sigma_{ow} \cos \theta = \sigma_{os} - \sigma_{ws}. \tag{8.4}$$

Here, σ_{ow} is the interface energy of the oil–water interface and σ_{os} and σ_{ws} are the energies of the oil–solid and water–solid interfaces, respectively. Hydrophilic or water-wet porous media, in which the water shows a greater affinity than oil to stick to the rock surface, are more widespread in nature than hydrophobic or oil-wet media. This explains why S_{or} usually is larger than S_{wr}. In a perfectly water-wet system, $\theta = 0$ so that water spreads evenly over the whole surface of the mineral grains. Likewise, in a perfectly oil-wet system, $\theta = 180°$, so that water forms spherical droplets at the solid surface.

8.1.3 Capillary Pressure

Because of the surface tension, the equilibrium pressure in two phases separated by a curved interface will generally be different. The difference in phase pressures is called the *capillary pressure*,

$$p_c = p_n - p_w, \tag{8.5}$$

and is always positive because the pressure in the non-wetting fluid is higher than the pressure in the wetting fluid. For a water-wet reservoir, the capillary pressure is therefore defined as $p_{cow} = p_o - p_w$, whereas one usually defines $p_{cog} = p_g - p_o$ in an oil–gas system where oil is the wetting phase.

The action of capillary pressures can cause liquids to move in narrow spaces, devoid of or in opposition to other external forces such as gravity. To illustrate this, we consider a thin tube immersed in a wetting and a non-wetting fluid, as shown in Figure 8.3. For the wetting fluid, adhesive forces between the solid tube and the liquid will form a concave meniscus and pull the liquid upward against the gravity force. It is exactly the same effect that causes water to be drawn up into a piece of cloth or paper dipped into water. In the non-wetting case, the intermolecular cohesion forces within the liquid exceed the adhesion forces between the liquid and the solid so that a convex meniscus is formed and drawn downwards relative to the liquid level outside of the tube. At equilibrium inside the capillary

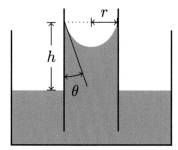 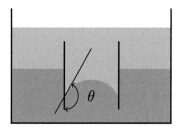

Figure 8.3 Capillary tubes for a wetting liquid and air (left) and a wetting and a non-wetting liquid (right). (The width of the capillary tube is exaggerated.)

tube, the upward and downward forces must balance each other. The force acting upward equals

$$2\pi r(\sigma_{as} - \sigma_{ls}) = 2\pi r \sigma \cos\theta,$$

where subscripts a and l refer to air and liquid, respectively. The capillary pressure is defined as force per unit area, or in other words,

$$p_c = \frac{2\pi r \sigma \cos\theta}{\pi r^2} = \frac{2\sigma \cos\theta}{r} \qquad (8.6)$$

The force acting downward can be deducted from Archimedes' principle as $\pi r^2 gh(\rho_l - \rho_a)$. By equating this with the action of the capillary pressure, we obtain

$$p_c = \frac{\pi r^2 gh(\rho_l - \rho_a)}{\pi r^2} = \Delta\rho gh. \qquad (8.7)$$

Void space inside a reservoir contains a large number of narrow pore throats that can be thought of as a bundle of nonconnecting capillary tubes of different radius. As we can see from the formulas developed thus far, the capillary pressure increases with decreasing tube radius for a fixed interface-energy difference between two immiscible fluids. Because the pore size is usually so small, capillary pressure will play a major role in establishing the fluid distribution inside the reservoir. To see this, consider a hydrocarbon phase migrating upward by buoyancy forces into a porous rock filled with water. If the hydrocarbon phase is to enter void space in the rock, its buoyancy force must exceed a certain minimum capillary pressure. The capillary pressure that is required to force the first droplet of oil into the rock is called the *entry pressure*. If we consider the rock as a complex assortment of capillary tubes, the first droplet will enter the widest tube, which according to (8.6) has the lowest capillary pressure. As the pressure difference between the buoyant oil and the resident water increases, oil will be able to enter increasingly smaller pore throats and hence reduce the water saturation. This means that there will be a relation between saturation and capillary pressure,

$$p_{cnw} = p_n - p_w = P_c(S_w), \qquad (8.8)$$

as illustrated in Figure 8.4. The slope of the curve is determined by the variability of the pore sizes. If all pores are of similar size, they will all be invaded quickly once we exceed the entry pressure and the curve will be relatively flat so that saturation decays rapidly with increasing capillary pressure. If the pores vary a lot in size, the decrease in saturation with increasing capillary pressure will be more gradual. As for the vertical distribution of fluids, we see that once the fluids have reached a hydrostatic equilibrium, the difference in densities between water and oil dictates the difference in phase pressures and hence oil saturation increases in the upward direction, which is also illustrated in the figure.

8.1 New Physical Properties and Phenomena   237

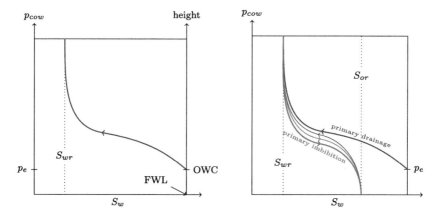

Figure 8.4 The left plot shows a capillary pressure curve giving the relation between capillary pressure p_{cow} and water saturation S_w. In addition, we have included how the capillary pressure and water saturation relate to the height above the free water level (FWL) for a system in hydrostatic equilibrium. Notice that the entry pressure p_e occurs at the oil–water contact (OWC), which is found above the FWL for systems with nonzero entry pressures. The right plot shows hysteretic behavior for repeated drainage and imbibition displacements.

Hysteresis, Drainage, and Imbibition

The argument earlier in this section was for the case of an invading non-wetting fluid displacing a wetting fluid. This type of displacement is called *drainage* to signify that the saturation of the wetting phase is decreasing in this type of displacement. The opposite case, called *imbibition*, occurs when a wetting fluid displaces a non-wetting fluid. As an example, let us assume that we inject water to flush out the oil in a pristine reservoir, in which the amount of connate water equals the irreducible water saturation. During this displacement, the water saturation will gradually increase as more water is injected. Hence, the oil saturation decreases until we reach the residual oil saturation, at which there is only *immobile oil* left. The displacement will generally not follow the same capillary curve as the primary drainage curve, as shown in the right plot in Figure 8.4. Likewise, if another drainage displacement takes place, starting from S_{or} or a larger oil saturation, this process will generally not follow the imbibition curve. The result is an example of what is called *hysteresis*, in which the behavior of a system depends both on the current state and its previous history. The hysteretic behavior can be explained by pore-scale trapping of oil droplets, by variations in the wetting angle between advancing and receding fluid at the solid interface, and by the fact that whereas the drainage process is controlled by the size of the widest non-invaded pore throat, the imbibition process is controlled by the size of the narrowest non-invaded pore.

As we can see from Figure 8.4, a relatively large fraction of the oil will be left behind in an immobile state after waterflooding. Several methods for enhanced oil recovery have been developed to mobilize this immobile oil, e.g., by injecting another fluid (e.g., CO_2 or gas) that mixes with immobile oil droplets so that a larger fraction of the oil can be washed out along with the invading fluid. In chemical and microbial methods, one adds chemical substances or small microorganisms to the injected fluids that alter the wetting properties inside the pores. Simulating these processes, however, will require more advanced models than those discussed herein; you can find more details in a textbook like [176].

Leverett J-Function

To use the relation between capillary pressure and saturation in practical modeling, it is convenient to express the capillary–saturation relationship $P_c(S_w)$ as an explicit or tabulated function. In the petroleum industry, one usually uses flooding experiments on core samples from the reservoir to develop empirical models based on observations of the relationship between average p_c and S_w values inside the core models. Each core sample will naturally generate a different capillary curve because of differences in pore-size distribution, porosity, and permeability. To normalize the measured data, it is common to use a so-called *Leverett J-function* [184], which takes the form

$$J(S_w) = \frac{P_c}{\sigma \cos\theta} \sqrt{\frac{K}{\phi}}. \tag{8.9}$$

Here, the surface tension σ and the contact angle θ are measured in the laboratory and are specific to a particular rock and fluid system. The scaling factor $\sqrt{K/\phi}$ is proportional to the characteristic, effective pore-throat radius. The function J can now be obtained as a (tabulated) function of S_w by fitting rescaled observed data (p_c and s_w) to a strictly monotone J-shaped function. Then, the resulting function is used to extrapolate capillary pressure data measured for a given rock to rocks that are similar, but have different permeability, porosity, and wetting properties. One cannot expect to find a J function that is generally applicable because the parameters that affect capillary pressure vary largely with rock type. Nevertheless, experience has shown that J-curves correlate well for a given rock type and in reservoir models it is therefore common to derive a J-curve for each specific rock type (facies)[1] that is represented in the underlying geological model.

To set up a multiphase flow simulation, we need the initial saturation distribution inside the reservoir. If we know the location of the OWC, the saturation higher up in the formation can be determined by combining (8.7) and (8.9)

$$S_w = J^{-1}\left(\frac{\Delta \rho g h}{\sigma \cos\theta} \sqrt{\frac{K}{\phi}}\right).$$

Other Relationships

In other application areas than petroleum recovery, it is common to use models that express the capillary pressure directly as an explicit function of the normalized (or effective) water saturation,

---

[1] By (litho)facies, we mean a part of the rock that is distinguishable by its texture, mineralogy, grain size, and the depositional environment that produced it.

$$\hat{S}_w = \frac{S_w - S_w^{\min}}{S_w^{\max} - S_w^{\min}}. \tag{8.10}$$

Here, $S_w^{\max}$ and $S_w^{\min}$ are the maximum and minimum values the saturation can attain during the displacement. For the primary drainage shown to the left in Figure 8.4, it is natural to set $S_w^{\max} = 1$ and $S_w^{\min} = S_{wr}$, whereas $S_w^{\max} = 1 - S_{or}$ and $S_w^{\min} = S_{wr}$ for all subsequent displacements. The following model was proposed by Brooks and Corey [54] to model the relationship between capillary pressure and water saturation in partially saturated media (i.e., in the vadoze zone where the two-phase flow consists of water and air; see Section 8.3.5):

$$\hat{S}_w = \begin{cases} (p_c/p_e)^{-n_b}, & \text{if } p_c > p_e, \\ 1, & p_c \le p_e. \end{cases} \tag{8.11}$$

Here, p_e is the entry pressure of air and $n_b \in [0.2, 5]$ is a parameter related to the pore-size distribution. Another classical model is the one proposed by van Genuchten [298]:

$$\hat{S}_w = \left(1 + (\beta_g p_c)^{n_g}\right)^{-m_g}, \tag{8.12}$$

where β_g is a scaling parameter related to the average size of pores and the exponents n_g and m_g are related to the pore-size distribution.

8.1.4 Relative Permeability

When discussing incompressible, single-phase flow, we saw that the only petrophysical parameter affecting how fast a fluid flows through a porous medium is the absolution permeability **K** that measures the capacity of the rock to transmit fluids, or alternatively the resistance the rock offers to flow. As described in Chapters 2 and 4, absolute permeability is an intrinsic property of the rock and does not depend on the type of fluid that flows through the rock. In reality, this is not true, mainly because of microscale interactions between rock and fluid may cause particles to move, pore spaces to be plugged, clays to swell when brought in contact with water, etc. Likewise, liquids and gases may not necessarily experience the same permeability, because gas does not adhere to the mineral surfaces in the same way as liquids do. This means that whereas the flow of liquids is subject to no-slip boundary conditions, gases may experience slippage that gives a pressure-dependent apparent permeability, which at low flow rates is higher than the permeability experienced by liquids. This is called the *Klinkenberg effect* and plays a substantial role for gas flows in low-permeable, unconventional reservoirs such as coal seams, tight sands, and shale formations. Herein, we will not consider reservoirs where these effects are pronounced and henceforth, we assume – as for the incompressible, single-phase flow models in Chapter 4 – that the absolute permeability **K** is an intrinsic quantity.

When more than one phase is present in the pore space, each phase α will experience an effective permeability $\mathbf{K}_\alpha^e$ that is less than the absolute permeability **K**. Looking at the conceptual drawing in Figure 8.1, it is easy to see why this is so. The presence of another phase will effectively present additional "obstacles," whose interfacial tension offers resistance to flow. Because interfacial tension exists between all immiscible phases, the sum of all the effective phase permeabilities will generally be less than one, i.e.,

$$\sum_\alpha \mathbf{K}_\alpha^e < \mathbf{K}.$$

To model this reduced permeability, we introduce a property called *relative permeability* [219], which for an isotropic medium is defined as

$$k_{r\alpha} = K_\alpha^e / K. \tag{8.13}$$

Because the effective permeability is always less than or equal to the absolute permeability, $k_{r\alpha}$ will take values in the interval between 0 and 1. For anisotropic media, the relationship between the effective and absolution permeability may in principle be different for each component of the tensors. However, it is still common to define the relative permeability as a scalar quantity postulated to be in the form

$$\mathbf{K}_\alpha^e = k_{r\alpha} \mathbf{K}. \tag{8.14}$$

Relative permeabilities will generally be functions of saturation, which means that for a two-phase system we can write

$$k_{rn} = k_{rn}(S_n) \quad \text{and} \quad k_{rw} = k_{rw}(S_w).$$

It is important to note that the relative permeabilities generally are nonlinear functions of the saturations, so that the sum of these functions at a specific location (with a specific composition) is not necessarily equal to one. As for the relationship between saturation and capillary pressure, the relative permeabilities are strongly dependent on the lithofacies. It is therefore common practice to associate a unique pair of curves with each rock type represented in the geological model. Relative permeabilities may also depend on pore-size distribution, fluid viscosity, and temperature, but these factors are usually ignored in models of conventional reservoirs.

In simplified models, it is common to assume that $k_{r\alpha}$ are monotone functions that assume unique values in $[0, 1]$ for all values $S_\alpha \in [0, 1]$, so that $k_{r\alpha} = 1$ corresponds to the case with fluid α occupying the entire pore space, and $k_{r\alpha} = 0$ when $S_\alpha = 0$; see the left plot in Figure 8.5. In practice, $k_{r\alpha} = 0$ occurs when the fluid phase becomes immobilized for $S_\alpha \leq S_{\alpha r}$, giving relative permeability curves as shown in the right plot in Figure 8.5. The preferential wettability can be deducted from the point where two curves cross. Here, the curves cross for $S_w > 0$, which indicates that the system is water-wet.

Going back to our previous discussion of hysteresis, we generally expect relative permeability to be different during drainage and imbibition. The drainage curve in Figure 8.6 corresponds to the primary drainage process discussed in connection with Figure 8.4, in which a non-wetting hydrocarbon phase migrates into a water-wetting porous medium completely saturated by water. After water has been drained to its irreducible saturation,

8.1 New Physical Properties and Phenomena 241

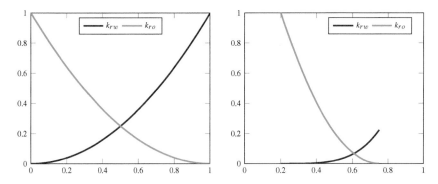

Figure 8.5 Illustration of relative permeabilities for a two-phase system. The left plot shows an idealized system with no residual saturations, while the right plot shows more realistic curves from a water-wet system.

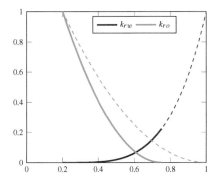

Figure 8.6 Illustration of relative permeability hysteresis, with drainage curves shown as thin, dashed lines and imbibition curves shown as thick lines.

water is reinjected to flush out the oil. In this particular illustration, the relative permeability of water k_{rw} exhibits no hysteretic behavior, whereas the imbibition and drainage curves deviate significantly for k_{ro}.

Two-Phase Relative Permeability

It is quite common to use simple analytic relationships to represent relative permeabilities. These are usually stated using the normalized or effective saturation $\hat{S}_w$ from (8.10). The simplest model possible is a pure power-law relationship, which is sometimes called a Corey model,

$$k_{rw} = \left(\hat{S}_w\right)^{n_w} k_w^0, \qquad k_{ro} = \left(1 - \hat{S}_w\right)^{n_o} k_o^0, \qquad (8.15)$$

where the exponents $n_w, n_o \geq 1$ and the constants k_α^0 used for end-point scaling, should be fit to measured data. Another popular choice is the use of the Brooks–Corey functions

$$k_{rw} = \left(\hat{S}_w\right)^{n_1+n_2 n_3}, \quad k_{ro} = \left(1 - \hat{S}_w\right)^{n_1}\left[1 - \left(\hat{S}_w\right)^{n_2}\right]^{n_3}, \qquad (8.16)$$

where $n_1 = 2$, $n_2 = 1 + 2/n_b$, and $n_4 = 1$ gives the Brooks–Corey–Burdine model and $n_1 = \eta$, $n_2 = 1 + 1/n_b$, and $n_3 = 2$ gives the Brooks–Corey–Mualem model. It is also possible to derive models that correspond with the van Genuchten capillary functions in (8.12). In the case that $m_g = 1 - 1/n_g$, this reads

$$\begin{aligned}k_{rw} &= \hat{S}_w^\kappa \big[1 - (1 - \hat{S}_w^{1/m_g})^{m_g}\big]^2, \\ k_{ro} &= (1 - \hat{S}_w)^\kappa \big[1 - \hat{S}_w^{1/m_g}\big]^{2m_g},\end{aligned} \qquad (8.17)$$

where the connectivity factor κ is a fitting parameter. This is called the van Genuchten–Mualem model, whereas the closed-form expressions obtained for $m_g = 1 - 2/n_g$,

$$\begin{aligned}k_{rw} &= \hat{S}_w^2 \big[1 - (1 - \hat{S}_w^{1/m_g})^{m_g}\big], \\ k_{ro} &= (1 - \hat{S}_w)^2 \big[1 - \hat{S}_w^{1/m_g}\big]^{m_g},\end{aligned} \qquad (8.18)$$

is called the van Genuchten–Burdine model.

Three-Phase Relative Permeability

Measuring relative permeability has traditionally been costly and complex. Recently, laboratory techniques have made great progress by using computer tomography and nuclear magnetic resonance (NMR) to scan the test cores where the actual phases are being displaced. Although standard experimental procedures exist for measuring relative permeabilities in two-phase systems, there is nevertheless still significant uncertainty concerning the relevance of the experimental values found, and it is difficult to come up with reliable data to be used in a simulator. This is mainly due to boundary effects. Particularly for three-phase systems, there is a lot of uncertainty in reported data (see the review by [29]) and it is not unfair to say that no reliable experimental technique exists. Thus, three-phase relative permeabilities are usually modeled using two-phase measurements, for which several theoretical models have been proposed. Most of them are based on an idea first proposed by Stone [282], and seek to combine sets of two-phase relative permeabilities to give three-phase relationships.

In Stone's original model, which was developed for water-wet porous media and is commonly referred to as the Stone I model, the idea was that the relative permeability of water and gas would be identical to the models measured in water/oil displacements and oil/gas displacements, respectively. In the following, I use the convention that k^{wo} denotes a relationship derived from water/oil displacements and k^{og} are relationships derived from oil/gas displacements. Hence, we have that

$$k_{rw}(S_w) = k_{rw}^{wo}(S_w) \qquad k_{rg}(S_g) = k_{rg}^{go}(S_g). \qquad (8.19)$$

The three-phase relative permeability of oil, on the other hand, depends nonlinearly on both water and gas saturations,

$$k_{ro}(S_w, S_g) = \frac{\hat{S}_o}{k_{ro}^{wo}(S_{wc})} \frac{k_{ro}^{wo}(S_w)}{1 - \hat{S}_w} \frac{k_{ro}^{og}(S_g)}{1 - \hat{S}_g}. \tag{8.20}$$

Here, we have introduced the following scaled quantities

$$\hat{S}_o = \frac{S_o - S_{om}}{1 - S_{wc} - S_{om}}, \quad \text{for } S_o \geq S_{om}$$

$$\hat{S}_w = \frac{S_w - S_{wc}}{1 - S_{wc} - S_{om}}, \quad \text{for } S_w \geq S_{wc}$$

$$\hat{S}_g = \frac{S_g}{1 - S_{wc} - S_{om}}.$$

Here, S_{wc} is the connate water saturation and S_{om} is the minimum oil saturation. To close the model, we can for instance use the expression for S_{om} given by Fayers and Matthews [109]

$$S_{om} = \alpha S_{orw} + (1 - \alpha) S_{org}, \quad \alpha = 1 - S_g/(1 - S_{wc} - S_{org}), \tag{8.21}$$

where S_{orw} and S_{org} denote the residual oil saturations for waterflooding and gasflooding, respectively.

Stone later proposed an alternative expression for the relative permeability of oil called the Stone II model [283]. A number of modifications and alternative formulations have been developed over the years; see, e.g., [29] for a review. MRST implements a different model that uses a linear relationship to interpolate the two-phase curves. This is the same model as the default choice in the commercial ECLIPSE simulator [271], and will be discussed in more detail in Section 11.3.

8.2 Flow Equations for Multiphase Flow

Having introduced the new physical parameters and key phenomena that characterize multiphase flow of immiscible fluids, we are in a position to develop mathematical models describing multiphase flow. To this end, we follow more or less the same steps as we did for single-phase flow in Section 4.2. Stating the generic equations describing multiphase flow is straightforward and will result in a system of partial differential equations that contains more unknowns than equations. The system must hence be extended with constitutive equations to relate the various physical quantities, as well as boundary conditions and source terms (see Sections 4.3.1 and 4.3.2) that describe external forces driving the flow. However, as indicated in the introduction of the chapter, the generic flow model has a complex mathematical character and contains delicate balances of various physical forces, whose characteristics and individual strengths vary a lot across different flow regimes. To reveal the mathematical character, but also to make models that are more computationally

8.2.1 Single-Component Phases

To develop a generic system of flow equations for multiphase flow, we use the fundamental principle of mass conservation. For N immiscible fluid phases that each consists of a single component, we write one conservation equation per phase,

$$\frac{\partial}{\partial t}(\phi \rho_\alpha S_\alpha) + \nabla \cdot (\rho_\alpha \vec{v}_\alpha) = \rho_\alpha q_\alpha. \tag{8.22}$$

Here, each phase can contain multiple chemical species, but these can be considered as a single component since there is no transfer between the phases so that their composition remains constant in time.

As for the flow of a single fluid, the primary constitutive relationship used to form a closed model is Darcy's law (4.2), which can be extended to multiphase flow by using the concept of relative permeabilities discussed above

$$\vec{v}_\alpha = -\frac{\mathbf{K}k_{r\alpha}}{\mu_\alpha}(\nabla p_\alpha - g\rho_\alpha \nabla z). \tag{8.23}$$

This extension of Darcy's law to multiphase flow is often attributed to Muskat and Wyckoff [219] and has only been rigorously derived from first principles in the case of two fluid phases. Equation (8.23) must therefore generally be considered as being phenomenological. Darcy's law is sometimes stated with the opposite sign for the gravity term, but herein we use the convention that g is a positive constant. Likewise, it is common to introduce phase mobilities $\lambda_\alpha = \mathbf{K} k_{r\alpha}/\mu_\alpha$ or relative phase mobilities $\lambda_\alpha = \lambda_\alpha \mathbf{K}$ to simplify the notation slightly. From the discussion in the previous section, we also have an additional closure relationship stating that the saturations sum to zero (8.1), as well as relations of the form (8.8) that relate the pressures of the different phases by specifying the capillary pressures as functions of the fluid saturations.

Most commercial reservoir simulators compute approximate solutions by inserting the multiphase Darcy equations (8.23) into (8.22), introducing functional relationships for capillary pressure and how ρ_α and ϕ depend on phase pressure, and then discretizing the resulting conservation equations more or less directly. If we use the discrete derivative operators from Section 4.4.2, combined with a backward discretization of the temporal derivatives, the resulting system of fully implicit, discrete equations for phase α reads

$$\frac{(\boldsymbol{\phi} S_\alpha \boldsymbol{\rho}_\alpha)^{n+1} - (\boldsymbol{\phi} S_\alpha \boldsymbol{\rho}_\alpha)^n}{\Delta t^n} + \operatorname{div}(\rho v)_\alpha^{n+1} = (\rho q)_\alpha^{n+1}, \tag{8.24a}$$

$$v_\alpha^{n+1} = -\frac{\mathbf{K} k_{r\alpha}}{\mu_\alpha^{n+1}}\left[\operatorname{grad}(p_\alpha^{n+1}) - g\rho_\alpha^{n+1} \operatorname{grad}(z)\right], \tag{8.24b}$$

Here, $\boldsymbol{\phi}, S_\alpha, p_\alpha \in \mathbb{R}^{n_c}$ denote vectors with one porosity value, one saturation, and one pressure value per cell, respectively, whereas v_α is the vector of fluxes for phase α per face, and so on. For properties depending on pressure and saturation, we have for simplicity not introduced any notation to distinguish whether these are evaluated in cells or at cell interfaces. The superscript refers to discrete times and Δt denotes the associated time step.

The main advantage of using this direct discretization is that it will generally give a reasonable approximation to the true solution as long as we are able to solve the resulting nonlinear system at each time step. To do this, we must pick as many unknowns as we have phases and perform some kind of linearization; we will come back in more detail later in the book, and discuss implementation of (8.24) in Chapter 11. Here, we simply observe that there are many ways to perform the linearization: we can choose the phase saturations as primary unknowns, the phase pressures, the capillary pressures, or some combinations thereof. How difficult it will be to solve for these unknowns will obviously depend on the coupling between the two equations and nonlinearity within each equation. The main disadvantages of discretizing (8.22) directly are that the general system conceals its mathematical nature and that the resulting equations are not well-posed in the case of an incompressible system. In Section 8.3, we will therefore go back to the continuous equations and analyze the mathematical nature of the resulting system for various choices of primary variables and simplifying assumptions.

8.2.2 Multicomponent Phases

In many cases, each phase may consist of more than one chemical species that are mixed at the molecular level and generally share the same velocity (and temperature). This type of flow differs from the immiscible case, since dispersion and Brownian motion will cause the components to redistribute if there are macroscale gradients in the mass fractions. The simplest way to model this is through a linear Fickian diffusion,

$$\vec{J}_\alpha^\ell = -\rho_\alpha S_\alpha \mathbf{D}_\alpha^\ell \nabla c_\alpha^\ell, \tag{8.25}$$

where c_α^ℓ is the mass fraction of component ℓ in phase α, ρ_α is the density of phase α, S_α is the saturation of phase α, and $\mathbf{D}_\alpha^\ell$ is the diffusion tensor for component ℓ in phase α. Likewise, the chemical species may interact and undergo chemical reactions, but a description of this is beyond the scope of this book.

For multicomponent, multiphase flow, we are, in principle, free to choose whether we state the conservation of mass for components or for fluid phases. However, if we choose the latter, we will have to include source terms in our balance equations that account for the transfer of components between the phases. This can be a complex undertaking and the standard approach is therefore to develop balance equations for each component. For a system of N fluid phases and M chemical species, the mass conservation for component $\ell = 1, \ldots, M$ reads

$$\frac{\partial}{\partial t}\left(\phi \sum_\alpha c_\alpha^\ell \rho_\alpha S_\alpha\right) + \nabla \cdot \left(\sum_\alpha c_\alpha^\ell \rho_\alpha \vec{v}_\alpha + \vec{J}_\alpha^\ell\right) = \sum_\alpha c_\alpha^\ell \rho_\alpha q_\alpha, \quad (8.26)$$

where $\vec{v}_\alpha$ is the superficial phase velocity and q_α is the source term. The system is closed in the same way as for single-component phases, except that we now also have to use that the mass fractions sum to zero, (8.3).

The generic system (8.23)–(8.26) can also describe miscible displacements in which the composition of the fluid phases changes when the porous medium undergoes pressure and saturation changes and one has to account for all, or a large majority, of the chemical species that are present in the flow system. For immiscible or partially miscible systems, it is common to introduce simplifications by lumping multiple species into pseudo-components, or even disregard that a fluid phase may be composed of different chemical species, as we will see next.

8.2.3 Black-Oil Models

The most common approach to simulate oil and gas recovery is to use the so-called black-oil equations, which essentially constitute a special multicomponent model with no diffusion among the components. The name refers to the assumption that the chemical species can be lumped together to form two *pseudo-components* at surface conditions, a heavy hydrocarbon component called "oil" and a light hydrocarbon component called "gas." At reservoir conditions, the two components can be partially or completely dissolved in each other depending on pressure and temperature, forming either one or two phases: a liquid oleic phase and a gaseous phase. In addition to the two hydrocarbon phases, the framework includes an aqueous phase that in the simplest models of this class is assumed to consist of only water. In more comprehensive models, the hydrocarbon components are also allowed to dissolve in the aqueous phase and the water component may be dissolved or vaporized in the two hydrocarbon phases. The composition of each hydrocarbon component, however, remains constant for all times.

We henceforth assume three phases (oleic, gaseous, and aqueous) and three components (oil, gas, and water). By looking at our model (8.23)–(8.26), we see that we so far have introduced 27 unknown physical quantities: 9 mass fractions c_α^ℓ and 3 of each of the following six quantities: ρ_α, S_α, $\vec{v}_\alpha$, p_α, μ_α, and $k_{r\alpha}$. In addition, the porosity will typically depend on pressure as discussed in Section 2.4.1. To determine these 27 unknowns, we have 3 continuity equations (8.26), an algebraic relation for the saturations (8.1), 3 algebraic relations for the mass fractions (8.3), and Darcy's law (8.23) for each of the 3 phases. Altogether, this constitutes only 10 equations. Thus, we need to add 17 extra closure relations to make a complete model.

The first five of these can be obtained immediately from our discussion in Sections 8.1.3 and 8.1.4, where we saw that the relative permeabilities $k_{r\alpha}$ are functions of the phase saturations. Likewise, capillary pressures are functions of saturation and can be used to relate phase pressures as follows:

8.2 Flow Equations for Multiphase Flow

$$p_o - p_w = P_{cow}(S_w, S_o), \qquad p_g - p_o = P_{cgo}(S_o, S_g).$$

The required functional forms are normally obtained from a combination of physical experiments, small-scale numerical simulations, and analytical modeling based on bundle-of-tubes arguments, etc. Viscosities are either assumed to be constant or can be established as pressure-dependent functions through laboratory experiments, which gives us another three equations. The remaining nine closure relations are obtained from mixture rules and PVT models that are generalizations of the equations of state discussed in Section 4.2.

By convention, the black-oil equations are formulated as conservation for the oil, water, and gas components. To this end, we employ a simple PVT model that uses pressure-dependent functions to relate fluid volumes at reservoir and surface conditions. Specifically, we use the *formation-volume factors* $B_\ell = V_\ell / V_{\ell s}$ that relate the volumes V_ℓ and $V_{\ell s}$ occupied by a bulk of component ℓ at reservoir and surface conditions, respectively. The formation-volume factors, and their reciprocal *shrinkage factors* $b_\ell = V_{\ell s}/V_\ell$, which I personally prefer to use for notational convenience, are assumed to depend on phase pressure. In dead-oil systems, the oil is at such a low pressure that it does not contain any gas or has lost its volatile components (which presumably have escaped from the reservoir). Neither of the components are therefore dissolved in the other phases at reservoir conditions. In live-oil systems, gas is dissolved in the oleic phase. When underground, hydrocarbon molecules are mostly in liquid phase, but gaseous components are liberated when the oil is pumped to the surface. The solubility of gas in oil is usually modeled through the pressure-dependent solution gas–oil ratio, $R_s = V_{gs}/V_{os}$, defined as the volume of gas, measured at standard conditions, that at reservoir conditions is dissolved in a unit of stock-tank oil. In condensate reservoirs, oil is vaporized in the gaseous phase, so that when underground, condensate oil is mostly a gas, but condenses into a liquid when pumped to the surface. The solubility of oil in gas is modeled as the pressure-dependent vaporized oil–gas ratio R_v, defined as the amount of surface condensate that can be vaporized in a surface gas at reservoir conditions. Using this notation,[2] the black-oil equations for a live-oil system reads,

$$\partial_t \left[\phi \left(b_o S_o + b_g R_v S_g \right) \right] + \nabla \cdot \left(b_o \vec{v}_o + b_g R_v \vec{v}_g \right) - \left(b_o q_o + b_g R_v q_g \right) = 0,$$

$$\partial_t \left(\phi b_w S_w \right) + \nabla \cdot \left(b_w \vec{v}_w \right) - b_w q_w = 0,$$

$$\partial_t \left[\phi \left(b_g S_g + b_o R_s S_o \right) \right] + \nabla \cdot \left(b_g \vec{v}_g + b_o R_s \vec{v}_o \right) - \left(b_g q_g + b_o R_s q_o \right) = 0.$$

To model enhanced oil recovery, we must extend the system with additional components representing chemical or microbial species that are added to the injected fluids to mobilize immobile oil by altering wettability and/or to improve sweep efficiency and hence the overall displacement of mobile oil. One may also model species dissolved in the resident fluids. Likewise, the black-oil equations may be expanded by an additional solid phase to account for salts and other minerals that precipitate during hydrocarbon recovery, and possibly also extended to include an energy equation that accounts for temperature effects.

---

[2] The convention in the petroleum literature is to use B_α rather than b_α.

We will return to a more detailed discussion of the general case of three-phase flow with components that may transfer between phases later in Chapter 11. In the rest of this chapter, we continue to study the special case of two single-component fluid phases with no interphase mass transfer.

8.3 Model Reformulations for Immiscible Two-Phase Flow

For the special case of two immiscible fluids, found in a wetting w and a non-wetting n phase, the general system of flow equations (8.22) simplifies to

$$\frac{\partial}{\partial t}(\phi S_w \rho_w) + \nabla \cdot (\rho_w \vec{v}_w) = \rho_w q_w,$$
$$\frac{\partial}{\partial t}(\phi S_n \rho_n) + \nabla \cdot (\rho_n \vec{v}_n) = \rho_n q_n. \tag{8.27}$$

In the following, we discuss various choices of primary variables and how these choices affect the mathematical structure of the resulting coupled system of nonlinear differential equations.

8.3.1 Pressure Formulation

If we choose the phase pressures p_n and p_w as primary unknowns, we must express the saturations S_n and S_w as functions of pressure. To this end, we assume that the capillary pressure has a unique inverse function $\hat{S}_w = P_c^{-1}(p_c)$ (see (8.8)) so that we can write

$$S_w = \hat{S}_w(p_n - p_w), \qquad S_n = 1 - \hat{S}_w(p_n - p_w). \tag{8.28}$$

Then, we can reformulate (8.27) as

$$\frac{\partial}{\partial t}(\phi \rho_w \hat{S}_w) + \nabla \cdot \left(\frac{\rho_w \mathbf{K} k_{rw}(\hat{S}_w)}{\mu_w} \left(\nabla p_w - \rho_w g \nabla z \right) \right) = \rho_w q_w,$$
$$\frac{\partial}{\partial t}(\phi \rho_n (1 - \hat{S}_w)) + \nabla \cdot \left(\frac{\rho_n \mathbf{K} k_{rn}(\hat{S}_w)}{\mu_n} \left(\nabla p_n - \rho_n g \nabla z \right) \right) = \rho_n q_n. \tag{8.29}$$

This system is unfortunately highly coupled and strongly nonlinear. The strong coupling comes from the fact that the difference in the primary variables $p_n - p_w$ enters the computation of $\hat{S}_w$ in the accumulation terms and also appears in the composite functions $k_{rw}(\hat{S}_w(\cdot))$ and $k_{rn}(1 - \hat{S}_w(\cdot))$ used to evaluate the relative permeabilities. As an example of the resulting nonlinearity, we can look at the van Genuchten model for capillary and relative permeabilities, (8.12) and (8.17). Figure 8.7 shows the capillary pressure and relative permeabilities as functions of S as well as the inverse of the capillary pressure and relative permeabilities as functions of $p_n - p_w$. Whereas the accumulation function is nonlinear, this nonlinearity is further accentuated when used inside the nonlinear relative permeability functions. The pressure formulation was used in the simultaneous solution scheme originally proposed by Douglas, Peaceman, and Rachford [86] in 1959, but has

8.3 Model Reformulations for Immiscible Two-Phase Flow 249

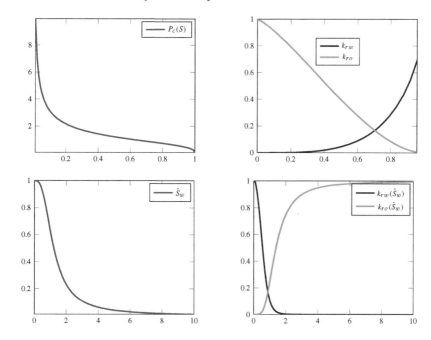

Figure 8.7 Capillary pressure and relative permeabilities as functions of S for the van Genuchten model for $\beta_g = 1$, $m_g = 2/3$, and $\kappa = 1/2$. The lower row shows saturation and relative permeabilities as function of $p_c = p_n - p_w$.

later been superseded by other formulations that reduce the degree of coupling and improve the nonlinear nature of the equation.

8.3.2 Fractional-Flow Formulation in Phase Pressure

The strong coupling and much of the nonlinearity seen in the previous formulation can be eliminated if we instead express the system in terms of one phase pressure and one phase saturation. A common choice is to use p_n and S_w, which gives the following system

$$\frac{\partial}{\partial t}(\phi S_w \rho_w) + \nabla \cdot \left(\frac{\rho_w \mathbf{K} k_{rw}}{\mu_w}\left(\nabla p_n - \nabla P_c(S_w) - \rho_w g \nabla z\right)\right) = \rho_w q_w,$$

$$\frac{\partial}{\partial t}(\phi(1-S_w)\rho_n) + \nabla\left(\frac{\rho_n \mathbf{K} k_{rn}}{\mu_n}\left(\nabla p_n - \rho_n g \nabla z\right)\right) = \rho_n q_n. \quad (8.30)$$

To further develop this system of equations, it is common to expand the derivatives to introduce rock and fluid compressibilities, as discussed for the single-phase flow equation in Section 4.2. First, however, we will look closer at the special case of incompressible flow and develop the *fractional-flow formulation*, which will enable us to further expose the mathematical structure of the system.

Incompressible Flow

For incompressible flow, the porosity ϕ only varies in space and the fluid densities ρ_α are constant. Using these assumptions, we can simplify the mass-balance equations to be on the form

$$\phi \frac{\partial S_\alpha}{\partial t} + \nabla \cdot \vec{v}_\alpha = q_\alpha. \tag{8.31}$$

To derive the fractional-flow formulation, we start by introducing the total Darcy velocity, which we can express in terms of the pressure for the non-wetting phase,

$$\begin{aligned}\vec{v} = \vec{v}_n + \vec{v}_w &= -\lambda_n \nabla p_n - \lambda_w \nabla p_w + (\lambda_n \rho_n + \lambda_w \rho_w) g \nabla z \\ &= -(\lambda_n + \lambda_w) \nabla p_n + \lambda_w \nabla p_c + (\lambda_n \rho_n + \lambda_w \rho_w) g \nabla z.\end{aligned} \tag{8.32}$$

We then add the two continuity equations and use the fact that $S_n + S_w = 1$ to derive a pressure equation without temporal derivatives

$$\nabla \cdot (\vec{v}_n + \vec{v}_w) = \nabla \cdot \vec{v} = q_n + q_w \tag{8.33}$$

If we now define the total mobility $\lambda = \lambda_n + \lambda_w = \lambda \mathbf{K}$ and total source $q = q_n + q_w$, insert (8.32) into (8.33), and collect all terms that depend on pressure on the left-hand side and all other terms on the right-hand side, we obtain an elliptic Poisson-type equation

$$-\nabla \cdot (\lambda \mathbf{K} \nabla p_n) = q - \nabla \big[\lambda_w \nabla p_c + (\lambda_n \rho_n + \lambda_w \rho_w) g \nabla z\big]. \tag{8.34}$$

This *pressure equation* (or flow equation) is essentially on the same form as the equation (4.10) governing single-phase, incompressible flow, except that both the variable coefficient and the right-hand side now depend on saturation through the relative mobility λ and the capillary pressure function $p_c = P_c$.

To derive an equation for S_w, we multiply each phase velocity by the relative mobility of the other phase, subtract the result, and use Darcy's law to obtain

$$\begin{aligned}\lambda_n \vec{v}_w - \lambda_w \vec{v}_n &= \lambda \vec{v}_w - \lambda_w \vec{v} \\ &= -\lambda_n \lambda_w \mathbf{K}(\nabla p_w - \rho_w g \nabla z) + \lambda_w \lambda_n \mathbf{K}(\nabla p_n - \rho_n g \nabla z) \\ &= \lambda_w \lambda_n \mathbf{K}\big[\nabla p_c + (\rho_w - \rho_n) g \nabla z\big].\end{aligned}$$

If we now solve for $\vec{v}_w$ and insert into (8.31) for the wetting phase, we obtain what is commonly referred to as the *saturation equation* (or transport equation)

$$\phi \frac{\partial S_w}{\partial t} + \nabla \cdot \big[f_w \vec{v} + f_w \lambda_n \mathbf{K} \Delta \rho g \nabla z\big] = q_w - \nabla \cdot \big(f_w \lambda_n \mathbf{K} P_c' \nabla S_w\big). \tag{8.35}$$

Here, $\Delta \rho = \rho_w - \rho_n$, and we have introduced the *fractional flow function* $f_w = \lambda_w/(\lambda_n + \lambda_w)$, which measures the fraction of the total flow that consists of the wetting fluid.

Equation (8.35) is parabolic and accounts for the balance of three different forces: viscous advection $f_w \vec{v}$, gravity segregation $(f_w \lambda_n) \mathbf{K} \Delta \rho g \nabla z$, and capillary forces $f_w \lambda_n \mathbf{K} P_c' \nabla S_w$. The first two terms both involve a first-order derivative and hence have a *hyperbolic* character, whereas the capillary term contains a second-order derivative

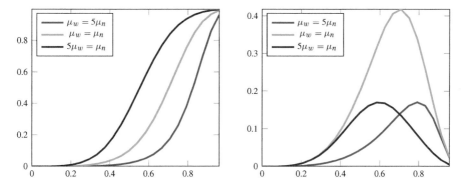

Figure 8.8 Fractional flow function f_w (left) and gravity segregation function $\lambda_n f_w$ for the van Genuchten–Mualem model (8.17) with $m_g = 2/3$, $\kappa = 1/2$, and $\Delta\rho > 0$.

and hence has a *parabolic* character. The negative sign in front of the capillary term is misleading; the overall term is positive since $P'_c < 0$ and hence gives a well-posed problem. Moreover, since λ_n is zero at the end point, the capillary term disappears at this point and the parabolic equation thus degenerates to being hyperbolic. This, in turn, implies that a propagating displacement front – in which a wetting fluid displaces a non-wetting fluid – has finite speed of propagation, and not the infinite speed of propagation that is typical for parabolic problems.

Figure 8.8 shows the functional form of the two hyperbolic terms for the van Genuchten–Mualem model (8.17). The fractional flow function f_w is monotonically increasing and has a characteristic S-shape. If the flow is dominated by viscous terms so that gravity and capillary forces are negligible, the flow will always be *cocurrent* with both phases flowing in the direction of $\vec{v}$ because the derivative of f_w is positive. The gravity segregation function, on the other hand, has a characteristic bell shape and describes the upward movement of the lighter phase and downward movement of the heavier phase. (The function is turned upside down if $\rho_w < \rho_n$.) If gravity is sufficiently strong, it may introduce *countercurrent flow*, with the fluid phases flowing in opposite directions. On a field scale, capillary forces are typically small and hence the overall transport equation will have a strong hyperbolic character. Neglecting the capillary term is generally a good approximation and gives a purely hyperbolic saturation equation. We will return to this model later and discuss a few examples in the special case of purely hyperbolic 1D flow.

The coupling between the elliptic pressure equation and the parabolic saturation equation is much weaker than the coupling between the two continuity equations in the formulation (8.29) with two pressures as primary unknowns. In the pressure equation (8.34), the coupling to saturation appears explicitly in the effective mobility, which makes up the variable coefficient in the Poisson problem, and on the right-hand side through the phase mobilities and the derivative of the capillary function. In (8.35), on the other hand, the saturation is only indirectly coupled to the pressure through the total Darcy velocity. In typical waterflooding scenarios, saturation variations are relatively smooth in most of the domain, except at the interface between the invading and the displaced fluids, and hence the pressure distribution is only mildly affected by the evolving fluid saturations. In fact, the pressure and transport equations are completely decoupled in the special case of linear relative permeabilities, no capillary pressure, and equal fluid densities and viscosities. These weak couplings are exploited in many efficient numerical methods, which use various forms of operator splitting to solve pressure and fluid transport in separate steps. This is also our method of choice for incompressible flow in MRST, and will be discussed in more details in Chapter 10.

The reformulation discussed above can easily be extended to N immiscible phases, and will result in one pressure equation and a (coupled) system of $N - 1$ saturation equations. It is also possible to develop fractional-flow formulations for multicomponent flow without mass exchange between the phases.

Compressible Flow

To develop a fractional-flow formulation in the compressible case, we start by expanding the accumulation term in (8.22) and then use fluid compressibilities to rewrite the derivative of the phase densities

$$\frac{\partial}{\partial t}(\phi \rho_\alpha S_\alpha) = \rho_\alpha S_\alpha \frac{\partial \phi}{\partial t} + \phi S_\alpha \frac{d\rho_\alpha}{dp_\alpha} \frac{\partial p_\alpha}{\partial t} + \phi \rho_\alpha \frac{\partial S_\alpha}{\partial t}$$

$$= \rho_\alpha S_\alpha \frac{\partial \phi}{\partial t} + \phi \rho_\alpha S_\alpha c_\alpha \frac{\partial p_\alpha}{\partial t} + \phi \rho_\alpha \frac{\partial S_\alpha}{\partial t}.$$

If we now insert this into (8.22), divide each equation by ρ_α, and sum the equations, we obtain

$$\frac{\partial \phi}{\partial t} + \phi c_n S_n \frac{\partial p_n}{\partial t} + \phi c_w S_w \frac{\partial p_w}{\partial t} + \frac{1}{\rho_n} \nabla \cdot (\rho_n \vec{v}_n) + \frac{1}{\rho_w} \nabla \cdot (\rho_w \vec{v}_w) = q_t, \quad (8.36)$$

where $q_t = q_n + q_w$. To explore the character of this equation, let us for the moment assume that capillary forces are negligible so that $p_n = p_w = p$ and that the spatial density variations are so small that we can set $\nabla \rho_\alpha = 0$. If we now introduce rock compressibility $c_r = d \ln(\phi)/dp$, (8.36) simplifies to

$$\phi c \frac{\partial p}{\partial t} - \nabla \cdot (\lambda \mathbf{K} \nabla p) = \hat{q}. \quad (8.37)$$

Here, we have introduced the *total compressibility* $c = (c_r + c_n S_n + c_w S_w)$ and a source function $\hat{q}$ that accounts both for volumetric sources and pressure variations with depth. Equation (8.37) is clearly parabolic, but simplifies to the elliptic Poisson equation (8.34) for incompressible flow ($c = 0$). Fluid compressibilities tend to decrease with increasing pressure, and we should therefore expect that the pressure equation has a strong elliptic character at conditions usually found in conventional reservoirs during secondary and tertiary production, in particular for weakly compressible systems. On the other hand, the pressure evolution will usually have a more pronounced parabolic character during primary

8.3 Model Reformulations for Immiscible Two-Phase Flow

depletion, in gas reservoirs, and in unconventional reservoirs, where **K** may be in the range of one micro Darcy or less.

These observations remain true also when capillary forces and spatial density variations are included. However, including these effects makes the mathematical structure of the equation more complicated. In particular, the pressure equation becomes a *nonlinear parabolic equation* if the spatial variations in density cannot be neglected. To see this, let us once again neglect capillary forces and consider the second-last term on the left-hand side of (8.36),

$$\frac{1}{\rho_n} \nabla \cdot (\rho_n \vec{v}) = \nabla \cdot \vec{v}_n + \vec{v}_n \cdot \frac{1}{\rho_n} \nabla \rho_n$$
$$= \nabla \cdot \vec{v}_n + \vec{v}_n \cdot (c_n \nabla p_n).$$

By using Darcy's law, we can transform the inner product between the phase flux and the pressure gradient to involve a quadratic term in ∇p_n or $\vec{v}_n$,

$$\vec{v}_n \cdot (c_n \nabla p_n) = -\lambda_n (\nabla p_n - \rho_n g \nabla z) \cdot (c_n \nabla p_n)$$
$$= c_n \vec{v}_n \cdot (-\lambda_n^{-1} \vec{v}_n + \rho_n g \nabla z).$$

The last term on the left-hand side of (8.36) has a similar form, and hence we obtain a nonlinear equation if at least one of the fluid phases has significant spatial density variations.

Going back to (8.36), we see that the equation contains both phase pressures and hence cannot be used directly alongside with a saturation equation. There are several ways to formulate a pressure equation so that it only involves a single unknown pressure. We can either pick one of the phase pressures as the primary variable, or introduce an average pressure $p_a = (p_n + p_w)/2$, as suggested in [246]. With p_n as the primary unknown, the full pressure equation reads

$$\phi c \frac{\partial p_n}{\partial t} - \left[\frac{1}{\rho_n} \nabla \cdot (\rho_n \lambda_n \nabla p_n) + \frac{1}{\rho_w} \nabla \cdot (\rho_w \lambda_w \nabla p_n) \right]$$
$$= q_n + q_w - \frac{1}{\rho_w} \nabla \cdot (\rho_w \lambda_w \nabla P_c) + c_w S_w \frac{\partial P_c}{\partial t}$$
$$- \frac{1}{\rho_n} \nabla \cdot (\rho_n^2 \lambda_n g \nabla z) - \frac{1}{\rho_w} \nabla \cdot (\rho_w^2 \lambda_w g \nabla z). \tag{8.38}$$

Likewise, with p_a as the primary unknown, we get

$$\phi c \frac{\partial p_a}{\partial t} - \left[\frac{1}{\rho_n} \nabla \cdot (\rho_n \lambda_n \nabla p_a) + \frac{1}{\rho_w} \nabla \cdot (\rho_w \lambda_w \nabla p_a) \right]$$
$$= q_t + \frac{1}{2} \left[\frac{1}{\rho_n} \nabla \cdot (\rho_n \lambda_n) - \frac{1}{\rho_w} \nabla \cdot (\rho_w \lambda_w) \right] \nabla P_c + \frac{1}{2} c_w S_w \frac{\partial P_c}{\partial t}$$
$$- \frac{1}{2} c_n S_n \frac{\partial P_c}{\partial t} - \frac{1}{\rho_n} \nabla \cdot (\rho_n^2 \lambda_n g \nabla z) - \frac{1}{\rho_w} \nabla \cdot (\rho_w^2 \lambda_w g \nabla z). \tag{8.39}$$

The same type of pressure equation can be developed for cases with more than two fluid phases. Also in this case, we can define a total velocity and use this to obtain $N-1$ transport equations in fractional-flow form.

The saturation equation for compressible flow is obtained exactly in the same way as in the incompressible case,

$$\frac{\partial}{\partial t}(\phi \rho_w S_w) + \nabla \cdot \left[\rho_w f_w (\vec{v} + \lambda_n \Delta \rho g \nabla z)\right]$$
$$= \rho_w q_w - \nabla \cdot (\rho_w f_w \lambda_n P_c' \nabla S_w). \tag{8.40}$$

However, whereas total velocity was the only quantity coupling the saturation equation to the fluid pressure in the incompressible case, we now have coupling also through porosity and density, which may both depend on pressure. This generally makes the compressible case more nonlinear and challenging to solve than the incompressible case.

8.3.3 Fractional-Flow Formulation in Global Pressure

The formulation in phase pressure discussed above has a relatively strong coupling between pressure and saturation in the presence of capillary forces. The strong coupling is mainly caused by the ∇p_c term on the right-hand side of (8.34). Let us therefore go back and see if we can eliminate this term by making a different choice of primary variables. To this end, we start by looking at the definition of the total velocity, which couples the pressure to the saturation equation:

$$\vec{v} = -\lambda_n \nabla p_n - \lambda_w \nabla p_w + (\rho_n \lambda_n + \rho_w \lambda_w) g \nabla z$$
$$= -(\lambda_n + \lambda_w)\nabla p_n + \lambda_w \nabla(p_n - p_w) + (\rho_n \lambda_n + \rho_w \lambda_w) g \nabla z$$
$$= -\lambda(\nabla p_n - f_w \nabla p_c) + (\rho_n \lambda_n + \rho_w \lambda_w) g \nabla z.$$

If we now introduce a new pressure variable p, called the *global pressure* [62], defined so that $\nabla p = \nabla p_n - f_w \nabla p_c$, we see that the total velocity can be related to the global pressure through an equation that looks like Darcy's law

$$\vec{v} = -\lambda(\nabla p - (\rho_w f_w + \rho_n f_n) g \nabla z). \tag{8.41}$$

By adding the continuity equations (8.31), and using (8.41), we obtain an elliptic Poisson-type equation for the global pressure

$$-\nabla \cdot \left[\lambda \mathbf{K}(\nabla p - (\rho_w f_w + \rho_n f_n) g \nabla z)\right] = q_t. \tag{8.42}$$

The advantages of this formulation is that it is very simple and highly efficient in the incompressible case, where pressure often is treated as an immaterial property and one is mostly interested in obtaining a velocity field for the saturation equation. The disadvantages are that it is not obvious how one should specify and interpret boundary conditions for the global pressure and that global-pressure values cannot be used directly in well models (of Peaceman type). The global-pressure formulation can also be extended to variable densities and to three-phase flow; the interested reader should consult [67, 65] and references therein. The global formulation is often used by academic researchers because of its simplicity, but is, to the best of my knowledge, hardly used for practical simulations in the industry.

8.3.4 Fractional-Flow Formulation in Phase Potential

The two fractional-flow formulations discussed above are generally well suited to study two-phase immiscible flow with small or negligible capillary forces. In such cases, flow is mainly driven through high-permeable regions, and flow paths are to a large extent determined by the permeability distribution, the shape of the relative permeability functions, and density differences (which determine the relative importance of gravity segregation). In other words, the flow is mainly governed by viscous forces and gravity segregation, which constitute the hyperbolic part of the transport equation, whereas capillary forces mostly contribute to adjust the width of the interface between the invading and displaced fluids. For highly heterogeneous media with strong contrasts in capillary functions, on the other hand, the capillary forces may have pronounced impact on flow paths by enhancing cross-flow in stratified media or reducing the efficiency of gravity drainage. Likewise, because capillary pressure is continuous across geological interfaces, differences in capillary functions between adjacent rocks of different type may introduce (strong) discontinuities in the saturation distribution.

An alternative fractional-flow formulation was proposed by Hoteit and Firoozabadi [136] to study heterogeneous media with strong contrasts in capillary functions. The formulation is similar to the ones discussed above, except that we now work with fluid potentials rather than fluid pressures. The fluid potential for phase α and the capillary potential are given by

$$\Phi_\alpha = p_\alpha - \rho_\alpha g z, \tag{8.43}$$

$$\Phi_c = \Phi_n - \Phi_w = p_c + (\rho_w - \rho_n)gz. \tag{8.44}$$

By manipulating the expression for the total velocity, we can write it as a sum of two new velocities, which each is given by the gradient of Φ_w and Φ_c, respectively:

$$\begin{aligned}\vec{v} &= -\lambda_n\big[\nabla p_n - \rho_n g \nabla z\big] - \lambda_w\big[\nabla p_w - \rho_n g \nabla z\big] \\ &= -\lambda_n\big[\nabla p_w + \nabla p_c - (\rho_n - \rho_w + \rho_w)g\nabla z\big] - \lambda_w\big[\nabla p_w - \rho_n g \nabla z\big] \\ &= -(\lambda_n + \lambda_w)\big[\nabla p_w - \rho_w g \nabla z\big] - \lambda_n\big[\nabla p_c + (\rho_w - \rho_n)g\nabla z\big] \\ &= -\lambda \nabla \Phi_w - \lambda_n \nabla \Phi_c = \vec{v}_a + \vec{v}_c.\end{aligned}$$

Here, we observe that

$$\vec{v}_w = -\lambda_w \nabla \Phi_w = -\frac{\lambda_w}{\lambda}\lambda \mathbf{K} \nabla \Phi_w, = f_w \vec{v}_a$$

so that the velocity $\vec{a}$ represents the same phase differential as the phase velocity of the wetting phase. The only exception is that in $\vec{v}_a$, the potential gradient is multiplied by the total mobility, which is a smoother and less varying function than the mobility λ_w of the water phase. This means that the saturation dependence of $\vec{v}_a$ is less than for the total velocity $\vec{v}$.

Proceeding analogously as above, we can derive the following coupled system for two-phase, immiscible flow:

$$-\nabla \cdot (\lambda \mathbf{K} \nabla \Phi_w) = q + \nabla \cdot (\lambda_n \mathbf{K} \nabla \Phi_c) \qquad (8.45)$$

$$\phi \frac{\partial}{\partial t} + \nabla \cdot (f_w \vec{v}_a) = q_w. \qquad (8.46)$$

Compared with the two other fractional-flow formulations, this system is simpler in the sense that capillary forces are only accounted for in the pressure equation. Equally important, the saturation equation is purely hyperbolic and only contains advective flow along $\vec{v}_a$. This makes the saturation much simpler to solve, since one does not have to resolve delicate balances involving nonlinear functions for gravity segregation and capillary forces. One can thus easily employ simple upwind discretizations and highly efficient solvers that have been developed for purely co-current flow, like streamline methods [79] or methods based on optimal ordering [220, 196]. On the other hand, the advective velocity $\vec{v}_a$ may vary significantly with time and hence represent a strong coupling between pressure and transport.

8.3.5 Richards' Equation

Another special case arises when modeling the vadose or unsaturated zone, which extends from the ground surface to the groundwater table, i.e., from the top of the earth to the depth at which the hydrostatic pressure of the groundwater equals one atmosphere. Soil and rock in the vadose zone will generally contain both air and water in its pores, and the vadose zone is the main factor that controls the movement of water from the ground surface to aquifers, i.e., the saturated zone beneath the water table. In the vadose zone, water is retained by a combination of adhesion and capillary forces, and in fine-grained soil one can find pores that are fully saturated by water at a pressure less than one atmosphere.

Because of the special conditions in the vadose zone, the general flow equations modeling two immiscible phases can be considerably simplified. First of all, we can expect that any pressure differences in air will be equilibrated almost instantaneously relative to water, since air typically is much less viscous than water. (At a temperature of 20 °C, the air viscosity is approximately 55 times smaller than the water viscosity.) Secondly, air will in most cases form a continuous phase so that all parts of the pore space in the vadose zone are connected to the atmosphere. If we neglect variations in the atmospheric pressure, the pore pressure of air can therefore be assumed to be constant. For simplicity, we can set the atmospheric pressure to be zero, so that $p_a = 0$ and $p_c = p_a - p_w = -p_w$. This, in turn, implies that the water saturation (and the water relative permeability) can be defined as functions of water pressure using one of the models for capillary pressure presented in Section 8.1.3. Inserting all of this into (8.27), we obtain

8.3 Model Reformulations for Immiscible Two-Phase Flow

$$\frac{\partial}{\partial t}(\phi \rho_w S_w(p_w)) + \nabla \cdot [\rho_w \lambda_{rw}(p_w)\mathbf{K}(\nabla p_w - \rho_w g \nabla z)] = 0.$$

If we further expand the accumulation term, neglect spatial gradients of the water density, we can divide the above equation by ρ_w to obtain the so-called *generalized Richards' equation*

$$C_{wp}(p_w)\frac{\partial p_w}{\partial t} + \nabla \cdot [\lambda_{rw}(p_w)\mathbf{K}\nabla(p_w - \rho g z)] = 0, \quad (8.47)$$

where $C_{wp} = c_w \theta_w + d(\theta_w)/dp_w$ is a storage coefficient and $\theta_w = \phi S_w$ is the water content. In hydrology, it is common to write the equation in terms of the pressure head

$$C_{wh}(h_w)\frac{\partial h_w}{\partial t} + \nabla[\kappa_w k_{rw}(h_w)\nabla(h_w - z)] = 0, \quad (8.48)$$

where $C_{wh} = \rho_w g C_{wp}$ and κ_w is the hydraulic conductivity of water (see Section 4.1 on page 113). If we further neglect water and rock compressibility, we obtain the pressure form of the classical Richards' equation [264]

$$C_{ch}(h_w)\frac{\partial h_w}{\partial t} + \nabla[\kappa_w k_{rw}(h_w)\nabla(h_w - z)] = 0, \quad C_{ch} = \frac{d\theta_w}{dh_w}. \quad (8.49)$$

The corresponding equation for the water flux

$$\vec{v}_w = -\kappa_s k_{rw}(h_w)\nabla(h_w - z) \quad (8.50)$$

was suggested earlier by Buckingham [55] and is often called the Darcy–Buckingham equation. Richards' equation can also be expressed in two alternative forms,

$$\frac{\partial \theta_w}{\partial t} + \nabla \cdot [\kappa_w k_{rw}(h_w)\nabla(h_w - z)] = 0,$$

$$\frac{\partial \theta_w}{\partial t} + \nabla[\mathbf{D}(\theta_w)\nabla \theta_w] = \kappa_w k_{rw}(\theta_w)\nabla z,$$

where $\mathbf{D}(\theta_w) = \frac{dh_w}{d\theta_w}\kappa_w k_{rw}(\theta_w)$ is the hydraulic diffusivity tensor. The latter, called the water-content form, is generally easier to solve numerically than the other forms, but becomes infinite for fully saturated conditions.

Various forms of Richards' equation are widely used to simulate water flow in the vadose zone. However, one should be aware that this class of models has important limitations. First of all, because air pressure is assumed to be constant, these models cannot describe air flow, and also assume that the air phase in the whole pore space is fully connected to the atmosphere. This may not be the case if the porous medium contains heterogeneities blocking the air flow, e.g., in the form of layers that are almost impermeable to air. Likewise, the derivation of the equations assumed that the mobility of air is infinite compared with water, which may not be the case if the relative permeability of air is much smaller than that of water. If this is not the case, one should expect significant discrepancies in simulation results obtained from Richards' equations and the general two-phase flow equations.

The official release of MRST does not yet have any implementation of Richards' equations, but work is underway to develop a new module [299].

8.4 The Buckley–Leverett Theory of 1D Displacements

To shed more light into the physical behavior of two-phase immiscible flow systems, we derive analytical solutions in a few simple cases. As part of this, you will also get a glimpse of the (wave) theory of hyperbolic conservation laws. This theory plays a very important role in the mathematical and numerical analysis of multiphase flow models, but is not described in great detail herein for brevity. If you want to learn more of this interesting topic, there are many good mathematical textbooks you can consult, e.g., [76, 134, 183, 293, 294]. Chapter 9 also introduces a few central concepts and briefly discusses development of specialized numerical methods.

8.4.1 Horizontal Displacement

As a first example, we consider incompressible displacement in a 1D homogeneous and horizontal medium, $x \in [0, \infty)$, with inflow at $x = 0$. In the absence of capillary forces, the two phase pressures are equal and coincide with the global pressure p. The pressure equation simplifies to

$$v'(x) = q, \qquad v(x) = -\lambda(x) p'(x).$$

If we assume that there are no volumetric source terms, but a constant inflow rate at $x = 0$, it follows by integration that $v(x)$ is constant in the domain $[0, \infty)$. Under these assumptions, the saturation equation (8.35) simplifies to,

$$\frac{\partial S}{\partial t} + \frac{v}{\phi} \frac{\partial f(S)}{\partial x} = 0. \tag{8.51}$$

Let us use this equation to study the propagation of a constant saturation value. To this end, we consider the total differential of $S(x,t)$,

$$0 = dS = \frac{\partial S}{\partial t} dt + \frac{\partial S}{\partial x} dx,$$

which we can use to eliminate $\partial S/\partial t$ from (8.51),

$$-\frac{\partial S}{\partial x} \left(\frac{dx}{dt} \right)\bigg|_{dS=0} + \frac{v}{\phi} \frac{df}{dS} \frac{\partial S}{\partial x} = 0.$$

This gives us the following equation for the path of a value of constant saturation

$$\left(\frac{dx}{dt} \right)\bigg|_{dS=0} = \frac{v}{\phi} \frac{df(S)}{dS}. \tag{8.52}$$

If we assume that f is a function of S only, it follows that states of constant saturation will propagate along straight lines, i.e., along paths given by

$$x(t) = x_0(t_0) + \frac{v}{\phi} \frac{df(S)}{dS} (t - t_0).$$

To make a well-posed problem, we must pose initial and boundary conditions for the saturation equation (8.51) in the form $S(x,0) = S_{0,x}(x)$ and $S(0,t) = S_{0,t}(t)$. For the special

8.4 The Buckley–Leverett Theory of 1D Displacements

case of a linear fractional flow function $f(S) = S$, which corresponds to a displacement with linear relative permeabilities and equal viscotities, the analytical solution is given on closed form

$$S(x,t) = \begin{cases} S_{0,x}(x - at), & x \geq at, \\ S_{0,t}(at - x), & x < at, \end{cases}$$

where $a = v/\phi$. Loosely speaking, the interpretation of this solution is that the saturation equation takes the initial and boundary data and transports them unchanged along the x-axis with a speed a or a speed equal $f'(S)v/\phi$ in the general case. Notice that to get a classical solution, the initial and boundary data must be continuous and matching so that $S_{0,x}(0) = S_{0,t}(0)$. This way of constructing a solution is called the *method of characteristics* and has been used a lot throughout the literature of porous media to study various kinds of displacement scenarios.

Next, let us consider the case of one immiscible fluid displacing another, assuming constant initial saturation S_0 and a constant injected saturation S_i, where S_i generally is different from S_0. Without lack of generality, we set $v = \phi = 1$ or alternatively introduce new time $t^* = tv/\phi$ to rescale the saturation equation so that v and ϕ disappear. This gives the classical *Buckley–Leverett problem*

$$\frac{\partial S}{\partial t} + \frac{\partial f(S)}{\partial S} = 0, \quad S(x,0) = S_0, \quad S(0,t) = S_i, \tag{8.53}$$

named after the authors [56] who first developed this type of analysis. Here, we see that all changes in saturation must originate from the jump in values from S_0 to S_i at the point $(x,t) = (0,0)$ and that these changes, which we henceforth refer to as waves, will propagate along straight lines. In (8.52), we saw that a wave corresponding to a constant S-value will propagate with a speed proportional to the derivative of the fractional flow function f. In most cases, $f(S)$ has a characteristic S-shape with one inflection point, i.e., a point at which $f'(S)$ has an isolated extremum. In essence, this means that saturation values near the extremum should travel faster than saturation values on both sides of the extremum. Thus, if we apply the characteristic equation (8.52) naively, we end up with a multivalued solution as illustrated in Figure 8.9, which is clearly not physical. Indeed, if several waves emanate from the same location in space and time, the speeds of these waves have to be nondecreasing in the direction of flow so that the faster waves move ahead of the slower waves.

We can avoid the unphysical behavior if we simply replace the multivalued solution by a discontinuity, as shown in Figure 8.9. To ensure that mass is conserved, the discontinuity can be constructed graphically so that the two shaded areas are of equal size. This, however, means that the solution cannot be interpreted in the classical sense, since derivatives are not guaranteed to exist pointwise. Instead, the solution must be defined by multiplying the saturation equation by a smooth function φ of compact support and integrating the result over time and space

$$\iint \left[\frac{\partial S}{\partial t} + \frac{\partial f(S)}{\partial x} \right] \varphi(x,t) \, dx \, dt = 0.$$

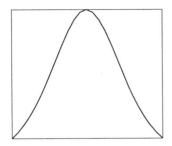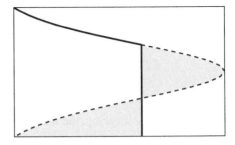

Figure 8.9 Construction of the self-similar Buckley–Leverett solution $S(x/t)$. The left plot shows $f'(S)$, whereas the right plot shows how the multivalued function given by a naive application of the characteristic equation (8.52) plotted as a dashed line is turned into a discontinuous solution plotted as a solid line by requiring that the shaded areas are of equal size.

We now integrate by parts and change the order of integration to transfer the derivatives to the smooth function φ,

$$\iint \left[S \frac{\partial \varphi}{\partial t} + f(S) \frac{\partial \varphi}{\partial x} \right] dx dt = 0. \tag{8.54}$$

A weak solution is then defined as any solution satisfying (8.54) for all smooth and compactly supported test functions φ.

In general, propagating discontinuities like the one in Figure 8.9 must satisfy certain conditions. To see this, let $x_d(t)$ denote the path of a propagating discontinuity, and pick two points x_1 and x_2 such that $x_1 < x_d(t) < x_2$. We first use the integral form of the saturation equation to write

$$f(S_1) - f(S_2) = -\int_{x_1}^{x_2} \frac{\partial f(S)}{\partial x} dx = \frac{d}{dt} \int_{x_1}^{x_2} S(x,t) \, dx, \tag{8.55}$$

and then decompose the last integral as follows:

$$\int_{x_1}^{x_2} S(x,t) \, dx = \lim_{\epsilon \to 0^+} \int_{x_1}^{x_d(t)-\epsilon} S(x,t) \, dx + \lim_{\epsilon \to 0^+} \int_{x_d(t)+\epsilon}^{x_2} S(x,t) \, dx.$$

Since the integrand is continuous in each of the integrals on the right-hand side of the equation, we can use the Leibniz rule to write

$$\frac{d}{dt} \int_{x_1}^{x_d(t)-\epsilon} S(x,t) \, dx = \int_{x_1}^{x_d(t)-\epsilon} \frac{\partial S}{\partial t} dx + S(x_d(t)-\epsilon, t) \frac{dx_d(t)}{dt}.$$

Here, the first term will vanish in the limit $x_1 \to x_d(t) - \epsilon$. By collecting terms, substituting back into (8.55), and taking appropriate limits, we obtain what is known as the *Rankine–Hugoniot condition*

$$\sigma(S^+ - S^-) = f(S^+) - f(S^-). \tag{8.56}$$

Here, $S^\pm$ denote the saturation values immediately to the left and right of the discontinuity and $\sigma = dx_d/dt$. Equation (8.56) describes a necessary relation between states on opposite

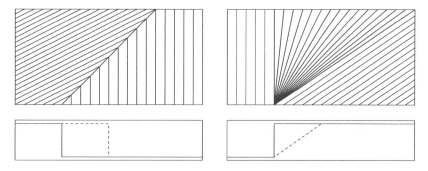

Figure 8.10 Illustration of self-sharpening (left) and spreading waves (right). The upper plots show the characteristics, with the characteristics belonging to the Riemann fan marked in blue. The lower plots show the initial solution (solid line) and the solution after some time (dashed line).

sides of a discontinuity, but does not provide sufficient conditions to guarantee that the resulting discontinuity is physically admissible. To this end, one must use a so-called *entropy condition* to pick out the physically correct solution. If the flux function is strictly convex with $f''(S) > 0$, or strictly concave with $f''(S) < 0$, a sufficient condition is provided by the *Lax entropy condition*,

$$f'(S^-) > \sigma > f'(S^+). \tag{8.57}$$

This condition basically states that the discontinuity must be a *self-sharpening* wave in the sense that saturation values in the interval between S^- and S^+ tend to come closer together upon propagation. Such a wave is commonly referred to as a *shock*, based on an analogy from gas dynamics. In the opposite case of a *spreading wave*, commonly referred to as a *rarefaction wave*, nearby states become more distant upon propagation, and for these continuous waves, the characteristic speeds are increasing in the flow direction. In Figure 8.9 this corresponds to the continuous part of the solution that decays towards the shock. For cases with convex or concave flux function, waves arising from discontinuous initial-boundary data will either be shocks or rarefactions, as illustrated in Figure 8.10. If the flux function is linear, or has linear sections, we can also have linearly degenerate waves for which the characteristic speeds are equal on both sides of the wave.

For cases where the flux function is neither strictly convex nor concave, one must use the more general *Oleinik entropy condition* to single out admissible discontinuities. This condition states that

$$\frac{f(S) - f(S^-)}{S - S^-} > \sigma > \frac{f(S) - f(S^+)}{S - S^+} \tag{8.58}$$

for all values S between S^- and S^+. For the classical displacement case with $S_0 = 0$ and $S_i = 1$, this condition implies that the correct saturation S^* at the shock front is given by

$$f(S^*)/S^* = f'(S^*), \tag{8.59}$$

i.e., the point on the fractional-flow curve at which the chord between the points $(0,0)$ and $(S^*, f(S^*))$ coincides with the tangent at the latter point.

If you are well acquainted with the theory of hyperbolic conservation laws, you will immediately recognize (8.53) as an example of a *Riemann problem*, i.e., a conservation law with a constant left and right state. Let these states be denoted S_L and S_R. If $S_L > S_R$, the solution of the Riemann problem is a self-similar function given by

$$S(x,t) = \begin{cases} S_L, & x/t < f_c'(S_L), \\ (f_c')^{-1}(x/t), & f_c'(S_L) < x/t < f_c'(S_R), \\ S_L, & x/t \geq f_c'(S_R), \end{cases} \quad (8.60)$$

which is often referred to as the *Riemann fan*. Here, $f_c(S)$ is the upper concave envelope of f over the interval $[S_R, S_L]$ and $(f_c')^{-1}$ the inverse of its derivative. Intuitively, the concave envelope is constructed by imagining a rubber band attached at S_R and S_L and stretched above f in between these two points. When released, the rubber band assumes the shape of the concave envelope and ensures that $f_c''(S) < 0$ for all $S \in [S_R, S_L]$. The case $S_L < S_R$ is treated symmetrically with f_c denoting the lower convex envelope. The petroleum literature usually attributes this construction to Welge [309]. Figure 8.10 shows examples of two such Riemann fans: a single shock in the left plot and a rarefaction fan in the right plot.

Example 8.4.1 *Let us first consider the Buckley–Leverett solution for a pure imbibition, for which the left and right states are $S_L = 1$ and $S_R = 0$. From the discussion above, we know that the self-similar solution of the associated Riemann problem is found by computing the upper concave envelope of the fraction flux function, which is linear in the interval $[0, S^*]$ and then coincides with $f(S)$ in the interval $[S^*, 1]$, where the saturation S^* behind the shock front is determined from (8.59). If we assume a simple case with relative permeabilities given by the Corey model (8.15) on page 241 with $n_w = n_o = 2$ and $\mu_w/\mu_o = M$, (8.59) reads*

$$\frac{S_M}{S_M^2 + M(1 - S_M)^2} = \frac{2M(1 - S_M)S_M}{\left(S_M^2 + M(1 - S_M)^2\right)^2}.$$

Here, we have used S_M to denote the front saturation $S^* = \sqrt{M/(M+1)}$ to signify that it depends on the viscosity ratio M. Figure 8.11 shows the Buckley–Leverett solution for three different viscosity ratios. In all three cases, the solution consists of a shock followed by a trailing continuous rarefaction wave. When the water viscosity is five times higher than oil ($M = 5$), the mobility of pure oil is five times higher than the mobility of pure water. Hence, we have a favorable *displacement in which the water front acts like a piston that displaces almost all the oil at once. In the opposite case with oil viscosity five times higher than water viscosity, we have an* unfavorable *displacement because the water front is only able to displace a small fraction of the oil. Here, the more mobile oil will tend to finger through the oil and create viscous fingers for displacements in higher spatial dimensions. We will come back to examples that illustrate this later in the book. In the opposite case of*

8.4 The Buckley–Leverett Theory of 1D Displacements

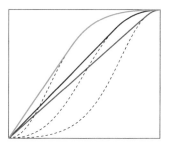
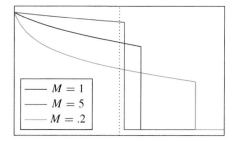

Figure 8.11 Buckley–Leverett solutions for a pure imbibition case with Corey relative permeabilities with exponent $n_w = n_0 = 2$ and mobility ratio $\mu_w/\mu_o = M$. The left plot shows the upper concave envelopes of the flux functions, while the right plot shows the self similar solution $S_w(x/t)$ with $x/t = 1$ shown as a dotted line.

a drainage process, it follows by symmetry that the solution will consist of a shock from 1 to $1 - S_M$ followed by a rarefaction wave from S_M to 0.

Instead of an infinite domain, let us place a producer some distance to the right of the origin. We then see that since the saturation is constant ahead of the shock, the amount of produced oil will be equal the amount of injected water until the first water reaches the producer. This means that the cumulative oil production has a linear slope until water breakthrough. After water has broken through, the well will be producing a mixture of oil and water and the slope of the cumulative oil production decreases.

Let us try to determine the amount of produced oil at water breakthrough. To this end, we can assume that water is injected at $x = 0$ and oil produced at $x = L$. At water breakthrough, the saturation at the producer is $S(x = L) = S^*$. The amount of produced oil equals the amount of water that has displaced it $L\bar{S}$, where $\bar{S}$ is the average water saturation defined as

$$\bar{S} = \frac{1}{L}\int_0^L S\,dx = \frac{1}{L}\Big[xS\Big]_{x=0}^{x=L} - \frac{1}{L}\int_1^{S^*} x\,dS$$
$$= S^* - \frac{t}{L}\int_1^{S^*}\left(\frac{df_c}{dS}\right)dS = S^* - \frac{t}{L}\big[f^* - 1\big].$$

The second equality follows from the total differential $d(xS) = x\,dS + S\,dx$, while the third equality follows from the equation (8.52) that describes the position $x(t)$ of a constant saturation value at time t. Finally, the water breakthrough occurs at time $t = L/f'(S^*)$, and hence we have that,

$$\frac{1 - f(S^*)}{\bar{S} - S^*} = f'(S^*).$$

We interpret this graphically as follows: S_w is the point at which the straight line used to define the front saturation S^* intersects the line $f = 1$ as shown in Figure 8.12.

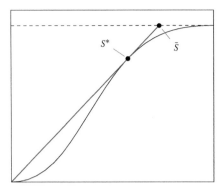

Figure 8.12 Graphical determination of front saturation and average water saturation at water breakthrough.

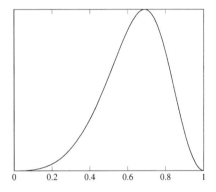

Figure 8.13 Fractional flow function for 1D gravity segregation for Corey relative permeabilities (8.15) with $n_w = 3, n_0 = 2$, and mobility ratio $\mu_w = 5\mu_o$ for $\rho_w > \rho_n$.

8.4.2 Gravity Segregation

Next, we consider a pure gravity displacement, for which the transport equation reads,

$$\frac{\partial S}{\partial S} + \frac{\partial}{\partial z}\left(\frac{\lambda_n \lambda_w}{\lambda_w + \lambda_n}\right) = 0, \tag{8.61}$$

where we, without lack of generality, have scaled away the constant $g\Delta\rho$. With a slight abuse of notation, we call the flux function for $g(S)$. This flux function will have a characteristic bell-shape, e.g., as shown in Figure 8.13, which points downward if $\rho_w > \rho_n$ and upward in the opposite case. We assume that the wetting fluid is most dense and consider the flow problem with one fluid placed on top of the other so that the fluids are separated by a sharp interface. If the lightest fluid is on top, we have that $S_L < S_R$, and from the discussion of (8.60) on page 262 we recall that the Riemann solution is found by computing the lower convex envelope of $g(S)$. Here, $g_c(S) \equiv 0$ and hence we have a stable situation. In the opposite case, $S_L > S_R$ and $g_c(S)$ is a strictly concave function for

8.4 The Buckley–Leverett Theory of 1D Displacements

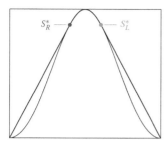 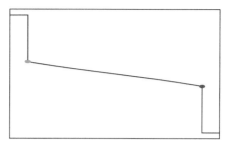

Figure 8.14 Buckley–Leverett solution for a gravity column with sealing top and bottom before the moving fluids have contacted the top/bottom of the domain. (Model: Corey relative permeabilities (8.15) with $n_w == n_0 = 2$, and mobility ratio $\mu_w = \mu_o$ and $\rho_w > \rho_n$.)

which $g'_c(0)' > 0 > g'_c(1)$. This is an unstable situation and the heavier fluid will start to move downward (in the positive z-direction) and the lighter fluid upward (in the negative z-direction) near the interface. Let us consider more details in a specific case:

Example 8.4.2 *We consider a gravity column inside a homogeneous sand body confined by a sealing medium at the top and the bottom. For simplicity, we scale the domain to be [0, 1], assume that the initial fluid interface is at $z = 0.5$, and use a Corey relative permeability model with $n_w = n_o = 2$ and equal viscosities.*

This choice of parameters gives a symmetric flux function $g(S)$, as shown in the left plot of Figure 8.14. Our initial condition with a heavy fluid on top of a lighter fluid is described by setting $S_L = 1$ in the top half of the domain and $S_R = 0$ in the bottom half of the domain. As explained above, the Riemann problem is solved by computing the upper concave hull of $g(S)$. The hull $g_c(S)$ is a linear function lying above $g(S)$ between 0 and S^*_R, follows the graph of $g(S)$ in the interval S^*_R to S^*_L, and is a linear function lying above $g(S)$ between S^*_L and 1. The point S^*_R is given by the solution to $g'(S^*_R)S^*_R = g(S^*_R)$. At the boundaries $x = 0, 1$, we have $g(S_L) = g(S_R) = 0$, and hence no-flow conditions as required. Initially, the solution thus consists of three different regions:

- *a single-phase region on top that only contains the heavier fluid ($S_L = 1$), which is bounded above by the lines $z = 0$ and below by the position of the shock $1 \to S^*_L$;*
- *a two-phase transition region that consists of a mixture of heavier fluid moving downward and lighter fluid moving upward; this corresponds to a centered rarefaction wave in which the saturation decays smoothly from S^*_L to S^*_R;*
- *a single-phase region at the bottom that only contains the light fluid. This region is bounded above by a shock $S^*_R \to 0$ and from below $z = 1$.*

As time passes, the two shocks that limit the single-phase regions of heavy and light fluids will move upward and downward, respectively, so that an increasing portion of the fluid column is in a two-phase state; see the right plot in Figure 8.14. Eventually, these shocks will reach the top and bottom of the domain. Because of the symmetry of the problem, it is sufficient for us to only consider one end of the domain, say $z = 0$. Since we have assumed

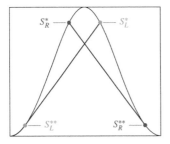 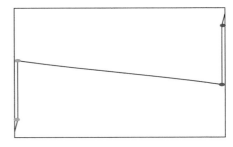

Figure 8.15 Buckley–Leverett solution for a gravity column with sealing top and bottom just after the light/heavy fluids started to accumulate at the top/bottom of the domain. To ensure no-flow, the solution is $S = 0$ at $z = 0$ and $S = 1$ at $z = 1$.

no flow across the top of the domain, the shock $1 \to S_L^*$ cannot continue to propagate upward. We thus get a Riemann problem, defined by right state S_L^* and a left state S_L, which could be any value satisfying $g(S_L) = 0$. The left state cannot be 1, since this would give back the same impermissible shock $1 \to S_L^*$. If the left value is $S_L = 0$ instead, i.e., a single-phase region with only light fluid forming on the top, the Riemann solution will be given by the concave envelope of $g(S)$ over the interval $[0, S_L^*]$ and hence will consist fully of waves with positive speed that propagate downward again. The left plot in Figure 8.15 shows the corresponding convex envelope, which we see gives a rarefaction wave 0 to S_L^{**} and a shock $S_L^{**} \to S_L^*$. As the leading shock propagates downward, it will interact with the upward-moving part of the initial rarefaction wave and form a shock, whose right state gradually decays towards $S = 0.5$. The solution at the bottom of the domain is symmetric, with accumulation of a single-phase region of the heavy fluid.

8.4.3 Front Tracking: Semi-Analytical Solutions

Whereas it was quite simple to find an analytical solution in closed form for simple Riemann problems like the ones discussed above, it becomes more complicated for general initial data or when different waves start to interact, as we observed when the two initial shocks reflected from the top and bottom of the gravity column. Generally, one must therefore resort to numerical discretizations in the form of a finite element, difference, or volume method. However, for one-dimensional saturation equations there is a much more powerful alternative, which we will introduce you to very briefly.

In the semi-analytical *front-tracking* method, the key idea is to replace the functions describing initial and boundary values by piecewise constant approximations. For a small time interval, the solution can then be found by piecing together solution of local Riemann problems, and consists of a set of constant states separated by shocks and rarefaction waves. Shocks correspond to the linear parts of the local convex/concave envelopes, whereas rarefaction waves correspond to the nonlinear parts of envelopes. If we further approximate the flux function by a piecewise linear function, all convex/concave envelopes are always

8.4 The Buckley–Leverett Theory of 1D Displacements

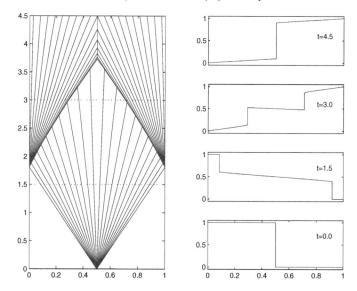

Figure 8.16 Solution for a gravity column with heavy fluid on top of a light fluid computed by front tracking for a model with Corey relative permeabilities with $n_w = n_o = 2$ and equal viscosities. The left column shows the wave pattern in the (z, t) plane, where each blue line corresponds to the propagation of a discontinuous state. The right column shows the solution $S(z, t)$ at four different times.

piecewise linear, and the solution of Riemann problems will consist only of shocks. In other words, the solution to the continuous, but approximate PDE problem consists entirely of constant states separated by discontinuities propagating at a constant speed given by the Rankine–Hugoniot condition (8.56) on page 260.

Each time two or more discontinuities "collide," we have a new Riemann problem that gives rise to a new set of discontinuities emanating out of the collision point. Computing local convex/concave envelopes and keeping track of propagating discontinuities can be formulated as numerical algorithms, and we end up with a semi-analytical method, which is unconditionally stable and formally first-order accurate. In fact, one can prove that for any finite spatial domain, there will be a finite number of shock collisions, and hence we can compute an approximate solution up to infinite time in a finite number of steps; see [134] for more details. In the following example, we use this method to study a few simple 1D cases.

Example 8.4.3 *We continue to study gravity segregation of a heavy fluid placed on top of a lighter fluid inside a sand box with sealing top and bottom. Figure 8.16 shows the solution computed by approximating the flux function with a piecewise linear function with 100 equally spaced segments. Initially, the solution contains two constant states $S_L = 1$ and $S_R = 0$, separated by a composite wave that consists of an upward-moving shock, a centered rarefaction wave, and a downward moving shock, as explained earlier* in Example 8.4.2. Notice that the approximate solution at time $t = 1.5$ is of the same form as shown in Figure 8.14, except that in the approximate front-tracking solution, the continuous rarefaction wave is replaced by a sequence of small discontinuities.

The right plot in Figure 8.15 illustrated the form of the exact solution just after the two initial shocks have reflected at the top and bottom of the domain. Figure 8.16 shows the solution after the reflected waves have propagated some time upward/downward. Here, we clearly see that these waves consist of a fast shock wave followed by a rarefaction wave. As the leading shock interacts with the centered rarefaction wave originating from $(0.5, 0)$, the difference between left and right states diminishes, and the shock speeds decay. Later, the two shocks collide and form a stationary discontinuity that will eventually become a stable steady state with the lighter fluid on top and the heavier fluid beneath.

In general, the transport in an inclined reservoir section will be driven by a combination of gravity segregation and viscous forces coming from pressure differentials. The next example studies one such case.

Example 8.4.4 (CO_2 storage) *We consider two highly idealized models of CO_2 storage in an inclined aquifer with a slow drift of brine either in the upslope or in the downslope direction. If the injection point lies deeper than approximately 800 meters below the sea level, CO_2 will appear in the aquifer as a supercritical liquid phase having much lower density than the resident brine. The injected fluid, which is commonly referred to as the CO_2 plume, will therefore tend to migrate in the upslope direction. As our initial condition, we assume (somewhat unrealistically) that CO_2 has been injected so that it completely fills a small section of the aquifer $x \in [\frac{1}{4}, \frac{3}{4}]$. To study the migration after injection has ceased, we can now use a fractional flow function of the form,*

$$F(S) = \frac{S^2 \mp 4S^2(1-S)^3}{S^2 + \frac{1}{5}(1-S)^3},$$

where the minus sign denotes upslope background drift and the plus sign downslope drift. Figure 8.17 shows the flux functions along with the envelopes corresponding to the two Riemann problems.

Let us first consider the case with upslope drift. At the tip of the plume ($x = \frac{3}{4}$), there will be a drainage process in which the more buoyant CO_2 displaces the resident brine. This gives a composite wave that consists of a shock $S_d^* \to 1$ that propagates upslope, followed by a centered rarefaction wave from S_d^* to S_d^{**} having wave speeds both upslope and downslope. The rarefaction is followed by a trailing shock wave $0 \to S_d^{**}$ that propagates in the downslope direction. Likewise, at the trailing edge of the plume, the resident brine will imbibe into CO_2, giving a wave consisting of a shock $S_i^* \to 0$ followed by a rarefaction wave, both propagating in the upslope direction. The two Riemann fans can be seen clearly in Figure 8.18, and will stay separated until $t \approx 0.18$, when the shock from the imbibition process collides with the trailing drainage shock. The result of this wave interaction is a new and slightly weaker shock that propagates increasingly faster in the upslope direction,

8.4 The Buckley–Leverett Theory of 1D Displacements

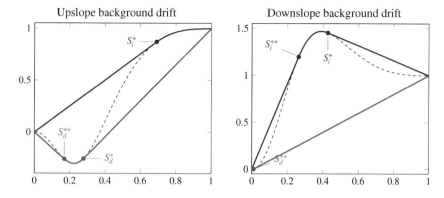

Figure 8.17 Flux functions for the CO_2 migration example. The dashed curves are the flux function $F(S)$, which accounts for a combination of gravity segregation and viscous background drift. The blue lines show the concave envelopes corresponding to imbibition at the left edge of the plume, while the red lines are the convex envelopes corresponding to drainage at the right edge the plume.

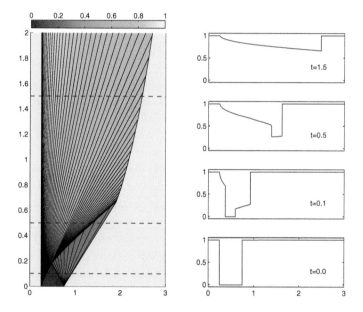

Figure 8.18 Solution for a conceptual model of CO_2 injected into an inclined 1D aquifer with a background upslope drift. Upslope direction is to the right. Dashed lines in the left plot signify the snapshots shown in the right column.

followed by a continuous rarefaction wave, which is an extension of the initial imbibition rarefaction. At time $t \approx 0.665$, the new shock wave has overtaken the tip of the plume, and the result of this interaction is that the upslope migration of the plume gradually slows down.

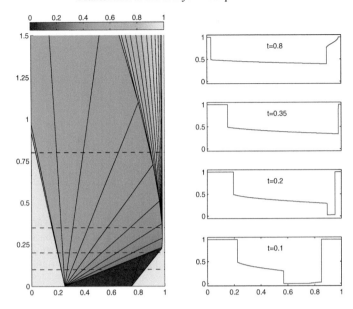

Figure 8.19 Solution for a conceptual model of CO_2 injected into an inclined 1D aquifer with a background downslope drift. Upslope direction is to the left.

If we remove the upslope drift, the initial imbibition process is no longer present, but the long-term behavior remains the same and consists of a leading shock wave followed by a rarefaction that slowly eats up and slows down the shock.

Figure 8.19 shows the opposite case with a downslope background drift. Here, the imbibition process at the left edge of the plume gives a shock $0 \to S_i^{**}$ that propagates in the downslope direction (to the right in the figure), a centered rarefaction wave from S_i^{**} to S_i^* with waves that propagate both in the downslope and upslope direction, followed by a shock $S_i^* \to 1$ that propagates upslope. The drainage at the right edge gives a strong shock $S_d^* \to 1$ that propagates downslope, followed by a weak rarefaction wave from 0 to S_d^*. At time $t \approx 0.25$ the downslope imbibition shock has overtaken both the rarefaction wave and the shock from the drainage process. The resulting wave interaction first forms a stationary shock with left state S_i^{**}. However, as the positive part of the imbibition rarefaction wave continues to propagate downslope, the left state gradually increases, and at some point the left state passes the value $S = \sqrt{1/10}$ for which $F(S) = 1$. When this happens, the stationary shock turns into a composite wave consisting of a shock and a trailing rarefaction wave that both propagate in the upslope direction.

Figure 8.20 shows the two migration cases over a longer time. Injecting CO_2 so that it must migrate upward against a downslope brine drift is clearly advantageous since it will delay the upslope migration of the gradually thinning plume.

The purpose of the previous example was to introduce you to the type of wave patterns that arise in hyperbolic saturation equations, and show that these can be relatively complicated even for simple initial configurations. While front-tracking is probably the best method

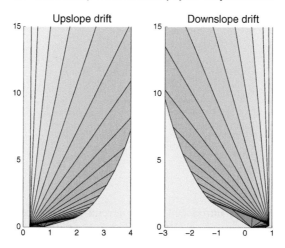

Figure 8.20 Long-time solutions for the two conceptual models of CO_2 storage.

you can find to study wave patterns in 1D hyperbolic saturation equations, it would not be the method of choice to study multiphase flow models in multiple spatial dimensions, unless we use a streamline method that transform the 3D transport problems to a family of 1D problems. In the next chapter, we therefore continue to discuss finite-volume methods that are more suitable for multidimensional reservoir simulation. In simulation of real reservoirs, one cannot hope to resolve the dynamics to the level of detail seen in the last example. To be able to compute shocks and other types of discontinuities, finite-volume (and finite-element) methods contain a certain amount of numerical dissipation that tends to smear any discontinuity. Likewise, strong heterogeneities and radial flow near wells can introduce orders of magnitude variations in Darcy velocities. This necessitates the use of small time steps or implicit temporal discretizations, which both tend to smear discontinuities and make it more difficult to resolve complex wave interactions in full detail.

We end our discussion of 1D solutions with a few remarks about the simulation of CO_2 storage, since this was touched upon conceptually in Example 8.4.4. In more realistic modeling of CO_2 storage, one would need a more detailed model that accounts for the effect that the lighter CO_2 plume will flow to the top of the formation and migrate upslope as a thin layer under the sealing caprock that bounds the aquifer from above. This can be done by either using a 3D or a 2D cross-sectional saturation equation, possibly coupled with a pressure equation. However, the most efficient approach to study large-scale, long-term CO_2 migration is to integrate the pertinent flow equations in the vertical direction to form a vertically averaged model that accounts for the vertical fluid distribution in an averaged sense. MRST has a large module, co2lab, that offers a wide variety of vertically averaged models and other types of reduced models suitable for modeling of large-scale CO_2 storage. A discussion of such models is beyond the scope of this book. I you are interested in more details, you should instead consult one of the papers describing the methods implemented in MRST-co2lab [19, 230, 227, 228, 232, 193, 21].

9
Discretizing Hyperbolic Transport Equations

So far in the book, we have introduced you to several basic discretizations: Section 4.4 introduced the two-point scheme for elliptic operators of the type $\nabla \cdot \mathbf{K}(\vec{x})\nabla$ and the upstream scheme for hyperbolic operators $\nabla \cdot \vec{v}$. Likewise, Section 7.2 introduced fully implicit discretizations for compressible single-phase flow, and discussed how to use a Newton–Raphson method to linearize and solve the resulting system of discrete equations. With some modifications, these techniques are all we need to describe the most widespread approach to simulate compressible, multiphase flow.

For incompressible multiphase flow, on the other hand, it is common to write the system as a pressure equation and one or more transport equations and use specialized discretization schemes for each subequations. Chapter 6 discussed consistent schemes for elliptic pressure equations. This chapter reviews some of the particular challenges that lie in discretizing transport equations. If you are not interested in this wider perspective and only wish to know the simplest possible approach to solving saturation equations, you can jump directly to Section 9.4 on page 286, which discusses the standard upstream mobility-weighting method implemented for general polyhedral grids in the `incomp` module of MRST.

9.1 A New Solution Concept: Entropy-Weak Solutions

In the first part of this chapter, we continue to discuss the homogeneous conservation law introduced in the previous chapter,

$$u_t + f(u)_x = 0, \qquad u(x,0) = u_0(x). \tag{9.1}$$

Here, u is some conserved quantity, which need not necessarily be a fluid saturation, and $f(u)$ is a generic flux function. Equation (9.1) usually arises from a more fundamental physical law on integral form,

$$\frac{d}{dt}\int_{x_1}^{x_2} u(x,t)\,dx = f\big(u(x_1,t)\big) - f\big(u(x_2,t)\big), \tag{9.2}$$

which states that the rate of change of quantity u within the interval $[x_1, x_2]$ equals the flux across the ends $x = x_1$ and $x = x_2$ of the interval.

9.1 A New Solution Concept: Entropy-Weak Solutions

In the previous chapter, we saw that solutions to (9.1) may develop discontinuities in finite time, even for smooth initial data if the flux function f is nonlinear. This means that the solution of (9.1) is usually understood in the weak sense,

$$\int_0^\infty \int_{\mathbb{R}} \left(u\varphi_t + f(u)\varphi_x\right) dt\,dx = \int_{\mathbb{R}} u_0(x)\varphi(x,0)\,dx. \tag{9.3}$$

Here, $\varphi(x,t)$ is a continuous and smooth test function that has *compact support* so that it vanishes outside a bounded region in the (x,t)-plane.

Solutions defined by the weak form (9.3) are not necessarily unique. The solution concept must therefore be extended to include additional admissibility conditions to single out the correct solution among several possible candidates satisfying the weak form. A classical method to obtain uniqueness is to add a regularizing second-order term to (9.1), giving a parabolic equation

$$u_t^\epsilon + f(u^\epsilon)_x = \epsilon u_{xx}^\epsilon,$$

which has unique, smooth solutions. The unique solution of the hyperbolic equation (9.1) is then defined as the limit of $u^\epsilon(x,t)$ as ϵ tends to zero. In models from fluid dynamics, such a second-order term can be proportional to the viscosity of the fluid, and the method is therefore called the *vanishing viscosity method*. For flow in porous media, the transport equations will already have a second-order term coming from capillary forces (see (8.35)). However, capillary forces are often neglected when studying flow dominated by global pressure gradients, or the capillary function may vary with rock type and this can introduce discontinuities in the phase saturations at the interface between different rock types.

Since the vanishing viscosity solutions $u^\epsilon(x,t)$ are smooth, it is possible to use techniques from classical analysis to prove the existence, uniqueness, and stability of the solution to (9.1). This was first done in a seminal work by Kružkov [173], which paved the road for the modern theory of nonlinear partial differential equation of a hyperbolic-parabolic type. Working with limits of viscous solutions is not very practical. Instead, we impose other admissibility conditions. We have already encountered the Lax and Oleinik entropy conditions, (8.57) and (8.58), which were used to single out permissible discontinuities. An alternative approach is to introduce a (convex) entropy function $\eta(u)$ and a corresponding entropy flux $\psi(u)$, and require that an admissible weak solution u must satisfy the entropy condition

$$\eta(u)_t + \psi(u)_x \leq 0, \tag{9.4}$$

which must be interpreted in the weak sense as

$$\int_0^\infty \int_{\mathbb{R}} \left(\eta(u)\varphi_t + \psi(u)\varphi_x\right) dt\,dx + \int_{\mathbb{R}} \eta(u_0(x))\varphi(x,0)\,dx \geq 0. \tag{9.5}$$

The solution $u(x,t)$ is called an *entropy weak solution* of (9.1) if it satisfies (9.5) for all $k \in \mathbb{R}$ and nonnegative test functions φ.

9.2 Conservative Finite-Volume Methods

You have already seen that using a finite-volume method made it simple to formulate a discretization on general polyhedral grids and ensured that the resulting method obeys the important property of mass conservation. Approximating the unknown function in terms of its cell averages has one additional advantage for hyperbolic conservation laws. Because the differential equation ceases to be pointwise valid in the classical sense once a discontinuity arises in $u(x,t)$, we should expect that pointwise approximations used in standard finite-difference methods will break down, since these rely on the assumption that the function is smooth in a small neighborhood where the discrete difference is taken. Instead of seeking solutions in a *pointwise* sense, the finite-volume approach seeks *globally defined solutions* of the integral form (9.2), which is more fundamental, represented in the form of discrete cell averages, as we have already seen in previous chapters.

To develop a finite-volume method for (9.1), we first define the sliding average,

$$\bar{u}(x,t) = \frac{1}{\Delta x}\int_{x-\frac{1}{2}\Delta x}^{x+\Delta x/2} u(\xi,t)\,d\xi. \tag{9.6}$$

We then associate these sliding averages with grid cells, $\{[x_{i-1/2}, x_{i+1/2}]\}$, where $x_{i\pm 1/2} = x_i \pm \frac{1}{2}\Delta x_i$, and set $u_i(t) = \bar{u}(x_i,t)$. By inserting $u_i(t)$ into the integral form of the conservation law (9.2), we obtain a semi-discrete version of (9.1),

$$\frac{d\bar{u}_i}{dt} = \frac{1}{\Delta x_i}\Big[f\big(u(x_{i-1/2},t)\big) - f\big(u(x_{i+1/2},t)\big)\Big]. \tag{9.7}$$

Alternatively, we may define $u_i^n = \bar{u}_i(t_n)$ for a set of discrete times t_n and then integrate (9.7) from t_n to t_{n+1} to derive a fully discrete version of (9.1),

$$u_i^{n+1} - u_i^n = \frac{1}{\Delta x}\int_{t_n}^{t_{n+1}} f(u(x_{i-1/2},t))\,dt - \frac{1}{\Delta x}\int_{t_n}^{t_{n+1}} f(u(x_{i+1/2},t))\,dt. \tag{9.8}$$

While (9.7) and (9.8) tell us how the unknown cell averages $u_i(t)$ and u_i^n evolve in time, they cannot be used directly *to compute* these cell averages, since we do not know the point values $u(x_{i\pm 1/2},t)$ needed to evaluate the integrand. To obtain these, we need to make additional assumptions and approximations, e.g., as discussed in Section 4.4.3 for the linear case. Nevertheless, (9.8) suggests that a viable numerical method for (9.1) should be of the form

$$u_i^{n+1} = u_i^n - r_i^n\big(F_{i+1/2}^n - F_{i-1/2}^n\big), \qquad r_i^n = \Delta t^n/\Delta x_i, \tag{9.9}$$

where $F_{i\pm 1/2}^n$ approximates the average flux over each cell interface,

$$F_{i\pm 1/2}^n \approx \frac{1}{\Delta t^n}\int_{t_n}^{t_{n+1}} f\big(u(x_{i\pm 1/2},t)\big)\,dt. \tag{9.10}$$

In Section 8.4 we showed that information propagates at a finite speed along so-called characteristics, which implies that the flux integral (9.10) will only depend on the solution $u(x,t_n)$ in a local neighborhood of the interface $x_{i\pm 1/2}$. This means, in turn, that $F_{i\pm 1/2}$ can be approximated in terms of a small collection of neighboring cell averages, i.e., $F_{i+1/2} = F(u^n_{i-p}, \ldots, u^n_{i+q})$, which we alternatively will write as $F_{i\pm 1/2} = F(u^n; i \pm \frac{1}{2})$.

Any numerical method written on the form (9.9) will be *conservative*. To see this, we multiply (9.9) by Δx_i and sum over i. The flux terms will cancel in pairs and we are left with

$$\sum_{i=-M}^{N} u^{n+1}_i \Delta x_i = \sum_{i=-M}^{N} u^n_i \Delta x_i - \Delta t^n \left(F^n_{N+1/2} - F^n_{-M-1/2} \right).$$

From this, it follows that the method is conservative on a finite domain, since the accumulation inside any interval balances the sum of the flux across the interval edges. To see that the same is true in an infinite domain, we must make some additional assumptions. If the initial solution $u_0(x)$ has bounded support, it follows that the two flux terms will cancel if we choose M and N sufficiently large. Hence, the scheme (9.9) is conservative in the sense that

$$\sum_i u^{n+1}_i \Delta x_i = \sum_i u^n_i \Delta x_i = \cdots = \int u_0(x)\, dx.$$

9.3 Centered versus Upwind Schemes

To provide important insight into the numerical solution of hyperbolic equation, we briefly outline two main classes of classical schemes.

9.3.1 Centered Schemes

The simplest approach to *reconstruct* the point values $u(x_{i\pm 1/2}, t)$ needed in (9.7) and (9.9), is to assume that $u(x,t) = u_i(t)$ inside each grid cell and average the flux values on opposite sides of the grid interface, i.e., set

$$F^n_{i\pm 1/2} = \tfrac{1}{2}\left[f(u^n_{i\pm 1}) + f(u^n_i) \right]. \tag{9.11}$$

This approximation is referred to as a *centered approximation*, which unfortunately gives a notoriously unstable scheme. We can add an artificial diffusion term, $\frac{\Delta x^2}{\Delta t} \partial^2_x u$, to stabilize, but then it is no longer possible to go back to the semi-discrete form (9.7), since the artificial diffusion will blow up in the limit $\Delta t \to 0$. Discretizing the artificial diffusion through standard centered differences gives the classical Lax–Friedrichs scheme

$$u^{n+1}_i = \frac{1}{2}\left(u^n_{i+1} + u^n_{i-1} \right) - \frac{1}{2} r \left[f(u^n_{i+1}) - f(u^n_{i-1}) \right]. \tag{9.12}$$

Alternatively, the scheme can be written in conservative form (9.9) using the numerical flux

$$F(u^n; i+1/2) = \frac{1}{2r}\left(u^n_i - u^n_{i+1} \right) + \frac{1}{2}\left[f(u^n_i) + f(u^n_{i+1}) \right]. \tag{9.13}$$

To ensure stability, we have to impose a restriction on the time-step through a *CFL condition*, named after Courant, Friedrichs, and Lewy, who wrote one of the first papers on finite-difference methods in 1928 [75]. The CFL condition states that the true domain of dependence for the PDE (9.1) should be contained in the domain of dependence for (9.9). For the Lax–Friedrichs scheme, this means that

$$\frac{\Delta t}{\Delta x} \max_u |f'(u)| \leq 1. \tag{9.14}$$

Under this condition, the scheme is very robust and will always converge, albeit painstakingly slow in many cases. The robustness is largely due to the added numerical diffusion, which tends to smear discontinuities. As an illustration, consider a stationary discontinuity satisfying $\partial_t u = 0$. Here, the approximate solution in each cell is defined as the arithmetic average of the cell averages in the neighboring cells, i.e., $u_i^{n+1} = \frac{1}{2}(u_{i+1}^n + u_{i-1}^n)$. From this, we see that the scheme will smear the discontinuity one cell in each direction per time step. By using a formal Taylor expansion on a single time step, it follows that the scheme has a truncation error of order two. In practice, we are more interested in the error at a fixed time, which we need to use an increasing number of time steps to reach as $\Delta x \to 0$ because of (9.14). This means that we must divide with Δt so that the error of the scheme is $\mathcal{O}(\Delta x)$ as $\Delta x \to 0$, and we say that the scheme is *formally first-order accurate*.

We can improve accuracy if we make a more accurate approximation to the integral defining $F_{i\pm1/2}^n$. Instead of evaluating the integral at the endpoint t_n, we can evaluate it at the midpoint $t_{n+1/2} = t_n + \frac{1}{2}\Delta t$. One can show that the corresponding point values can be *predicted* with acceptable accuracy by the Lax–Friedrichs scheme on a grid with half the grid spacing. This gives a second-order, predictor-corrector scheme called the (Richtmeyer two-step) Lax–Wendroff method [178]

$$\begin{aligned} u_{i+1/2}^{n+1/2} &= \tfrac{1}{2}(u_i^n + u_{i+1}^n) - \tfrac{1}{2}r\big[f(u_{i\pm1}^n) - f(u_i^n)\big], \\ u_i^{n+1} &= u_i - r\big[f(u_{i+1/2}^{n+1/2}) - f(u_{i-1/2}^{n+1/2})\big], \end{aligned} \tag{9.15}$$

which is stable under the same CFL condition (9.14) as the Lax–Friedrichs scheme. The corresponding numerical flux reads

$$F(u^n; i + 1/2) = f\left(\tfrac{1}{2}(u_i^n + u_{i+1}^n) - \tfrac{1}{2}r\big[f(u_{i+1}^n) - f(u_i^n)\big]\right). \tag{9.16}$$

Both the Lax–Friedrichs and the Lax–Wendroff scheme can either be interpreted as a finite-difference or as a finite-volume scheme, and by only looking at the formulas written in terms of u_i^n, it is not possible to determine which formulation is in use. However, the underlying principles are fundamentally different. Finite-difference methods evolve a discrete set of point values by using discrete differences to approximate the differential operators in (9.1). Finite-volume methods evolve *globally defined solutions* given by (9.2) and *realize* them in terms of a discrete set of cell averages. This latter perspective is the key to modern so-called high-resolution methods [183, 294], which have proved to be very successful.

Let us now see how we can implement these two schemes compactly in MATLAB. When computing on a bounded domain, one generally has to modify the stencil of the scheme in the cells next to the domain boundary to account for boundary conditions.

9.3 Centered versus Upwind Schemes

A widely used trick to avoid this is to pad the domain with a layer of *ghost cells*, i.e., extra cells that are not really part of the domain and whose states can be set manually to impose the desired boundary conditions.

```
function u=lxf(u0,cfl,dx,T,flux,df,boundary)
u = u0; t = 0.0;
dt = cfl*dx/max(abs(df(u0)));
i = 2:numel(u0)-1;   % do not compute on the ghost cells
while (t<T)
  if t+dt>T, dt = T-t; end
  t=t+dt;
  u = boundary(u);   % set values in the ghost cells
  f = flux(u);
  u(i) = 0.5*(u(i+1)+u(i-1)) - 0.5*dt/dx*(f(i+1)-f(i-1));
  dt = cfl*dx/max(abs(df(u)));
end
```

Here, u0 gives the initial data, cfl and dx are the CFL number and the spatial discretization parameter, T is the desired end time, and flux, df, and boundary are handles to functions that compute the flux f and its derivative and set appropriate states in the ghost cells. Notice, in particular, how we use the vector i to avoid computing in the ghost cells that pad our domain. To get the Lax–Wendroff scheme, we introduce an auxiliary array U and then simply replace the second last line by

```
U(i) = 0.5*(u(i)+u(i+1)) - 0.5*r*(f(i+1)-f(i));
U = boundary(U);  f = flux(U);
u(i) = u(i) - r*(f(i)-f(i-1));
```

9.3.2 Upwind or Godunov Schemes

The centered schemes introduced in the previous subsection are black-box schemes that can be utilized without any particular knowledge about the flux function apart from an estimate of its maximum and minimum derivative (or eigenvalues for systems of equations). The reason for this is that these schemes utilize information from both sides of the grid interface when computing approximations to the flux integral in (9.10). In many cases we can do better, since we know more of how the solution behaves at the interface. If $f'(u) > 0$, for instance, we know that all characteristics are positive, and hence we can use the solution from the left side of the grid interface to evaluate the integral in (9.10). With a constant reconstruction inside each cell, this means that $F_{i+1/2} = f(u_i^n)$ is the *exact* value of the integral, provided that the solution at time t_n is constant and equal u_i^n in grid cell i. The resulting scheme,

$$u_i^{n+1} = u_i^n - r\big[f(u_i^n) - f(u_{i-1}^n)\big], \tag{9.17}$$

is called an *upwind scheme*, which you have already encountered in Chapters 4 and 7. Similarly, if $f'(u) \leq 0$, the upwind scheme takes the form

$$u_i^{n+1} = u_i^n - r[f(u_{i+1}^n) - f(u_i^n)]. \tag{9.18}$$

In either case, the upwind scheme is a two-point scheme that, instead of using a centered difference, uses a one-sided difference in the *upwind* direction, i.e., in the direction from which the waves (or information) is coming. This type of upwind differencing is the underlying design principle for so-called *Godunov schemes*, named after the author of the seminal paper [118]. If we assume a constant reconstruction so that $u(x,t^n) \equiv u_i^n$ inside grid cell number i, the evolution of $u(x,t)$ can be decomposed into a set of local Riemann problems

$$\partial_t v + \partial_x f(v) = 0, \qquad v(x,0) = \begin{cases} u_i^n, & x < x_{i+1/2}, \\ u_{i+1}^n, & x \geq x_{i+1/2}, \end{cases} \tag{9.19}$$

which are of the same type as those we have already accounted in Section 8.4. Each of these Riemann problems admits a self-similar solution $v_{i+1/2}(x/t)$, which can be constructed by introducing a local convex or concave envelope[1] $f_c(u; u_i^n, u_{i+1}^n)$ of the flux function f over the interval $[u_i^n, u_{i+1}^n]$, as discussed in (8.60). The cell averages can then be correctly evolved a time step Δt forward in time by (9.9) if we use $v_{i\pm1/2}(0)$ – or a good approximation thereof – to evaluate f in the flux integral (9.10). The Riemann fan will generally expand in each direction at a constant speed, and using $v_{i\pm1/2}(0)$ to evaluate the integrand of (9.10) is only correct until the first wave from the neighboring Riemann problem reaches the interface. This implies a time-step restriction of the form

$$\max_{i,n} |f_c'(\cdot; u_i^n, u_{i+1}^n)| \frac{\Delta t^n}{\Delta x_i} \leq 1, \tag{9.20}$$

which we recognize as a sharper version of the CFL condition (9.14). The resulting scheme is formally first-order accurate, but will generally be much less diffusive than the centered Lax–Friedrichs scheme.

Let us now go back to the general transport equation (8.35), which is hyperbolic if $P_c' \equiv 0$. Here, the fractional flow function $f(S)$ has a characteristic S-shape and its derivative is therefore always positive. This means that if gravity forces are negligible, like in the horizontal displacements we studied in Section 8.4.1, the correct Godunov scheme would be either (9.17) or (9.18), depending upon the sign of the flux $\vec{v}$. For the gravity segregation problem considered in Section 8.4.2, on the other hand, the flux function $g(S) = \lambda_w \lambda_n/(\lambda_w + \lambda_n)$ has derivatives of both signs and we would generally have to solve the Riemann problem (9.19) like we did in Figure 8.14 on page 265 to apply the Godunov scheme. The same is generally true for flow involving a combination of viscous and gravity forces, as illustrated in Figure 8.17. Having to solve a Riemann problem each time we need to evaluate the flux $F_{i\pm1/2}^n$ may seem rather cumbersome, and fortunately there is a rather elegant and physically motivated fix to this. Instead of upwinding the flux itself, we can use the phase fluxes to evaluate the relative mobilities in the upstream direction

---

[1] The notation $f(\cdot; a, b)$ signifies that the functional form of the function f depends on the parameters a and b.

9.3 Centered versus Upwind Schemes

$$\lambda_\alpha(S)^u\big|_{x_{i+1/2}} = \begin{cases} \lambda_\alpha(S_i^n), & \text{if } v_\alpha > 0, \\ \lambda_\alpha(S_{i+1}^n), & \text{otherwise.} \end{cases} \qquad (9.21)$$

and then use these upwind values to evaluate both the S-shaped viscous flux $f_w(S)$ and the bell-shaped gravity flux $\lambda_n(S) f_w(S)$. Here, we have used the phase flux v_α, which in the multidimensional case can be computed as $(\nabla p_\alpha - g\rho_\alpha \nabla z)\cdot \vec{n}_i$, where $\vec{n}_i$ is the normal to the interface. The resulting method, called the explicit *upstream mobility-weighting* scheme, is usually the method of choice for porous media application because of its accuracy and simplicity. In [49] it was shown that the method also satisfies the mathematical properties necessary to get a consistent and convergent scheme.

9.3.3 Comparison of Centered and Upwind Schemes

The classical schemes introduced so far in this section all have certain weaknesses and strengths that are common for many other schemes as well. In short, first-order schemes tend to smear discontinuous solutions, whereas second-order schemes introduce spurious oscillations. To highlight these numerical artifacts, we consider two different cases; you can find complete source codes for both examples in the `nummet` directory of the `book` module. The directory also contains two examples of high-resolution methods, which to a large extent cure these problems. Very cursorily explained, the idea of such a scheme is to introduce a nonlinear function that enables the scheme to switch from being high-order on smooth solutions to being first-order near discontinuities.

Example 9.3.1 *Consider linear advection on the unit interval with periodic boundary conditions*

$$u_t + u_x = 0, \qquad u(x,0) = u_0(x), \qquad u(0,t) = u(1,t).$$

The advantage of using periodic data is that we know that the exact solution at time $t=n$ for an integer number n equals $u_0(x)$. As initial data $u_0(x)$ we choose a combination of a smooth squared sine wave and a double step function,

$$u(x,0) = \sin^2\left(\frac{x-0.1}{0.3}\pi\right)\chi_{[0.1,0.4]}(x) + \chi_{[0.6,0.9]}(x).$$

Using the `lxf` *solver outlined in Section 9.3.1, we can set up the case as follows, where the function* `periodic` *simply sets* `u(1)=u(end-1)` *and* `u(end)=u(2)`.

```
dx = 1/100;
x  = -.5*dx:dx:1+.5*dx;
u0 = sin((x-.1)*pi/.3).^2.*double(x>=.1 & x<=.4);
u0((x<.9) & (x>.6)) = 1;

f  = @(u) u;        % flux function
df = @(u) 0*u+1;
uf = lxf(u0, .995, dx, 20, f, df, @periodic);
```

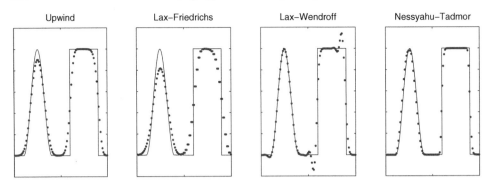

Figure 9.1 Approximate solutions at time $t = 20.0$ for the linear advection equation $u_t + u_x = 0$, with periodic boundary conditions computed by three classical schemes and a high-resolution scheme on a grid with 100 cells and CFL number 0.995.

Figure 9.1 shows approximate solutions after 20 periods ($t = 20$) computed by the upwind, Lax–Friedrichs, and Lax–Wendroff methods and the high-resolution Nessyahu–Tadmor scheme [224] on a grid with 100 cells for $\Delta t = 0.995 \Delta x$. Both the centered schemes clearly give unacceptable resolution of the solution profile. The first-order Lax–Friedrichs scheme smears both the smooth and the discontinuous part of the advected profile. The second-order Lax–Wendroff scheme preserves the smooth profile quite accurately, but introduces spurious oscillations at the two discontinuities. This behavior is representative for classical schemes. The upwind method represents a reasonable compromise in the sense that it does not smear the smooth wave and the discontinuities as much as Lax–Friedrichs, and does not introduce oscillations like Lax–Wendroff. The Nessyahu–Tadmor scheme gives accurate resolution of both the smooth and the discontinuous parts of the profile, thereby combining the best of the low and high-order schemes.

The advection equation can serve as a conceptual model for more complex cases. As we saw in Figures 8.9 and 8.10, a displacement front will be self-sharpening in the sense that if the leading discontinuity is smeared, smeared states behind the displacement front will travel faster than the discontinuity (since they have characteristics pointing into it) and will hence "catch up." Likewise, smeared states ahead of the front will travel slower and be overrun by the front, since these states have characteristics that point into the front. For linear waves, the characteristics of any smeared state will run parallel to the discontinuity and hence there will be no self-sharpening to counteract the numerical dissipation inherent in any numerical scheme. Linear waves are therefore more susceptible to numerical dissipation than nonlinear shock waves.

In water-based EOR methods, for instance, active chemical or biological substances are added to modify the physical properties of the fluids or/and the porous media at the interface between oil and water. The resulting displacement processes are governed by complex interplays between the transport of chemical substances and how these substances affect the flow by changing the properties of the fluids and the surrounding rock. These

9.3 Centered versus Upwind Schemes   281

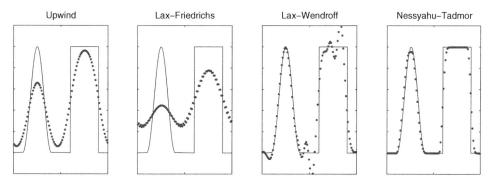

Figure 9.2 Approximate solutions at time $t = 1.0$ for the linear advection equation $u_t + u_x = 0$, with periodic boundary conditions computed by three classical schemes on a grid with 100 cells and time step of 0.5 relative to the stability limit.

property changes are often very nonlinear and highly sensitive to threshold parameters that determine sharp transitions between regions of very different behavior. The transport of chemical substances is largely linear and hence more affected by numerical diffusion. A simulation may therefore fail to resolve the local displacement process if the chemical fronts are smeared out or contain overshoots. As a result, unresolved simulation can therefore lead to misleading predictions of injectivity and recovery profiles.

Example 9.3.2 *For many systems, the linear waves travel significantly slower than the leading shock wave, and hence will be computed with a smaller effective CFL number. In Figure 9.2, we have rerun the experiment assuming that the linear waves travel at half the speed of a leading shock wave. Even after one period, the first-order schemes have introduced significantly more smearing, and the oscillations for Lax–Wendroff are much worse. The high-resolution scheme, on the other hand, maintains its resolution much better.*

The two previous examples are admittedly idealized cases compared with what we see in reservoir simulation, but illustrate very well the potential pitfalls of classical schemes. In the next example, we consider a more relevant case.

Example 9.3.3 *Let us revisit the classical Buckley–Leverett profile studied in Example 8.4.1, i.e., consider the following equation,*

$$S_t + \left(\frac{S^2}{S^2 + (1-S)^2}\right)_x = 0, \quad S(0,t) = 1, \quad S(x,0) = \begin{cases} 1, & x < 0.1, \\ 0, & x \geq 0.1. \end{cases}$$

Figure 9.3 shows the solution at time $t = 0.65$ computed by the same four schemes. The first-order schemes compute qualitatively correct approximations of both the leading shock wave and the trailing rarefaction wave. As expected, Lax–Friedrichs introduces significant smearing of both the shock and the kink. The upwind scheme delivers acceptable accuracy. Lax–Wendroff fails completely to capture the correct structure of the composite: not only does it introduce oscillations behind the displacement front, but the propagation speed is

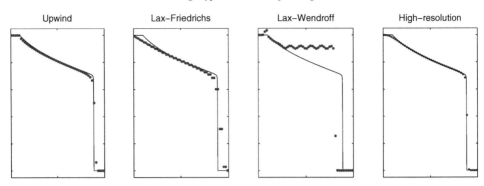

Figure 9.3 Approximate solutions at time $t = 0.65$ for the Buckley–Leverett problem on a grid with 50 cells computed by three classical schemes.

also incorrect. The reason is that once overshoots are introduced near the leading discontinuity, the only way to maintain a mass-conservative solution is if the numerically computed wave (with overshoots) travels slower than the true wave. Likewise, the scheme introduces an overshoot at the kink where the solution has discontinuous derivative. These are serious deficiencies typical of higher-order classical schemes. The high-resolution scheme computes a qualitatively correct solution and overall delivers the best accuracy. On the other hand, it is more computationally costly and would be outperformed by the upwind scheme if we compare accuracy versus computational cost.

COMPUTER EXERCISES

9.3.1 Try to set $\Delta x = \Delta t$ so that the CFL number equals one exactly for the periodic advection problem in Example 9.3.1. Can you explain what you observe? Why is using a CFL number identical to one not that interesting from a practical point of view?

9.3.2 Implement a Godunov solver that works correctly for strictly convex or strictly concave flux functions. Hint: try to find a formula that gives the correct self-similar Riemann solution along the line $x/t = 0$.

9.3.3 Implement the single-point upstream mobility-weighting scheme (9.21) and use it to simulate the 1D cases discussed in Examples 8.4.2–8.4.4.

9.3.4 Extend the schemes introduced in this section to the two-dimensional conservation law $u_t + f(u)_x + g(u)_y = 0$, and implement a solver that can run a 2D analogue of the periodic advection problem in Example 9.3.1. Try to make the changes in the existing solver as small as possible. Hint: In 2D, the Lax–Friedrichs scheme reads,

$$u_{ij}^{n+1} = \tfrac{1}{4}\left(u_{i+1,j}^n + u_{i-1,j}^n + u_{i,j+1}^n + u_{i,j-1}^n\right)$$
$$- \tfrac{1}{2}r\left[f(u_{i+1,j}^n) - f(u_{i-1,j}^n)\right] - \tfrac{1}{2}r\left[g(u_{i,j+1}^n) - g(u_{i,j-1}^n)\right].$$

9.3.5 The nummet folder of the book module also contains another high-resolution scheme, the semi-discrete central-upwind scheme Kurganov et al. [174]. Use the supplied scripts to also familiarize yourself with this scheme.

9.3.4 Implicit Schemes

So far, we have only considered schemes based on explicit temporal discretization, since this is the most common approach in the literature on numerical methods for hyperbolic conservation laws. In principle, we could also have used *implicit* temporal discretization, as we have already encountered in Chapter 7. Going back to (9.8), we could approximate the integral using point values at t_{n+1} rather than at time t_n, which would give us a numerical method on the form,

$$u_i^{n+1} + r_i^n \left(F_{i+1/2}^{n+1} - F_{i-1/2}^{n+1} \right) = u_i^n. \tag{9.22}$$

To get a specific scheme, we must specify how the cell averages $u_{i-p}^{n+1}, \ldots, u_{i+q}^{n+1}$ are used to evaluate the numerical flux $F_{i+1/2}^{n+1}$. The standard approach in reservoir simulation is to use the upstream mobility-weighting approximation from (9.21) with S^n replaced by S^{n+1}.

In the explicit method (9.9), the flux terms were written on the right-hand side to signify that they are given in terms of the known solution u_i^n. In (9.22), the numerical flux terms appear on the left-hand side to signify that they depend on the unknown cell averages u_i^{n+1}. Equation (9.22) hence represents a coupled system of *nonlinear* discrete equations, which typically must be solved by use of a Newton–Raphson method as discussed in Section 7.1. Computing a single time step of an implicit method is typically significantly more costly than computing an explicit update, since the latter involves fewer evaluations of the flux function and its derivatives. On the other hand, the fully implicit discretization (9.22) is *unconditionally stable* in the sense that there is no stability condition (CFL condition) limiting the size of the time step. In principle, this means that the additional computational cost can be offset by using much larger time steps. In practice, however, viable time-step sizes are limited by numerical errors (as we will see shortly) and by the convergence of the numerical method used to solve the nonlinear system. For Newton–Raphson-type methods, in particular, the convergence may be quite challenging if the flux function contains inflection points/lines separating regions of different convexity; see e.g., [147, 306, 213] for a more detailed discussion.

For 1D horizontal displacements, we have already seen that the flux function has a characteristic S-shape with positive derivatives. Like in the explicit case, we can therefore use one-sided values to evaluate the flux integrals, giving the following scheme,

$$u_i^{n+1} + r_i^n \left[f\left(u_i^{n+1}\right) - f\left(u_{i-1}^{n+1}\right) \right] = u_i^n.$$

The result is a triangular nonlinear system with one nonzero band below the diagonal. For cases like the Buckley–Leverett displacement considered in Example 9.3.3, we can

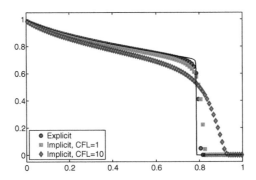

Figure 9.4 Approximate solutions at time $t = 0.65$ for the Buckley–Leverett problem on a grid with 100 cells computed by the single-point upstream mobility-weighting scheme with explicit and implicit time discretization.

therefore solve the nonlinear system very efficiently using a substitution method. That is, we start at the cell next to the inflow boundary,

$$u_1^{n+1} + r_1^n f(u_1^{n+1}) = u_1^n + r_1^n f_L,$$

where f_L is the known inflow. This single nonlinear equation can be solved robustly for an arbitrarily large Δt by use of a classical bracketing method that tracks the end points of an interval containing the unknown root. Once u_1^{n+1} has been computed, we can move to the next cell, and solve a similar scalar equation,

$$u_2^{n+1} + r_2^n f(u_2^{n+1}) = u_2^n + r_2^n f(u_1^{n+1}),$$

and so on. The same strategy can, in fact, be applied to the multidimensional case under certain assumptions guaranteeing that the system has a similar cocurrent flow property; see [220] for more details.

Example 9.3.4 *Let us revisit the setup from Example 9.3.3 and compare the explicit (9.9) and the implicit (9.22) discretizations. Instead of implementing the 1D implicit scheme, we rely on the general transport solvers from the* incomp *module in MRST, which will be discussed in more detail in the next chapter. Figure 9.4 shows solutions computed by the explicit scheme with a unit CFL number and by the implicit scheme with CFL numbers 1 and 10. At a unit CFL number, the explicit scheme resolves the displacement front somewhat sharper than the implicit scheme. When the CFL number is increased ten times for the implicit scheme, the numerical smearing is significant and will lead to the prediction of a (much) too early water breakthrough.*

To explain the difference in numerical smearing, let us do some simple numerical analysis. However, rather than studying the nonlinear Buckley–Leverett model, we go back to the simple advection equation, $u_t + au_x = 0$, and consider the schemes discussed earlier in this section as finite-difference schemes rather than finite-volume schemes. The classical way of analyzing numerical schemes is to first compute the truncation error, which is obtained by inserting the exact solution into the numerical scheme, assume a smooth solution, and use Taylor expansions to express the various terms around a common point (x,t). In addition, we also use the fact that $u_t = -au_x$ and that Δt and Δx are proportional. Starting with the explicit scheme,

$$\begin{aligned} 0 &= \frac{u(x,t+\Delta t) - u(x,t)}{\Delta t} + a\frac{u(x,t) - u(x-\Delta x,t)}{\Delta x} \\ &= u_t(x,t) + \tfrac{1}{2}\Delta t\, u_{tt}(x,t) + \mathcal{O}(\Delta t^2) + au_x(x,t) - \tfrac{1}{2}a\Delta x\, u_{xx}(x,t) + \mathcal{O}(\Delta x^2) \\ &= u_t + \tfrac{1}{2}\Delta t\, a^2 u_{xx} + \mathcal{O}(\Delta t^2) + au_x - \tfrac{1}{2}a\Delta x\, u_{xx} + \mathcal{O}(\Delta x^2) \\ &= u_t + au_x - \tfrac{1}{2}a\big[\Delta x - a\Delta t\big]u_{xx} + \mathcal{O}(\Delta x^2). \end{aligned}$$

In other words, the explicit scheme will up to order $\mathcal{O}(\Delta x^2)$ solve a *parabolic* equation. This equation is only well-posed if the coefficient in front of the second-order term is negative, implying that $a\Delta t/\Delta x \leq 1$, which is exactly the same requirement as in the CFL condition. We also notice that the higher the CFL number we use, the less numerical diffusion we introduce in the approximate solution and in the special case that $a\Delta t = \Delta x$, the scheme has no numerical diffusion.

For the implicit scheme, we only need to expand the u_{i-1}^{n+1} term since we have already computed the expansion for u_i^{n+1} for the explicit scheme,

$$\begin{aligned} u_{i-1}^{n+1} &= u(x-\Delta x, t+\Delta t) \\ &= u(x-\Delta x,t) - a\Delta t\, u_x(x-\Delta x,t) + \tfrac{1}{2}(a\Delta t)^2 u_{xx}(x-\Delta x,t) + \mathcal{O}(\Delta t^3) \\ &= u - \Delta x u_x + \tfrac{1}{2}\Delta x^2 u_{xx} + \mathcal{O}(\Delta x^3) - a\Delta t\, \partial_x\big[u - \Delta x u_x + \mathcal{O}(\Delta x^2)\big] \\ &\quad + \tfrac{1}{2}(a\Delta t)^2 \partial_x^2\big[u + \mathcal{O}(\Delta x)\big] + \mathcal{O}(\Delta t^3). \end{aligned}$$

Collecting terms, we have that

$$\begin{aligned} 0 &= \frac{u(x,t+\Delta t) - u(x,t)}{\Delta t} + a\frac{u(x,t) - u(x-\Delta x,t)}{\Delta x} \\ &= u_t + au_x - \tfrac{1}{2}a\big[\Delta x + a\Delta t\big]u_{xx} + \mathcal{O}(\Delta x^2). \end{aligned}$$

This means that the implicit scheme also will solve a parabolic equation with accuracy $\mathcal{O}(\Delta x^2)$. In this case, the coefficient in front of the u_{xx} term is always negative and the scheme is therefore stable regardless of the choice of Δt. However, this also means that the numerical diffusion *increases* with increasing time steps.

One can also conduct a similar analysis in the nonlinear case and arrive at similar conclusions. This in turn implies an interesting observation if we look back at our discussion of linear waves on page 280. For an explicit scheme, slow waves are computed with *more* numerical diffusion than fast waves. For an implicit scheme, on the other hand, slow waves are computed with *less* numerical diffusion than fast waves. Explicit schemes are therefore best suited for systems having relatively small differences in wave speeds, whereas implicit schemes are better suited for systems with large differences in wave speeds. In a typical reservoir setting, wave speeds are large in the near-well region, of medium size in high-flow regions away from the wells, and low near or in stagnant zones. It is therefore reasonable to expect that implicit schemes are best suited for simulating real reservoir systems. However, explicit schemes are still used, mainly because of their simplicity and in connection with higher-resolution discretizations.

COMPUTER EXERCISES

9.3.1 Implement the implicit upwind scheme in 1D and verify against the solver from the `incomp` module.

9.3.2 Use the single-point upstream mobility-weighting scheme (9.21) to extend your implicit solver to also account for countercurrent flow and use it to simulate the 1D cases discussed in Examples 8.4.2–8.4.4. Compare with the corresponding explicit scheme.

9.4 Discretization on Unstructured Polyhedral Grids

So far, this chapter has introduced you to various types of numerical methods that can be used to discretize saturation equations. The methods discussed, and some of their high-resolution extensions included in the book module, are all reasonably simple to implement on regular Cartesian and rectilinear grids and on grids consisting entirely of simplices (triangles in 2D and tetrahedrons in 3D). However, as you have seen multiple times throughout the book, realistic grid models are rarely that simple. For the baseline solvers in MRST we have therefore chosen to go with the most robust choice of them all, namely the explicit or implicit version of the single-point upstream mobility weighting scheme. This section therefore discusses in more detail how to formulate this scheme on unstructured, polyhedral grids for a general transport equation

$$\phi \frac{\partial S}{\partial t} + \nabla \cdot \vec{H}(S) = 0. \tag{9.23}$$

Here, the flux function $\vec{H}$ includes viscous flow driven by pressure gradients, as well as gravity segregation and capillary forces,

$$\vec{H}(S) = \frac{\lambda_w}{\lambda_w + \lambda_n} \vec{v} + \frac{\lambda_w \lambda_n \mathbf{K}}{\lambda_w + \lambda_n} \left(\Delta \rho \, \vec{g} + \nabla P_c(S) \right)$$
$$= \vec{H}_f(S) + \vec{H}_g(S) + \vec{H}_c(S). \tag{9.24}$$

To discretize this equation, we start from a multi-dimensional analogue of the integral form (9.8) of the conservation law,

$$S_i^{n+1} - S_i^n = \frac{1}{\phi_i |\Omega_i|} \sum_k \int_{t_n}^{t_{n+1}} \int_{\Gamma_{ik}} \vec{H}\big(S(\vec{x},t)\big) \cdot \vec{n}_{i,k} \, ds \, dt, \tag{9.25}$$

9.4 Discretization on Unstructured Polyhedral Grids

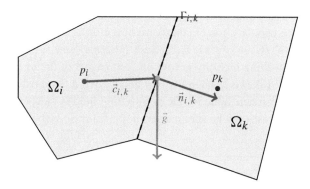

Figure 9.5 Two cells used to define the finite-volume discretization of the general two-phase transport equation (9.23). Here, the interface $\Gamma_{i,k}$ between the two cells Ω_i and Ω_k has normal vector $\vec{n}_{i,k}$ and area $A_{i,k}$.

defined over a general grid cell Ω_i as illustrated in Figure 9.5. We do not want to develop a higher-order discretization, and hence it is sufficient to evaluate the time integral at one of its end-point, so that (9.25) simplifies to

$$S_i^{n+1} - S_i^n = \frac{\Delta t}{\phi_i |\Omega_i|} \int_{\Gamma_{ik}} \vec{H}(S(\vec{x}, t_m)) \cdot \vec{n}_{i,k} \, ds, \qquad m = n, n+1. \tag{9.26}$$

Before we describe the discretization of the three flux functions H_f, H_g, and H_c, let us quickly recap how we discretize the viscous flux $\vec{v} = -\mathbf{K}\nabla p$ in the corresponding pressure equation. Referring to Figure 4.10 on page 133, the flux reads

$$v_{i,k} \approx A_{i,k} \mathbf{K}_i \frac{(p_i - \pi_{i,k})\vec{c}_{i,k}}{|\vec{c}_{i,k}|^2} \cdot \vec{n}_{i,k} = T_{i,k}(p_i - \pi_{i,k})$$
$$= \left[T_{i,k}^{-1} + T_{k,i}^{-1}\right]^{-1}(p_i - p_k),$$

where the first line gives the one-sided formulation computed entirely from quantities associated with cell i, and the second line gives the formulation that also couples to the neighboring cell k. From this expression, it follows naturally that the capillary term should be discretized as follows:

$$A_{i,k} \nabla P_c(S) \cdot \vec{n}_{i,k} \approx \left[T_{i,k}^{-1} + T_{k,i}^{-1}\right]^{-1} \left[P_c(S_i) - P_c(S_k)\right] = P_{i,k}(S). \tag{9.27}$$

Likewise, we can define a "gravity flux" g_{ik} that is independent of saturation,

$$g_{ik} = \left[g_{i,k}^{-1} + g_{k,i}^{-1}\right]^{-1}, \qquad \begin{array}{l} g_{i,k} = (\Delta \rho)|_{\Omega_i} (\mathbf{K}_i \vec{g}) \cdot \vec{n}_{i,k}, \\ g_{k,i} = (\Delta \rho)|_{\Omega_k} (\mathbf{K}_k \vec{g}) \cdot \vec{n}_{k,i}. \end{array} \tag{9.28}$$

With this, all we need to do is to find the correct upstream value λ_α^u for each of the two phases, and then the overall approximation of the interface flux reads:

$$H_{ik} = \frac{\lambda_w^u}{\lambda_w^u + \lambda_n^u} v_{ik} + \frac{\lambda_w^u \lambda_n^u}{\lambda_w^u + \lambda_n^u} \left[g_{ik} + P_{ik}\right]. \tag{9.29}$$

To determine the upstream directions, we need to compare the Darcy flux v_{ik} and the gravity flux g_{ik}. (If there are capillary forces, these are henceforth assumed to be added to the gravity flux, i.e., $g_{ik} + P_{ik} \to g_{ik}$.) If v_{ik} and g_{ik} have the same signs, we know that the correct choice of the wetting mobility is from the left (right) of the interface if the fluxes are positive (negative). Likewise, when the two fluxes have different signs, we know that the correct choice of upstream value for the non-wetting fluid is independent of the actual mobility values. These cases can be summarized in the following tabular:

sign(v_{ik})	sign(g_{ik})	λ_w^u	sign(v_{ik})	sign(g_{ik})	λ_n^u
≥ 0	≥ 0	$\lambda_w(S_L)$	≥ 0	≤ 0	$\lambda_n(S_L)$
≤ 0	≤ 0	$\lambda_w(S_R)$	≤ 0	≥ 0	$\lambda_n(S_R)$

If neither of these cases are fulfilled, we need to check the sign of the phase fluxes to determine the correct upstream direction. In practice, we do not check the phase fluxes themselves, but rather the phase fluxes divided by the fractional flow functions as these quantities are less expensive to compute. That is, we check the sign of $v_{ik} + g_{ik}\lambda_n$ and $v_{ik} - g_{ik}\lambda_w$ and use this to pick the correct upstream value.

For most cases in reservoir simulation, we also have source terms that drive flow, so that (9.23) is replaced by an inhomogeneous equation

$$\phi\frac{\partial S}{\partial t} + \nabla \cdot \vec{H}(S) = \max(q,0) + \min(q,0)f(S).$$

In the next chapter, we describe how the resulting transport solvers are implemented in MRST, and discuss how to combine them with `incompTPFA`, or one of its consistent alternative, in a sequential solution procedure to solve incompressible, multiphase flow on general polygonal and polyhedral grids.

10
Solvers for Incompressible Immiscible Flow

The multiphase flow equations introduced in Chapter 8 can describe very different flow behavior depending upon what are the dominant physical effects. During the formation of petroleum reservoirs, fluid movement is primarily driven by buoyancy and capillary forces, which govern how hydrocarbons migrate upward and enter new layers of consolidated sediments. The same effects are thought to dominate long-term geological carbon storage if the buoyant CO_2 phase continues to migrate upward in the formation long after injection has ceased. In recovery of conventional hydrocarbon resources, on the other hand, the predominant force is viscous advection caused by pressure differentials. Here, pressure disturbances will in most cases propagate much faster through the porous medium than the material waves that transport fluid phases and chemical components. This is one of the main reasons why solving multiphase flow equations turns out to be relatively complicated. In addition, we have all the other difficulties already encountered for single-phase flow in Chapters 5–7. The variable coefficients entering the flow equations are highly heterogeneous with orders of magnitude variation and complex spatial patterns involving a wide range of correlation lengths. The grids used to describe real geological media tend to be highly complex, having unstructured topologies, irregular cell geometries, and orders of magnitude aspect ratios. Flow in injection and production wells takes place on small scales relative to the reservoir and hence needs to be modeled using approximate analytical expressions, and so on.

This chapter will teach you how to solve multiphase flow equations in the special case of incompressible rock and immiscible and incompressible fluids. As we saw in Chapter 8, the system of PDEs can then be reformulated so that it consists of an elliptic equation for fluid pressure and one or more transport equations. These transport equations are generally parabolic, but have a strong hyperbolic character (see Section 8.3). Since the pressure and saturations equations have very different mathematical characteristics, it is natural to solve them in consecutive substeps. Examples of such methods include the classical IMPES method [276, 284] (see also [73] and references therein), the adaptive-implicit method (AIM) [280], and the sequentially implicit method [307]. Operator splitting is also used in streamline simulation [79] and in recent multiscale methods [195].

Incompressible models are best suited for fluid systems consisting mainly of liquid phases such as in waterflooding of oil reservoirs that either do not contain gas components or are well below the bubble point. Such systems often have weak coupling between pressure and fluid transport, which can be exploited by sequential solution procedures. For systems with a high gas content, significant compressibility effects, strong coupling between different types of flow mechanisms, or large differences in time constants, one generally has to use compressible flow models and fully implicit solvers. Nonetheless, also in this case the mixed elliptic-hyperbolic character of the model equations plays a key role in developing efficient preconditioning strategies [304, 305]. The combination of incompressible models and sequential solution procedures is very popular in academia and for research purposes, since it provides a simple means to develop more clean-cut model equations that still have many of the salient features for multiphase flow.

10.1 Fluid Objects for Multiphase Flow

In Chapter 5, we discussed the basic data objects entering a flow simulation. When going from a single-phase to a multiphase flow model, the most prominent changes take place in the fluid model. It is this model that generally will tell your solver how many phases are present and how these phases affect each other when flowing together in the same porous medium. We therefore start by briefly outlining a few fluid objects that implement the basic fluid behavior discussed in Chapter 8.

To describe an incompressible flow model, we need to know the viscosity and the constant density of each fluid phase, as well as the relative permeabilities of the fluid phases. If the fluid model includes capillary forces, we also need one or more functions that specify the capillary pressure as function of saturation. The most basic multiphase fluid object in MRST implements a simplified version of the Corey model (8.15)

```
fluid = initSimpleFluid('mu' , [    1,   10]*centi*poise           , ...
                        'rho', [1014, 859]*kilogram/meter^3, ...
                        'n'  , [    2,    2]);
```

Here, the residual saturations S_{wr} and S_{nr} are assumed to be zero and the end-points are scaled to unity, so that $k_{rw} = (S_w)^{n_w}$ and $S_{rn} = (1 - S_w)^{n_n}$. To recap from Chapter 5, the fluid object offers the following interface to evaluate the petrophysical properties of the fluid:

```
mu = fluid.properties();      % gives mu_w and mu_n
[mu,rho] = fluid.properties(); %   .... plus rho_w and rho_n
```

New to multiphase flow is the `relperm` function, which takes a single fluid saturation or an array of fluid saturations as input and outputs the corresponding values of the relative permeabilities. To plot the relative permeability curves of the `fluid` object, we can use the following code:

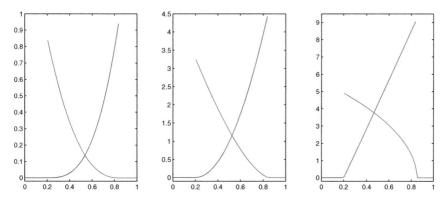

Figure 10.1 Corey relative permeabilities (left) and their first and second derivatives (middle and right) constructed by the `initCoreyFluid` function.

```
s=linspace(0,1,20)';
kr = fluid.relperm(s);
plot(s,kr(:,1),'-s',s,kr(:,2),'-o');
```

The `relperm` function can also return the first and second derivatives of the relative permeability curves when called with two or three output arguments.

The `incomp` module also implements the general Corey model with end-point scaling k_α^0 and nonzero residual saturations S_{wr} and S_{nr}.

```
fluid = initCoreyFluid('mu' , [   1,   10]*centi*poise      , ...
                      'rho', [1014, 859]*kilogram/meter^3, ...
                      'n'  , [   3,  2.5]                    , ...
                      'sr' , [ 0.2,  .15]                    , ...
                      'kwm', [   1,  .85]);
```

Figure 10.1 shows the relative permeabilities and their first and second derivatives for this particular model.

Whether the flow equations incorporate capillary pressure is specified by the fluid object. The `incomp` module implements two different capillary-pressure models, a simple linear relationship of the form $P_c(S) = C(1 - S)$ and the Leverett J-function scaling (8.9). Both models take the same input as `initSimpleFluid` and are generated the following functions

```
fluid = initSimpleFluidPc( .., 'pc_scale', 2*barsa);
fluid = initSimpleFluidPc( .., 'rock', rock,
       'surf_tension',10*barsa*sqrt(100*milli*darcy/0.1));
```

```
fluid =
   properties: @(varargin)properties(opt,varargin{:})
   saturation: @(x,varargin)x.s
      relperm: @(s,varargin)relperm(s,opt,varargin{:})
           pc: @(state)pc_funct(state,opt)
```

You may notice that the capillary function pc is evaluated using a *state* object and not a saturation. This may seem awkward, but provides a simpler interface to P_c functions that depend on rock properties, like Leverett-J.

COMPUTER EXERCISES

10.1.1 Modify the Corey model so that it also can output the residual saturations and the end-point scaling values.

10.1.2 Implement the Brooks–Corey (8.16) and the van Genuchten models (8.17) and (8.18).

10.1.3 Extend the models to also include the capillary functions (8.11) and (8.12).

10.2 Sequential Solution Procedures

To solve the two-phase, incompressible model, we rely entirely on the fractional-flow formulation developed in Section 8.3.2. As you may recall, in this formulation, the flow equations consists of an elliptic pressure equation

$$\nabla \cdot \vec{v} = q, \qquad \vec{v} = -\lambda\bigl(\nabla p_n - f_w \nabla P_c - (\rho_w f_w + \rho_n f_n) g \nabla z\bigr) \qquad (10.1)$$

and a parabolic transport equation

$$\phi \frac{\partial S_w}{\partial t} + \nabla \cdot \bigl[f_w\bigl(\vec{v} + \lambda_n(\Delta \rho g \nabla z + \nabla P_c)\bigr)\bigr] = q_w. \qquad (10.2)$$

Here, the capillary pressure $p_c = p_w - p_n$ is assumed to be a known function P_c of the wetting saturation S_w, and the transport equation becomes hyperbolic whenever P_c' is zero.

In the standard sequential solution procedure, the system (10.1)–(10.2) is evolved in time using a set of discrete time steps Δt_i. Let us assume that p, $\vec{v}$, and S_w are all known at time t and that we want to evolve the solution to time $t + \Delta t$. At the beginning of the time step, we first assume that the saturation S_w is fixed. This means that the parameters λ, f_w, and f_n in (10.1) become functions of the spatial variable $\vec{x}$ only. We then use the resulting Poisson-type equation to update pressure p_n and Darcy velocity $\vec{v}$. Next, we hold $\vec{v}$ and p_n fixed while (10.2) is evolved a time step Δt to define an updated saturation $S_w(\vec{x}, t + \Delta t)$. This saturation is then held fixed when we update p_n and $\vec{v}$ in the next time step, and so on.

Some authors refer to this solution procedure as an *operator splitting* method, since the solution procedure effectively splits the overall solution operator of the flow model into two parts that are evolved in consecutive substeps. Likewise, some authors refer to the sequential solution procedure as IMPES, which is short-hand for *implicit pressure, explicit saturation*. Strictly speaking, using the name IMPES is only correct if the saturation evolution is approximated by a single time step of an explicit transport solver. The size of the splitting step Δt is then restricted by the CFL condition of the explicit scheme. In many cases, the overall flow system does not have stability requirements that necessitate such a restriction on Δt. Indeed, by writing the flow model in the fractional-flow formulation, we have isolated the parts of the system that have stability restrictions to the hyperbolic saturation equation. The elliptic pressure equation, on the other hand, describes (smooth) solutions resulting from the instant redistribution of pressure in a system with infinite speed of propagation. For this equation, we can therefore in principle use as large time step as we want.

As long as we ensure that the evolving discontinuities and sharp transitions are propagated in a stable manner in the saturation equation, our only concern when choosing the size of the splitting step should be to control or minimize the *splitting error* introduced by accounting for pressure and transport in separate substeps. The fractional-flow formulation underlying our operator splitting was developed to minimize the coupling between saturation and pressure. For incompressible flow models, the effect that dynamic changes in the saturation field have on pressure is governed entirely by the total mobility $\lambda(S)$, which in many cases is a function that locally has small and relatively smooth variation in time. For this reason, you can typically use splitting steps that are significantly larger than the CFL restriction from the hyperbolic part of the saturation equation and still accurately resolve the coupling between pressure and saturation. In other words, for each pressure update, the saturation can be updated by an explicit solver using multiple saturation substeps, or by an implicit solver using either a single or multiple saturation substeps. If necessary, you can also iterate on the splitting steps.

10.2.1 Pressure Solvers

The pressure equation (10.1) for incompressible, multiphase flow is time dependent. This time dependence comes as the result of three factors:

- $\mathbf{K}/\mu$ is replaced by the total mobility $\lambda(S_w)$, which depends on time through the saturation $S_w(\vec{x}, t)$,
- the constant density ρ is replaced by a saturation-dependent quantity $\rho_w f_w(S_w) + \rho_n f_n(S_w)$, and
- the new source term $q - \nabla \lambda_w(S_w) \nabla P_c(S_w)$ depends on saturation.

Nevertheless, once S_w is held fixed in time, all three quantities become functions of $\vec{x}$ only, and we hence end up again with an elliptic Poisson-type equation having the same spatial variation as in (4.10) on page 117. Hence, we can either use the two-point scheme introduced in Section 4.4.1 or one of the consistent discretization methods from Chapter 6, *mutatis mutandis*. The incompTPFA solver discussed in Chapter 5 and the incompMimetic and incompMPFA solvers from Chapter 6 are implemented so that they solve the pressure equation for a general system of m incompressible phases. Whether this system has one or more phases is determined by the fluid object and the reservoir state introduced in Section 5.1.2. We will therefore not discuss the pressure solvers in more detail.

10.2.2 Saturation Solvers

Apart from the time loop, which we have already encountered in Chapter 7, the only remaining part we need is a solver for the transport equation (10.2) that implements the discretizations we introduced in Section 9.4. Summarized, this can be written in the following residual form, for each cell Ω_i

$$\mathcal{F}_i(s,r) = s_i - r_i + \frac{\Delta t}{\phi_i |\Omega_i|} \left[\sum_k H_{ik}(s) - \max(q_i, 0) - \min(q_i, 0) f(S_i) \right]. \quad (10.3)$$

Here, s and r are cell-averaged quantities and subscript i refers to the cell the average is evaluated in. The sum of the interface fluxes for cell i

$$H_i(s) = \sum_k \frac{\lambda_w^u(s_i, s_k)}{\lambda_w^u(s_i, s_k) + \lambda_n^u(s_i, s_k)} \Big[v_{ik} + \lambda_n^u(s_i, s_k)(g_{ik} + P_{ik}) \Big]. \quad (10.4)$$

is computed using the single-point, upstream mobility-weighting scheme discussed on page 287, whereas the fractional flow function f in the source term is evaluated from the cell average of S in cell Ω_i. The explicit scheme is given as $S^{n+1} = S^n - \mathcal{F}(S^n, S^n)$ and the implicit scheme follows as a coupled system of discrete nonlinear equations if we set $\mathcal{F}(S^{n+1}, S^n) = 0$. In the following, we discuss the inner workings of these solvers in more detail.

Explicit Solver

The `incomp` module offers the following explicit transport solver

```
state = explicitTransport(state, G, tf, rock, fluid, 'mech1', obj1, ..)
```

which evolves the saturation given in the `state` object a step `tf` forward in time. The function requires a complete and compatible model description consisting of a grid structure `G`, petrophysical properties `rock`, and a fluid model `fluid`. For the solver to be functional, the `state` object must contain the correct number of saturations per cell and an *incompressible* flux field that is consistent with the global drive mechanisms given by the `'mech'` argument (`'src'`, `'bc'`, and/or `'wells'`) accompanied by correctly specified objects `obj`, as discussed in Sections 5.1.3–5.1.5.

In practice, this means that the *input value* of `state` must be the *output value* of a previous call to an incompressible solver like `incompTPFA`, `incompMPFA`, or `incompMimetic`. In addition, the function takes a number of optional parameters that determine whether the time steps are prescribed by the user or to be automatically computed by the solver. The solver can also ignore the Darcy flux and work as a pure gravity segregation solver if the optional parameter `onlygrav` is set to true. Finally, the solver will issue a warning if the updated saturation value is more than `satwarn` outside the interval $[0,1]$ of physically meaningful values (default value: `sqrt(eps)`).

The explicit solver involves many of the same operations and formulas used for the spatial discretizations as the implicit solver. To avoid duplication of code we have therefore introduced a private helper function

```
[F,Jac] = twophaseJacobian(G, state, rock, fluid, 'pn1', pv1, ..)
```

that implements the residual form (10.3) and its Jacobian matrix $J = d\mathcal{F}$, returned as two function handles, F and Jac. With this, the key lines of the explicit saturation solver read

```
F = twophaseJacobian(G, state, rock, fluid, 'wells', opt.wells, ..);
s = state.s(:,1);
t = 0;
while t < tf,
    dt  = min(tf-t, getdt(state));
    s(:) = s - F(state, state, dt);
    t   = t + dt;
    s   = correct_saturations(s, opt.satwarn);
    state.s = [s, 1-s];
end
```

Here, the function getdt implements a CFL restriction on the time step by estimating the maximum derivative of each function used to assemble interface fluxes and source terms (you can find details in the code). The function correct_saturations ensures that the computed saturations stay inside the interval of physically valid states. If this function issues a warning, it is highly likely that your time step exceeds the stability limit, or something is wrong with your fluxes or setup of the model.

Implicit Solver

The implicit solver has the same user interface and parameter requirement as the explicit solver

```
state = implicitTransport(state, G, tf, rock, fluid, 'mech1', obj1, ..)
```

In addition, there are optional parameters controlling the Newton–Raphson method used to solve for S^{n+1}. To describe this method, we start by writing the residual equations (10.3) for all cells in vector form

$$F(s) = s - S + \frac{\Delta t}{\phi |\Omega|} \big[H(s) - Q^+ - Q^- f(s) \big] = 0. \tag{10.5}$$

Here, s is the unknown state at time tf and S is the known state at the start of the time step. As you may recall from Section 7.1, the Newton–Raphson linearization of an equation like (10.5) can be written as

$$0 = F(s_0 + \delta s) \approx F(s_0) + \nabla F(s_0) \delta s,$$

which naturally suggests an iterative scheme in which the approximate solution $s^{\ell+1}$ in the $(\ell+1)$-th iteration is obtained from

$$J(s^\ell) \delta s^{\ell+1} = -F(s^\ell), \qquad s^{\ell+1} \leftarrow s^\ell + \delta s^{\ell+1}. \tag{10.6}$$

Here, $\delta s^{\ell+1}$ is called the *Newton update* and J is the Jacobian matrix. The `incomp` module was implemented before automatic differentiation was introduced in MRST, and hence the Jacobian is computed analytically through the following expansion

$$J(s) = \frac{dF}{ds}(s) = 1 + \frac{\Delta t}{\phi|\Omega|}\left[\frac{dH}{ds}(s) - Q^{-}\frac{df}{ds}(s)\right],$$

$$\frac{dH}{ds} = \frac{dH}{d\lambda_w}\frac{d\lambda_w}{ds} + \frac{dH}{d\lambda_n}\frac{d\lambda_n}{ds} + f_w\lambda_n\frac{dP}{ds},$$

$$\frac{dH}{d\lambda_w} = \frac{f_w}{\lambda}\left[v + \lambda_n(g+P)\right], \qquad \frac{dH}{d\lambda_n} = -\frac{f_w}{\lambda}\left[v - \lambda_w(g+P)\right].$$

In general, we are not guaranteed that the resulting values in the vector $s^{\ell+1}$ lie in the interval [0, 1]. To ensure physically meaningful saturation values, we can introduce a *line-search* method, which uses the Newton update to define a *search direction* $p^{\ell} = \delta s^{\ell+1}$ and tries to find the value α that minimizes $h(\alpha) = F(s^{\ell} + \alpha\, p^{\ell})$. We may now either solve $h'(\alpha) = 0$ exactly, or use an *inexact line-search method* that only asks for a sufficient decrease in h. In the implicit solver discussed herein, we have chosen the latter approach and use an unsophisticated method that reduces α in a geometric sequence. The following code should give you the idea:

```
function [state, res, alph, fail] = linesearch(state, ds, target, F, ni)
   capSat = @(sat) min(max(0, sat),0);
   [alph,i,fail] = deal(0,0,true);
   sn = state;
   while fail && (i < ni),
      sn.s(:,1) = capSat(state.s(:,1) + pow2(ds, alph));
      res  = F(sn);
      alph = alph - 1;  i = i + 1;
      fail = ~(norm(res, inf) < target);
   end
   alph = pow2(alph + 1); state.s = sn.s;
```

Here, F is a function handle to the residual function F. The number of trials `ni` in the line-search method is set through the optional parameter `'lstrails'`, whereas the target value is set as the parameter `'resred'` times the residual error upon entry. Default values for `'lstrails'` and `'resred'` are 20 and 0.99, respectively.

The implicit discretization is stable in the sense that there exists a solution S^{n+1} for an arbitrarily large time increment Δt. Unfortunately, there is no guarantee that we will be able to find this solution using the line-search method described previously. If the time step is too large, the Newton method may simply compute search directions that do not point us toward the correct solution. To compensate for this, we also need a mechanism that reduces the time step if the iteration does not converge and then uses a sequence of shorter time step to reach the prescribed time `tf`. First of all, we need to define what we mean by convergence. This is defined by the optional `'nltol'` parameter, which sets the absolute tolerance ϵ (default value 10^{-6}) on the residual $\|\mathcal{F}(S^{n+1}, S^n)\|_\infty \leq \epsilon$. In addition, we use a parameter `'maxnewt'` that gives the maximum number of iteration steps (default value 25) the method can take to reach a converged solution. The following code gives the essence of the overall algorithm of the iterative solver, as implemented in the helper function `newtonRaphson2ph`

```
mints   = pow2(tf, -opt.tsref);
[t, dt] = deal(0.0, tf);
while t < tf && dt >= mints,
    dt = min(dt, tf - t);
    redo_newton = true;
    while redo_newton,
        sn_0 = resSol; sn = resSol; sn.s(:) = min(1,sn.s+0.05);
        res  = F(sn, sn_0, dt);
        err  = norm(res(:), inf);
        [nwtfail, linfail, it]  = deal(err>opt.nltol,false,0);
        while nwtfail && ~linfail && it < opt.maxnewt,
            J  = Jac(sn, sn_0, dt);
            ds = -reshape(opt.LinSolve(J, reshape(res', [], 1)), ns, [])';
            [sn, res, alph, linfail] = update(sn, sn_0, ds, dt, err);
            it   = it + 1;
            err  = norm(res(:), inf);
            nwtfail = err > opt.nltol;
        end
        if nwtfail,
            % Chop time step in two, or use previous successful dt
        else
            redo_newton = false;
            t = t + dt;
            % If five successful steps, increase dt by 50%
        end
    end
    resSol = sn;
end
```

The algorithm has two optional parameters: `'tsref'` with default value 12 gives the number of times we can halve the time step, whereas `'LinSolve'` is the linear solver, which defaults to `mldivide`. Beyond this, the best way to find more details about the solver is to read the code.

10.3 Simulation Examples

You have now been introduced to all the functionality you need to solve incompressible, multiphase flow problems. It is therefore time to start looking into the qualitative behavior of such systems and the typical multiphase phenomena you may encounter in practice. The examples presented in the following are designed to highlight individual effects, or combinations of effects, and may not always be fully realistic in terms of physical scales, magnitude of the parameters and effects involved, etc. We will also briefly look at the structure of the discrete systems arising in the implicit transport solver. The last section of the chapter discusses numerical errors resulting from specific choices of discretizations and solution strategies.

We have already seen that there are essentially three effects that determine the direction of a single-phase flow field. The first is *heterogeneity* (i.e., spatial variations in permeability) that affects the local magnitude and direction of the flow. The second effect is introduced by *drive mechanisms*, such as wells and boundary conditions, that determine where fluids flow to and from. However, the further you are from the location of a well or a boundary condition, the less effect it will have on the flow direction. The last effect is gravity.

Multiphase flow is more complicated, since the fluid dynamics is now also affected by the viscosity and density ratios of the fluids present, as well as by relative permeability and capillary pressure. These effects will introduce several challenges. If the displacing fluid is more mobile than the resident fluid, it will tend to move rapidly into this fluid, giving weak shock fronts and long rarefaction waves. For a homogeneous reservoir, this will result in early breakthrough of the displacing fluid, and slow and incremental recovery of the resident fluid. In a heterogeneous reservoir, you may also observe *viscous fingering*, which essentially means that the displacing fluid moves unevenly into the resident fluid. This is a self-reinforcing effect that causes the fingers to move farther into the resident fluid.

Gravity segregation, on the other hand, will force fluids having different density to segregate, and lead to phenomena such as *gravity override*, in which a lighter fluid moves quickly on top of a denser fluid. This is a problem in many recovery methods relying on gas injection and for geological storage of CO_2. Finally, we have capillary effects, which tend to spread out the interface between the invading and displaced fluids. When combined with heterogeneity, these effects are generally difficult to predict without detailed simulations. Sometimes, they work in the same direction to aggravate sweep and displacement efficiency, but can also counteract each other to cancel undesired behavior. Gravity and capillary forces, for instance, may both reduce viscous fingering that would otherwise give undesired early breakthrough.

This section discusses most of the phenomena just outlined in more detail. In most cases, we only consider a single effect at the time. Throughout the examples, you will also learn how to set up various types of simulations using MRST. As a rule, for brevity we will not discuss complete codes, and when reading the material you should therefore take the time to also read the accompanying codes found in the `in2ph` directory of the `book` module. You will gain much more insight if you run these codes and try to modify them to study the effect of different parameters and algorithmic choices. I also strongly encourage you to do as many of the computer exercises as possible.

10.3.1 Buckley–Leverett Displacement

As a first example, let us revisit the 1D horizontal setup from Example 9.3.4 on page 284, in which we compared explicit and implicit transport solvers for the classic Buckley–Leverett displacement profile arising when pure water is injected into pure oil. The following code sets up a slightly rescaled version of the problem, computes the pressure solution, and then uses the explicit transport solver to evolve the saturations forward in time

```
G     = computeGeometry(cartGrid([100,1]));
rock  = makeRock(G, 100*milli*darcy, 0.2);
fluid = initSimpleFluid('mu' , [1, 1].*centi*poise, ...
                       'rho', [1000, 1000].*kilogram/meter^3, 'n', [2,2]);
bc    = fluxside([], G, 'Left',   1, 'sat', [1 0]);
bc    = fluxside(bc, G, 'Right', -1, 'sat', [0 1]);
hT    = computeTrans(G, rock);
rSol  = initState(G, [], 0, [0 1]);
rSol  = incompTPFA(rSol, G, hT, fluid, 'bc', bc);
rSole = explicitTransport(rSol, G, 10, rock, fluid, 'bc', bc, 'verbose',true);
```

The explicit solver uses 199 time steps to reach time 10. Let us try to see if we can do this in one step with the implicit solver:

```
[rSoli, report] = ...
   implicitTransport(rSol, G, 10, rock, fluid, 'bc', bc, 'Verbose', true);
```

This corresponds to running the solver with a CFL number of approximately 200. From the output in Figure 10.2 we see that this is not a big success. With an attempted time step of $\Delta t = 10$, the solver only manages to reduce the residual by a factor 2.5 within the allowed 25 iterations. Likewise, when the time step is halved to $\Delta t = 5$, the solver still only manages to reduce the residual one order of magnitude within the 25 iterations. When the time step is halved once more, the solver converges in 20 iterations in the first step and then in 9 iterations in the next 3 substeps. This time-step chopping is not very efficient: Altogether, more than half of the iterations (50 out of 97) were wasted trying to compute time steps that would not converge. The explicit solver avoids this problem since the time step is restricted by a CFL condition, but requires significantly more time steps.

To overcome the problem with wasted iterations, we can explicitly subdivide the pressure step into multiple saturation steps:

```
rSolt = rSol;
for i=1:n
   rSolt = implicitTransport(rSolt, G, 10/n, rock, fluid, 'bc', bc);
end
```

Figure 10.3 reports the approximate solutions and the overall number of iterations used by the implicit solver with n equally spaced substeps. The solver typically needs more iterations during the first time steps when the displacement front is relatively sharp. As the simulation progresses, the shock is smeared across multiple cells, which contributes to weaken the nonlinearity of the discrete system and hence reduce the number of required iterations. This explains why the reported number of iterations is not an exact multiple of the number of time steps. We also see that to get comparable accuracy as the explicit

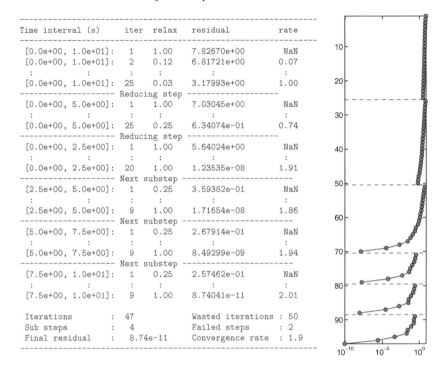

```
Time interval (s)     iter  relax   residual        rate
--------------------------------------------------------
[0.0e+00, 1.0e+01]:    1    1.00    7.82670e+00     NaN
[0.0e+00, 1.0e+01]:    2    0.12    6.81721e+00     0.07
     :                 :     :         :             :
[0.0e+00, 1.0e+01]:   25    0.03    3.17993e+00     1.00
------------------ Reducing step ------------------------
[0.0e+00, 5.0e+00]:    1    1.00    7.03045e+00     NaN
     :                 :     :         :             :
[0.0e+00, 5.0e+00]:   25    0.25    6.34074e-01     0.74
------------------ Reducing step ------------------------
[0.0e+00, 2.5e+00]:    1    1.00    5.64024e+00     NaN
     :                 :     :         :             :
[0.0e+00, 2.5e+00]:   20    1.00    1.23535e-08     1.91
------------------- Next substep ------------------------
[2.5e+00, 5.0e+00]:    1    0.25    3.59382e-01     NaN
     :                 :     :         :             :
[2.5e+00, 5.0e+00]:    9    1.00    1.71654e-08     1.86
------------------- Next substep ------------------------
[5.0e+00, 7.5e+00]:    1    0.25    2.67914e-01     NaN
     :                 :     :         :             :
[5.0e+00, 7.5e+00]:    9    1.00    8.49299e-09     1.94
------------------- Next substep ------------------------
[7.5e+00, 1.0e+01]:    1    0.25    2.57462e-01     NaN
     :                 :     :         :             :
[7.5e+00, 1.0e+01]:    9    1.00    8.74041e-11     2.01

Iterations      :  47           Wasted iterations :  50
Sub steps       :   4           Failed steps      :   2
Final residual  :  8.74e-11     Convergence rate  : 1.9
--------------------------------------------------------
```

Figure 10.2 Results from running the `implicitTransport` solver on a 1D Buckley–Leverett displacement problem with CFL number 200. Left: screen output, where several lines have been deleted for brevity. Right: convergence history for the residual, with cumulative iteration number increasing from top to bottom.

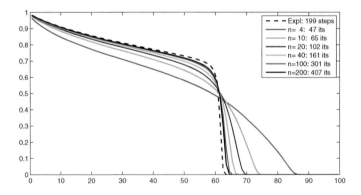

Figure 10.3 Approximate solutions computed by the explicit transport solver and the implicit transport solver with n time steps.

transport solver, the implicit solver needs to use at least 40 time steps, which amounts to more than 160 iterations. In this case, there is thus a clear advantage of using the explicit transport solver if we want to maximize accuracy versus computational cost.

10.3.2 Inverted Gravity Column

In the next example, we revisit the inverted gravity column from Example 8.4.3 on page 267 with a light fluid at the bottom and a heavier fluid at the top. We change the setup slightly so that the fluids are representative for supercritical CO_2 and brine found at conditions that would be plausible when storing CO_2 in a deep saline aquifer. The following is the essence of the simulator (plotting commands are not included for brevity):

```
gravity reset on
G     = computeGeometry(cartGrid([1, 1, 40], [1, 1, 10]));
rock  = makeRock(G, 0.1*darcy, 1);
fluid = initCoreyFluid('mu' , [0.30860, 0.056641]*centi*poise, ...
        'rho', [975.86,686.54]*kilogram/meter^3, ...
        'n', [2,2], 'sr', [.1,.2], 'kwm',[.2142,.85]);
hT = computeTrans(G, rock);
xr = initResSol(G, 100.0*barsa, 1.0); xr.s(end/2+1:end) = 0.0;
xr = incompTPFA(xr, G, hT, fluid);
dt = 5*day; t=0;
for i=1:150
   xr = explicitTransport(xr, G, dt, rock, fluid, 'onlygrav', true);
   t = t+dt;
   xr = incompTPFA(xr, G, hT, fluid);
end
```

In Example 8.4.3 the fluids had the same viscosity and hence moved equally fast upward and downward. Here, supercritical CO_2 is much more mobile than brine and will move faster to the top of the column than brine moves downward. Hence, whereas the CO_2 reaches the top of the column after 250 days, it takes more than 400 days before the first brine has sunk to the bottom. After approximately two years, the fluids are clearly segregated and separated by a sharp interface; see Figure 10.4.

In the simulation, we used relatively small splitting steps (150 steps of 5 days each) to march the transient solution towards steady state. Looking at Figure 10.5, which shows the vertical pressure distribution every fiftieth day (i.e., for every tenth time step), we see that the pressure behaves relatively smoothly compared with the saturation distribution. It is therefore reasonable to expect that we could get away with using a smaller number of splitting steps in this particular case. How many splitting steps do you think we need?

Before leaving the problem, let us inspect the discrete nonlinear system from the implicit transport solver in some detail. Figure 10.6 contrasts the sparsity pattern at two instances in time to that of the 1D horizontal Buckley–Leverett problem. From the discussion in Section 9.3.4, we know that the latter only has a single nonzero band below the diagonal and hence can be solved more robustly if we instead of using Newton's method, use a nonlinear substitution method with a bracketing method for each single-cell problem [220]. Since the lighter fluid moves upward and the heavier fluid moves downward during gravity segregation, there will be nonzero elements above and below the diagonal in the

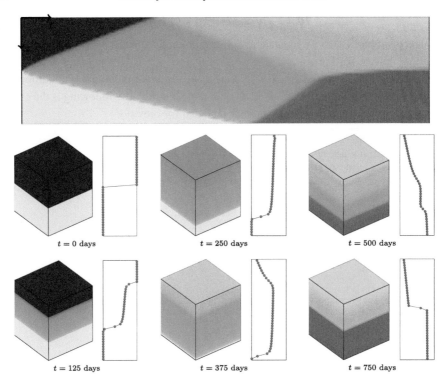

Figure 10.4 Simulation of an inverted gravity column where pure CO_2 initially fills the bottom half and brine the upper half of the volume. The upper plot shows $S(z,t)$ with yellow color signifying pure CO_2 and blue color signifying pure brine. At equilibrium, the CO_2 at the top contains some irreducible water and the brine at the bottom contains residual CO_2, which thus can be considered as safely trapped within the immobile brine.

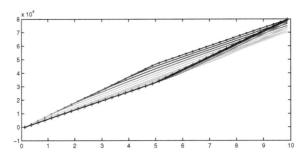

Figure 10.5 Pressure distribution at every tenth time step as function of depth from the top of the gravity column, with blue color and dotted markers indicating initial time and red color and cross markers indicating end of simulation.

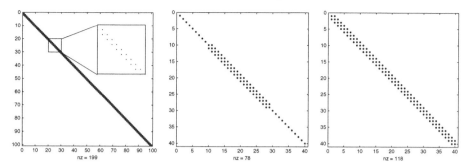

Figure 10.6 Sparsity pattern for the 1D Buckley–Leverett problem (left) and the inverted gravity column after 125 days (middle) and 390 days (right)

two-phase region and at the interface between the two phases. In the figure, we also see how the sparsity pattern changes as the two-phase region expands upward and downward from the initial interface. Since the nonlinear system is no longer triangular, a substitution method cannot be used, but each linearized system can be solved efficiently using the Thomas algorithm, which is a special $\mathcal{O}(n)$ Gaussian elimination method for tridiagonal systems. Knowing the sparsity pattern of the problem is a key to efficient solvers. The mldivide solver (i.e., A\b) in MATLAB performs an analysis of the linear system and picks efficient solvers for triangular, tridiagonal, and other special systems. To confirm that the optimal solver is indeed used, you can type spparms('spumoni',2) before you run the example.

10.3.3 Homogeneous Quarter Five-Spot

To gain more insight into the simulation of multiphase displacement processes, we consider the classical confined quarter five-spot test case discretized on a 128×128 grid. As you may recall from Section 5.4.1, this test case consists of one quarter of a symmetric pattern of four injectors surrounding a producer (or vice versa), repeated to infinity in each direction. We neglect capillary and gravity forces and assume a simplified Corey model with exponent 2.0 and zero residual saturation, i.e., $k_{rw} = S^2$ and $k_{ro} = (1 - S)^2$. We start by setting the viscosity to 1 cP for both fluid phases, giving a unit mobility ratio. The injector and producer operate at fixed bottom-hole pressure, giving a total pressure drop of 100 bar across the reservoir. Since we have assumed incompressible flow, equal amounts of fluid must be produced from the reservoir so that injection and production rates sum to zero. The total time is set such that 1.2 pore volumes of fluid will be injected if the initial injection is maintained throughout the whole simulation. However, the actual injection rate will depend on the total resistance to flow offered by the reservoir, and hence vary with time when the total mobility varies throughout the reservoir as a result of fluid movement. (Remember that the total mobility is less in all cells containing two fluid phases.) The simulation code

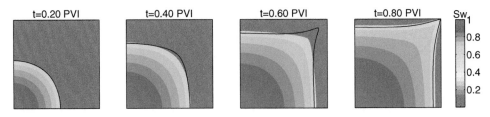

Figure 10.7 Evolution of a two-phase displacement front for a homogeneous quarter five-spot case with water injected into oil. The single line shown in each plot is a contour at value $t/(2\sqrt{2}-2)$ PVI for the time-of-flight field computed from the corresponding single-phase problem (i.e., with $\lambda \equiv 1$).

follows the same principles as outlined in the two examples just discussed; you can find details in the file `quaterFiveSpot2D.m`.

Figure 10.7 shows how the displacement profile resulting from injection of water into the oil-filled reservoir expands circularly near the injector. As the displacement front propagates into the reservoir, it gradually elongates along the diagonal and forms a finger that extends towards the producer. As a result, water breaks through in the producer long time before the displacement front has managed to sweep the stagnant regions near the northwest and southeast corners. The evolution of the saturation profile is the result of two different multiphase effects.

To better understand these effects, it is instructive to transform our 3D transport equation into streamline coordinates. Since the flow field is incompressible, we can use (4.44) to write $\vec{v} \cdot \nabla = \phi \frac{\partial}{\partial \tau}$ so that (10.2) transforms to a family of 1D transport equations, one along each streamline,

$$\frac{\partial S}{\partial t} + \frac{\partial f_w(S)}{\partial \tau} = \frac{q_w}{\phi}. \qquad (10.7)$$

Hence, the first flow effect is exactly the same Buckley–Leverett displacement as we saw in Section 10.3.1, except that it now acts along streamlines rather than along the axial directions. It is therefore tempting to suggest that to get a good idea of how a multiphase displacement will evolve, we can solve a single-phase pressure equation for the initial oil-filled reservoir, compute the resulting time-of-flight field, and then map the 1D Buckley–Leverett profile (8.60) computed from (8.53) onto time-of-flight. How accurate this approximation is depends on the coupling between the saturation and pressure equations.

With linear relative permeabilities and unit mobility ratio, there is no coupling between pressure and transport, and mapping 1D solutions onto time-of-flight thus produces the correct solution. In other cases, changes in total mobility will modify the total Darcy velocity and hence the time-of-flight. To illustrate this, let us compare the propagation of the leading shock predicted by the full multiphase simulation and our simplified streamline analysis. For fluids with a viscosity ratio $\mu_w/\mu_n = M$, it follows by solving $f'(S) = f(S)/S$ that the leading shock of the Buckley–Leverett displacement profile moves at a speed $M/(2\sqrt{M+1}-2)$ relative to the Darcy velocity, shown as a single black line for each snapshot in Figure 10.7. Compared with our simplified streamline analysis, the movement of the injected water is retarded by the reduced mobility in the two-phase region behind the leading displacement front.

In most simulations, the primary interest is to predict well responses. To extract these, we introduce the following function:

```
function wellSol = getWellSol(W, x, fluid)
mu = fluid.properties();
wellSol(numel(W))=struct;
for i=1:numel(W)
  out  = min(x.wellSol(i).flux,0); iout = out<0;   % find producers
  in   = max(x.wellSol(i).flux,0); iin  = in>0;    % find injectors
  lamc = fluid.relperm(x.s(W(i).cells,:))./mu;     % mob in completed cell
  fc   = lamc(:,1)./sum(lamc,2);                   %
  lamw = fluid.relperm(W(i).compi)./mu;            % mob inside wellbore
  fw   = lamw(:,1)./sum(lamw,2);                   %
  wellSol(i).name = W(i).name;
  wellSol(i).bhp  = x.wellSol(i).pressure;
  wellSol(i).wcut = iout.*fc + iin.*fw;
  wellSol(i).Sw   = iout.*x.s(W(i).cells,1) + iin.*W(i).compi(1);
  wellSol(i).qWs  = sum(out.*fc) + sum(in.*fw);
  wellSol(i).qOs  = sum(out.*(1-fc)) + sum(in.*(1-fw));
end
```

Inside the simulation loop, this function is called as follows

```
[wellSols, oip] = deal(cell(N,1), zeros(N,1));
for n=1:N
   x = incompTPFA(x, G, hT, fluid, 'wells', W);
   x = explicitTransport(x, G, dT, rock, fluid, 'wells', W);
   wellSols{n} = getWellSol(W, x, fluid);
   oip(n)      = sum(x.s(:,2).*pv);
end
```

The loop also computes the oil in place at each time step. Storing well responses in a cell array may seem unnecessarily complicated and requires a somewhat awkward construction to plot the result

```
t = cumsum(dT);
plot(t, cellfun(@(x) x(2).qOs, wellSols));
```

The call to `cellfun` passes elements from the cell array `wellSols` to an anonymous function that extracts the desired field qOs containing the surface oil rate of the second well (the producer). Each element represents an individual time step. The reason for using cell arrays is to provide compatibility with the infrastructure developed for compressible models of industry-standard complexity. Here, use of a cell array provides the flexibility needed to process much more complicated well output. As a direct benefit, we can use a GUI developed for visualizing well responses:

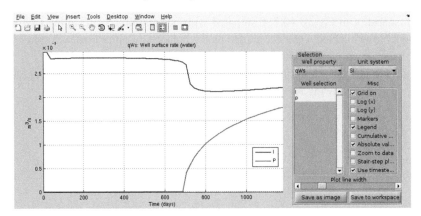

Figure 10.8 Graphical user interface `plotWellSols` from the `ad-core` module for plotting computed well responses. The GUI does not work in GNU Octave.

```
mrstModule add ad-core
plotWellSols(wellSols,cumsum(dt))
```

This brings up a plotting window as shown in Figure 10.8, where we have chosen to visualize the surface water rate for injector and producer.

To better understand the deviation between the ideal and the actual recovery, we can look at the well responses and total mass balance in more detail, as shown in Figure 10.9. We see that the oil rate drops immediately as water enters the system in the first time step and then decays slowly until water breaks through in the producer around time $t = 0.7$. Since this well now produces a mixture of water and oil, the oil rate decays rapidly as the smoothed displacement front enters the well, and then decays more slowly when the inflow of water is determined by the trailing rarefaction wave. The left plot shows the cumulative oil production computed in two different ways: (i) using the oil rate from the well solution, and (ii) measured as the difference between initial and present oil in place. Up to water breakthrough, the two estimates coincide. After water breakthrough, the production estimated from the well solution will be slightly off, since it for each time step uses a simplified approximation that multiplies the size of the time step with the total flow rate computed *at the start* of the time step and the fractional flow in the completed cell *at the end of the time step*.

COMPUTER EXERCISES

10.3.1 Repeat the experiment with wells controlled by rate instead of pressure. Do you observe any differences and can you explain them?

10.3.2 Repeat the experiment with different mobility ratios and Corey exponents.

10.3.3 Can you correct the computation of oil/water rates?

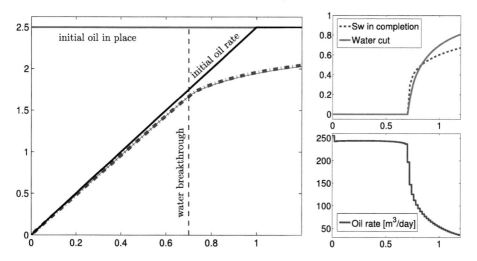

Figure 10.9 Well responses computed for the homogeneous quarter five-spot test. The left plot shows the cumulative oil production computed from the well solution shown as a thin line compared with the amount of extracted oil derived from a mass-balance computation (initial oil in place minus current oil in place) shown as a thick dashed line. The lower-right plot shows oil rate and the upper-right plot shows water saturation and water cut (fractional flow) in the well perforation.

10.3.4 Heterogeneous Quarter Five-Spot: Viscous Fingering

In the previous example we studied imbibition in a homogeneous medium, which resulted in symmetric displacement profiles. However, when a displacement front propagates through a porous medium, the combination of viscosity differences and permeability heterogeneity may introduce viscous fingering effects. In general, the term *viscous fingering* refers to the onset and evolution of instabilities at the interface between the displacing and displaced fluid phases. Fingering can arise because of viscosity differences between two phases or as a result of viscosity variations within a single phase that, for instance, contains solutes. In the laboratory, viscous fingering is usually studied in so-called Hele–Shaw cells, which consist of two flat plates separated by a tiny gap. The plates can be completely parallel, or contain small-scale variations (rugosity) to emulate a porous medium. When a viscous fluid confined in the space between the two plates is driven out by injecting a less viscous fluid (e.g., dyed water injected into glycerin), beautiful and complex fingering patterns can be observed. I recommend a search for "Hele–Shaw cell" on YouTube.

Figure 10.10 shows results of three different simulations of a quarter five-spot on a square domain represented on a uniform 60×120 grid with petrophysical properties sampled from the topmost layer of the SPE 10 data set. The wells operate at a fixed rate, corresponding to the injection of one pore volume over a period of 20 years. To reach a final time, we use 200 pressure steps and the implicit transport solver. As in the previous example, the fluid is assumed to obey a simple Corey fluid model with quadratic relative permeabilities and no residual saturations. (Complete code for the example is found in `viscousFingeringQ5.m`.)

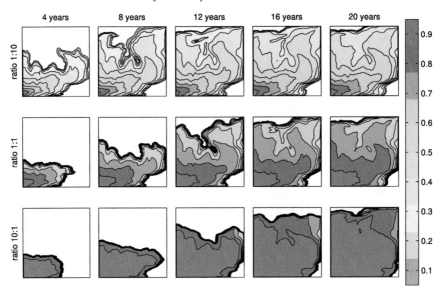

Figure 10.10 Quarter five-spot solutions on a subsample from the first layer of the SPE 10 model for three different viscosity ratios $\mu_w : \mu_n$.

If the viscosity of the injected fluid is (significantly) less than that of the resident fluid, we get an *unfavorable displacement* in which the displacing phase forms a weak shock front that will "finger" rapidly through the less mobile phase that initially fills the medium. Once a finger develops, it will create a preferential flow direction for the injected phase, which causes the finger to extend towards the producer, following the path of highest permeability. In the opposite case of a (significantly) more viscous fluid being injected into a less viscous fluid, one obtains a strong front that acts almost like a piston and creates a very *favorable and stable displacement* with a leading front that has much fewer buckles than in the unfavorable case. Not only does this front have better local displacement efficiency (i.e., it can push out more oil), but the areal sweep is also better. The unit viscosity case is somewhere in between the two, having a much better local displacement efficiency than the unfavorable case, but almost the same areal sweep at the end of the simulation.

Figure 10.11 reports well responses for the three simulations. Because water is injected at a fixed rate, the oil rate will remain constant until water breaks through in the producer. This happens after 1,825 days in the unfavorable case, after 4,050 days for equal viscosities, and after 6,300 days in the favorable case. As discussed in the previous example, the decay in oil rate depends on the strength of the displacement front and will hence be much more abrupt in the favorable mobility case, which has an almost piston-like displacement front. On the other hand, by the time the favorable case breaks through, the unfavorable case has reached a water cut of 82%. Water handling is generally expensive and in the worst case the unfavorable case might shut down before reaching the end of the 20-year production period.

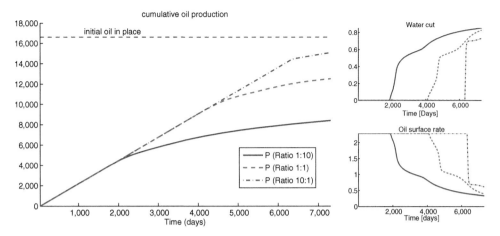

Figure 10.11 Well responses computed for the heterogeneous quarter five-spot with different mobility ratios $\mu_w : \mu_n$.

Next, we look at the sparsity of the transport equation. In the homogeneous case, all fluxes point in the positive axial directions and the Jacobian matrix is thus lower triangular. With heterogeneous permeability, or another well pattern, this unidirectional flow property is no longer present and the Jacobian matrix will have elements above and below the diagonal; see Figure 10.12. However, if we look at the transformation to streamline coordinates (10.7), it is clear that we still have unidirectional flow along streamlines. This means that the Jacobian matrix can be permuted to triangular form by performing a *topological sort* on the flux graph derived from the total Darcy velocity. In MATLAB, you can try to permute the matrix to upper-triangular form by use of the Dulmage–Mendelsohn decomposition:

```
[p,q] = dmperm(J); Js = J(p,q);
```

The result is shown in the middle plot in Figure 10.12. This permutation is similar to what is done inside MATLAB's linear solver `mldivide`. Since we can permute the Jacobian matrix to triangular form, we can also do the same for the nonlinear system, and hence apply a highly efficient nonlinear substitution method [220] as discussed for 1D Buckley–Leverett problems in Section 9.3.4. The same applies also for 3D cases as long as long as capillary forces are neglected and we have purely co-current flow. Countercurrent flow can be introduced by gravity segregation, as we saw in Section 10.3.2, or if the flow field is computed by one of the consistent discretization schemes from Chapter 6, which are generally not monotone.

It is also possible to permute the discretized system to triangular form by performing a potential ordering [175], as shown to the right in Figure 10.12:

```
[~,i] = sort(x.pressure); Jp = J(i,i);
```

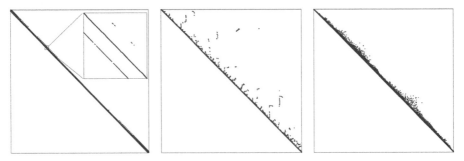

Figure 10.12 The left plot shows the sparsity structure for the Jacobian matrix for the heterogeneous quarter five-spot with viscosity ratio 1:10. The middle plot shows the sparsity structure after a topological sort, whereas the right plot shows the sparsity structure after potential ordering.

Implementing this type of nonlinear substitution methods is unfortunately not very efficient in MATLAB and should be done using a compiled language. In MRST, we therefore mainly rely on the intelligence built into `mldivide` to give us the required computational performance.

A remark at the end: in a real case, *injectivity* would obviously be a decisive factor, i.e., how high pressure is required to ensure a desired injection rate without fracturing the formation, or vice versa, which rate one would obtain for a given injection pressure below the fracturing pressure. This is not accounted for in our earlier discussion; for illustration purposes we tacitly assumed that the desired injection rate could be maintained.

COMPUTER EXERCISES

10.3.1 Repeat the experiments with wells controlled by pressure, fixed water viscosity, and varying oil viscosity. Can you explain the differences you observe?

10.3.2 Run a systematic study that repeats the quarter five-spot simulation from the previous exercise for each of the 85 layers of the SPE 10 model. Plot and compare the resulting production curves. (You can run the experiment for a single mobility ratio to save computational time.)

10.3.3 Run the same type of study with 100 random permeability fields, e.g., as generated by the simplified `gaussianField` routine from Section 2.5.2.
Alternatively, you can use any kind of drawing program to generate a bitmap and generate a channelized permeability as follows

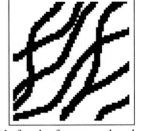

```
K = ones(G.cartDims)*darcy;
I = imread('test.pbm');
I = flipud(I(:,:,1))';
K(I) = milli*darcy;
```

This can easily be combined with different random fields for the foreground and background permeability.

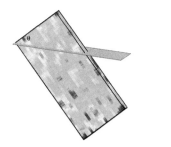 

Figure 10.13 Illustration of the sloping sandbox used for the buoyancy example and how it is simulated by rotating the gravity vector. (Color: Gaussian porosity field.)

10.3.5 Buoyant Migration of CO_2 in a Sloping Sandbox

In Section 10.3.2, we considered the buoyant migration of supercritical CO_2 inside a vertical column. Here, we extend the problem to three spatial dimensions and simulate the upward movement of CO_2 inside a sloping sandbox with sealing boundaries. The rectangular sandbox has dimensions $100 \times 10 \times 200$ m^3, and we consider two different petrophysical models: homogeneous properties or Gaussian porosity with isotropic permeability given from a Carman–Kozeny transformation similar to (2.6). The sandbox is rotated around the y-axis so that the top surface makes an angle of inclination θ with the horizontal plane. Instead of rotating the grid so that it aligns with the aquifer geometry, we will rotate the coordinate system by rotating the gravity vector an angle θ around the y-axis; see Figure 10.13. The rotation is introduced as follows:

```
R = makehgtform('yrotate',-pi*theta/180);
gravity reset on
gravity( R(1:3,1:3)*gravity().' );
```

MRST defines the gravity vector as a persistent, global variable, which by default (for historical reasons) equals $\vec{0}$. The second line sets $\vec{g}$ to the standard value (pointing downward vertically) before we perform the rotation.

To initialize the problem, we assume that CO_2, which is lighter than the resident brine, fills up the model from the bottom and to a prescribed height,

```
xr = initResSol(G, 1*barsa, 1);
d  = gravity() ./ norm(gravity);
dc = G.cells.centroids * d.';
xr.s(dc>max(dc)-height) = 0;
```

For accuracy and stability, the time step is ramped up gradually as follows,

```
dT = [.5, .5, 1, 1, 1, 2, 2, 2, 5, 5, 10, 10, 15, 20, ...
      repmat(25,[1,97])].*day;
```

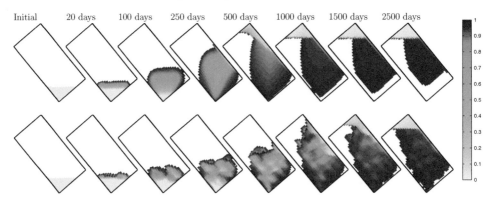

Figure 10.14 Buoyant migration of CO_2 in a sandbox with sealing boundaries. The upper plots show the homogeneous case and the lower plots the case with Gaussian heterogeneity. In the plots, only cells containing some CO_2 are colored.

to reach a final simulation time of 2,500 days. The remaining code is similar to what was discussed in earlier examples; details can be found in buoyancyExample.m.

For the homogeneous case, the buoyant CO_2 plume will initially form a cone shape as it migrates upward and gradually drains the resident brine. After approximately 175 days, the plume starts to accumulate as a thin layer of pure CO_2 under the sloping east face of the box. This layer will migrate quickly up towards the topmost northeast corner of the box, which is reached after approximately 400 days, This corner forms a structural trap that will gradually be filled as more CO_2 migrates upward. The trapped CO_2 forms a diffused and curved interface (see the plots at 500 and 1,000 days), but as time passes, the interface becomes sharper and flatter. During the same period, brine will imbibe into the trailing edge of the CO_2 plume and gradually form a layer of pure brine at the bottom. After approximately 1,000 days, the only CO_2 left below the interface in the northeast corner is found at small saturation values and will therefore migrate very slowly upward.

The heterogeneous case follows much of the same pattern, except that the leading drainage front will finger into high-permeability regions of the sandbox. Low-permeability cells, on the other hand, will retard the plume migration. Altogether, we see a significant delay in the buoyant migration compared with the homogeneous case. Since the permeability is isotropic, the plume will still mainly migrate upward. This should not be expected in general. Many rocks have significantly lower permeability in the vertical direction or consist of strongly layered sandstones containing mud drapes or other thin deposits that inhibit vertical movement between layers. For such cases, one can expect a much larger degree of lateral movement.

COMPUTER EXERCISES

10.3.1 To gain more insight into the flow physics of a buoyant phase, you should experiment more with the buoyancyExample script. A few examples:

- Set $\theta = 88$ and initial height to 10 meters for the heterogeneous case.
- Set $\theta = 60°$ and impose an anisotropic permeability with ratio 0.1:1:5 to mimic a case with strong layering.
- In the experiments so far, we used an unrealistic fluid model without residual saturations. Replace the fluid model by a more general Corey model having residual saturations (typical values could be 0.1 or 0.2) and possibly also endpoint scaling. How does this affect the upward plume migration? (Hint: in addition to the structural trapping at the top of the formation, you will now have residual trapping.)

10.3.2 Rerun the experiment with capillary forces included, e.g., by using `initSimpleFluidJfunc` with parameters set as in its documentation.

10.3.3 Replace the initial layer of CO_2 at the bottom of the reservoir by an injector that injects CO_2 at constant rate before it is shut in. For this, you should increase the spatial and temporal scales of the problem.

10.3.4 Gravity will introduce circular currents that destroy the unidirectional flow property we discussed in Section 10.3.4. To investigate the sparsity of the discretized transport equations, you can set a breakpoint inside the private function `newtonRaphson2ph` used by the implicit transport solver and use the `plotReorder` script from the book module to permute the Jacobian matrix to block-triangular form. Try to stop the simulation in multiple time steps to investigate how the sparsity structure and degree of countercurrent flow change throughout the simulation.

10.3.6 Water Coning and Gravity Override

Water coning is a production problem in which water (from a bottom drive) is sucked up in a conical shape towards a producer. This is highly undesirable since it reduces the hydrocarbon production. As an example, we consider a production setup on a sector model consisting of two different rock types separated by a fully conductive, inclined fault as shown in Figure 10.15.

A vertical injector is placed in the low-permeability stone ($K = 50$ md and $\phi = 0.1$) to the east of the fault, whereas a horizontal producer is perforated along the top of the more permeable rock ($K = 500$ md and $\phi = 0.2$) to the west of the fault. The injector operates at a fixed bottom-hole pressure of 700 bar and the producer operates at a fixed bottom-hole pressure of 100 bar. To clearly illustrate the water coning, we consider oil with somewhat contrived properties: density 100 kg/m^3 and viscosity 10 cP. The injected water has density 1,000 kg/m^3 and viscosity 1 cP. Both fluids have quadratic relative permeabilities. The large density difference was chosen to ensure a bottom water drive, while the high viscosity was chosen to enhance the coning effect. Complete source code can be found in `coningExample.m` in the book module.

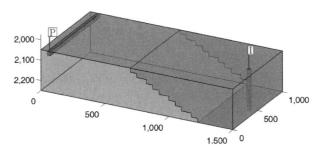

Figure 10.15 Sector model used to demonstrate water coning. Blue color indicates low permeability and red color high permeability.

The idea of using a vertical injector is that the lower completions will set up a bottom water drive in the good rock to the west of the fault, whereas the upper completions will provide volumetric sweep of the low-quality rock to the east of the fault. The water front from the lowest perforations penetrates through to the better zone west of the fault after approximately 40 days and then gradually builds up a water tongue that moves westward more rapidly along the bottom of the reservoir as seen in the two plots in the upper row of Figure 10.16. The advancing water front reaches the far west side of the reservoir after approximately 670 days and forms a cone that extends upward towards the horizontal producer. After 810 days, the water front breaks through in the midsection of the producer, and after 1,020 days, the whole well is engulfed by water. If the well had been instrumented with intelligent inflow control devices, the operator could have reduced the flow rate through the midsection of the well to try to delay water breakthrough. As the water is sucked up to the producer, it gradually forms a highly conductive pathway from the injector to the producer as seen in the snapshot from time 1,500 days shown at the bottom-left of Figure 10.16. A significant fraction of the injected water will therefore cycle through the water zone without contributing significantly to sweep any unproduced oil. Cycling water like this contributes to significantly increase the energy consumption and the operational costs of the production operation and is generally not a good production strategy.

Looking at the well responses in Figure 10.17, we first of all observe that the initial injection rate is very low because of the high viscosity of the resident oil. Thus, we need high injection pressure to push the first water into the reservoir. Once this is done, the injectivity increases steadily as more water contacts and displace a fraction of the oil. Because the reservoir rock and the two fluids are incompressible, increased injection rates give an equal increase in oil production rates until water breaks through after approximately 800 days. Since there is no heterogeneity to create pockets of bypassed oil, and residual saturations are zero, we will eventually be able to displace all oil by continuing to flush the reservoir with water. However, the oil rate drops rapidly after breakthrough, and increasing amounts of water need to be cycled through the reservoir to wash out the last parts of the remaining oil. By 4,500 days, the recovery factor is 73% and the total injected and produced water amount to approximately 2.6 and 1.8 pore volumes, respectively.

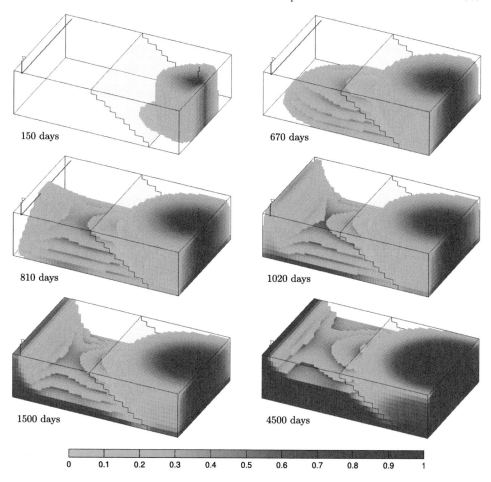

Figure 10.16 Evolution of the displacement profile for the water-coning case.

Another related problem is that of gravity override, in which a less dense and more mobile fluid flows preferentially above a denser and less mobile fluid. To illustrate this multiphase flow phenomenon, we consider a reservoir consisting of two horizontal zones placed on top of each other. Light fluid of high mobility is injected into the lower zone by a vertical well placed near the east side. Fluids are produced from a well placed near the west side, perforated in the lower zone only; see Figure 10.18. Densities of the injected and resident fluids are assumed to be 700 kg/m^3 and 1,000 kg/m^3, respectively. To accentuate the phenomenon, we assume that both fluids have quadratic relative permeabilities, zero residual saturations, and viscosities 0.1 cP and 1.0 cP, respectively. In the displacement scenario, 0.8 pore volumes of the light fluid are injected at constant rate. For simplicity, we refer to this fluid as water and the resident fluid as oil.

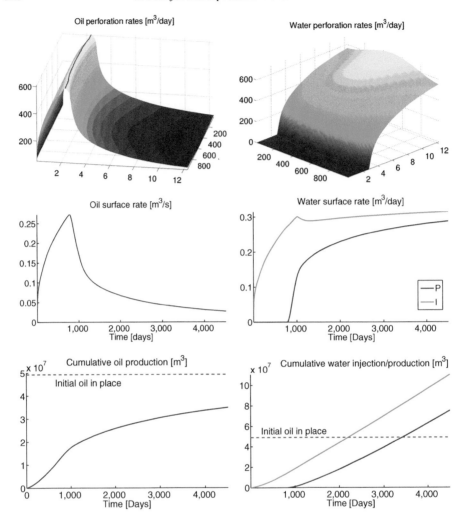

Figure 10.17 Well responses for the simulation of water coning. The top plots show rates in each perforation of the horizontal producer; the red lines indicate water breakthrough. The middle plots show total surface rates, whereas the lower plots show cumulative oil production and cumulative water injection and production.

We consider two different scenarios: one with high permeability in the upper zone and low permeability in the lower zone, and one with low permeability in the upper zone and high permeability in the lower zone. To simplify the comparison, the injection rate is the same in both cases. Figure 10.19 compares the evolving displacement profiles for the two cases. In both cases, buoyancy quickly causes the lighter injected fluid to migrate into the upper zone. With high permeability at the top, the injected fluid accumulates under the sealing top and flows fast towards the west boundary in the upper layers, as seen

10.3 Simulation Examples

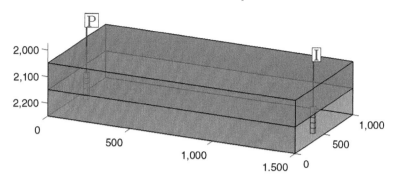

Figure 10.18 Setup for the case used to illustrate gravity override. Blue and red color indicate different permeabilities. Perforated cells are colored green.

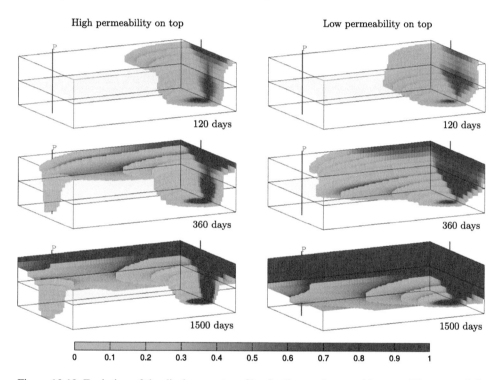

Figure 10.19 Evolution of the displacement profiles for the gravity-override setup. The reservoir is viewed from a position below and to the southeast of the reservoir. Cells containing only the resident fluid are not plotted.

in the upper-left plot of Figure 10.19. Looking at Figure 10.20, you may observe that since the production well is pressure-controlled and perforated in the low-permeability zone, the perforation rates will increase toward the top, i.e., the closer the perforation lies to the high-permeability zone above. As the leading displacement front reaches the producer

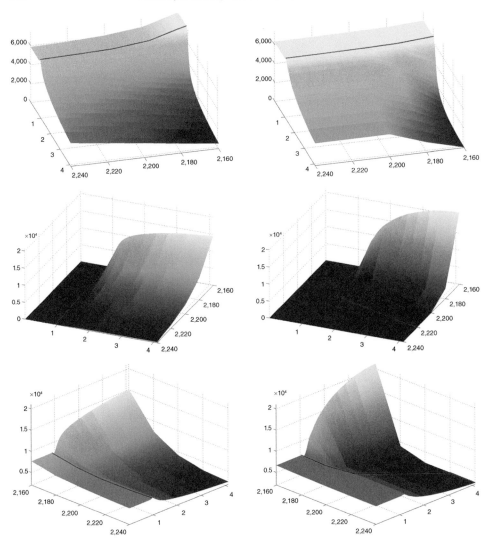

Figure 10.20 Perforation rates for the gravity-override case: oil (top), water (middle), and total rate (bottom). For the oil rate, the red lines indicate peak production, while they indicate water breakthrough in the plots of water and total rate. The left column reports the case with a high-permeability upper zone, and the right column the opposite case.

near the west boundary, it is sucked down toward the open perforations and engulfs them almost instantly, as seen after 360 days in the middle-left plot of Figure 10.19 and the red line in the lower-left plot of Figure 10.20. This causes a significant drop in oil rate and a corresponding increase in water rates towards the top of the well, which lies closer to the flooded high-permeability zone. The oil rate is also reduced in the lowest perforations, but

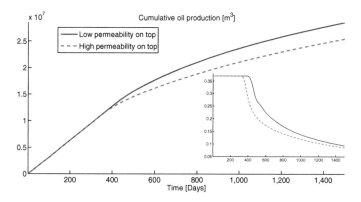

Figure 10.21 Cumulative oil production for the gravity-override case. The inset shows surface oil rates.

because these grid layers do not produce much of the injected fluid, the drop in oil rate is an effect of the internal adjustment of total perforation rates along the well, because mobility is so much higher near the top of the well. Indeed, some time after water breakthrough the oil rate is higher in the lower than in the upper perforations.

Also with a low-permeability zone on top, buoyancy will cause the injected fluid to migrate relatively quickly into the upper zone. However, because this case has better direct connection between the injector and producer through the high-permeability lower zone, the displacement front will move relatively uniformly through all layers of the top zone and the upper layer of the lower zone. The leading part of the displacement front is now both higher and wider and has swept a larger part of the upper zone by the time it breaks through in the producer. Unlike in the first case, the displacing fluid will not engulf the whole well, but only break through in the three topmost perforations. However, after breakthrough, most of the production comes from the topmost perforation, and the total rate in the three lowest perforations quickly drops below 20% of the rate in topmost perforation.

Finally, looking at cumulative oil production and surface oil rates reported in Figure 10.21, we see that low permeability on top not only delays the water breakthrough, but also enables us to maintain a higher oil rate for a longer period. This case is more economically beneficial than with high permeability on the top. To increase the recovery of the latter case, one possibility would be to look for a substance to inject with the displacing fluid to reduce its mobility in the upper zone.

10.3.7 The Effect of Capillary Forces – Capillary Fringe

As you may recall from Section 8.1.3, the pressure in a non-wetting fluid is always greater than the pressure in the wetting phase. In a reservoir simulation model, the capillary pressure – defined as $p_c = p_n - p_w$ in a two-phase system – has the macroscale effect of determining the local saturation distribution at the interface between the wetting and non-wetting fluid, or in other words, there is a relation $p_c = P_c(S)$ between capillary pressure and saturation. This is seen in two ways: In a system that is initially in hydrostatic equilibrium, capillary forces enforce a smooth, vertical transition in saturation upward from a (horizontal) fluid contact. This transition is often referred to as the *capillary fringe*, and will be discussed in more detail in this section. The second effect is that capillary forces will redistribute fluids slightly near a dynamic displacement front so that this is not a pure discontinuity, as assumed in the hyperbolic models discussed in Chapter 9, but rather a smooth wave. For many field and sector models, the characteristic width of the transition zone is small compared to the typical grid size, and hence capillary forces can be safely neglected. In other cases, capillary forces have a significant dampening effect on the tendency for viscous fingering and are therefore crucial to include in the simulation model.

Returning to the formation of a capillary fringe near a fluid interface, let z_i denote the depth of the contact between the wetting and non-wetting fluid, and let $p_{n,i}$ and $p_{w,i}$ denote the phase pressures at this depth. The phase pressures and the capillary pressure are then given by

$$p_w(z) = p_{w,i} + g\rho_w(z - z_i), \qquad p_n(z) = p_{n,i} + g\rho_n(z - z_i)$$
$$p_c(z) = p_{c,i} + g\Delta\rho(z - z_i), \qquad (10.8)$$

where $\Delta\rho = \rho_n - \rho_w$ is density difference and $p_{c,i}$ is capillary pressure at z_i. This pressure is the capillary pressure necessary to initiate displacement of the wetting fluid by the non-wetting fluid and is called the *entry pressure*. It follows from (10.8), that the total height of the capillary fringe is given by $p_{c,n}/g\Delta\rho$. Figure 10.22 illustrates the concept of a capillary fringe for a case with zero residual saturations.

If we know the phase contact z_i, we can find the saturation directly by first computing the capillary force as function of depth using (10.8) and then using the capillary pressure

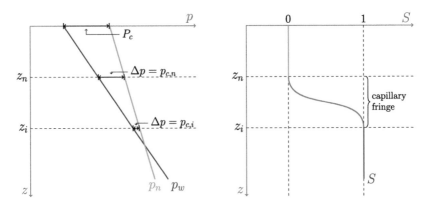

Figure 10.22 Diagrams showing phase and capillary pressures (left) and saturation (right) as function of depth. Here, z_i is the depth of the contact between the non-wetting and the wetting fluid, and z_n is the depth of pure non-wetting fluid.

function $P_c(S)$ to invert for saturation. Alternatively, if we only know the volume of the fluids, we can use the incompressible solvers to determine the hydrostatic fluid distribution. To illustrate, we consider a 100×100 m^2 vertical cross-section represented on a 20×40 grid. We assume a fluid system with CO_2 and brine having the same basic properties as in Section 10.3.2. With a density difference of approximately 290 kg/m^3, the capillary fringe corresponding to a capillary pressure of 1 bar has a height of 35 m. The relationship between saturation and capillary pressure depends on permeability and porosity, and to model this, we use the Leverett J-function (8.9). The initSimpleFluidJfunc implements a simplified Corey-type fluid in which $J(S) = 1 - S$. This gives

$$P_c(S) = \sigma \sqrt{\frac{\phi}{K}}(1-S).$$

The surface tension σ is usually specific to the fluid. To illustrate capillary raise, we choose σ such that median rock properties give a capillary pressure of one bar,

```
fluid = initSimpleFluidJfunc('mu' , [0.30860, 0.056641]*centi*poise, ...
    'rho', [ 975.86,  686.54]*kilogram/meter^3, ...
    'n' , [     2,       2], ...
    'surf_tension',1*barsa/sqrt(mean(rock.poro)/(mean(rock.perm))),...
    'rock',rock);
```

We set initial data such that the rock is filled with half a pore volume of wetting fluid at the bottom and half a pore volume of non-wetting fluid at the top, and then simulate the system forward in time until steady-state is reached. The time-loop is set up with a gradual ramp-up of the time step to increase the stability of the solution procedure (we will come back to the choice of time step in Section 10.4.1),

```
dt = dT*[1 1 2 2 3 3 4 4 repmat(5,[1,m])]*year;
dt = [dt(1).*2.^[-5 -5:1:-1], dt(2:end)];
s  = xr.s(:,1);
for k = 1 : numel(dt),
    xr = incompTPFA(xr, G, hT, fluid);
    xr = implicitTransport(xr, G, dt(k), rock, fluid);
    t  = t+dt(k);
    if norm(xr.s(:,1)-s,inf)<1e-4, break, end;
end
```

We consider two different permeability setups. In the first case, the permeability field varies linearly from 50 md in the west to 400 md in the east. The porosity is assumed to be constant. When the system is released from the artificial initial state with a sharp interface at $z = 50$, the non-wetting fluid starts draining downward into the wetting phase, whereas the wetting phase starts imbibing upward into the non-wetting phase. After approximately 3.5 years, the system reaches the steady state shown in the middle plot in the upper row of Figure 10.23. Here, we say that steady-state is reached when the saturation difference between two consecutive time steps is less than 10^{-4} measured in the L^∞ norm. Because

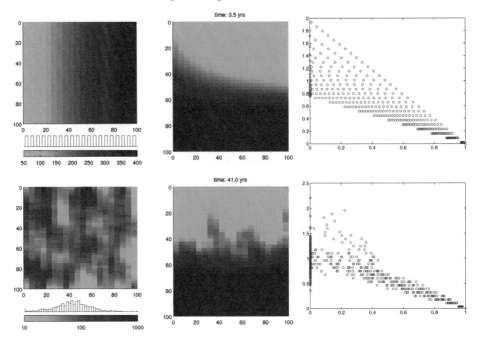

Figure 10.23 Capillary fringe for two different permeability fields: permeability increasing linearly from west to east (top), and lognormal permeability (bottom).

permeability is homogeneous in the vertical direction, the steady-state saturation decreases linearly upward from pure wetting to pure non-wetting fluid. The height of the capillary fringe is much higher in west where the permeability is low, since the capillary scales as $1/\sqrt{K}$, and lower in the east where the permeability is high. In the plot of capillary pressure versus saturation to the upper-right in Figure 10.23, you may also be able to identify the twenty different lines corresponding to the twenty columns of homogeneous permeability in the grid.

The second case has random petrophysical parameters with a permeability field that is related to a Gaussian porosity field through a Carman–Kozeny relationship; see the lower-left plot in Figure 10.23. The permeability values span three orders of magnitude, from 1 md to 1 darcy, which in turn gives a wider span in time constants than in the case with linear permeability. Because of the heterogeneities, the imbibing wetting phase will be sucked higher up and sideways into regions of lower permeability. If you run the script `capillaryColumn` yourself, you will see that the high-permeability regions surrounding the initial sharp interface reach equilibrium within a few years, whereas the rise is much slower in the low-permeability regions. The column next to the east boundary, in particular, is the last to reach steady state after approximately 41 years. At steady state, the fringe extends above the top of the reservoir section in the east-most column. Moreover, because permeability is heterogeneous, the saturation at steady state is no longer monotone in the vertical (and horizontal) direction.

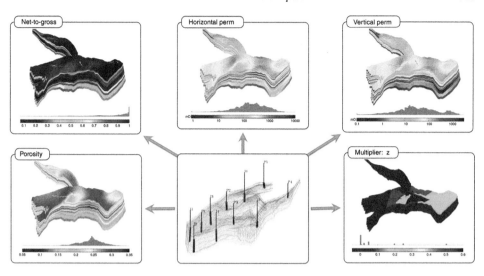

Figure 10.24 The Norne test case with grid and petrophysical data from the simulation model of the real field. The well pattern is artificial and has nothing to do with how the real field is operated.

10.3.8 Norne: Simplified Simulation of a Real-Field Model

Having worked with highly idealized models so far in this chapter, it is now time to look at a more realistic model. For this, we will use the grid geometry and petrophysical properties from a simulation model of the Norne field from the Norwegian Sea. More details about Norne and the reservoir geometry were given in Section 3.3.1. Figure 10.24 shows the petrophysical properties as well as a well-pattern that was chosen somewhat haphazardly for illustration purposes. We notice that the permeability is anisotropic and heterogeneous, with a clear layered structure. This layered structure is also reflected in the histograms, which show several modes. (Such histograms are discussed in more detail in Sections 2.5.3 and 2.5.5 for the SPE 10 and SAIGUP models.) The lateral permeability has four orders of magnitude variations, whereas the vertical permeability is up to two orders lower and has five orders of magnitude variations. In addition, the vertical communication is further reduced by a *multiplier field* (MULTZ keyword), which contains large regions having values close to zero in the middle layers of the reservoir. The porosities span the interval [0.094, 0.347], but since the model has a net-to-gross field to model that a portion of the cells may consist of impermeable shale, the effective porosity is much smaller in some of the cells. For a model like this, we thus cannot expect to be able to use the explicit transport solver and must instead rely on the implicit solver. (Note also that the incomp module contains a similar example with synthetic petrophysical properties; see incompExampleNorne2ph.)

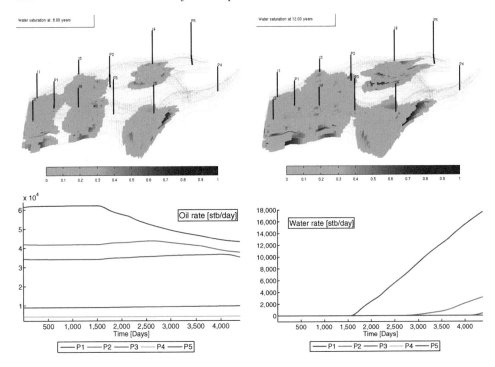

Figure 10.25 Incompressible two-phase simulation for the Norne model. The upper plots show snapshots of the solution (after 6 and 12 years, respectively) and the lower plots show oil and water surface rates for all producers.

Setting up the model proceeds as discussed previously; you can consult the runNorne Simple script in the book module for full details. The only difference is how to account for the transmissibility multipliers. This is done as follows:

```
hT = computeTrans(G, rock, 'Verbose', true);
tmult = computeTranMult(G, grdecl);
hT = hT.*tmult;
```

Here, the Eclipse input structure grdecl contains data for the MULTZ keyword. The second call takes the MULTZ values associated with cells and assigns a corresponding reduction value between 0 and 1 to all half-faces. We then multiply the half-transmissibilities hT by tmult to get the reduced transmissibility. It is important that these multipliers are assigned *before* computing the intercell transmissibilities. Altogether, approximately 6.5% of the half-faces have reduced transmissibility.

Figure 10.25 shows results of a simulation of a scenario in which water is injected into oil. The oil has a five times higher viscosity and hence the injected water will form an unstable displacement with a weak displacement front similar to the case discussed in Section 10.3.4. The main displacement takes place in the region that involves injectors I1 to I5 and producers P1–P3. The last two producers are located in regions that are poorly connected to the rest of the reservoir where the injectors are placed, and hence contribute less to the overall production. This is particularly true for producer P4, which most likely is completely misplaced. As we have seen in previous examples, once water breaks through in a well (primarily in P1 and P2), the oil rate decays significantly.

The main purpose of the example lies in the actual code and not in the results it produces. Discussing the simulation results beyond this point is therefore somewhat futile since the fluid system and the well pattern have limited relevance for the real reservoir, which contains a three-phase oil–gas–water system that is modeled by the compressible black-oil equations that will be discussed in the next chapter. The main takeaway message is that the incompressible solvers can be applied to models that have the geometrical and petrophysical complexity seen in real reservoir models.

COMPUTER EXERCISES

10.3.1 To get more acquainted with the multiphase incompressible solvers and see their versatility, you should return to a few examples presented earlier in the book and try to set them up as multiphase test cases:

- Consider the reservoir in Exercise 3.1.3 and place one injector to the south and two producers symmetrically along the northern perimeter. Simulate the injection of one pore volume of water into an oil.
- Consider the test case with non-rectangular reservoir geometry in Figure 5.6 on page 164 and set up a simulation that injects one pore volume from the flux boundary. How would you compute the flux out of the pressure-controlled boundary?
- Pick any of the faulted grids generated by the `simpleGrdecl` routine as shown in Figure 3.31 on page 97 and place an injector in one fault block and a producer in the other and simulate the injection of half a pore volume of water.

10.3.2 Consider a rectangular reservoir with two wells (see Figure 3.37) and compare solutions computed with three different grids: a uniform coarse grid, a uniform fine grid, and a coarse grid with radial well refinement.

10.3.3 Try to study the Norne model in more detail.

- Are the multipliers important for the simulation result?
- Is gravity important or can it be neglected?
- Can you come up with a better recovery strategy, i.e., improved placement and control strategy for wells?
- Do you get very different solutions if you use a consistent solver? (Hint: Although multipliers can be incorporated into these solvers, as described in [231], this is not part of the public implementation and for this comparison you should therefore neglect the MULTZ keyword.)

10.4 Numerical Errors

There are several errors involved in the computations in the previous section. First of all, numerical discretization errors obviously arise when approximating a continuous differential equation by a set of discrete finite-volume equations. For single-phase, incompressible flow, errors were purely spatial. These errors will decrease with decreasing size of the grid as long as the spatial discretization is consistent. However, as we saw in Chapter 6, the standard two-point scheme is not consistent unless the grid is strictly K-orthogonal, and the incompTPFA pressure solver can in general be expected to produce errors for anisotropic permeabilities and skewed grids. When using a sequential method to solve multiphase flow equations, there will also be *temporal errors* arising from three different factors: discretization errors arising when temporal derivatives in the transport equations are discretized by finite differences, amplifications of spatial errors with time, and errors introduced by the operator splitting underlying the sequential solution procedure. In this section, we briefly discuss the two last error types in more detail.

10.4.1 Splitting Errors

When using a sequential solution procedure, the total velocity is computed from the fluid distribution at the start of each time step. This means that the effect of mobility on the flow paths is frozen in time, and for each time step appears as if we solved a single-phase flow problem with reduced permeability in all parts of the domain that contain more than one fluid phase. Within a single splitting step the transport solver will thus only resolve the dynamic effect of mobility along each flow path, but will not account for the fact that mobility changes reduce the effective permeability along each flow path or move the flow paths themselves. This introduces a time lag in the simulation, which may lead to significant errors in the propagation of displacement fronts if the splitting steps are chosen too large.

Homogeneous Quarter Five-Spot

To illustrate this, we can revisit the homogeneous quarter five-spot from Section 10.3.3 and study the self-convergence of approximate solutions defined on a fixed grid as the number of splitting steps increases. Figure 10.26 shows approximate solutions at time $t = 0.6$ PVI (pore volumes injected) computed with 4^ℓ steps for $\ell = 0, \ldots, 3$. With a single pressure step, the displacement front coincides with the time line from the single-phase flow field, since the pressure computation only sees the initial oil saturation. The only exception is a certain smearing introduced by the spatial and temporal discretizations of the explicit scheme used to compute the transport step. The retardation effect that oil has on the invading water is better accounted for as the number of splitting steps increases, and hence the splitting solution gradually approaches the correct solution. Table 10.1 reports the self-convergence towards a reference solution computed on the same grid with 256 splitting steps. To estimate the convergence rate we assume that the error scales like $\mathcal{O}(\Delta t^r)$. If the

Table 10.1 *Self-convergence for the homogeneous quarter five-spot computed on a* 128×128 *grid with n splitting steps relative to a reference solution computed with 256 time steps after 0.6 PVI. The errors are reported in relative norms.*

n	saturation L^1-error	rate	pressure L^2-error	rate
1	5.574e−02	–	4.950e−03	–
2	4.368e−02	0.35	5.140e−04	3.27
4	2.778e−02	0.65	1.554e−04	1.73
8	1.445e−02	0.94	4.524e−05	1.78
16	6.389e−03	1.18	1.226e−05	1.88
32	2.394e−03	1.42	2.998e−06	2.03
64	7.869e−04	1.61	5.990e−07	2.32

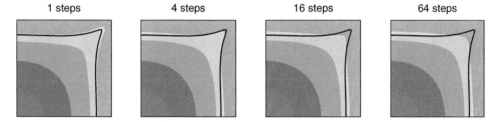

Figure 10.26 Quarter five-spot solution at time 0.6 PVI computed on a uniform 128×128 grid with the explicit transport solver and different number of splitting steps. The solid lines are time lines at $t = 0.6a$ for a single-phase displacement.

solution having error E_2 is computed using twice as many time steps as the solution having error E_1, the corresponding convergence rate is

$$r = \log(E_1/E_2)/\log(2).$$

Pressure is smooth and will therefore converge faster than saturation, which is a discontinuous quantity and hence will have much larger errors. The convergence for high n values is exaggerated since we are measuring self-convergence toward a solution computed with the same method, but with a larger number of steps. The code necessary to run this experiment is found in `splittingErrorQ5hom.m` in the `in2p` directory of the `book` module.

Heterogeneous Quarter Five-Spot

As pointed out earlier, the coupling between the pressure and transport equations depends on the variation in total mobility $\lambda(S)$ throughout the simulation. If variations are small and smooth, the two equations will remain loosely coupled, and relatively large time steps can

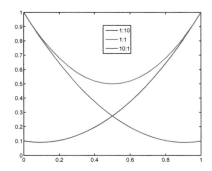

Figure 10.27 Total mobility for three fluid models with different viscosity ratios.

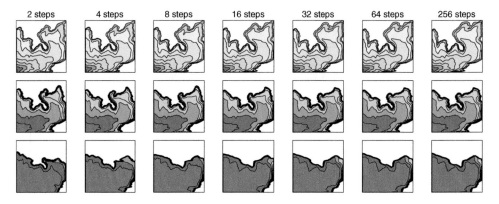

Figure 10.28 Self convergence with an increasing number of equally spaced splitting steps to reach time 0.5 PVI for a quarter five-spot setup on a subsample from the first layer of the SPE 10 model for three different viscosity ratios $M = \mu_w/\mu_n$ (top: M = 1/10, middle: M = 1, bottom: M = 10). Saturation profiles are shown at a time scaled by $v(1)/v(M)$, where $v(M)$ is the characteristic wave speed of the displacement front with viscosity ratio M.

be allowed without seriously decaying solution accuracy. On the other hand, when λ has large variations over the interval $[0, 1]$, pressure and transport are more tightly coupled, and we cannot expect to be able to use large splitting steps. To illustrate this, we revisit the setup with three different fluid models used to study viscous fingering in Section 10.3.4. From the plot of total mobilities in Figure 10.27 it is obvious that both the unfavorable (1:10) and the favorable (10:1) mobility cases have stronger coupling between saturation and pressure than the case with equal viscosities.

Figure 10.28 shows the self convergence of the saturation profiles with respect to the number of equally spaced time steps used to reach time 0.5 PVI. To isolate the effect of splitting errors and avoid introducing excessive numerical smearing in the solutions with few splitting steps, we have subdivided the transport steps into multiple steps so that the implicit solver uses the same step length and hence introduces the same magnitude of

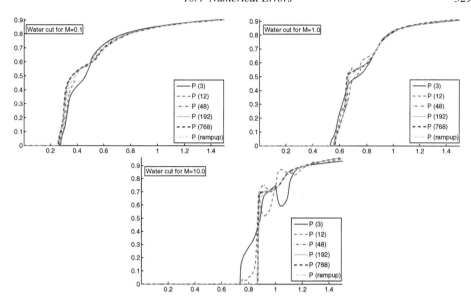

Figure 10.29 Water cut in the producer for various number of splitting steps for the heterogeneous quarter five-spot setup from Figure 10.28.

numerical smearing in all simulations. From the figure it is clear that even with very few splitting steps, the sequential solution method manages to capture the qualitatively correct behavior for all viscosity ratios. As expected, the discrepancies in solutions with few and many time steps are larger for the favorable and unfavorable cases than for unit viscosity ratio.

To investigate how the size of the splitting steps affects the *quantitative* behavior of the approximate solutions, we reran the same experiments up to 1.5 PVI. Figure 10.29 reports water cut in the producer for all three fluid models. Starting with the unfavorable case, we see that the water cut has a dent. This is a result of the secondary finger that initially extends along the western edge, making contact with the main finger and hence contributing to a more rapid incline in water production. All curves, except the one using only three time steps, follow the same basic trend. The main reason is that small variations in the saturation profile will not have a large effect on the water cut since the displacement front is so weak. With three splitting steps only, the main dent is a result of the second pressure update. For unit viscosity ratio, the production curves are still close, but here we notice a significant incline in the curve computed with 12 steps after the pressure update at 0.75 PVI. The 12-step and 48-step curves also show non-monotone behavior at 0.625 PVI and 0.69 PVI, respectively. Similar behavior can be seen for the favorable case, but since this case has an almost piston-like displacement front, the lack of monotonicity is significantly amplified. For comparison, we have also included a simulation with 96 splitting steps, in which the first two first steps have been replaced by ten smaller splitting steps that gradually ramp up to the constant time step. These profiles, and similar profiles run with 48 steps, are monotone, which suggests that the inaccuracies in the evolving saturation profiles are introduced early in the simulation when the profile is rapidly expanded by high fluid velocities in the near-well region. In our experience, using such a ramp-up is generally advisable to get more well-behaved saturation profiles. You can find the code for this experiment in `splittingErrorQ5het.m` in the `in2p` directory of the `book` module.

Capillary-dominated flow

The sequential solution procedure discussed in this chapter is as a general rule reasonably well-behaved for two-phase scenarios where the fluid displacement is dominated by the hyperbolic parts of the transport equation, i.e., by viscous forces (pressure gradients) and/or gravity segregation. If the parabolic part of the solution dominates, on the other hand, a sequential solution procedure will struggle more, in particular when computing fluid equilibrium govern by a delicate balance between gravity and capillary forces. To illustrate this, we revisit the computation of capillary fringe from Section 10.3.7. If you look carefully in the accompanying code, you will see that we compute the two cases with a time step that is ten times larger for the Gaussian case than for the case with linear permeability. The time steps (and the initial ramp-up sequence) were chosen by trial and error and are close to what appears to be the stability limit. If one, for instance, increases the final time steps by 150% for the case with linear permeability, the simulation will not converge but instead ends up predicting an oscillatory interface, as illustrated in Figure 10.30. Similar problems may arise, e.g., when simulating structural trapping of CO_2 using vertical equilibrium models, in which gravity gives rise to a parabolic term that plays the same role as capillary pressure in the upscaled flow equations; see e.g., [227] for more details.

10.4.2 Grid Orientation Errors

As you may recall from the discussion on page 287, the saturation-dependent mobility at the interface between two grid blocks is usually approximated by single-point, upstream mobility weighting. Like the TPFA method, the resulting scheme only accounts for information on opposite sides of a cell interface and does not take any transverse transport effects into account. It is therefore well known that this method also suffers from grid orientation errors, especially when applied to unfavorable displacements, as we will see in the following example. In passing, we also note that to evaluate gravity and capillary-pressure terms, the transport solvers use two-point approximations similar to what is used in the TPFA method. This may introduce additional errors, but we will not discuss these in detail herein.

Homogeneous Quarter Five-Spot

When the single-point transport solver is combined with the classical TPFA pressure solver, computed displacement fronts tend to preferentially move along the axial directions of the

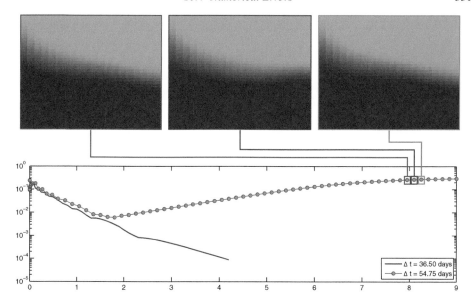

Figure 10.30 Unstable solution computed by a sequential solution with too large splitting step. The lower plot shows difference in L^∞ norm between consecutive time steps – which is used as convergence criterion by the solver – for a convergent solution with $\Delta t = 36.5$ days and for a divergent solution with $\Delta t = 54.75$ days. The upper plot shows three consecutive solutions for the divergent simulation.

grid, i.e., in the direction of the normal vectors of the cell faces. This will lead to grid orientation effects like those discussed earlier in Chapter 6, even if the resulting grids are K-orthogonal. To illustrate this, we compare and contrast solutions of the standard quarter five-spot setup with a rotated setup in which the grid is aligned with the directions between injectors and producers, as illustrated in Figure 10.31. This test problem was first suggested by Todd et al. [290] and has later been used by many other authors to study grid orientation errors in miscible displacements [321, 255, 279], which are particularly susceptible to this type of truncation error. To avoid introducing too much diffusion when using few time steps, we use the explicit transport solver. (See `runQ5DiagParal` for complete setup of the two cases and `gridOrientationQ5` for the following experiments.)

For the standard setup, the combination of a single-point transport solver and a two-point pressure solver overestimates the movement into the stagnant regions along the x and y axes and underestimates the diagonal movement in the high-flow direction between injector and producer along the diagonal. In the rotated setup, the grid axes follow the directions between the wells and the solvers will hence tend to overestimate flow in the high-flow zone and underestimate flow toward the stagnant zones. The upper row in Figure 10.32 shows that this effect is pronounced for the unfavorable displacement ($M = 0.1$), evident with equal viscosities ($M = 1$), and hardly discernible for the favorable displacement ($M = 10$). For equal viscosities, the difference between the two grids can be almost

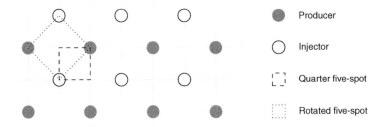

Figure 10.31 Well setup for the quarter five-spot comparison. Displacement fronts have preferential movement parallel to the axial directions and hence the rotated setup will predict earlier breakthrough than the original setup.

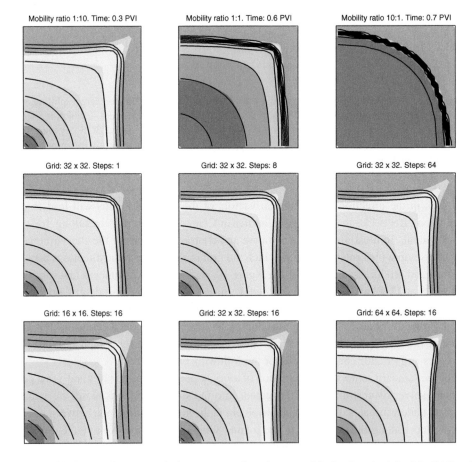

Figure 10.32 Quarter five-spot solutions computed on the rotated (colors) and original (solid lines) geometry for a 32×32 grid. The top row shows solutions computed with 16 steps for different mobility ratios. The middle plot shows convergence for mobility ratio 1:10 with respect to time step and the lower row with respect to grid resolution.

10.4 Numerical Errors

eliminated by increasing the number of splitting steps for a fixed Δx and/or by increasing the grid resolution provided that a certain number of time steps are used. In the unfavorable case, the plots in the middle row of Figure 10.32 show that increasing the number of time steps changes both solutions in the same direction, but does not necessarily reduce their difference. However, since both the original and the rotated grid are regular, the grid orientation effects will diminish if we increase the grid resolution.

A possible remedy to the behavior we just observed is to replace the single-point scheme by a multidimensional upwind scheme (see e.g., [155]) or a modern high-resolution scheme. Such methods are not yet part of the public MRST release, but work is currently in progress [190].

Symmetric Well Pattern on a Skew Grid

We have already seen several times that since the TPFA scheme cannot approximate transverse fluxes that are parallel to grid interfaces, the `incompTPFA` solver will introduce grid orientation errors for anisotropic permeabilities and grids that are not K-orthogonal. In Chapter 6, we showed that we can significantly reduce, but not completely eliminate, these errors if we replace TPFA by a consistent scheme. In this example, we will revisit one of the cases discussed in Chapter 6 to investigate if we see the same reduction for two-phase flow.

The computational setup consists of a horizontal 400×200 m^2 sector model. Water is injected from at the midpoint of the northern perimeter and fluids are produced from two wells located 50 m from the southeast and southwest corners, respectively. Since the well pattern is symmetric within a confined domain and the petrophysical parameters are homogeneous and isotropic, the true displacement profile will also be symmetric. The grid, however, is skewed and compressed towards the southeast corner. This will induce a preferential flow direction towards the southeast producer. Seeing that grid orientation effects are more pronounced for unstable displacements, we use the same configuration as in the previous example with a viscosity ratio $M = 10$, which gives a very mobile, weak displacement front that will tend to finger rapidly into the resident oil. Figure 10.33 shows a snapshot of the pressure and saturation profile after 250 days along with water cuts and cumulative oil production over the whole 1,200-day simulation period. The flow field computed with the `incompTPFA` pressure solver exhibits the same lack of symmetry as seen in Figure 6.5 on page 187. The result is a premature breakthrough in the southeast producer and delayed breakthrough in the southwest producer. With a mimetic finite-difference (MFD) solver, the water-cut curves are much closer and less affected by grid orientation errors. Moreover, since all wells operate under pressure control, we see that the total oil production predicted by TPFA is significantly less than for the MFD solver.

As a second example, we use the same grid to describe a vertical cross-section, in which we inject water from two horizontal injectors at the bottom of the reservoir and produce fluids from a horizontal producer at the top of the reservoir. Producers and injectors operate under the same pressure control as for the horizontal reservoir section. Figure 10.34 shows

334 Solvers for Incompressible Immiscible Flow

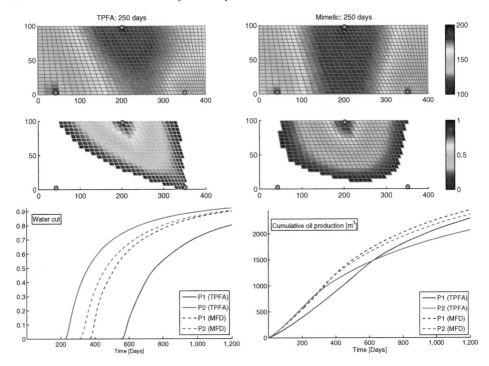

Figure 10.33 Simulation of a symmetric flow problem in a horizontal, homogeneous domain represented on a skew grid.

simulation results with the TPFA and MFD solvers. The density difference between the injected and resident fluids is 150 kg/m^3, and hence gravity tends to oppose the imbibing water front in a way that accentuates the grid orientation effects for both solvers. For comparison, Figure 10.35 reports the simulations performed on a regular Cartesian grid. Both schemes produce symmetric, but slightly different displacement profiles and the match in production profiles is largely improved compared with the skew grid.

Altogether, the two examples presented in this section hopefully show you that you not only need to take care when designing your grid, but should also be skeptic to simulations performed by a single method or a single choice of time steps. A good piece of advice is to conduct simulations with more than one scheme, different time-step selection, and, to the extent possible, different grid types to get an idea of how numerical errors influence your results. To further investigate grid orientation effects, consider any of the following computer exercises.

COMPUTER EXERCISES

10.4.1 Repeat the test case with the original/rotated quarter five-spot using one of the consistent solvers from Chapter 6 to compute the pressure. Do you see any differences?

10.4 Numerical Errors 335

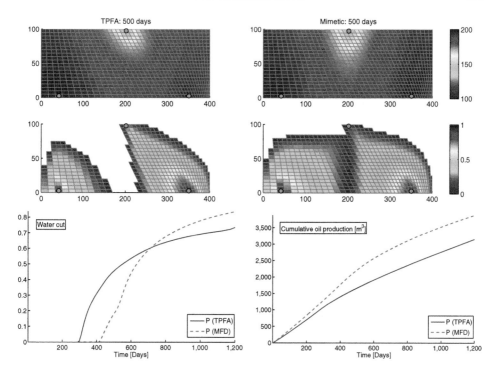

Figure 10.34 Simulation of a symmetric flow problem in a vertical, homogeneous domain represented on a skew grid with two horizontal injectors at the bottom and a horizontal producer at the top.

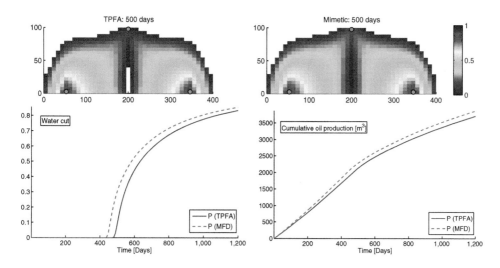

Figure 10.35 Simulation of a symmetric flow problem in a vertical, homogeneous domain computed on a regular Cartesian grid.

10.4.2 Set up a flow problem to compare solutions computed on the extruded Delaunay or Voronoi grids shown in Figure 3.32 on page 97.

10.4.3 Set up a quarter five-spot test using the grids shown in Figure 3.44 on page 108.

10.4.4 The `CaseB4` test case shown in Figure 3.20 on page 80 represents the same geology with two different grid formats, using either a deviated pillar grid or a stair-stepped grid. Both grids are sampled at two different resolutions. Set up a flow problem with four wells, two injectors and two producers, one in each corner of the reservoir. Run simulations on all four grids and compare production curves. You can either use homogeneous permeability or a layered permeability with homogeneous properties within each layer.

10.4.5 Pick any of the models in the `bedModels1` or `bedModel2` data sets and run two-phase simulations injecting one pore volume from south to north and from west to east. Investigate splitting and grid orientation effects.

11
Compressible Multiphase Flow

The black-oil equations constitute the industry-standard approach to describe compressible three-phase flow. Black-oil models generally have stronger coupling between fluid pressure and the transport of phases/components than the two-phase, incompressible flow models discussed in the previous chapter. For this reason it is common to use a fully coupled solution strategy, in which the whole system of equations is discretized implicitly and all primary unknowns are solved for simultaneously. This chapter introduces you to the underlying physics and describes the various rock-fluid and PVT properties that enter these models. We also explain the basics of how these models are discretized and implemented in MRST. Our implementation will rely heavily on the discrete operators discussed earlier in the book. As a gentle introduction to the numerics of the black-oil equations, we pick up again the simple AD solver from Chapter 7 and extend it to compressible two-phase flow without mass transfer between the phases. The code is applicable to general unstructured grids in 2D and 3D, but it is still a far stretch from a full-fledged reservoir simulator capable of simulating real assets, as discussed in Chapter 12.

11.1 Industry-Standard Simulation

A commercial reservoir simulator is a complex software that involves a wide variety of flow models, computational algorithms, and a lot of subtle tricks to ensure robust behavior. Covering all of these features, as well as the many different scenarios for which a simulator can be used, is well beyond the scope of this chapter. Instead, I will try to provide you with a minimum of insight into the key principles and show how to set up your first black-oil simulation in MRST using the AD-OO framework (this framework is described in more detail in Chapter 12). Let us start by a brief overview before we discuss the different constituents in more detail.

Flow Models and Phase Behavior

Reservoir fluids are complex mixtures of many chemical components that distribute differently among the fluid phases present in the reservoir depending upon temperature and pressure. The composition and characteristics of hydrocarbon fluids vary largely from one reservoir to another, but may also have large variations across reservoir compartments. As we have seen already, the black-oil model groups the chemical components into two pseudo-components reflecting the phase they belong to at surface conditions. At reservoir conditions, each component can be present in both a liquid oleic and a gaseous phase. We motivate and briefly explain the phase diagrams used to describe the phase distribution of the pseudo-components, but an in-depth discussion of the underlying physics and the many different forms these diagrams can take is outside the scope of the book.

The black-oil equations can also model processes involving other chemical species. In enhanced oil recovery, various forms of chemical and biological substances are added to the injected fluid to increase sweep efficiency and improve local displacement. You can model such systems by adding extra fluid components to the three phases. MRST offers an EOR module with basic models for polymer and surfactant flow; see [30] for more details, as well as a module for simulating solvents. We have already discussed CO_2 storage multiple times. Here, the buoyant and supercritical phase formed when CO_2 is injected sufficiently deep into the subsurface can dissolve into the resident brine. The behavior of the CO_2–brine system can thus be modeled as a hydrocarbon system in which brine plays the role as oil and CO_2 the role as gas. Extensive support has been developed in MRST to describe the long-term migration of buoyant CO_2 in large-scale aquifer systems, in which fully implicit solution methods developed for the black-oil equations play a key role. You can find more details in the thesis by Andersen [19].

Grouping the hydrocarbons into two pseudo-components does not describe all recovery processes with sufficient accuracy. *Compositional simulation* models the reservoir fluids as a system with M components and describes their phase behavior by an equation of state. The basic conservation equations for such systems was given in Section 8.2.2 and typical examples of equations of state were briefly reviewed as part of our discussion of single-phase flow in Section 4.2. As of release 2017b, MRST offers a module for compositional simulation. We will not discuss this at all herein; you can find preliminary details in the thesis by Møyner [212]. Likewise, a new module for geochemistry was also added in the same release, which, in particular, offers non-isothermal aqueous speciation, surface chemistry, redox chemistry, as well as solid and gas phase equilibrium.

Wells and Production Control

In previous chapters, wells have always been completed in a single cell or along straight lines following the axial directions, and have either been controlled by constant pressure or constant fluid rate. Real wells are significantly more complex and generally follow highly deviated trajectories and possibly consist of multiple branches. Many subtleties go into well modeling. In particular, special care is needed in the solution algorithm to handle cross-flow, in which fluids may flow into one part of the wellbore and out of another.

To ensure that each well operates under realistic engineering constraints, reservoir simulators offer a comprehensive set of control options that can be applied to fluid rates and bottom-hole or tubing-head pressures. A typical control strategy consists of setting a target rate and constraints on the pressure the well can sustain before the rate must be reduced. Well control may thus switch depending upon the computed pressures and flow rates once the well fails to meet one of its constraints. In so-called history mode, observed fluid rates at reservoir conditions are set as controls to reproduce production history as closely as possible. Controls also include economic constraints, which force wells to be completely shut in or recompleted if their production falls below a lower economic limit, and gas lift or other mechanisms introduced to help lift the well fluids to the surface when the oil or liquid rates fall below a specified limit. This will strongly affect the bottom-hole pressure in individual wells.

Control strategies can also be imposed on groups of wells, or on the field as a whole. Typical strategies here either involve guide controls, in which total rates are distributed to individual rates according to their production potential, or prioritization schemes that turn wells on in decreasing order of priority. Group and field controls may also involve various restrictions in terms of total pump and lift capacities, reinjection of produced formation gas or water, limitations in surface facilities, etc. Altogether, the resulting control strategies may involve a lot of quite intricate logic. MRST contains data structures for representing surface facilities, but only a limited number of models and control options have so far been implemented in the public version. I therefore consider a more detailed discussion of control strategies to be outside the scope of this book. Examples of third-party implementations with MRST are discussed in two master theses [141, 115].

Many wells also have advanced completions and devices for inflow control. In *multisegment* well models, the well trajectory is discretized into a series of segments that can be linked together into branches. Each branch can, in turn, contain sub-branches. A segment describes how the flow rate between two points (nodes) relates to the pressure drop between the same two points. Nodes can be located at different positions along the wellbore or on opposite sides of devices that restrict the flow, such as chokes and valves. Each node has an associated volume, and mass-conservation equations are solved between the nodes to describe pressure drops and wellbore storage effects. As of release 2017a, MRST contains a general multisegment model that allows branches of segments to be defined in the tubing and in the void space (annulus) between the inner and outer tubing. The branches can be joined at multiple points in a flexible manner to form a general network. Altogether, this enables accurate description of inflow devices and the near-well region, as well as complex cross-flow patterns that may arise in multilateral wells. An in-depth discussion of multisegment well modeling is outside the scope of this book, but as is always the case with MRST, you are free to consult the code to understand the necessary details.

Data Input

To model real assets, a simulator needs comprehensive and flexible input functionality to specify the setup of all ingredients that constitute the simulation model. The industry-standard approach is to define an input language consisting of a number of keywords the user can utilize to construct a model of a computation (i.e., a state machine). The detailed syntax will vary from one simulator to another, but the general setup is conceptually the same. You have already encountered the ECLIPSE input format [270, 33] multiple times throughout the book. Over the years, this software has amassed options for simulating a wide variety of physical processes, and the input language currently has several thousand keywords, many of which have multiple available options. There are also similar file formats that can be used to store simulation results so that these can be reloaded and used to continue a previous simulation.

So far, we have only discussed how ECLIPSE input describes grids and petrophysical properties, and even for this purpose there are a large number of keywords we have not discussed. In general, an ECLIPSE input deck consists of eight different sections, some of which are optional:

RUNSPEC – required section, includes a description of the simulation, including parameters such as name of the simulation case, grid dimensions, phases and components present, number of wells, etc.

GRID – required section, describes basic grid geometry/topology and petrophysical properties (porosity, permeability, net-to-gross).

EDIT – optional section describing user-defined changes that are applied to the grid data *after* these have been processed to modify pore volume, cell centers, transmissibilities, local grid refinements, etc.

PROPS – required section describing fluid and rock properties such as relative permeabilities, capillary pressure, and PVT specification as a function of fluid pressures, saturations, and compositions.

REGIONS – optional section defining how the computational grid can be split into multiple regions for calculating PVT properties, relative permeabilities and capillary pressures, initial conditions, and fluids in place.

SOLUTION – required section specifying how the model is to be initialized: calculated from specified depths of fluid contacts, read from a restart file (output) from a previous run, or specified for each grid cell.

SUMMARY – optional section that enables the user to specify the reservoir responses (well curves, field-average pressure, etc.) to be written to a summary file after each time step.

SCHEDULE – required section, defines wells and how they are to be operated, specifies time-step selection and tolerances for the solvers, controls the output of quantities defined over the grid (pressures, saturations, fluxes, etc.).

The sections *must come in this prescribed order* in the input file, but the keywords within each section can be given in arbitrary order. Altogether, the input data file can be seen as a kind of scripting language that not only enables you to input data to the simulator, but also provides control of the time dependency and how the simulator will run a prescribed set of time steps.

In the following, we borrow the naming convention from ECLIPSE and refer to the way a simulation is to be run as a *schedule*. In the simplest case, a schedule consists of a number of prescribed time steps and one control per well. However, schedules representing real assets tend to be much more complicated, as can be inferred from the previous discussion of wells.

Let me also make two additional remarks: Since several of the keywords, and in particular those in the GRID section, may contain huge amounts of data, the input format offers an INCLUDE keyword allowing input data to be separated into multiple files. Moreover, although MRST aims to support the most widely used keywords in the required sections, there are many keywords that are *not* supported. In particular, no effort has been put into supporting the flexible, yet complex, options for user-specified postprocessing of geological models in the EDIT section.

The following text will give some details about ECLIPSE input [270, 33], but only when this is prudent to understand how you can set up the various part of a black-oil fluid description in MRST. If you are interested in a more thorough introduction, I recommend the lecture notes by Pettersen [250], which are much to the point, freely available, and complementary to the discussion in my book. An early version of the ECLIPSE 100 User Course [269] can also be found online.

More Advanced Solution Methods

As evidenced by our discussion, discrete flow models found in reservoir simulators are significantly more complex than the incompressible, two-phase flow equations considered in previous chapters. For small models with relatively simple phase behavior and solution trajectories, it is straightforward to linearize the flow equations and solve them with standard linear algebra, much in the same way as we did for single-phase flow in Section 7.2. However, for more complex phase behavior, there are many features complicating the solution process, like the need to switch primary unknowns from one cell to another when phases appear or disappear. Likewise, you may need extra postprocessing steps to record historic solution trajectories to model hysteretic behavior of relative permeabilities and capillary pressure, and so on. As the complexity increases for each of the three constituents of the reservoir model (geology model, fluid model, well model), one really sees the benefit of using automatic differentiation to perform the necessary linearizations.

All these model complications translate to increased nonlinearity in the discretized system, and a simple nonlinear Newton–Raphson solver cannot be expected to always converge unless we instrument it with inexact line-search methods and methods for time-step control, as briefly introduced in Section 10.2.2. Likewise, there is a need for correct scaling of the residual equations, as well as more advanced linear solvers, e.g., a preconditioned Krylov subspace method like the flexible general minimal residual (GMRES) method, combined with specialized preconditioning strategies aimed at lessening the impact of the largely different mathematical characteristics of the various sub-equations. Classic choices of single-stage preconditioners include the incomplete lower/upper (ILU) family or the nested factorization method [24], which is still the default method in ECLIPSE. Another popular choice is the two-stage constrained-pressure residual (CPR) method [304, 305]. Here, the first step computes an approximate solution of a reduced-pressure matrix that is appropriately scaled and selected so that one can use, e.g., a very efficient algebraic multigrid solver [285, 119]. The second step then applies a broadband smoother like ILU to the entire system. Section 12.3.4 briefly outlines such preconditioners and the exact solution strategy used in MRST to obtain robust simulation of real cases. Full details can, as always, be found in the code.

Details of the MRST Implementation

To a large degree, MRST has many of the model complexities discussed thus far implemented in its black-oil type solvers. Our first successful attempt at implementing black-oil simulators relied on discrete operators for discretization and automatic differentiation and used the type of procedural programming discussed so far in the book. As more complex behavior was introduced, the code gradually became more unwieldy, and at a certain stage, we realized that a major restructuring was required to make the code easier to extend and maintain. The result was the object-oriented AD-OO framework [170, 212, 30], which will be discussed in more detail in the next chapter. The key idea of the AD-OO framework is to separate the implementation of physical models (i.e., conservation equations and constitutive relationships), discrete operators, nonlinear solvers and time-stepping, and assembly and solution of the linear system. By using a combination of classes, structures, and functions for generic access, details from within each of these contexts can be hidden from the user/developer and only exposed if really needed. As a result, it is now much simpler to implement generic algorithmic components or extend existing models with more comprehensive flow physics.

This introduces a pronounced divide in MRST between simulators written for incompressible and compressible flow. It also represents a pedagogical challenge. I have solved it by relying on the simplified codes written in a procedural form to introduce key concepts that are also found in the object-oriented framework, albeit implemented in a somewhat different form. Once the basic concepts have been introduced, I switch to using the AD-OO framework when presenting simulation cases with black-oil models. This means, in particular, that the codes used to run black-oil simulations will not be presented at the same level of detail as in previous chapters of the book, but all examples have complete source codes in the `ad-blackoil` or `book` modules.

11.2 Two-Phase Flow without Mass Transfer

For a system consisting of two immiscible phases, each with a fixed chemical composition and no mass transfer between the phases, the basic flow model consists of the conservation equations (8.22) and the standard multiphase extension of Darcy's law, (8.23). This if often referred to as a *dead oil* model, and is representative for reservoirs with relatively thick or residue oil that has lost its light and volatile hydrocarbon components, or for oils containing a constant amount of dissolved gas because pressure is so high that gas cannot appear as a separate phase at reservoir conditions (more about this later).

If we for simplicity neglect capillary pressure and introduce discrete approximations for all derivatives, the discrete flow equations read,

$$\frac{(\boldsymbol{\phi} S_\alpha \boldsymbol{\rho}_\alpha)^{n+1} - (\boldsymbol{\phi} S_\alpha \boldsymbol{\rho}_\alpha)^n}{\Delta t^n} + \mathrm{div}(\boldsymbol{\rho v})_\alpha^{n+1} = (\boldsymbol{\rho q})_\alpha^{n+1}, \tag{11.1a}$$

$$\boldsymbol{v}_\alpha^{n+1} = -\frac{\boldsymbol{K} k_{r\alpha}^{n+1}}{\mu_\alpha^{n+1}}\bigl[\mathrm{grad}(p^{n+1}) - g\boldsymbol{\rho}_\alpha^{n+1}\mathrm{grad}(z)\bigr]. \tag{11.1b}$$

11.2 Two-Phase Flow without Mass Transfer

There are only two issues that are conceptually different from the corresponding single-phase model (7.2) we studied in Chapter 7: first of all, we now have two phases and must ensure that all saturation-dependent quantities are averaged correctly, as discussed in the previous chapter. Secondly, wells must be treated slightly differently, since the wellbore now may contain a mixture of two fluid phases. We shall put this issue off for now, and return to it later.

A Simple Compressible Two-Phase Simulator

To develop our first simulator, we consider a simple rectangular $200 \times 200 \times 50$ m^3 reservoir, initially filled with oil with oil in the upper 2/3 and by water in the bottom 1/3 of the volume. Water is injected in the lower southwest corner and fluids are produced from the upper northeast corner. To mimic the effect of injection and production wells, we manipulate the flow equations to ensure reasonable inflow/outflow for the perforated cells. The resulting code, called `twoPhaseAD.m`, can be found in the book module and follows essentially the same steps as the single-phase code in Section 7.2. Setup of the geological model is essentially the same and we skip the details.

For the fluid model, we assume that the water phase is slightly compressible with quadratic relative permeabilities, whereas the oil phase is lighter, more compressible, and has cubic relative permeabilities

```
[muW, muO ] = deal(1*centi*poise, 5*centi*poise);
[cw,  co  ] = deal(2e-6/psia,     1e-5/psia);
[krW, krO ] = deal(@(S) S.^2,     @(S) S.^3);
```

The reference densities at 200 bar are 1014 kg/m^3 for water and 850 kg/m^3 for oil. Since the capillary pressure is assumed to be zero, we can compute the initial pressure distribution exactly as for single-phase flow on page 206. Transmissibilities and discrete operators are independent of the flow equation and are defined as in Section 7.2.2, except that we now also have to define the upwind operator:

```
upw  = @(flag, x) flag.*x(N(:, 1)) + ~flag.*x(N(:, 2));
```

The setup of the simulation loop follows along the same path as in Section 7.2.4. We start by initializing the unknown variables and introduce indices so that we that later can easily extract the individual variables from the global vector of unknowns,

```
[p, sW]   = initVariablesADI(p0, sW0);
[pIx, sIx] = deal(1:nc, (nc+1):(2*nc));
```

Next, we initialize the necessary iteration variables and create a simulation schedule containing a gradual ramp-up of the time step. Then, the main loop is more or less identical to the one shown on page 210. The only difference is how we compute the discrete equations. We start by computing all cell-based properties

```
% Densities and pore volumes
[rW,rW0,r0,r00] = deal(rhoW(p), rhoW(p0), rhoO(p), rhoO(p0));
[vol, vol0]     = deal(pv(p), pv(p0));

% Mobility: Relative permeability over constant viscosity
mobW = krW(sW)./muW;
mobO = krO(1-sW)./muO;
```

In the flux terms, we approximate the product of density and mobility at the interface between cells using upwind-mobility weighting. To this end, we first define the difference in phase pressures across each cell interface,

```
dp  = grad(p);
dpW = dp-g*avg(rW).*gradz;
dpO = dp-g*avg(rO).*gradz;
```

which we then use to determine the upstream direction for each phase and pick cell values from the upstream side

```
vW = -upw(double(dpW) <= 0, rW.*mobW).*T.*dpW;
vO = -upw(double(dpO) <= 0, rO.*mobO).*T.*dpO;
```

We now have all terms we need to compute the residual equations in each cell,

```
water = (1/dt(n)).*(vol.*rW.*sW - vol0.*rW0.*sW0) + div(vW);
oil   = (1/dt(n)).*(vol.*rO.*(1-sW) - vol0.*rO0.*(1-sW0)) + div(vO);
```

The injector is set to inject 0.1 pore volume of water over a period of 365 days. We model the injection by a source term having constant mass rate equal the total volumetric rate times the surface density of water

```
water(inIx) = water(inIx) - inRate.*rhoWS;
```

For the producer, we just fake plausible behavior; that is, we replace the oil equation by an equation that equates the cell pressure to the specified bottom-hole pressure and set the residual equation for water equal to S_w,

```
oil(outIx)   = p(outIx) - outPres;
water(outIx) = sW(outIx);
```

These conditions will ensure that fluids flow into the perforated cell.

Structure of the Linearized System

As before, the linearized system can be assembled into a global system matrix by concatenating the two ADI vectors,

```
eqs = {oil, water};   eq = combineEquations(eqs{:}); % i.e., cat(eqs{:});
```

11.2 Two-Phase Flow without Mass Transfer

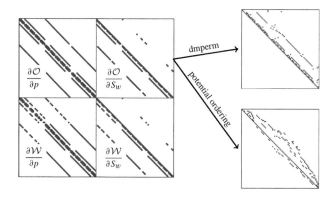

Figure 11.1 Linearized equations for the compressible, two-phase system without mass transfer between phases. The discrete equations have a 2 × 2 block structure corresponding to the derivatives of the water ($\mathcal{W}$) and oil ($\mathcal{O}$) equations with respect to the two primary variables p and S_w.

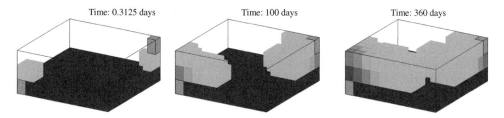

Figure 11.2 Saturation profiles for the simple two-phase problem without mass transfer between the phases. The plots only show cells in which $S_w > 10^{-3}$.

Let us inspect this system in some detail on a small 5 × 5 × 3 grid. A large number of the coefficients in the residual equations are zero in the first iteration steps since the pressure differential induced by the wells has not yet started to cause significant movement of fluids. We therefore run five of the ramp-up steps, bringing us to time five days, before we inspect the linearized system.

In the linearized system in Figure 11.1 we recognize the sparsity pattern of a standard seven-point scheme in matrix blocks that have derivatives with respect to p. Blocks with S_w-derivatives can be reordered to obtain the triangular structure we discussed in Section 10.3.4, which is typical for hyperbolic problems. Indeed, we have already seen that we can rewrite this compressible system as a parabolic pressure equation and a hyperbolic transport equation in the absence of capillary forces.

Figure 11.2 clearly shows that the suggested production strategy might not be optimal. Already in the first time step, water is sucked up from the bottom layer and towards the producer. However, by injecting in the water layer, rather than in the oil zone, we ensure that oil is not pushed down into the water zone. Maybe you can think of a better design to produce this reservoir? You can for instance try to move the injector to the upper southwest corner and repeat the simulation.

We leave the setup of wells and instead investigate what effect compressibility has on the system by comparing solutions in the compressible and the incompressible case. If we reset all the compressibilities to zero and rerun the code, we get the following error message:

```
Warning: Matrix is close to singular or badly scaled.
Results may be inaccurate.
RCOND =  2.977850e-23.
```

This is usually reason to worry, and we will get back to this issue shortly. However, let us first look at the computed results in terms of oil production, pressure in cell perforated by the injector, and field-average pressure. Oil production can be measured from a simple mass-balance calculation. To this end, we need the initial oil in place, which we compute as part of the initialization.

```
ioip = sum(pv(p0).*(1-sW0).*rhoO(p0));
```

To extract injector pressure, average field pressure, and oil in place at the end of each time step we postprocess the array of structures used to store the computed solutions,

```
t   = arrayfun(@(x) x.time, sol)/day;
ipr = arrayfun(@(x) x.pressure(inIx), sol)/barsa;
fpr = arrayfun(@(x) sum(x.pressure)/G.cells.num, sol)/barsa;
oip = arrayfun(@(x) sum(pv(x.pressure).*(1-x.s).*rhoO(x.pressure)), sol);
```

Notice here the combination of `arrayfun`, which enables us to operate on each element of an array, and anonymous functions used to extract and process the fields belonging to each array element. Figure 11.3 reports these reservoir responses. Both the pressure in the perforated injector cell and the field-average pressure shown in the left plot exhibit

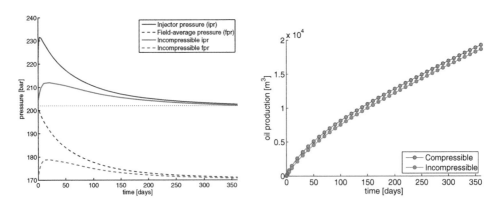

Figure 11.3 Reservoir responses for the first two-phase example without mass transfer. The left plot shows the pressure in the perforated cell and field-average pressure (FPR) and the right plot shows total field oil production (FOPT). (The dotted line indicates initial FPR.)

significantly different behavior in the compressible and incompressible models. In the compressible case, we see that the injector pressure[1] is significantly higher since some of the injected energy is used to compress the oil and expand the pore space rather than only pushing fluids, as in the incompressible case. At the same time, the reduced pressure at the producer will cause the compressible oil to expand. Hence, we also observe a somewhat higher total field-oil production. Looking at the FPR, we see that this is high initially in the compressible case, but decays gradually as more fluids are produced. We also observe the same trend in the incompressible case, but here we have an initial transient period during which the pressure first decreases and then increases slightly before we observe the expected gradual decay. Given that we do not have a real well model, it is somewhat futile to put too much attempt into interpreting this behavior.

Numerical Challenges

Let us return to the numerical difficulties of the problem. Figure 11.4 reports the number of iterations observed along with a plot of how the reciprocal condition number changes with time for the two simulations. The more well-conditioned the system is, the closer the reciprocal condition number should be to unity. All numbers are very small in our case, indicating quite ill-conditioned systems. In particular, during the first four iterations steps, the reciprocal condition numbers for the incompressible case are of the order 10^{-37} to 10^{-24}, which means that this system is effectively singular. Nevertheless, the direct solver manages to compute a solution that can be used to continue the nonlinear iteration toward convergence. This is a well-known problem in reservoir simulation; fully implicit formulations are not suited for solving incompressible flow problems. A common trick to avoid singular matrices is therefore to add a little compressibility to the problem to make it less ill-posed. In our example, we could, for instance, assume incompressible fluids but set the rock compressibility to 10% of the value used in the compressible simulation without visibly affecting the pressure and production curves reported in Figure 11.3.

Also in the compressible case, the condition numbers are not very good. From a pragmatic point of view, you may think that this is not a serious problem since the direct solver in MATLAB has no problem to solve the system. Nevertheless, the problem will get worse if the reservoir has strong heterogeneity. Moreover, when moving to larger systems, we would eventually have to utilize an iterative solver, which in most cases needs a suitable preconditioner to deliver satisfactory convergence, as we will discuss later.

To summarize: This simple example has introduced you to the typical structure of a compressible, multiphase problem. You have seen that the elliptic and hyperbolic characteristics discussed in previous chapters indeed can be observed in the discrete system. We have also demonstrated that there are significant differences in compressible and incompressible systems and that these generally require different solution strategies; hence the divide in MRST between incompressible and compressible solvers. We will return to more numerical

---

[1] Referring back to the Peaceman well model (4.35) on page 128, this pressure corresponds to p_R and will be lower than the bottom-hole pressure p_{bh} inside the wellbore.

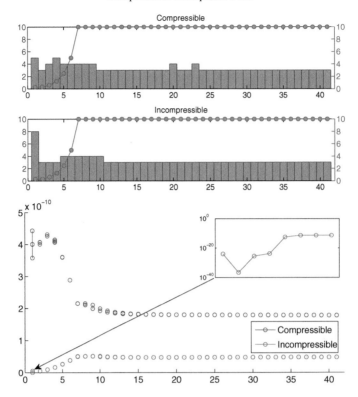

Figure 11.4 Numerical stability for the two-phase problem. In the upper plot, the bars show the number of iterations to convergence for each of the 41 time steps; time-step lengths (in units days) are shown as a line plot. The lower plot shows estimates of the variation in reciprocal condition numbers for the linearized system.

examples later in the chapter, but now it is time for a more in-depth discussion of the fluid and phase behavior of three-phase models.

11.3 Three-Phase Relative Permeabilities

When extending our two-phase simulator to three phases, we have to determine how to model saturation-dependent quantities like capillary pressure and relative permeabilities. In principle, one could imagine that all three phases could flow simultaneously in all parts of the pore volume and that relative permeability and capillary pressure generally were functions of two independent saturations. It is more common to assume that the flow is essentially two-phase, with the third phase acting as a passive background phase that restricts the flow volume and provides resistance to flow but does not contribute actively to the flow. Effectively, this means that the relative permeabilities of gas and water are functions of gas and water saturation, respectively, as we saw in Section 8.1.4. (The same applies to capillary pressure.) A commercial simulator like ECLIPSE offers several different options [271] for combining two-phase models to obtain three-phase relative permeabilities, including both the Stone I [284] and II [283] models. Common for all options is that they rely on interpolation of tabulated input. The various options are classified into families that either use one-dimensional or two-dimensional input. This offers great flexibility, but may also slow down the simulation if you are not careful when you specify the input tables. Fully implicit simulators implemented with the AD-OO framework in the `ad-blackoil` module utilize fluid objects, as discussed in Sections 5.1.1 and 10.1 to evaluate relative permeabilities and capillary pressures as general functions of phase saturations. It is up to you whether these functions should be based on analytic expressions or tabulated input, and the default dependence on saturation only can easily be overridden should you need to include other variable dependencies.

For convenience, the `ad-props` module offers a generic interface for constructing fluid objects based on input in ECLIPSE format. So far, we have only implemented the most common family of input parameters and the default option for interpolating three-phase oil relative permeability from two-phase curves. Nonetheless, it should not be a very difficult exercise to extend this implementation to include other families of input options. The rest of this section describes the default model and its implementation in the `ad-props` module in more detail.

11.3.1 Relative Permeability Models from ECLIPSE 100

The default model for calculating three-phase oil relative permeability in ECLIPSE 100 is based on the same key idea as the Stone I model (8.19)–(8.20) on page 243, but where the Stone I model assumes a nonlinear relationship to interpolate the two-phase curves, the default model employs a simpler linear relationship, which according to [271] avoids problems of negative values and poor conditioning. The model makes the key assumption that water and gas are completely segregated within each cell, giving the fluid distribution illustrated in Figure 11.5. This means that each cell is divided into two zones (here, S_w, S_o, and S_g denote the average saturations):

- in the upper gas zone, only oil and gas flow, and the oil saturation is S_o, the water saturation is S_{wc}, and the gas saturation is $S_g + S_w - S_{wc}$;
- in the lower water zone, only oil and water flow, and the oil saturation is S_o, the water saturation is $S_g + S_w$, and the gas saturation is 0.

Then, the oil relative permeability is interpolated linearly from two-phase relative permeabilities

$$k_{ro}(S_g, S_w) = \frac{S_g\, k_{rog}(S_o)}{S_g + S_w - S_{wc}} + \frac{(S_w - S_{wc})\, k_{row}(S_o)}{S_g + S_w - S_{wc}}, \quad (11.2)$$

with $S_o = 1 - S_w - S_g$. Here, k_{rog} is the oil relative permeability for a system with oil, gas, and connate water, and k_{row} is the oil relative permeability for a system with oil and gas

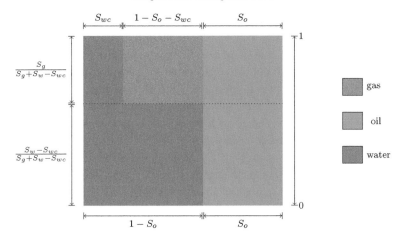

Figure 11.5 Fluid distribution inside each cell assumed by the default method for computing three-phase oil relative permeability in MRST; adapted from [271].

only, both tabulated as a function of oil saturation. This model is implemented as default within MRST, too.

To create a fluid object, we must read and process data from the PROPS section of ECLIPSE input files. For this, we use the `deckformat` module, which offers general functionality for reading and processing input decks. The most important function is the reader,

```
deck = readEclipseDeck(fn);
```

which reads the input file `fn` and constructs a structure `deck` with one field for each of the sections in the input file, except EDIT and SUMMARY. In passing, we mention that the `deckformat` module also offers several functions to construct objects for grids, rock properties, and wells, as well as functionality for reading and processing ECLIPSE *output* files. Hence, you can use MRST to visualize output from ECLIPSE simulations, and it is also possible to use existing ECLIPSE runs to initialize simulation cases in MRST.

To understand the input, we must look in more detail at the input format of the PROPS section. All data are given as capital letter keywords, followed by numbers and terminated by a single /. As an example, let us take an excerpt from the file BENCH_SPE1.DATA, which is supplied with MRST and describes input for the SPE 1 benchmark case [241]

```
PROPS    ===============================================================
-------- THE PROPS SECTION DEFINES THE REL. PERMEABILITIES, CAPILLARY
-------- PRESSURES, AND THE PVT PROPERTIES OF THE RESERVOIR FLUIDS
----------------------------------------------------------------

-- ROCK COMPRESSIBILITY
--    REF. PRES   COMPRESSIBILITY
ROCK
       14.7         3.0E-6            /
```

11.3 Three-Phase Relative Permeabilities

Here, all lines starting with a double-dash (--) are interpreted as comments and the only data contained in this part of the file are the two parameters necessary to describe the compressibility of the rock, i.e., a reference pressure (in units bar) and a compressibility (in units 1/bar). For this particular input file, the relative permeabilities are given by the two keywords SWOF and SGOF:

```
SWOF
--   SWAT              KRW                 KRO             PCOW
  0.1200000000                    0      1.000000000000    0
  0.1210000000      0.000000011363636    1.000000000000    0
  0.1400000000      0.000000227272727    0.997000000000    0
  0.1700000000      0.000000568181818    0.980000000000    0
  0.2400000000      0.000001363636364    0.700000000000    0
  0.3199999990      0.000002272727261    0.350000004375    0
  0.3700000000      0.000002840909091    0.200000000000    0
  0.4200000000      0.000003409090909    0.090000000000    0
  0.5200000000      0.000004545454545    0.021000000000    0
  0.5700000000      0.000005113636364    0.010000000000    0
  0.6200000000      0.000005681818182    0.001000000000    0
  0.7200000000      0.000006818181818    0.000100000000    0
  0.8200000000      0.000007954545455    0.000000000000    0
  1.0000000000      0.000010000000000                 0    0
/
```

The four columns represent water saturation (SWAT), relative permeability of water (KRW), relative permeability of oil (KRO), and oil–water capillary pressure (PCOW). After processing, the deck.PROPS field will have one subfield containing each of these three data

```
deck.PROPS =
      SWOF: {[14x4 double]}
      SGOF: {[15x4 double]}
         :
      ROCK: [1.0135e+05 4.3511e-10 NaN NaN NaN NaN]
```

The careful reader may observe that each of the SWOF and SGOF fields represent the input data as a cell containing an array. The reason is that the relative permeability input may contain multiple arrays of different length corresponding to different regions of the reservoir. Each array will then be represented as a cell and the whole function as a cell array.

We can then create a fluid object by calling initDeckADIFluid from ad-props

```
fluid = initDeckADIFluid(deck);
```

The essential part of initDeckADIFluid is a generic loop that traverses the fields of deck.PROPS and tries to pass each of them onto a constructor function written specifically for each field

```
props = deck.PROPS;
fns   = fieldnames(props);
for k = 1:numel(fns)
    fn  = fns{k};
    if doAssign(fn)
        asgn = str2func(['assign',fn]);
        try
```

```
            fluid = asgn(fluid, props.(fn), reg);
         catch  %#ok
            warning(msgid('Assign:Failed'), ...
               'Could not assign property ''%s''.', fn)
         end
      end
end
```

The second line extracts a list with the names of all fields present, the fifth line checks the field name against a list of field names that either have not been implemented yet or should for some reason not be processes. The sixth line makes a function pointer by appending the field name to the string "assign," in our case creating function pointers to `assignROCK`, `assignSWOF`, and `assignSGOF`. These functions are then called in the eight line, with the fluid object `fluid` and the appropriate fluid data as input. (The third input argument, `reg`, represents region information.) Let us look at the implementation of rock compressibility. If we for simplicity disregard region information, the model for rock compressibility is set up as follows:

```
function f = assignROCK(f, rock, reg)
  [cR, pRef] = deal(rock(2),rock(1));
  f.cR = cR;
  f.pvMultR = @(p)(1 + cR.*(p-pRef));
end
```

Although rock compressibility, strictly speaking, is not a *fluid* relationship, it has for historic reasons been bundled with the saturation and pressure-dependent fluid and rock-fluid relationships since it represents a purely pressure-dependent quantity and will enter the flow equations in the same way as the other elements of the fluid object. Notice also that with the generic setup in `initDeckADIFluid`, all you must do to support a new keyword MYKEY, is to implement the function `assignMYKEY` and place it in the right directory of the ad-props module.

Assigning relative permeabilities is somewhat more involved. If we disregard code lines having to do with the region information (i.e., that the model can use different relative permeability functions in different sub-regions of the grid), the constructor takes the following form

```
function f = assignSWOF(f, swof, reg)
  f.krW  = @(sw, varargin) krW(sw, swof, reg, varargin{:});
  f.krOW = @(so, varargin) krOW(so, swof, reg, varargin{:});
  f.pcOW = @(sw, varargin) pcOW(sw, swof, reg, varargin{:});
  swcon  = cellfun(@(x)x(1,1), swof);
  f.sWcon = swcon(1);
end
```

That is, we add three function pointers to the fluid object implementing the relative permeability of water, the relative permeability from an oil–water displacement, and the capillary pressure of between oil and water. Let us take the `krOW` function as an example,

```
function v = krOW(so, swof, reg, varargin)
  satinx = getRegMap(so, reg.SATNUM, reg.SATINX, varargin{:});
  T = cellfun(@(x) x(:,[1,3]), swof, 'UniformOutput', false);
  T = extendTab(T);
  v = interpReg(T, 1 - so, satinx);
end
```

The first line sets up a map that lets us select the correct saturation functions if there are multiple regions. The next line extracts the first and the third column of the input data, i.e., the SWAT and KRO columns. Here, we use `cellfun` in combination with an anonymous function to extract individual elements in the same way as we used `arrayfun` to extract simulation results from the `sol` array of structures in the previous section. The saturation values in the relative permeability tables may not necessarily span the whole unit interval, and in the third line we add a constant extrapolation to ensure that we can use T to extrapolate for all legal saturation values. The last line calls an interpolation function that uses region information to pick the correct relative permeability function from the cell array and then calls an appropriate interpolation function. If T had been a simple array, the last line would have amounted to

```
F = griddedInterpolant(T(:,1), T(:,2) , 'linear', 'linear');
v = F(1-so);
```

The functions `krW` and `pcOW` are implemented exactly the same way, with the obvious modifications, and so are the functions in `assignSGOF`.

11.3.2 Evaluating Relative Permeabilities in MRST

By default, fluid objects constructed by `initDeckADIFluid` only contain functions for evaluating the tabulated two-phase systems found in the input file through functions `krW`, `krOW`, `krG`, and `krOG`. For convenience, the ad-props module offers the function `assignRelPerm` that implements the three-phase relative permeability model described in the beginning of the section. You can run `assignRelPerm` to instrument your fluid object with a generic interface to evaluate relative permeabilities. However, before this function is called, you must have run the necessary routines to assign data tables and functions for interpolating these data to the fluid object f, or alternatively have implemented your own fluid object that offers the same functions `krW`, `krOW`, etc. Notice also that if you later update any of the two-phase data in the fluid object, `assignRelPerm` must be run again to ensure correct evaluation.

The `assignRelPerm` function inspects the fields that are present in the fluid object and assigns the correct relative-permeability functions accordingly

```
function f = assignRelPerm(f)
  if ~isfield(f, 'krOG')      % two-phase water-oil
    f.relPerm = @(sw, varargin)relPermWO(sw, f, varargin{:});
  elseif ~isfield(f, 'krOW')  % two-phase oil-gas
    f.relPerm = @(sg, varargin)relPermOG(sg, f, varargin{:});
  else                        % three-phase
    f.relPerm = @(sw, sg, varargin)relPermWOG(sw, sg, f, varargin{:});
  end
end
```

The logic is quite simple: If there is no oil–gas data (krOG) for a three-phase system available, we must have a water–oil system. If there is no oil–water data (krOW) for a three-phase system available, we must have an oil–gas system. Otherwise, all three phases are present and we must set up a full three-phase model. The relative permeabilities for the two-phase water–oil system are set up as follows:

```
function [krW, krO] = relPermWO(sw, f, varargin)
  krW = f.krW(sw, varargin{:});
  if isfield(f, 'krO')
    krO = f.krO(1-sw, varargin{:});
  else
    krO = f.krOW(1-sw, varargin{:});
  end
end
```

Here, the default is to look for a two-phase table called krO for the oil relative permeability. If this is not present, the user must have supplied an oil–water table from a three-phase model, and we use this instead. The oil–gas system is treated analogously. Finally, for the three-phase model:

```
function [krW, krO, krG] = relPermWOG(sw, sg, f, varargin)
  swcon = f.sWcon;
  swcon = min(swcon, double(sw)-1e-5);

  d    = (sg+sw-swcon);
  ww   = (sw-swcon)./d;
  krW  = f.krW(sw, varargin{:});

  wg   = 1-ww;
  krG  = f.krG(sg, varargin{:});

  so   = 1-sw-sg;
  krow = f.krOW(so, varargin{:});
  krog = f.krOG(so, varargin{:});
  krO  = wg.*krog + ww.*krow;
```

11.3.3 The SPE 1, SPE 3, and SPE 9 Benchmark Cases

In 1981, Aziz S. Odeh [241] launched an initiative for the independent comparison of the state-of-the-art reservoir simulators for 3D black-oil models. To represent a good cross-section of solution methods used throughout the industry, seven companies were selected, each participating with their own simulator. This type of comparison was well received by both the industry and the academic community. Over the next two decades, the Society of Petroleum Engineering (SPE) published a series of nine additional studies, which are today known as the ten SPE comparative solution projects. In each study, a number of oil companies, software providers, research institutes, universities, and consultants participated on a voluntarily basis, reporting their best attempt to solve a specified problem. Altogether, these ten problems constitute a rich source of test problems that can be used to verify and validate various multiphase models and solution strategies. You have already seen data from the tenth project used multiple times throughout this book.

In this example, we use input decks for the SPE 1 [241], SPE 3 [158], and SPE 9 [159] comparison projects to demonstrate how you can use MRST to set up and compare different fluid models. If you have not already done so, you can download the necessary files by use of MRST's data set manager:

```
mrstDatasetGUI;
```

This will place the data sets on a standard path, so that they can be read and processed using the following set of commands (with the obvious modifications for SPE 3 and SPE 9):

```
mrstModule add deckformat ad-core ad-props
pth  = getDatasetPath('spe1');
fn   = fullfile(pth, 'BENCH_SPE1.DATA');
deck = readEclipseDeck(fn);
deck = convertDeckUnits(deck);
f    = initDeckADIFluid(deck);
f    = assignRelPerm(f);
```

This produces a fluid object with a number of fields (here for SPE 1):

```
    krW: @(sw,varargin)krW(sw,swof,reg,varargin{:})
   krOW: @(so,varargin)krOW(so,swof,reg,varargin{:})
   pcOW: @(sw,varargin)pcOW(sw,swof,reg,varargin{:})
  sWcon: 0.1200
    krG: @(sg,varargin)ireg(Tkrg,sg,varargin{:})
   krOG: @(so,varargin)ireg(Tkro,sgas(so),varargin{:})
   pcOG: @(sg,varargin)ireg(Tpcog,sg,varargin{:})
       :
relPerm: @(sw,sg,varargin)relPermWOG(sw,sg,f,varargin{:})
```

Here, we recognize relative permeability of water, oil relative permeability of an oil–water system, oil–water capillary pressure, connate water injection, etc. The last function is a

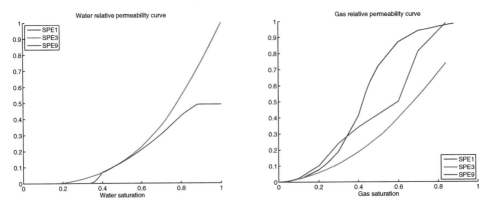

Figure 11.6 Relative permeabilities for water and gas for the fluid models from three of the SPE Comparative Solution Projects.

generic interface for evaluating relative permeabilities of all phases in the three-phase system. Let us look at the two-phase relative permeabilities krW and krG, which we evaluate by a call on the form

```
s    = linspace(0,1,51)';
krW = f.krW(s);
```

Figure 11.6 shows the resulting curves (extrapolated krG values are not plotted for saturations $S_w < S_{wc}$). SPE 1 is represented as a three-phase system in this particular input file, but it is really a two-phase gas–oil problem that describes gas injection into an undersaturated oil, i.e., an oil that can dissolve more gas. (We discuss this sort of PVT behavior in the next section.) The water relative permeability is therefore virtually zero over the whole saturation interval. For the SPE 3 model, both relative permeability curves have monotone derivatives similar to what we have seen in the analytic models discussed in Section 8.1.4. This is not the case for the gas relative permeability of SPE 1, which has an S-shape similar to the fractional flux functions we have seen in two-phase systems. For the SPE 9 model, the relative permeability curves have significant kinks, and this will contribute to increase the nonlinearity of the discretized flow equations.

Figure 11.7 shows the corresponding relative permeability curves for two-phase oil–water and gas–oil systems and the resulting interpolated three-phase oil relative permeability curve. The plots were generated as follows:

```
[sw, sg] = meshgrid(linspace(0,1,201)); so=1-sw-sg;
[~, krO, ~] = f.relPerm(sw(:),sg(:));
krO = reshape(krO,size(sw)); krO(sw+sg>1)=nan;

[mapx,mapy] = ternaryAxis('names',{'S_w';'S_g';'S_o'});
contourf(mapx(sw,sg,so), mapy(sw,sg,so), krO, 20)
```

11.3 Three-Phase Relative Permeabilities

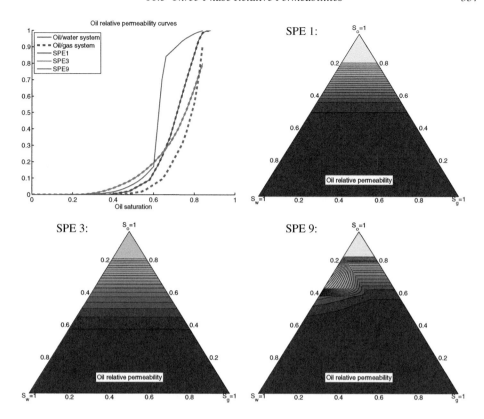

Figure 11.7 Three-phase relative permeabilities of oil for three of the SPE Comparative Solution Projects generated by linear interpolation of two-phase curves from oil–water and oil–gas displacements (upper left plot).

The `krOW` and `krOG` curves representing two-phase displacements coincide for both the SPE 1 and SPE 3 data sets, and hence the interpolated `krO` surface is smooth and monotone along straight lines in phase space. For SPE 9, the two-phase curves are not only significantly different, but the `krOW` curve has a very steep gradient. As a result, we observe kinks and it is also possible to identify straight lines in the phase diagram along which `krO` is non-monotone; see e.g.,

```
sw = linspace(0,.6,21); sg = linspace(0,.4,21);
[~,krO] = f.relPerm(sw',sg');
```

As is warned about in [270], abrupt changes in relative permeability values over small saturation intervals can be the source of significant convergence problems when solving the discretized flow equations and may cause the simulator to slow down. Like ECLIPSE, MRST honors the input data and does not attempt to overcome convergence problems by smoothing kinks or retabulating the input data.

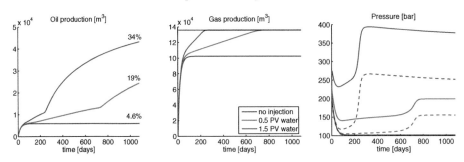

Figure 11.8 Conceptual three-phase simulation with no mass transfer between the phases. The reservoir is initially filled with gas on top, oil in the middle, and water at the bottom. Fluids are produced from the middle layer. The plots show results of simulations with no water injection and with injection of 0.5 and 1.5 pore volumes of water into the bottom layer. The plots to the right show average pressure in the injector cell (solid lines) and FPR (dashed lines).

11.3.4 A Simple Three-Phase Simulator

Having established how to model three-phase relative permeabilities, it is a simple exercise to extend the simple simulator from Section 11.2 to three phases under the assumption of no mass transfer between the phases. This assumption is not very realistic, but is introduced to isolate and illustrate certain aspects of multiphase flow. To exemplify, we consider a setup similar to the one discussed in Section 11.2, with the upper 1/3 initially filled with gas ($S_g = 1 - S_{wc}$), the middle 1/3 filled with oil ($S_o = 1 - S_{wc}$), and the bottom 1/3 is filled with water. Fluids are produced from a cell at the northeast corner of the middle section by setting the pressure in this cell to be 100 bar, which is approximately 100 bar lower than the initial average reservoir pressure. We change the fluid model so that water is twice as viscous as oil, and assume that relative permeabilities follow the SPE 3 fluid model.

When the producer starts depleting the reservoir pressure, the compressed gas will expand so that a large fraction of it is driven out of the reservoir through the oil layer below. Gas is more mobile than oil and will flow past the oil rather than displace it. Hence, only a minor fraction of the oil will be recovered. To increase oil recovery, we inject water into the water zone at the lower southwest corner. Code to simulate this case is given in `threePhaseAD.m` in the book module. We run a comparison of no water injection and injection of 0.5 and 1.5 pore volumes of water

```
pvi=0; threePhaseAD; pvi=.5; threePhaseAD; pvi=1.5; threePhaseAD
```

Figure 11.8 reports observed hydrocarbon production and pressure behavior. Without water injection, the reservoir pressure is quickly depleted and reaches a steady state after approximately 100 days, and out of the volumes initially in place, we recover 75%, but only 4.6% of the oil. By injecting 0.5 pore volumes of water, we are able to maintain reservoir pressure and hence get higher oil production in the period until all the gas is depleted and the injected water breaks through in the producer. This happens after approximately 740 days. Towards

11.4 PVT Behavior of Petroleum Fluids

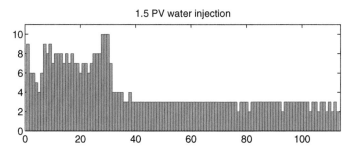

Figure 11.9 Number of nonlinear iterations for the conceptual three-phase simulation.

the end of this period, the pressure increases significantly as the water front approaches the producer and then stabilizes at a plateau. When gas disappears and oil is the most mobile phase, the oil production increases abruptly, but then decays slowly as a result of gradually increasing water cut in the producer. Similar behavior can be observed when injecting 1.5 pore volumes of water over the same time period. However, since injection rates are three times higher, water breakthrough occurs much earlier and we also observe pressures that are much higher than the initial field pressure.

If we try to repeat the simulation with a total injection of two pore volumes, the simulator fails to converge within 15 iterations in the 23rd time step, right before the water front breaks through in the producer. We also observe a similar increase in the number of iterations towards water breakthrough for the other water injection cases; see Figure 11.9. It is well known that Newton's method may experience convergence problems if saturations (or compositions) change too much during a single time step. We have already discussed this issue in Section 10.2.2, where we also introduced a reactive strategy for chopping of the time step.

COMPUTER EXERCISES

11.3.1 Visualize the structure of the linearized three-phase system in the same way as in Figure 11.1. Try to permute the system so that the numbering in the vector of unknowns first runs over variables and then over cells.

11.3.2 What happens if you set water to be incompressible?

11.3.3 Extend the simulator in `threePhaseAD.m` to include the reactive time-step control discussed in Section 10.2.2 and try to extend it so that it also uses changes in saturation as a criterion to chop time steps. Does this enable you to run a simulation with more than two pore volumes injected?

11.4 PVT Behavior of Petroleum Fluids

Our assumptions of no mass transfer between phases and that each fluid phase has constant compressibility, but otherwise constant fluid properties, represent gross oversimplifications.

Pressure and temperature will not only affect the compressibility and viscosity of each fluid phase, but also determine which phases are present and how the different chemical components in the reservoir fluids distribute among the aqueous, oleic, and gaseous phases at reservoir and surface conditions. Hydrocarbon fluids can come in a wide variety of forms, including the bitumen (semisolid rock form of crude oil) of the Athabasca oil province in Canada and heavy (dead) oils containing almost no light hydrocarbon components. We will not discuss such cases in any detail but instead focus on hydrocarbon fluids that have a significant gas component at surface conditions. That is, medium oils containing lighter components that are liberated as free gas at surface conditions; volatile oils that shrink significantly when liberating free gas; rich gases that condense liquid at reduced pressure; wet gases that condensate fluids at lower temperatures; and dry gases that mainly consist of light hydrocarbons and stay in single-phase vapor form.

11.4.1 Phase Diagrams

We have previously discussed equations of state modeling the relationship among pressure, volume (density), and temperature. For reservoir fluids it is more common to use *phase diagrams*, which in this setting are a type of chart showing pressure and temperature conditions at which thermodynamically distinct phases occur and coexist at equilibrium. This section discusses such diagrams for reservoir fluids.

Single-Component Substances

As a starting point, however, let us look at a typical phase diagram for a substance consisting of a single component. To explain such diagrams, we first look at *Gibbs' phase rule* for a system without chemical reactions. This rule states that the number of independent degrees of freedom F that must be specified to determine the intensive state of a system is given by

$$F = 2 + n_c - n_p,$$

where n_c is the number of components and n_p is the number of phases. For a single-component system ($n_c = 1$), we have $F = 3 - n_p$, and hence there can at most be three phases. A state with three phases corresponds to zero degrees of freedom and the phases must therefore coexist in a single point called the *triple point*, at which pressure and temperature are uniquely determined. If two phases coexist, there is one degree of freedom, and pressure can be chosen freely if temperature is fixed, and vice versa. When the substance is in a single-phase state, both pressure and temperature can be chosen independently.

Figure 11.10 illustrates the resulting phase diagram. In the diagram, the three single-phase regions corresponding to solid, liquid, or vapor phase are separated by the *sublimation*, *melting*, and *vaporization* curves. Along these curves, two phases coexist in equilibrium. Across the curves, various properties of the substance (compressibility, viscosity,

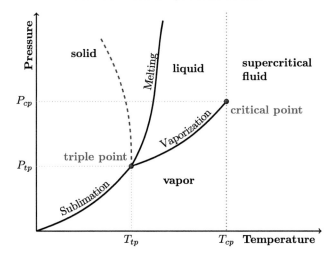

Figure 11.10 Conceptual phase diagram for a single-component substance. The dashed blue line illustrates the melting curve for water, which unlike most other fluids has negative slope.

etc.) change abruptly and have discontinuous derivatives with respect to temperature and pressure. At the triple point, all phases exist in equilibrium. For temperatures below the *critical point* T_{tp}, the substance is in vapor phase if the pressure is below the sublimation curve and in solid state for pressures above. For pressures and temperatures along the sublimation curve, vapor and solid are in equilibrium. For temperatures between the triple point and the critical point ($T_{tp} < T < T_{cp}$), the substance is in vapor state if the pressure is below the vaporization curve, in liquid state if the pressure is above the vaporization curve, and in solid state if the pressure is above the melting curve.

The liquid–vapor curve terminates at the critical point. The critical temperature T_{cp} is the temperature at, and above, which vapor of the substance cannot be liquefied, no matter how much pressure is applied. Critical temperatures depend on the size of the molecule and hence vary a lot for hydrocarbons, from low temperatures for small molecules to high temperatures for large molecules. For water, the critical temperature is 374°C, whereas carbon dioxide has a much lower critical temperature of 31.2°C. Likewise, the critical pressure P_{cp} is the pressure required to liquefy a gas at its critical temperature, or in other words, the pressure above which distinct liquid and gaseous phases do not exist. The critical pressure of water is 217.7 atm and for CO_2 it is 73.0 atm. For temperatures above T_{cp}, the substance is said to be in a *supercritical state*, i.e., in a state in which the liquid and gaseous phases are indistinguishable. This is the state in which CO_2 is usually found when stored in a deep geological formation (approximately below 800 m).

For most substances, the melting curve separating liquid and solid phase has a positive derivative with respect to temperature; the denser the molecules of the substance are brought together, the higher temperature is required to break them apart to form a liquid state. Water is one exception to this rule.

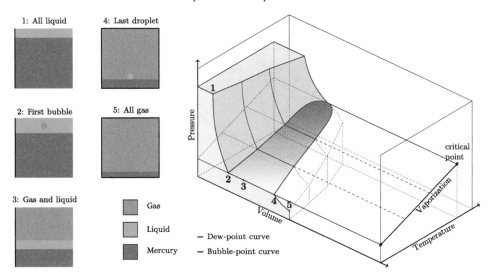

Figure 11.11 PVT behavior of a single-component hydrocarbon substance. The left figures illustrate measurement of pressure-volume behavior in a high-pressure cell at fixed temperature. Mercury is connected to a pump that can inject or extract this fluid to control pressure. Mercury has high surface tension and will hence be non-wetting for most container surfaces like glass. The right plot shows the measurements plotted in a PVT diagram, with the bubble-point and vaporization curves projected onto a pressure-temperature diagram.

For hydrocarbon recovery, we are primarily interested in the *vaporization curve*. To better understand the phase behavior for multicomponent systems, we first briefly explain an experimental setup for determining this curve using a *high-pressure* cell, as illustrated in Figure 11.11. Initially, the cell is filled with liquid hydrocarbon at the top and mercury at the bottom. As mercury is gradually removed, the liquid hydrocarbon expands and the pressure drops. This continues until the first bubble of gas comes out of the solution as we reach the *bubble point* (point 2 in the figure). Beyond this point, the gas phase expands the volume rapidly while keeping the pressure constant (point 3 in the figure). Eventually, we reach the , where the last liquid droplet vaporizes (point 4 in the figure). As more mercury is removed, the gas expands while the pressure decreases. If we repeat the experiments for multiple temperatures, we get the diagram shown in the right plot. By projecting the locus of bubble points, which we henceforth refer to as the *bubble-point curve*, and the locus of dew points, which is called the *dew-point curve*, we obtain the *vaporization curve* in the pressure-temperature diagram.

Binary Substances

Hydrocarbon reservoir fluids are made up of a wide range of chemical substances, and hence their phase diagrams will be very complicated. Gibbs' phase rule obviously gives us little guidance since the number of actual phases occuring in the reservoir is much less

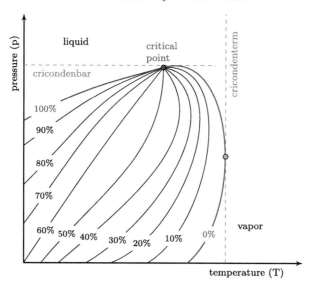

Figure 11.12 A pressure-temperature diagram for a binary system of fixed composition. Bubble-point curve is shown in blue and dew-point curve shown in red. The lines inside the two-phase envelope represent constant percentages of liquid. The diagram is specific to a given chemical composition of the binary mixture and will change if this composition changes, e.g., if components are extracted at different rates.

than the maximum number of phases that may exist. In black-oil models, the hydrocarbon fluids are considered to be a two-component system. (Notice that this distinction between oil and gas components is only relevant when the reservoir is in a state where both can coexist.) For binary substances, the number of degrees of freedom for a two-phase system is $F = 2$, and hence two phases could in principle exist for arbitrary combinations of temperature and pressure, but this is not the case in practice. To describe the reservoir fluids, we utilize a bubble-point/dew-point projection similar to the one we studied for the single-component fluid above. This projection is defined for a constant mole fraction of one of the components, and hence the phase diagram will depend on the composition of the binary mixture. In the projected plot, the bubble-point and dew-point curves no longer coincide along a single vaporization curve. Once the bubble point is reached, the expansion is accompanied by a pressure decrease, since the gaseous phase has a different chemical composition than the oleic phase. The projection hence produces a temperature and pressure envelope, inside which the binary mixture forms two phases as shown in Figure 11.12.

At pressures and temperatures above the bubble-point curve, the binary mixture forms a liquid phase. Likewise, a gaseous phase is formed for all points located below or to the right of the dew-point curve. The bubble-point and dew-point curves meet in the *critical point*, where the phase compositions and all phase properties are identical. Notice that the critical point does not necessary coincide with the maximum temperature of the two-phase envelope. The maximum temperature is called the *cricondentherm* and the maximum pressure is called the *cricondenbar*. The critical point represents a mathematical discontinuity in the system, and the phase behavior is generally difficult to define near this point, meaning that it is not obvious whether a (slight) pressure decline will cause the formation of bubbles of vapor or droplets of liquid.

11.4.2 Reservoir Types and Their Phase Behavior during Recovery

In petroleum literature, the phase diagrams discussed in the previous subsection are sometimes used to classify reservoirs according to the type of phase behavior that takes place during (primary) production. To motivate this classification, and the choice of physical variables and closure relationships to be discussed in the next section, we start by a schematic of a typical production setup for a depleting reservoir as shown in Figure 11.13.

When a well is opened to start production of hydrocarbons, the pressure inside the reservoir will start to decline. Reservoirs are thermally inert, and unless other fluids having significantly different temperature are injected into the reservoir, the temperature of the reservoir fluids will remain (almost) constant at the value given by the geothermal gradient of the area. Inside the reservoir, the extraction of hydrocarbons therefore takes place along a vertical line in the pressure-temperature diagram. Also when fluids are injected to displace the resident hydrocarbons, temperature effects are in many cases so small that the temperature can be considered as constant (near the producing well) in the reservoir model. This means that various types of reservoirs can be classified by the location of the

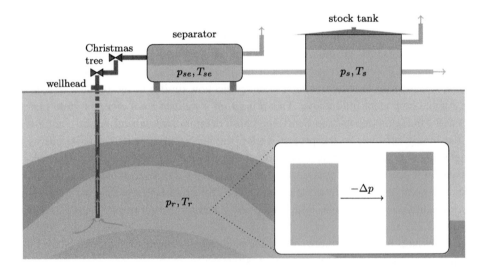

Figure 11.13 Illustration of pressure and temperature changes resulting in liberation and expansion of gas and shrinkage of oil during production of a hydrocarbon reservoir containing both oil and gas. Stock-tank conditions vary but will roughly be 1 atm and 20°C.

11.4 PVT Behavior of Petroleum Fluids

initial reservoir pressure and temperature relative to the bubble-point/dew-point envelope. If the temperature of the reservoir is less than the critical temperature ($T_r < T_c$) of the hydrocarbon mixture, the reservoir is said to be an *oil reservoir*. Conversely, if the temperature is larger than the critical temperature ($T_r > T_c$), the reservoir is said to be a *gas reservoir*. In both cases, the reservoir may contain hydrocarbons in both liquid and vapor form, depending on the type of the reservoir.

As the hydrocarbons leave the reservoir and flow to the surface, they undergo a series of changes that reduce the pressure and temperature to surface conditions. These changes generally cause liberation of gas, expansion of the gaseous phase, and shrinkage of the liquid oleic phase. Pressure reductions come in two types: in *flash liberation*, gas comes out of solution as a result of a sudden pressure drop, and then stays confined with the remaining oil. In *differential liberation*, gas is liberated from the oil as pressure is gradually decreased and is continuously removed from the oil. How the reservoir fluids contained in the well stream are brought from reservoir pressure and temperature to surface condition can have a large effect on how much oil and gas is recovered from a given quantity of reservoir fluids. Traditionally, oil has had much higher economic value than gas, which often has been considered as a waste product. It has therefore been highly important to design a production system that maximizes the stock-tank oil yield.

There are many mechanisms that induce pressure reduction: friction and gravity forces acting in the opposite direction of the flow, loss of kinetic energy due to expansion, contraction in the fluid area, flow through constrictions, etc. At the surface, the fluids pass through the *Christmas tree* mounted on top of the well head, whose main purpose is to control the flow of reservoir fluids out of the well. The Christmas tree has isolation valves to stop the fluid flow for maintenance or safety purposes, and flow lines and chokes that control the flow into the downstream production system. Inside the production system, the fluids can continue into one or more pressure vessels (separators) used to separate the produced fluids into oil, gas, and water components, and possibly remove gas and/or water. The oil component may continue to a stock tank for temporary storage, inside which further gas liberation may take place.

We already know that the pressure drop inside the reservoir and the pressure drop from the reservoir to the wellbore both depend on flow rate; this follows from Darcy's law and the Peaceman well model, respectively. Similar relationships between pressure drop and flow rate can be developed for the other parts of the flow system from the bottom hole to the stock tank; that is, pressure loss in the tubing, in the Christmas tree and flow lines, across the choke, and in the separator units. A few examples are presented in Section 11.7.2. These relationships are used by production engineers to optimize the hydrocarbon recovery. For a more detailed discussion, you can consult a textbook on so-called *nodal analysis* [145].

Gas Reservoirs

All reservoirs in which the initial temperature is higher than the critical temperature are called gas reservoirs. Methane is by far the dominant hydrocarbon component in gas reservoirs. If the initial temperature is higher than the cricondentherm, the hydrocarbon mixture

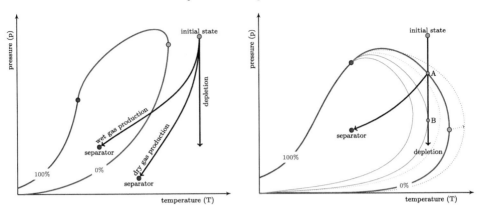

Figure 11.14 Phase diagrams for gas reservoirs. In a single-phase dry/wet gas reservoir (left), the temperature is above the cricondentherm. In a retrograde gas condensate reservoir (right), the temperature is above the critical temperature, but below the cricondentherm. During retrograde depletion below the dew-point curve, the composition of the mixture will change inducing a change in the diagram indicated by the dotted lines.

will be in a single-phase gaseous state and will continue to be so during depletion. As the reservoir is depleted, the composition of the produced fluids do not change, but the hydrocarbon mixture may enter the two-phase envelope on its way to the surface, depending upon the cooling process, as illustrated in the left plot of Figure 11.14.

If a gas reservoir produces liquid hydrocarbons at the surface (condensate), the reservoir is said to be a *condensate gas*, *wet gas*, or *rich gas* reservoir. Conversely, if the temperature–pressure path followed during production does not penetrate the two-phase envelope, all hydrocarbons produced at the surface will be in gas phase and the reservoir is said to be a *dry gas* reservoir. Dry gas reservoirs rarely contain hydrocarbons with carbon number five or higher and typically consist of C_1, C_2, N_2, CO_2, and H_2S. No hydrocarbon separation is needed at the surface since the well stream consists of gas only, but separators/dehydrators may be needed to remove produced/condensed water. Although wet-gas reservoirs mainly consist of methane, they also contain both intermediate hydrocarbons (ethane to hexane) as well as a significant fraction of hydrocarbons with carbon number seven or higher. The condensation of liquid hydrocarbons at the surface is a result of cooling, but whether the well stream is brought to surface conditions in one big flash or through a series of smaller flashes does not have a large effect on the relative amounts of gas and liquid observed at the surface.

Unlike dry/wet gas reservoirs, *dew-point* or *retrograde gas condensate* reservoirs contain a significant fraction of heavier hydrocarbon components (C_{7+}). The right plot in Figure 11.14 shows such a hydrocarbon mixture, which is initially in (an indeterminate) vapor state and will continue to be so as the reservoir is depleted toward point A on the dew-point curve. During this initial decline, the composition of the produced fluid remains constant. Below point A, a liquid oleic phase containing heavier hydrocarbons will condense out of the vapor as a dew or fog and continue to do so until the process reaches point *B* of maximum liquid percentage. This process is called retrograde, since one normally would see vaporization and not condensation when pressure decays.

The condensed liquid is immobile at low saturations and generally less mobile than the vapor phase. Hence, more light hydrocarbons will be recovered from the reservoir, giving an increased gas–oil ratio at the surface. This changes the composition of the hydrocarbon mixture, increasing the fraction of heavier hydrocarbons, which in turn shifts the two-phase envelope toward the right, as indicated by the dotted lines in the figure. This aggravates the loss of heavier hydrocarbons, which are retained as immobile droplets inside the rock pores. If production is continued, the system will sooner or later reach a point of maximum liquid percentage, below which the condensed liquid will start to vaporize. This improves the mobility of the liquid components and leads to a decrease in the surface gas–oil ratio.

For a dew-point reservoir, the exact design of how the reservoir fluids are brought to surface conditions can have a large impact on the amount of stock-tank oil and gas. This may call for a complex network of separators in which separated vapor or liquid is reinjected into the liquid–vapor mixture to better distribute heavier hydrocarbon components to the stock-tank oil and lighter components to the final gas stream.

In closing, we remark that the distinction between condensate and oil is somewhat artificial and has more to do with how the liquid hydrocarbons are recovered than their chemical composition. Liquid hydrocarbons in the stock tank are called oil if they originate from a liquid oleic state in the reservoir and condensate when they have been condensed from a gaseous state.

Bubble-Point Oil Reservoirs

A reservoir having initial temperature below the cricondentherm is characterized as an oil reservoir. Unlike the dead oil considered in Section 11.2, which either contained a constant amount of dissolved gas or had lost all its components, the oleic phase in bubble-point reservoirs consists of a *live oil* containing dissolved gas, which may be released at lower pressures inside the reservoir and at the surface. This type of reservoir is also called a *solution-gas* reservoir, and will be the topic for most of the simulation examples discussed later in the chapter. Although somewhat artificial and imprecise, it is common to subdivide oil reservoirs into two categories: *black-oil* reservoirs are mainly made up of a variety of hydrocarbons, including heavy and non-volatile components (C_{7+}). *Volatile* reservoirs contain fewer heavy components and have more intermediate hydrocarbon molecules (ethane to hexane) and the oleic phase therefore tends to shrink dramatically when gas is liberated. Black-oil reservoirs tend to have much less shrinkage. The main property that distinguishes the two categories, however, is the presence of vaporized oil in the gaseous phase, which is assumed to be negligible in black-oil reservoirs and significant in volatile reservoirs. This might be a bit confusing since we later will use *black-oil equations* to model both types of reservoirs. The reason is most likely historic. Traditionally, the largest and most profitable reservoirs in the world have predominantly been of black-oil type, and the black-oil equations were developed to model these. Volatile reservoirs are typically found at

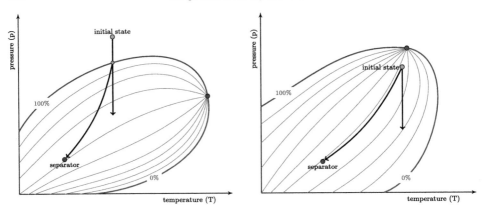

Figure 11.15 Phase diagram for bubble-point oil reservoirs. The left plot shows a black-oil reservoir with undersaturated initial oil and the right plot a volatile oil reservoir.

greater depths, and as exploration goes deeper, more reservoirs of this type have been discovered and entered production. Volatile reservoirs contain less oil per volume, but since the reservoir fluids are more mobile because of lower oil viscosity and higher dissolved-gas content, this type of reservoir can still be more economically attractive than a black-oil reservoir.

If the pressure is above the bubble-point curve as shown in the left plot of Figure 11.15, the hydrocarbon mixture of the oil reservoir is in a liquid state usually referred to as *undersaturated*, since all the available gas has been dissolved in the oleic phase. However, when the first liquid is brought to the surface, it liberates free gas. As production continues, the pressure will decline toward the bubble-point curve. Below the bubble point, gas will be released from the oleic phase as vapor bubbles that gradually form a free gaseous phase. Because the oil cannot hold all the available gas, we say that it is in a *saturated* state. When the gaseous phase exceeds its residual saturation, it will start to flow toward the producer. At the surface, the gas thus consists of both expanded free gas and gas liberated from the oil during production. Oftentimes, surface facilities have to limit the gas rate, which in turn can reduce the oil rate below an economic level unless secondary mechanisms like water injection are applied to maintain the reservoir pressure and push the oleic phase toward the producer (as we saw for the dead oil system in Section 11.3).

The volatile reservoir shown to the right in Figure 11.15 is initially in an saturated state inside the two-phase envelope. The reservoir fluids thus consist of a liquid oleic phase which exists in equilibrium with a gaseous phase containing gas condensate. Because the gaseous phase has a different composition and is lighter than the oleic phase, the gaseous phase will form a gas zone or gas cap overlying the liquid oil zone and has a phase diagram that is largely different from that of the oleic phase. Equilibrium between the two phases implies that the oleic phase must be at its bubble point, whereas the gaseous phase must be at its dew point and may either exhibit retrograde or non-retrograde behavior, as illustrated

11.4 PVT Behavior of Petroleum Fluids

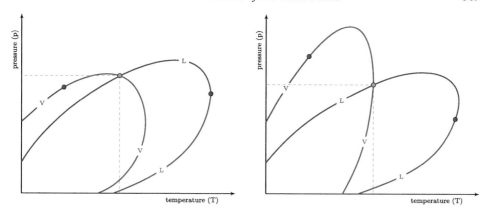

Figure 11.16 Phase diagrams for a reservoir at saturated conditions so that liquid oil is in equilibrium with a gas cap containing vaporized oil. The gaseous phase will exhibit retrograde behavior in the diagram to the left and non-retrograde behavior in the diagram to the right.

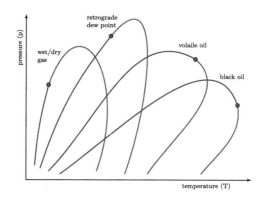

Figure 11.17 Sketch of phase diagrams for different types of hydrocarbon reservoirs.

in Figure 11.16. Notice also that volatile reservoirs are not very different from retrograde gas condensate reservoirs for temperatures close to the critical temperature.

We end our discussion about reservoir classification by presenting two comparisons of the different reservoir types. Figure 11.17 shows a schematic with the different phase diagrams plotted relative to each other. This sketch is by no means accurate, but illustrates how the phase diagrams move to the right when the hydrocarbon mixture contains (a larger fraction of) heavier hydrocarbon components. Table 11.1 reports a number of characteristic properties for different reservoir types.

Table 11.1 *Petroleum fluids and their characteristics at initial conditions or bubble-point pressure (adapted from [177]).*

	Oils			Gases	
Characteristic	Heavy oils and tars	Black oils	Volatile oils	Gas condensates	Wet and dry gases
Mol. weight	210+	70–210	40–70	23–40	<23
Oil color in stock tank	Black	Brown to light green	Greenish to orange	Orange to clear	Clear
Stock-tank °API	5–15	15–45	42–55	45–60	45+
$C_{7}+$ fraction (mol%)	>50	35–50	10–30	1–6	0–1
Reservoir temperature °F	90–200	100–200	150–300	150–300	150–300
Saturation pressure [psia]	0–500	300–5,000	3,000–7,500	1,500–9,000	—
Initial B_o [RB/STB]	1.0–1.1	1.1–1.5	1.5–3.0	3.0–20.0	20+
Initial R_s [scf/STB]	0–200	200–900	900–3,500	3,500–30,000	30,000+
Initial R_v [STB/MMscf†]	0	0–10	10–200	50–300	0–50
Max vol% liquid‡	100	100	100	0–45	0

† At bubble-point pressure.
‡ At constant composition expansion of reservoir fluid.

11.4.3 PVT and Fluid Properties in Black-Oil Models

The phase diagrams discussed in the previous section are useful to classify reservoir fluids and visualize the pressure-temperature path from reservoir to surface conditions. They aid the reservoir engineer in deciding the depletion process, designing top-side processing facilities (separators, etc.), and devising appropriate strategies for improved oil recovery (water injection or reinjection of gas to maintain pressure, etc.). Phase diagrams are also useful when setting up a fluid sampling program to determine parameters describing PVT behavior in flow models.

As we have seen already in Section 8.2.3 on page 246, the fluid and PVT description in a black-oil model tries to mimic how the gas and oil pseudo-components at surface conditions partition into the gaseous and oleic phases at reservoir conditions, as illustrated in Figure 11.13 and discussed in detail in the previous section. To this end, the composition and density of the oil and gas components are defined at surface conditions (where they easily can be measured), and then the following three parameters are used to determine the density of the oleic and gaseous phases at reservoir conditions:

11.4 PVT Behavior of Petroleum Fluids

B_o – The oil formation-volume factor is defined as the volume of oil in liquid phase that must be recovered from the reservoir to produce one volume unit of stock-tank oil; in other word, $B_o = V_o/V_{os}$. A reservoir volume V with porosity ϕ and oil saturation S_o thus contains $\phi V S_o/B_o$ stock-tank volumes of oil. In SI units, the formation-volume factor is a dimensionless quantity, but in oilfield units it is usually reported in units reservoir barrels per stock-tank barrels [bbl/STB], where 1 bbl equals 0.1589873 m³. Some authors prefer the reciprocal quantity, $b_o = 1/B_o$, i.e., the factor by which one reservoir volume of oil shrinks (or expands) when brought to surface conditions. MRST uses shrinkage factors rater than formation-volume factors.

B_g – The gas formation-volume factor is defined similarly, as the volume of gaseous phase that must be recovered from the reservoir to produce a unit volume of gas at the surface, i.e., $B_g = V_g/V_{gs}$. It is common to express B_g in units reservoir barrels per standard cubic feet [bbl/scf]. MRST uses the reciprocal expansion factor $b_g = 1/B_g$.

R_s – The solution gas–oil ratio is the volume of gas at standard conditions that dissolves into a unit stock-tank volume of oil at reservoir conditions and is usually reported in units standard cubic feet per stock-tank barrel [scf/STB]. A reservoir volume V with porosity ϕ and oil saturation S_o thus contains $\phi V S_o R_s/B_o$ surface volumes of liberated gas.

This description assumes that the reservoir is in instantaneous phase equilibrium. In addition to these three parameters, we also need the bubble point p_b of the oil, a similar formation-volume factor for water, as well as the dynamic viscosities of the gaseous, oleic, and aqueous phase as function of pressure (and temperature). To model volatile-oil, retrograde dew-point, and wet-gas reservoirs, one also need to model the amount of vaporized oil present in the gaseous phase:

R_v – The vaporized oil–gas ratio (or volatile oil–gas ratio) is defined as the stock-tank volume of condensate oil per unit volume of free gas at surface conditions produced from the gaseous phase at reservoir conditions.

The formation-volume factors and gas–oil and oil–gas ratios all depend on reservoir pressure and temperature and are determined by laboratory experiments (or simulations) reproducing the sequence of differential (and flash) liberations that most likely take place during production of the reservoir. Since reservoirs are almost thermally inert, it is usually sufficient to express these quantities as function of pressure and not temperature.

In the following, we discuss various fluid properties and relationships in some more detail, and in most cases try to mention what are plausible ranges of their values. Readers interested in a more in-depth discussion should consult a textbook on reservoir fluids like e.g., [207, 316, 177] or www.petrowiki.org.

Equation of State: Fluid Densities

Using the quantities defined earlier, we can write the following equations of state that relate the densities of the aqueous, oleic, and gaseous phases (denoted by capital letters W, O, G) at reservoir conditions to the densities of the oil, water, and gas pseudo-components (denoted by o, w, g) at surface conditions

$$\rho_W = \frac{\rho_{ws}}{B_w} = b_w \rho_{ws} \tag{11.3}$$

$$\rho_O = \frac{\rho_{os} + R_s \rho_{gs}}{B_o} = b_o(\rho_{os} + R_s \rho_{gs}) \tag{11.4}$$

$$\rho_G = \frac{\rho_{gs} + R_v \rho_{os}}{B_g} = b_g(\rho_{gs} + R_v \rho_{os}) \tag{11.5}$$

Water density is almost invariant to pressure changes. The oleic phase behaves very differently above and below the bubble point. Above the bubble point, it behaves like a standard fluid in the sense that the density increases with increasing pressure. Below the bubble point, gas dissolution will in most cases dominate compression effects, so that density decreases with increasing pressure as a result of dissolved gas that increases the fraction of lighter hydrocarbon components. The gaseous phase is strongly affected by fluid compression so that density increases with pressure.

API and Specific Gravity

In our discussion of phase behavior, we repeatedly talked about light and heavy hydrocarbon components in the mixture. In petroleum literature, it is common to use the American Petroleum Institute gravity as a measure of how heavy or light a petroleum liquid is. The API gravity is defined as

$$\gamma_{API} = \frac{141.5}{\gamma_\ell} - 131.5, \qquad \gamma_\ell = \frac{\rho_\ell}{\rho_w}, \tag{11.6}$$

where γ_ℓ is the specific gravity of liquid ℓ to water, i.e., the ratio of the density of liquid to the density of water. Bubble-point pressure p_b, oil viscosity μ_o, and the solution gas–oil ratio R_s, all depend strongly on oil gravity. It also has an indirect effect on oil compressibility and oil formation-volume factors, since these depend on R_s.

The API gravity is specified as degrees on a hydrometer (°API) and is defined so that any liquid having API gravity greater than 10° is lighter and floats on water, whereas liquids with values less than 10° are heavier and sink. Using this scale, crude oil is commonly classified as follows:

- light oil: API gravity higher than 31.1°, i.e., less than 870 kg/m^3
- medium oil: API gravity between 22.3 and 31.1°, i.e., 870–920 kg/m^3
- heavy oil: API gravity below 22.3°, i.e., 920–1,000 kg/m^3
- extra heavy oil: API gravity below 10.0°, i.e., greater than 1,000 kg/m^3

Typical API gravity values are also reported in the overview of fluid characteristics for different reservoirs in Table 11.1.

11.4 PVT Behavior of Petroleum Fluids

For gas, the specific gravity is measured relative to air and is defined as

$$\gamma_g = \frac{\rho_g}{\rho_{air}} = \frac{M_g}{M_{air}},$$

where M denotes the molar mass and M_{air}= 28.94 kg/kmol. Specific gas gravity is one of the parameters that determine bubble-point pressure, viscosity, and compressibility of gas, as well as the solution gas–oil ratio. Typical values range from 0.55 for dry gas to approximately 1.5 for wet sour gas. Petroleum gases have gravity approximately equal 0.65.

Reservoir water is saline and hence has a specific gravity greater than unity, but has no impact on compressibility, formation-volume factor, and viscosity.

Compressibility and Formation-Volume Factor of Dry Gas

You may recall from Section 4.2 that the equation of state of an ideal gas is given as $pV = nRT$, where n is the number of moles of gas and R is the universal gas constant. The literature is full of equations of state for real gases, but in petroleum engineering it is common to use a so-called *compressibility equation of state* to describe dry gas

$$pV = ZnRT, \qquad Z = \frac{V_{real}}{V_{ideal}}. \tag{11.7}$$

The *compressibility factor* or *gas deviation factor* Z measures how much the real gas volume deviates from that of an ideal gas at the same temperature. The gas compressibility factor typically varies between 0.8 and 1.2. For constant temperature and moderate pressures, the gas molecules are close enough to exert an attraction on each other that causes the actual gas volume to be less than that of an ideal gas. As p approaches 0, the molecules move more apart and the gas behaves like an ideal gas. At higher pressures, repulsive forces will tend to force modules more apart and the real gas therefore has a higher volume than an ideal gas. The same type of equation of state can be used also for wet gas and mixtures if Z is replaced by a pseudo function that depends on the composition of the mixture. For volatile oils and retrograde-condensate gases it is more common to use a more advanced equation of state like one of the cubic equations discussed in Section 4.2. For a dry gas, the formation-volume factor can be derived from (11.7),

$$B_g = \frac{p_s}{Z_s T_s} \frac{ZT}{p}, \tag{11.8}$$

where subscript s refers to standard conditions. This varies throughout the world, and can be e.g., 14.65 psia and 60°F or 1 atm and 25°C.

The isothermal compressibility of the gas is defined as the fractional change of volume with respect to pressure for constant temperature,

$$c_g = -\frac{1}{V}\left(\frac{\partial V}{\partial p}\right)_T = -\frac{1}{B_g}\left(\frac{\partial B_g}{\partial p}\right)_T. \tag{11.9}$$

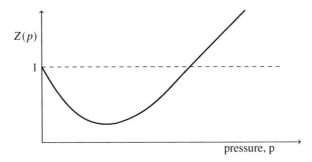

Figure 11.18 Characteristic shape of gas compressibility factor Z as function of pressure for fixed temperature.

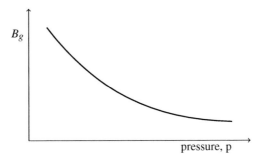

Figure 11.19 Characteristic shape of gas formation-volume factor as function of pressure for fixed temperature and chemical composition.

Inserting the compressible equation of state, it follows that

$$c_g = -\frac{1}{V}\left[\frac{nRT}{p}\left(\frac{\partial Z}{\partial p}\right)_T - \frac{ZnRT}{p^2}\right] = \frac{1}{p} - \frac{1}{Z}\left(\frac{\partial Z}{\partial p}\right)_T.$$

At low pressures, $\partial Z/\partial p$ is negative and hence the compressibility of a real gas is larger than that of an ideal gas. At high pressures, $\partial Z/\partial p$ is positive and hence the compressibility is lower than for an ideal gas.

Gas Viscosity

Dynamic gas viscosities typically range from 0.01 to 0.03 cP and are generally difficult to determine experimentally. It is therefore common to use correlations to determine values. At low pressures, gas molecules are far apart and can easily move past each other. As pressure increases the molecules are forced closer to each other and hence the viscosity of the gas increases. In MRST, gas viscosities are either given as simple explicit formulas (as seen so far) and correlations [316], or as tabulated input read from an ECLIPSE simulation deck (as we will see later).

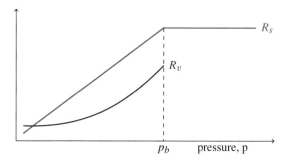

Figure 11.20 Characteristic shape of the solution gas–oil ratio and vaporized oil–gas ratio as functions of pressure for fixed temperature and chemical composition.

Bubble-Point Pressure

Bubble-point pressure is one of the key parameters in the phase behavior of reservoir fluids since it marks the limit between the liquid and two-phase liquid–vapor region for hydrocarbon fluids. In the previous section we outlined briefly how bubble-point pressure can be measured when studying fluid expansion as function of pressure and temperature. The literature is full of correlations that can be used to determine the bubble-point pressure all on the form $p_b = f(T, \gamma_{API}, \gamma_g, R_s)$; see e.g., [177] or www.petrowiki.org for an overview. None of these are implemented in MRST. Instead, the software follows the convention from leading commercial simulators and requires tabulated B and R_s factors, in which p_b is only given implicitly.

Solution Gas–Oil Ratio and Vaporized Oil–Gas Ratio

When discussing dissolved gas, we are in reality referring to light hydrocarbon molecules that appear in gaseous phase at the surface and in liquid oleic phase at reservoir conditions. For a given temperature, the quantity of light molecules present in the oleic phase is limited only by the pressure and the number of light molecules present. For pressures below the bubble point, the oleic phase will release some free gas if the pressure is lowered slightly, and hence we say that the phase is *saturated*. Above the bubble point, the oleic phase is able to dissolve more gas than what is available and is said to be *undersaturated*. Altogether, this means that a plot of the quantity of dissolved gas in the oleic phase as function of pressure increases linearly up to the bubble point and then remains constant; see Figure 11.20. The vaporized oil–gas ratio is only defined for pressures below the bubble point.

Oil Formation-Volume Factor

Most measures of fluid production has traditionally taken place at the surface. It has therefore been customary to introduce the formation volume-factor to relate volumes measured at surface conditions to volumes at reservoir conditions. The oil formation-volume factor is affected by *three* mechanisms: The most important is the liberation of gas when bringing liquid oil from the reservoir to the surface, which will cause a decrease in the liquid volume,

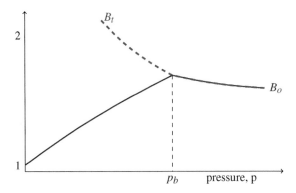

Figure 11.21 Characteristic shape of the oil and total formation-volume factor as functions of pressure for fixed temperature and chemical composition.

so that the B_o factor always assumes values greater or equal 1; the more gas that is dissolved, the larger the B_o factor. At reservoir conditions, increased pressure introduces two additional effects: fluid compression and swelling caused by dissolution of gas. The latter is dominant and below the bubble point, the formation-volume factor therefore increases with increasing pressures. Above the bubble point, there is no more gas to be dissolved, and hence fluid compression causes a gradual decrease in B_o; see Figure 11.21.

The compressibility of oil is defined in the same way as for gas. For undersaturated oil, we therefore have that

$$c_o = -\frac{1}{B_o}\frac{\partial B_o}{\partial p}, \qquad B_o = B_{ob}\exp(c_o(p_b - p)), \quad p > p_b, \qquad (11.10)$$

where B_{ob} is the oil formation-volume factor at the bubble point.

Total Formation-Volume Factor

To better understand the behavior for saturated oil, we introduce the total or two-phase formation-volume factor, defined as the ratio between the total hydrocarbon volume at reservoir conditions and the volume of oil or gas at standard conditions. For simplicity, let us first consider the case of a black oil with a surface oil volume V_o, i.e., a system with dissolved gas but no vaporized oil. At the bubble point, the reservoir fluids inside the reservoir are found as a single oleic phase having a volume $V_o B_{ob}$ and containing $V_o R_{sb}$ volumes of dissolved oil. If we lower the pressure slightly, the system will enter the two-phase region, so that the oleic phase liberates an amount of gas that at the surface would amount to a volume $V_o(R_{sb} - R_s)$. At reservoir conditions, this corresponds to a volume $B_g(R_{sb} - R_s)$ of free gas. Hence, the total formation-volume factor with respect to oil reads,

$$B_{to} = B_o + B_g(R_{sb} - R_s).$$

Because of the expanding gas, the B_{to} function can have significantly larger values at lower pressures than at the bubble point.

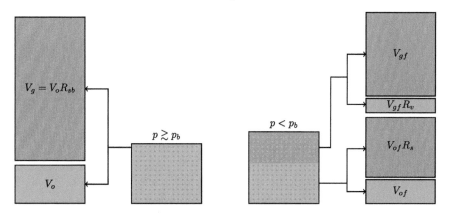

Figure 11.22 Illustration of the volumes involved in a volatile oil model.

For a retrograde gas condensate reservoir, we can start at the dew point and carry out a similar argument to obtain

$$B_{tg} = B_g + B_o(R_{vd} - R_v),$$

where R_{vd} is the vaporized oil–gas ratio at the dew point.

For volatile oils, things get a bit more complicated as Figure 11.22 illustrates. If no water is present, the pore volume V_ϕ is filled by the oleic and gaseous phase that together contain an amount V_o of oil and V_g of gas. Let V_{of} be the stock-tank volume of free oil and V_{gf} the surface volume of free gas. Then, we have

$$V_\phi = V_{of} B_o + V_{gf} B_g.$$

For pressures below the bubble point, oil and gas will appear in both the oleic and gaseous phase, and hence $V_o = V_{of} + V_{gf} R_v$ and $V_g = V_{gf} + V_{of} R_s$. Combining these two and solving for V_{gf} gives $V_{gf} = (V_g - V_o R_s)/(1 - R_v R_s)$. Inserting into the equation above and simplifying, we have

$$V_\phi = \frac{(B_o - R_s B_g) V_o + (B_g - R_v B_o) V_g}{1 - R_s R_v}. \tag{11.11}$$

Just above the bubble point, the hydrocarbon is in liquid phase and hence all gas is resolved in the oil, i.e., $V_g = V_o R_{sb}$. By definition, $V_\phi = B_{to} V_o$. Inserting into the above equation, dividing by V_o, and reorganizing terms, we obtain the following expression for the two-phase formation-volume factor

$$B_{to} = \begin{cases} B_o, & p \geq p_b, \\ \frac{B_o(1 - R_{sb} R_v) + B_g(R_{sb} - R_s)}{1 - R_v R_s}, & p < p_b. \end{cases} \tag{11.12}$$

The two-phase formation-volume factor for retrograde gas condensate can be derived similarly as

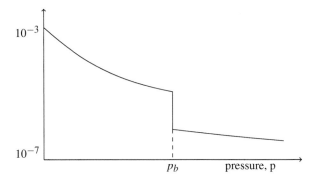

Figure 11.23 Isothermal oil compressibility as function of pressure for fixed temperature and chemical composition.

$$B_{tg} = \begin{cases} B_g, & p \geq p_d, \\ \dfrac{B_g(1-R_{vd}R_s)+B_o(R_{vd}-R_v)}{1-R_v R_s}, & p < p_d. \end{cases} \quad (11.13)$$

Isothermal Oil Compressibility

Isothermal oil compressibility is defined as the relative change in oil volume per change in pressure. It follows from the earlier discussion that oil compressibility has different behavior above and below the bubble point. For an undersaturated oleic phase above the bubble point, the compressibility is defined in the same way as for a gas, i.e., $c_o = (-1/B_o)(\partial B_o/\partial p)_T$, and is a primary drive mechanism. At the bubble point, gas comes out of solution and causes a sharp increase in compressibility. Below the bubble point, the compressibility becomes a much stronger function of pressure as the oil volume now depends on expansion/compression of both liquid oil and dissolved gas

$$c_o = -\frac{1}{B_o}\left(\frac{\partial B_o}{\partial p}\right)_T + \frac{1}{B_o}\frac{B_g - B_o R_v}{1 - R_s R_v}\left(\frac{\partial R_s}{\partial p}\right)_T. \quad (11.14)$$

As a drive mechanism, however, oil compressibility is dominated by the compressibility of gas that has come out of solution.

Oil Viscosity

We have already seen in Section 7.3 that the viscosity of a fluid is affected by both temperature and pressure, with temperature being the most dominant for a fluid of constant composition. Viscosity is also related directly to the type and size of molecules the fluid is made up of. Above the bubble point, the oleic phase has a constant chemical composition and the lower the pressure is, the easier the molecules can move past each other, since they are further apart. Hence, the viscosity increases almost linearly with pressure. Below the bubble point, the oleic phase changes chemical composition as more lighter molecules are removed from the liquid when gas is liberated. The fraction of larger molecules with complex shapes will gradually become larger, causing the viscosity to increase. One order of magnitude

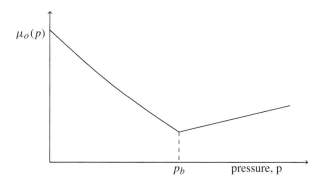

Figure 11.24 Oil viscosity as function of pressure for fixed temperature and chemical composition.

increase in viscosity from bubble point to low reservoir pressure is not uncommon, and this largely reduces the mobility of the remaining oil.

Water Properties

The aqueous phase is usually a brine that contains a number of different dissolved salts and hence has specific density greater than unity. Water may also contain certain amounts of dissolved gas, but this is usually ignored in standard simulation models since the solubility factor in water is orders of magnitude lower than in the liquid oleic phase.

Water viscosity decreases with increasing temperature and increases with increasing pressure, but is generally a very weak function of pressure. Typical values at reservoir conditions range from 0.5 to 1 cP. Water is only slightly compressible at reservoir conditions, with values in the range from 1×10^{-6} to 9×10^{-6} psi^{-1}, which implies that the water formation-volume factor is usually close to unity.

11.5 Phase Behavior in ECLIPSE Input Decks

Use of tabulated data from the PROPS section of ECLIPSE input is the primary input mechanism to set up simulation models for the simulators implemented in the ad-blackoil, ad-eor, and other derived modules. As we saw earlier on page 350, functionality for reading these input files, parsing and processing them to create specific fluid objects is offered by the deckformat and ad-props modules.

To make the book as self-contained as possible, this section outlines the necessary ECLIPSE keywords and show a few examples of specific phase behavior for the first, third, and ninth SPE benchmark cases. For a more comprehensive discussion of the input format, you should consult the lecture notes of Pettersen [250], or the documentation of ECLIPSE [270, 269] or OPM Flow [33]. The following presentation is inspired by these sources.

Dead Oil

Dead oils are assumed to always stay above the bubble point and hence behave like a standard fluid that is compressed and becomes more viscous at higher pressures. Most dead oils contain a constant amount of dissolved gas that is liberated as the oleic phase is brought to the surface. There are two different ways to describe dead oil in an ECLIPSE input deck:

PVDO – specifies the properties of an undersaturated oil above the bubble point in terms of a table of formation-volume factors and viscosities versus monotonically increasing pressure values. The table is attached to specific (groups of) cells, and the keyword can contain multiple tables if the reservoir contains several zones with distinct oils. The number of tables is therefore given by the NTPVT element of the TABDIMS keyword from the RUNSPEC section. The constant amount of dissolved oil and the associated bubble-point pressure should be specified using the RSCONST keyword. The following example is taken from [250]:

```
PVDO
--  Po      Bo       mu_o
    27.2    1.012    1.160
    81.6    1.004    1.164
   136.0    0.996    1.167
   190.5    0.988    1.172
   245.0    0.9802   1.177
   300.0    0.9724   1.181
   353.0    0.9646   1.185  /
RSCONST
--  Rs      Pbp
   180     230  /
```

Here, the pressures Po and Pbp are given in units bar, the formation-volume factor Bo and the gas–oil ratio Rs are dimensionless, and the viscosity mu_o has unit cP.

PVCDO – describes an oleic phase with constant compressibility by specifying the five constants in the formulas

$$B_o(p) = B_{or} e^{-c_o(p-p_r)}, \qquad (B_o\mu_o)(p) = B_{or}\mu_{or} e^{-(c_o-c_\mu)(p-p_r)}.$$

Here, p_r is reference pressure, B_{or} reference formation-volume factor, c_o compressibility, μ_{or} reference viscosity, and c_μ "viscosibility." In addition, one must specify RCONST and a maximum pressure PMAX. The following example is taken from [269] (with pressures given in units psi):

```
PVCDO
--  P       Bo      Co      muO     Cmu
    2500   1.260   6E-6    0.5    1E-6  /
RSCONST
--  GOR     Pb
   0.656   2500 /
PMAX
   4500 /
```

Live Oil

Undersaturated oil has a well-defined bubble point that can be determined from the temperature and the oil's chemical composition. Saturated oils, on the other hand, change chemical composition as gas is dissolved or liberated. To model the phase behavior of a live oil, you need to describe both how the bubble point changes with changing composition of the oil, as well as the properties in the undersaturated region for each unique composition. In a black-oil model, the fluid composition of the oleic phase is represented in terms of the solution gas–oil ratio, and all we need to discriminate saturated from undersaturated states is a relationship between R_s and the bubble point p_b. In ECLIPSE input, this is represented by tabulating $p_b(R_s)$. For each value of $p_b(R_s)$, the dead-oil behavior in the undersaturated region is prescribed in the same way as described for dead-oil models. Hence, the input language offers two keywords for describing live oils:

PVTO – consists of NTPVT tables that each give the properties above and below the bubble point. Each table contains from two to NRPVT records that specify PVT data for given R_s values. The records consist of at least four numbers specifying R_s, bubble point pressure p_b, formation-volume factor B_{ob} for saturated oil at the bubble point, and viscosity μ_{ob} at the bubble point. However, each record can also be composed of from two to NPPVT rows of data organized in three columns that describe oil pressure (increasing values), formation-volume factor (decreasing values), and viscosity for undersaturated oil for the specified value of R_s. These extra data *must* be specified for the maximum R_s value inside each table but can optionally be given for any R_s value. We present a specific example of this input format for the SPE 9 model on page 383.

Notice that ECLIPSE interprets the input data by linear interpolation of the reciprocal values of B_o and $(\mu_o B_o)$ between data points.

PVCO – specifies properties above and below the bubble point in compressibility form; that is, instead of specifying the undersaturated behavior in terms of a tabular, it is described as for the PVCDO keyword. The data thus consists of NTPV tables that each may have up to NRPVT rows. The following example is from [269]:

```
PMAX
  6000 /
PVCO
--Pb        Rs      Bo        Vo       Co    Cvo
  1214.7   0.137   1.17200   1.97000   1E-5  0
  1414.7   0.195   1.20000   1.55600   1*    0
  1614.7   0.241   1.22100   1.39700   1*    0
  1814.7   0.288   1.24200   1.28000   1*    0
  2214.7   0.375   1.27800   1.09500   1*    0
  2614.7   0.465   1.32000   0.96700   1*    0
  3014.7   0.558   1.36000   0.84800   1*    0
  3414.7   0.661   1.40200   0.76200   1*    0
  3814.7   0.770   1.44700   0.69100   1*    0
/
```

Notice, in particular the use of 1* to indicate a default value that should be computed by linear interpolation. Since the table only gives one value in the first row, this value should be copied to all the next rows.

Dry Gas

Dry gases are found either below the dew point or (well) to the left of the critical point. Their phase behavior is equivalent to that of dead oils and is described as a tabular of formation-volume factor and viscosity as function of pressure, or by a Z-factor and viscosity as function of pressure:

PVDG – specifies NTPVT tables that each consists of three columns and up to NPPVT rows giving gas pressure (increasing monotonically downwards), gas formation factor, and gas viscosity. A specific example of this input format is presented for the SPE 9 model on the next page.

PVZG – specifies properties using Z-factors. The data consists of NTPV tables that each consists of two records. The first gives the reference temperature used to convert Z-values into formation-volume factors (see [11.8] on page 373), and the second consists of from 2 up to NPPVT rows giving gas pressure, Z-factor, and gas viscosity. Pressures and Z-values should increase monotonically down the table. The lowest pressure value should correspond to the dew point. The gas may contain a constant amount of vaporized oil, specified by the RVCONST keyword, which also gives the dew-point pressure.

Wet Gas

Wet or rich gases with vaporized (condensate) oil are similar to live oils in the sense that the phase behavior is different for saturated and undersaturated states. The composition of the wet gas is represented by vaporized oil–gas ratio R_v. To discriminate between saturated and undersaturated states, you must tabulate the maximum amount of vaporized oil as function of pressure, i.e., $R_v(p_g)$. The corresponding keyword is:

PVTG – describes properties of wet gas with vaporized oil. The format is similar to the PVTO keyword used to describe live oil, and consists of from two to NRPVT records that specify PVT data (R_v, B_g, and μ_g) at the saturated state for given p values. Each record can also be composed of from two to NPPVT rows of three-column data that specify R_v, B_g, and μ_g for undersaturated states. These extra data *must* be specified for the highest pressure data in each of the NTPVT tables.

We shown an example of this input format on page 387 when discussing the PVT behavior of the SPE 3 benchmark.

Miscellaneous Properties

Water properties are described with the keyword PVTW, which follows the same format as the PVCDO keyword for dead oil. In addition to the PVT behavior, you also need to prescribe the density or specific gravity of each fluid at surface condition, which is defined by the keywords DENSITY or GRAVITY. The PROPS section of the input deck should also describe how the pore volume depends on pressure, either by the ROCK keyword that specifies a reference pressure and rock compressibility, or by the ROCKTAB keyword that tabulates porosity (and permeability) multipliers as function of pressure.

Example: SPE 1 and SPE 9 – Live Oil With Dry Gas

To exemplify live oil and dry gas models, we revisit the SPE 1 and SPE 9 benchmarks discussed on page 355. The SPE 9 benchmark describes a live oil with dry gas specified as follows (the setup for SPE 1 is similar):

```
-- LIVE OIL (WITH DISSOLVED GAS)            -- DRY GAS (NO VAPORIZED OIL)
PVTO                                        PVDG
-- Rs      Pbub    Bo       Vo              --     Pg      Bg        Vg
   .0      14.7    1.0000   1.20 /                 14.7    178.08    .0125
   .165    400.    1.0120   1.17 /                 400.    5.4777    .0130
   .335    800.    1.0255   1.14 /                 800.    2.7392    .0135
   .500    1200.   1.0380   1.11 /                 1200.   1.8198    .0140
   .665    1600.   1.0510   1.08 /                 1600.   1.3648    .0145
   .828    2000.   1.0630   1.06 /                 2000.   1.0957    .0150
   .985    2400.   1.0750   1.03 /                 2400.   0.9099    .0155
  1.130    2800.   1.0870   1.00 /                 2800.   0.7799    .0160
  1.270    3200.   1.0985    .98 /                 3200.   0.6871    .0165
  1.390    3600.   1.1100    .95 /                 3600.   0.6035    .0170
  1.500    4000.   1.1200    .94                   4000.   0.5432    .0175 /
           5000.   1.1189    .94 /
/                                           DENSITY
                                            --  Oil      Water    Gas
PVTW                                            44.98    63.01    0.0702 /
--Pref    Bw      Comp    Vw    Cv
  3600.   1.0034  1.0E-6  0.96  0.0 /
```

The model is specified in field units, so that densities are given as lb/ft^3 and not kg/m^3 and pressure values are given in unit psi. For the live oil, we see that undersaturated behavior is described for the largest R_s value. We do not repeat the statements required to load these model, but simply assume that we have created a processed deck structure with the necessary input data and created a fluid object f. Once the units have been converted to the SI system and the input data have been processed to construct a fluid object, the phase behavior of the oil is represented by the following functions and constants:

```
   b0: @(po,rs,flag,varargin) b0(po,rs,pvto,flag,reg,varargin{:})
  mu0: @(po,rs,flag,varargin) mu0(po,rs,pvto,flag,reg,varargin{:})
rsSat: @(po,varargin) rsSat(po,pvto,reg,varargin{:})
rhoOS: 720.5105
```

MRST has functions for computing the reciprocal expansion/shrinkage factors rather than the usual formation-volume factors. We also note that R_s is called rsSat rather than Rs. The reason is that dissolved gas will be an unknown in our system of flow equations and hence we change the name of the *parameter R_s* that describes the maximum dissolved gas amount for a given pressure to avoid confusion.

Looking at the functions in more detail, we see that the b0 function takes three primary arguments: oil pressure, the amount of dissolved oil, and a flag that indicates whether a given element in po and rs represents a saturated or an undersaturated state. In the underlying functions, pvto is an object that holds the data of the PVTO keywords, whereas reg represents region information that picks the right table, should the model have multiple oils. The implementation of the BO, bO, muO, and rSat functions can be found in the file assignPVTO.m and consists of calls to two utility functions that extract region information, which is subsequently used to interpolate the correct data tables. For saturated oil, we use a simple linear interpolation between the data points, whereas the interpolation for undersaturated oil is set up to preserve compressibility and viscosibility if only one undersaturated data curve is given, or to interpolate compressibility and viscosibility if multiple curves are given. I will not go through the interpolation functions in detail; these have been optimized for speed and are only easy to understand if you are a black-belt MATLAB programmer.

Instead, we can plot the various relationships. (You can find the necessary code in the showSPEfluids script in the book module.) We start by extracting the input data from the deck structure:

```
pvto = deck.PROPS.PVTO{1};
rsd  = pvto.key([1:end end]);
pbp  = pvto.data([pvto.pos(1:end-1)' end],1);
Bod  = pvto.data([pvto.pos(1:end-1)' end],2);
muOd = pvto.data([pvto.pos(1:end-1)' end],3);
```

Here, we have only extracted the data points representing saturated states as well as the undersaturated data for the maximum R_s value, which is the minimal data requirement and coincides with what is given in the input file in our case. To simplify subsequent interpolation, the input processing in MRST adds an extra data point inside the undersaturated region for each data point, but we ignore these in our simplified data extraction.

Let us take the SPE 9 model as an example: The input data has eleven data records for R_s values 0, 0.165, 0.335,...,1.5. The last record also contains formation-volume factor and viscosity at pressure 5,000 psi inside the undersaturated region. The three pairs of pressure, B_o, and μ_o values specified for $R_s = 1.5$ define values for compressibility and viscosibility. Assuming that these are constant inside the whole undersaturated region, the input routine computes one additional value at pressure values $p = p_{bp} + 1{,}000$ psi, so that pvto.data becomes a 22×3 array. If we need R_s values for *all* data points represented in the deck structure, including extra points that have been added in the undersaturated region, we can use the following call:

```
rsd = pvto.key(rldecode((1:numel(pvto.key))',diff(pvto.pos),1));
```

To evaluate the fluid model, we make a mesh of pressure and dissolved gas states and then determine whether each state represents a saturated or undersaturated state by comparing with the interpolated R_s value for the corresponding pressure:

```
[RsMax,pMax] = deal(max(rsd), max(pbp));
[rs,p]       = meshgrid(linspace(10,RsMax-10,M), linspace(0,pMax,N));
Rs           = reshape(f.rsSat(p(:)), N, M);
```

11.5 Phase Behavior in ECLIPSE Input Decks

```
isSat      = rs >= Rs;
rs(isSat)  = Rs(isSat);
Bo         = reshape( f.BO (p(:), rs(:), isSat(:)), N,M);
muO        = reshape( f.muO(p(:), rs(:), isSat(:)), N,M);
```

Figure 11.25 shows plots of R_s as function of pressure as well as plots of B_o and μ_o as function of pressure and amount of dissolved gas. For comparison, we have overlain the plots for SPE 1 and SPE 9, which describe a gas-injection scenario and a case with

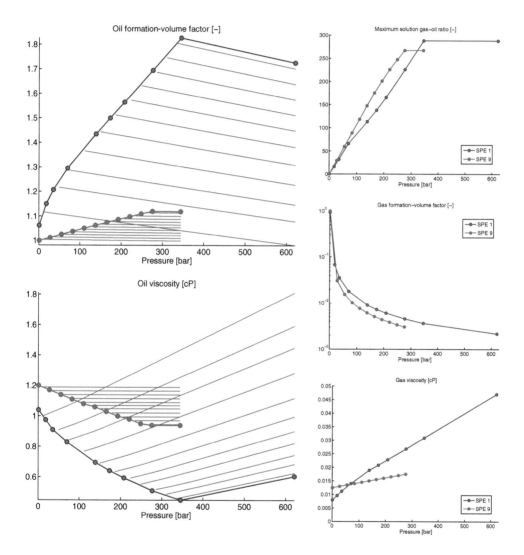

Figure 11.25 Description of phase behavior for the live-oil and dry-gas models in the SPE 1 and SPE 9 benchmarks, shown as blue and green lines, respectively. In the B_o and μ_o plots, the red lines correspond to undersaturated states.

water injection, respectively. There is a pronounced difference in the phase properties of the two models: Whereas the SPE 1 oil shows strong compressibility and viscosibility effects both for the saturated and undersaturated states, the undersaturated SPE 9 oil is almost incompressible and has constant viscosity. In part, this is also indicated by the surface densities of oil and gas, which are 786.5 kg/m^3 and 0.9698 kg/m^3 for SPE 1 and 720.5 kg/m^3 and 1.1245 kg/m^3 for SPE 9. The SPE 1 model has a larger fraction of both lighter and heavier hydrocarbon components and therefore exhibits larger variations in formation-volume factor and viscosity than SPE 9.

For completeness, let us also look at the gaseous and aqueous phases. The behavior of dry gas is represented by the following functions and constants (here with values for SPE 9):

```
   bG: @(pg,varargin) ireg(TbG,pg,varargin{:})
   muG: @(pg,varargin) ireg(TmuG,pg,varargin{:})
 rhoGS: 1.1245
```

The constant arguments TBG, TbG, and TmuG to the functions are data objects containing the tabulated formation-volume factor, its reciprocal expansion factor, and the viscosity from the PVDG keyword, respectively. The function name `ireg` is a shorthand for the `interpReg` function we encountered when discussing computation of three-phase relative permeabilities on page 353.

The aqueous phase is represented as a weakly compressible fluid (here for SPE 9),

```
    cW: 1.4504e-10
  muWr: 9.6000e-04
    bW: @(pw,varargin) bW(pw,pvtw,reg,varargin{:})
   muW: @(pw,varargin) muW(pw,pvtw,reg,varargin{:})
 rhoWS: 1.0093e+03
```

To be consistent with ECLIPSE, we use a Taylor expansion to approximate the weakly compressible functions

$$B_w(p) = B_{wr} e^{-c_w(p-p_r)} = B_{wr} e^{-\xi} \approx \frac{B_{wr}}{1 + \xi + \frac{1}{2}\xi^2}.$$

At this point, I encourage you to consult the tutorial `fluidInspectionExample` from the `ad-core` module, which introduces a graphical interface for fluid models from the AD-OO framework that enables you to interactively produce many of the same plots as shown here and in Section 11.3. Creating models and launching the GUI only requires three lines of code:

```
[G, rock, fluid, deck] = setupSPE9();          % or setupSPE1(), setupSPE3()
spe9 = selectModelFromDeck(G, rock, fluid, deck);
inspectFluidModel(spe9)
```

Example: SPE 3 – Rich Gas With Retrograde Condensation

The SPE 3 benchmark was designed to compare the ability of compositional simulators to match PVT data for a problem with gas cycling in a rich-gas reservoir with retrograde condensation. The input deck distributed with MRST is a translation of the problem into a black-oil framework, with which one can only hope to describe the pertinent flow physics to a limited extent. The resulting model consists of a dead oil, described with the PVDO keyword, and a rich gas described by the PVTG keyword. Here, we only show excerpts of the specification of the rich gas:

```
-- 'Gas Pressure'  'Gas OGR'    'Gas FVF'    'Gas Visc'
-- Units: psia     stb /Mscf    rb /Mscf     cp
PVTG
       1214.7000   0.0013130    2.2799       0.0149
                   0            2.2815       0.01488/
       1814.7000   0.00353      1.4401       0.01791
                   0.001313     1.4429       0.01782
                   0            1.4445       0.01735 /
       2414.7000   0.01102      1.0438       0.02328
       :
       :
       3449.3322   0.0642       0.7783       0.0403
                   0.0612       0.7777       0.0395
                   0.0567       0.7769       0.03892
                   0.0454       0.7747       0.03748
                   0.0331       0.7723       0.03594
                   0.01102      0.7681       0.03325
                   0.00353      0.7666       0.03236
                   0.001313     0.7662       0.0321
                   0            0.7660       0.03194 /
/
```

The maximum amount of vaporized oil increases as function of pressure, and the input table uses an increasing number of data points within each record, starting with two points for the first pressure value and ending with nine values for the highest pressure value. The data points are quite unevenly scattered in (p_g, R_v) space, as you can see from the plot of R_v and μ_g versus p_g in Figure 11.26. You should therefore be wary of potential numerical inaccuracies in the numerical interpolation. (The showSPEfluids script in the book module gives the full code used to generate the plots in the figure.)

Looking at the qualitative behavior of the gas properties shown in Figure 11.26, we see that as pressure increases, more oil can be vaporized into the gas and this will increase the viscosity of the gaseous phase. The data points along the vertical lines traverse the phase envelope for fixed pressure, from the blue line corresponding to a saturated state with full vaporization (i.e., maximal supplied R_v value) to the red line representing the dew point, at which the vapor phase contains no vaporized oil ($R_v = 0$). Vaporization of oil also affects the gas formation-volume factor, but since the amount of vaporized oil is so small, the resulting variations in volume across the phase envelope for a fixed pressure are so small that they are only visible when zooming in on the plot. Interestingly, we observe that B_g decreases with increasing vaporization for pressures below 220 bar, but *increases* with increasing vaporization for pressures above 220 bar. However, the density of

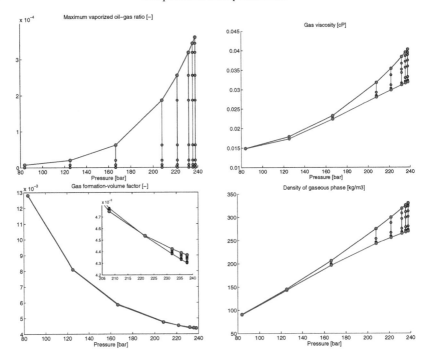

Figure 11.26 Description of phase behavior for the rich-gas system of the SPE 3 benchmark problem. Vertical lines represent transition across the phase envelope for fixed pressure, with the blue lines representing max R_v and red lines $R_v = 0$.

the gaseous phase increases monotonically with increasing oil vaporization, as one would expect physically.

The figure does not report properties of the dead oil, but these behave as expected, with B_o decreasing and μ_o increasing with increasing pressure.

11.6 The Black-Oil Equations

In this section, we discuss the flow equations for black-oil reservoir model, which we presented in Section 8.2.3, in more detail. We go through the equations, component by component, and discuss the various special cases and present how to discretize the various terms using discrete differentiation and averaging operators. As described in Section 11.4, the family of black-oil models assume isothermal conditions inside the reservoir, so that the PVT behavior is modeled as a function of pressure only. The resulting models work best when operating in single-phase hydrocarbon regions far from the critical point, but can also be used in two-phase hydrocarbon regions as long as the variation in composition is relatively small.

11.6.1 The Water Component

The discussion in this chapter has primarily focused on the gas and oil components contained in the gaseous and oleic hydrocarbon phase. Reservoirs will nonetheless always contain aqueous fluids present as connate water within the hydrocarbon-bearing rock layers. The reason is that the reservoir rock was fully or partially saturated with water before hydrocarbons migrated upwards from a deeper source rock. Since most rocks are fully or partially water-wet, the invading hydrocarbons would not be able to fully displace the resident water. Oftentimes, hydrocarbon-bearing rocks will have extensive underlying aquifer that contribute pressure support and fluid inflow that should be accounted for in the reservoir model.

As a rule, aqueous reservoir fluids are saline. The most common dissolved salt is sodium chloride, but the aqueous phase typically contains a wide range of compositions and concentrations of other inorganic chemical species. These species are not accounted for in basic black-oil type models, which simply assume that the resident brine and injected water is a single, chemically inert component that only is present in the aqueous phase. (You can consult Trangenstein and Bell [295] for an alternative model, in which gas is allowed to dissolve into the aqueous phase.) The corresponding flow equations read

$$\partial_t\left(\phi b_w S_w\right) + \nabla \cdot \left(b_w \vec{v}_w\right) - b_w q_w = 0, \tag{11.15}$$

with Darcy's law for the phase velocity

$$\vec{v}_w = -\frac{\mathbf{K} k_{rw}}{\mu_w}(\nabla p_w - g\rho_w \nabla z).$$

These equations are discretized exactly in the same way as (11.1) on page 342, except for the obvious modifications resulting from ρ_w having been replaced by b_w. Let us now look at how you could implement the discrete flow equations in MRST. We have already discussed most of the tools you need. As before, we assume that the fluid properties are collected in a fluid object called `fluid`. For convenience, the `ad-core` module collects all the discrete differentiation and averaging operators in a special structure, which we here will refer to by the short-hand s. We start by computing the necessary fluid properties:

```
bW      = fluid.bW(p0);
rhoW    = bW.*fluid.rhoWS;
rhoWf   = s.faceAvg(rhoW);
mobW    = krW./fluid.muW(p0);
pvMultR = fluid.pvMultR(p0);
```

Here, `faceAvg` is the same as the `avg` operator we have used previously, but has been renamed to specify explicitly that this is an averaging operator that should be used to compute average values at cell interfaces. Notice that we evaluate the pressure-dependent properties using oil pressure and not water pressure. In general, there are several ways you can choose the primary unknowns in the black-oil models. Herein, we have chosen to follow the conventions from ECLIPSE and use oil pressure p_o and gas and water saturations, S_w and S_g, as our primary unknowns. (We will get back to the choice of primary variables later in this section.)

The water flux across cell faces will, as before, be computed using single-point upstream-mobility weighting. To this end, we must first compute the water pressure in each grid cell,

```
pW = pO
if isfield(fluid, 'pcOW') && ~isempty(sW)
   pW = pW - fluid.pcOW(sW);
end
```

Then, we can compute the water flux from Darcy's law

```
dpW  = s.Grad(pW) - rhoWf.*gdz;
upcw = (double(dpW)<=0);
vW   = -s.faceUpstr(upcw, mobW).*T.*dpW;
bWvW = s.faceUpstr(upcw, bW).*vW;
```

The function `faceUpstr` implements the upstream operator we previously denoted by `upw`. With this, we have all quantities we need to compute the homogeneous water equation on residual form

```
water = (s.pv/dt).*( pvMult.*bW.*sW - pvMult0.*bW0.*sW0 ) + s.Div(bWvW);
```

Here, quantities ending with 0 refer to variables evaluated at the beginning of the time step and the vector `s.pv` contains pore volumes in each cell at reference pressure. The code lines in this subsection are in essence what is implemented in the flow solvers of the `ad-blackoil` module. However, if you look at the code in e.g., `equationsBlackOil`, you will see that it is more complicated since we in general need to also include (multisegment) well models, source terms, and boundary conditions. Likewise, we must account for various user-defined multipliers that modify transmissibilities and mobilities, enable solution of adjoint equations for computing sensitivities, and store computed results; more about this in the next chapter.

11.6.2 The Oil Component

In the basic black-oil model, the oil component is only present in the liquid oleic phase, and we hence get the following flow equation:

$$\partial_t(\phi b_o S_o) + \nabla \cdot (b_o \vec{v}_o) - b_o q_o = 0, \tag{11.16}$$

with Darcy's law for the phase velocity

$$\vec{v}_o = -\frac{\mathbf{K} k_{rw}}{\mu_o}(\nabla p_o - g\rho_o \nabla z).$$

11.6 The Black-Oil Equations

As for the water component, we start by computing the necessary fluid properties. How we compute density, shrinkage factor, and viscosity depends upon whether the oil phase contains dissolved gas or not. If the system contains no dissolved gas, the parameters only depend on oil pressure and can be computed as follows:

```
bO  = fluid.bO(pO);
if isfield(fluid, 'BOxmuO')
   muO = fluid.BOxmuO(pO).*bO;
else
   muO = fluid.muO(pO);
end
rhoO = bO.*fluid.rhoOS;
```

As pointed out on page 384, the convention in ECLIPSE 100 input is that tabulated formation-volume factors and viscosity for oil should not be interpolated independently, but instead be interpolated as their reciprocal product, $1/(\mu_o B_o)$. MRST signifies input data that require this particular interpolation by a special function BOxmuO in the fluid object. If this function is not present, the viscosity and the shrinkage factors are computed independently. If oil also contains dissolved gas, the same properties are evaluated as follows:

```
bO  = fluid.bO(pO, rs, isSat);
muO = fluid.muO(pO, rs, isSat);
rhoO = bO.*(rs*fluid.rhoGS + fluid.rhoOS);
```

Here, rs is a vector holding the gas–oil ratio and isSat is a Boolean vector that tells whether each element of pO and rs represents a saturated or an undersaturated state. We will get back to how rs is computed in Section 11.6.4. With density, shrinkage factor, and viscosity given, we can proceed to compute the oleic phase flux bOvO using upstream-mobility weighting as described for the aqueous phase in Section 11.6.1. Altogether, this gives us the homogeneous residual equation

```
oil = (s.pv/dt).*( pvMult.*bO.*sO - pvMult0.*bO0.*sO0 ) + s.Div(bOvO);
```

If oil is allowed to vaporize into the gaseous phase, the conservation equation for the oil component reads

$$\partial_t\left[\phi\left(b_o S_o + b_g R_v S_g\right)\right] + \nabla \cdot \left(b_o \vec{v}_o + b_g R_v \vec{v}_g\right) - \left(b_o q_o + b_g R_v q_g\right) = 0. \quad (11.17)$$

We therefore need to compute the properties and the phase flux for the gaseous phase before we can compute the homogeneous residual oil equation:

```
oil = (s.pv/dt).*( pvMult.* (bO.* sO  + rv.* bG.* sG) - ...
      pvMult0.*(bO0.*sO0 + rv0.*bG0.*sG0) ) + s.Div(bOvO + rvbGvG);
```

Notice, in particular, that we use single-point upstream-mobility weighting to evaluate the vaporized oil–gas ratio in the term representing flux of oil vaporized in the gas phase rvbGvG. This is completely analogous to how we evaluate the shrinkage factors for oil and water at the interface between two neighboring cells.

11.6.3 The Gas Component

The gas component can generally be present both in the gaseous and the liquid oleic phases. The corresponding conservation equation reads

$$\partial_t \left[\phi \left(b_g S_g + b_o R_s S_o \right) \right] + \nabla \cdot \left(b_g \vec{v}_g + b_o R_s \vec{v}_o \right) - \left(b_g q_g + b_o R_s q_o \right) = 0 \qquad (11.18)$$

with Darcy's law for the phase velocity

$$\vec{v}_g = -\frac{\mathbf{K} k_{rg}}{\mu_g} (\nabla p_g - g \rho_g \nabla z).$$

Like for oil, the computation of the gas properties differs for systems with and without vaporized oil:

```
if vapoil
    bG   = fluid.bG(p0, rv, isSat);
    muG  = fluid.muG(p0, rv, isSat);
    rhoG = bG.*(rv*fluid.rhoOS + fluid.rhoGS);
else
    bG   = fluid.bG(p0);
    muG  = fluid.muG(p0);
    rhoG = bG.fluid.rhoGS;
end
```

We also need to compute the phase pressure to determine the phase velocity:

```
pG = p0;
if isfield(fluid, 'pcOG') && ~isempty(sG)
    pG = pG + fluid.pcOG(sG);
end
```

With these quantities given, we can proceed as above to compute the remaining terms necessary to evaluate the accumulation and flux terms in the homogeneous residual equation

```
gas = (s.pv/dt).*( pvMult.* (bG.* sG  + rs.* b0.* s0) - ...
      pvMult0.*(bG0.*sG0 + rs0.*b00.*s00 ) ) + s.Div(bGvG + rsb0v0);
```

11.6.4 Appearance and Disappearance of Phases

With all the details presented thus far, it should be a surmountable task to extend the simple two-phase, dead-oil implementation discussed in Section 11.2 to a similar simplified simulator for black-oil models. In particular, dissolved gas in the liquid oleic phase and/or vaporized oil in the gaseous phase is not too difficult to include as long as each hydrocarbon phase either stays in a saturated state or an unsaturated state throughout the whole simulation. For saturated phases, it is common to use one phase pressure (p_o in MRST) and the

11.6 The Black-Oil Equations

water and gas saturation (S_w and S_g) as independent variables since both the solution gas–oil ratio R_s and the vaporized oil–gas ratio R_v can be computed as a function of pressure. If oil is in an undersaturated state, there is no free gas available and the flow problem is essentially a two-phase flow problem (assuming that water is present). However, such cases can also be simulated as a three-phase problem if we set R_s as a primary variable. Likewise, we can choose R_v as a primary variable for cases with undersaturated gas.

Somewhat more intricate details are needed when the hydrocarbon phases switch between saturated and undersaturated states so that phases may appear or disappear during the simulation. As you have seen already, MRST introduces the solution gas–oil ratio rs as an extra variable alongside Sg for cases with dissolved gas. In addition, we need a status flag in each cell that tells which of the two is the primary unknown. When oil is in a saturated state, Sg is the primary variable and rs is a secondary variable that can be determined from the pressure. If the computed gas saturation in a cell is zero or falls below, we know that the cell has switched from a saturated to an undersaturated state. In cells containing undersaturated oil, there is no free gas available and we solve for rs as a primary variable to track the amount of gas flowing in and out of each cell with the oil. Once a new rs value is computed, we need to check whether rs>fluid.rsSat(p0), which means that we have a transition from undersaturated to saturated state and hence need to increase Sg to the correct amount of liberated gas.

Switching of variables may unfortunately introduce instabilities in the nonlinear solution process, and in practice we do not allow states to cross from a saturated/undersaturated state and well into the undersaturated/saturated region. We rather stop the update at the interface, i.e., set rs to rsMax=rsSat(p0) and Sg=0, and chop the update to improve robustness of the Newton iteration loop. Cases with transitions between saturated and undersaturated gas are treated analogously. The following code summarizes the logic just explained to choose the primary variable x representing gas:

```
% Assumption: rs = rsSat for saturated cells, likewise for rv
% status 1 (oil, no gas): x = rs, sg = 0,     rv = rvMax
% status 2 (gas, no oil): x = rv, sg = 1-sw, rs = rsMax
% status 3 (oil and gas): x = sg, rs = rsMax, rv = rvMax
watOnly    = sW > 1- sqrt(eps);

if ~vapoil, oilPresent = true;  else, oilPresent = or(sO > 0, watOnly); end
if ~disgas, gasPresent = true;  else, gasPresent = or(sG > 0, watOnly); end

status = oilPresent + 2*gasPresent;
if ~disgas, st1 = false; else, st1 = status==1; end
if ~vapoil, st2 = false; else, st2 = status==2; end
st3 = status == 3;

x = st1.*rs + st2.*rv + st3.*sG;
```

11.7 Well Models

Earlier in the book, we have discussed well models for single-phase flow. Section 4.3.2 introduced the basic inflow-performance relationship, which states that the flow rate in or out from a well is proportional to the pressure difference between the wellbore and the reservoir, $q = J(p_R - p_{wb})$. This section describes well models for multiphase flow in more detail. In particular, we discuss wells with multiple connections and explain how to handle cross flow and introduce a more comprehensive family of models that use a network of 1D flow models to describe flow in horizontal and multilateral wells.

11.7.1 Inflow-Performance Relationships

The inflow-performance relationship for black-oil models is written in terms of volumetric production rates for each fluid phase α at stock-tank conditions

$$q_{\alpha,i} = J_i \lambda_{\alpha,i} \left(p_i - p_{bh} - H_{wi} \right). \tag{11.19}$$

Here, J_i is the productivity index (sometimes called connection transmissibility factor or R-factor) for connection number i, $\lambda_{\alpha,i}$ is phase mobility at the connection, p_i is pressure in the perforated grid block, p_{bh} is bottom-hole pressure, and H_{wi} is pressure drop from the well's datum point to connection i.

The standard Peaceman-type expression for the productivity index of a vertical well penetrating a Cartesian grid block reads

$$J = \frac{\theta K h}{\ln(r_e/r_w) + S}.$$

Here, Kh is effective permeability times net thickness of the connection, r_w is wellbore radius, and r_e is the equivalent radius. We have previously discussed expressions for K and r_e for various special cases in Section 4.3.2. The parameter θ gives the angle the wellbore is connected to the reservoir block, with $\theta = 2\pi$ when the well is perforated in the center of the cell, $\theta = \pi$ when the well is perforated along a cell face, and $\theta = \pi/2$ when the well is perforated along a corner edge; see Figure 11.27. The *skin factor S* accounts for the difference between ideal pressure drawdown predicted by Darcy's law and actual drawdown resulting from formation changes in the near-well region; S also includes the constant 3/4 from pseudo-steady flow. The typical range of skin factors is $-6 < S <$

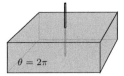
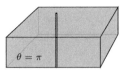

Figure 11.27 The parameter θ measures the angular section of the wellbore that is connected to the reservoir and depends on where the well is placed inside a grid block.

Figure 11.28 Illustration of crossflow in a well; blue and green colors represent different fluid phases.

100. Negative values represent wells stimulated by hydraulic fracturing or other means to improve the permeability in the near-well zone, whereas large positive values represent wells suffering from severe formation damage caused, e.g., by various forms of clogging induced by drilling fluids, fines migration, precipitation of soluble salts, etc.

The mobility $\lambda_{\alpha,i}$ is computed differently for injection and production wells. For *producing connections*, the mobility of a free phase of gas, oil, or water, is given as

$$\lambda_{\alpha,i} = \frac{(b_\alpha k_{r\alpha})_i}{\mu_{\alpha,i}}, \quad (11.20)$$

where the relative permeability $k_{r\alpha}$, the shrinkage/expansion factor b_α, and the phase viscosity μ_α are computed using pressure and saturation values from the grid block perforated by connection number i. If the produced gaseous or oleic phases contain vaporized oil or dissolved gas, respectively, the mobilities account for this as follows

$$\lambda_O = \lambda_o + R_v \lambda_g, \qquad \lambda_G = \lambda_g + R_s \lambda_o. \quad (11.21)$$

For *injecting connections*, MRST computes the mobility as the sum over all phases

$$\lambda_{\alpha,i} = y_\alpha b_\alpha \sum_\beta \left(\frac{k_{r\beta}}{\mu_\beta}\right)_i. \quad (11.22)$$

where y_α is the volume fraction of phase α in the mixture *inside the wellbore*. This means that the injectivity is a function of the total mobility of the connected grid block.

Wells connected to multiple grid blocks may sometimes experience *crossflow* as illustrated in Figure 11.28. This means that fluids flow into some connections and out of other connections. Injectivities must therefore be modified so that they represent the mixture of the fluids inside the wellbore. To this end, we first compute inflow from producing connections at reference (surface/stock-tank) conditions and use this to compute the *average* wellbore mixture at reference conditions. Then, we compute the average volumetric mixture at the conditions of each injection connection and use this to define the correct mobility. A similar average wellbore mixture is used to calculate the pressure drop between the well's datum point and each connection.

11.7.2 Multisegment Wells

Many wells are designed so that the reservoir fluids do not flow directly from the sandface and into the wellbore. Oftentimes, the wellbore (production casing) lies inside another casing that has been cemented in place inside the drilled hole. The void *annular* space between

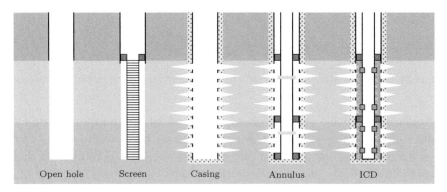

Figure 11.29 Conceptual illustration of different types of well completions.

the casing and the wellbore is limited by sealing packers, which limit the annular space into sections, and sleeves that provide flow paths between the annulus and the wellbore. One can also have various types of screens that aim to reduce/eliminate the inflow of solid particles that would cause erosion inside the wellbore. Simple versions of *sand screens* are made up of ribs, rods, or wire filters, possibly filled with gravel, whereas more modern ones consist of high-permeable, resin-coated gravel packed between two layers of wire-wrap filter media (*gravel pack*). In addition, wells may have various components that can partially or fully choke the inflow to prevent unwanted fluids to enter the wellbore. *Inflow control devices* (ICDs) are passive or autonomous devices installed along the completion interval to optimize inflow along the length of the wellbore. These devices are designed to exert higher back pressure at high flow rates and can be used to equalize inflow from high and low-permeable zones, or delay water or gas breakthrough across selected intervals of horizontal wells. *Inflow control valves* serve the same purpose and are controlled from the surface through electric or hydraulic signals.

To further increase complexity, modern wells typically have directional profiles designed to intersect multiple targets with a single wellbore. Such wells can have multiple production zones, or consist of multiple deviated or horizontal branches (*multilaterals*). To increase the productivity of a well, it is common to install *downhole pumps* that lower the wellbore flowing pressure so that fluids can be drawn from the reservoir at a higher rate. If the downhole pressure is too low, it may be necessary to apply *artificial gas lift*, in which gas is injected downhole to lower the density of the flowing fluids and use gas expansion to help the produced fluids to the surface. Most oil wells require artificial lift at some point to ensure that they keep flowing when their bottom-hole pressure decreases. Wells may also incorporate equipment for downhole compression and separation of fluid phases, which may be used to separate gas (or water) so that it can be reinjected downhole. Downhole processing can also be introduced to reduce the size and weight of surface or seafloor facilities or to overcome bottlenecks in existing surface processing systems.

All these different components will induce pressure drawdown and/or changes in the composition of the flowing fluids. To model this complex behavior, one may use a so-called *multisegment well model*, which can be thought of as a flow network consisting of a set of nodes connected by a set of segments (edges). Each node n has an associated volume and can store fluids, and fluids can flow along each segment connecting two nodes as described by a set of one-dimensional flow equations.

Continuous Flow Equations

The one-dimensional flow equations can either be formulated using phase velocities $\vec{v}_\alpha$ and volume fractions y_α, or component *mixture velocities* $\vec{v}_c^m$ and mass fractions x_c for the gas, oil, and water components. Capillary pressure is negligible inside the wellbore, hence we can write the conservation equations for volumetric quantities as follows:

$$\partial_t(b_w y_w) + \partial_x(b_w v_w) - b_w q_w = 0,$$
$$\partial_t(b_o y_o + b_g R_v y_g) + \partial_x(b_o v_o + b_g R_v v_g) - b_o q_o - b_g R_v q_g = 0, \quad (11.23)$$
$$\partial_t(b_g y_g + b_o R_s y_o) + \partial_x(b_g v_g + b_o R_s v_o) - b_g q_g - b_o R_s q_o = 0.$$

The phase velocities do not relate to pressure gradients through Darcy's law, but rather through some nonlinear relation modeling friction, acceleration, inflow devices, valves, and so on. We can write this relationship as

$$\nabla p - g\rho \nabla z - G(v, \rho, \mu) = 0, \quad (11.24)$$

where ρ, v, and μ are the mixture density, velocity, and viscosity, respectively, and G represents the (nonlinear) part of the pressure drop that cannot be explained by a Darcy relationship. The equivalent set of equations on mass form reads,

$$\partial_t(x_c \rho) + \nabla \cdot \vec{v}_c^m - q_c^m = 0, \quad c \in \{g, o, w\}. \quad (11.25)$$

Here, q_c^m is the mass source term of component c.

Discrete Well Equations

To define the discrete multisegment well models, we utilize a vector notation analogous to the discrete operators we introduced in Section 4.4.2 for flow equations in the porous medium. Where we previously had two mappings F between cells and their bounding cell faces and C_1, C_2 between faces and adjacent cells, we now have similar mappings S from nodes to all connected segments and N_1, N_2 from segments to the adjacent nodes; see Figure 11.30. Given these mappings, the *discrete gradient operator* reads

$$\mathtt{grad}(v)[s] = p[N_2(s)] - p[N_1(s)],$$

and similarly the *discrete divergence operator* reads

$$\mathtt{div}(v)[n] = \sum_{s \in S(n)} v[s] \mathbf{1}_{\{n = N_1(s)\}} - \sum_{s \in S(n)} v[s] \mathbf{1}_{\{n = N_2(s)\}}.$$

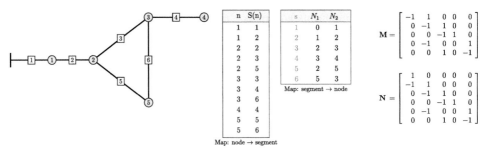

Figure 11.30 Example of well topology and corresponding discrete operators.

These operators can both be represented as sparse matrices, almost like the D matrix. However, the top segment of a well needs special treatment, since the top node corresponds to the reference depth at which the well's bottom-hole pressure is defined. Hence, instead of calculating a pressure drop across the top segment, we impose a boundary condition on the top node corresponding to the correct control of the well. This means that we write the operators as $\mathrm{grad}(p) = Mp$ and $\mathrm{div}(v) = -N^T v$, where M equals N except that the first row is removed. Discrete averaging operators are defined analogously. Given these operators, the discrete flow equations are defined as follows: If we use the mass form, the conservation equation in each node reads

$$\frac{V}{\Delta t}\left[x_c \rho - (x_c \rho)^0\right] + \mathrm{div}(v_c^m) - q_c^m = 0, \tag{11.26}$$

where superscript 0 refers to values at the previous time step. The equation for the pressure drop reads,

$$\mathrm{grad}(p) - g\,\mathrm{upw}(\rho)\mathrm{grad}(z) - G\big(v^m, \mathrm{upw}(\rho), \mathrm{upw}(\mu)\big) = 0. \tag{11.27}$$

Here, $\mathrm{grad}(z)$ is the discrete gradient of the node depth, i.e., the depth difference between neighboring nodes. Note that (11.27) is not defined for the topmost segment and we need to specify a boundary or control equation at the top to close the coupled system (11.26)–(11.27).

Pressure-Drop Relations

The function G in (11.24) can be used to impose almost any pressure drop. For general network topologies, however, we assume that $\partial G/\partial v^m$ is nonzero. As an example, the pressure drop over a *simple valve* can be modeled as

$$\Delta p = \frac{\rho u_c^2}{2 C_v^2}, \tag{11.28}$$

where u_c is the velocity in the constriction and C_v is the discharge coefficient. Another example is the *frictional pressure drop* along the wellbore, which may have a significant effect on the behavior of (long) horizontal wells. In ECLIPSE 100 [271], this pressure loss is given by

$$\Delta p = \frac{2fL}{D}\rho u^2. \quad (11.29)$$

Here, f is the Fanning friction factor, L is the length of the segment, D is the corresponding tubing diameter, ρ is the mixture density, and u is the mixture velocity. For a mixture mass flux v^m or mixture volume flux v, the velocity u is given by

$$u = \frac{v^m}{\rho\pi(D/2)^2} = \frac{v}{\pi(D/2)^2}. \quad (11.30)$$

For annular pressure drop, $(D/2)^2$ is replaced by $(D_o/2)^2 - (D_i/2)^2$, where D_o and D_i are the diameters of the outer and inner tubing, respectively. The Fanning friction factor depends on the Reynolds number $\text{Re} = \rho u D/\mu$ and is calculated as

$$\sqrt{f^{-1}} = -3.6\log_{10}\left[\frac{6.9}{\text{Re}} + \left(\frac{e}{3.7D}\right)^{10/9}\right], \quad (11.31)$$

for $\text{Re} > 4{,}000$, and $f = 16/\text{Re}$ in the laminar region for $\text{Re} < 2{,}000$. In between, one interpolates between the corresponding end values for Re equal 2,000 and 4,000. MRST uses the same approach, but also allows use of (11.31) for all velocities. The coefficient e gives roughness of the tubing.

11.8 Black-Oil Simulation with MRST

Altogether, we have discussed a great number of details that must be brought together to make a simulator. In addition, there are still a number of other important topics like preconditioning and details of boundary conditions, time chopping, nonlinear solution strategy, and so on, that I have *not* discussed in detail. As stated in the introduction of the chapter, I consider many of these details to be outside the scope of the book and I hence feel that time is right to switch the presentation to the AD-OO framework, which implements all the necessary ingredients entering industry-standard reservoir simulators. I therefore end the chapter by first briefly outlining how you can use the AD-OO framework to simulate the SPE 1 benchmark model and then discuss some limitations and potential pitfalls that you, as a user of MRST and other reservoir simulators, should be aware of. The next chapter will give a more systematic overview of the AD-OO framework and discuss more simulation examples.

11.8.1 Simulating the SPE 1 Benchmark Case

You have already encountered the fluid model from the SPE 1 benchmark [241] multiple times in this chapter. The benchmark describes a problem with gas injection in a small $10 \times 10 \times 3$ reservoir with a producer and an injector placed in diagonally opposite corners. The porosity is uniform and equals 0.3, whereas the permeability is isotropic with values

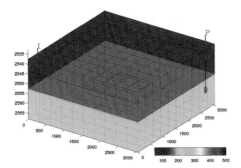
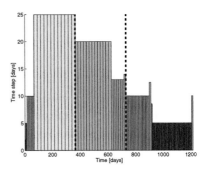

Figure 11.31 Horizontal permeability and well positions for the SPE 1 benchmark case. The injector is perforated in the upper and the producer in the lower layer. The right plot shows the time steps in the simulation schedule.

500, 50, and 200 md in the three layers with thickness 20, 30, and 50 ft; see Figure 11.31. The reservoir is initially undersaturated with constant pressure in each layer and a uniform mixture of water ($S_w = 0.12$) and oil ($S_o = 0.88$) with no initial free gas ($S_g = 0.0$) and a constant solution gas–oil ratio ($R_s \approx 226.2$) throughout the model. Although the problem is really a two-phase gas–oil problem describing gas injection into an undersaturated oil, the input file describes a three-phase problem with almost stationary water. The original problem was posed to study ten years of production, but herein we only report and compare solutions for the first 1,216 days.

We have already discussed the steps to load the ECLIPSE input data and create the necessary objects representing grid, petrophysical and fluid properties in detail earlier in the book. For completeness, however, we include them all here:

```
gravity reset on
deck  = readEclipseDeck(fullfile(getDatasetPath('SPE1'), 'BENCH_SPE1.DATA'));
deck  = convertDeckUnits(deck);
G     = computeGeometry(initEclipseGrid(deck));
rock  = compressRock(initEclipseRock(deck), G.cells.indexMap);
fluid = initDeckADIFluid(deck);
```

In addition, we must set up an initial state. The SOLUTION section in the input file specifies how the reservoir should be initialized. Here, we hard-code the fluid equilibrium state for simplicity:

```
[k, k] = ind2sub([prod(G.cartDims(1:2)), G.cartDims(3)], G.cells.indexMap);
p0     = [329.7832774859256 ; ...   % Top layer
          330.2313357125603 ; ...   % Middle layer
          330.9483500720813 ];      % Bottom layer
p0     = convertFrom(p0(k), barsa);
s0     = repmat([ 0.12, 0.88, 0.0 ], [G.cells.num, 1]);
rs0    = repmat( 226.1966570852417 , [G.cells.num, 1]);
rv0    = 0; % dry gas

state  = struct('s', s0, 'rs', rs0, 'rv', rv0, 'pressure', p0);
```

11.8 Black-Oil Simulation with MRST

The first statement creates a partition vector containing the layer index, which we use to lookup in the vector of equilibrium pressures values. All the statements just presented are collected in a utility function `setupSPE1` in the `ad-blackoil` module.

The next step is to create an object representing the whole *simulation model* with description of reservoir geometry, petrophysical and fluid parameters, drive mechanisms, and parameters determining the correct flow equations

```
model = selectModelFromDeck(G, rock, fluid, deck);
```

Technically, the `model` object is an instance of a *class* defined using `classdef` and includes both properties and methods that can operate on input parameters and the data contained within the object. Classes can also define events and enumerations, but these are not used in AD-OO. If you are not familiar with the concept of class objects, you can think of `model` as a standard `struct` containing both data members and function pointers. In this particular case, the input data will create a three-phase, black-oil model with dissolved gas but no vaporization of oil. The following is a list of a few of the data members:

```
ThreePhaseBlackOilModel with properties:
             disgas: 1
             vapoil: 0
              fluid: [1x1 struct]
               rock: [1x1 struct]
              water: 1
                gas: 1
                oil: 1
            gravity: [0 0 9.8066]
       FacilityModel: [1x1 FacilityModel]
          operators: [1x1 struct]
                  G: [1x1 struct]
```

The fields `G`, `rock`, and `fluid` are the standard MRST structures we have used earlier to represent grid, petrophysics, and fluid properties. The `operators` structure contains transmissibilities, pore volume, and discrete operators and is the same structure we referred to as s in the previous section. `FacilityModel` is an instance of another class designed to describe models of wells and surface facilities. Last but not least, the object contains five flags telling which of the three phases are present and whether gas is allowed to dissolve in oil and oil is allowed to vaporize into gas.

We have already encountered the flag `vapoil` in the code that computed gas properties on page 392. Likewise, `disgas` and `vapoil` were used to determine the correct primary unknown for the gas phase on page 393. These, and all the other code pieces for computing flow equations on residual form presented in Section 11.6, are excerpts from the function `equationsBlackOil` and a number of helper functions and have been simplified and reorganized for pedagogical purposes. The AD-OO framework never calls `equationsBlackOil` directly, but instead accesses it indirectly through a generic member function called `getEquations`. This way, the parts of the framework that implement the time loop and the nonlinear solver does not need to know any specifics of the model as long as we know that the model exists and that calling `getEquations` will compute the corresponding flow equations on residual form with automatic differentiation activated for all primary unknowns. In addition to the parameters just outlined, the model contains a number of other parameters that control the nonlinear solver, chopping of time steps, and output of results. More details about different model classes, time loops, nonlinear and linear solvers, are given in Chapter 12.

The next thing we must do is create a *simulation schedule*, which essentially consists of a set of time steps and a specification of the drive mechanisms that will be present during each of these time steps. These are described in the SCHEDULE section of the input deck, which for SPE 1 declares two different wells:

```
-- WELL SPECIFICATION DATA
--
--   WELL GROUP LOCATION  BHP   PI
--   NAME NAME    I  J   DEPTH DEFN
WELSPECS
    'P'  'G'   10 10    8400 'OIL' /
    'I'  'G'    1  1    8335 'GAS' /
/
```

```
-- COMPLETION SPECIFICATION DATA
--
-- WELL -LOCATION-  OPEN/ SAT CONN WELL
--  NAME  I  J K1 K2 SHUT TAB FACT DIAM
COMPDAT
    'P'  10 10 3  3  'OPEN' 0  -1  0.5 /
    'I'   1  1 1  1  'OPEN' 1  -1  0.5 /
/
```

The WELSPECS keyword shown to the left introduces the wells, defines their name, the position of the well head, the reference depth for the bottom holes, and the preferred phase. The keyword COMPDAT shown to the right, specifies how each well is completed, i.e., in which cells each well can inject/produce fluids. The next we need to define is how the wells are controlled during the simulation:

```
-- PRODUCTION WELL CONTROLS
--   WELL OPEN/  CNTL   OIL   WATER  GAS   LIQU   RES   BHP
--   NAME SHUT   MODE   RATE  RATE   RATE  RATE   RATE
WCONPROD
    'P'  'OPEN' 'ORAT' 20000  4*                        1000 /
/

-- INJECTION WELL CONTROLS
--   WELL INJ   OPEN/  CNTL   FLOW
--   NAME TYPE  SHUT   MODE   RATE
WCONINJE
    'I'  'GAS' 'OPEN' 'RATE' 100000 100000 50000/
/
```

The specification says that the producer should be controlled by oil rate, whereas the injector should inject gas and be controlled by rate. To turn this input into a data object describing the simulation schedule in a form that MRST understands, we issue the following command:

```
schedule = convertDeckScheduleToMRST(model, deck);
```

which parses the deck structure and creates the following data structure:

```
schedule =                schedule.step =              schedule.control =
    step: [1x1 struct]        control: [120x1 double]       W: [2x1 struct]
    control: [1x1 struct]         val: [120x1 double]      bc: []
                                                          src: []
```

The schedule has two data members: a list of steps and a list of controls. The controls specify a set of *targeted*[2] drive mechanisms (wells, boundary conditions, and source terms) that each is unique and will be applied during at least one step. Each step consists of a time-step length (`val`) and the number (`control`) of the corresponding control. In our case, there is only one control consisting of the injection and production wells described previously (see Figure 11.31), which will be active during each of the 120 time steps. The control steps are given by the keyword TSTEP, which can be repeated to specify the full schedule (see Figure 11.31):

```
TSTEP                                       -- INITIAL STEP: 1 DAY, MAXIMUM: 6 MONTHS
1.0 2*2.0 2*5.0 5*10.0 11*25.0              TUNING
 /                                          1    182.5  /
TSTEP                                       1.0 0.5 1.0E-6 /
  25.0                                      /
 /
```

The TUNING keyword gives parameters for time-step control, time truncation and convergence control (here: mass-balance error $\leq 10^{-6}$), and control of Newton and linear iterations (not given here). This keyword is ignored by MRST.

To run the schedule, we must select a nonlinear solver, which is implemented as a class in the AD-OO framework:

```
nls = NonLinearSolver('useLinesearch', true);
```

The NonLinearSolver class implements an abstract framework for nonlinear solvers, which uses the Newton linearization (or some other mechanism like a fix-point step) provided by the model to iterate towards a solution. It is capable of time-step selection and chopping based on convergence rates and can be extended via subclassing with alternative linear solvers and time-step classes. Convergence is handled by the model. The NonLinearSolver simply responds based on what the model reports in terms of convergence to ensure some level of encapsulation. In the current example, we initialize the solver class to use line-search, as discussed in Section 10.2.2. The excerpts from the `nls` object

```
NonLinearSolver with properties:
          maxIterations: 25
        maxTimestepCuts: 6
           LinearSolver: [1x1 BackslashSolverAD]
        timeStepSelector: [1x1 SimpleTimeStepSelector]
           useLinesearch: 1
                verbose: 0
                      :
```

show that the nonlinear solver will at most use 25 iterations, be allowed to cut the time step six times before giving up. The solver will use the BackslashSolverAD linear solver and SimpleTimeStepSelector to select time steps, use line-search as specified, run in non-verbose mode so that no details about convergence are reported to screen, and so on.

---

[2] This is an important distinction: the schedule may for instance request that a given well is operated according to a prescribed rate (*rate control*). However, if this rate cannot be achieved without violating physical limits on the bottom-hole pressure, the simulation will switch to using a bottom-hole control instead.

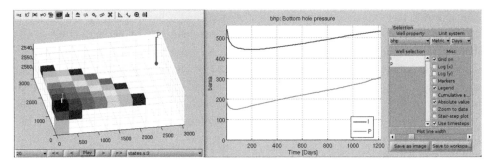

Figure 11.32 Graphical user interfaces for plotting reservoir data and well responses. The left plot shows the saturation of free gas after 20 time steps. The right plot shows the variation in bottom-hole pressure in the injector and producer throughout the simulation.

We are now ready to run the full simulation. We specify that we wish output of well solutions, reservoir states, and a report with details about time steps, convergence, etc.:

```
[wellSols, states, report] = ...
    simulateScheduleAD(state, model, schedule, 'nonlinearsolver', nls);
```

```
Solving timestep 001/120:                      -> 1 Day
Solving timestep 002/120: 1 Day                -> 3 Days
   :                        :                      :
Solving timestep 120/120: 3 Years, 111 Days    -> 3 Years, 121 Days
*** Simulation complete. Solved 120 control steps in 103 Seconds .. ***
```

To inspect the solutions, we issue the following commands, which will bring up the two graphical user interfaces shown in Figure 11.32. First, the plot of well responses

```
plotWellSols(wellSols, report.ReservoirTime)
```

and then the plot of reservoir states

```
figure;
plotToolbar(G, states)
plotWell(G, schedule.control(1).W,'radius',.5)
axis tight, view(-10, 60)
```

Let us study the dynamics of the reservoir in some detail. Figure 11.33 shows how the bottom-hole pressure in the injector and the solution gas–oil ratio inside the reservoir evolve as functions of time. Before production starts, the reservoir is in an undersaturated state, shown as a green marker in the plots of rs as function of p0. Initially, the injector needs a relatively high bottom-hole pressure (bhp) to force the prescribed gas rate into the reservoir. Because the reservoir is undersaturated, the injected gas will dissolve in the oil phase, so that the viscosity decreases. A lower bhp is thus required to maintain the injection rate. As the injection continues, cells near the well will gradually become saturated so that free gas is liberated and starts accumulating at the top of the reservoir. Similarly, some free

11.8 Black-Oil Simulation with MRST

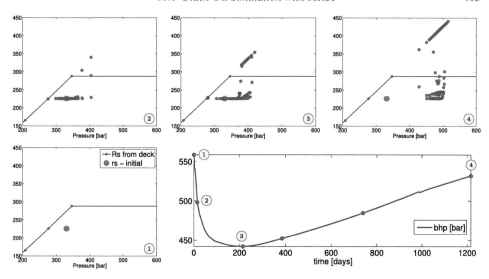

Figure 11.33 Reservoir dynamics of the SPE 1 case during the first three years. The lower-right plot shows the bottom-hole pressure of the injector as function of time. The other plots show the solution gas–oil ratio as function of pressure for all cells in the reservoir at four different instances in time.

gas will be liberated near the producer because of the pressure drawdown. As more gas is injected into the reservoir, more cells transition from an undersaturated to a saturated state, giving an increasing plume of free gas that expands out from the injector along the top of the reservoir (see the left plot in Figure 11.32). The total injection rate is higher than the production rate, and hence the pressure in the reservoir will increase, and after approximately 200 days, we see that the injector bhp starts to increase again to maintain the specified rate.

11.8.2 Comparison against a Commercial Simulator

To build confidence in our simulator, we can verify the simulation in the previous subsection against ECLIPSE. To compare the two, we first extract surface rates and bottom-hole pressures from MRST as plain arrays:

```
[qWs, qOs, qGs, bhp] = wellSolToVector(wellSols);
```

We also extract time steps and determine indices for injector and producer:

```
T    = convertTo(cumsum(schedule.step.val), year);
inj  = find([wellSols{1}.sign] == 1);
prod = find([wellSols{1}.sign] == -1);
```

Output from ECLIPSE is usually stored in so-called *summary/restart files*, which can be used to reinitialize a simulation so that it can be continued. The `deckformat` module in MRST offers useful functionality for reading and interpreting such (binary) output files (see

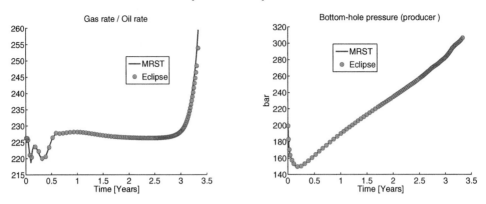

Figure 11.34 Comparison of well responses for the SPE 1 benchmark simulated by MRST and ECLIPSE 100.

`readEclipseSummaryFmt` and `readEclipseSummaryUnFmt`). To simplify the comparison for the SPE 1 benchmark, the data set supplied with MRST includes a `*.mat` file with selected well responses obtained by simulating the exact same input deck with ECLIPSE 100. These data are stored as a MATLAB structure, which also contains a few functions for extracting the raw data. We load the summary

```
load SPE1_smry
ind = 2:size(smry.data,2);
Te  = smry.get(':+:+:+:+', 'YEARS', ind);
```

Here, we have ignored the first entry, which contains zero values. Figure 11.34 shows a comparison of producer gas–oil ratio and bottom-hole pressure in the producer. These are extracted as follows:

```
eGOR  = convertFrom(smry.get('PRODUCER', 'WGOR', ind), 1000*ft^3/stb)';
eBHPp = convertFrom(smry.get('PRODUCER', 'WBHP', ind), psia)';
eBHPi = convertFrom(smry.get('INJECTOR', 'WBHP', ind), psia)';

mGOR  = qGs(:,prod)./qOs(:,prod);
mBHPp = bhp(:,prod);
mBHPi = bhp(:,inj);
```

As we see from the figure, there is (almost) perfect match between MRST and ECLIPSE, which is what we should expect, given that the two simulators implement (almost) exactly the same numerical methods.

11.8.3 Limitations and Potential Pitfalls

Numerical flow models based on the black-oil equations and their like are complex creatures and should be treated cautiously to ensure that the results they produce make sense.

A commercial simulator usually has a large number of precautions built into it, so that it on one hand will refuse to simulate unphysical situations, but on the other hand manages to give an answer even though the user gives incomplete or inconsistent data. (You may also recall our discussion of time-step chops implemented to ensure convergence when faced with severe nonlinearity, which potentially *could* be the result of an inadequate model or incorrect parameters.) Being aware of limitations and potential pitfalls is important, in particular when using a prototyping tool like MRST that has very few precautions built into it, but also when using mature commercial tools, whose alluring robustness may sometimes be deceptive. In this section, I will try to review a few potential issues and comment on possible precautions.

When Can Black-Oil Models be Applied?

Any model is no better than its assumptions, and it is important to know what the limitations of a model are so that it is not used for the wrong purposes. Black-oil models can, in principle, be applied to simulate all the five types of reservoirs outlined in Section 11.4.2, as long as these are produced by depletion or waterflooding. As a general rule of thumb, you should be more precautious when applying black-oil models to gas injection processes that involve significant changes in the composition of the oil and gas pseudo-components during the displacement. Vaporization and retrograde condensation in rich gases are examples of such processes, and here you should be careful to check the validity of black-oil models against more comprehensive compositional models. To ensure that the black-oil description is sufficiently accurate, the recovery process should satisfy the following characteristics [269]:

- The recovery process should be isothermal.
- The path taken by the reservoir fluids should be far from the critical point.
- The amount of liberated gas or condensate dropout should be a small part of the hydrocarbon in place.
- The remaining hydrocarbon composition should not change significantly when gas is liberated or vaporized condensate drops out.

If these conditions are not satisfied, you should consider using a compositional simulator.

Consistent PVT Parameters

PVT parameters used in black-oil models were originally derived from laboratory analysis of (recombined) reservoir fluids, but in recent years it has become more common to derive the necessary pressure-dependent parameters by simulating differential liberation experiments by use of compositional models with appropriate equations of state. Regardless of their source, the derived PVT model must satisfy certain physical properties; see [73]. First of all, the total compressibility of the system must be non-negative, i.e.,

$$c_t = S_w c_w + S_o c_o + S_g c_g + c_r \geq 0,$$

where c_r is the rock compressibility and c_α is the isothermal compressibility of phase α,

$$c_w = -\frac{B'_w}{B_w}, \qquad c_o = -\frac{B'_o}{B_o} + \frac{(B_g - B_o R_v) R'_s}{B_o(1 - R_s R_v)},$$

$$c_g = -\frac{B'_g}{B_g} + \frac{(B_o - B_g R_s) R'_v}{B_g(1 - R_s R_v)}.$$

Since this property has to be valid for all saturations, it follows that each of the isothermal compressibilities must also be nonnegative, which gives the following three consistency conditions:

$$B'_w \leq 0,$$
$$(1 - R_s R_v) B'_o \leq (B_g - R_v B_o) R'_s,$$
$$(1 - R_s R_v) B'_g \leq (B_o - R_s B_g) R'_v. \qquad (11.32)$$

To ensure that the partial volumes in (11.11) are also positive, the following additional conditions must be satisfied:

$$R_s R_v < 1, \qquad B_w > 0, \qquad B_g - R_v B_o > 0, \qquad B_o - R_s B_g > 0. \qquad (11.33)$$

As we have seen in Section 11.4, B_o, R_s, and R_v are normally increasing functions with positive derivatives, whereas B_g and B_w are decaying functions with negative derivatives.

Use of Tabulated Data

Use of tabulated data, as discussed earlier in Section 11.5, is very common in industry-grade simulation models. This means that when the simulator requires values between data points, these need to be interpolated. Likewise, values requested outside the range of tabulated data need to be extrapolated. Algorithmic choices of how data are interpolated and extrapolated are specific to the individual simulator, but can have a significant impact on both the accuracy and stability of the simulator, and the user should therefore be aware of what methods are used.

The default choice in ECLIPSE is to use linear interpolation and zeroth order extrapolation for rock-fluid properties, so that the simulator picks the value of the first or last tabulated point rather than using trend interpolation based on the closest points. This means that *derivatives* of tabulated quantities are piecewise constants and effectively are computed using finite-differences between tabulated points. PVT properties are also interpolated within their range of definition, but properties like R_s and R_v may also be extrapolated. In MRST, we have chosen to follow the interpolation conventions from ECLIPSE in the ad-props module to be able to produce simulation results that are as identical as possible when using the same discretizations and solvers. This raises several issues you should be aware of:

- First of all, you should make sure that the data points cover the value range of the simulation. For PVT properties, this usually means that data should be defined for higher pressures than what are initially found in the reservoir, since pressure can increase during the simulation as a result of fluid injection. In ECLIPSE 100, you can use the keyword EXTRAPMS to force the simulator to issue a warning whenever it extrapolates outside of tabulated data. At the time of writing, such a safeguard is not yet implemented in MRST.

- To reduce numerical errors from the numerical interpolation, make sure that data points are sufficiently dense and distributed so that interpolation errors do not vary too much in magnitude throughout the parameter range.
- Try to avoid discontinuities, strong kinks, and other abrupt changes in tabulated properties unless these are dictated by physics. (See, e.g., the relative permeabilities for SPE 9 in Figures 11.6 and 11.7.) If necessary, introduce extra data points near strong nonlinearities and local extremal points to avoid introducing unphysical oscillations and too-large values for derivatives. This will generally improve the convergence of the nonlinear solver and reduce the number of (unnecessary) time-step chops.
- On the other hand, introducing too many points will make table lookup more expensive and contribute to slow down your simulation. For MRST, you should be particularly beware of multiple regions. Property evaluation is vectorized over all cells to make it as efficient as possible, but with multiple relperm or PVT regions, each region must be processed sequentially in a for loop, and this can add a significant computational overhead if the number of regions is large.
- In certain cases, ECLIPSE interpolates functions of the input data or functional *combinations* of multiple properties, rather than the tabulated values. The primary example is the PVT data for oil. Here, ECLIPSE 100 interpolates reciprocals of B_o and $\mu_o B_o$, whereas ECLIPSE 300 interpolates reciprocals of B_o, but interpolates μ_o directly. Using different interpolations may have a significant effect on simulation results.

To give you an idea of how things can go wrong, let us revisit two of the SPE benchmark models, starting with SPE 3. To sample the model, we make a 101×21 regular mesh of (p, R_v) values covering the rectangular domain $[p_0, p_1] \times [0, R_v(p_1)]$ for $p_0 = 80$ and $p_1 = 280$ bar. In the input data, properties are only defined for pressures in the interval $[83.75, 237.82]$ bar, and hence we need to use both interpolation and extrapolation to compute values. Figure 11.35 reports the computed gas viscosities for values toward the upper end of the pressure interval. Here, the left plot clearly shows that the viscosity is not necessarily a monotone function of pressure for fixed R_v. More important, the right plot shows that we get incorrect behavior with highly unrealistic viscosities for R_v values less than $R_v(p)$ once the pressure exceeds maximum data values by more than a few percent.

Figure 11.36 shows that the PVT model in SPE 1 P violates the consistency condition $B_o - R_s B_g > 0$ in (11.33) for high pressure values, as observed previously by Coats [73]. By convention, the formation-volume factor B_o of oil is computed by interpolating the shrinkage factor b_o rather than the tabulated B_o values. In the tabulated data, b_o declines with increasing R_s values and will hence cross zero for sufficiently high values. This leads to blowup and negative values for B_o. This behavior is only observed for (p, R_s) values that are larger than what is to be expected in practical computations, but it is still interesting to note this artifact.

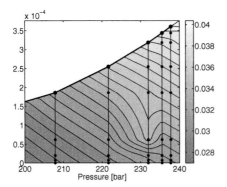 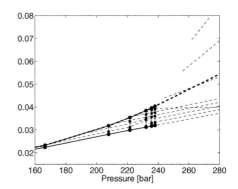

Figure 11.35 Interpolated and extrapolated gas viscosity for the SPE 3 fluid model. The left plot shows contours of μ_g over parts of the (p, R_v) domain, with data points shown as filled circles. The right plot shows μ_g as function of pressure only, with dashed lines corresponding to constant R_v values.

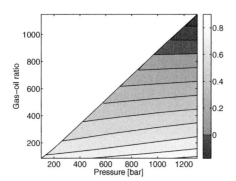 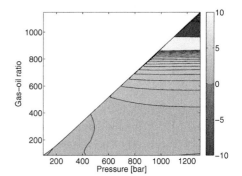

Figure 11.36 Inconsistency in the PVT data of the SPE 1 benchmark giving negative partial volumes. The left plot shows the interpolated shrinkage factor b_o, whereas the right plot shows the resulting values of $B_o - R_s B_g$. To emphasize the blowup and the negative values, the color scale has been cropped to $[-10, 10]$, but the actual values are up to two orders larger.

Initialization and Equilibration

In all examples studied so far in the book, the initial saturation has either been a constant, or the fluid phases have been uniformly segregated throughout the reservoir. In a real reservoir, the initial fluid distribution will usually be more complicated, as we saw for the heterogeneous capillary fringe in Section 10.3.7. Setting the correct initial state is essential to obtain accurate estimates of fluids in place and predict how the reservoir will behave under production. ECLIPSE input offers two ways of setting up the initial state:

- you can either input individual cell values of the initial state using the keywords PRESSURE, SWAT, SGAS, and RS or PBUB in an ECLIPSE input file, or
- you can calculate the initial state equilibration, i.e., determine it on the basis of hydrostatic equilibrium.

11.8 Black-Oil Simulation with MRST

At the time of writing, MRST's support of initialization is somewhat unfinished. All cells in a grid can be assigned specific pressure and saturation values given in a deck by the function `initStateDeck(model,deck)`. The function also performs a full equilibration initialization, proceeding in two steps

```
regions = getInitializationRegionsDeck(model, deck);
state0 = initStateBlackOilAD(model, regions);
```

The first call extracts the different subregions from the REGIONS section of the input deck, whereas the second call computes vertical equilibrium inside each region. This part of the software will be unified in the near future.

Let us discuss this initialization in some more detail. In the previous chapter we saw how variations in rock properties can cause variations in capillary pressures. Observed fluid contacts (gas–oil and oil–water) generally show vertical variation and will not always be planar. To determine the free fluid contacts, we must know the phase pressures inside the reservoir. The level of free gas is then defined as the depth at which the gas and oil pressures coincide, i.e., the depth at which the oil–gas capillary pressure is zero. The level of free water is defined analogously.

Typical input data to an equilibration calculation consist of pressure at a datum depth, gas–oil and oil–water contacts, capillary pressure at each contact, and gas–oil ratio (R_s) or bubble-point pressure (p_b) as a function of depth. Given these data, it is, in principle, a straightforward task to determine the fluid distribution, which generally will consist of five different zones:

1. A gas cap, in which gas is the only continuous phase ($S_g = 1 - S_{wi}$)

$$\frac{dp_g}{dz} = g\rho_g, \quad p_o = p_g - P_{cog}(S_g), \quad p_w = p_o - P_{cow}(S_{wi}).$$

2. An oil–gas transition zone, in which both gas and oil are continuous:

$$\frac{dp_g}{dz} = g\rho_g, \quad \frac{dp_o}{dz} = g\rho_o, \quad p_w = p_o - P_{cow}(S_{wi}).$$

The saturations are $S_w = S_{wi}$, $S_g = P_{cog}^{-1}(p_g - p_o)$, and $S_o = 1 - S_g - S_{wi}$.

3. An oil zone, in which oil is the only continuous phase ($S_o = 1 - S_{wi}$):

$$\frac{dp_o}{dz} = g\rho_o, \quad p_g = p_o + P_{cog}(0), \quad p_w = p_o - P_{cow}(S_{wi})$$

4. An oil–water transition zone, in which both oil and water are continuous:

$$\frac{dp_o}{dz} = g\rho_o, \quad \frac{dp_w}{dz} = g\rho_w, \quad p_g = p_o + P_{cog}(0).$$

The saturations are $S_g = 0$, $S_w = P_{cow}^{-1}(p_o - p_w)$, and $S_o = 1 - S_w$.

5. A water zone, in which only water is present:

$$\frac{dp_w}{dz} = g\rho_w, \quad p_o = p_w + P_{cow}(1), \quad p_g = p_o - P_{cog}(0).$$

In practice, however, the equilibration procedure is complicated by the fact that the fluid contacts may vary between different compartments in the reservoir and that capillary pressure curves may vary from one cell to the next. Notice also that numerical artifacts and inaccuracies may arise if the grid is not aligned with the fluid contacts; see e.g., [251].

Once your model is initialized, it is a good precaution to run a simulation with no external drive mechanisms applied to see if the reservoir contains any movable fluids, i.e., check whether the initial phase and pressure distributions remain invariant in time.

12
The AD-OO Framework for Reservoir Simulation

In Chapter 7 we showed that combining a fully implicit formulation with automatic differentiation makes it simple to extend basic flow models with new constitutive relationships, extra conservation equations, new functional dependencies, and so on. By using numerical routines and vectorization from MATLAB, combined with discrete differential and averaging operators from MRST, these equations can be implemented in a very compact form close to the mathematical formulation. Writing a simulator as a single script, like we did for single-phase flow in Chapter 7 or for two-phase flow without phase transfer in Section 11.2, has the advantage that the code is self-contained, quick to implement, and easy to modify as long as you work with relatively simple flow physics. However, as the complexity of the flow models and numerical methods increases, and more bells and whistles are added to the simulator, the underlying code will inevitably become cluttered and unwieldy.

When you research new computational methods for reservoir simulation, it is important to have a flexible research tool that enables you to quickly test new ideas and verify their performance on a large variety problems, from idealized and conceptual cases to full simulation setups. The AD-OO framework was introduced to structure our fully implicit simulators and enable us to rapidly implement new proof-of-concept codes also for more complex flow physics. Altogether, the AD-OO framework offers many of the features found in commercial simulators. Understanding its design and all the nitty-gritty details of the actual implementation will obviously provide you with a lot of valuable insight into practical reservoir simulation. If you are not interested in code details, you can read Section 12.1 and then jump directly to Section 12.4, which presents a few simulation examples so as to outline the functionality and technical details you need to set up your own black-oil simulation both with and without the use of ECLIPSE input. The sections in between give a somewhat detailed discussion of the most central classes and member functions to demonstrate the general philosophy behind the framework, but do not discuss in detail how boundary conditions, source terms, and wells are implemented. The purpose of these sections is to provide you with sufficient detail to start developing new simulators with new numerics or other types of flow physics.

12.1 Overview of the Simulator Framework

Looking back at the procedural implementations presented so far in the book, we have introduced a number of data objects to keep track of all the different entities that make up a simulation model:

- a *state object* holding the unknown pressures, saturations, concentrations, inter-cell fluxes, and unknowns associated with wells;
- the *grid structure* G giving the geometry and topology of the grid;
- a structure rock representing the petrophysical data: primarily porosity and permeability, but net-to-gross as well;
- a sparse *transmissibility vector* (or inner-product matrix), possibly including multipliers that limit the flow between neighboring cells;
- a structure fluid representing the *fluid model*, having a collection of function handles that can be queried to give fluid densities and viscosities, evaluate relative permeabilities, formation volume factors, etc.;
- additional structures that contain the global *drive mechanisms*: wells, volumetric source terms, boundary conditions;
- and an optional structure that contains *discrete operators* for differentiation and averaging.

In a procedural code, these data objects are passed as arguments to solvers or functions performing subtasks like computing residual equations. As an example, all incompressible flow solvers in the incomp family require the same set of input parameters: reservoir state, grid and transmissibilities, a fluid object, and optional parameters that describe the drive mechanisms. The transport solvers, on the other hand, require state, grid, fluid, rock properties, and drive mechanisms. Solvers for the fully implicit equation system require the union of these quantities to set up the residual equations. An obvious simplification would be to collect all data describing the reservoir and its fluid behavior in a model object, which also implements utility functions to access model behavior. Given a reservoir state, for instance, we can then query values for physical variables with a syntax of the form:

```
[p, sW, sG, rs, rv] = ...
    model.getProps(state, 'pressure', 'water', 'gas', 'rs', 'rv');
```

As we saw in the previous chapter, the criteria by which we choose s_g, r_s, and r_v as primary reservoir variable will vary from cell to cell depending on the fluid phases present. Ideally, the model object should keep track of this and present us with the correct vector of unknowns for all cells. It is also natural that the model object should be able to compute residual flow equations (and their Jacobian) by a call like

```
eqs = model.getEquations(oldState, newState, dT);
```

With a generic interface like this, you can develop nonlinear solvers that require no specific or very limited knowledge of the physical model. A next step would be to separate the

12.1 Overview of the Simulator Framework

time-stepping strategy and the linear solver from the nonlinear solvers, so that we easily can switch between different methods or replace a generic method by a tailor-made method. Moreover, a critical evaluation of the many simulation examples developed with the procedural approach shows that large fractions of the scripts are devoted to input and output of data, plotting results, etc. Likewise, implementing variations of the same model either results in code duplication or a large number of conditionals that tend to clutter the code.

A standard approach to overcome these difficulties is to use object orientation. The object-oriented, automatic-differentiation (AD-OO) framework in MRST is tailor-made to support rapid prototyping of new reservoir simulators based on fully or sequentially implicit formulations. An essential idea of the framework is to develop general model and simulator classes, and use inheritance to develop these into simulators for specific equations. The framework splits

- *physical models* describing the porous medium and the fluids flowing in it,
- *nonlinear solvers* and *time-stepping* methods,
- *linearization* of discrete equations formulated using generic discretization, averaging, and interpolation operators as discussed previously, and
- *linear solvers* for the solution of the linear system

into different *numerical contexts*. This way, we only expose needed details and enable more reuse of functionality that has already been developed.

Figure 12.1 illustrates how these contexts may appear in the main loop of a simulator. The left part of the figure shows an unsophisticated Newton loop of the type used for single-phase flow in Chapter 7, which consists of an outer loop running the time steps, and an inner loop performing the Newton iteration for each time step. The right part shows a more advanced nonlinear solver that utilizes some kind of time-control mechanism to run

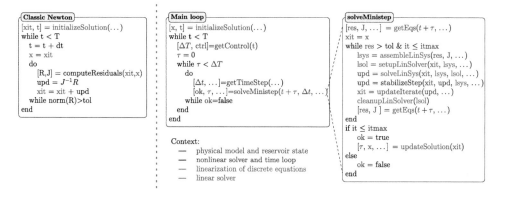

Figure 12.1 Time loop of a simulator for a simple Newton solver (left) and for a more sophisticated solver with stabilization and time-step control (right). For the latter, the iteration has been divided into various numerical contexts.

the targeted *control steps* of a simulation schedule. (Usually, the simulator should report the solution after each such step.)

In this setup, the *physical model* is responsible for initializing the reservoir state. The *nonlinear solver* implements the mechanism that selects and adjusts the time steps and performs a nonlinear iteration for this time step. Inside this nonlinear iteration, the *physical model* computes residual equations defined over all cells and possibly also cell faces and discrete entities in the well representation (connections or nodes and segments). When the residual equations are evaluated, the corresponding Jacobian matrix containing the linearized residual equations is computed implicitly by the AD library. This produces a collection of matrix blocks that each represents the derivative of a specific residual equation with respect to a primary variable. The *linearization* context assembles the matrix blocks into an overall Jacobian matrix for the whole system, and if needed, eliminates certain variables to produce a reduced linear system. The *linear solver* sets up appropriate preconditioners and solves the linear system for iteration increments, while the *nonlinear solver* performs necessary stabilization, e.g., in the form of an inexact line-search algorithm. Finally, the *physical model* updates reservoir states with computed increments and recomputes residuals to be checked against prescribed tolerances by the *nonlinear solver*.

Figure 12.2 shows how a single nonlinear solve with time-step estimation is realized in terms of classes and structures in the AD-OO framework. The `ad-core` module utilizes MRST's core functionality to implement a generic nonlinear solver, a set of generic time-step selectors, and a set of linear solvers. The nonlinear solver does not know any specific details of the physical model, except that it is able to perform the generic operations we just outlined; we will come back to model classes in the next section. The time-step selectors can implement simple heuristic algorithms that limit how much solutions can change during an iterative step, e.g., as in the Appleyard and modified Appleyard chop [271] used in commercial simulators. The linear solver class provides an interface for

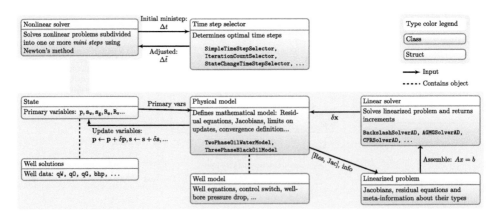

Figure 12.2 Data diagram for a single nonlinear solve with time-step estimation in the AD-OO framework.

various solvers including MATLAB's default backslash solver as well as a state-of-the-art, constrained pressure residual (CPR) preconditioner [304, 305, 119], which can be combined with efficient algebraic multigrid solvers such as the aggregation-based method found in [238], or the more recent AMGCL library [81]. Assembly of the linearized system is relegated to a special class that stores meta-information about the primary variables and the residual equations, i.e., whether they are reservoir, well, or control equations. The linear solver class employs this information to help setting up preconditioning strategies that exploit special structures in the problem.

Example 12.1.1 (1D Buckley–Leverett) *To illustrate the use of the nonlinear solver, let us revisit the classical Buckley–Leverett problem from Example 9.3.4 on page 284 and Section 10.3.1. For completeness, we show all code lines necessary to set up this problem in the AD-OO framework (which you also can find in* adBuckleyLeverett1D.m *in the* ad-core *module). First, we construct the reservoir and water–oil fluid data with constant shrinkage factors:*

```
mrstModule add ad-core ad-props ad-blackoil
G     = computeGeometry(cartGrid([50, 1, 1], [1000, 10, 10]*meter));
rock  = makeRock(G, 1*darcy, .3);
fluid = initSimpleADIFluid('phases', 'WO', 'n', [2 2]);
```

We then create a two-phase black-oil model (from ad-blackoil*) and set up the initial state:*

```
model  = TwoPhaseOilWaterModel(G, rock, fluid);
state0 = initResSol(G, 50*barsa, [0, 1]);
state0.wellSol = initWellSolAD([], model, state0);
```

Notice that the water–oil model assumes that the state object contains a subfield holding well solutions. Since there are no wells in our case, we instantiate this field to be an empty structure having the correct subfields. Finally, we set up the correct drive mechanisms: constant rate at the left and constant pressure at the right end.

```
injR = sum(poreVolume(G,rock))/(500*day);
bc   = fluxside([], G, 'xmin', injR, 'sat', [1, 0]);
bc   = pside(bc, G, 'xmax', 0*barsa, 'sat', [0, 1]);
```

Having set up the problem, the next step is to instantiate the nonlinear solver:

```
solver = NonLinearSolver();
```

By default, the solvers employs MATLAB's standard backslash operator to solve the linear problems. The time-step selector employs a simplified version of the time-chop strategy outlined in Section 10.2.2 to run the specified control steps: time steps are chopped in two if the Newton solver requires more than 25 iterations, but no attempt is made to increase successful substeps during a control step. The setup is the same as for the SPE 1 test case 11.8.1 on page 399, except that we then used inexact line search to stabilize the

Newton updates. Line search is not needed here since the fluid behavior of the Buckley–Leverett problem is much simpler and does not involve transitions between undersaturated and saturated states.

There are several ways we can use the nonlinear solver class. One alternative is to manually set up the time loop and only use the `solveTimestep` *function of the nonlinear solver to perform one iteration:*

```
[dT, n]  = deal(20*day, 25);
states   = cell(n+1, 1);  states{1} = state0;
solver.verbose = true;
for i = 1:n
    states{i+1} = solver.solveTimestep(states{i}, dT, model, 'bc', bc);
end
```

The results are stored in a generic format and can be plotted with the `plotToolbar` *GUI discussed in the previous chapter. Alternatively, we can construct a simple simulation schedule and call the generic simulation loop implemented in* `ad-core`*:*

```
schedule = simpleSchedule(repmat(dT,1,25), 'bc', bc);
[~,sstates] = simulateScheduleAD(state0, model, schedule);
```

This loop has an optional input argument that sets up dynamic plotting after each step in the simulation, e.g., to monitor how the well curves progress during the simulation, or to print out extra information to the command window. Here, we use this hook to set up a simple graphical user interface that visualizes the progress of the simulation and enables you to modify the time steps interactively, dump the solution to workspace, and stop the simulation and continue running it in debug mode. The GUI may in turn call any or both of the two GUIs shown in Figure 11.32 to plot reservoir states and well solutions. We request 1D plotting of the water saturation, and that we turn off plotting of wells, since the example does not have any:

```
fn = getPlotAfterStep(state0, model, schedule, ...
    'plotWell', false, 'plotReservoir', true, 'field', 's:1', ...
    'lockCaxis',true, 'plot1d', true);

[~,sstates,report] = ...
    simulateScheduleAD(state0, model, schedule,'afterStepFn', fn);
```

Figure 12.3 shows a snapshot of the GUI and the saturation profile just before the displacement front reaches the right end of the domain. The first control step requires 7 iterations, whereas the next 15 steps only use 4 iterations, giving a total of 67 iterations, or an average of 4.19 iterations for the first 16 control steps. The GUI also reports the total runtime so far and uses the iteration history to estimate the total runtime. A total runtime of 7 seconds for such a small model may seem high and can mainly be attributed to the GUI, plotting, and computational overhead introduced by the AD-OO framework. This overhead is significant for small models, but decreases in importance as the size of the model increases.

12.1 Overview of the Simulator Framework

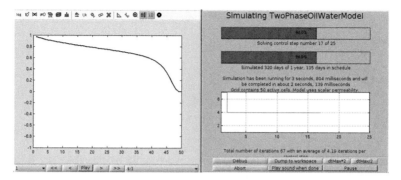

Figure 12.3 The default graphical user interface in the generic simulation loop reports simulation progress and enables simple computational steering.

For large models it may not be feasible to run interactive sessions and store all simulation results in memory. The solver also has a hook that enables you to control how the computed states are stored. To control the storage, we use a so-called result handler, *which behaves almost like a cell array, whose content can either be stored in memory, on disk, or both places. This is very useful if you have long simulations you want to run in batch mode, or if you expect that the simulator may fail or be interrupted before the simulation is finished and you want to store all completed time steps. Let us set up a handler for storing computed states to disk*

```
handler = ResultHandler('writeToDisk', true, 'dataFolder', 'ad-bl1d');
```

We can then pass the handler to the simulator as an optional argument

```
simulateScheduleAD(state0, model, schedule, 'outputHandler', handler);
```

The simulator will now store each state as a separate MATLAB-formatted binary file in a subdirectory ad-bl1d *somewhere on a standard path relative to where you have installed MRST. By default, the files are called* state1.mat, *and so on. The name can be changed through the option* 'dataPrefix', *whereas the parameter* 'dataDirectory' *lets you change the base directory relative to which the files are stored. To maintain results in memory, you need to set the option* 'storeInMemory'.

You can now use the result handler as a standard cell array, and you need not really know where the results are stored. You can for instance plot the results through the same GUI we used in the previous chapter

```
plotToolbar(G, handler, 'field', 's:1','lockCaxis', true, 'plot1d', true);
```

If you at a later time wish to retrieve the data, you simply create a new handler and will then have access to all data stored on disk. The following example shows how to extract every second time step:

```
handler = ResultHandler('dataFolder', 'ad-bl1d');
m = handler.numelData();
states = cell(m, 1);
for i = 1:2:m
   states{i} = handler{i};
end
```

You can delete the stored data through the call `handler.resetData()`.

To make the AD-OO simulator framework as versatile as possible, we have tried to develop a modular design, so that any simulator consists of a number of individual components that can easily be modified or replaced if necessary. Figure 12.4 shows how various classes, structures, and functions work together in the generic `simulateScheduleAD` function. The function takes physical model, initial state, and schedule as input, and then loops through all the control steps, updating time steps and drive mechanisms, calling the nonlinear solver, and performing any updates necessary after the nonlinear solver has converged (to track hysteretic behavior, etc.). In the next two sections, we discuss the main elements shown in Figure 12.4.

12.2 Model Hierarchy

The `mrst-core` module implements two basic model classes: `PhysicalModel` and `ReservoirModel`. Their main purpose is to provide generic templates for a wide variety of specific flow models. As such, these two generic model classes do not offer any flow equations and hence cannot be used directly for simulation. Instead, they provide generic quantities and access mechanisms, which we will discuss in more detail shortly. We can then develop specific models by expanding the generic models with specific flow equations and fluid and rock properties. The `ad-core` module also contains classes implementing well models: either simple Peaceman wells of the type we discussed in Sections 4.3.2, 5.1.5, and 11.7.1, or advanced multisegment well models from Section 11.7.2 that give more accurate representation of deviated, horizontal, and multilateral wells with advanced completions and inflow control devices.

The `ad-blackoil` module implements black-oil models, as discussed in Chapter 11, utilizing functionality from the `deckformat` module to read ECLIPSE input decks and from `ad-props` to construct models for rock-fluid and PVT properties; see Figure 12.5. The resulting simulators can simulate industry-standard black-oil models and compute adjoint gradients and sensitivities.

Sequential solvers, like those discussed in Chapter 10, can often be significantly faster than fully implicit solvers, especially for where the total velocity changes slowly during the simulation. Such solvers are also a natural starting point when developing, e.g., multiscale methods [215, 216]. The `blackoil-sequential` module implements sequential (fully implicit) solvers for the same set of equations as in `ad-blackoil`, based on a fractional

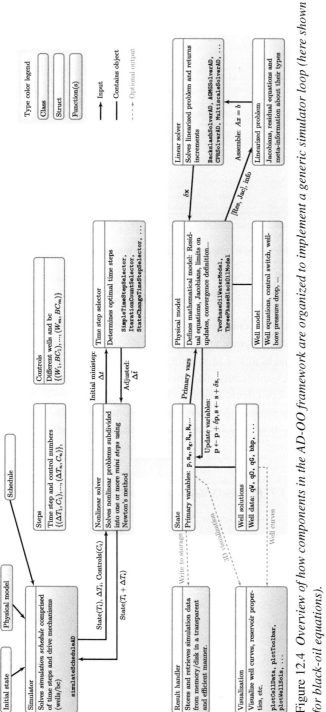

Figure 12.4 *Overview of how components in the AD-OO framework are organized to implement a generic simulator loop (here shown for black-oil equations).*

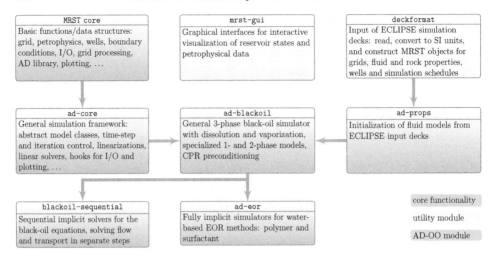

Figure 12.5 Modules used to implement black-oil models in the AD-OO framework.

flow formulation wherein pressure and transport are solved as separate steps and thus are represented as separate models.

The `ad-eor` module [30] implements extensions of the black-oil equations to water-based EOR methods like polymer and surfactant flooding and the `co2lab` module [19] uses inheritance from the `black-oil` module to implement specialized simulators for CO_2 sequestration. The black-oil family also includes modules for solvents and routines for solving optimal control problems based on the AD-OO framework. In addition, MRST contains other AD-OO modules for *compositional* simulators [212], geomechanics, and coupled geomechanics–flow simulations that do inherit from `ad-blackoil`, and more modules are in the making. These are all (unfortunately) outside the scope of this book.

12.2.1 PhysicalModel – Generic Physical Models

At the lowest level of the model hierarchy, we have a generic physical model that defines basic properties to be inherited by other models and model classes. All models in the AD-OO framework are interpreted in a discrete sense and are assumed to consist of a set of model equations on residual form. The primary properties of a physical model are a structure holding *discrete operators* used to define the model equations and a *nonlinear tolerance* that defines how close the residual values must be to zero before the model is fulfilled. Actual operators are *not* implemented in the `PhysicalModel` class and must be specified in derived classes or explicitly set by the user for concrete instances of `PhysicalModel`. These operators are typically defined over a *grid*, and the `ad-core` module offers the function `setupOperatorsTPFA` for setting up the two-point operators we have discussed in previous chapters. In addition to the optional grid property, the class contains a flag signifying if the model equations are linear (and hence can be solved by a single matrix inversion), and a flag determining how verbose the member functions should be in their output

```
classdef PhysicalModel
   properties
      operators
      nonlinearTolerance
      G
      verbose
      stepFunctionIsLinear
   end
   methods
      function model = PhysicalModel(G, varargin)
         model.nonlinearTolerance = 1e-6;
         model.verbose = mrstVerbose();
         model = merge_options(model, varargin{:});
         model.G = G;
         model.stepFunctionIsLinear = false;
      end
   end
end
```

Notice that all class names in MRST start with a capital letter to distinguish them from functions, which always start with a lowercase letter.

Model Equations and Physical Properties

The basic class has several member functions that implement generic tasks associated with a model. The first, and most important, is to evaluate residual equations and Jacobians, which is done by the following function:

```
[problem, state] = model.getEquations(state0, state, dt, forces, vararg)
```

Here, problem is an instance of the LinearizedProblem class that implements what we referred to as the *linearization numerical context* in the simulator loop in Figure 12.1. This class has data structures to represent the linearized system for given states and member functions that correctly assemble the full linear system from residuals and Jacobian block matrices stored for the individual sub-equations making up the flow model; we discuss this in more detail in Section 12.3. Input argument forces represents driving forces that are currently active. These may vary from one step to the next and are specified by the control part of the schedule; see e.g., Figure 12.4. Active driving forces are set by

```
forces = model.getDrivingForces(control);
```

The function creates a cell array that lists the forces on the same form as for the incompressible solvers. What are the valid forces for a model (with reasonable defaults) can be queried

```
validForces = model.getDrivingForces()
```

In addition, the class has utility functions for interacting with the state structure in a consistent manner. This includes querying values for one or more specific variables/properties, setting or capping specific values, or correctly incrementing values

```
p     = model.getProp (state, 'pressure');
[p,s] = model.getProps(state, 'pressure', 's');
state = model.setProp (state, 'pressure', 5);
state = capProperty   (state, 's', 0, 1)
state = model.incrementProp(state, 'pressure', 1);
```

These operations are generic and hence implemented in the PhysicalModel class. However, the specific function arguments will only work for instances of ReservoirModel or derived classes that specify the variables pressure and s. You can check whether a variable is defined or not as follows

```
[fn, index] = model.getVariableField(name)
```

which produces the field name and column index that will extract the correct variable from the state object. The function produces an error when called with any name for a PhysicalModel, since this class does not implement any variables.

Linear Updates to States

The second purpose of the basic model class is to correctly update states as part of a linear or nonlinear solution process. First of all, the function

```
[state, report] = model.stepFunction(state, state0, dt, forces, ...
                                     linsolver, nonlinsolver, itno, varargin)
```

implements the generic behavior of a single nonlinear update; that is, it executes the steps of the inner while loop of the right-most box in Figure 12.1. Here, nonlinsolver is an instance of the NonLinearSolver class, which we briefly introduced on on page 399, and linsolver is an instance of a similar *linear* solver class; more detail are given in Section 12.3. The second output, report, is a standardized step report that provides information such as the CPU time used by the linear solver, setup of the stabilization step, convergence status, size of residual, etc. There are also functions to extract the computed update, update the state, check whether it is converged or not, check whether the controls have changed, as well as a hook enabling a state to be updated after it has converged, e.g., to model hysteretic behavior.

Let us in particular look at the function for updating the state with a given increment. The following code updates the pressure value from 10 to 110:

```
state = struct('pressure', 10);
state = model.updateStateFromIncrement(state, 100, problem, 'pressure')
```

We can also restrict the maximum relative changes, e.g., to be at most 10%

```
state = model.updateStateFromIncrement(state, 100, problem, 'pressure', .1)
```

which sets pressure to 11. Relative limits such as these are important when working with tabulated and nonsmooth properties in a Newton-type loop, as the initial updates may be far outside the reasonable region of linearization for a complex problem. On the other hand, limiting the relative updates can delay convergence for smooth problems with analytic properties and will, in particular, prevent zero states from being updated, so use with care.

Quality Assurance

A model must also be able to check its on validity. The function

```
model.validateModel(forces)
```

where `forces` is an optional argument, validates that a model is suitable for simulation. If missing or inconsistent parameters can be fixed automatically, an updated model is returned. Otherwise, an error should occur. This function does not have meaning for a `PhysicalModel`, but should be implemented in derived classes. Likewise, there is a function to validate the state for use with the model

```
model.validateState(state)
```

The function should check that required fields are present and of the right dimensions, and if missing fields can be assigned default values, return `state` with the required fields added. If reasonable default values cannot be assigned, a descriptive error should be thrown telling the user what is missing or wrong (and ideally how to fix it). Also this function must be implemented in derived classes. The only function whose behavior *is* implemented in `PhysicalModel` is

```
model.checkProperty(state, 'pressure', [nc, 1], [1, 2]);
```

here called for a model of the derived `ReservoirModel` class to check that the pressure field has nc elements along its first dimension and one element along the second dimension.

Adjoint Equations

One of the reasons for introducing the AD-OO framework was to simplify the solution of so-called adjoint equations to compute parameter gradients and sensitivities. (How this is done is outside the scope of this book.) The base class therefore contains an interface for computing adjoint equations:

```
[problem, state] = model.getAdjointEquations(state0,state,dt,forces,vararg)
```

Solving adjoint equations boils down to running a backward simulation with a set of linearized equations. `PhysicalModel` also offers a generic update function `solveAdjoint` that computes a single step for the corresponding adjoint equations.

12.2.2 ReservoirModel – Basic Reservoir Models

The next class in the hierarchy has properties to represent entities found in most reservoir models: fluid model, petrophysical properties, indicators for each of the basic three phases (aqueous, gaseous, and oleic), variables representing phase saturations and chemical components, and a model of production facilities

```
classdef ReservoirModel < PhysicalModel
properties
   fluid, rock, gravity, FacilityModel
   water, gas, oil

   % Iteration parameters
   dpMaxRel, dpMaxAbs, dsMaxAbs, maximumPressure, minimumPressure,
   useCNVConvergence, toleranceCNV, toleranceMB
   :
```

Notice in particular that the reservoir model declares absolute and relative tolerances on how much pressures and saturations can change during an iteration step as well as lower and upper bounds on the pressure. In addition, the class declares that convergence of nonlinear iterations can be measured in two different ways, either by mass balance (MB) or by scaled residuals (CNV); more details will follow in Section 12.3.

To construct an instance of the class, we first construct an instance of a `PhysicalModel` and then proceed to declare the additional properties that are part of a reservoir model. The following is a somewhat simplified excerpt:

```
function model = ReservoirModel(G, varargin)
   model = model@PhysicalModel(G);
   model.rock  = varargin{1}; model.fluid = varargin{2};
   [model.water, model.gas, model.oil] = deal(false);
   model.dpMaxRel = inf;    model.dpMaxAbs = inf;
   model.dsMaxAbs = .2;
   :
   model.operators = setupOperatorsTPFA(G, model.rock, ..)
end % call for construction: ReservoirModel(G, rock, fluid, ...)
```

By default, the reservoir model class does not contain any active phases; these must be specified in derived classes. However, we do declare that discrete operators should be of the two-point type. We also declare limits on how much pressures and saturations may change from one iteration step to the next: the default is that changes in pressure can be arbitrarily large, whereas saturations are only allowed an absolute change of 0.2.

Physical Variables and Active Phases

Physical variables known to a particular class are declared by the following function:

```
function [fn,ix] = getVariableField(model, name)
   switch(lower(name))
     case {'pressure' , 'p'}
        ix = 1; fn = 'pressure' ;
     case {'s', 'sat', 'saturation'}
        ix = ':'; fn = 's' ;
     case {'so' , 'oil'}
        ix = find(strcmpi(model.getSaturationVarNames, name));
        fn = 's' ;
     :
```

The purpose of this function is to provide a convenient and uniform way to access physical variables without knowing their actual storage in the state object. Let us consider saturation as an example. Saturation is stored in the field `state.s`, but this vector may have one, two, or three entries per cell, depending upon the number of phases present. To access individual phase saturations, the class implements two additional utility functions:

```
function vars = getSaturationVarNames(model)
   vars = {'sw', 'so', 'sg'};
   vars = vars(model.getActivePhases());
end
function isActive = getActivePhases(model)
   isActive = [model.water, model.oil, model.gas];
end
```

Here, we see that oil saturation is stored in the second column of `state.s` for models containing water, but in the first column for a model without water. Oil saturation can now be extracted as follows:

```
[fn, ix] = model.getVariableField('so');
so = state.(fn)(:, ix);
```

This may seem unnecessary complicated, but enables access to variables in a uniform way for all models so that the code continues to work if we later decide to change the name of the saturation variable, or implement models with more than three phases to represent precipitated solids, emulsified phases, etc.

Altogether, the reservoir model declares the possible presence of variables for pressure; saturations for the aqueous, oleic, and gaseous phases; temperature; and well solutions, but does not declare variables for dissolved gas–oil ratio and vaporized oil–gas ratios, as these are specific to black-oil models. There are also a number of other functions for getting the active phases and components and correctly storing saturations, mobilities, face fluxes, densities, shrinkage factors, and upstream indices in the state object.

Fluid Behavior

The `ReservoirModel` class implements a generic method for evaluating relative permeabilities:

```
function varargout = evaluateRelPerm(model, sat, varargin)
   active = model.getActivePhases();
   nph = sum(active);
   varargout = cell(1, nph);
   names = model.getPhaseNames();

   if nph > 1
      fn = ['relPerm', names];
      [varargout{:}] = model.(fn)(sat{:}, model.fluid, varargin{:});
   elseif nph == 1
      varargout{1} = model.fluid.(['kr', names])(sat{:}, varargin{:});
   end
```

The function is a general interface to the `relPermWO`, `relPermOG`, `relPermWG`, and `relPermWOG` functions discussed in Section 11.3, which in turn are interfaces to the more basic `krW`, `krO`, `krG`, `krOG`, and `krOW` functions of fluid objects in MRST. The generic reservoir model neither implements nor offers any interface for PVT behavior, since this is very different in black-oil models and models relying on an equation of state. However, there is a function for querying surface densities of all phases.

Global Driving Forces

The basic physical model was aware of driving forces but did not implement any. `ReservoirModel` has gravity as a property and can evaluate the gravity vector and compute a gravity gradient. In addition, the class specifies the potential presence of boundary conditions, source terms, and wells

```
function forces = getValidDrivingForces(model)
   forces = getValidDrivingForces@PhysicalModel(model);
   forces.W   = [];
   forces.bc  = [];
   forces.src = [];
end
```

Wells are special because each well is represented by another model, i.e., an instance of a derived class of `PhysicalModel`. Instances of such models are stored in the `FacilityModel` property of the reservoir model, which in principle can contain models of all production facilities used to bring hydrocarbons from the reservoir to the surface. New well instances are created by the following function whenever the well controls change

```
function [model, state] = updateForChangedControls(model, state, forces)
  model.FacilityModel = model.FacilityModel.setupWells(forces.W);
  state.wellSol = initWellSolAD(forces.W, model, state);
  [model,state] = updateForChangedControls@PhysicalModel(model, state, forces);
end
```

The generic operation of adding the effect of wells to a system of equations by adding the corresponding source terms and augmenting the system with additional model equations for the wells is done in the `insertWellEquations` function. The same holds for source terms and boundary conditions. If you are more interested, you should consult the code for further details.

Updating States and Models

The computation of linearized increments in a reservoir model is inherited from the `PhysicalModel` class without any special adaptions or extensions, since it follows the exact same algorithm as in a generic Newton loop. On the other hand, the way these increments are used to update reservoir states differs from other types of models and hence requires special implementation. The function `updateState` splits state variables into four different categories: well variables, saturation variables, pressure, and optional remaining variables. Updating well variables is quite involved and will not be discussed herein; as always, you can find all necessary details in the code.

Saturation variables are characterized by the fact that they should sum to unity so that pore space is completely filled by fluid phases. If our model contains n_{ph} phases, the linearized increments will contain $n_{ph} - 1$ saturation increments. The n_{ph}'th increment is set as the negative sum of all the other increments; i.e., values added to the first $n_{ph} - 1$ variables must be subtracted from the saturation of the last phase so that the total increment over the n_{ph} phases is zero. The increments are then passed to the `updateStateFromIncrement` function to compute new saturations satisfying limits on absolute and relative updates. To ensure physically correct states, the updated saturations are then cropped to the unit interval, and then we renormalize saturations in any cells containing capped values, so that $S_\alpha \leftarrow S_\alpha / \sum_\alpha S_\alpha$.

We update pressure by first updating according to relative/absolute changes and then capping values to fulfill limits on minimum and maximum pressure. Any remaining variables are updated with no limits on changes or capping.

Validating Models and States

You may recall that `PhysicalModel` defined functions for validating models and states but did not implement any specific behavior. The reservoir model checks that the model instance contains a facility model, and if not, instantiates one such object with default values:

```
function model = validateModel(model, varargin)
   if isempty(model.FacilityModel)
      model.FacilityModel = FacilityModel(model);
   end
   if nargin > 1
      W = varargin{1}.W;
      model.FacilityModel = model.FacilityModel.setupWells(W);
   end
   model = validateModel@PhysicalModel(model, varargin{:});
```

The validation function can also be called with a structure containing drive mechanisms as argument. In this case, we extract the well description and pass it on to the facility model so that this model can initialize itself properly. The function then calls the validation function inherited from its parent class (`PhysicalModel`), which in the current implementation does nothing. Because the reservoir model does not contain any concrete equations and parameters, further consistency checks of model equations and petrophysical, fluid, and PVT properties must be performed in derived classes or implemented explicitly by the user. However, the model should be able to check a reservoir state and make sure that this state is compatible with the model:

```
function state = validateState(model, state)
   % Check parent class
   state = validateState@PhysicalModel(model, state);
   active = model.getActivePhases();
   nPh = nnz(active);
   nc = model.G.cells.num;
   model.checkProperty(state, 'Pressure', [nc, 1], [1, 2]);
   if nPh > 1
      model.checkProperty(state, 'Saturation', [nc, nPh], [1, 2]);
   end
   state = model.FacilityModel.validateState(state);
```

To be consistent, the state should contain one unknown pressure and the correct number of phase saturations in each cell. Once we have ensured that the reservoir states meets the requirements, the facility model should check that the reservoir state contains the correct data structures for representing well states. Neither of these operations check whether the *actual values* represented in the reservoir state are physically meaningful.

12.2.3 Black-Oil Models

The `ThreePhaseBlackOilModel` class is the base class in the `ad-blackoil` module and is derived from the `ReservoirModel` class. Unlike the two previous abstract classes, this base class is a *concrete class* designed to represent a general compressible, three-phase, black-oil model with dissolved gas and vaporized oil. The model therefore declares two new properties, `disgas` and `vapoil`, to signify the presence of dissolved gas and

12.2 Model Hierarchy

vaporized oil, respectively. The corresponding data fields `rs` and `rv` are defined in the `getVariableField` function. We also extend `validateState` to check that the reservoir state contains a `rs` and/or a `rv` field if dissolution and/or vaporization effects are active in a specific model instance. The class also defines the maximum absolute/relative increments allowed for r_s and r_v. By default, the constructor sets up a dead-oil model containing the aqueous, gaseous, and oleic phases, but with no dissolution and vaporization effects.

Unlike the generic classes from `ad-core`, `ThreePhaseBlackOilModel` implements concrete equations. The member function `getEquations` calls the `equationsBlackOil`, which generates linearized equations according to how the general black-oil model is configured. We have already discussed the essential code lines contained in this function, but as stated in Chapter 11, the actual code is structured somewhat differently and contains more conditionals and safeguards to ensure that discrete residuals are computed correctly for all combinations of drive mechanisms and possible special cases of the general model.

In addition to implementing flow equations, the class extends `updateState` to implement variable switching when the reservoir state changes between saturated and undersaturated states. There is also a new member function that computes correct scaling factors to be used by CPR-type preconditioners, which we outline on page 444.

The `ad-blackoil` module also implements three special cases of the general black-oil model: a single-phase water model and two-phase oil–water and oil–gas models. Since most functionality is inherited from the general model, these special cases are implemented quite compactly, for instance:

```
classdef TwoPhaseOilWaterModel < ThreePhaseBlackOilModel
properties
end
methods
  function model = TwoPhaseOilWaterModel(G, rock, fluid, varargin)
    model = model@ThreePhaseBlackOilModel(G, rock, fluid, varargin{:});
    model.oil = true;
    model.gas = false;
    model.water = true;
    model = merge_options(model);
  end
```

Residual equations are evaluated by the function `equationsOilWater`, which is a simplified version of the `equationsBlackOil` function in which all effects relating to gas, dissolution, and vaporization have been removed.

```
function [problem, state] = ...
          getEquations(model, state0, state, dt, drivingForces, varargin)
   [problem, state] = ...
       equationsOilWater(state0, state, model, dt, drivingForces, varargin{:});
end
```

For completeness, let us quickly go through the essential parts of this function, disregarding source terms, boundary conditions, and computation of adjoints. We start by getting driving forces and discrete operators and setting names of our primary variables,

```
W = drivingForces.W;
s = model.operators;
primaryVars = {'pressure', 'sW', wellVarNames{:}};
```

Then, we can use the reservoir and facility models to extract values for the primary unknowns:

```
[p, sW, wellSol] = model.getProps(state, 'pressure', 'water', 'wellsol');
[p0, sW0, wellSol0] = model.getProps(state0, 'pressure', 'water', 'wellSol');
[wellVars, wellVarNames, wellMap] = ...
        model.FacilityModel.getAllPrimaryVariables(wellSol);
```

Once this is done, we compute the residual equations more or less exactly the same way as discussed in Section 11.6. That is, we first compute fluid and PVT properties in all cells. Then, we compute gradient of phase potentials, apply suitable multipliers, and average properties correctly at the interface to compute intercell fluxes.

```
sO  = 1 - sW;
[krW, krO] = model.evaluateRelPerm({sW, sO});
[pvMult, transMult, mobMult, pvMult0] = getMultipliers(model.fluid, p, p0);
krW = mobMult.*krW; krO = mobMult.*krO;
T = s.T.*transMult;
    :
```

We now have all we need to compute the residual equations

```
water = (s.pv/dt).*( pvMult.*bW.*sW - pvMult0.*bW0.*sW0 ) + s.Div(bWvW);
oil   = (s.pv/dt).*( pvMult.*bO.*sO - pvMult0.*bO0.*sO0 ) + s.Div(bOvO);
```

In all codes discussed so far, we have included wells by explicitly evaluating the well equations and adding the corresponding source terms to the residual equations in individual cells before concatenating all the residual equations. Here, however, we first concatenate the reservoir equations and then use a generic member function from the `ReservoirModel` class to insert the necessary well equations

```
eqs   = {water, oil};
names = {'water', 'oil'};
types = {'cell', 'cell'};
[eqs, names, types, state.wellSol] = ...
    model.insertWellEquations(eqs, names, types, wellSol0, wellSol,
                    wellVars, wellMap, p, mob, rho, {}, {}, dt, opt);
```

Finally, we construct an instance of the linearized problem class

```
problem = LinearizedProblem(eqs, types, names, primaryVars, state, dt);
```

We will discuss this class in more detail in Section 12.3.1.

12.2.4 Models of Wells and Production Facilities

The AD-OO framework has primarily been developed to study hydrocarbon recovery, CO_2 storage, geothermal energy, and other applications in which fluids are produced from a porous rock formation and brought up to the surface, or brought down from the surface and injected into a rock formation. By default, all *reservoir* models are therefore assumed to have an entity representing this flow communication between the surface and the subsurface. If you only want to describe flow in the porous medium, you can set the well model to be void, but it must still be present.

Facility Models

The well models discussed earlier in the book apply to a single well (see Section 11.7). Wells are sometimes operated in groups that are subject to an overall control. The corresponding well models must hence be coupled. As an example, wells can be coupled so that gas or water produced from one well is injected into another well as displacing fluid. Wells from different reservoirs may sometimes also produce into the same flowline or surface network and hence be subject to overall constraints that couple the reservoir models. Hence, there is a need to represent a wider class of *production and injection facilities*.

Instead of talking of a well model, the AD-OO framework therefore introduces a general class of *facility models* that can represent different kinds of facilities used to enhance and regulate the flow of reservoir fluids from the sandface to the well head or stock tank. The `FacilityModel` class is derived from the `PhysicalModel` class and is a general *container class* that holds a collection of submodels representing individual wells. In principle, the `FacilityModel` container class can also incorporate models for various types of topside production facilities, but at the time of writing, no such models are implemented in MRST.

Peaceman-Type Well Models

The inflow performance relation, $q = J(p_R - p_{wb})$, discussed in Sections 4.3.2 and 11.7.1, gives an unsophisticated description of injection and production wells, and is generally best suited for simple vertical wells, in which fluids inside the wellbore can be assumed to be in hydrostatic equilibrium. Apart from the well index (and possibly a skin factor), no attempt is made to model the complex flow physics that may take place in various types of well completions. The flow in and out of the wellbore is regulated by controlling either the surface flow rate or the bottom-hole pressure (bhp). In the AD-OO framework, this type of well model is implemented in the `SimpleWell` class, which is derived from the `PhysicalModel` class.

Multisegment Well Models

The `MultisegmentWell` class is also derived from the `PhysicalModel` class and implements the more general concept of multisegment wells, which we briefly introduced in Section 11.7.2. In these models, the flow of fluids from the sandface to the well head is described in terms of a set of 1D flow equations on a general flow network. In ECLIPSE [270], the network must assume the form of a tree (i.e., be a directed graph). The implementation in MRST is more general and allows for complex network topologies involving circular dependencies among the nodes.

A Word of Caution about the Implementation

Implementing the correct behavior of wells and facilities involves quite complicated logic and a lot of intricate details that quickly make the code difficult to understand. A detailed discussion of the facility and well classes is therefore outside the scope of this book. As usual, you can find all necessary details in the actual code, but let me add a small warning: To retain the same computational efficiency as the rest of the simulator classes, the well classes have been optimized for computational efficiency. The reason is that well and facility models typically have orders-of-magnitude fewer unknowns than the number of states defined over the reservoir grid. The vectorized AD library in MRST is primarily designed to perform well for long vectors and large sparse matrices and is not necessarily efficient for scalars and short vectors. To keep the computational efficiency when working with individual wells, it has proved necessary to meticulously distinguish between variables that need automatic differentiation and variables that do not. As a result, the code contains explicit casting of AD variables to standard variables, which in my opinion has reduced readability.

In other words: please go ahead and read the code, but be warned that it may appear less comprehensible than other parts of the AD-OO framework.

12.3 Solving the Discrete Model Equations

In this section, we go through the three different types of classes the AD-OO framework uses to solve the discrete equations, motivate their design, and briefly describe their key functionality. If you are interested in developing your own solution algorithms and solvers, the classes define generic information about the discrete equations and interfaces to the rest of the simulator you can use as a starting point.

12.3.1 Assembly of Linearized Systems

The main purpose of the `LinearizedProblem` class is to assemble the individual matrix blocks computed by the AD library invoked within a model class into a sparse matrix that represents the linearization of the whole reservoir model. To this end, the class contains the residual equations evaluated for a given specific state along with meta-information about the equations and the primary variables they are differentiated with respect to. We refer to this as a *linearized problem*. This description can be transformed into a linear system and solved using a subclass of the `LinearSolverAD` class to be discussed in Section 12.3.4, provided that the number of equations matches the number of primary variables. The class also contains the following data members:

```
classdef LinearizedProblem
properties
    equations, types, equationNames
    primaryVariables
    A, b
    state
    dt, iterationNo, drivingForces
end
```

The cell array `equations` contains the residuals evaluated at the reservoir state represented in the `state` object. The residuals can be real numbers, but are most typically AD objects. The cell array `types` has one string per equation indicating its type. (Common types: `'cell'` for cell variables, `'well'` for well equations, etc.) These types are available to any linear solver and can be used to construct appropriate preconditioners and solver strategies. The cell arrays `equationNames` and `primaryVariables` contain the names of equations and primary variables; these are used, e.g., by the linear solvers and for reporting convergence. The sparse matrix A represents the linear system and the vector b holds the right-hand side. For completeness, we also store the time step, the iteration number, and the driving forces, which may not be relevant for all problems.

The linear system corresponding to a linearized problem can be extracted by the member function

```
[A, b] = problem.getLinearSystem();
```

The function will check whether the linear system already exists, and if not, it will assemble it using the function:

```
function problem = assembleSystem(problem)
    if isempty(problem.A)
        iseq = cellfun(@(x) ~isempty(x), problem.equations);
        eqs = combineEquations(problem.equations{iseq});
        if isa(eqs, 'ADI')
            problem.A = -eqs.jac{1};
            problem.b = eqs.val;
        else
            problem.b = eqs;
        end
    end
end
```

In our previous AD solvers, the equations were always AD objects, and we could assemble the linearized system simply by concatenating the AD variables vertically. In a general model, we cannot guarantee that all residual equations are AD objects, and hence the vertical concatenation is implemented as a separate function we can overload by specialized implementations. If the residual equations are given as real numbers, this function simply reads

```
function h = combineEquations(varargin)
  h = vertcat(varargin{:});
end
```

In this case, we cannot construct a linear system since no partial derivatives are available. If the residual equations are represented as AD objects, the linear system is, as before, assembled from the Jacobian block matrices stored for each individual (continuous) equation. The AD library from `mrst-core` implements an overloaded version of `combineEquations`, which ensures that AD objects are concatenated correctly and also enables concatenation of AD objects and standard doubles.

How the linear equations are solved, will obviously depend on the linear solver. The simplest approach is to solve all equations at once. However, as we will come back to in Section 12.3.4, it may sometimes be better to eliminate some of the variables and only solve for a subset of the primary variables. To this end, the `LinearizedProblem` class has member functions that use a block-Gaussian method to eliminate individual variables

```
[problem, eliminatedEquation] = problem.eliminateVariable(name)
```

where `name` corresponds to one of the entries in `problem.equationNames`. Likewise, you can eliminate all variables that are not of a specified type

```
[problem, eliminated] = reduceToSingleVariableType(problem, type)
```

Using this function, you can for instance eliminate all equations that are not posed on grid cells. There is also a function that enables you to recover the increments in primary variables corresponding to equations that have previously been eliminated. In addition, the `LinearizedProblem` class contains member functions for appending/prepending additional equations, reordering the equations, computing the norm of each residual equation, querying indices and the number of equations, querying the number of equations of a particular type, as well as a number of utility functions for sanity checks, clearing the linear system, and so on.

12.3.2 Nonlinear Solvers

The `NonLinearSolver` class is based on a standard Newton–Raphson formulation and is capable of selecting time steps and cutting them if the nonlinear solver convergences too slowly. To modify how the solver works, you can either develop your own subclass and/or combine the existing solver with modular linear solvers and classes for time-step selection. You have already seen some of the properties of the class in Section 11.8.1 on page 399. The following is an almost complete declaration of the class and its data members

12.3 Solving the Discrete Model Equations

```
classdef NonLinearSolver < handle
properties
   identifier
   maxIterations, minIterations
   LinearSolver, timeStepSelector, maxTimestepCuts
   useRelaxation, relaxationParameter, relaxationType, relaxationIncrement
   minRelaxation, maxRelaxation
   useLinesearch, linesearchReductionFn
   linesearchReductionFactor, linesearchDecreaseFactor, linesearchMaxIterations
   linesearchConvergenceNames, linesearchResidualScaling
   enforceResidualDecrease, stagnateTol
   errorOnFailure, continueOnFailure
```

The nonlinear solver must call a linear solver and hence has a `LinearSolver` class object; the default is the standard backslash solver. Likewise, we need time-step selection, which in the default implementation does not offer any advanced heuristics for computing optimal time steps. Furthermore, there are limits on the number of times the control steps can be subdivided and the maximum/minimum number of iterations within each subdivided step. If `useRelaxation` is true, the computed Newton increments should be relaxed by a factor between 0 and 1. The `relaxationParameter` is modified dynamically if `useRelaxation` is true and you should in general not modify this parameter unless you really know what you are doing. Relaxation can be any of the following three choices

```
x_new = x_old + dx;                     % relaxationType = 'none'
x_new = x_old + dx*w;                   % relaxationType = 'dampen'
x_new = x_old + dx*w + dx_prev*(1-w);   % relaxationType = 'sor'
```

where `dx` is the Newton increment and `w` is the relaxation factor. Similarly, there are various parameters for invoking and controlling line search.

When `enforceResidualDecrease` is invoked, the solver will abort if the residual does not decay more than the stagnation tolerance during relaxation. If `errorOnFailure` is not enabled, the solver will return even though it did not converge. Obviously, you should never rely on non-converged results, but this behavior may be useful for debugging purposes.

Use of Handle Classes

You may have observed that `NonLinearSolver` is implemented as a subclass of the handle class. All classes defined so far have been so-called *value classes*, meaning that whenever you copy an object to another variable or pass it to a function, MATLAB creates an independent copy of the object and all the data the object contains. This means that if you want to change data values of a class object inside a function, you need to return the class as an output variable. Classes representing nonlinear solvers do a lot of internal bookkeeping inside a simulator and hence need to be passed as argument to many functions. By making this class a *handle class*, the class object is passed by reference to functions, so that any changes occuring inside the function will also take effect outside the function without having to explicitly return the class object as output argument.

Computing a Control Step

We have already encountered the primary member function

```
[state, report, ministates] = solver.solveTimestep(state0, dT, model)
```

whose purpose is to solve a specified control step. Recall from our discussion of nonlinear Newton-solvers in Section 10.2.2 that each control step may be subdivided into several substeps (ministeps) to ensure stable computations and proper convergence. Each ministep will involve of one or more *nonlinear iterations*. Each of these may consist of one matrix inversion if we use a direct linear solver or one or more *linear iterations* if we use an iterative solver.

Whether the solver will move forwards using a single time step or multiple substeps depends on the convergence rate and what the time-step selector predicts to be a feasible time step for the given state. For the solveTimestep function to work, the model object must contain a valid implementation of the stepFunction member function declared in the PhysicalModel class. Driving forces that are active for the particular model must be passed as optional arguments in the same way as for the incompressible solvers.

Upon completion, the function will return the reservoir state after the time step and a report structure containing standard information like iteration count, convergence status, and any other information passed on by model.stepFunction. The cell array ministates contains all ministeps used to advance the solution a total time step dT. The class also has member functions for computing these ministeps, applying line search, stabilizing Newton increments, and implementing checks for stagnation or oscillating residuals. You have already been exposed to the essential ideas of these operations; the implementation in the NonLinearSolver class contains more safeguards against possible errors, has more flexible hooks for pluggable linear solvers and time-step selection schemes, as well as more bells and whistles for monitoring and reporting the progress of the solver.

Measures of Nonlinear Convergence

Whether the nonlinear solution process has converged or not is determined by the PhysicalModel class; the NonLinearSolver class simply responds based on what the model reports in terms of convergence. There are several different ways convergence can be measured. The AD-OO framework currently implements the same two measures of convergence as in ECLIPSE [271].

The first measure is *mass balance*. When summing residual equations over all cells in the grid, all intercell fluxes cancel because the flux *out of* a cell across an internal face has the same magnitude but opposite sign of the corresponding flux *into* the neighboring cell. The sum of residuals therefore equals the difference between the net mass accumulation of the phase minus the net influx from wells, source terms, and open boundaries, which is the definition of the *mass-balance error*. By convention, these errors are scaled to make them problem-independent,

$$\text{MB}_\alpha = \Delta t \bar{B}_\alpha \left(\sum_i \mathcal{R}_{\alpha,i} \Big/ \sum_i \Phi_i \right), \tag{12.1}$$

where $\bar{B}_\alpha$ is the formation-volume factor of phase α evaluated at average pressure, $\mathcal{R}_{\alpha,i}$ is the residual equation evaluated in cell i, and Φ_i is the pore volume of that cell. The result can be interpreted as mapping from masses to saturations at field conditions. We say the solution process is converged when the all scaled residuals are less than the `toleranceMB` tolerance declared in the `ReservoirModel` class.

The second test computes the *maximum normalized residual* defined as

$$\text{CNV}_\alpha = \Delta t \bar{B}_\alpha \max_i \left| \frac{\mathcal{R}_{\alpha,i}}{\Phi_i} \right| \tag{12.2}$$

and says that the solution is converged when this quantity is less than `toleranceCNV` for all three phases. Default values for the MB and CNV tolerances are the same as in ECLIPSE 100, i.e., 10^{-7} and 10^{-3}, respectively.

12.3.3 Selection of Time-Steps

As indicated in the previous subsection, appropriate selection of time step is crucial to ensure low runtimes and good convergence in the nonlinear solver. We have already encountered several selection schemes that use experience from previous time steps to select optimal time steps, e.g., by halving the time step if the nonlinear solver fails to converge, or increasing the time step back once we have managed to successfully run a certain number of chopped steps. Questions to be asked:

- How difficult was it to converge the previous time step? As a simple rule of thumb, a single iteration is usually considered as too easy, two to three iterations is easy, whereas more than ten iterations indicate problems with time step or model.
- How does a specified control step compare with the last successful step?
- For a ministep: how far are we away from the end of the control step?
- How does the required number of iterations for the last successful step compare with any iteration targets?
- Which mechanism determined the size of the previous time step?

Heuristics based on these and similar questions are implemented in the time-selector classes of MRST, which like `NonLinearSolver` are handle classes:

```
classdef SimpleTimeStepSelector < handle
properties
   history, maxHistoryLength
   isFirstStep, isStartOfCtrlStep
   previousControl, controlsChanged, resetOnControlsChanged
   firstRampupStep, firstRampupStepRelative
   maxTimestep, minTimestep
   maxRelativeAdjustment, minRelativeAdjustment
   stepLimitedByHardLimits
```

The class implements a generic member function `pickTimestep`,

```
dt = selector.pickTimestep( dtPrev, dt, model, solver, statePrev, state)
```

which is called by the `NonLinearSolver` class to determine the next time step. The logic of this function is as follows: We start by checking whether this is the first step, or if the well controls have changed and we want to induce a gradual ramp-up

```
if selector.controlsChanged && ...
    (selector.resetOnControlsChanged || selector.isFirstStep);
        dt = min(dt, selector.firstRampupStepRelative*dt);
        dt = min(dt, selector.firstRampupStep);
        selector.stepLimitedByHardLimits = true;
end
```

The default behavior is to not use such a ramp-up. This means that the default value of the relative adjustment is set to unity, the lower limit on the first ministep attempted after controls have changed is set to infinity, and the reset flag is disabled. The time step is then passed on to the function

```
dt = selector.computeTimestep( dt, dtPrev, model, solver, statePrev, state)
```

which will try to compute an optimal adjustment, using some sort of heuristics based on the stored time-step and iteration history. (By default, this consists of 50 entries.) The base class offers no heuristics and returns an unchanged time step. The `IterationCountTimeStepSelector` subclass tries to select a time step that will maintain the number of nonlinear iterations as close as possible to a prescribed target, whereas `StateChangeTimeStepSelector` attempts to ensure that certain properties of the state change at prescribed target rates during the simulation. This can often be a good way of controlling time steps and minimizing numerical error if good estimates of the error are known; see [58] for more details.

The last step of the selection algorithm is to ensure that the computed time step does not change too much compared with the previous time step, unless we are at the start of a new control step. The default choice of tolerances is to allow relative changes of at most a factor two. Likewise, all time steps are required to be within certain lower and upper bounds (which by default are 0 and ∞, respectively). Alternative heuristics for choosing time steps can be introduced by implementing a new subclass of the simple selector base class or one of its derived classes.

12.3.4 Linear Solvers

The linear solver classes in AD-OO implement methods for solving linearized problems. All concrete solvers are subclasses of the following handle class:

```
classdef LinearSolverAD < handle
 properties
    tolerance, maxIterations
    extraReport, verbose
    replaceNaN, replacementNaN, replaceInf, replacementInf
 end
```

The intention of the class is to provide a general interface to all types of linear solvers, both direct and iterative. Hence, the class declares both a tolerance for the linear solver (default: 10^{-8}) and an upper limit on the number of linear iterations (default: 25), even though these do not make sense for direct solvers. The flag extraReport determines the amount of reporting; turning it on may consume a lot of memory for problems with many unknowns.

The main interface to linear solvers is through the following member function, which assembles and solves a linear system:

```
function [dx, result, report] = solveLinearProblem(solver, problem, model)
    problem = problem.assembleSystem();
    ⋮
    timer = tic();
    [result, report] = solver.solveLinearSystem(problem.A, problem.b);
    [result, report] = problem.processResultAfterSolve(result, report);
    report.SolverTime = toc(timer);
    ⋮
    dx = solver.storeIncrements(problem, result);
```

The only place this function is called is within the generic stepFunction implemented in the PhysicalModel class. The code excerpts above show the main steps of solving a linear problem: first, the linear system is assembled by the problem object, which is an instance of the LinearizedProblem class discussed in Section 12.3.1. Next, we call the main member function solveLinearSystem, which is nonfunctional in the base class and must be implemented in concrete subclasses. The linear solution is then passed back to the LinearizedProblem class for potential postprocessing. Postprocessing is not necessary for the models discussed herein, but the hook is provided for generality. Linear solvers may sometimes produce infinite values or not-a-number, which the linear solver class can replace by other values if the replaceInf and/or replaceNaN flags are set; this is done by two simple if statements. Finally, we extract the results from the solution vector and store them in the cell array dx, which has one increment entry per primary variable in the linearized problem.

In addition to the two member functions for solving linearized problems and linear systems, the LinearSolverAD class implements various utility functions that may be utilized by concrete linear solvers. This includes interfaces for setup and cleanup of linear solvers as well as a generic function that can reduce the problem by eliminating some of the variables and a similar generic function for recovering increments for eliminated variables from the linear solution. There is also a special function that reduces a problem to cell variables only, solves it, and then recovers the eliminated variables afterwards. Last, but not least, the `solveAdjointProblem` function solves backward problems for computing adjoints.

MATLAB's Standard Direct Solver

By far, the simplest approach is to use the standard direct solver. Use of this solver is implemented very compactly:

```
classdef BackslashSolverAD < LinearSolverAD
   methods
       function [result, report] = solveLinearSystem(solver, A, b) %#ok
           result = A\b;
           report = struct();
       end
```

The resulting solver is very robust and should be able to solve almost any kind of linear system that has a solution. On the other hand, black-oil models have several unknowns per cell and can have many nonzero matrix entries because of unstructured grid topologies. Sparse direct solvers do generally not scale very well with the number of unknowns, which means that both memory consumption and runtime increase rapidly with model sizes. MATLAB's direct solvers should therefore only be used for models having at most a few tens of thousands of unknowns.

GMRES with ILU Preconditioner

To get a more scalable solver, you need to use an iterative method. Iterative methods used today to solve large-scale, sparse systems are usually of the Krylov subspace type that seek to find an approximation to the solution of the linear system $Ax = b$ in the Krylov space $\mathcal{K}$ formed by repeatedly applying the matrix A to the residual $r_0 = b - Ax_0$, where x_0 is some initial guess:

$$\mathcal{K}_\nu = \mathcal{K}_\nu(A, r_0) = \text{span}\{r_0, Ar_0, A^2 r_0, \ldots, A^{\nu-1} r_0\}.$$

One widely used example is the generalized minimal residual (GMRES) method, which seeks the approximate solution as the vector x_ν that minimizes the norm of the ν'th residual $\|r_\nu\| = \|b - Ax_\nu\|$. Krylov vectors are often linearly dependent, and hence GMRES employs a modified Gram–Schmidt orthogonalization, called Arnoldi iteration, to find an orthonormal basis for $\mathcal{K}_\nu$.

The GMRES iteration is generally robust and works well for nonsymmetric systems of linear equations. Iterative solvers are nonetheless only efficient if the condition number of the matrix is not too high. Black-oil models often give linear systems that have quite bad condition numbers. To remedy this, it is common to use a *preconditioner*. That is, instead of solving $Ax = b$, we first construct a matrix B that is inexpensive to invert. Expanding, we have $b = AB^{-1}Bx = (AB^{-1})y$, which is less expensive to solve if AB^{-1} has better condition number than A.

12.3 Solving the Discrete Model Equations

Herein, we apply an incomplete LU factorization (ILU) method. The true LU factors for a typical sparse matrix can be much less sparse than the original matrix, and in an incomplete factorization, one seeks triangular matrices L and U such that $LU \approx A$. When used as a preconditioner, it is common to construct L and U so that these matrices preserve the sparsity structure of the original matrix, i.e., that we have no fill-in of new nonzero elements. The result is called ILU(0).

The GMRES_ILUSolverAD class implements a GMRES solver with ILU(0) preconditioning by use of the built-in gmres and ilu solvers from MATLAB. As for the direct solver, the implementation is quite compact:

```
function [result, report] = solveLinearSystem(solver, A, b)
    nel = size(A, 1);
    if solver.reorderEquations
        [A, b] = reorderForILU(A, b);
    end
    [L, U] = ilu(A, solver.getOptsILU());
    prec = @(x) U\(L\x);
    [result, flag, res, its] = ...
        gmres(A, b, [], solver.tolerance, min(solver.maxIterations, nel), prec);
    report = struct('GMRESFlag', flag, 'residual', res, 'iterations', its);
end
```

Notice that we permute the linear system to ensure nonzero diagonals to simplify the construction of the ILU preconditioner. Notably, this utility is useful whenever well equations are added, since these may not have derivatives with respect to all well controls.

Well equations generally consist of a mixture of different variables and can hence be badly scaled. Control equation on pressure gives residuals having magnitudes that are typically $\mathcal{O}(10^7)$ since all equations are converted to SI units in MRST. Rate equations, on the other hand, give residuals that typically are $\mathcal{O}(10^{-5})$. To be on the safe side, we overload the solver for linearized problems by a new function that eliminates all equations that are not posed on grid cells by use of a block-Gaussian algorithm, solve the resulting problem by the GMRES-ILU(0) method, and then recover the eliminated variables

```
function [dx, result, report] = solveLinearProblem(solver, problem, model)
    keep = problem.indexOfType('cell');
    [problem, eliminated] = solver.reduceToVariable(problem, keep);
    problem = problem.assembleSystem();

    timer = tic();
    [result, report] = solver.solveLinearSystem(problem.A, problem.b);
    report.SolverTime = toc(timer);

    dxCell = solver.storeIncrements(problem, result);
    dx = solver.recoverResult(dxCell, eliminated, keep);
```

We eliminate well equations (and other non-cell equations) by use of member functions from the LinearSolverAD base class, which in turn relies on block-Gaussian algorithms implemented in the `LinearizedProblem` class. This algorithm is not specific to the GMRES-ILU(0) and in the actual code the content of this function is implemented as a member function of the `LinearSolverAD` class.

Constrained Pressure Residual (CPR)

The state-of-the-art approach for solving linear systems arising from black-oil type models is to use a so-called constrained pressure residual (CPR) method [305, 119]. To explain the method, we start by writing the equation for the Newton update in block-matrix form

$$-\begin{bmatrix} J_{pp} & J_{ps} \\ J_{sp} & J_{ss} \end{bmatrix} \begin{bmatrix} \Delta x_p \\ \Delta x_s \end{bmatrix} = \begin{bmatrix} R_p \\ R_s \end{bmatrix}.$$

Here, we have decomposed the unknown Newton increment in primary unknowns (p), which typically consist of pressure variables, and secondary unknowns (s) that consist of saturations and mass fractions. The Jacobian matrix blocks read $(J_{m,\ell})_{i,j} = \partial R_{\ell,i}/\partial x_{m,j}$, where the indices m, ℓ run over the variable types $\{p, s\}$ and i, j run over all cells.

Instead of applying an iterative solver to the full linear system, we first form an (almost) elliptic equation for the primary unknowns. This equation can be achieved by modifying the fully implicit system to the form

$$-\begin{bmatrix} J_{pp}^* & J_{ps}^* \\ J_{sp} & J_{ss} \end{bmatrix} \begin{bmatrix} \Delta x_p \\ \Delta x_s \end{bmatrix} = \begin{bmatrix} R_p^* \\ R_s \end{bmatrix},$$

where the off-diagonals of the coupling matrix J_{ps}^* are small, so that $J_{pp}^* \Delta x_p = R_p^*$ resembles an incompressible pressure equation. Because this CPR pressure equation is almost elliptic, we can use on of the many efficient solvers developed for Poisson-type equations to compute an approximate increment Δx_p in an inexpensive manner. This *predictor stage* is used in combination with a *corrector stage*, in which we iterate on the entire system using the approximated pressure update as our initial guess.

The solver implemented in the `CPRSolverAD` class employs an IMPES-like reduction (plus dynamic row sums from [119]), in which we determine appropriate weights w_c so that the accumulation terms of the residual equations cancel

$$\sum_c w_c A_c = 0, \quad A_{c,i} = \left(\phi \sum_\alpha b_\alpha S_\alpha x_{c,\alpha}\right)_i^{n+1} - \left(\phi \sum_\alpha b_\alpha S_\alpha x_{c,\alpha}\right)_i^n.$$

Here, $x_{c,\alpha}$ gives the volume of component c at standard conditions present in a unit stock-tank of phase α at reservoir conditions, i.e., $x_{o,o} = 1$, $x_{o,g} = R_v$, and so on. The (almost) decoupled CPR pressure equation is then taken to be $R_p^* = \sum_c w_c R_c$. Notice that if components are in one phase only, the weight of component c is simply $1/b_\alpha$, where α is the phase of component c. An advantage of this decoupling strategy is that the resulting pressure block J_{pp}^* is fairly symmetric. Moreover, the decoupling can be done at a nonlinear level, before the Jacobians are computed, and is thus less prone to errors.

Aggregation-Based Algebraic Multigrid

The pressure equation in the CPR solver can obviously be solved using a standard direct solver. However, a more scalable approach is to use an iterative multilevel method. To present the idea, we consider only two levels: the fine level on which we seek the unknown solution x and a coarser level on which we seek another unknown vector $\tilde{x}$ having much fewer elements. To move between the two levels, we have a prolongation operator P and a restriction operator Q. Then, the iterative method can be written as follows:

$$x^* = x_\nu + S(b - Ax_\nu),$$
$$x_{\nu+1} = x^* + PA^{-1}Q(b - Ax^*).$$

Here, S is a so-called smoother, which typically consists of a few steps of an inexpensive iterative solver (like Jacobi's method or ILU(0)). For a multilevel method, this procedure is performed recursively on a sequence of coarser levels. In classical methods, the coarser levels are chosen by geometric partitions of the original grid. The state-of-the-art approach is to select the coarse levels algebraically and build the operator P by analyzing the matrix coefficients of A. The restriction operator is then chosen as the transpose of the interpolation matrix, $Q = P^T$, and the coarse matrix is $A_c = P^T A P$.

The AGMGSolverAD class is an interface to the aggregation-based algebraic multigrid (AGMG) solver [238, 15], which builds prolongation operators using recursive aggregates of cells with constant interpolation so that the operator on each level is as sparse as possible to avoid that the coarse matrices A_c become more dense the more one coarsens. AGMG was originally released as free open-source, but recent versions are only freely available for academic users. The main advantage of AGMG, beyond a simple MATLAB interface, is that it has low setup cost and low memory requirements. In my experience, it is easy to integrate and seems to work very well with MRST. For the black-oil equations, this solver should either be used as a preconditioner for the CPR solver, or as a pressure solver for the sequentially implicit method implemented in the blackoil-sequential module.

AGMG can also be run in two stages: a setup stage that builds the coarse hierarchies and the prolongation operators, and a solution stage that employs the prolongation operators to solve a specific problem. This way, you can hope to reduce computational costs by reusing the setup stage for multiple subsequent solves. This behavior can be set up as follows:

```
function [result, report] = solveLinearSystem(solver, A, b)
  cleanAfter = false;
  if ~solver.setupDone
     solver.setupSolver(A, b); cleanAfter = true;
  end
  if solver.reuseSetup
     fn = @(A, b) agmg(A, b, [], solver.tolerance, ...
                             solver.maxIterations, [], [], 1);
  else
     fn = @(A, b) agmg(A, b, [], solver.tolerance, solver.maxIterations);
  end
```

```
    [result, flag, relres, iter, resvec] = fn(A, b);
    if cleanAfter, solver.cleanupSolver(A, b); end
end
function solver = setupSolver(solver, A, b, varargin)
    if solver.reuseSetup
        agmg(A,[],[],[],[],[],[], 1); solver.setupDone = true;
    end
end
```

For brevity, we have dropped a few lines that create the full convergence report. This example hopefully shows that it is not very difficult to set up an external solver with MRST if the solver has a standard MATLAB interface.

Interface to External Linear Solvers

Recently, we added a new module called linearsolver to MRST, whose purpose is to offer a uniform interface to other external solvers on the form:

```
x = solver(A, b, tolerance, iterations)
```

So far, the module implements support for AMGCL [81], which is a recent and highly efficient header-only C++ library for solving large sparse linear systems with algebraic multigrid methods [296, 285]. AMGCL can either be used as a multigrid solver or as a smoother, and offers a variety of basic building blocks that can be combined in different ways:

- *smoothers*: damped Jacobi, Gauss–Seidel, ILU(0), ILU(k), sparse approximate inverse, and others;
- *coarsening methods*: aggregation, smoothed aggregation, smoothed aggregation with energy minimization, and Ruge–Stüben; and
- *Krylov subspace solvers*: conjugated gradients, stabilized biconjugate gradients, GMRES, LGMRES, FGMRES, and induced dimension reduction.

To use AMGCL as a CPR solver, you can, e.g., set it up as follows:

```
lsolve = AMGCL_CPRSolverAD('block_size', 2, ... ,
                          'maxIterations', 150, 'tolerance', 1e-3);
lsolve.setCoarsening('aggregation')
lsolve.setRelaxation('ilu0');
lsolve.setSRelaxation('ilu0');
lsolve.doApplyScalingCPR = true;
lsolve.trueIMPES = false;
:
```

At the time of writing, AMGCL is not part of the official MRST release and must be downloaded from a software repository [81], compiled, and integrated via a MEX interface. Initial tests show that AMGCL is significantly faster than AGMG, and installing it is definitely worth the effort if you want to use MRST on models with hundreds of thousand unknowns.

ECLIPSE Solver: Orthomin and Nested Factorization

The default solver in ECLIPSE is a Krylov subspace method that relies on a set of search vectors defined so that each new vector $q_{\nu+1}$ is orthogonal to all previous vectors $q_0, q_1, \ldots, q_\nu$. Assuming a preconditioner B and an initial guess x_0, we start the iteration process by setting $q_0 = B^{-1} r_0$, where $r_0 = b - Ax_0$. The next approximate solution is then defined as

$$x_{\nu+1} = x_\nu + \omega_\nu B^{-1} q_\nu,$$

which inserted into the linear system gives the residual

$$r_{\nu+1} = r_\nu - \omega_\nu A B^{-1} q_\nu = r_\nu - \omega_\nu \hat{q}_\nu.$$

The relaxation parameter ω_ν is optimized by minimizing $\|r_{\nu+1}\|_2$. The resulting *orthomin* method is a special version of the *conjugate gradient* method. For completeness, let us derive the remaining steps. The optimum value for $\omega_{\nu+1}$ is determined by setting

$$0 = \frac{\partial}{\partial \omega_{\nu+1}} \|r_{\nu+1}\|_2 = \frac{\partial}{\partial \omega_{\nu+1}} \left[r_\nu \cdot r_\nu - 2\omega_\nu r_\nu \cdot \hat{q}_\nu + (\omega_\nu)^2 \hat{q}_\nu \cdot \hat{q}_\nu \right]$$

$$= -2 r_\nu \cdot \hat{q}_\nu + 2\omega_\nu \hat{q}_\nu \cdot \hat{q}_\nu \quad \longrightarrow \quad \omega_\nu = \frac{r_\nu \cdot \hat{q}_\nu}{\hat{q}_\nu \cdot \hat{q}_\nu},$$

The search vectors are set to lie in the space spanned by $r_{\nu+1}$ and all the previous search vectors, i.e.,

$$q_{\nu+1} = r_{\nu+1} + \sum_{\ell=0}^{\nu} \alpha_\nu^\ell q_\ell,$$

subject to the constraint that $\hat{q}_{\nu+1}$ is orthogonal to all $\hat{q}_k$ for $k \leq \nu$, i.e.,

$$0 = \hat{q}_{\nu+1} \cdot \hat{q}_\ell = \left[AB^{-1} r_{\nu+1} + \sum_{\ell=0}^{\nu} \alpha_\nu^\ell \hat{q}_\ell \right] \cdot \hat{q}_k.$$

Using the fact that $\hat{q}_\ell \cdot \hat{q}_k$ for $\ell, k \leq \nu$ and $\ell \neq k$, we can easily show that

$$\alpha_\nu^\ell = -\frac{(AB^{-1} r_{\nu+1}) \cdot \hat{q}_\ell}{\hat{q}_\ell \cdot \hat{q}_\ell}.$$

This completes the definition of the *orthomin* method.

For grids having a rectangular topology, it is possible to develop a particularly efficient preconditioner called *nested factorization* [24]. As we have seen previously, MRST uses a *global* numbering of the unknowns, in which the numbering of unknowns first runs over cells and then over variables; this to utilize highly efficient vectorization from MATLAB. In a simulator written in a compiled language, it is usually more efficient to use a *local* numbering that first runs over variables and then over cells. With a standard TPFA and single-point upwind-mobility discretization, the resulting system will have a block-heptagonal form,

$$A = L_3 + L_2 + L_1 + D + U_1 + U_2 + U_3$$

where D is a diagonal matrix whose elements are $m \times m$ matrices representing the unknowns per cell. The block matrices L_i and U_i are lower and upper bands representing connections to neighboring cells in logical direction $i = 1, 2, 3$. Experience shows that the numbering should be chosen so that direction 1 corresponds to the direction of highest transmissibilities, which usually is the z-direction; see [271], whose presentation we henceforth follow. The key idea of nested factorization is now to utilize this special structure to recursively define a set of preconditioners, whose action can be computed iteratively by solving block-tridiagonal systems. These can be solved very efficiently by Thomas's algorithm that consists of two substitution sweeps, one forward and one backward. The factorization is built iteratively using the following definition

$$B = (P + L_3)P^{-1}(P + U_3), \tag{12.3}$$

$$P = (T + L_2)T^{-1}(T + U_2), \tag{12.4}$$

$$T = (G + L_1)G^{-1}(G + U_1). \tag{12.5}$$

With a slight abuse of notation, we should understand the terms on the right-hand side as values from the previous iteration. The matrix G is a block-diagonal matrix, which for a fully implicit discretization is defined as,

$$G = D - L_1 G^{-1} U_1 - \mathrm{colsum}(L_2 T^{-1} U_2) - \mathrm{colsum}(L_3 P^{-1} U_3). \tag{12.6}$$

This definition ensures that although the solution may not yet be correct, it will have no mass-balance errors. Alternatively, in the IMPES and AIM methods, we can construct G using row sum rather than column sum to ensure accurate pressure.

Let us now look at how the solution proceeds: Starting at the outmost level, we see that to solve $By = r$, we need to solve $(P + L_3)(I + P^{-1}U_3)y = r$. This can be decomposed as follows

$$\hat{y} = P^{-1}(r - L_3 \hat{y}), \qquad y = \hat{y} - P^{-1}U_3 y$$

If $\hat{y}$ is known in the first plane of cells, solving these two equations reduces to a forward elimination and a backward substitution, since the products $L_3 \hat{y}$ and $U_3 y$ only involve values along planes where the solution is known. In this procedure, we need to invert P and compute vectors $z = P^{-1}u$ along planes of cells. Applying the same idea to (12.4), we obtain

$$\hat{z} = T^{-1}(u - L_2 \hat{z}), \qquad z = \hat{z} - T^{-1}U_2 z,$$

which we can solve by a forward elimination and a backward substitution provided we know the solution along a first line of cells in each plane. Here, we must compute $w = T^{-1}v$ on each line, which again can be decomposed and solved efficiently as follows:

$$\hat{w} = T^{-1}(v - L_1\hat{w}), \qquad w = \hat{w} - T^{-1}U_1 w,$$

provided we know G along the lines. From the structure of (12.6), it follows that we can construct G one cell at a time: Starting from a known inverse block G^{-1} in a cell, we can compute the contribution $L_1 G^{-1} U_1$ to the next cell in direction 1, and so on. Once G^{-1} is known along a line in direction 1, we can compute the contribution of $\operatorname{colsum}(L_2 T^{-1} U_2)$ using the forward-elimination, backward-substitution algorithm for T^{-1}. When all the lines in a plane are known, we can compute the contributions from $\operatorname{colsum}(L_3 P^{-1} U_3)$ to the next plane in a similar manner. The method may look laborious, but is highly efficient. Unfortunately, it is not applicable to fully unstructured grids.

12.4 Simulation Examples

You have now been introduced to all key components that make up the AD-OO framework. In the following, we demonstrate in more detail how to use the AD-OO framework to perform simulations with or without the use of structured input decks on the ECLIPSE format.

12.4.1 Depletion of a Closed/Open Compartment

The purpose of this example[1] is to demonstrate how to set up a simulator from scratch without the use of input files. To this end, we consider a 2D rectangular reservoir compartment with homogeneous properties filled by a single-phase fluid. The reservoir fluid is drained by a single producer at the midpoint of the top edge. The compartment is either closed (i.e., sealed by no-flow boundary conditions along all edges), or open with constant pressure support from an underlying, infinite aquifer, which we model as a constant-pressure boundary condition.

We start by making the geological model and fluid properties

```
G     = computeGeometry(cartGrid([50 50],[1000 100]));
rock  = makeRock(G, 100*milli*darcy, 0.3);
pR    = 200*barsa;
fluid = initSimpleADIFluid('phases','W', ... % Fluid phase: water
                           'mu',  1*centi*poise, ... % Viscosity
                           'rho', 1000,          ... % Surface density [kg/m^3]
                           'c',   1e-4/barsa,    ... % Fluid compressibility
                           'cR',  1e-5/barsa));  ... % Rock compressibility
```

This is all we need to set up a single-phase reservoir model represented by `WaterModel`, which is a specialization of the general `ReservoirModel` implemented in `ad-core`. The only extra thing we need to do is to explicitly set the gravity direction. By default, gravity

---

[1] Complete code: `blackoilTutorialOnePhase.m` in ad-blackoil.

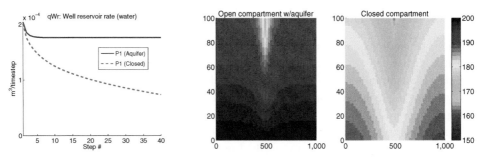

Figure 12.6 Drainage from a single reservoir compartment that is either sealed or supported by an infinite aquifer from below.

in MRST is a three-component vector that points in the positive z-direction. Here, we set it to a two-component vector pointing in the negative y-direction.

```
gravity reset on
wModel = WaterModel(G, rock, fluid,'gravity',[0 -norm(gravity)]);
```

As always, we also need to specify the drive mechanisms by use of the standard data structures for wells and boundary conditions:

```
wc = sub2ind(G.cartDims, floor(G.cartDims(1)/2), G.cartDims(2));
W = addWell([], G, rock, wc,         ...
        'Type', 'bhp', 'Val', pR - 50*barsa, ...
        'Radius', 0.1, 'Name', 'P1','Comp_i',1,'sign',1);
bc = pside([],G,'South',200*barsa,'sat',1);
```

We then create two simulation schedules, one with well and constant pressure on the south boundary, and the other with well and no-flow conditions on all boundaries:

```
schedule1 = simpleSchedule(diff(linspace(0,5*day,41)), 'bc', bc, 'W', W);
schedule2 = simpleSchedule(diff(linspace(0,5*day,41)), 'W', W);
```

The pristine reservoir is in vertical equilibrium, which we compute as a steady-state solution of the flow equation subject to boundary conditions only

```
state.pressure = ones(G.cells.num,1)*pR;
state.wellSol  = initWellSolAD([], wModel, state);
nonlinear = NonLinearSolver;
state = nonlinear.solveTimestep(state, 10000*day, wModel, 'bc', bc);
```

We can the compute each flow solution as follows:

```
[wellSols1, states1] = simulateScheduleAD(state, wModel, schedule1);
```

Figure 12.6 shows the solution of the two problems. Once the well is turned on, the drawdown at the wellbore will create an elongated pressure pulse that moves fast downward and expands slowly outward. The open compartment quickly reaches a steady-state in which the flow rate into the well equals the influx across the bottom boundary. In the closed case, the pressure inside the compartment is gradually depleted towards the perforation pressure, which in turn causes the production rate to decline towards zero.

12.4.2 An Undersaturated Sector Model

The next example[2] continues in the same vein as the previous: we consider a $1,000 \times 1,000 \times 100$ m^3 rectangular sector model filled with undersaturated oil and compare and contrast the difference in production when the boundaries are closed or held at constant pressure. The phase behavior of the oil and dissolved gas is assumed to follow that of the SPE 1 benchmark, but unlike in the original setup, water is assumed to be mobile, following a simple quadratic relative permeability.

```
[~, ~, fluid, deck] = setupSPE1(); fluid.krW = @(s) s.^2;
```

Permeability is homogeneous and isotropic, and saturation and pressure are for simplicity assumed to be uniformly distributed inside the reservoir:

```
[s0, p0] = deal([0.2, 0.8, 0], 300*barsa);
state0 = initResSol(G, p0, s0);
state0.rs = repmat(200, G.cells.num, 1);
model = ThreePhaseBlackOilModel(G, rock, fluid, 'disgas', true);
```

The reservoir is produced from a single vertical well, placed in the center, operating at a fixed oil rate, but with a lower bottom-hole pressure (bhp) limit:

```
ij = ceil(G.cartDims./2);
W = verticalWell([], G, rock, ij(1), ij(2), [], ...
                'val', -0.25*sum(model.operators.pv)/T, ...
                'type', 'orat', 'comp_i', [1, 1, 1]/3, ...
                'sign', -1, 'name', 'Producer');
W.lims.bhp = 100*barsa;
```

The AD-OO framework assumes no-flow boundary conditions by default, and hence we do not need to specify boundary conditions to simulate closed boundaries. For the other case, we assume that the pressure and fluid composition at the boundary are held at their initial values to model additional pressure support, e.g., from a gas cap or an aquifer with highly mobile water. Although such boundary conditions have been used often in the past, there is in reality no physical mechanism that would be able to maintain constant boundary conditions in a reservoir, so this is a gross simplification that should not be applied when modeling real cases.

---

[2] Complete code: blackoilSectorModelExample.m in ad-blackoil

```
bc = []; sides = {'xmin', 'xmax', 'ymin', 'ymax'};
for side = 1:numel(sides)
   bc = pside(bc, G, sides{side}, p0, 'sat', s0);
end
```

We also need to specify the R_s value on all boundaries. For this purpose, MRST employs a dissolution matrix, which can either be specified per cell interface, or for all interfaces. Here, the fluid composition is the same on the whole boundary, so we only specify a single matrix

```
bc.dissolution = [1, 0, 0;... % Water fractions in phases
                  0, 1, 0; ...% Oil fractions in phases
                  0, 200, 1]; % Gas fractions in phases
```

To define specific dissolution factors on n individual cell faces, we would have to set up an $n \times 3 \times 3$ array.

Turning on the well induces a fast pressure transient that lasts a few days. We do not try to resolve this, but gradually ramp up the time step, as discussed earlier in the book, to improve the initial stability of the simulation:

```
dt = rampupTimesteps(T, 30*day);
```

Specifically, the routine splits the simulation horizon T into a number of time step of length 30 days (the last step may be shorter if T is not a multiple of the specified time step). It then uses a geometric sequence 1./2.^[n n:-1:1] to further subdivide the first time step (default value of n is 8). We can now set up and simulate the two schedules in the same way as discussed in the previoius subsection

```
schedule = simpleSchedule(dt, 'W', W, 'bc', bc);
[ws, states] = simulateScheduleAD(state0, model, schedule);
```

Figure 12.7 shows bottom-hole pressure and surface oil rate for the two simulation, i.e., the quantities used to control the production well. After the initial transient, the case with constant pressure boundary settles at a steady state with constant bhp and oil rate maintained at its target value. With closed boundaries, there is no mechanism to maintain the reservoir pressure, and hence the bottom-hole pressure starts to gradually decay. After 870 days, the bhp drops below its lower limit of 100 bar, and the well switches control from oil rate to bhp. As production continues, the reservoir pressure will gradually be depleted until the point where there is not sufficient pressure difference between the reservoir and the well to drive any flow into the wellbore; this happens after approximately 4.5 years. Grossly speaking, qualitative behavior of the pressure is similar to what we observed in the previous single-phase example.

Figure 12.8 reports cumulative production of oil, gas, and water at surface conditions, as well as instantaneous production rates of the oleic, gaseous, and aqueous phases at

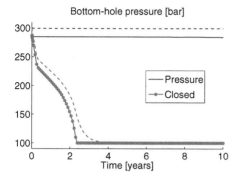
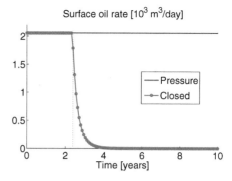

Figure 12.7 Bottom-hole pressure and surface oil rate for the undersaturated sector model; dashed lines show field-average pressure. With closed boundaries, the well control switches from oil surface rate to bottom-hole pressure after 870 days (shown as dotted lines).

reservoir conditions. As the reservoir pressure is gradually lowered, some cells drop below the bubblepoint and free gas is liberated, starting near the wellbore, where the pressure drop is largest, and gradually moving outward as shown in Figure 12.9.

Oil is only produced from the oleic phase. When free gas is liberated, the fraction of oil in the oleic phase increases and the oil reservoir rate decays slightly because the reservoir volume of the oleic phase required to produce a constant surface rate of oil decreases. The liberated gas is very mobile and causes an accelerating pressure drop as more gas starts flowing into the wellbore. Gas production at the surface thus increases dramatically, because gas is now produced both from the oleic and the gaseous phase. Gas can sometimes be sold and contribute to generate revenue, but the operator may also have to dispose of it by reinjecting it into the reservoir to maintain reservoir pressure. Unlike in the original SPE 1 case, the aqueous phase is mobile. As gas is liberated, the mobility of oil is reduced and we also see a significant increase in water production. Topside facilities usually have limited capacity for handling produced water, and the whole production may be reduced or shut down if the water rate exceeds the available capacity. Here, it might have been advantageous if the well was instrumented with autonomous inflow devices that would choke back water and gas and only let oil flow through. We will return to this in the next example.

As the well switches to bhp control, all rates fall rapidly. For oil in particular, the additional recovery after the well switches is very limited: at 870 days, the cumulative oil production has reached 93% of what is predicted to be the total production after 10 years. The liberated gas (and the water) will continue to flow longer, but also here the production ceases more or less after 4.5 years.

This idealized example is merely intended to demonstrate use of the AD-OO framework and illustrate basic flow mechanisms and makes no claim of being representative of real hydrocarbon reservoirs.

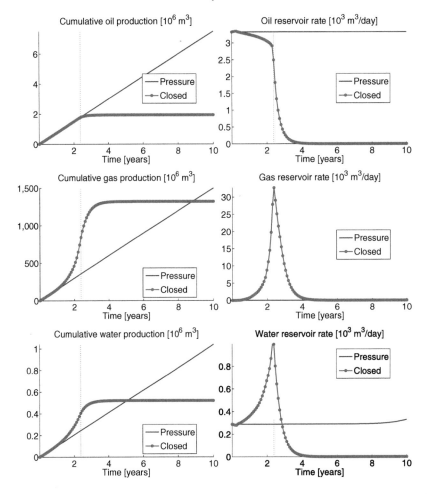

Figure 12.8 Production profiles for the undersaturated sector model. The left column shows cumulative production of the oil, gas, and water components at surface conditions, whereas the right column shows instantaneous phase rates at reservoir conditions.

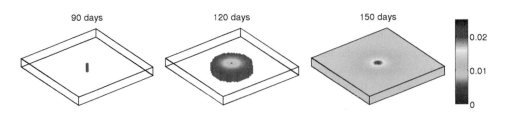

Figure 12.9 Liberation of free gas as the drawdown from the well causes the pressure in the undersaturated oil to drop below the bubblepoint. (The plot shows $S_g > 5\text{e-}3$.)

COMPUTER EXERCISES

12.4.1 Try to simulate the initial pressure transient, check how long it lasts, and whether it behaves as in the previous example.

12.4.2 The simulation discussed in this section did not include any consideration of production relative to the amount of fluids initially in place. Try to include this to determine the recovery factor for the closed case. Can you also use this to say something about the influx across the boundaries in the case with pressure boundaries.

12.4.3 With constant pressure boundaries, the water rate starts to increase after approximately seven years. Can you explain this behavior?

12.4.4 Gravity was not included in this example. Can you set up multilayer simulations to investigate if gravity has a significant effect on the result?

12.4.3 SPE 1 Instrumented with Inflow Valves

This example[3] demonstrates the use of multisegment wells in MRST. The default well model in MRST assumes that flow in the well takes place on a very short time-scale compared to flow in the reservoir and thus can be accurately modeled as being instantaneous, following the linear inflow performance relationship discussed in Section 11.7.1. Such models are not adequate for long wellbores that have significant pressure drop due to friction, or for wells having valves or more sophisticated inflow control devices. The multisegment well class in MRST supports transient flow inside the well and enables a more fine-grained representation of the well itself that allows the effects of friction, acceleration, and nonlinear pressure drops across valves and other flow constrictions to be included; see Section 11.7.2.

To illustrate, we revisit the SPE 1 case from Section 11.8.1 and replace the vertical producer by a horizontal well completed in six cells with valves between the production tubing and each sandface connection. We consider two models, a simple model assuming instantaneous flow with a standard inflow relationship and hydrostatic pressure along the wellbore, and a more advanced model that accounts for pressure drop across the valves and frictional pressure drop along the horizontal wellbore; see Figure 12.10.

First, we load the description of the fluid model and the simulation schedule from an ECLIPSE input deck:

```
[G, rock, fluid, deck, state] = setupSPE1();
[nx, ny] = deal(G.cartDims(1),G.cartDims(2));

model = selectModelFromDeck(G, rock, fluid, deck);
model.extraStateOutput = true;
schedule = convertDeckScheduleToMRST(model, deck);
```

---

[3] Complete source code: multisegmentWellExample.m in ad-blackoil

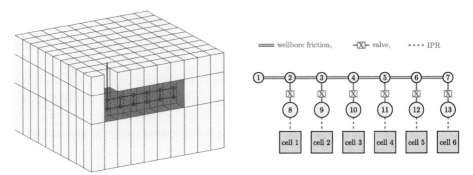

Figure 12.10 Multisegment well for the modified SPE 1 case. Nodes 1–7 represent the wellbore, whereas nodes 8–13 represent the void outside of the valves that separate the wellbore from the sandface.

The model selector examines which phases are present and whether the deck prescribes dissolved gas or vaporized oil, and then determines the specific black-oil subclass that is most suitable for simulating the deck. As we already know, the SPE 1 benchmark describes a three-phase model with dissolved gas but no vaporized oil. The model will thus be an instance of the general `ThreePhaseBlackOilModel` class. Likewise, the schedule converter will parse the specification of all wells and create a suitable schedule object, which here has 120 steps and a single control.

To specify the well models, we first initialize the production well as a standard well structure completed in the six given grid cells:

```
c = nx*ny + (2:7)';
prod0 = addWell([], G, rock, c, 'name', 'prod', ...
          'refDepth', G.cells.centroids(1,3), ...
          'type', 'rate', 'val', -8e5*meter^3/day);
```

We then define the N_1, N_2 mappings for the twelve segments, as well as the mapping from network nodes to reservoir cells represented in the well structure

```
topo = [1 2 3 4 5 6 2 3  4  5  6  7
        2 3 4 5 6 7 8 9 10 11 12 13]';
%       |   tubing  |     valves    |
cell2node = sparse((8:13)', (1:6)', 1, 13, 6);
```

The `cell2node` matrix has dimension equal the number of nodes times the number of reservoir cells perforated by the well. We must also specify the segment lengths and diameters and the node depths and volumes

```
lengths = [300*ones(6,1); nan(6,1)];
diam    = [.1*ones(6,1); nan(6,1)];
depths  = G.cells.centroids(c([1 1:end 1:end]), 3);
vols    = ones(13,1);
```

With this, we have all the information we need to convert the data structure of the simple well into a corresponding data structure describing a multisegment network model:

```
prodMS = convert2MSWell(prodS, 'cell2node', cell2node, 'topo', topo, ...
                       'G', G, 'vol', vols, 'nodeDepth', depths, ...
                       'segLength', lengths, 'segDiam', diam);
```

Finally, we must set up the flow model for each segment. The first six segments are modeled using the wellbore-friction model (11.29)–(11.31) with roughness $e = 10^{-4}$ m and discharge coefficient $C_v = 0.7$. The next six segments represent valves, whose pressure losses are modeled by use of (11.29) with mixture mass flux. The valves have thirty openings, each with a nozzle diameter of 25 mm. We then set up the flow model as function of velocity, density, and viscosity

```
[wbix, vix]  = deal(1:6, 7:12);
[roughness, nozzleD, discharge, nValves] = deal(1e-4, .0025, .7, 30);

prodMS.segments.flowModel = @(v, rho, mu)...
    [wellBoreFriction(v(wbix), rho(wbix), mu(wbix), ...
                     prodMS.segments.diam(wbix), ...
                     prodMS.segments.length(wbix), roughness, 'massRate'); ...
     nozzleValve(v(vix)/nValves, rho(vix), nozzleD, discharge, 'massRate')];
```

The injector is a simple well modeled by the standard inflow performance relationship, i.e., a Peaceman type well model,

```
inj = addWell([], G, rock, 100, 'name', 'inj', 'type', 'rate', 'Comp_i', ...
    [0 0 1], 'val', 2.5e6*meter^3/day,'refDepth', G.cells.centroids(1,3));
```

To compare the effect of instrumenting the production well with inflow valves, we run two different schedules: The first is a baseline simulation in which the producer is treated as a simple well with instantaneous flow and one node per contact between the well and the reservoir:

```
schedule.control.W = [inj; prodS];
[wellSolsSimple, statesSimple] = simulateScheduleAD(state, model, schedule);
```

For the second simulation, we must combine the simple injector and multisegment producer into a facility model since the two wells are represented with different types of well models

```
W = combineMSwithRegularWells(inj, prodMS);
schedule.control.W = W;
model.FacilityModel = model.FacilityModel.setupWells(W);
```

This is normally done automatically by the simulator, but here we do it explicitly on the outside to view the output classes. These classes are practical if we want to incorporate per-well adjustments to tolerances or other parameters. The model classes and the well structures read

```
SimpleWell with properties:                         MultisegmentWell with properties:

                                                                signChangeChop: 0
                      W: [1x1 struct]                                         W: [1x1 struct]
          allowCrossflow: 1                             allowCrossflow: 1
        allowSignChange: 0                             allowSignChange: 0
    allowControlSwitching: 1                         allowControlSwitching: 1
                dpMaxRel: Inf                                   dpMaxRel: Inf
                dpMaxAbs: Inf                                   dpMaxAbs: Inf
                dsMaxAbs: 0.2000                                dsMaxAbs: 0.2000
                VFPTable: []                                    VFPTable: []
                operators: []                                   operators: [1x1 struct]
        nonlinearTolerance: 1.0000e-06                  nonlinearTolerance: 1.0000e-06
                      G: []                                           G: []
                verbose: 0                                      verbose: 0
    stepFunctionIsLinear: 0                         stepFunctionIsLinear: 0

W is structure with fields:                         W is structure with fields:
                cells: 100                                      cells: [6x1 double]
                type: 'rate'                                    type: 'bhp'
                val: 28.9352                                    val: 25000000
                  r: 0.1000                                       r: 0.1000
                    :                                                :
            cell2node: []                                   cell2node: [13x6 double]
              connDZ: []                                      connDZ: [6x1 double]
                isMS: 0                                         isMS: 1
                nodes: []                                       nodes: [1x1 struct]
            segments: []                                    segments: [1x1 struct]
```

The main difference in the two classes is that the `SimpleWell` model does not have any operators defined. Likewise, this model does not have any representation of cell to node mappings and structures representing the nodes and segments in the W structure. However, what MRST uses to distinguish simple and multisegment wells is the flag `isMS`, which is set to false for the injector and true for the producer. For completeness, let us also look at the data structures representing nodes and segments for the instrumented producer

```
segments:                                           nodes:
            length: [12x1 double]                           depth: [13x1 double]
        roughness: [12x1 double]                            vol: [13x1 double]
              diam: [12x1 double]                           dist: [13x1 double]
        flowModel: [function_handle]
              topo: [12x2 double]
    deltaDistance: [12x1 double]
```

The crucial element here is the function handle `flowModel` in the segments, which is used to assemble the well equations.

Figure 12.11 shows bhp in the producer and injector, along with field-average pressure predicted by the two different well models. Resistance to flow when the flow is constricted through the nozzles of the valves and by friction along the tubing causes a significantly higher pressure drop from the sandface to the datum point in the wellbore. Hence, the bhp must be significantly lower for the multisegment model, compared with the simple well model to sustain the same surface production rate[4]. Looking at the injector bhp, we see that a relatively high pressure is required initially to force the first gas into the reservoir. As soon as the first gas has entered the reservoir, the total mobility in the near-well region increases

---

[4] Notice that this comparison is somewhat exaggerated. In reality, one would likely have adjusted either the well index or the skin factor in the simple well model to account for some of this pressure loss. On the other hand, such representative values can only be found after one has observed some production history.

12.4 Simulation Examples

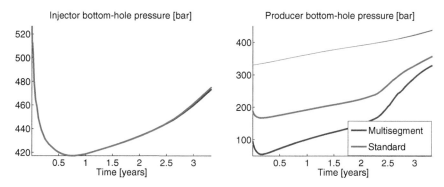

Figure 12.11 Comparison of bottom-hole pressures in the injector (left) and the producer (right) predicted for the modified SPE 1 model with a simple well model and a well model accounting for pressure drop across valves and caused by friction along the horizontal wellbore. Thin lines in the right plot show average-field pressure.

and the injection pressure decays. With some gas accumulated, the total flow resistance of the pore space from injector to producer will decrease, and hence the bhp starts to increase in the producer after approximately 60 days. On the other hand, the injected gas is not able to displace a similar amount of reservoir fluids from the reservoir because of lower density and higher compressional effects, and hence the reservoir pressure will gradually increase. This effect is observed in both models.

After approximately 20 months, the first gas has reached a neighboring cell of the producer and will start to engulf the cells above and gradually be drawn down into the perforated cells; see Figure 12.12. This engulfing process causes a significant decay in the oil rate and a steeper incline in producer bhp. The rate decay is slightly delayed in the multisegment model, but the effect on the total oil production is a mere 1.3% increase (which still could represent significant income).

Figure 12.13 shows difference in pressure between the wellbore and the perforated cells for both well models. The simple well model assumes instantaneous hydrostatic fluid distribution, and since the producer is completely horizontal, the pressure is identical in all the six points where the wellbore is open to the reservoir. The instrumented well has a significantly larger pressure drawdown near its heel, which also lies furthest away from the injector and the advancing gas front. Both models use the same inflow performance relationship and hence the drawdown outside of the valve in the multisegment model will on average be approximately the same as for the simple well model when both wells are run on rate controls.

COMPUTER EXERCISES

12.4.5 Our setup had no lower limit on the bhp. What happens if bhp in the producer is not allowed to drop below 100 bar or 150 bar?

12.4.6 Rerun the two simulations with producer controlled by a bhp of 250 bar. Can you explain differences in the producer and the bhp of the injector?

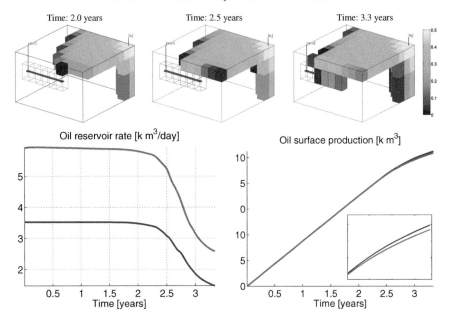

Figure 12.12 Displacement front engulfing the horizontal producer causes decay in oil production. Because of different pressure in the wellbore, the volumetric reservoir rates is largely different in the two models, even though surface rates are almost identical.

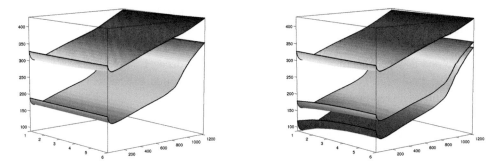

Figure 12.13 Pressure in perforated cells (top) and wellbore (bottom) as function of time for the simple well model (left) and the multisegment well model (right). In the latter plot, the middle surface represents pressure outside of the valves.

12.4.4 The SPE 9 Benchmark Case

In our last example,[5] we set up a more realistic simulation case, show how to use CPR preconditioning together with an algebraic multigrid solver, and compare our simulation results against ECLIPSE to verify that the AD-OO black-oil simulator is correct. To this

---

[5] Complete source code: blackoilTutorialSPE9.m in ad-blackoil

12.4 Simulation Examples

end, we consider the model from the SPE 9 comparative solution project [159], which was posed in the mid-1990s to compare contemporary black-oil simulators. The benchmark consists of a water-injection problem in a highly heterogeneous reservoir described by a $24 \times 25 \times 15$ regular grid with a 10-degree dip in the x-direction. By current standards, the model is quite small, but contains several salient features that must be properly handled by any black-oil simulator aiming to solve real problems. On one hand, the fluid and PVT model have certain peculiarities; we have already discussed these models in detail in Sections 11.3.3 and 11.5, respectively. In particular, the discontinuity in the water–oil capillary pressure curve is expected to challenge the Newton solver when saturations change significantly. Other challenges include strong heterogeneity, transition from undersaturated to saturated state, abrupt changes in prescribed production rates, and dynamic switching between rate and bhp control for the producers. Altogether, this makes the case challenging to simulate.

Petrophysical Model and Initial Conditions

The basic setup of the grid, petrophysical parameters, fluid model, and initial state is more or less identical to that of the SPE 1 model discussed in Section 11.8.1. For convenience, this is implemented in a dedicated function:

```
[G, rock, fluid, deck, state0] = setupSPE9();
model    = selectModelFromDeck(G, rock, fluid, deck);
schedule = convertDeckScheduleToMRST(model, deck);
```

The reservoir rock is highly heterogeneous, albeit not as bad as the SPE 10 benchmark. The model consists of 14 unique layers of varying thickness. The permeability is isotropic and follows a lognormal distribution with values spanning more than six orders of magnitude; see Figure 12.14. The porosity field has no variation within each layer. One of the high-permeability layers is represented by two cells in the vertical direction, whereas two of the other layers have the same porosity, altogether giving 12 unique porosity values.

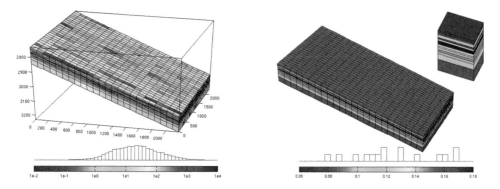

Figure 12.14 Petrophysical data for the SPE 9 model: the left plot shows absolute permeability in units md and the right plot shows porosity.

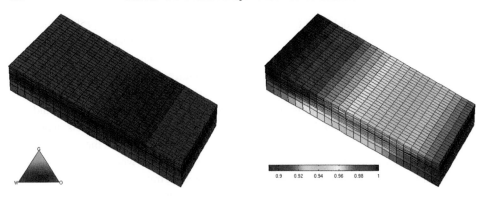

Figure 12.15 Initial fluid distribution inside the reservoir. The left figure shows the saturation, which lies on the two-phase O–W axis in the ternary diagram, whereas the right plot shows the degree of saturation defined as $r_s/R_s(p)$.

Initially, the reservoir does not contain any free gas, but a significant amount of gas is dissolved into the oleic phase. The dissolved amount r_s is constant throughout the whole model, except for three individual cells. Some of the dissolved gas may be liberated as free gas if the reservoir pressure drops below the bubblepoint during production. As we know, the solution gas–oil ratio R_s varies with pressure, which in turn increases with depth. At the top of the reservoir, the oleic phase is fully saturated ($r_s = R_s$), whereas the degree of saturation is only 89% at the bottom of the reservoir ($r_s = 0.89 R_s$). We should therefore expect that some free gas will form at the top of the reservoir as pressure drops once the wells start producing.

Wells and Simulation Schedule

The reservoir is produced from 25 vertical wells that initially operate at a maximum rate of 1,500 STBO/D (standard barrels of oil per day). Pressure is supported by a single water injector, operating at a maximum rate of 5,000 STBW/D (standard barrels of water per day) and a maximum bhp of 4,000 psi at reference depth. Between days 300 and 360, the production rate is lowered to 100 STBO/D, and then raised again to its initial value until the end of simulation at 900 days. Throughout the simulation, most of the wells switch from rate control to pressure control.

The simulation schedule consists of three control periods: days 0–300, days 300–360, and days 360–900, with a ramp-up of time steps at the beginning of the first and third period. All 26 wells are present during the entire simulation. The injector is injecting a constant water rate, while the producers are set to produce a target oil rate, letting bhp and gas/water production vary. In addition to the rate targets, the producers are limited by a lower bhp value of 1,000 psi, which could, e.g., signify the lowest pressure we can have in the well and still be able to lift reservoir fluids to the surface. The prescribed limits are given in the `lims` field of each well. For producer number two, this would be `schedule.control(1).W(22).lims`:

12.4 Simulation Examples

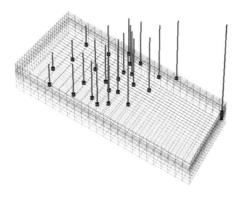
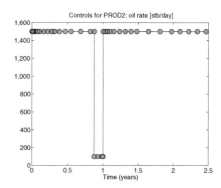

Figure 12.16 Well placement in the reservoir (left) and simulation schedule with rate targets for the producers (right).

```
Well limits for PROD2:
   orat: -0.0028
   wrat: -Inf
   grat: -Inf
   lrat: -Inf
    bhp: 6.8948e+06
```

By convention, the oil rate has negative sign for a producer. The pressure control will be activated if the wellbore pressure falls below the lower limit.

Customizing Linear Solver and Time-Step Control

The model has 9,000 active cells, which for a three-phase problem corresponds to a linear system with 27,000 cell unknowns; the total number of unknowns is slightly higher because of the 26 wells. Models of this size can be solved with the standard direct solvers in MATLAB on most computers, but runtimes can be reduced if we instead use a CPR preconditioner. To solve the elliptic pressure subsystem, we use the algebraic multigrid solver AGMG, with MATLAB's standard direct solver as potential backup, if AGMG is not installed on the computer:

```
try
  mrstModule add agmg
  pressureSolver = AGMGSolverAD('tolerance', 1e-4);
catch
  pressureSolver = BackslashSolverAD();
end
linsolve = CPRSolverAD('ellipticSolver', pressureSolver, ...
                      'relativeTolerance', 1e-3);
```

We also adjust the default time-step control slightly to ensure stable simulation by tightening the restriction on relative changes pressure and absolute changes in saturation

```
model.dpMaxRel  = .1;
model.dsMaxAbs  = .1;
```

Beyond this, we make no modifications to the standard configuration of the simulator, which employs robust defaults from ECLIPSE.

Running the Simulation

The simulation can now be started by the following statements:

```
model.verbose = true;
[wellsols, states, reports] =...
    simulateScheduleAD(state0, model, schedule,'LinearSolver', linsolve);
```

We give the schedule with well controls and control time steps. The simulator may use other time steps internally, but it will always return values at the specified control steps. Turning on verbose mode ensures that we get an extensive report of the nonlinear iteration process for each time step:

```
Solving timestep 01/35:                          -> 1 Day
Well INJE1: Control mode changed from rate to bhp.
=================================================================================================
| It # | CNV_W     | CNV_O     | CNV_G     | MB_W      | MB_O      | MB_G      | waterWell| oilWells | gasWells | closure  |
=================================================================================================
|    1 | 2.82e-02  | 1.10e+00  | 5.00e-03  | 1.18e-05  | 2.00e-04  | 1.58e-06  | 6.81e-04 | 5.00e-02 | 1.73e-01 |*0.00e+00 |
|    2 | 1.14e-01  | 7.25e-02  | 5.47e-02  | 5.17e-06  | 1.48e-04  | 8.86e-04  | 4.40e-03 | 2.20e-03 | 1.24e+00 | 1.47e+06 |
Well PROD26: Control mode changed from orat to bhp.
|    3 |*6.37e-04  | 2.38e-02  | 6.68e-02  |*7.78e-09  | 2.24e-04  | 1.13e-03  |*6.04e-07 | 4.58e-05 | 6.02e-02 |*1.73e-18 |
|    4 | 1.89e-03  | 8.12e-03  | 2.55e-02  | 7.23e-06  | 4.74e-05  | 3.56e-04  |*5.09e-06 | 1.85e-04 | 5.16e-02 | 5.67e+05 |
|    5 |*6.04e-04  | 1.69e-03  |*6.50e-04  | 1.32e-06  | 3.47e-06  | 1.90e-06  |*5.17e-07 | 1.92e-05 | 4.85e-03 |*8.67e-19 |
|    6 |*6.31e-04  | 1.46e-03  |*6.72e-04  | 1.16e-06  | 4.47e-06  | 1.74e-06  |*2.28e-10 | 1.56e-05 | 3.87e-03 |*0.00e+00 |
|    7 |*9.16e-05  |*3.79e-04  |*2.20e-04  |*7.16e-08  | 3.82e-07  |*9.41e-09  |*3.58e-11 |*2.41e-07 | 6.84e-05 |*0.00e+00 |
|    8 |*7.96e-05  |*2.45e-04  |*1.08e-04  |*5.50e-08  | 2.40e-07  |*7.63e-08  |*2.87e-14 |*3.82e-11 |*1.10e-08 |*0.00e+00 |
|    9 |*1.37e-05  |*5.70e-05  |*3.43e-05  |*2.15e-09  |*1.20e-09  |*1.22e-09  |*1.79e-14 |*7.86e-11 |*2.40e-08 |*0.00e+00 |
=================================================================================================
Solving timestep 02/35: 1 Day                    -> 2 Days
```

Here, we recognize the mass-balance (MB) errors and the maximum normalized residuals (CNV) discussed on page 438 for water, oil, and gas. The next three columns report residuals for the inflow performance relationship defined per perforations, whereas the last column reports residual for the control equations defined per well. A star means that the residual is below the tolerance. In the first time step, mass-balance error for oil was the last one to drop below the prescribed value of 10^{-9}. Nine iterations is a quite high number and indicates that there are significant nonlinearity in the residual equations, primarily caused by initial transient effects. The next seven control steps all converge within 6 iterations. Between days 80 and 100, liberated gas reaches the water zone and the next two steps now require 7 and 12 iterations. Large jumps in the iteration count can typically be traced back to interaction of different displacement fronts, liberation of gas or other abrupt phase transitions, and/or steep/discontinuous changes in the well controls or other parameters affecting the global drive mechanisms.

Switching of Well Controls

Figure 12.17 shows which control is active during each of the 35 control steps. Initially, all wells are set to be rate controlled. The targeted injection rate of 5,000 STBW/day is much

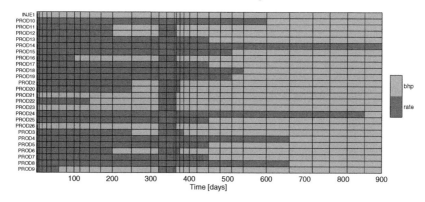

Figure 12.17 Well controls active during the simulation of the SPE 9 benchmark.

higher than what can be sustained with reasonable injection pressures. The injector thus switches immediately from rate to bhp control during the first time step and stays there throughout the whole simulation. Likewise, producer PROD26 and producers PROD21 and PROD23 switch from rate to bhp control during the first and second control step, respectively. This switching is reported to screen during simulation, and in the report on the facing page, we see that the injector switches during the first iteration and producer PROD26 after the second iteration. The overall picture is that producers are mostly rate controlled in the beginning and mostly controlled by bhp at the end; the only exception is PROD14, which runs rate control the whole period. This general switch from rate to bhp control is a result of the average field pressure dropping during the simulation as fluids are removed from the reservoir. We also notice the middle period when rates are reduced by a factor of ten. A smaller drawdown is required to sustain this rate, and hence all producers are able to stay above the bhp limit.

Analysis of Simulation Results

This reservoir is an example of a combined drive, in which two different mechanisms contribute to push hydrocarbons towards the producers. The first is injection of water below the initial oil–water contact, which contributes to maintain reservoir pressure and push water into the oil zone to displace oil. Ideally, this would have been a piston-like displacement that gradually pushed the oil–water contact upward, but because of heterogeneity and nonlinear fluid mobilities, water displaces oil as a collection of "fingers" extending from the oil–water contact and towards the producers; see Figure 12.18. Because the injector operates at fixed pressure, the injection rate increases monotonically as the reservoir is gradually depleted.

The second drive mechanism is liberation of free gas, which starts near the producers where the pressure drawdown is largest. As pressure continues to decay, more gas is liberated at the top of the formation where the pristine oil phase is closest to a saturated state. The liberated gas forms a gas cap that gradually grows downward and contributes to push more gas into the wells and reduce the production rates of oil; see the left column in

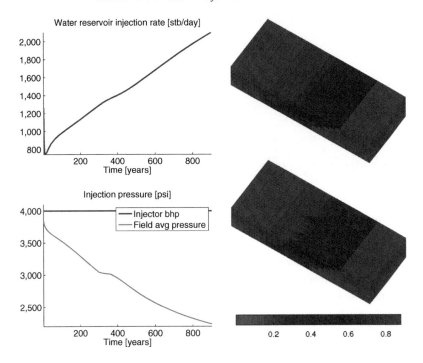

Figure 12.18 Water drive contributing to enhanced production. The left column shows surface injection rate (top) and bhp compared with the average field pressure (bottom). The right column shows the bottom water drive, with initial water saturation at the top and water saturation after 900 days at the bottom.

Figure 12.19. Any produced and free gas must come from the gas initially dissolved in oil, since no gas has been injected. Figure 12.19 reports the amount of dissolved gas initially and after 900 days.

Figure 12.20 shows production from the 25 wells. First of all, we note that there is significant spread in the recovery from the individual wells, with some wells that experience declining oil rates almost immediately and deviate from the upper production envelope that corresponds to the prescribed rate targets. These are wells at which free gas appears already during the first step. Since the free gas is more mobile and occupies a larger volume, it will effectively choke back the production of the oleic phase so much that we see a *decay* in the total surface gas rates from these wells. Some of the other wells are able to maintain the prescribed oil rate for a longer period. These wells experience increasing surface gas rates until the oil rate is reduced at 300 days. Other wells are of intermediate type and first experience increasing and then decaying gas rates. Most wells only produce small amounts of water, and for several of those who produce significant amounts of water, the breakthrough comes after the period with reduced oil rates has ended. The only exception in PROD17, which produces almost two orders of magnitude more water than the next

12.4 Simulation Examples 467

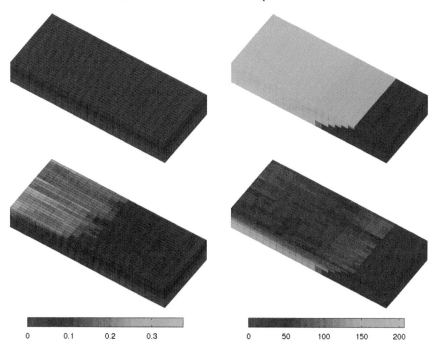

Figure 12.19 Liberation of gas. The left column shows initial gas saturation at the top and gas saturation after 900 days at the bottom. The right column shows the amount of gas dissolved in the oleic phase.

producer. (PROD17 is not shown in the lower-left plot in Figure 12.20.) We note in passing that the plot of bhp confirms our previous discussion of switching well controls.

Verification Against ECLIPSE

The default solvers in `ad-blackoil` have been implemented as to reproduce ECLIPSE as closely as possible. This does not necessarily mean that ECLIPSE always computes the *correct* result, but given the major role this simulator plays in the industry, it is very advantageous to be able to match results, so that this can be used as a baseline when investigating the performance of new models or computational methods. We have already shown that MRST matches ECLIPSE very well for the SPE 1 benchmark. To build further confidence in the simulator, we also show that MRST matches for the SPE 9 benchmark, given identical input. To this end, we use the same functionality from the `deckreader` module for reading *restart* files as we used in Section 11.8.2:

```
addir = mrstPath('ad-blackoil');
compare = fullfile(addir, 'examples', 'spe9', 'compare');
smry = readEclipseSummaryUnFmt(fullfile(compare, 'SPE9'));
```

We first get the number of report steps and the time at which these are computed

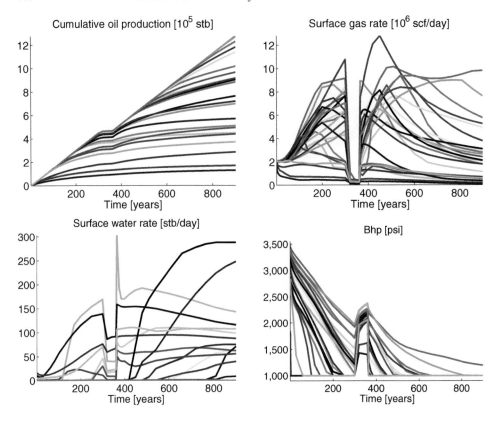

Figure 12.20 Fluid production from the 25 producers.

```
compd = 1:(size(smry.data, 2));
Tcomp = smry.get(':+:+:+:+', 'YEARS', compd);
```

Let us for instance compare the bhp of PROD13. We extract the corresponding data as follows:

```
comp = convertFrom(smry.get('PROD13', 'WBHP', compd), psia)';
mrst = getWellOutput(wellsols, 'bhp', 'PROD13');
```

and subsequently plot well responses using standard MATLAB commands; see Figure 12.21. The results computed by the two simulators are not identical. If you run the example, you will see that the output from ECLIPSE contains 37 steps whereas MRST reports 35 steps. The reason is that MRST does not include the initial state in its output and that ECLIPSE has a somewhat different time-step algorithm that adds one extra time step during the ramp-up after the target oil rates have been increased back to their original value at 360 days. Nevertheless, the match between the two computed solutions appears to

12.4 Simulation Examples

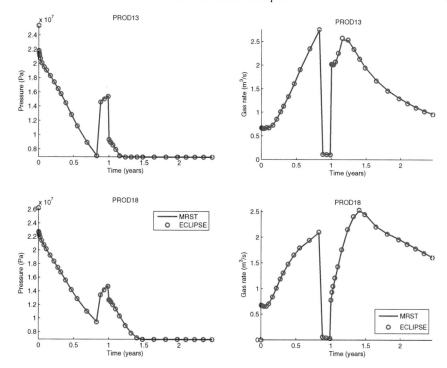

Figure 12.21 Verification of the AD-OO black-oil solver from `ad-blackoil` against ECLIPSE. The left column shows bhp for two different producers, whereas the right column shows surface gas rates.

be almost perfect. Similar matches are also seen for other types of well responses and all the remaining 24 wells.

Bao et al. [30] report similar good match for a set of polymer injection cases including non-Newtonian fluid rheology, In addition, we have run a number of cases in-house verifying that the AD-OO framework provides excellent match with commercial simulators. This includes challenging compositional cases as well as black-oil cases. Altogether, this shows that the software is a good starting point for case studies. However, when working with black-oil models, you should expect that you may stumble upon features that are not yet implemented in MRST, but this is also the point where you can contribute by adding new features.

COMPUTER EXERCISES

12.4.7 Set up an animation of water and gas to better understand how the aqueous and gaseous phases displace the oleic phase.

12.4.8 As pointed out on page 466, well PROD17 produces orders of magnitude more water than the other wells. How would shutting this well in from the start influence the overall field production.

12.4.9 Run the simulation using CPR with AGMG or MATLAB's direct solver and compare with just using the direct solver for the whole system. Are there any noticeable differences in simulation times? Try to install AMGCL and use functionality from the `linearsolver` module discussed on page 446 to see if you can reduce runtime even further.

12.4.10 In some cases, produced gas cannot be sold but must be reinjected. Can you set up such a simulation based on the case discussed in this section? Where would you place the gas injector? (Hint: Operations involving reinjection generally require complex topside logic to determine correct rates for the reinjected fluid. Such logic is not yet available in MRST. As a first approach, you can just assume that the gas is readily available so that whatever amount of gas produced in step n can be injected in step $n + 1$.)

12.5 Improving Convergence and Reducing Runtime

I hope that this and the previous chapters have given you a better understanding of reservoir simulation, and also provided you with sufficient detail about MRST so that you can use the software to perform interesting simulations. Although the software is primarily designed for flexibility and generality, the black-oil simulators implemented by use of the AD-OO framework offer such a comprehensive set of features that they can be used to simulate quite complicated models. Out of the box, MRST may not always be as fast as a simulator written in a compiled language, but with a bit customized setup you can easily make the simulators much more competitive. A key factor to improve computational efficiency of MRST is to replace MATLAB's default direct solver by a modern iterative solver with appropriate multilevel preconditioning as discussed in Section 12.3.4. With such capabilities on board, MRST will typically be a few times slower than ECLIPSE or comparable research simulators like OPM Flow for moderately sized models with $\mathcal{O}(10^5-10^6)$ unknowns as long as many-core or multi-node parallelism is not utilized.

As you explore other models than those discussed in the previous section, you will sooner or later run into convergence problems and/or cases for which the computational performance is not satisfactory. I end this chapter by briefly discussing how to address such problems. First, I provide a few general recommendations that apply equally well to commercial simulators like ECLIPSE, and then end the section with a few issues that are specific to MRST.

Tweaking Control Parameters for Solvers

Like many other things in the AD-OO framework, default values for the various parameters that control the behavior of the `NonLinearSolver` class largely follow what is suggested for ECLIPSE. These default values have been developed based on many years of experience on a large variety of simulation models and will work well in most cases. You can observe improved performance of your simulation by tuning parameters in specific cases, but it is generally *not recommended* that you change the default values unless you

12.5 Improving Convergence and Reducing Runtime

know what you are doing. In particular, it is rarely necessary to *weaken* the convergence tolerances. When convergence problems occur in the nonlinear iterations, the culprit usually lurks in the time-step selection or the model itself, and it is too late to try to fix the problem by adjusting the convergence control. Contrary to what you may think, faster overall convergence can often be observed by *tightening* convergence controls rather than weakening them since this may prevent inaccuracies earlier in the simulation, which otherwise may accumulate from one time step to the next and cause severe errors later in the simulation.

As general advice: the best way to improve convergence and computational performance is in most cases to choose a more robust linear solver and/or time-stepping scheme. In some cases, the cause of poor convergence and long runtimes may also lie in the model parameters, e.g., as discussed in Section 11.8.3.

Selection of Control Steps

How you set up your simulation schedule will largely determine the efficiency of your simulation. In many cases, the number of control steps and their individual lengths are determined by external factors. These can be requirements of monthly or quarterly reports for prediction runs, or the times of data points for history-matching runs. In other cases, there are no specific requirements, and the schedule can be chosen more freely. The following guidelines may be useful for improving convergence and general computational performance:

- Make sure that the limit on the maximum time step is compatible with the specified control steps. A classical mistake is to set the maximum time step to be 30 days, and design a simulation schedule that reports results at the end of each month. As we all know, every second month has 31 days ...

- For cases with complex flow physics like changes between undersaturated and saturated states, flow reversals, abrupt changes in drive mechanisms, etc., it may be advantageous to reduce long time steps since reductions in iteration count often will compensate for the increased number of time steps. As an example, using a gradual ramp-up of time steps initially and after major changes in well controls and other drive mechanisms will ensure that transients are better resolved, which in turn means that the simulation runs more smoothly.

- In some cases, you may be asking for shorter control steps than you actually need. If, for instance, each control step consists of only one time step and/or the nonlinear solver converges in very few iterations (say, one to three), it may be advantageous to *increase* the length of the control steps to give the time-step selector more freedom to choose optimal time steps.

For simulations based on ECLIPSE input decks, you should be particularly wary of the TUNING keyword, which is easy to misuse so that it interferes adversely with the standard mechanisms for time-step selection. (This keyword is not used by MRST.)

Issues Specific to MRST/MATLAB

In implementing MRST, we have tried to utilize MATLAB and its inherent vectorization as effectively as possible. However, maintaining both flexibility and efficiency is not always possible, and there are a few sticky points in the implementation of the AD-OO framework that may contribute to increased computational time. For vectors with a few hundred or thousand unknowns, the computational overhead in the AD library and other parts of the software is significant. The cost of linearization and assembly of linear systems can therefore completely overwhelm the runtime of the linear solver for small problems. This computational overhead should in principle, and will also in many cases, gradually disappear with larger problem sizes. However, this is not always the case.

The first factor that may contribute to maintain a significant overhead also for larger models, is our handling of wells. Wells are generally be perforated in multiple cells throughout the grid, and to assemble their models, we need to access and extract small subsets of large sparse matrices. Such access, which some refer to as *slicing*, is not as efficient in MATLAB as one could hope, particularly not in older versions of the software. To be able to handle different types of well models like the SPE 1 case with inflow valves discussed in Section 12.4.3, the `FacilityModel` class requires a somewhat heterogeneous memory access, which we have not yet figured out how to implement efficiently in MATLAB. Using this class may therefore introduce significant overhead. For cases where all wells assume vertical equilibrium (and a Peaceman-type inflow performance relationship), you can instead use a *special facility class* called `UniformFacilityModel` that has been optimized to utilize uniform memory access. Once you have created a reservoir model, you can convert the well description to this type by the following call:

```
model.FacilityModel = UniformFacilityModel(model);
```

For the SPE 9 case discussed in Section 12.4.4, the 26 wells perforated 80 out of the 9,000 cells in the grid. Adding this transformation led to a 25% reduction in runtime on my computer.

A second factor comes from the way MRST's vectorized AD library is constructed. By default, this library always constructs a sparse matrix to represent the derivatives (Jacobian) of a vector unknown. Constructing and multiplying such matrices has a significant cost that should be avoided whenever possible. For cases with many independent primary variables, many of the Jacobian sub-matrices tend to be be diagonal. You can represent a set of diagonal matrices as a smaller, dense matrix, for which elementary operations are highly efficient. At the time of writing, work is in progress to develop more *efficient ADI backends* that utilize diagonal structures (and other known matrix patterns) to reduce construction, multiplication, and assembly costs. You can instrument you models by a specific AD back end through an assignment of the form:

```
model.AutoDiffBackend = DiagonalAutoDiffBackend;
```

I expect that improved support for more efficient backends will continue to develop in future versions of the software.

Last, but not least, *plotting and creating new axes* is slow in MATLAB and has in my experience become even slower after the new graphics engine ("HG2") was introduced in R2014b. Many of the examples presented in the book have plotting integrated into simulation loops. For brevity and readability I have not optimized this part, and instead explicitly cleared axes and replotted grids and cell values whenever the plot needs to be updated. To reduce runtime, you should instead delete and recreate individual graphics handles, or simply manipulate the color data of existing graphics objects, rather than drawing the whole scene anew.

Part IV

Reservoir Engineering Workflows

13
Flow Diagnostics

It is rarely sufficient to study well responses and pressure and saturation distributions to understand the communication patterns in a complex reservoir model. To gain a better qualitative picture, you typically want understand from what region a given producer drains. To what region does a given injector provide pressure support? Which injection and production wells are in communication? Which parts of the reservoir affect this communication? How much does each injector support the recovery from a given producer? Do any of the wells have backflow? What is the sweep and displacement efficiency within a given drainage, sweep, or well-pair region? Which regions are likely to remain unswept? Likewise, you may also want to perform what-if and sensitivity analyses to understand how different parameters and their inherent uncertainty affect reservoir responses.

Simulating a full reservoir model containing a comprehensive description of geology, reservoir fluids, flow physics, well controls, and coupling to surface facilities is a computationally demanding task that may take hours or even days to complete. This limits your ability to study parameter variations and is at odds with the philosophy of modern reservoir characterization techniques, which often generate hundreds of equiprobable realizations to quantify uncertainty in the characterization. In this chapter, we therefore introduce a set of simple techniques, referred to as *flow diagnostics*, which you can use to delineate volumetric communications and improve your understanding of how flow patterns in the reservoir are affected by geological heterogeneity and respond to engineering controls. In their basic setup, these techniques only assume that you have a geocellular model of the reservoir and rely on the solutions of single-phase, incompressible flow problems, as discussed in Chapters 4 and 5. However, you can equally well employ the techniques to analyze representative flow field extracted from multiphase simulations of the type discussed in Chapters 10–12.

My definition of flow diagnostics is *a family of simple and controlled numerical flow experiments that are run to probe a reservoir model, establish connections and basic volume estimates, and quickly provide a qualitative picture of the flow patterns in the reservoir and quantitative measures of the heterogeneity in dynamic flow paths.* You can also use these methods to compute quantitative information about the recovery process in settings somewhat simpler than what would be encountered in an actual field. Using flow diagnostics, you can rapidly and iteratively perturb simulation input and evaluate the resulting changes in volumetric connections and communications to build an understanding of cause and effects in your model. The following presentation is inspired by [274] and [218], who developed the concept of flow diagnostics based on standard finite-volume discretizations for time-of-flight and tracer partitions [221, 98]. Similar ideas have previously been used within streamline simulation [79] for ranking and upscaling [140, 27, 278], identifying reservoir compartmentalization [127], rate optimization [287, 244, 144], and flood surveillance [32].

Flow diagnostics is inexpensive to compute, and I generally recommend that you use these methods to prescreen your models before you start to conduct more comprehensive multiphase simulations. MRST already offers a few simple interactive GUIs to this end, and I expect that similar capabilities will appear in commercial tools in the not too distant future. Interactive tools are not easy to present in book form, and if you want to see these work in practice, you should try out some of the examples that follow MRST, or study the exercises presented throughout this chapter. Much of the same ideas can also be used to postprocess multiphase flow simulations; such capabilities are currently available in the OPM ResInsight viewer (http://resinsight.org). Early research [218, 171] indicates that flow diagnostics offers a computationally inexpensive complement (or alternative) to full-featured multiphase simulations to provide flow information in various reservoir management workflows, e.g., as inexpensive objection functions for well placement and rate optimization, or as a means to perform what-if and sensitivity analyzes in parameter regions surrounding preexisting simulations.

13.1 Flow Patterns and Volumetric Connections

You have already been introduced to the basic quantities that lie at the core of flow diagnostics in Chapters 4 and 5: as you probably recall from Section 4.3.3, we can derive time lines that show how heterogeneity affects flow patterns for an instantaneous velocity field $\vec{v}$ by computing

- the *forward time-of-flight*, defined by

$$\vec{v} \cdot \nabla \tau_f = \phi, \qquad \tau_f|_{\text{inflow}} = 0, \tag{13.1}$$

  which measures time it takes a neutral particle to travel to a given point in the reservoir from the nearest fluid source or inflow boundary; and
- the *backward time-of-flight*, defined by

$$-\vec{v} \cdot \nabla \tau_b = \phi, \qquad \tau_b|_{\text{outflow}} = 0, \tag{13.2}$$

  which measures the time it takes a neutral particle to travel from a given point in the reservoir to the nearest fluid sink or outflow boundary.

The sum of the forward and backward time-of-flight at a given point in the reservoir gives the total *residence time* of an imaginary particle as it travels from the nearest fluid source or inflow boundary to the nearest fluid sink or outflow boundary.

Studying iso-contours of time-of-flight gives you a time map for how more complex multiphase displacements may evolve under fixed well and boundary conditions and reveals more information about the flow field than pressure and velocities alone. We illustrated this already in Chapter 5, where Figure 5.3 on page 160 showed time lines for a quarter five-spot flow pattern, and the total residence time was used to distinguish high-flow regions from stagnant regions. Likewise, in Figure 5.11 on page 173 we used time-of-flight to identify non-targeted regions for a complex field model; that is, regions with high τ_f values that were likely to remain unswept and hence would be obvious targets to investigate for placement of additional wells. From time-of-flight, you can derive various measures of dynamic heterogeneity, as we will see in the next section, or compute proxies of economical measures such as net-present value for multiphase flow models; see [218, 171].

In a similar manner, we can determine all points in the reservoir that are influenced by a given fluid source or inflow boundary by solving the following equation to compute the *volumetric injector partition*:

$$\vec{v} \cdot \nabla c_i = 0, \qquad c_i|_{\text{inflow}} = 1. \tag{13.3}$$

To understand what this equation represents, let us think of an imaginary painting experiment in which we inject a tracer dye (a massless, non-diffusive ink) of a unique color at each fluid source or point on the inflow boundary we want to trace the influence from. The ink will start flowing through the reservoir and paint every point it contacts. Eventually, the fraction of different inks that flows past each point in the reservoir reaches a steady state, and by measuring these fractions, we can determine the extent to which each different ink influences a specific point and find the *influence region* of an inflow point. Likewise, to determine how much the flow through each point is influenced by a fluid sink or point on the outflow boundary, we reverse the flow field and solve similar equations for producer (or outflow) partitions,

$$-\vec{v} \cdot \nabla c_p = 0, \qquad c_p|_{\text{outflow}} = 1. \tag{13.4}$$

To summarize, flow diagnostics requires three computations: (i) solution of a pressure equation, or extraction of fluxes from a previous computation, to determine the bulk fluid movement; (ii) solution of a set of steady advection equations (13.3)–(13.4) to partition the model into influence or volumetric flow regions; and (iii) solution of time-of-flight equations (13.1)–(13.2) to give averaged time lines that describe the flow within each region. This chapter describes how you can combine and process tracer partitions and time-of-flight to gain more insight into flow patterns and volumetric connections in the reservoir. In more advanced methods, we also use *residence-time distributions* over all flow paths and analogies with tracer analysis [277, 139] for more accurate analysis of flow paths.

13.1.1 Volumetric Partitions

If each source of inflow (outflow) is assigned a unique "color" value, the resulting color concentration should in principle produce a *partition of unity* in all parts of the reservoir that are in communication with the sources of inflow (outflow). In practice, you may not be able to obtain an exact partition of unity because of numerical errors. Based on the inflow/outflow partitions, we can further define:

- *drainage regions* – each such region represents the reservoir volume that eventually will be drained by a given producer (or outflow boundary), provided that the current flow field $\vec{v}$ prevails until infinity;
- *sweep regions* – each such region represents the reservoir volume that eventually will be swept by a given injector (or inflow boundary) if the current flow conditions remain forever;
- *well pairs* – injectors and producers in communication with each other;
- *well-pair regions* – regions of the reservoir in which the flow between a given injector and producer takes place.

When visualizing multiple drainage or sweep regions in 3D, it is common to assign a unique region to each cell determined by a majority vote over all "colors" that affect the cell. Well pairs are determined by finding all injectors whose concentration is positive in one of the well completions of a given producer (or vice versa). Well-pair regions are found by intersecting injector and producer partitions, possibly followed by a majority vote. Well allocation factors will be discussed in more detail in Section 13.1.3 and in Chapter 15, in conjunction with single-phase upscaling.

Our default choice would be to assign a unique "color" to each injector and producer, but you can also subdivide wells into multiple segments and trace the influence of each segment separately. You can, for instance, use this to determine if a (horizontal) well has cross-flow, so that fluid injected in one part of the well is drawn back into the wellbore in another part of the well, or fluids produced in one completion are pushed out again in another.

We have already encountered `computeTimeOfFlight` for computing time-of-flight in Section 5.3. The `diagnostics` module additionally offers

```
D = computeTOFandTracer(state, G, rock, 'wells', W, ...)
```

that computes forward and backward time-of-flight, injector and producer coloring, as well as sweep and drainage regions in one go for models with flow driven by wells. These quantities are represented as fields in the structure D:

- `inj` and `prod` give the indices for the injection and production wells in the well structure W;
- `tof` is a $2 \times n$ vector with τ_f in its first and τ_b in its second column;
- `itracer`/`ptracer` contain color concentrations for the injectors and producers;
- `ipart` and `ppart` hold the partitions resulting from a majority vote.

Obviously, you can associate similar quantities with boundary conditions and/or source terms. (MRST does not yet fully implement this.) Well pairs are identified by analyzing the

13.1 Flow Patterns and Volumetric Connections

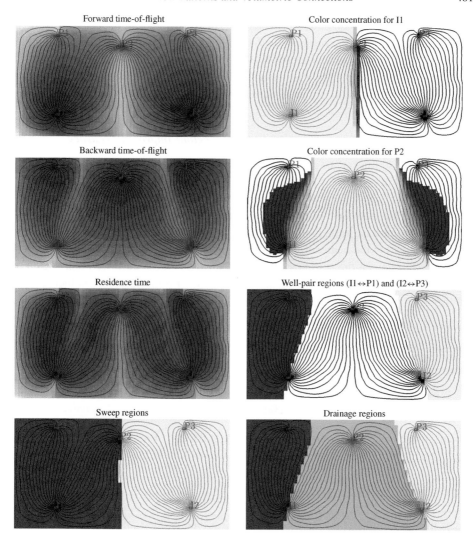

Figure 13.1 Time-of-flight, color concentration, and various types of volumetric delineations for a simple case with two injectors and three producers. Source code necessary to generate the plots is given in the `showDiagnostBasic` tutorial in the `book` module.

`itracer` and `ptracer` fields. The following function also computes the pore volume of the region of the reservoir associated with each pair:

```
WP = computeWellPairs(G, rock, W, D)
```

Figure 13.1 shows time-of-flight, color concentrations, volumetric partitions, and a few combinations thereof for a simple flow problem with two injectors and three production wells. There are many other ways you can combine and plot these basic quantities to reveal volumetric connection and provide enhanced insight into flow patterns. You can, for instance, combine sweep regions with time-of-flight to provide a simple forecast of *contacted volumes* or to visualize how displacement fronts move through the reservoir. To this end, you would typically also include estimates of characteristic wave-speeds from multiphase displacement theory as discussed in Section 8.4.

13.1.2 Time-of-Flight Per Partition Region: Improved Accuracy

It is important to understand that the time-of-flight values computed by a finite-volume method like the one discussed in Section 4.4.3 are *volume-average values*, which cannot be compared directly with the *point values* you obtain by tracing streamlines, as described in Section 4.3.3. Because time-of-flight is a quantity defined from a global line integral (see (4.39)), the point-wise variations can be large inside a single grid cell, in particular near flow divides and in the near-well regions, where both high-flow and low-flow streamlines converge. To improve the accuracy of the time-of-flight within each well-pair region, we can use the color concentrations to recompute the time-of-flight values for influence region number k,

$$\vec{v} \cdot \nabla \left(c_i^k \tau_f^k \right) = c_i^k \phi. \tag{13.5}$$

In MRST, this is done by passing the option computeWellTOFs to the time-of-flight solver, using a call that looks something like

```
T = computeTimeOfFlight(state, G, rock, 'wells', W, ...
                'tracer',{W(inj).cells},'computeWellTOFs', true);
```

which appends one extra column for each color at the end of return parameter T. You can also use a higher-order discretization, as discussed in [221, 261].

13.1.3 Well Allocation Factors

Apart from a volumetric partition of the reservoir, one is often interested in knowing the fraction of the inflow to a given producer that can be attributed to each of the injectors, or conversely, how the push from a given injector is distributed to the different producers. We refer to this as *well allocation factors*. These factors can be further refined so that they also describe the cumulative flow from the toe to the heel of the well. By computing the cumulative flux along the wellbore and plotting this flux as a function of the distance from the toe (with the flux on the x-axis and distance on the y-axis) we get a plot that is reminiscent of the plot from a *production logging tool*.

To formally define well allocation factors, we use the notation from Section 4.4.2, so that $x[c]$ denotes the value of vector x in cell c. Next, we let c_n^i denote the injector color concentration associated with well (or segment) number n, let c_m^p denote the producer concentration associated with well number m, q the vector of well fluxes, and $\{w_k^n\}_k$ the cells in which well number n is completed. Then, the cumulative factors are defined as

13.2 Measures of Dynamic Heterogeneity

$$a_{nm}^i[w_\ell^n] = \sum_{k=1}^{\ell} q[w_k^n]\, c_m^p[w_k^n],$$
$$a_{mn}^p[w_\ell^m] = \sum_{k=1}^{\ell} q[w_k^m]\, c_n^i[w_k^m]. \tag{13.6}$$

The total well allocation factor equals the cumulative factor evaluated at the heel of the well. Well allocation factors computed by computeWellPair are found as two arrays of structs, WP.inj and WP.prod, which give the allocation factors for all the injection and production wells (or segments) accounted for in the flow diagnostics. In each struct, the array alloc gives the a_{nm} factors, whereas influx or outflux that cannot be attributed to another well or segment is represented in the array ralloc.

13.2 Measures of Dynamic Heterogeneity

Whereas primary recovery can be reasonably approximated using averaged petrophysical properties, secondary and tertiary recovery are strongly governed by the intrinsic variability in rock properties and geological characteristics. This variability, which essentially can be observed at all scales in the porous medium, is commonly referred to as *heterogeneity*. As we have seen in previous chapters, both the rock's ability to store and transmit fluids are heterogeneous. However, it is the heterogeneity in permeability that has the most pronounced effect on flow patterns and volumetric connections in the reservoir. The importance of heterogeneity has been recognized from the earliest days of petroleum production, and over the years a number of static measures have been proposed to characterize heterogeneity, such as flow and storage capacity, Lorenz coefficient, Koval factor, and Dykstra–Parson's permeability variation coefficient, to name a few; see, e.g., [176] for a more comprehensive overview.

In the following, we discuss how some of the static heterogeneity measures from classical sweep theory can be reinterpreted in a dynamic setting if we calculate them from the time-of-flight (and color concentrations) associated with an instantaneous flow field [278]. Static measures describe the spatial distribution of permeability and porosity, and large static heterogeneity means that there are large (local) variations in the rock's ability to store and transmit fluids. *Dynamic heterogeneity measures*, on the other hand, describe the distribution of flow-path lengths and connection structure, and large heterogeneity values show that there are large variations in travel and residence times, which again tends to manifest itself in early breakthrough of injected fluids. Experience has shown that these measures, and particularly the dynamic Lorenz coefficient, correlate very well with forecasts of hydrocarbon recovery predicted by more comprehensive flow simulations. They can thus be used as effective flow proxies in various reservoir management workflows; see [278, 274, 218].

13.2.1 Flow and Storage Capacity

Dynamic heterogeneity measures are computed from the total residence time, which in turn is computed as the sum of forward and backward time-of-flight. To define dynamic flow and

Figure 13.2 Streamtube analogue used to define dynamic flow and storage capacity. Colors illustrate different average porosities.

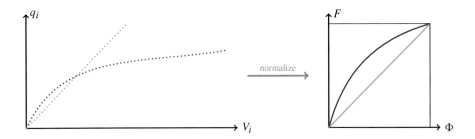

Figure 13.3 Construction of the F–Φ diagram. The plot to the left shows flow rates q_i plotted as function of streamtube volumes V_i for a homogeneous displacement (green) and for a heterogeneous displacement (blue). The right plot shows the corresponding F–Φ diagrams, where the flow rates and the streamtube volumes have been normalized.

storage capacity, we can think of the reservoir as a set of N non-communicating volumetric flow paths (streamtubes) that each has a volume V_i, a flow rate q_i, and a residence time $\tau_i = V_i/q_i$. The streamtubes are sorted so that their residence times are ascending, $\tau_1 \leq \tau_2 \leq \cdots \leq \tau_N$; see Figure 13.2. Inside each streamtube, we assume a piston type displacement; think of a blue fluid pushing a red fluid from the left to the right in the figure. We then define the normalized flow capacity F_i and storage capacity Φ_i by,

$$\Phi_i = \sum_{j=1}^{i} V_j \Big/ \sum_{j=1}^{N} V_j, \qquad F_i = \sum_{j=1}^{i} q_j \Big/ \sum_{j=1}^{N} q_j. \qquad (13.7)$$

Here, Φ_i is the volume fraction of all streamtubes that have broken through at time τ_i and F_i represent the corresponding fractional flow, i.e., the fraction of the injected fluid to the total fluid being produced. These two quantities can be plotted in a diagram as shown in Figure 13.3. From this diagram, we can also define the fractional recovery curve defined as the ratio of inplace fluid produced to the total fluid being produced; that is $(1 - F)$ plotted versus dimensionless time $t_D = d\Phi/dF$ measured in units of pore volumes injected.

To understand how the F–Φ diagram can be seen as a measure of dynamic heterogeneity, we first consider the case of a completely homogeneous displacement, for which all streamtubes will break through at the same time τ. This means that, $(\Phi_i - \Phi_{i-1})/(F_i - F_{i-1}) \propto V_i/q_i$ is constant, which implies that $F = \Phi$, since both F and Φ are normalized quantities. Next, we consider a heterogeneous displacement in which all the streamtubes have the same flow rate q. Since the residence times τ_i form a monotonically increasing sequence, we have that $\{V_i\}$ will also be monotonically increasing. In general, $F(\Phi)$ is a concave function, where the steep initial slope corresponds to high-flow regions giving early breakthrough, whereas the flat trailing tail corresponds to low-flow and stagnant regions.

In the continuous case, we can (with a slight abuse of notation) define the storage capacity as

$$\Phi(\tau) = \int_0^\tau \phi(\vec{x}(s))\,ds, \tag{13.8}$$

where $\vec{x}(\tau)$ represents all streamlines whose total travel time equals τ. By assuming incompressible flow, we have that pore volume equals the flow rate times the residence time, $\phi = q\tau$, and hence we can define the flow capacity as

$$F(\tau) = \int_0^\tau q(\vec{x}(s))\,ds = \int_0^\tau \frac{\phi(\vec{x}(s))}{s}\,ds. \tag{13.9}$$

From this, we can define *normalized, dynamic* flow and storage capacities by

$$\hat{\Phi}(\tau) = \frac{\Phi(\tau)}{\Phi(\infty)}, \qquad \hat{F}(\tau) = \frac{F(\tau)}{F(\infty)}.$$

Henceforth, we only discuss the normalized quantities and for simplicity we drop the hat symbol. To compute these quantities for a discrete grid model, we would have to first compute a representative set of streamlines, associate a flow rate, a pore volume, and a total travel time to each streamline, and then compute the cumulative sums as in (13.7).

Next, we consider how these concepts carry over to our flow diagnostics setting with time-of-flight computed by a finite-volume method and not by tracing streamlines. Let pv be an $n \times 1$ array containing pore volumes of the n cells in the grid and tof be an $n \times 2$ array containing the forward and backward time-of-flights. We can now compute the cumulative, normalized storage capacity Phi as follows:

```
t       = sum(tof,2);       % total travel time
[ts,ind] = sort(t);          % sort cells based on travel time
v       = pv(ind);           % put pore volumes in correct order
Phi     = cumsum(v);         % cumulative sum
vt      = full(Phi(end));    % total volume of region
Phi     = [0; Phi/vt];       % normalize to units of pore volumes
```

We can compute the flow rate for each cell directly as the ratio between pore volume and residence time if we assume incompressible flow. With this, the computeFandPhi calculates the cumulative, normalized flow capacity F as follows

```
q  = v./ts;           % back out flux based on incompressible flow
ff = cumsum(q);       % cumulative sum
ft = full(ff(end));   % total flux computed
F  = [0; ff/ft];      % normalize and store flux
```

The result of the above calculation is that we have two sequences Φ_i and F_i that are both given in terms of the residence time τ_i. If we sort the points (Φ_i, F_i) according to ascending values of τ_i, we obtain a sequence of discrete points that describe a parametrized curve in 2D space. The first end point of this curve is at the origin: If no fluids have entered the domain, the cumulative flow capacity is obviously zero. Likewise, full flow capacity is reached when the domain is completely filled, and since we normalize both Φ and F by their value at the maximum value of τ, this corresponds to the point (1,1). Given that both F and Φ increase with increasing values of τ, we can use linear interpolation to define a continuous, monotonic, increasing function $F(\Phi)$.

13.2.2 Lorenz Coefficient and Sweep Efficiency

The *Lorenz coefficient* is a popular measure of heterogeneity and is defined as the difference in flow capacity from that of an ideal piston-like displacement:

$$L_c = 2 \int_0^1 \bigl(F(\Phi) - \Phi\bigr) d\Phi. \tag{13.10}$$

In other words, the Lorenz coefficient is equal twice the area under the $F(\Phi)$ curve and above the line $F = \Phi$, and has values between zero for homogeneous displacement and unity for an infinitely heterogeneous displacement; see Figure 13.4. Assuming that the

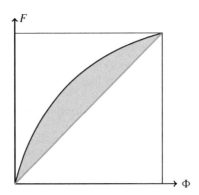

Figure 13.4 The definition of Lorenz coefficient from the F–Φ diagram. The green line represents a completely homogeneous displacement in which all flow paths have equal residence times. Te blue line is a heterogeneous displacement with significant variation in residence times. The Lorenz coefficient is defined as two times the gray area.

13.2 Measures of Dynamic Heterogeneity

flow and storage capacity are given as two vectors F and Phi, the Lorenz coefficient can be computed by applying a simple trapezoid rule

```
v  = diff(Phi,1);
Lc = 2*(sum((F(1:end-1)+F(2:end))/2.*v) - .5);
```

This is implemented in the computeLorenz function in the diagnostics module.

We can also use the F–Φ diagram to compute the *volumetric sweep efficiency* E_v, which measures how efficient injected fluids are used. Here, E_v denotes the volume fraction of inplace fluid that has been displaced by injected fluid, or equivalently, the ratio between the volume contacted by the displacing fluid at time t and the volume contacted at time $t = \infty$. In our streamtube analogue, only streamtubes that have not yet broken through will contribute to sweep the reservoir. Normalizing by total volume, we thus have

$$E_v(t) = \frac{q}{V} \int_0^t \left[1 - F\big(\Phi(\tau)\big)\right] d\tau$$

$$= \frac{qt}{V} - \frac{q}{V} \int_0^t F(\tau)\, d\tau = \frac{qt}{V} - \frac{q}{V}\left[F(t)t - \int_0^F \tau\, dF\right]$$

$$= \frac{t}{\bar{\tau}}\big(1 - F(t)\big) + \frac{1}{\bar{\tau}} \int_0^\Phi \bar{\tau}\, d\Phi = \Phi + (1 - F)\frac{d\Phi}{dF} = \Phi + (1-F)t_D.$$

The third equality follows from integration by parts, and the fourth equality since $\bar{\tau} = V/q$ and $\tau\, dF = \bar{\tau}\, d\Phi$. Here, the quantity $d\Phi/dF$ takes the role as dimensionless time. Prior to breakthrough, $E_v = t_D$. After breakthrough, Φ is the volume of fully swept flow paths, whereas $(1 - F)t_D$ is the volume of flow paths being swept.

The implementation in MRST is quite simple and can be found in the utility function computeSweep. Starting from the two arrays F and Phi, we first remove any flat segments in F to avoid division by zero

```
inz        = true(size(F));
inz(2:end) = F(1:end-1)~=F(2:end);
F          = F(inz);
Phi        = Phi(inz);
```

Then, dimensionless time and sweep efficiency can be computed as follows:

```
tD = [0; diff(Phi)./diff(F)];
Ev = Phi + (1-F).*tD;
```

Figure 13.5 shows the F–Φ and the sweep diagram for the simple example with two injectors and three producers from Figure 13.1. For this particular setup, the Lorenz coefficient is approximately 0.25, which indicates that we can expect a mildly heterogeneous displacement with some flow paths that break through early and relatively small stagnant regions. From the diagram of the sweep efficiency, we see that 70% of the fluids in place can be produced by injecting one pore volume. By injecting two additional pore volumes almost all the in-place fluid can be produced.

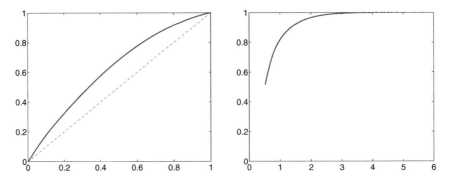

Figure 13.5 F–Φ and sweep diagram for the simple case with two injectors and three producers, which has a Lorenz coefficient of 0.2475.

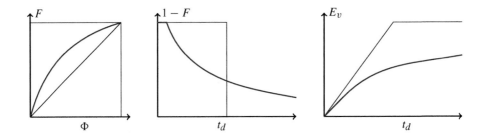

Figure 13.6 The three basic flow diagnostics curves: F–Φ diagram, fractional recovery curve, and sweep efficiency. All quantities Φ, F, E_v, and t_D are dimensionless; t_D is given in terms of pore volumes injected (PVI).

Altogether, we have defined three different curves that can be derived from the residence/travel time. The curves shown in Figure 13.6 are visually intuitive and emphasize different characteristics of the displacement:

- The F–Φ curve is useful for assessing the overall level of displacement heterogeneity. The closer this curve is to a straight line, the better is the displacement.
- The fractional recovery curve emphasizes early-time breakthrough behavior and can have utility as a proxy for fractional recovery of the fluid in place.
- Sweep efficiency highlights the behavior after breakthrough and has utility as a proxy for recovery factor.

The curves can be defined for the field as a whole, or be associated with sector models, individual swept volumes, well-pair regions, and so on.

Whereas visually intuitive information is useful in many workflows, others need measures defined in terms of real numbers. In our work, we have primarily used the Lorenz

13.3 Residence-Time Distributions

coefficient, but dynamic analogues of other classical heterogeneity measures can be defined in a similar fashion. For instance, the dynamic Dykstra–Parsons' coefficient is defined as

$$V_{DP} = \frac{(F')_{\Phi=0.5} - (F')_{\Phi=0.841}}{(F')_{\Phi=0.5}}, \quad (13.11)$$

where $\Phi = 0.5$ corresponds to the mean storage capacity, while $\Phi = 0.841$ is the mean value plus one standard deviation. Likewise, we can define the dynamic flow heterogeneity index,

$$F_{HI} = F(\Phi^*)/\Phi^*, \qquad F'(\Phi^*) = 1. \quad (13.12)$$

One can show that $(\frac{dF}{d\Phi})_i = \frac{\bar{\tau}}{\tau_i}$, where $\bar{\tau} = V/q = \sum V_i / \sum q_i$ is the average residence time for all the streamtubes. These heterogeneity measures have not yet been implemented in the `diagnostics` module.

COMPUTER EXERCISES

13.2.1 Implement Dykstra–Parsons' coefficient (13.11) and the flow heterogeneity index (13.12) as new functions for the diagnostics module.

13.2.2 Use time-of-flight values defined per influence region as discussed in Section 13.1.2 on page 482 to implement refined versions of the dynamic heterogeneity measures.

13.2.3 Compute heterogeneity measures for each well pair in the model with two injectors and three producers. (Original source code: `showDiagnostBasics`.) Are there differences between the different regions?

13.2.4 Try to make the displacement shown in Figures 13.1 and 13.5 less heterogeneous by moving the wells and/or by changing the relative magnitude of the injection/production rates.

13.2.5 Define a set of multiphase flow simulations cases, e.g., one five-spot simulation for each layer of the SPE 10 model (see Section 13.4.2). Which of the heterogeneity measures above has the strongest correlation with the total hydrocarbon recovery from multiphase simulations?

13.3 Residence-Time Distributions

You have previously seen how we can compute point-wise values of time-of-flight (and influence regions) by tracing streamlines or compute cell-averaged values by solving a steady, linear transport problem using a finite-volume method. This section sheds more light into the finite-volume approach and discusses how we can overcome some of its potential limitations. The discussion follows [171] closely and presents functionality that is new in MRST 2018b.

We assume incompressible flow to simplify the discussion and start by writing the finite-volume discretization of the time-of-flight equation (13.1) in vector form as $V\tau = \phi$. Here, τ is the vector of unknown time-of-flight values, ϕ is the vector of pore volumes per cell, and V is the upstream flux matrix corresponding to the operator $\vec{v}\cdot\nabla$. Consider a cell j with total influx v_j^i, and let c_b^j denote the vector of *backward* tracer concentrations corresponding to an imaginary experiment in which a tracer is injected in cell j and allowed to flow in the reverse direction of v. Let χ_j denote the characteristic vector, which equals one in cell j and zero in all other cells. We can then write the discrete equations for the forward time-of-flight τ and the backward influence region c_b^j as follows

$$V\tau_f = \phi \quad \text{and} \quad V^T c_b^j = \chi_j v_j^i. \tag{13.13}$$

From this, we obtain the following expression for the time-of-flight value in cell j

$$\tau_f[j] = \chi_j^T V^{-1} \phi = \left(\frac{1}{v_j^i} c_b^j\right)^T \phi. \tag{13.14}$$

Hence, $\tau_f[j]$ equals the pore volume of the backward influence region of cell j divided by the influx into the cell. If cell j lies close to a producer, the backward influence region typically contains flow paths with very different residence times. This implies that the forward time-of-flight value computed by the finite-volume method represents the average of a distribution that potentially can have (very) large variance. A similar argument applies to the backward time-of-flight. The averaging will introduce a systematic bias when we use averaged time-of-flight values to compute the dynamic heterogeneity measures discussed in the previous section. This bias may be acceptable in many applications, e.g., if we use dynamic heterogeneity measures to rank different model realizations or as a simple reduced-order model to predict recovery of secondary oil recovery [144, 218]. On the other hand, residence times computed from time-of-flight values will in most cases *overestimate* the time to breakthrough in heterogeneous displacements.

For a better description of the dynamic heterogeneity of a reservoir model, we can consider the *distribution* of time-of-flight for each cell. This is particularly interesting for cells perforated by production wells, since pointwise time-of-flight values in these cells describe the residence times, or time to breakthrough, for individual flow paths. To determine this *residence-time distribution* (RTD), we start by considering the linear transport equation

$$\phi \frac{\partial c}{\partial t} + \vec{v} \cdot \nabla c = 0, \quad c|_{\Gamma_i} = \delta(t), c(\vec{x},0) = 0, \tag{13.15}$$

which describes the transport of a unit pulse through a domain Ω from an inflow boundary Γ_i to an outflow boundary Γ_o. (The inflow and outflow will obviously be through wells in most reservoir models, but in the following I refer to these as boundaries for mathematical convenience.) For each point $\vec{x}$, the normalized time-of-flight distribution $\mathcal{T}(\cdot;\vec{x})$ is now simply defined by the Dirac function

$$\mathcal{T}(t;\vec{x}) = c(\vec{x},t) = \delta(t - \tau(\vec{x})). \tag{13.16}$$

At the outflow, the normalized RTD is given as

$$\mathcal{T}_o(t) = \frac{1}{F_o} \int_{\Gamma_o} c\, \vec{v} \cdot \vec{n}\, ds, \quad F_o = \int_{\Gamma_o} \vec{v} \cdot \vec{n}\, ds. \tag{13.17}$$

13.3 Residence-Time Distributions

It follows from the definition of the Dirac distribution that $\int \mathcal{T}_o(t)\,dt = 1$. From the RTD, we obtain flow capacity and storage capacity as follows [277]

$$F(t) = \int_0^t \mathcal{T}_o(s)\,ds, \qquad \Phi(t) = \frac{F_o}{\Phi_o} \int_0^t s\, \mathcal{T}_o(s)\,ds. \tag{13.18}$$

As before, both quantities are normalized so that $F(\infty) = \Phi(\infty) = 1$. From this definition, it also follows that the mean value of $\mathcal{T}_o(t)$ corresponds to the time $\bar{t} = \Phi_o/F_o$ it takes to inject one pore volume (1 PVI).

To compute discrete approximations to the time-of-flight and RTD, we introduce a finite-volume approximation,

$$\frac{d\boldsymbol{c}}{dt} + \boldsymbol{M}\boldsymbol{c} = 0, \quad \boldsymbol{c}(0) = \boldsymbol{c}_0 = \frac{\boldsymbol{q}_i}{\phi}. \tag{13.19}$$

Here, $\boldsymbol{M}$ is the discretization of the linear operator $\phi^{-1}\vec{v}\cdot\nabla$ defined in the same way as for the time-of-flight equation, and $\boldsymbol{q}_i$ is the vector of inflow flow rates. In the `diagnostics` module, we use a backward Euler method to solve (13.19) directly. This method is robust and can take large time steps, but is not very accurate.

Alternatively, you can write the solution formally in terms of matrix exponentials $\boldsymbol{c}(t) = \exp(-t\boldsymbol{M})\boldsymbol{c}_0$. Let $\boldsymbol{q}_o$ be the vector that collects the flux across all outflow boundaries of individual cells, $\mathbf{1}$ be a vector with value one in each cell, and Φ_o be the total pore volume drained by the outflow boundary Γ_o. We then represent the discrete counterparts of (13.16)–(13.17) as follows

$$\boldsymbol{\mathcal{T}}(t)[j] = \boldsymbol{\chi}_j^T \exp(-t\boldsymbol{M})\boldsymbol{c}_0 \quad \text{and} \quad \mathcal{T}_o(t) = \boldsymbol{q}_o^T \exp(-t\boldsymbol{M})\boldsymbol{c}_0/\boldsymbol{q}_o^T\mathbf{1}. \tag{13.20}$$

The vector $\boldsymbol{\mathcal{T}}$ contains time-of-flight distributions in all cells (with value $\boldsymbol{\mathcal{T}}[j]$ in cell j) and $\mathcal{T}_o$ is the RTD over Γ_o. You can then employ a rational Padé approximation to evaluate the action of the matrix exponential,

$$\boldsymbol{\mathcal{T}}(t+\Delta t) = \exp(-\Delta t \boldsymbol{M})\boldsymbol{\mathcal{T}}(t) \approx P(-\Delta t \boldsymbol{M})\, Q(-\Delta t \boldsymbol{M})^{-1}\boldsymbol{\mathcal{T}}(t), \tag{13.21}$$

for suitable polynomials P and Q. One example would be to use the first-order polynomials $P(x) = 1 + x/2$ and $Q(x) = 1 - x/2$ to reduce fill in, i.e.,

$$\boldsymbol{\mathcal{T}}(t+\Delta t) = \left(\boldsymbol{I} - \tfrac{1}{2}\Delta t\, \boldsymbol{M}\right)\left(\boldsymbol{I} + \tfrac{1}{2}\Delta t\, \boldsymbol{M}\right)^{-1}\boldsymbol{\mathcal{T}}(t).$$

For each successive value you compute, you must solve a linear system and perform a matrix multiplication. The matrix $\boldsymbol{I} + \boldsymbol{M}$ has the same sparsity as the discretization matrix for the time-of-flight equation; i.e., is triangular, possibly after a permutation. Each linear solve is thus highly efficient, but the time step is restricted by a CFL condition because of the explicit part of the discretization coming from the polynomial $P(x)$. In our experience, using backward Euler is thus more efficient.

The next two examples illustrate the difference between averaged values and distributions and how these differences impact measures of dynamic heterogeneity.

Example 13.3.1 *We consider a single horizontal layer from SPE 10 with an injector along the south boundary and a producer along the north boundary. Assume we have computed a flow solution with fluxes represented in* state, *cell-wise pore volumes in* pv, *time-of-flight and tracer values represented in the structure* D *discussed on page 480, and well pair information in* WP. *The following routine computes the propagating pulse using (13.19) with a backward Euler time discretization and extracts the outflux observed in the producer as function of time:*

```
rtd = computeRTD(state, G, pv, D, WP, W);
```

The solid lines in Figure 13.7 show an unnormalized version of $T_o(t)$ as function of time for two layers; here the integral equals the total allocation. The leading pulse for the Tarbert layer is spread out and has a small secondary hump. For Upper Ness, the pulse breaks through earlier and is more focused because of high-permeable channels connecting the

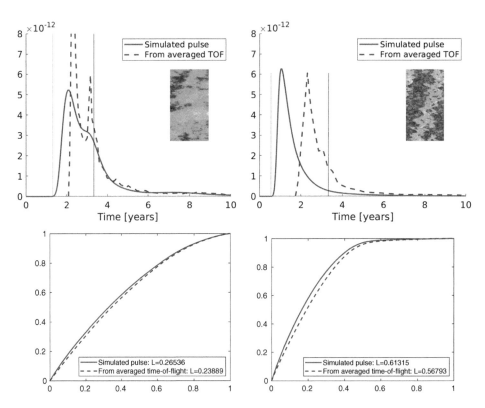

Figure 13.7 Residence-time distributions and F–Φ diagrams for two permeability field sampled from the SPE 10 benchmark. Solid lines are computed by tracing a unit pulse through the reservoir to determine the distribution at the outlet. Dashed lines are obtained by solving the time-of-flight equation by a finite-volume method and backing out data for representative flow paths through each of the cells of the model. Thin solid and dotted lines represent the mean of the distribution, i.e., the time it takes to inject one pore volume, and the time to breakthrough for the fastest flow path.

south and north boundaries. The mean of each distribution equals *1 PVI* by construction. This may not be apparent from the plots since the distributions have very long tails, particularly in the channelized case.

We can also estimate this distribution from the average residence times $\tau_r = \tau_f + \tau_b$, obtained by solving the forward and backward time-of-flight equations. We start by using the relationship $f_j = \phi[j]/\tau_r[j]$ to back out the flux

```
D = computeTOFandTracer(state, G, rock, 'wells', W, ..)
f = poreVolume(G,rock)./sum(D.tof,2);
```

Plotting f against τ_r values sorted in ascending order and normalized by *1 PVI* gives a discrete *(point)* description of how a given total flux through the permeable medium will distribute itself over a finite number of flow paths. This distribution is highly irregular, and to get a continuous distribution we sort the τ_r values into small bins, sum the total flux within this bin, and compute an averaged flux density value over the whole length of the bin. This is implemented in the function

```
RTD = estimateRTD(pv, D, WP);
```

Using averaged time-of-flight values introduces a significant delay in the breakthrough time, in particular for Upper Ness. The two types of distributions also suffer from different types of numerical errors: Tracing a pulse by a finite-volume method preserves flux allocation and not pore volume and can also contain significant temporal smearing. Backing out a distribution from averaged time-of-flight values preserves total volume but not flux allocation. This is taken into account by the following function, which computes F–Φ diagrams from a RTD,

```
[f, phi] = computeFandPhi(rtd);
```

Figure 13.7 also compares the resulting F–Φ diagram with similar diagrams computed directly from τ_r and pore volume. As expected, the diagrams computed from the RTD are more concave and the corresponding Lorenz coefficients are somewhat larger than when computed from averaged time-of-flight values. However, the differences are not very large, and the measures computed from averaged time-of-flight values carry a systematic underestimation bias and can thus still be robustly used to rank and discriminate different cases.

Notice that we do *not* report F–Φ diagrams derived from *estimated* RTDs. The binning and averaging used in the estimation process introduces a strong smoothing that generally obfuscates dynamic heterogeneity measures. Hence, we strongly advise against computing any heterogeneity measure from *estimated* RTDs; these distributions should only be used for illustrations like Figure 13.7.

Example 13.3.2 *We repeat the same setup as in the previous example, but now with an injector placed slightly off-center along the south perimeter and two producers placed symmetrically at the northern perimeter. Figure 13.8 reports residence distributions and F–Φ diagrams for each of the two injector pairs. In both cases, producer P2 is connected*

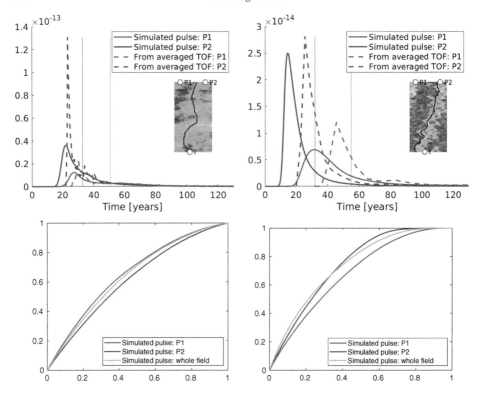

Figure 13.8 Residence-time distributions and F–Φ diagrams for two permeability field sampled from the SPE 10 benchmark with one injector and two producers.

to the injector through more permeable rocks than producer P1, even though the physical distance from I to P2 is larger than to P1. The fastest flow paths thus break through earlier in P2 than in P1, and overall, producer P2 allocates a significantly larger portion of the flux than P1.

Next, we look at the dynamic heterogeneity. For Tarbert, the (I,P2) well pair allocates approximately twice as much flux and a 30% larger flow region than (I,P1). The majority of the flow paths in the (I,P2) have small residence times and dynamically, this region is less heterogeneous than (I,P1), whose RTD contains a heavier tail. Hence, the F–Φ curve of P1 lies above P2 and has a higher Lorenz coefficient. For the reservoir as a whole, these two effects balance each other: In the beginning, the produced tracer comes predominantly from fast flow paths connected to P2, whereas the last tracer volumes produced come predominantly from slow flow paths connected to P1. Hence, fast flow paths from P2 ensure that the F–Φ curve of the whole reservoir lies below the curve of P1, whereas contributions from the heavier tail of P1 ensure that the overall curve lies above that of P2.

For Upper Ness, 50% more flux is allocated to P2 even though the associated flow region is smaller than for P1. In the (I,P1) region, all flow paths traverse regions of both high and

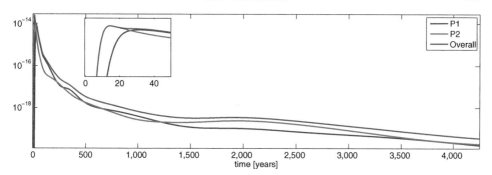

Figure 13.9 Residence-time distributions for the Upper Ness case shown on a semi-logarithmic axis. (One PVI corresponds to 41 years.)

low permeability, which means that this region is more homogeneous than the (I,P2) region, which contains a large fraction of flow paths that (almost) exclusively traverse cells with high permeability. Hence, the F–Φ curve of P2 lies above P1. Explaining the F–Φ curve of the whole reservoir is significantly more difficult. Figure 13.9 shows the RTD on a semi-logarithmic axis. During the first 20 years, tracer production is predominantly from P2. Producer P1 has the largest contribution from year 27 to year 1220, and then P2 contributes most for the rest of the simulated period. Obviously, this exercise has nothing to do with real reservoir engineering but serves to illustrate how tiny details on (unrealistically) long time scales affect measures like F–Φ diagrams.

Distributions of residence times are used in the study of chemical reactors and tracer tests [277, 139]. In [218] we demonstrate how you can convolve these distributions with representative, 1D displacement profiles to develop reduced-order models that, e.g., enable you to quickly estimate and discriminate improvements in macroscopic and microscopic displacement efficiency for polymer flooding.

13.4 Case Studies

The use of flow diagnostics is best explained through examples. In this section we therefore go through several cases and demonstrate various ways flow diagnostics can be used to enhance our understanding of flow patterns and volumetric connections, tell us how to change operational parameters such as well placement and well rates to improve recovery, and so on.

13.4.1 Tarbert Formation: Volumetric Connections

As our first example, we consider a subset of the SPE 10 data set consisting of the top twenty layers of the Tarbert formation; see Section 2.5.3. We modify the original inverted five-spot well pattern by replacing the central injector by two injectors that are moved a

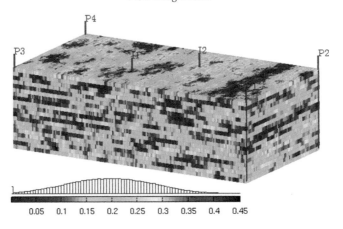

Figure 13.10 Porosity and well positions for a model consisting of subset of the Tarbert formation in Model 2 from the 10th SPE Comparative Solution Project

short distance from the model center (see Figure 13.10), assume single-phase incompressible flow, and solve the corresponding flow problem. You can find complete description of the setup in showWellPairsSPE10.m in the book module.

Given the geological model represented in terms of the structures G and rock, the wells represented by W, and the reservoir state including fluxes by rS, we first compute the time-of-flight and volumetric partitions:

```
D = computeTOFandTracer(rS, G, rock, 'wells', W);
```

This gives us the information we need to partition the volume into different drainage and sweep volumes. The simplest way to do this for the purpose of visualization is to use a majority vote over the injector and producer partitions to determine the well that influences each cell the most as shown in Figure 13.11. The result of this majority vote is collected in D.ppart and D.ipart, respectively, and the essential commands to produce the two upper plots in Figure 13.11 are:

```
plotCellData(G,D.ipart, ..);
plotCellData(G,D.ppart,D.ppart>1, ..);
```

Since there are two injectors, we would expect to only see two colors in the to plot of sweep regions. However, there is also a small blue volume inside the triangular section bounded by I1, P1, and P2, which corresponds to an almost impermeable part of the reservoir that will not be swept by any of the injectors. The well pattern is symmetric and for a homogeneous medium we would therefore expect that the two pressure-controlled injectors would sweep symmetric volumes of equal size. However, the Tarbert formation is highly heterogeneous and hence the sweep regions are irregular and clearly not symmetric. The injection rate of I2 is approximately six times that of I1 because the wells are completed in cells with very different permeability, hence I2 will sweep a much larger region than I1. In particular, we

13.4 Case Studies

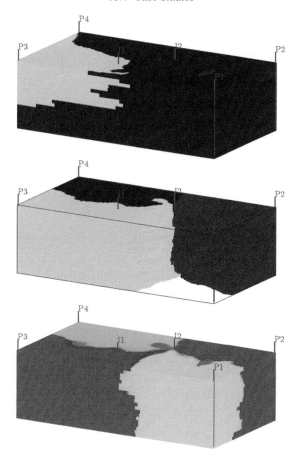

Figure 13.11 Sweep (top) and drainage regions (middle) determined by a majority vote over injector and producer partitions, respectively, for the Tarbert model. The bottom plot shows producer influence region without majority vote, with gray color signifying cells that affect multiple producers.

see that I2 is the injector that contributes most to flooding the lower parts of the region near P3, even though I1 is located closer. Looking at the drainage and sweep regions in conjunction, it does not seem likely that I1 will contribute significantly to support the production from wells P1 and P2 unless we increase its rate.

When using a majority vote to determine drainage and sweep regions, we disregard that individual cells may be influenced by more than one well. To better visualize the influence regions, we can blend in a gray color in all cells in which more than one color has nonzero concentration, as shown in the bottom plot of Figure 13.11. The plot is generated by the following call:

```
plotTracerBlend(G, D.ppart, max(D.ptracer, [], 2), ..);
```

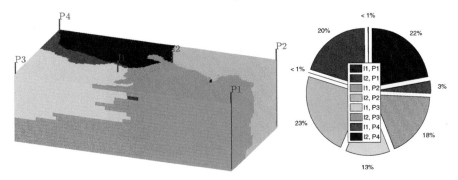

Figure 13.12 Well-pair regions and associated fraction of the total pore volume for the upper 20 layers of the Tarbert formation.

Having established the injection and producer partitions, we can identify well pairs and compute the pore volumes of the region associated with each pair:

```
WP = computeWellPairs(rS, G, rock, W, D);
pie(WP.vols, ones(size(WP.vols)));
legend(WP.pairs,'location','Best');
```

To visualize the volumetric regions, we compute the tensor product of the injector and producer partitions and then compress the result to get a contiguous partition vector with a zero value signifying unswept regions:

```
p = compressPartition(D.ipart + D.ppart*max(D.ipart))-1;
plotCellData(G,p,p>0,'EdgeColor','k','EdgeAlpha',.05);
```

The result shown in Figure 13.12 confirms our previous observations of the relative importance of I1 and I2. Altogether, I1 contributes to sweep approximately 16% of the total pore volume, shown as the light red and the yellow regions in the 3D plot.

It is also interesting to see how these volumetric connections affect the fluxes in and out of wells. To this end, we should look at the cumulative well allocation factors, which are defined as the cumulative flux in/out of a well from the bottom to the top perforation of a vertical well, or from toe to heel for a deviated well. We start by computing the flux allocation manually for the two injectors (outflow fractions are given from well head and downward and hence must be flipped):

```
for i=1:numel(D.inj)
    subplot(1,numel(D.inj),i); title(W(D.inj(i)).name);
    alloc = cumsum(flipud(WP.inj(i)).alloc,1);
    barh(flipud(WP.inj(i).z), alloc,'stacked'); axis tight
    lh = legend(W(D.prod).name,4);
    set(gca,'YDir','reverse');
end
```

13.4 Case Studies   499

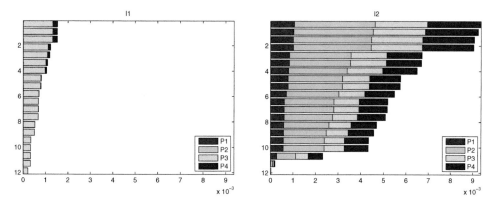

Figure 13.13 Well allocation factors for the two injectors of the Tarbert model.

Figure 13.13 shows the resulting bar plots of the cumulative allocation factors. These plots confirm and extend the understanding we have developed by studying volumetric connections: I1 will primarily push fluids towards P3. Some fluids are also pushed towards P4, and we observe that there is almost no outflow in the top three perforations where the rock has low quality. Injection from I2, on the other hand, contributes to uphold the flux into all four producers. We also see that the overall flux is not well balanced. Producer P1 has significantly lower inflow than the other three. Alternatively, we can use plotWellAllocationPanel(D, WP) to compute and visualize the well allocation factors for all the wells in the model, as shown in Figure 13.14. (This function is only useful for a small number of wells.)

When your simulation case contains only a few well pairs, you may get significantly more information about the flow patterns in the reservoir if you subdivide the wells into multiple segments (sets of completions). To exemplify, let us look more closely at the performance of the different completions along the well paths. To this end, we divide the completion intervals into bins and assign a corresponding set of pseudo-wells for which we recompute flow diagnostics. As an example, we split the completions of I1 into three bins and the completions of I2 into four bins.

```
[rSp,Wp] = expandWellCompletions(rS,W,[5, 3; 6, 4]);
Dp = computeTOFandTracer(rSp, G, rock, 'wells', Wp);
```

Figure 13.15 shows the majority-voted sweep regions for the four segments of I2. To better see the various sweep regions, the 3D plot is rotated 180 degrees compared with the other 3D plots of this model. To obtain the figure, we used the following key statements:

```
plotCellData(G, Dp.ipart,(Dp.ipart>3) & (Dp.ipart<8),...);
WPp = computeWellPairs(rSp, G, rock, Wp, Dp);
avols = accumarray(WPp.pairIx(:,1),WPp.vols);
pie(avols(4:end));
```

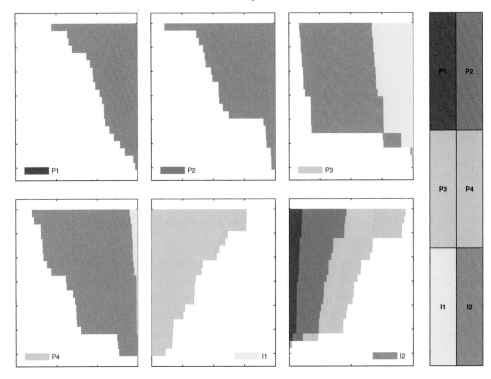

Figure 13.14 Normalized well allocation factors for all wells of the Tarbert model.

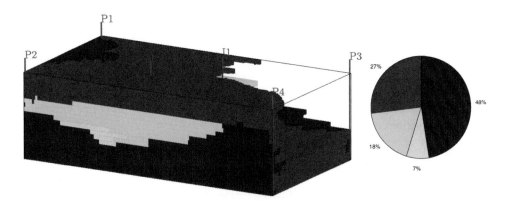

Figure 13.15 Majority-voted sweep regions for the Tarbert case with I2 of divided into four segments that each are completed in five layers of the model. (Notice that the view angle is rotated 180 degrees compared with Figure 13.11.) The pie chart shows the fraction of the total sweep region that can attributed to each segment of the well.

Notice, in particular, that fluids injected in the lowest segment is the major contributor in almost half of the well's total sweep region.

13.4.2 Heterogeneity and Optimized Well Placement

In this example, we first compute the Lorenz coefficient for all layers of the SPE 10 model subject to an inverted five-spot well pattern. We then pick one of the layers and show how we can balance the well allocation and improve the Lorenz coefficient and the areal sweep by moving some of the wells to regions with better sand quality.

To compute Lorenz coefficient for all layers in the SPE 10 model, we first define a suitable $60 \times 220 \times 1$ grid covering a rectangular area of $1{,}200 \times 2{,}200 \times 2$ ft^3. Then, we loop over all the 85 layers using the following essential lines:

```
for n=1:85
   rock = getSPE10rock(1:cartDims(1),1:cartDims(2),n);
   rock.poro = max(rock.poro, 1e-4);
   T    = computeTrans(G, rock);

   for w = 1:numel(wtype),
      W = verticalWell(..);
   end
   rS = incompTPFA(initState(G, W, 0), G, T, fluid, 'wells', W);

   D      = computeTOFandTracer(rS, G, rock, 'wells', W, 'maxTOF', inf);
   [F,Phi] = computeFandPhi(poreVolume(G,rock), D.tof);
   Lc(n)  = computeLorenz(F,Phi);
end
```

Transmissibility is recomputed inside the loop because the permeability changes for each new layer. Likewise, we regenerate the well objects to ensure correct well indices when updating the petrophysical data. You find the complete source code in the script computeLorenzSPE10 in the book module.

Figure 13.16 reports Lorenz coefficients for all layers when the injector and the four producers are controlled by bottom-hole pressure. The relatively large dynamic heterogeneity and the variation among individual layers within the same formation is a result of our choice of well controls. The actual injection and production rates achieved with pressure-controlled wells are very sensitive to each well being perforated in a region of good sand quality. This is almost impossible to ensure when using fixed well positions for all layers of SPE 10, and hence pressure-controlled wells will generally accentuate heterogeneity effects. To see this, you should modify the script and rerun the case with equal rates in all four producers.

Our overall setup and somewhat haphazard well placement is not representative of a real reservoir, but serves well to illustrate our next point. Since the Lorenz coefficient generally is quite large, most of the cases would have suffered from early breakthrough if we were

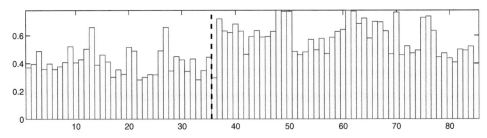

Figure 13.16 Lorenz coefficient for the 85 horizontal layers of the SPE 10 model subject to an inverted five-spot pattern with injector and producers controlled by bottom-hole pressure. The dashed line shows the border between the Tarbert and Upper Ness layers.

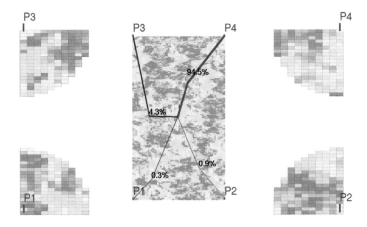

Figure 13.17 Well configuration and flux allocation for the four well pairs with initial well configuration for Layer 61. (Red colors are good sands, while blue colors signify sands of low permeability and porosity.)

to use this initial well placement for multiphase fluid displacement. Let us therefore pick one of the layers and see if we can try to improve the Lorenz coefficient and hence also the sweep efficiency. Figure 13.17 shows the well allocation and the sand quality near the producers for Layer 61, which is the layer giving the worst Lorenz coefficient. The flux allocation shows that we have a very unbalanced displacement pattern where producer P4 draws 94.5% of the flux from the injector and producers P1 and P2 together only draw 1.2%. This can be explained by looking at the sand quality in our reservoir. Producer P1 is completed in a low-quality sand and will therefore achieve low injectivity if all producers operate at the same bottom-hole pressure. Producers P2 and P3 are perforated in cells with better sand, but are both completely encapsulated in regions of low sand quality. On the other hand, producer P4 is connected to the injector through a relatively contiguous region of high-permeable sand.

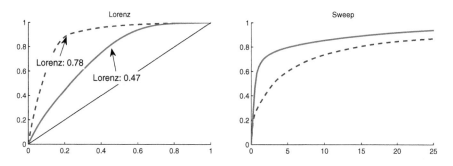

Figure 13.18 F–Φ and sweep diagrams before (dashed line) and after (solid line) producers P1 to P4 have been been moved to regions with better sand quality.

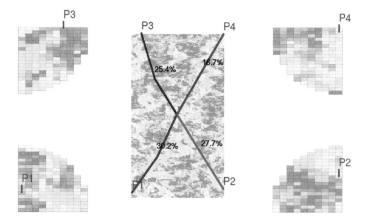

Figure 13.19 Well configuration and flux allocation for the four well pairs after all producers have been moved to regions with better sand quality.

The largely concave F–Φ diagram shown to the left in Figure 13.18 testifies that the displacement is strongly heterogeneous and is characterized by large differences in the residence time of different flow paths, or in other words, suffers from early breakthrough of the displacing fluid. Hence, we must inject large amounts of the displacing fluid to recover the hydrocarbons from the low-quality sand; this can be seen from the weakly concave sweep diagram to the right in Figure 13.18. Altogether, we should expect a very unfavorable volumetric sweep from this well placement.

To improve the displacement, we can try to move each producer to a better location. To this end, we should look for cells in the vicinity of each well that have better sand quality (higher porosity and permeability) *and* are connected to the injector by more contiguous paths of good quality sands. Figure 13.19 shows the well allocation after we have moved P1 to P3. Producer P4 was already in a good location and is only shifted one cell. The result is a much more balanced well allocation, which is confirmed by the rates reported in Table 13.1. (Notice also that the overall reservoir rate has increased slightly.) Figure 13.18

Table 13.1 *Volumetric flow rates in units 10^{-5} m^3/s for each of the wells in Layer 61 of the SPE 10 model. (These are obviously far from rates seen in real reservoirs.)*

Placement	P1	P2	P3	P4	I
Initial	−0.0007	−0.0020	−0.0102	−0.2219	0.2348
Improved	−0.0719	−0.0660	−0.0604	−0.0398	0.2381

shows how the F–Φ diagram has become significantly less concave and that the sweep diagram has become more concave. This testifies that the variation in residence times associated with different flow paths is much smaller. We should thus expect a more efficient and less heterogeneous volumetric sweep.

Dynamic heterogeneity measures correlate surprisingly well with recovery for water-flooding processes [278, 144, 259, 260, 218], and are generally easy to use as a guide when searching for optimal or improved well placement. Because of their low computational cost, the measures can be used as part of a manual, interactive search or combined with more rigorous mathematical optimization techniques. In the `diagnostics` module you can find examples that use flow diagnostics in combination with adjoint methods to determine optimal well locations and set optimal injection and production rates. The interested reader can find more details about optimization based on flow diagnostics in [218].

Computer exercises

13.4.1 Repeat the experiment using fixed rates for the producers (and possibly also for the injector). Can you explain the differences that you observe?

13.4.2 Compute Lorenz coefficients from RTDs instead and compare.

13.4.3 Use the interactive diagnostic tool introduced in the next section to manually adjust the bottom-hole pressures (or alternatively the production rates) to see if you can improve the sweep even further.

13.4.4 Can you devise an automated strategy that uses flow diagnostics to search for optimal five-spot patterns?

13.4.5 Cell-averaged time-of-flight values computed by a finite-volume scheme can be quite inaccurate when interpreted pointwise, particularly for highly heterogeneous media. To investigate this, pick one of the fluvial layers from SPE 10 and use the `pollock` routine from the `streamline` module to compute time-of-flight values on a 10×10 subsample inside a few selected cells. Alternatively, you can use the finite-volume method on a refined grid.

13.5 Interactive Flow Diagnostics Tools

The ideas behind most flow diagnostics techniques are relatively simple to describe and their computation is straightforward to implement. The real strength of these techniques nevertheless lies in their visual appeal and the ability for rapid user interaction. Together, MATLAB and MRST provide a wide variety of powerful visualization routines you can use to visualize input parameters and simulation results. The `diagnostics` module supplies additional tools for enhanced visualization, but using a script-based approach to visualization means that you each time need to write extra code lines to manually set color map and view angle or display various additional information such as legends, colorbars, wells, and figure titles.

We have mostly omitted these extra code lines in our previous discussion, but if you go in and examine the accompanying scripts, you will see that a large fraction of the code lines focus on improving the visual appearance of plots. Such code is repetitive and should ideally not be exposed to the user of flow diagnostics. More importantly, a script-based approach gives a static view of the data and offers limited capabilities for user interaction, apart from zooming, rotating, and moving the displayed data sets. Likewise, a new script must be written and executed each time we want to look a new plot that combines various types of diagnostic data, e.g., to visualize time-of-flight or petrophysical values within a given color region, use time-of-flight to threshold the color regions, etc.

Preprocessing: Predicting Fluid Movement

To simplify the user interface to flow diagnostics, we have integrated most of the flow diagnostics capabilities discussed earlier in the chapter into a graphical tool that enables you to interact more directly with your data set

```
interactiveDiagnostics(G, rock, W);
```

The interesting part of this GUI discussion is not the exact layout of the user interface, which likely will change in future versions of MRST, but the different quantities it can produce.

The GUI has primarily been written for preprocessing to predict future fluid movement and uses the standard two-point incompressible flow solver for a single-phase fluid with density 1,000 kg/m^3 and viscosity 1 cP to compute a representative flow field, which is then fed to the function `computeTOFandTracer` to compute the basic flow diagnostics quantities. Once the computation is complete, the graphical user interface is launched; see Figure 13.20: At the time of writing, this consists of a *plotting window* showing the reservoir model and a *control window*, which contains a set workflow tabs and menus that enable you to use flow diagnostics to explore volumetric connections, flow paths, and dynamic heterogeneity measures.

The control window has three different tabs. The "Region selection" tab is devoted to displaying the various kinds of volumetric regions discussed in Section 13.1.1. The "Plots" tab lets you compute F–Φ diagram, Lorenz coefficient, and the well allocation

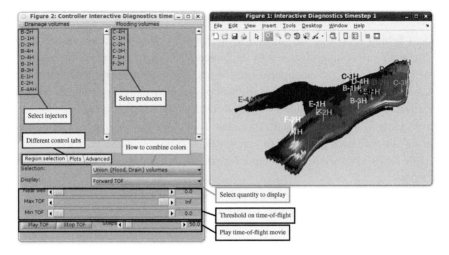

Figure 13.20 A graphical user interface to flow diagnostics. The plotting window shows forward time-of-flight, which is the default value displayed upon startup. In the control window, we show the "Region selection" tab that lets you select which quantity to show in the plotting window, select which wells to include in the plot, specify how to combine color concentrations to select volumetric regions, and as well set maximum and minimum time-of-flight values to crop the volumetric regions.

panels shown in Figure 13.14. From this tab, you can also bring up a dialog box to edit the well settings and recompute a new flow field and the resulting flow diagnostics (unless the parameter `computeFlux` is set to false). In the "Advanced" tab, you can control the appearance of the 3D plot by selecting whether to display grid lines, well-pair information, and well paths, set 3D lighting and transparency value, etc. This tab also allows you to export the current volumetric subset as a vector of boolean cell indicators.

Regions to be displayed in the plotting window are specified by selecting a set of active wells and by choosing how the corresponding colorings should be combined. That is, for a given set of active wells, you can either display all cells that have nonzero concentration for the injector *or* producer colors, only those cells that have nonzero concentration for injector *and* producer colors, only cells with nonzero concentration of producer colors, or only cells with nonzero concentrations for injector colors. To select a set of active wells, you can either use the list of injectors/producers in the control window, or mark wells in the plotting window. In the latter case, the set of active wells consist of the well you selected plus all other wells this well communicates with. Selecting a well in the 3D view also brings up a new window that displays a pie chart of the well allocation factors and a graph that displays cumulative allocation factors per connection. Figure 13.21 shows this type of visualization for a model where we also have access to a set of states from a full multiphase simulation. In this case, we invoke the GUI by calling

```
interactiveDiagnostics(G, rock, W, 'state', state, 'computeFlux', false);
```

13.5 Interactive Flow Diagnostics Tools

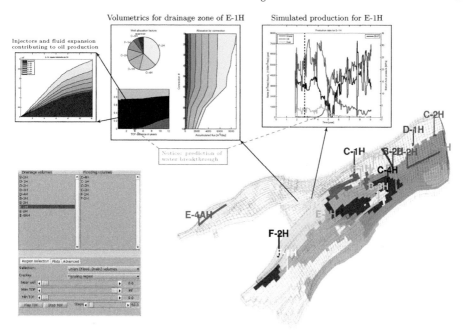

Figure 13.21 Flow diagnostics used to analyze volumetric connections and predict future inflow of fluids to well E-1H in a three-phase, black-oil flow simulation of the Norne field. For comparison, the figure also reports the *simulated* production from E-1H, where the dashed vertical line represent present time.

so that the fluxes used to compute time-of-flight and color concentrations are extracted from the given reservoir state given in `state`. Left-clicking on producer E-1H brings up a plot of the well allocation together with a plot of the fluid distribution displayed as function of the backward time-of-flight from the well. To further investigate the flow mechanism, we can mark each fluid and get a plot of how various injectors and fluid expansion in the reservoir contribute to push the given fluid toward the producer. In Figure 13.21 we have expanded the GUI by another function that plots the simulated well history of each well. This history confirms what we see in the backward time-of-flight analysis of the drainage zone: a few years into the future, water arrives at the producer (blue line), and there is a marked decay in oil and gas production (red and green lines).

The GUI also offers functionality to load additional cell-based data sets that can be displayed in the 3D plot:

```
interactiveDiagnostics(G, rock, W, celldata);
interactiveDiagnostics(G, rock, W, celldata, 'state', state);
```

and has some functionality for postprocessing simulations with multiple time steps, but this is beyond the scope of the current presentation.

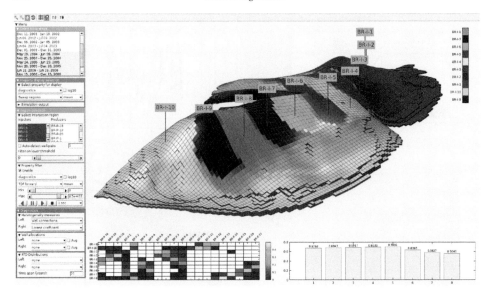

Figure 13.22 Variation in sweep regions over eight time report steps from an ECLIPSE simulation of the Brugge benchmark case. Clear color indicate regions swept by a single injector in all time steps, whereas grayish colors indicate regions being swept by more than one injector over the time interval.

Postprocessing: Enhanced Visualization of Simulation Results

As of release 2018b, MRST also offers another user interface dedicated entirely to *postprocessing* simulation data

```
PostProcessDiagnostics()
PostProcessDiagnostics('filename')
```

Figures 13.22–13.24 illustrate how flow diagnostics is used to enhance the understanding of the fluid communication for a full ECLIPSE simulation of the Brugge benchmark case [249].

Figure 13.22 shows the average sweep regions over eight report steps for all injectors, limited outward by an upper bound on the forward time-of-flight. Regions swept by a single injector over the whole simulation period are shown as clear colors, whereas diffuse or grayish colors represent regions swept by multiple injectors. You can use such plots to analyze how sweep or drainage regions change over time. The matrix plot to the lower left reports communication strength between each well pair, with rows representing injectors and columns injectors. The lower right plot shows how the dynamic heterogeneity measured by the Lorenz coefficient changes over the eight time steps. By selecting a single injector or producer, you can break up this analysis to see the dynamic heterogeneity of individual sweep, drainage, or well-pair regions, measured in terms of $F-\Phi$ diagrams or Lorenz coefficients.

13.5 Interactive Flow Diagnostics Tools

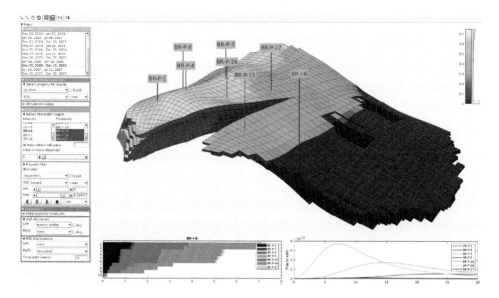

Figure 13.23 Sweep region of injector BR-I-6 for the Brugge benchmark case. The lower plots show flux allocation for the injector and the RTD for a passive tracer tracer released from the injector.

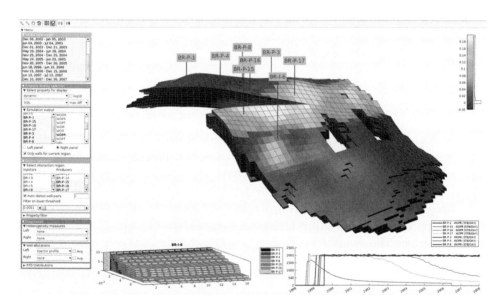

Figure 13.24 Saturation changes within the sweep region of injector BR-I-6 for the Brugge benchmark case. The lower-right plot reports flux allocation every half year for the injector from toe to heel (top to bottom in the plot). The lower-right plot reports water production rates for all connected producers.

Figure 13.23 shows oil saturation inside the sweep region of BR-I-6. From the flux allocation in the lower-left plot, we see that this injector primarily supports BR-P-16, BR-P-4, and BR-P-1 with the current set of drive mechanisms. Producer BR-P-15 lies closer than any other wells, but has been shut in (see lower-right plot in Figure 13.24) and is only shown since significant fluid volumes have been going to this producer earlier in the simulation. Also BR-P-17 lies closer to BR-I-6 than the other producers with large flux allocation, but as you can see from Figure 13.22, this well lies in the middle of BR-I-5's sweep region and is predominantly supported by that injector. The lower-right plot reports the first 30 years of an RTD analysis and shows that a passive tracer injected now from BR-I-6 will first break through in BR-P-16, and then in BR-P-4 and BR-P-1. This gives time lines for displacement fronts originating from BR-I-6.

Figure 13.24 reports accumulated saturation changes inside the sweep region over the whole simulation period. We clearly see how the displacement profile has engulfed producers BR-P-15 to 17 and caused a decline in production rates. In particular, BR-P-15 hardly managed to stay on plateau and started its decline less than half a year after it started operating.

The types of flow diagnostics discussed previously are also available in ResInsight, which is an open-source tool for postprocessing of simulation results on ECLIPSE format. This software is part of the OPM Initiative and currently functions as an outlet of research ideas from MRST towards professional use in industry.

In the rest of the chapter, we use the preprocessing GUI to study two reservoir models. The accompanying scripts do not reproduce all figures directly, but specify the manual actions you need to reproduce many of the figures.

13.5.1 Synthetic 2D Example: Improving Areal Sweep

The first example considers a slightly modified version of the setup from Figure 13.1. We have introduced two low-permeable barriers, moved the two injectors slightly to the south, and switched all wells from rate to pressure control; see Figure 13.25. Table 13.2 reports specific values for the well controls under the label "base case."

Our previous setup had a relatively symmetric well pattern in which producers P1 and P3 were supported by injectors I1 and I2, respectively, whereas producer P2 was supported by both injectors. This symmetry is broken by the two barriers, and now injector I1 also provides significant support for producer P3. You can infer this from the plot of producer partition: the gray areas between the magenta (P2) and red (P3) regions signify parts of the reservoir that are drained by both producers. The relatively large gray area southeast of I1 indicates that this injector supports both P2 and P3. There is also a gray area southwest of injector I2, but since this is less pronounced, we should only expect that only a small portion of P2's inflow can be attributed to I2. To confirm this, you can load the model in the interactive viewer (see the script `interactiveSimple`), and click the names for each individual producer to bring up a pie chart reporting the corresponding flux allocation.

13.5 Interactive Flow Diagnostics Tools

Table 13.2 *Well controls given in terms of bottom-hole pressure [bar] for a simple 2D reservoir with five initial wells (I1, I2, P1, P2, P3) and one infill well (I3).*

	I1	I2	I3	P1	P2	P3	Lorenz
Base case	200	200	—	100	100	100	0.2730
Case 1	200	200	—	100	130	80	0.2234
Case 2	200	200	—	100	130	80	0.1934
Case 3	200	220	140	100	130	80	0.1887

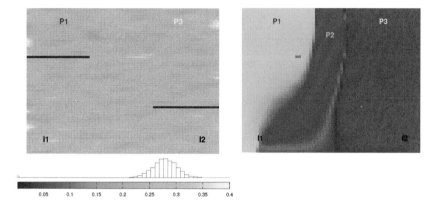

Figure 13.25 A simple 2D reservoir with two injectors and three producers. The left plot shows porosity and the right plot the corresponding producer partition.

To figure out to what extent this is a good well pattern or not, we start by looking at how a displacement front would propagate if the present flow field remains constant. For a displacement front traveling with a unit speed relatively to the Darcy velocity given by the flux field, the region swept by the front at time t consists of all cell for which $\tau_f \leq t$. Using the interactive GUI shown in Figure 13.20, you can either show swept regions by specifying threshold values manually, or use the "Play TOF" button to animate how the displacement front advances through the reservoir. Figure 13.26 shows four snapshots of such an advancing front. We notice, in particular, how the sealing fault to the northwest of the reservoir impedes the northbound propagation of the displacement front from I1 and leads to early breakthrough in P2. There is also a relatively large region that remains unswept after the two displacement fronts have broken through in all three producers.

As a more direct alternative to studying snapshots of an imaginary displacement front, we can plot the residence time, i.e., the sum of the forward and backward time-of-flight, as shown in the upper-left plot of Figure 13.27. Here, we use a nonlinear gray-map to more clearly distinguish high-flow zones (dark gray) from stagnant regions (white) and other

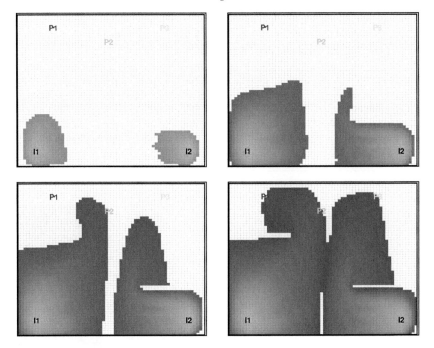

Figure 13.26 Evolution of an imaginary displacement front illustrated by thresholding time-of-flight iniside the sweep regions for the base case.

regions of low flow (light gray). In the figure, we see that wells I1 and P2 are connected by a high-flow region, which explains the early breakthrough we observed in Figure 13.26. The existence of high-flow regions can also be seen from the F–Φ diagram and the Lorenz coefficient of 0.273.

The interactive diagnostic tool has functionality that lets you modify the well controls and if needed, add new wells or remove existing ones. We now use this functionality to try to manually improve the volumetric sweep of the reservoir, much in the same way as we did for a layer of SPE 10 in Section 13.4.2. We start by reducing the high flow rate in the region influenced by I1 and P2. That is, we increase the pressure in P2 to, say, 130 bar to decrease the inter-well pressure drop. The resulting setup, referred to as "Case 1," gives more equilibrated flow paths, as you can see from the Lorenz coefficient and the upper-right plot in Figure 13.27.

Case 2 further decreases the pressure in P3 to 80 bar to increase the flow in the I2–P3 region. As a result of these two adjustments to the well pressures, we have reduced the stagnant region north of P2 and also diminished the clear flow-divide that extended from the south of the reservoir to the region between P2 and P3. To also sweep the large unswept region east of P3, we can use infill drilling to introduce a new well to the southeast of this region, just north of the sealing fault. Since the new well is quite close to the

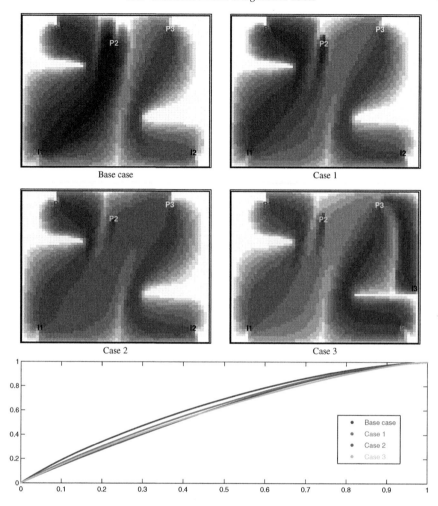

Figure 13.27 In Cases 1 and 2, well controls have been manually adjusted from that of the base case (see Table 13.2) to equilibrate total travel time throughout the reservoir. Case 3 includes infill drilling of an additional injector. The bottom plot shows the corresponding F–Φ curves.

existing producer, we should assign a relatively low pressure to avoid introducing too high flow rates.

In Case 3, we let the well operate at 140 bar and at the same time we have increased the pressure in I2 to 220 bar. Figure 13.28 shows snapshots of the advancing front at the same instances in time as was used in Figure 13.26 for the base case. Altogether, the well configuration of Case 3 gives significantly improved areal sweep and reduces the Lorenz coefficient to 0.189. We can therefore expect this configuration to give better displacement if the setups were rerun with a multiphase simulator.

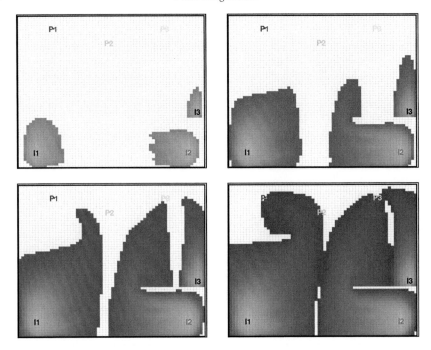

Figure 13.28 Evolution of an imaginary displacement front illustrated by thresholding time-of-flight iniside the sweep regions for Case 3.

You can find a more detailed description of how you should use the interactive GUI to perform the experiments just described in the interactiveSimple.m script of the book module. I encourage you to use the script to familiarize yourself with interactive flow diagnostics. Are you able to make further improvements?

13.5.2 SAIGUP: Flow Patterns and Volumetric Connections

The previous example was highly simplified and chosen mainly to illustrate the possibilities that lie in the interactive use of flow diagnostics. In this example, we take a closer look at the volumetric connection in the SAIGUP example from Section 5.4.4 on page 169. We start by rerunning the example script to set up the simulation model and compute a flow field:

```
saigupWithWells; close all
clearvars -except G rock W state
```

Henceforth, we only need the geological model, description of the wells, and the reservoir state. We clear all other variables and close all plots produced by the script. We can pass the reservoir state to the interactive GUI and use it in pure postprocessing mode:

```
interactiveDiagnostics(G, rock, W, 'state', state, 'computeFlux', false);
```

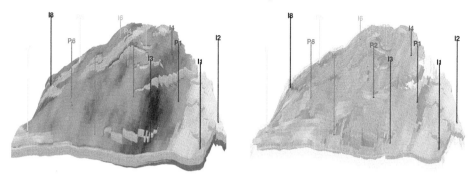

Figure 13.29 Horizontal permeability ($\log_{10} K_x$) for the SAIGUP model. The left plot shows the full permeability field, while the right plot only shows the permeability in cells that have a total residence time less than 100 years. (The reservoir is plotted so that the north–south axis goes from left to right in the figure.)

In this mode, we are not able to edit any well definitions and recompute fluxes.

In Figure 5.11 on page 173, we saw that, even though injectors and producers are completed in all 20 grid layers of the model, there is almost no flow in the bottom half of the reservoir. A more careful inspection shows that there is almost no flow in the upper layers in most of the reservoir either because the best sand quality is found in the upper-middle layers of the reservoir. Figure 13.29 confirms this by comparing the permeability in the full model with the permeability in cells having a residence time less than 100 years.

Because each injector is controlled by a total fluid rate, large fluid volumes are injected in completions connected to good quality sand, while almost no fluid is injected into zones with low permeability and porosity. You can see this in Figure 13.30, which shows overall and cumulative wellallocation factors for four of the injectors. (These plots are conceptually similar to what you would obtain by running a production logging tool for all these wells.) Injectors I3 and I4 are completed in the southern part of the reservoir, and here low-quality sand in the bottom half of the reservoir leads to almost negligible injection rates in completions 11–20. In a real depletion plan, the injectors would probably not have been completed in the lower part of the sand column. Injector I5 located to the west in the reservoir is completed in a column with low permeability in the top four and the bottom layer, high permeability in Layers 6–9, and intermediate permeability in the remaining layers. Hence, almost no fluid is injected through the top four completions, which hence are redundant. Finally, injector I6 is completed in a column with poor sand quality in the top three layers, high permeability in Layers 4–9, and intermediate permeability in the remaining layers.

Figure 13.30 also shows the flux allocation for all well pairs in the reservoir. Each curved line corresponds to a injector–producer connection, the line color signifies the producer, and the percentage signifies the fraction of the total flux from each injector that goes to the different producers. We have truncated the connections to only pairs that correspond

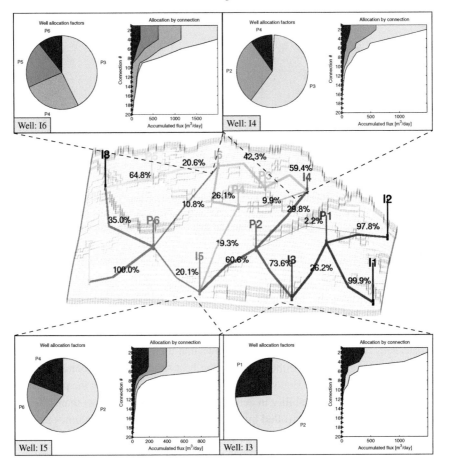

Figure 13.30 Flux allocation for all well pairs of the SAIGUP model and cumulative well-allocation factors from bottom to top layer for injectors I3, I4, I5, and I6.

to at least 1% of the flux. This explains why not all fractions sum up to unity. Let us take injector I4 as an example. Figure 13.31 shows two plots of the influence region for this injector. From the well allocation plots, we have already seen that I4 is connected to producers P2 to P4. Almost all the completions of these wells lie inside the region colored by I4. Producer P1, on the other hand, is only completed in a single cell inside the colored region, and this cell is in the top layer of the reservoir where the sand quality is very poor. It is therefore not clear whether P1 is actually connected to I4 or if this weak connection is a result of inaccuracies in the computation of color concentrations. Since we only use a first-order discretization, the color concentrations can suffer from a significant amount of numerical smearing near flow divides, which here is signified by blue-green colors.

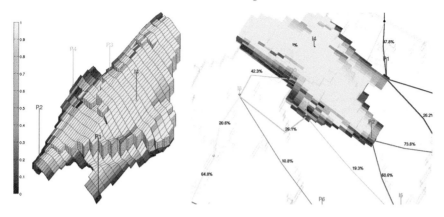

Figure 13.31 Plot of the influence region for well I4.

COMPUTER EXERCISES

13.5.1 Change all the injectors to operate at a fixed bottom-hole pressure of 300 bar. Does this significantly change the flow pattern in the reservoir? Which configuration do you think is best?

13.5.2 Use the flow diagnostic tool to determine well completions in the SAIGUP model that have insignificant flow rate, eliminate these well completions, and rerun the model. Are there any apparent changes in the flux allocation and volumetric connections?

13.5.3 Consider the model given in `makeAnticlineModel.m`. Can you use flow diagnostics to suggest a better well configuration?

13.5.4 Go back to the SPE 9 model from Section 12.4.4. Would it make sense to use flow diagnostics to analyze the flow patterns of this model? If so, how would you do it?

14
Grid Coarsening

Over the last decades, methods for characterizing subsurface rock formations have improved tremendously. This has, together with a dramatic increase in computational power, enabled the industry to build increasingly detailed and complex models to account for heterogeneous structures on different spatial scales. Using gridding techniques similar to the ones outlined in Chapter 3, today one can easily build complex geological models consisting of multiple millions of cells to account for most of the features seen in typical reservoirs. In most cases, geocellular models used for reservoir characterization contain more geological layers and model fine-scale heterogeneity with higher resolution than what is used for flow simulations.

Through parallelization and use of massively parallel computers it is possible to simulate fluid flow on grid models with up to a billion cells [84, 85, 240], but such simulations require expensive infrastructure and a very high power budget and are rarely seen in practice. Contemporary high-fidelity models seem to be in the range of a few million cells, whereas the majority of asset models have ten times fewer cells, since engineers usually want to spend available computational power on more advanced flow physics or on running a large number of model realizations instead of a few highly resolved ones. To obtain computationally tractable simulation models, it is therefore common to develop reduced models through some kind of upscaling (homogenization) procedure that removes spatial detail from the geological description. Typically, a coarser model is developed by identifying regions consisting of several cells and then replacing each region by a single, coarse cell with homogeneous properties that represent the heterogeneity inside the region in some averaged sense. We discuss such upscaling methods in more detail in Chapter 15.

14.1 Grid Partitions

You can obviously generate coarse grids in the same way as the original by simply specify a lower spatial resolution. However, this approach has two obvious disadvantages: first of all, if the original grid has complex geometry (and topology), it is very challenging to preserve the exact geometry of the fine grid for an arbitrary coarse resolution. Second, except in simple cases, there is generally not a one-to-one mapping between cells in the fine and coarse grids. Herein, we therefore choose a different approach. *In MRST, a coarse grid always refers to a grid that is defined as a partition of another grid, which is referred to as the fine grid.*

Tools for partitioning and coarsening of grids are found in two different modules of MRST: The `coarsegrid` module defines a basic grid structure for representing coarse grids and supplies simple routines for partitioning grids with an underlying Cartesian topology. The `agglom` module offers tools for defining flexible coarse grids that adapt to geological features and flow patterns, e.g., as discussed in [125, 124, 194, 187]. Coarse partitions also form a basis for contemporary multiscale methods [195] and are used extensively in the various multiscale modules of MRST (`msrsb`, `msmfem`, `msfv`, etc). In this chapter, we first discuss functionality for generating and representing coarse grids found in the `coarsegrid` module and then briefly outline some of the more advanced functions found in the `agglom` module. In an attempt to distinguish fine and coarse grids, we henceforth refer to fine grids as consisting of *cells*, whereas coarse grids are said to consist of *blocks*.

Coarse grids in MRST are represented by a structure that consists entirely of topological information stored in the same fields as for the general grid structure introduced in Section 3.4. As a naming convention, we use CG to refer to a coarse-grid structure and G to refer to the usual (fine) grid. A coarse grid is always related to a fine grid in the sense that

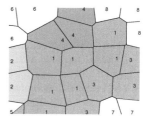

- each cell in the fine grid G belongs to one, and only one, block in the coarse grid CG;
- each block in CG consists of a *connected* subset of cells from G; and
- CG is defined by a *partition vector* p defined such that $p(i) = \ell$ if cell i in G belongs to block ℓ in CG.

This concept is quite simple, but has proved to be very powerful in defining coarse grids that can be applied in a large variety of computational algorithms. We will come back to the details of the grid structure in Section 14.2. First, let us discuss how to define partition vectors in some detail, as this is more useful from a user perspective than understanding the details of how the CG structure is implemented.

To demonstrate the simplicity and power of using partition vectors to define coarse grids, we go through a set of examples. You can find complete codes for all the following examples in `showPartitions.m` from the book module.

14.1.1 Uniform Partitions

For all grids having a logically Cartesian topology, i.e., grids that have a valid field `G.cartDims`, we can use the function `partitionUI` to generate a relatively uniform partition that consists of the tensor product of a load-balanced linear partition in each index direction. As an example, let us partition a 7×7 fine grid into a 2×2 coarse grid:

```
G = cartGrid([7,7]);
p = partitionUI(G, [2,2]);

plotCellData(G, p, 'EdgeColor', 'y');
outlineCoarseGrid(G, p, 'k');
axis tight off,
caxis([.5 max(p)+.5]);
colormap(lines(max(p)));
set(colorbar,'YTick',1:max(p));
```

The call to `partitionUI` returns a vector with one element per cell taking one of the integer values $1, 2, 3, 4$ that represent the four blocks. Since seven is not divisible by two, the coarse blocks do not have the same size but consist of 4×4, 3×3, 4×3, and 3×4 cells. To better distinguish different blocks in the plot, we have used `outlineCoarseGrid(G, p)` to find and plot all faces in G whose neighboring cells have different values of p.

The same procedure can, of course, also be applied to partition any grid in 3D that has a Cartesian topology. As an example, we consider a simple box geometry:

```
G = cartGrid([10,10,4]);
p = partitionUI(G, [3,3,2]);

plotCellData(G, p, 'Edgecolor', 'w');
outlineCoarseGrid(G, p, ...
     'EdgeColor','k','lineWidth',4);
colormap(colorcube(max(p)))
view(3);  axis off
```

Here, we have used the `colorcube` colormap, which is particularly useful for visualizing partition vectors, since it contains as many regularly spaced colors in RGB color space as possible. The careful reader will also observe that the arguments to `outlineCoarseGrid` changes somewhat for 3D grids.

14.1.2 Connected Partitions

All you need to partition a grid is a partition vector. This vector can be given by the user, read from a file, generated by evaluating a geometric function, or given as the output of some user-specified algorithm. As a simple example of the latter, let us partition the box model $[-1, 1] \times [-1, 1]$ into nine different blocks using the polar coordinates of the cell centroids. The first block is defined as $r \leq 0.3$, whereas the remaining eight are defined by segmenting $4\theta/\pi$:

14.1 Grid Partitions

```
G = cartGrid([11, 11],[2,2]);
G.nodes.coords = ...
    bsxfun(@minus, G.nodes.coords, 1);
G = computeGeometry(G);
c = G.cells.centroids;
[th,r] = cart2pol(c(:,1),c(:,2));
p = mod(round(th/pi*4)+4,4);
p(r<.3) = max(p)+1;
```

In the second-to-last line, the purpose of the modulus operation is to avoid wrap-around effects as θ jumps from $-\pi$ to π.

The human eye should be able to distinguish nine different coarse blocks in the plot above, but the partition does unfortunately not satisfy all the criteria we prescribed on page 519. Indeed, as you can see from the colorbar, the partition vector only has five unique values and thus corresponds to five blocks, according to our definition of p: cell i belongs to block ℓ if $p(i) = \ell$. Hence, what the partition describes is one connected block at the center surrounded by four *disconnected blocks*. To determine whether a block is connected or not, we will use some concepts from graph theory. We first form a local undirected graph (or a network) in which nodes are cells and edges are cell faces connecting two cells with the same partition value. A block is then said to be disconnected if the graph has multiple *connected components*. A connected component is defined as the subgraph in which any two nodes are connected to each other through a path. In other words, a coarse block is disconnected if there exists at least one pair of cells that cannot be connected by a continuous path in the local grid graph. To get a partition satisfying our requirements, we must split the four disconnected blocks. This is done by the following call:

```
q = processPartition(G, p);
```

which splits disconnected components that have the same p-value into separate blocks and updates the partition vector accordingly. In the current case, Blocks 1–4 will be split in two, whereas Block 5 remains unchanged, giving the nine blocks separated by solid lines in the figure.

Let us quickly outline how this routine works. We start by identifying all cells having the same partition value; the left plot in Figure 14.1 shows these cells for the first block. The grid graph is the same that we used to construct the discrete divergence and gradient operators, with the edge list given in terms of the two columns $C_1(f)$ and $C_2(f)$ from G.faces.neighbors that specify the cells connected by face f. The only exception is that we now have to exclude any connections between cells having different p-values. From this list, we can construct a local *adjacency matrix*. This symmetric matrix is defined so that all cells corresponding to nonzero entries in a single row (or column) are directly connected. In the two adjacency matrices shown in Figure 14.1, we have used different color for cells

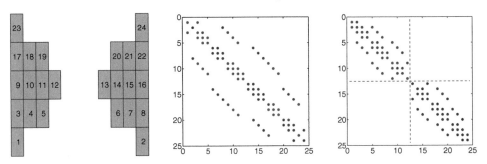

Figure 14.1 Partition of a disconnected block. The left plot shows all cells with the same p-value. The middle plot shows the adjacency matrix based on the original numbering, whereas the right plot shows the adjacency matrix after a Dulmage–Mendelsohn permutation has separated the connected components.

belonging to each of the two components for clarity. The middle plot shows that cell 1 is connected to cell 3; cell 2 is connected to cell 8; cell 3 is connected to cells 1, 4, and 9; and so on. To find the connected components, we use a Dulmage–Mendelsohn permutation. Disconnected components will then appear as diagonal blocks in the permuted adjacency matrix; see the right plot in Figure 14.1.

Graph operations like this can generally be used to adapt the partition to features in the geological model. As an example, the processing routine can also take an additional parameter `facelist` that specifies a set of faces across which the connections will be removed before processing:

```
q = processPartition(G, p, facelist)
```

Using this functionality one can, for instance, prevent coarse blocks from crossing faults inside the model.

14.1.3 Composite Partitions

In many cases, it may be advantageous to create partition vectors by combining more than one partition principle. As an example, consider a heterogeneous medium consisting of two different facies (rock types), one with high permeability and one with low, that each form large contiguous regions. If we now let the coarse blocks respect the facies boundaries, we can assign each block a homogeneous property and avoid upscaling. Within each facies, we can further use a rectangular partition generated by `partitionCartGrid`, which is simpler and less computationally expensive than `partitionUI` but only works correctly for a grid having a fully intact logically Cartesian topology with no inactive cells, no cells that have been removed by `removeGrid`, and so on. The following code illustrates the principle:

```
G = cartGrid([20, 20], [1 1]);
G = computeGeometry(G);

% Facies partition
f  = @(c) sin(4*pi*(c(:,1)-c(:,2)));
pf = 1 + (f(G.cells.centroids) > 0);

% Cartesian partition
pc = partitionCartGrid(G.cartDims, [4 4]);

% Alternative 1:
[b,i,p] = unique([pf, pc], 'rows' );
% Alternative 2:
q = compressPartition(pf + max(pf)*pc);
```

The example also shows two alternative techniques for combining different partitions. The first alternative collects the partitions as columns in an array A. The call [b,i,p]=unique(A,'rows') will return b as the unique rows of A so that b=A(i) and A=b(p). Hence, p will be a partition vector that represents the unique intersection of all the partitions collected in A. In the second method, we treat the partition vectors as multiple subscripts, from which we compute a linear index. This index is not necessarily contiguous. Most routines in MRST require that partition vectors are contiguous to avoid having to treat special cases arising from noncontiguous partitions. To fulfill this requirement, we use the function compressPartition that renumbers a partition vector to remove any indices corresponding to empty grid blocks. The two alternatives have more or less the same computational complexity, and which alternative you choose in your implementation is largely a matter of what you think will be easiest to understand for others.

Altogether, the examples presented so far in this chapter explain the basic concepts of how you can create partition vectors. Before we go on to explain details of the coarse-grid structure and how to generate this structure from a given partition vector, we show one last and a bit more fancy example. To this end, we create a cup-formed grid, partition it, and then visualize the partition using a nice technique. To generate the cup-shaped grid, we use the fictitious-domain technique we previously used to generate the ellipsoidal grid in Figure 3.4 on page 61.

```
x = linspace(-2,2,41);
G = tensorGrid(x,x,x);
G = computeGeometry(G);
c = G.cells.centroids;
r = c(:,1).^2 + c(:,2).^2+c(:,3).^2;
G = removeCells(G, (r>1) | (r<0.25) | (c(:,3)<0));
```

Assume that we wish to partition this cup model into 100 coarse blocks. We could, for instance, try to use partitionUI to impose a regular $5 \times 5 \times 4$ partition. Because of the

Figure 14.2 The partition of the cup-formed grid visualized with `explosionView`.

fictitious method, a large number of the ijk indices from the underlying Cartesian topology will correspond to cells that are not present in the actual grid. Imposing a regular Cartesian partition on such a grid typically gives block indices in the range $[1, \max(p)]$ that do not correspond to any cells in the underlying fine grid. In this particular case, only 79 out of the desired 100 blocks correspond to a volume within the grid model. To see this, we use the function `accumarray` to count the number of cells for each block index and plot the result as a bar chart:

```
subplot(2,1,1);
p = partitionUI(G,[5 5 4]);
bar(accumarray(p,1)); shading flat

q = compressPartition(p);
subplot(2,1,2);
bar(accumarray(q,1)); shading flat
set(gca,'XLim',[0 100]);
```

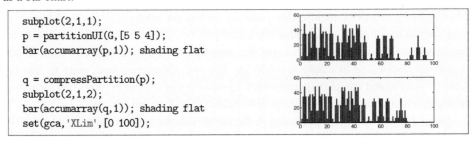

Figure 14.2 shows the partition obtained after we have compressed the partition vector. To clearly distinguish the different blocks, we have useed `explosionView` to create an *explosion view*, which is a useful technique for visualizing coarse partitions.

14.2 Coarse Grid Representation in MRST

When working with coarse grids, it is not always sufficient to only know the partition, i.e., which cells belong to which blocks. It can often be more advantageous to treat the coarse partition as a general polyhedral grid and have access to the inherent topology and possibly also the geometry. In MRST, we have chosen a compromise and use a coarse-grid structure that represents topology explicitly and geometry implicitly. This choice is motivated by flow solvers, which we previously have seen can be posed on an arbitrary grid graph provided each node has associated pore volume and depth value and each connection has an associated transmissibility.

14.2 Coarse Grid Representation in MRST

Given a grid structure G and a partition vector p, we generate a structure CG representing the coarse grid by the following call:

```
CG = generateCoarseGrid(G, p)
```

The coarse-grid structure consists entirely of topological information stored in the same way as described in Section 3.4 for G: The fields cells and faces represent the coarse blocks and their connections. As a result, we can use CG seamlessly with many of the standard solvers in MRST. Unlike the original grid structure, however, CG does not represent the geometry of the coarse blocks and faces explicitly and does therefore not have a nodes field. The geometry information is instead obtained from the parent grid G and the partition vector p, copies of which are stored in the fields parent and partition, respectively.

The structure, CG.cells, that represents the coarse blocks consists of the following mandatory fields:

- **num:** the number N_b of blocks in the coarse grid.
- **facePos:** an indirection map of size [num+1,1] into the faces array, which is defined completely analogously as for the fine grid. Specifically, the connectivity information of block i is found in the submatrix

    faces(facePos(i): facePos(i+1)-1, :)

    You can now compute the number of connections of each block using the statement diff(facePos).

- **faces:** an $N_c \times 2$ array of connections associated with a given block. Specifically, if faces(i,1)==j, then connection faces(i,2) is associated with block number j. To conserve memory, only the second column is actually stored in the grid structure. The first column can be reconstructed by a call to rldecode. Optionally, one may append a third column that contains a tag inherited from the parent grid.

In addition, the cell structure can contain the following optional fields, which typically are added by a call to coarsenGeometry, assuming that the corresponding information is available in the parent grid:

- **volumes:** an $N_b \times 1$ array of block volumes
- **centroids:** an $N_b \times d$ array of block centroids in $\mathbb{R}^d$

The face structure, CG.faces, consists of the following mandatory fields:

- **num:** the number N_c of global connections in the grid.
- **neighbors:** an $N_c \times 2$ array of neighboring information. Connection i is between blocks neighbors(i,1) and neighbors(i,2). One of the entries in neighbors(i,:), but not both, can be zero, to indicate that connection i is between a single block (the nonzero entry) and the exterior of the grid.

- `connPos, fconn`: packed data-array representation of the coarse → fine mapping. Specifically, the elements `fconn(connPos(i):connPos(i+1)-1)` are the connections in the parent grid (i.e., rows in `G.faces.neighbors`) that constitute coarse-grid connection `i`.

In addition to the mandatory fields, `CG.faces` has optional fields that contain geometry information and typically are added by a call to `coarsenGeometry`:

- `areas`: an $N_c \times 1$ array of face areas.
- `normals`: an $N_c \times d$ array of accumulated area-weighted, directed face normals in $\mathbb{R}^d$.
- `centroids`: an $N_c \times d$ array of face centroids in $\mathbb{R}^d$.

Like in G, the coarse grid structure also contains a field `CG.griddim` that is used to distinguish volumetric and surface grids, as well as a cell array `CG.type` of strings describing the history of grid constructor and modifier functions used to define the coarse grid.

As an illustrative example, let us partition a 4×4 Cartesian grid into a 2×2 coarse grid. This gives the following structure:

```
CG = 
         cells: [1x1 struct]
         faces: [1x1 struct]
     partition: [16x1 double]
        parent: [1x1 struct]
       griddim: 2
          type: {'generateCoarseGrid'}
```

with the `cells` and `faces` fields given as

```
CG.cells =                          CG.faces =
       num: 4                              num: 12
   facePos: [5x1 double]            neighbors: [12x2 double]
     faces: [16x2 double]             connPos: [13x1 double]
                                        fconn: [24x1 double]
```

Figure 14.3 shows relations between entities in the coarse grid and its parent grid. For instance, we see that block number one consists of cells one, two, five, and six because these are the rows in `CG.partition` that have value equal one. Likewise, we see that because `CG.faces.connPos(1:2)=[1 3]`, coarse connection number one is made up of two cell faces that correspond to faces number one and six in the parent grid because `CG.faces.fconn(1:2)=[1 6]`, and so on.

14.2.1 Subdivision of Coarse Faces

In the discussion so far, we have always assumed that there is only a single connection between two neighboring coarse blocks and that this connection is built up of a set of cell faces corresponding to all faces between pairs of cells in the fine grid that belong to the two different blocks. While this definition is useful for many workflows like in

14.2 Coarse Grid Representation in MRST

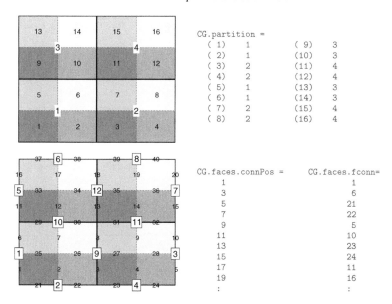

Figure 14.3 The relation between blocks in the coarse grid and cells in the parent grid (top) and between connections in the coarse grid and faces from the parent grid (bottom).

standard upscaling methods, there are also problems for which one may want to introduce more than one connection between neighboring blocks. To define a subdivision of coarse faces, we once again use a partition vector with one scalar value per face in the fine grid, defined completely analogous to vectors used for the volumetric partition. Assuming that we have two such partition vectors, pv describing the *volumetric* partition and pf describing the partition of cell faces, the corresponding coarse grid is built through the call:

```
CG = generateCoarseGrid(G, pv, pf);
```

In my experience, the simplest way to build a face partition is to compute it from an ancillary volumetric partition using the routine:

```
pf = cellPartitionToFacePartition(G, pv)
```

which assigns a unique, nonnegative integer for each pair of cell values occurring in the volumetric partition vector pv, and hence constructs a partitioning of all faces in the grid. Fine-scale faces that are not on the interface between coarse blocks are assigned zero value.

As an illustration, we continue the example from page 523. We first partition the 8×8 fine grid into a 2×2 coarse grid and then use facies information to subdivide faces of the coarse grid so that each coarse connection has a a given combination of facies values on opposite sides of the interface:

```
G = computeGeometry(cartGrid([8, 8], [1 1]));
f = @(c) sin(3*pi*(c(:,1)-c(:,2)));
pf = 1 + (f(G.cells.centroids) > 0);
plotCellData(G, pf,'EdgeColor','none');

pv = partitionCartGrid(G.cartDims, [2 2]);
pf = cellPartitionToFacePartition(G,pf);
pf = processFacePartition(G, pv, pf);
CG = generateCoarseGrid(G, pv, pf);
CG = coarsenGeometry(CG);
cmap = lines(CG.faces.num);
for i=1:CG.faces.num
    plotFaces(CG,i,'LineWidth',6,'EdgeColor', cmap(i,:));
end
text(CG.faces.centroids(:,1), CG.faces.centroids(:,2), ...
    num2str((1:CG.faces.num)'),'FontSize',20,'HorizontalAlignment', 'center');
```

As for the volumetric partition, we require that each interface that defines a connection in the face partition consists of a connected set of cell faces. That is, it must be possible to connect any two cell faces belonging to given interface by a path that only crosses edges between cell faces that are part of the interface. To ensure that all coarse interfaces are connected collections of fine faces, we have used the routine processFacePartition, which splits disconnected interfaces into one or more connected interfaces.

The same principles apply also in 3D, here illustrated for a rectangular block with a rectangular cut-out:

```
G = computeGeometry(cartGrid([20 20 6]));
c = G.cells.centroids;
G = removeCells(G, ...
    (c(:,1)<10) & (c(:,2)<10) & (c(:,3)<3));
plotGrid(G); view(3); axis off
```

We introduce a volumetric partition and a face partition:

```
p = partitionUI(G,[2, 2, 2]);
q = partitionUI(G,[4, 4, 2]);
CG = generateCoarseGrid(G, p, ...
    cellPartitionToFacePartition(G,q));

plotCellData(CG,(1:max(p))','EdgeColor','none');
plotFaces(CG,1:CG.faces.num,...
    'FaceColor' , 'none' , 'LineWidth' ,2);
view(3); axis off
```

The structure CG also contains lookup tables for mapping blocks and interfaces in the coarse grid to cells and faces in the fine grid. To illustrate, we visualize one connection of a subdivided coarse face that consists of several fine faces, along with the fine cells that belong to the neighboring blocks:

```
face  = 66;
sub   = CG.faces.connPos(face):CG.faces.connPos(face+1)-1;
ff    = CG.faces.fconn(sub);
neigh = CG.faces.neighbors(face,:);

show = false(1,CG.faces.num);
show(boundaryFaces(CG)) = true;
show(boundaryFaces(CG,neigh)) = false;
plotFaces(CG, show,'FaceColor',[1 1 .7]);
plotFaces(G, ff, 'FaceColor', 'g')
plotFaces(CG,boundaryFaces(CG,neigh), ...
    'FaceColor','none','LineWidth', 2);
plotGrid(G, p == neigh(1), 'FaceColor', 'none', 'EdgeColor', 'r')
plotGrid(G, p == neigh(2), 'FaceColor', 'none', 'EdgeColor', 'b')
```

14.3 Partitioning Stratigraphic Grids

You can also apply the principles outlined in the previous section to stratigraphic grids. To demonstrate this, we coarsen two corner-point models of industry-standard complexity: the sector model of the Johansen aquifer introduced in Section 2.5.4 and the SAIGUP model from Section 2.5.5. We also apply a few more advanced partition methods to the sector model with intersecting faults from Section 3.3.1. Full details are given in the scripts coarsenJohansen, coarsenSAIGUP, and coarsenCaseB4.

14.3.1 The Johansen Aquifer

The Johansen models were originally developed to study a potential site for geological storage of CO_2 injected as a supercritical fluid deep in the formation. In Section 2.5.4, we saw that the heterogeneous NPD5 sector model contains three formations: the Johansen sandstone delta bounded above by the Dunlin shale and below by the Amundsen shale. These three formations have distinctively different permeabilities (see Figure 2.15 on page 46) and play a very different roles in the sequestration process. The Johansen sandstone has relatively high porosity (and permeability) and is the container in which the CO_2 is to be kept. The low-permeability Dunlin shale acts as a seal that prevents the CO_2 from escaping back to the sea bottom, and we generally expect that the buoyant CO_2 phase will accumulate as a thin plume that migrates upward under the caprock in the up-dip direction.

To accurately simulate the up-dip migration under the top seal, it is highly important to preserve the correct interface between the Johansen sandstone and the Dunlin shale in the coarse model. (In general, it is a good advice to avoid creating coarse blocks containing large media contrasts, which would otherwise adversely affect upscaling accuracy.) We

Table 14.1 *Permeability values used to distinguish the different formations in the NPD5 sector model of the Johansen formation.*

Dunlin	Johansen	Amundsen
$K \leq 0.01$ mD	0.1 mD $< K$	0.01 mD $< K \leq 0.1$ mD

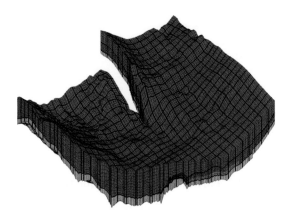

Figure 14.4 A $4 \times 4 \times 11$ coarsening of the NPD5 model of the Johansen aquifer that preserves the Amundsen, Dunlin, and Johansen formations.

therefore coarsen the three formations separately, using permeability values K as indicator as shown in Table 14.1. Likewise, to correctly resolve the formation and migration of the thin plume, it is essential that the grid has as high vertical resolution as possible. Unless we use a vertically integrated model (as in MRST co2lab [19]), we would normally only reduce the lateral resolution, say by a factor of four in each lateral direction. Here, however, we first use only a single block in the vertical direction inside each formation to more clearly demonstrate how the coarsening can adapt to the individual formations.

Assuming that the grid G and the permeability field K have been initialized properly as described in Section 2.5.4, the coarsening procedure reads

```
pK = 2*ones(size(K)); pK(K<=0.1) = 3; pK(K<=0.01)= 1;
pC = partitionUI(G, [G.cartDims(1:2)/4 1]);
[b,i,p] = unique([pK, pC], 'rows');
p = processPartition(G,p);
CG = generateCoarseGrid(G, p);
plotCellData(G,log10(K),'EdgeColor','k','EdgeAlpha',.4); view(3)
outlineCoarseGrid(G,p,'FaceColor','none','EdgeColor','k','LineWidth',1.5);
```

Figure 14.4 shows the coarse grid obtained by intersecting the partition vector pC, which has only one block in the vertical direction, with the partition vector pK that represents the

14.3 Partitioning Stratigraphic Grids

different formations. In regions where all formations are present, we get three blocks in the vertical direction. In other regions, only the Dunlin and Amundsen shales are present and we hence have two blocks in the vertical direction.

The aquifer model contains one major and several minor faults. As a result, 1.35% of the cells in the original grid have more than six neighbors. Coarsening a model with irregular geometry (irregular perimeter, faults, degenerate cells, etc.) uniformly in index space will in most cases give many blocks with geometry that deviates quite a lot from being rectangular. The resulting coarse grid thus contains a larger percentage of unstructured connections than the original fine model. For this particular model, 20.3% of the blocks have more then six coarse faces. If we look at a more realistic coarsening that retains the vertical resolution of the original model, 16.5% of the blocks have more than six neighboring connections. This model is obtained if we repeat the earlier construction using

```
pC = partitionUI(G, G.cartDims./[4 4 1]);
```

Figure 14.5 shows six different coarse blocks sampled from the top grid layer of the Dunlin shale. Block number one is sampled from a part of the aquifer perimeter that does not follow the primary grid directions and thus has irregular geometry. The other five blocks contain (parts of) a fault and will therefore potentially have extra connections to blocks in grid layers below. Despite the irregular geometry of the blocks, the coarse grid can be used directly with most of the solvers discussed earlier in the book. In our experience, the quality of the coarse solution is generally more affected by the quality of the upscaling of the petrophysical parameters (see Chapter 15) than by the irregular block geometry. In fact, irregular blocks that preserve the geometry of the fine-scale model respect the layering and connections in the fine-scale geology and therefore often give more accurate results than a coarse model with more regular blocks if upscaled correctly.

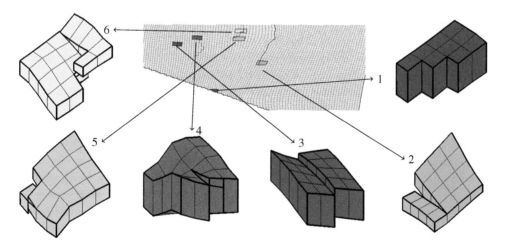

Figure 14.5 Six coarse blocks sampled from the top grid layer of the Dunlin formation in a $4 \times 4 \times 1$ coarsening of the NPD5 sector model of the Johansen formation.

14.3.2 The SAIGUP Model

The SAIGUP model, introduced in Section on page 97, has six user-defined rock types (also known as saturation regions; see Figure 2.20) that are used to specify different rock-fluid behavior (relative permeability and/or capillary pressure functions). Depending upon the purpose of the reduced model, we may want to preserve these rock types using the same type of technique as described in the previous example. This has the advantage that if each coarse block is made up of one rock type only, we would not have to upscale the rock-fluid properties. On the other hand, this typically leads to coarse grids with (highly) irregular block geometries and large variations in block volumes. To illustrate this point, we start by partitioning the grid uniformly into $6 \times 12 \times 3$ coarse blocks in index space:

```
p = partitionUI(G,[6 12 3]);
```

This introduces a partition of all cells in the logical $40 \times 120 \times 20$ grid, including cells that are inactive. To get a contiguous partition vector, we remove blocks that contain no active cells, and then renumber the vector. This reduces the total number of blocks from 216 to 201. Some of the blocks may contain disconnected cells because of faults and other nonconformities, and we must therefore postprocess the grid in physical space and split each disconnected block into a new set of connected sub-blocks:

```
p = compressPartition(p);
p = processPartition(G,p);
```

The result is a partition with 243 blocks that each consists of a set of connected cells in the fine grid. Figure 14.6 shows an explosion view of the individual coarse blocks. Whereas all cells in the original model are almost exactly the same size, the volumes of the coarse blocks span almost two orders of magnitude. In particular, the irregular boundary near the crest of the model introduces small blocks consisting of only a single fine cell in the lateral direction. Large variations in block volumes will adversely affect any flow solver if we later run a flow simulation on the coarsened model. To get coarse blocks with a more even size distribution, we therefore pick the smallest blocks and merge them with the neighbor that has the largest block volume. We repeat this process until the volumes of all blocks are above a prescribed lower threshold.

The merging algorithm is quite simple: we compute block volumes, select the block with the smallest volume and merge this block with one of its neighbors. Next, we update the partition vector by relabeling all cells in the block with the new block number and compress the partition vector to get rid of empty entries. Finally, we regenerate a coarse grid, recompute block volumes, pick the block with the smallest volume in the new grid, and repeat the process. In each iteration, we plot the selected block and its neighbors:

14.3 Partitioning Stratigraphic Grids

Figure 14.6 Logically Cartesian partition of the SAIGUP model. The plot to the left shows an explosion view of the individual blocks colored with `colorcube`. The bar graph to the right shows the volumes in units [m^3] for each of the blocks in the partition.

```
blockVols = CG.cells.volumes;
meanVol   = mean(blockVols);
[minVol, block] = min(blockVols);
while minVol<.1*meanVol

   % Find all neighbors of the block
   clist = any(CG.faces.neighbors==block,2);
   nlist = reshape(CG.faces.neighbors(clist,:),[],1);
   nlist = unique(nlist(nlist>0 & nlist~=block));
   plotBlockAndNeighbors(CG, block, ...
       'PlotFaults', [false, true], 'Alpha', [1 .8 .8 .8]);

   % Merge with neighbor having largest volume
   [~,merge] = max(blockVols(nlist));

   % Update partition vector
   p(p==block) = nlist(merge);
   p = compressPartition(p);

   % Regenerate coarse grid and pick the block with the smallest volume
   CG = generateCoarseGrid(G, p);
   CG = coarsenGeometry(CG);
   blockVols = CG.cells.volumes;
   [minVol, block] = min(blockVols);
end
```

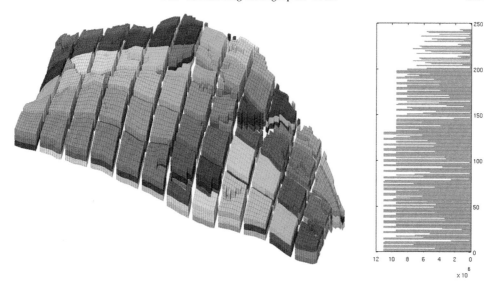

To find the neighbors of a given block, we first select all connections that involve block number `block`, which we store in the logical mask `clist`. We then extract the indices of the blocks involved in these connections by using `clist` to index the connection list `CG.faces.neighbors`. The block cannot be merged with the exterior or itself, so values 0 and `block` are filtered out. A pair of blocks may generally share more than one connection, so `unique` is used to remove multiple occurrences of the same block number.

When selecting which neighbor a block should be merged with, there are several points to consider from a numerical point of view. We typically want to keep blocks as regular and uniformly sized as possible, make sure that the cells inside each new block are well connected, and limit the number of new connections we introduce between blocks. Figure 14.7 shows several of the small blocks and their neighbors. In the algorithm shown in the most recent code box, we have chosen a simple merging criterion: each block is merged with the neighbor having the largest volume. In iterations one and three, the blue blocks are merged with the cyan blocks, which is probably fine in both cases. However, if the algorithm later wants to merge the yellow blocks, using the same approach may not give good results as shown in iteration ten. Here, it is natural to merge the blue block with the cyan block that lies in the same geological layer (same K index) rather than merging it with the magenta block that has the largest volume. The same problem is seen in iteration number four, where the blue block is merged with the yellow block, which is in another geological layer and only weakly connected to the blue block. As an alternative, we could choose to merge with the neighbor having the smallest volume, in which case the blue block would be merged with the yellow block in iterations one and three, with the magenta block in iteration four, and with the cyan block in iteration six. This would tend to create blocks consisting of cells from a single column in the fine grid, which may not be what we want. Another solution would be to merge across the coarse face having the largest transmissibility (sum of fine-scale transmissibilities).

Altogether, we see that there are several issues we need to take into consideration and these could potentially lead to mutually conflicting criteria. This obviously makes it difficult to devise a good algorithm that merges blocks in the grid in a simple and robust manner. A more advanced strategy could, for instance, include additional constraints that penalize merging blocks belonging to different layers in the coarse grid. Likewise, you may want to avoid creating connections that only involve small surface areas. However, such connections may already be present in the coarsening we start from, as illustrated by the yellow and cyan blocks in the lower-right plot in Figure 14.7, which were created when partitioning the grid uniformly in index space.

The basic `processPartition` routine only checks that there is a connection between all cells inside a block. A more sophisticated routine could obviously also check the face areas associated with each connection, or the magnitude of the associated transmissibility, and consider different parts of the block to be disconnected if the area or transmissibility that connects them is too small. As a step in this direction, we could consider the face area when we postprocess the first uniform partition, e.g., do something like the following,

14.3 Partitioning Stratigraphic Grids

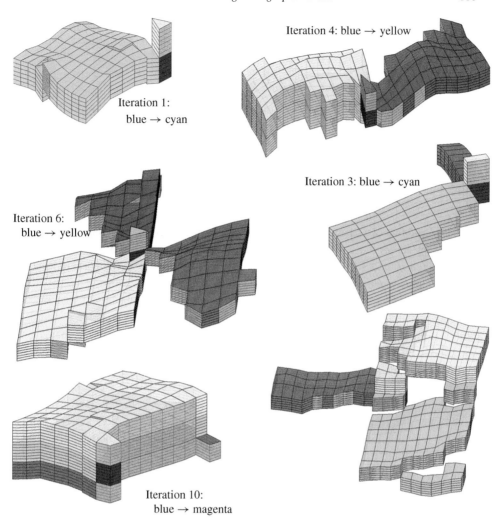

Figure 14.7 Examples of small blocks (in blue) being merged with one of their neighbors shown in semitransparent color with fault faces in gray.

```
p = partitionUI(G,[6 12 3]);
p = compressPartition(p);
p = processPartition(G, p, G.faces.areas<250);
```

This increases the number of blocks in the final grid obtained after merging small blocks from 220 to 273, but avoids constructing blocks looking like the yellow and cyan block in the lower-right plot in Figure 14.7. On the other hand, this approach obviously involves a threshold parameter that will vary from case to case and has to be set by an expert. The result may also be very sensitive to the choice of this parameter. To see this, you can try to redo the initial partition with threshold values 300 and 301 for the face areas.

14.3.3 Near Well Refinement for CaseB4

In this example[1], we consider the smallest $36 \times 48 \times 12$ pillar grid from CaseB4 (see Section 3.3.1) describing a reservoir section with two intersecting dip-slip faults. The reservoir is known to have four distinct geological layers with k indices 1, 2–7, 8–11, and 12. We start by making a standard load-balanced partition in index space:

```
p0 = partitionUI(G, [7 9 1]);
pf = processPartition(G, p0, find(G.faces.tag==1));
```

In the second line, we identify all fault faces and ask the processing routine to remove any connections across these faces when examining whether any of the 63 original blocks are disconnected or not. Blocks that are fully penetrated by a fault will contain at least two connected components and will hence be split. After splitting, the partition has 79 blocks. Although not needed for this model, it is quite simple to also split blocks that are only partially penetrated by faults; see [187] for more details. The disadvantage of imposing hard constraints like fault faces on the partitioning process is that it tends to increase the variation in block sizes, as shown in Figure 14.8.

Pressure gradients and flow rates are usually much larger near wells than inside the reservoir. How accurate we are able to capture the flow solution in the near-well region strongly influences the accuracy of the overall simulation. It may therefore be desirable to have higher grid resolution in the near-well zone than inside the reservoir. To refine blocks near the two wells, we use `refineNearWell` from the `coarsegrid` module, which takes a set of points and partitions these into cylindrical sectors around a single point in the xy-plane. The first well is placed in the corner of the reservoir section and we partition the perforated well block into five radial sections. The width of the radial sections is set to decay as the logarithm of the radial distance:

```
[wc,p] = deal( W(i).cells(1), pf);     % Pick well cell in top layer
wpt    = G.cells.centroids(wc,:);      % Center point of refinement
cells  = (p == p(wc));                 % All cells in block
pts    = G.cells.centroids(cells,:);   % Points to be repartitioned
out    = refineNearWell(G.cells.centroids(cells,:), wpt, ...
             'angleBins', 1, 'radiusBins', 4, ...
             'logbins', true, 'maxRadius', inf);
p(cells) = max(p) + out;               % Insert new partition in block
p = compressPartition(p);              %   and compress vector
```

The second well is perforated in the middle of a coarse block. It is therefore natural to refine both in the radial and the angular directions. The number of blocks in the angular

---

[1] Complete code: coarsenCaseB4.m in the book module.

14.3 Partitioning Stratigraphic Grids

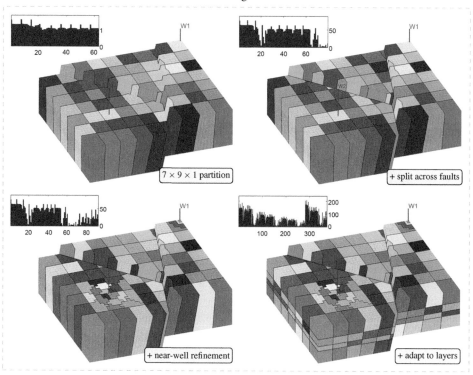

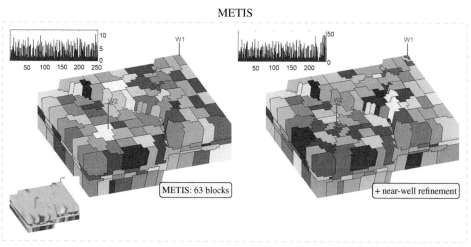

Figure 14.8 Partitioning of the CaseB4 pillar grid. The rectangular partition starts by a call to `partitionUI`, followed by the splitting of blocks across faults, near-well refinement by `refineNearWell`, and finally, vertical partition following geological layers. The METIS partition starts by a graph partitioning with transmissibility as edge weights followed by a near-well refinement. The large plots show how the partition is gradually refined by including new partition principles. The small plots show the variation in block volumes relative to the volume of the smallest block.

direction can be set for each radial section. Here, we therefore let the number of blocks increase outward, starting with a single block in the innermost layer, and then four, six, and nine blocks in the next three layers. To be able to place such a refinement, we repartition not only the well block but also the eight surrounding neighbors. To find these blocks, we construct a temporary coarse grid structure and use the topology from CG to construct an adjacency matrix similar to the one we discussed on page 522.

```
% Adjacency matrix
CG = generateCoarseGrid(G, pw);
N  = getNeighbourship(CG);
A  = getConnectivityMatrix(N,true,CG.cells.num);

% Find all axial/diagonal neighbors
rblk = zeros(CG.cells.num,1);
rblk(pwv)=1;
rblk((A*A*rblk)>1) = 1;
cells = rblk(pw)>0;
```

$r = $ (grid with 1)

$Ar = $ (grid with 1; 1 1 1; 1)

$A^2 r = $ (grid with 1; 2 2 2; 1 2 5 2 1; 2 2 2; 1)

We start by defining an indicator vector rblk equal one in the well block and zero in all other blocks. Multiplying by the adjacency matrix A will set the value in each block equal the sum over the block and its face-neighbors. After one multiplication, rblk therefore has ones in all the face-neighbors, and after two multiplications all blocks surrounding the initial block should have value larger than one. Next, we reset rblk to one in all blocks with A*A*rblk > 1 and index the resulting vector by the partition vector pw to extract all cells inside these blocks. These cells can be repartitioned as for the first well:

```
out    = refineNearWell(pts, wpt, 'angleBins', [1,4,6,9],...
         'radiusBins', 4, 'logbins', true, 'maxRadius', inf);
```

Altogether, this increases the number of coarse blocks to 93. Finally, we can impose vertical refinement that follows the four geological layers. Assuming the indices of these layers are given as an indirection map, this partition is constructed as follows:

```
% Here: layers = [1 2 8 12 13]
pK = rldecode((1:numel(layers)-1)',diff(layers));
[~,~,k]=gridLogicalIndices(G);
pk = compressPartition((pK(k)-1)*max(pw)+pw);
pk = processPartition(G, pk);
```

The resulting partition shown in Figure 14.8 has 372 blocks and significant variation in block sizes. This does not necessarily make the coarse grid ill-suited for flow simulations, but in a more careful implementation it would have been natural to postprocess the grid to get rid of small blocks away from the near-well regions that can be merged with their neighbors.

As an alternative to the regular partition in index space, we can use a standard software like METIS [154], which offers state-of-the-art partitioning of graphs and meshes based on multilevel recursive-bisection, multilevel k-way, and multi-constraint partitioning algorithms. METIS generally tries to make blocks that are as uniform as possible and at the same time have as few connections as possible. If we let fine-scale transmissibilities measure connection strengths between cells for the edge-cut minimization algorithm, the software tries to construct blocks without crossing large permeability contrasts. The `coarsegrid` module implements a wrapper that greatly simplifies the task of calling METIS. The following call gives a partition with approximately the same number of blocks as in the structural partition with four vertical layers, just discussed:

```
mrstModule add incomp
hT = computeTrans(G, rock);
p0 = partitionMETIS(G, hT, 7*9*4, 'useLog', true);
```

The partitioning function uses the standard `incompTPFA` solver from the `incomp` module to set up a fine-scale discretization matrix that describes how cells are connected. Here, we take the logarithm of the transmissibilities before the connection strength is constructed. Experience shows that this gives better partitions when transmissibilities vary several orders of magnitude locally. The resulting partition is shown in the lower-left plot in Figure 14.8. Compared with the rectangular partition, the grid blocks are more irregular and have more connections and larger variation in block sizes. Also for this partition, we can introduce near-well refinement exactly as described in this subsection. The first well falls within a smaller grid block and hence gives slightly different local blocks. For the second well, our detection of neighboring blocks selects a larger region to repartition and hence `refineNearWell` is able to generate a refinement that looks more circular compared our first partition method.

A technical note: To inform the system where to find METIS, we use a global variable `METISPATH` that can be set in your `startup_user` function. If you, like me, use Linux and have installed METIS in, e.g., `/usr/local/bin`, you add the following line to your `startup_user.m` file:

```
global METISPATH; METISPATH = fullfile('/usr','local','bin');
```

COMPUTER EXERCISES

14.3.1 Rerun CaseB4 with the stair-stepped model instead. Why does the code not work? (Hint: check the number of active cells.) Can you make a partition that splits blocks properly on opposite sides of faults?

14.3.2 The code example also contains a set up that tries to force METIS to split blocks across faults, which does not work well. Can you develop an alternative strategy to ensure that fault faces are also preserved in the METIS partition? (Hint: try to partition each fault block separately.)

14.3.3 Try to implement the merging technique from the SAIGUP example to reduce the large difference in cell volumes. (Hint: make sure that you do not merge any blocks in the near-well regions.)

14.4 More Advanced Coarsening Methods

The previous section outlined two main types of partitioning methods. Systematic graph partitioning algorithms like those implemented in METIS are widely used and very robust, but do not necessarily give you the desired control over the coarsening process. In particular, it can be quite difficult to rigorously formulate the desired coarsening as a set of cost functionals and constraints that are well posed and computationally tractable. More ad hoc and straightforward approaches that explicitly impose a structured partition in index space or use features from the model as partition vectors (Section 14.1.3) leave you in full control, but can easily lead to undesired artifacts like large variations in block volumes as we saw for SAIGUP and CaseB4. For large and complex models one cannot rely on visual quality control and manual repair of individual blocks. A significant body of research has therefore been devoted to develop automated or semiautomated coarsening procedures that incorporate certain features of the geology or flow physics.

Most methods reported in the literature try to generate a coarser and more optimal grid consisting of standard grid blocks with spatially varying resolution. The size of each grid block is usually determined by equilibrating a density measure or by introducing a background indicator whose variation is minimized within blocks and maximized between blocks. In particular, it is common to impose constraints on the geometrical shape of the new grid blocks (Delaunay/Voronoi) and the degree to which they align with the cells in the original grid [257, 47, 100]. Density measures and indicators can be defined using geological quantities like permeability [116, 160]; flow-based quantities like fluid velocity [101], vorticity [201], or streamlines [301, 72, 59, 310, 126]; local a priori error measures [162]; or statistical or a posteriori goal-oriented error indicators that measure how a particular point influences the error in production responses or other predefined quantities. Flow-based coarsening has been shown to be a powerful approach in combination with upscaling, and has been developed both for structured and unstructured grids. The basic goal of flow-based griding [93] is to introduce higher resolution in regions of high flow and coarser resolution in regions of lower flow.

An alternative approach is to generate coarse blocks by *agglomerating* cells that are similar in a sense prescribed by the user. This gives coarse blocks with complex polyhedral forms that follow the geometries of the original fine grid. An early approach in this direction [3] was aimed at coarse-scale discretization of transport equations and suggested to group cells according to the magnitude of the velocity (or flux) field so that each coarse block consists of a collection of cells through which the flow has approximately the same magnitude. Later research has shown that the original *nonuniform coarsening method* is just a special case of a much wider set of heuristic methods for producing coarse partitions that adapt to various types of geological or flow constraints. These methods have been developed for three different purposes: (i) to increase the accuracy of multiscale pressure solvers by adapting to fine-scale geological features [5, 6, 222, 194, 17, 191]; (ii) to define partitions that preserve rock types or relperm regions and thereby reduces the need for upscaling; and (iii) to find flow-adapted partitions suitable for coarse-scale transport solvers [3, 125, 124].

These methods can, in turn, be incorporated into a more general framework that tries to impose various types of geological/flow constraints in a hierarchical manner [187]. Similar aggregation-based methods have been applied to fractured media [153, 138]. The `agglom` module implements a set of modular components that can be combined in different ways to create various types of agglomerated partitions as outlined in the next section.

14.5 A General Framework for Agglomerating Cells

This section outlines the general agglomeration framework implemented in the `agglom` module. The agglomeration process can be viewed as governed by three rules:

- a neighbor definition that gives the *permissible directions* one can search for new cells to agglomerate;
- a set of indicator functions that determine the *feasible directions* among all the permissible directions;
- one or more indicators to determine how far the agglomeration should proceed, i.e., indicators that can be used to locally determine the size of each grid block.

Permissible directions are given by the grid topology and will in the default setup consist of all face neighbors. However, the permissible directions can also be extended to include cells sharing a common edge/node or cells whose centroids lie within a prescribed geometric distance. Likewise, the permissible directions can be restricted locally so that blocks cannot be agglomerated across faults or boundaries between different facies types, initialization regions, saturation regions, etc. The *feasible directions* are defined through a set of cell-based indicator functions that may contain geological or petrophysical features, flow-based quantities, or general user-set expert knowledge. One can also impose additional rules that prevent the total indicator function from growing too large within blocks, try to regularize the outline of the blocks, minimize aspect ratios. Altogether, this gives a very flexible framework that can be used to construct many different types of grids.

To realize this framework, the `agglom` module implements a set of source functions that create new partition vectors, and a set of filter functions that take one or more partitions as input and create a new partition as output. Examples of filter operations include combining/intersecting multiple partition vectors; performing sanity checks and modifications to ensure that no blocks are disconnected or contained within other blocks; or modifying the partition vector by merging small blocks or splitting large blocks.

14.5.1 Creating Initial Partitions

You have already seen several examples of topological partitions, e.g., generated by `partitionUI` from the `coarsegrid` module. Another way to create initial partitions is to segment indicator functions representing the heterogeneity of the medium. Examples of such indicator functions include the logarithm of permeability, velocity magnitude,

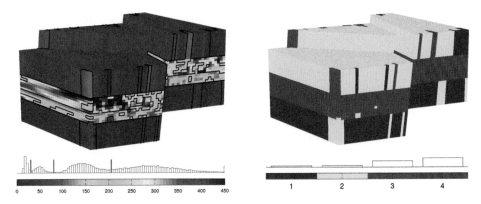

Figure 14.9 Segmentation with permeability as indicator for CaseB4 with four lognormally distributed layers (left). The right plot shows the resulting bins.

time-of-flight or residence time, vorticity, etc. Assuming we have an indicator function I that takes a single value in each cell of a grid G, the indicator can be segmented into bins by calling

```
p = segmentIndicator(G, I, bins)
```

Constructing indicator functions and making initial partitions will typically involve some kind of expert knowledge, and hence the MRST module does not offer specific routines. However, the tutorials contain several examples that should be instructive. To illustrate this function, let us revisit the CaseB4 example from Section 14.3.3 and see if we can use this function with permeability as an indicator to automatically detect the four geological layers in the model. The permeability is lognormal within each layer. The histogram of permeability values in Figure 14.9 has four distinct peaks, but the different layers are not clearly separated. We thus pick limits for the bins in the local minima of the distribution (K is given in md):

```
p = segmentIndicator(G, K, [0 30 80 205 inf], 'split', false);
```

Here, we have asked the function to only output the segmented bins. These bins, shown to the right in Figure 14.9, do not follow the layers exactly since permeability values are not unique to a single layer. If it is important to preserve the layering exactly, we would be better off by using the k index for the initial partition. In other cases like the SPE 10 model, all we have to bin different rock types are the permeability and/or porosity values.

14.5.2 Connectivity Checks and Repair Algorithms

Partitions defined by segmenting indicator functions usually give bins that consist of multiple connected components. These can be split by the function processPartition as discussed in Section 14.1.2. This function is called by default from within

14.5 A General Framework for Agglomerating Cells

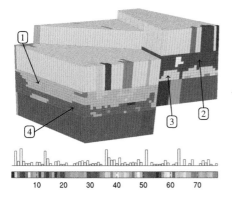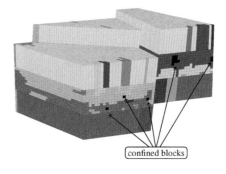

Figure 14.10 The left plot shows coarse blocks from the initial segmentation of CaseB4; the histogram of cells per block has logarithmic y-scale. Blocks 1,2 and 3,4 are disconnected components from layers two and three, respectively. The right figure shows isolated blocks that are completely confined within another block.

segmentIndicator, and if we drop the optional 'split' flag from the call, the initial bins are split so that they form blocks satisfying MRST's basic requirements for a coarse grid. Figure 14.10 shows that the resulting partition consists of six large coarse blocks and a number of smaller ones. Layers two and three are not fully connected across all faults and the corresponding bins have therefore each been split in two large blocks (1 and 2, and 3 and 4 in the plot). The remaining small blocks are the result of permeability variations inside the layers.

Blocks that are entirely confined within other blocks only have a single interface. When the resulting grid is used for flow simulation, the only flow that can pass the interface between the block and its surrounding neighbor must be caused by source terms or compressible effects. This tends to create numerical artifacts unless the block contains a well. For incompressible flow, in particular, the corresponding block interface will act as an internal no-flow boundary. An additional sanity check should therefore be performed to detect such blocks:

```
cb = findConfinedBlocks(G, p);
[cb,p] = findConfinedBlocks(G,p);
```

The first call just outputs a list of confined blocks, whereas the second also merges these blocks with their surrounding neighbor. The right plot in Figure 14.10 illustrates confined blocks for CaseB4.

The detection algorithm is somewhat simplified and the function cannot handle nested cases properly. In the case of recursively confined blocks, only the outermost and the innermost blocks are listed in cb. To find a more general solution, we can look to algorithms for detecting *biconnected components* from graph theory. Let me explain the idea by a simple example: Figure 14.11 shows a 7×7 grid partitioned into six coarse blocks together

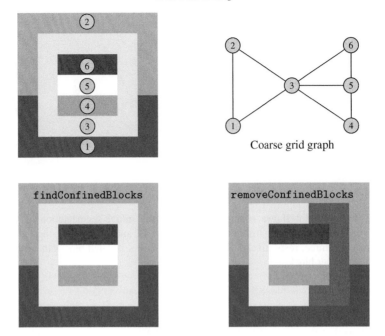

Figure 14.11 Detection and repair of recursively confined blocks by use of functionality from the MATLAB Boost Graph Library. The simple findConfinedBlocks function fails to detect and repair cases where a set of blocks are confined within another block.

with the corresponding coarse-grid graph. This graph is not biconnected: node 3 belongs to two connected components (1–2–3 and 3–4–5–6), and if this node is removed, the graph becomes disconnected. The function

```
pn = removeConfinedBlocks(G, p)
```

merges single confined blocks with their surrounding blocks. In addition, it uses functionality from the MATLAB Boost Graph Library[2] to determine whether a graph is biconnected or not and to find the corresponding components of the graph. If the graph is not biconnected, any grid block that completely surrounds a set of other blocks is split in two (along a plane orthogonal to the x-axis through the block center). The resulting partition is not necessarily singly connected and may have to be processed by a further call to processPartition.

14.5.3 Indicator Functions

The agglomeration framework relies on two indicator functions, which we for simplicity refer to as a *volume indicator* and a *flow indicator*. We assume that each indicator function

---

[2] This software is freely available under a BSD License. The matlab_bgl module has a function downloadMBGL that downloads and installs this library. This is one of the few examples of external dependencies in MRST.

takes value $I(c_i)$ in cell c_i and that these values can be interpreted as densities, i.e., that they are positive, additive, and normalized by the cell volume $|c_i|$. As for the cells, we associate indicator values $I(B_j)$ to each block, defined as the volumetric average of the cell indicator values within the block (or as the arithmetic average if I is not a density).

The volume indicator is only used to determine block volumes and is either set to be unity for bulk volumes and ϕ for pore volumes. The flow indicator can in principle be any user-defined quantity, but typical examples include quantities derived from permeability, time-of-flight or residence time, velocity or flux, vorticity, and so on (see discussion in [124]). Error indicators or sensitivities can also be used as flow indicators. Specific examples are given later.

To understand the role of these two indicators, let us look at four general principles you can use to control how cells are grouped together:

1. The variation of the flow indicator should be minimized inside each block. That is, each block should be as homogeneous as possible so that cells with high and low permeabilities, large and small fluxes, high and low travel times, and so on, are separated in different blocks. Likewise, cells from different facies, deposition environments, flow units, initialization regions, relperm/capillary regions, and so on, should preferably not be agglomerated into the same block. Algorithmically, this means that the flow indicator is used to pick the most feasible among the permissible neighbors of a block when adding a new cell to a growing block.
2. The integral of the flow indicator should be equilibrated over blocks. You can use this principle refine resolution in regions with high flow, low residence time, high error or sensitivity indicator, and so on. Full equilibration will in general require rigorous optimization. Algorithms in the `agglom` framework instead try to ensure that the integral of the flow indicator is within prescribed upper bounds. These bounds determine when to stop agglomerating more cells into a block, or alternatively, determine when blocks are too big and must be split.
3. The size of each block should be with certain upper and lower bounds. Algorithmically, upper limits on block sizes are imposed by the upper bound on the flow indicator, whereas lower bounds are imposed through the volume indicator when determining the blocks that should potentially be merged with their neighbors.
4. Blocks should not have too-large aspect ratios and be as regular as possible. Regularity is determined in part by the source functions generating initial partitions, and in part by filter functions that split or merge individual blocks. Controlling regularity may be difficult in ad hoc algorithms, and if regularity is important, you may be better off using METIS.

14.5.4 Merge Blocks

We have already seen how manually constructed partitions can give grids with very large variations in block sizes. Variations in block sizes can be even more pronounced if the initial partition is created by segmenting an indicator function. One method to remedy the problem would be to loop through all blocks in the partition and try to merge each small blocks with one of its neighbors. Our simple implementation in the SAIGUP example from Section 14.3.2 was not very efficient as it required generation of a new coarse grid structure in each iteration. Our choice of merging with the neighbor having largest volume was also somewhat ad hoc.

In the agglomeration framework, we use the volume indicator I_v to measure the size of cells and blocks with $I_v \equiv 1$ for bulk volume and $I_v = \phi$ for pore volume. (The latter definition can also be modified to include net-to-gross or other factors that affect the pore volume.) Let $G = \{c_i\}_{i=1}^n$ denote the whole grid consisting of n cells that each has a bulk volume $|c_i|$. Then, block $B_j = \{c_i \mid p(c_i) = j\}$ should be merged with a neighbor if it violates the following lower bound

$$I_v(B_j)|B_j| \geq n_l \bar{I}_v, \qquad \bar{I}_v = \frac{1}{n}\sum_{i=1}^n I_v(c_i)|c_i|, \qquad (14.1)$$

where $n_l \geq 1$ is a prescribed constant and $I_v(B_j)|B_j| = \sum_{c_i \in B_j} I_v(c_i)|c_i|$. In other words, block B_j should be merged if its integrated indicator value is less than n_l times the average volume indicator value for all cells. To determine which block to merge with, we use the flow indicator I_f and merge block B_j with the block among block B_j's permissible neighbors $\mathcal{N}(B_j)$ that has the closest indicator value, i.e.,

$$\tilde{B} = \operatorname{argmin}_{B \subset \mathcal{N}(B_j)} |I_f(B) - I_f(B_j)|. \qquad (14.2)$$

This is implemented in the function

```
q = mergeBlocks(p, G, Iv, Ifl, NL, 'static_partition', ps)
```

which takes a partition p, a volume indicator Iv, and a positive flow indicator Ifl and merges all blocks that violate (14.1). In doing so, each merged block is agglomerated into the neighboring block that has the closest I_f value. Optionally, the merging operation can be set to respect a static partition p2 so that no block ends up having different ps values.

The problem with this method is that it does not impose any control on the upper size of the merged blocks. In particular, merging with the nearest flow indicator tends to work against the principle that flow indicators should be equilibrated across all the grid blocks. The idea[3] of the flow indicator is that it should enable us to aggressively coarsen the grid in regions of low flow or regions that has limited impact on the overall simulation result, while keeping higher resolution in regions of high flow.

To provide upper control on the size of the blocks, we introduce a second requirement, this time on the flow indicator:

$$I_f(B)|B| \leq n_u \bar{I}_f, \qquad (14.3)$$

---

[3] In principle, the flow indicator need not have anything to do with flow; I still think it is important that you know the underlying philosophy so that you can design indicators so that the merging/refinement process works as you desire.

14.5 A General Framework for Agglomerating Cells 547

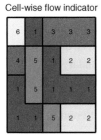
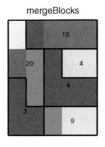
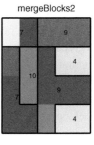

Figure 14.12 Comparison of the two merging algorithms for a simple 5×4 grid. The left plot shows the initial partition and the flow indicator I_f per cell. The middle and right plot shows partitions after merging, with numbers giving the integrated flow indicator, which according to (14.3) should not exceed 7.5.

where n_u is a prescribed parameter and $\bar{I}_f$ is the flow indicator averaged over all cells, defined exactly the same way as $\bar{I}_v$. The second merging function

```
q = mergeBlocks2(p, G, Iv, Ifl, NL, NU)
q = mergeBlocks2(p, G, Iv, Ifl, NL, NU, 'nblock', NB, 'cfac', cfac)
```

not only provides a more efficient implementation of the merging process, but also imposes (14.3) as a *soft constraint*, meaning that the routine tries its best to avoid violating this condition, but if no other options are available, it merges blocks even if the flow indicator of the new block becomes too high. It is also possible to specify an optional relative factor cfac at which (14.3) is turned into a *hard constraint*. Likewise, the parameter NB gives an upper bound on the number of cells that can be agglomerated into a single block. (Notice also that this routine does *not* support static partitions.)

Figure 14.12 compares the two methods for a small 5×4 grid, called with the parameters

```
p1 = mergeBlocks(p, G, ones(n,1), I, 2);
p2 = mergeBlocks2(p, G, ones(G.cells.num,1), I, 2, 3);
```

Here, mergeBlocks identifies four single-cell blocks that are merged with their neighbor with closest value: $4\to 6$, $5\to(2,2)$, $(4,6)\to(5,5)$, and $1\to(3,3,3)$. The result is three new blocks that all violate (14.3). The alternative routine merges $4\to(1,1,1)$ and $6\to 1$, which both are acceptable. For cell with value 5, the only option is to merge with $(1,1,1,1)$, even if this violates (14.3).

14.5.5 Refine Blocks

The next natural step would now be to refine blocks that violate the upper bound (14.3). You can do this by invoking any of the two calls

```
p = refineBlocks(p, G, Ifl, NU, @refineUniform, 'cartDims',[Nx Ny Nz])
p = refineBlocks(p, G, Ifl, NU, @refineGreedy)
```

The first call refines blocks by encapsulating the corresponding cell indices in a rectangular bounding box and then performing a load-balanced partition inside this box. The second line uses a greedy algorithm that starts by picking an arbitrary cell c_0 on the block boundary and then agglomerates a new block from the cell on the block boundary that is furthest away from the first cell. Once this new block is large enough, the routine once again picks the cell among the remaining ones that is furthest away from c_0 and starts to agglomerating a second block, and so on. The greedy algorithm comes in four different versions:

- refineGreedy – grows a new block inward from the block boundary by adding one entire ring of permissible neighbors in each iteration [3]. The permissible level-one neighbors are all cells sharing a face with existing cells in the growing block (optional parameter 'nlevel'=1). The default behavior ('nlevel'=2) is to also include cells that share (at least) two faces with existing cells in the block or level-one cells in the ring.
- refineGreedy2 – improved algorithm that only adds parts of the ring of level-one or level-two neighbors to honor the upper bound (14.3) on the flow indicator.
- refineGreedy3 – sorts cells in the neighbor ring according to number of faces shared with the growing block. Permissible neighbors are defined by the function neighboursByNodes and includes all cells that share a face or a node with cells in the growing block. Computing this neighborship is expensive for large grids.
- refineGreedy4 – sorts cells in the ring of neighbors according to difference in Ifl value, using the same permissible neighbors as refineGreedy3.

To illustrate how these functions operate for permissible level-one neighbors only, we consider a 4×4 grid with uniform flow indicator $I_f \equiv 1$. The functions are run with the following parameters:

```
p = refineGreedy(I,G,I,4,'nlevel',n);
    :
p = refineUniform(I,G,I,4,'CartDims',[2,2]);
```

where I is a vector of all ones. Figure 14.13 shows that the uniform partition gives four blocks, as expected. All greedy algorithms pick $c_0 = 1$ and start agglomerating blocks from cell 16. The simplest greedy algorithm, refineGreedy, only adds face neighbors and hence tends to create diamond-shaped blocks like the second block,

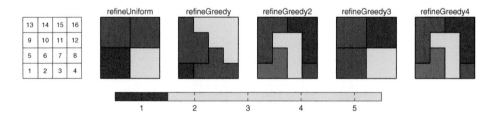

Figure 14.13 Refinement algorithms using level-one permissible neighbors on a 4×4 grid with uniform flow indicator.

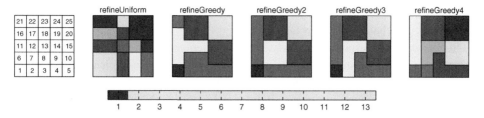

Figure 14.14 Refinement algorithms using level-two permissible neighbors (which is the default setting) on a 5 × 5 grid with uniform flow indicator.

which has six cells (8,11,12,14,15,16) and thus violates (14.3). After agglomerating three blocks, the algorithm is only left with cell c_0 and terminates. For `refineGreedy2` and `refineGreedy4`, the first step agglomerates cells (12,15,16). In the next step, cells (8,11,14) are possible merge candidates, out of which 8 is chosen since it has the smallest index. After the algorithm has agglomerated five blocks, there are no remaining cells in the original block and the algorithm terminates. The empty block is removed by a call to `compressPartition`, and blocks two to six are renumbered as blocks one to five. The third algorithm, `refineGreedy3`, selects cell 11 in the second step since this cell shares one more face with cells (12,15,16) than the other two candidates. Once the first block is formed, the greedy algorithms selects the remaining cell with maximum distance to c_0 as the next seed (cell 4 for the first algorithm, cell 11 for the second and fourth, and cell 8 for the third.)

Figure 14.14 shows a similar test case with level-two neighbors and a 5 × 5 grid. Here, the uniform refinement splits the block into four blocks of size 3 × 3, 2 × 3, 3 × 2, and 2 × 2. The first three violate (14.3) and are thus further split twice in each direction. To understand the irregular block shapes for the greedy algorithm, we can consider agglomeration of the fourth block, which starts by cells (4,5). Cell 3 is the only permissible neighbor in step two, whereas the third step has three candidates (2,7,8). All three are added by the first algorithm, whereas `refineGreedy2` chooses the cell with lowest index, which coincidentally gives a regular block, albeit with a larger aspect ratio. Why the other two algorithms choose cell 8 is a random effect caused by how MATLAB picks extremal values for a vector with identical entries.

The greedy refinement algorithms apply equally well to general polygonal/polyhedral grids with unstructured topologies. Uniform Cartesian refinement, on the other hand, relies on an ijk numbering and thus primarily works for grids with rectangular topology. However, the same idea can be generalized to unstructured grids if we instead of relying on topology use the cell centroids to sample from a rectangular subdivision of each block's geometric bounding box (see function `sampleFromBox`). This type of refinement is implemented in

```
p = refineRecursiveCart(p, G, I, NU, cartDims)
```

The function does not always produce the exact same results as `refineBlocks` with uniform refinement, but usually runs faster.

Refinement and merging are complementary operations that in principle can be applied in any order. A large number of tests nonetheless show that it is typically better to first merge small blocks, and then refine large ones, and improved grids can often be obtained if the merging-refinement process is iterated a few times. Notice also that the greedy algorithms tend to accentuate irregularity of blocks and should not be used uncritically.

14.5.6 Examples

We end by a few examples that show how the algorithmic components discussed in this section can be used to generate various types of coarse grids.

Example 14.5.1 (CaseB4) *As you may recall from Section 14.3.3, we first partitioned the CaseB4 grid in the ij coordinates and then split the resulting coarse blocks across the three faults. Figure 14.15 reports block sizes before and after splitting. To get rid of the smallest blocks, we can merge all blocks with a pore volume less than 20% of the mean block volume, which is approximately 50 times the mean cell volume. Applying one of the merging algorithms directly to the partition risks merging blocks across faults. We avoid this if we create a new vector that partitions the model into three fault blocks and use this to constrain the merging algorithm:*

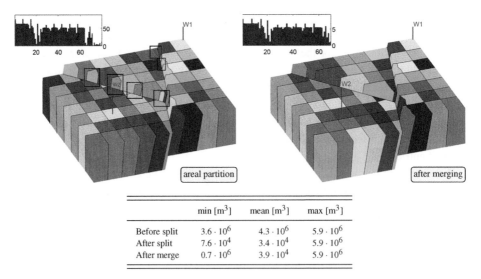

	min [m^3]	mean [m^3]	max [m^3]
Before split	$3.6 \cdot 10^6$	$4.3 \cdot 10^6$	$5.9 \cdot 10^6$
After split	$7.6 \cdot 10^4$	$3.4 \cdot 10^4$	$5.9 \cdot 10^6$
After merge	$0.7 \cdot 10^6$	$3.9 \cdot 10^4$	$5.9 \cdot 10^6$

Figure 14.15 Areal partition for CaseB4 before and after merging of blocks whose pore volume is less than 50 times the average cell volume ($1.3 \cdot 10^4$ m^3). The tabular shows pore volumes in various stages of the algorithm.

```
Iv = poreVolume(G, rock)./G.cells.volumes;
pm = mergeBlocks(pf, G, Iv, Iv, 50, 'static_partition', ...
    processPartition(G,ones(G.cells.num,1),find(G.faces.tag==1)));
```

Notice, that we here use the same indicator for both volume and flow, which generally would cause the algorithm to merge with blocks having similar average porosity values. After merging, the ratio between the smallest and the largest block is decreased by a factor ten, and we see that the algorithm has merged small blocks with neighbors on the correct side of the faults. It is important to merge small blocks before *we refine around wells to avoid merging any of the refined blocks in the near-well region.*

There are many different ways one can define flow indicators, for instance:

- From Darcy's law, we know that if a region is subject to a constant pressure gradient, the local magnitude of flow is proportional to permeability. Pressure gradients are rarely constant throughout a reservoir, but permeability can still be used as a simple a priori indicator to separate potential local regions of high and low flow before any flow solution is computed.
- Given a flow solution, it is natural to use velocity magnitude in each cell as a flow indicator. For a finite-volume method, this quantity must be reconstructed from the intercell fluxes (see discussion of non-Newtonian flow in Section 7.4), and alternatively one can use the sum of the absolute values of the fluxes as a measure that is simpler to compute. In a certain sense, flux/velocity can be considered as *local* indicators, since they only measure relative variations between individual cells and do not account for the actual flow paths.
- To get a *global* flow indicator that also takes representative flow paths into account, we can use time-of-flight or residence time.

As all three quantities tend to vary several orders of magnitude, we use their logarithm instead as suitable indicators:

```
iK = log10(rock.perm(:,1)); iK = iK - min(iK) + 1;
v  = sqrt(sum(faceFlux2cellVelocity(G, state.flux).^2, 2));
iV = log10(v); iV = iV - min(iV) + 1;
iT = -log10(Tf+Tb); iT = iT - min(iT) + 1;
```

assuming that `state` holds a representative flow solution and that `Tf`,`Tb` are the corresponding forward and backward time-of-flight values. Instead of residence time, we could have used the product of the two time-of-flight values. Let us illustrate these indicators by an example.

Example 14.5.2 (Adaptive coarsening) *Consider a five-spot pattern posed on Layer 25 from the SPE 10 model. For improved readability of plots, we only consider a 120×60 excerpt, initially partitioned into a 6×3 coarse grid. With a uniform flow indicator, we would get a coarse grid with 5×5 cells per block if we apply a 2×2 refinement twice.*

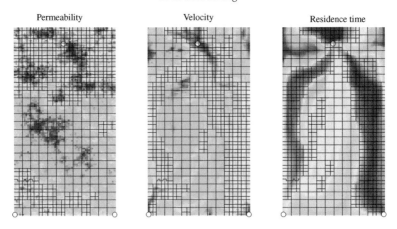

Figure 14.16 Recursive coarsening of a 120 × 60 excerpt of a five-spot well setup on SPE 10, Layer 25. The region is first partitioned uniformly into 6 × 3 coarse grids, and then refined recursively by a factor 2 × 2 by different flow indicators with $n_u = 25$ in (14.3).

With a heterogeneous flow indicator, we should expect to see local variations in the grid resolution. Figure 14.16 confirms this. All the initial blocks are refined twice, so that no grid block has more than 5 × 5 cells. Indicator iK *adds one extra refinement to all blocks having high permeabilities, since these would experience high flow if all cells in the model were subject to the same pressure drop. However, not all these blocks experience high flow for the specific five-spot flow pattern, and similarly there will be significant flow through many blocks with low(er) permeability because the injector and the two producers in the southeast/southwest corners lie on opposite sides of the low-permeability belt in the middle of the model.*

High-flow connections between the injector and the two producers are detected much better by the velocity indicator iV. *Unlike the indicator* iT *based on time-of-flight, the velocity indicator does not take into account the cumulative resistance to flow along flow paths and may therefore fail to refine blocks that have low local flow velocities but still lie along a flow path with low residence time. On the other hand, if a partition is built using a lot of specific flow information, it might be less suitable when applied to simulate radically different flow patterns.*

Example 14.5.3 (NUC algorithm) *The original nonuniform coarsening (NUC) algorithm [3] starts by segmenting a velocity indicator and then uses a sequence of merge-refine-merge operations as illustrated in Figure 14.17. Here, we have used the upper half of the model from the previous example, and constructed the NUC partition by the following sequence of operations:*

```
[volI,flwI] = deal( poreVolume(G, rock)./G.cells.volumes, iV);
ps  = segmentIndicator(G, flwI, 5);
pm1 = mergeBlocks(ps, G, volI, flwI, 20);
pr  = refineGreedy(pm1, G, flwI, 30);
pm2 = mergeBlocks(pr, G, volI, flwI, 20);
```

Figure 14.17 Application of the original nonuniform coarsening method of [3] for a 60×60 excerpts of a five-spot setup on Layer 25 of SPE 10.

The rightmost plot in the figure is obtained by replacing the merge and refinement operations by mergeBlocks2 *and* refineGreedy2, *respectively. Although this type of partition originally was developed to coarsen transport equations, it is currently implemented as part of an industry-standard tool for single-phase upscaling [187] with time-of-flight as flow indicator.*

The agglom module contains several tutorial examples that give a more in-depth discussion of the NUC family of coarsening methods, and also includes an example of how the techniques in this section can be used to generate *dynamic* grid coarsening that adapts to evolving displacement fronts.

COMPUTER EXERCISES

14.5.1 How would you merge small blocks *after* the near-well refinement is introduced in CaseB4 to avoid merging these refined blocks?

14.5.2 Replace the merging in the SAIGUP example (Section 14.3.2) by mergeBlocks. Which flow indicator would you choose? Can you change the merging algorithm so that it can use the SATNUM field as flow indicator? (Hint: to this end you should use a majority vote rather than a volume-weighted average to get representative flow indicator per block.)

14.5.3 Rerun the NUC example using the other greedy algorithms with level-1 and level-2 neighbors and permeability and time-of-flight indicators. Try to use the NUC algorithm on other examples, e.g., CaseB4.

14.5.4 A greedy agglomeration strategy may not always be optimal: Assume that you have a coarse block with integrated indicator 410, and that the upper bound is 400. The algorithm then tries to agglomerate two blocks with indicators 400 and 10, respectively. Can you make a function that instead tries to split the indicator more evenly among the new blocks?

14.6 Multilevel Hierarchical Coarsening

Geocellular models are realizations of a deeper geological understanding that include structural, stratigraphic, sedimentologic, and diagenetic aspects. On the way to generate a grid model and populating it with petrophysical properties, the reservoir may have been divided into different units, flow zones, environments of deposition, layering, and lithographic facies that each represents characteristics of the rock and how it was formed. The geocellular models we have encountered so far in the book have mainly consisted of a grid populated with a set of petrophysical properties like permeability, porosity, and net-to-gross. Some geocellular model may also contain geological indicators such as facies or rock type used to generate petrophysical properties. We have also seen that reservoirs can be subdivided into various types of regions to model spatial dependence in relative permeability models (PVTNUM), PTV behavior (PVTNUM), rock compressibility (ROCKTAB), and equilibration regions (EQLNUM).

The purpose of coarsening a grid is usually to develop one or more reduced models that only contain the most essential properties that affect fluid flow. It is well known that structural and stratigraphic frameworks have the largest impact on flow patterns, and preserving key concepts of these frameworks all the way to flow simulation is crucial to reliably predict flow patterns in the reservoir [48]. Likewise, it is important to preserve initial fluid contacts. Unfortunately, the volumetric partitions imposed by basic geological characterization are oftentimes lost in geostatistical algorithms and rarely made available as cell or face properties in the geocellular models. As an example, consider the distinction between high-permeability sand channels and low-permeability shales and coal in the Upper Ness formation from SPE 10. Likewise, think of CaseB4, where we in Figure 14.9 somewhat unsuccessfully tried to delineate four different geological layers based on permeability. To simplify future creation of coarse models, I therefore recommend that as much as possible of the basic geological characterization is preserved in terms of cellular indicators even after the geocellular model has been populated with petrophysical properties.

The idea of *hierarchical multilevel coarsening* is to define the various geological characterizations and regions just discussed as partition vectors, give them individual priority, and apply them recursively. By varying the number of features included, one can create a hierarchy of coarse partitions of increasing resolution. This approach can be combined with the agglomeration framework from the previous section to separate models into flow-dependent compartments representing high-flow and low-flow zones, or zones that are close to or far away from wells. A detailed discussion of this approach is outside the scope of this book. Instead, I give two illustrative examples.

Example 14.6.1 *The first example is a conceptual 40×40 model that has three different geological features: flow unit, lithofacies assemblage (LFA), and two intersecting fractures. The two first are represented by cell indicators* pu *and* pl, *whereas faults are represented as an explicit list of faces,* faults. *When combined, the three geological properties subdivide the reservoir into 15 different regions; see plots in the upper row of Figure 14.18. Flow unit and LFA are used to populate the geocellular model with stochastic realizations of the petrophysical properties. The lower-left plot in Figure 14.18 shows a flow indicator* I *derived from permeability, which has mean value approximately equal two in LFA 1 and*

14.6 Multilevel Hierarchical Coarsening 555

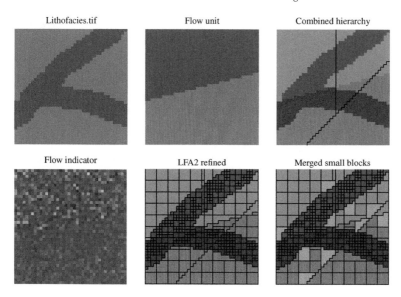

Figure 14.18 Multilevel hierarchical partition for a conceptual geological model with three geological properties, flow unit, lithofacies assemblage, and faults, which combined subdivide the reservoir into fifteen regions. The second row shows flow indicator, hierarchical 2×2 refinement, and the same refinement after blocks with less than four cells in LFA 1 have been merged (shown in a red shade).

ten in LFA 2. We use this flow indicator to introduce a hierarchical Cartesian subdivision with higher resolution in the second LFA[4] :

```
pc = ones(G.cells.num,4);
for i=1:4
    pc(:,i) = partitionCartGrid(G.cartDims,[5 5].*2^(i-1));
end
p = applySuccessivePart(processPartition(G,pu,faults),G, I, 8, [pl pc]);
```

That is, we first introduce a basic partition that splits the flow units and fault blocks, and then use the flow indicator to further partition the grid according to lithofacies and Cartesian blocks when the accumulated block indicator exceeds eight times the average cell value. This subdivision will create small blocks also inside the first lithofacies. To merge blocks with less than four cells, we can use `mergeBlocksByConnections`, *which merges blocks according to connection strength between fine-scale cells. The function does not merge blocks that have negative connection strength, and hence we set negative connection strength inside LFA 2 and between cells that belong to different regions in the geological hierarchy:*

---

[4] Complete code: `showHierCoarsen.m` in book.

Figure 14.19 Conceptual illustration of a reservoir model with aggressive coarsening in the aquifer zone, modest coarsening above the initial water contact, and original resolution for cells with low residence time.

```
qb = [0; processPartition(G,pl+2*(pu-1), faults)];
ql = [0; pl];
T = 2*(qb(G.faces.neighbors(:,1)+1)==qb(G.faces.neighbors(:,2)+1))-1;
T(ql(G.faces.neighbors(:,1)+1)==2) = -1;
pm = mergeBlocksByConnections(G, p, T, 4);
```

The lower-right plot in Figure 14.18 shows the result. This example is admittedly simple, but similar principles can be applied to industry-standard models [187].

Example 14.6.2 *In the second example, we consider a conceptual model of an anticline reservoir that overlies an aquifer. We use a hierarchical approach to aggressively coarsen cells below the initial water contact, which are assumed to be water-filled throughout the whole simulation. For the cells above the water contact, we apply a more modest coarsening. To capture the initial displacement between injectors and producers, we pick all cells with residence time less than twice the median residence time of all cells perforated by wells. Cells with low residence time are part of the high-flow zones and are thus kept at their original resolution. In addition, all blocks in the coarsest partition that are perforated by a well are refined back to their original resolution to ensure that we accurately capture flow in the near-well zone. Figure 14.19 shows the resulting model. You can find complete source code in the script* showAnticlineModel.m *in the* book *module.*

14.7 General Advice and Simple Guidelines

Over the years, I have made the general and perhaps somewhat disappointing observation that making a good coarse grid is more an art than an exact science. In particular, I do not believe black-box approaches are able to automatically select the optimal coarsening. As a result, MRST does not provide well-defined workflows for coarsening, but rather offers tools that (hopefully) are sufficiently flexible to support you in combining your creativity, physical intuition, and experience to generate good coarse grids.

14.7 General Advice and Simple Guidelines

Coarse partition are often used as input to a flow simulation. I conclude the chapter by suggesting a few guidelines that may contribute to decrease numerical errors introduced by coarsening:

1. Keep the grid as simple as possible: Use regular partitions for mild heterogeneities and cases where saturation/concentration profiles are expected to be regular.
2. Try to keep the number of connections between coarse-grid blocks as low as possible to minimize the bandwidth of the discretized systems, and avoid having too many small blocks, which otherwise would increase the dimension of the discrete system and adversely affect its stability without necessarily improving the accuracy significantly.
3. Use your understanding of the reservoir to judge what are the important features to preserve when coarsening the model. In doing so, you should think of what are primary geological features that control pressure communication and the main flow directions.
4. Make sure you have sufficient *vertical resolution* if you study systems with large density differences between the fluid phases (e.g., gas injection or CO_2 sequestration) to be able to capture the vertical segregation and the lighter fluid's tendency to override the other fluid(s).
5. Explore alternative partitioning strategies and compare the results.
6. Use relatively robust methods like METIS with transmissibility weights or regular partitions if you have limited understanding of which features affect the flow patterns.
7. Use basic geological features when available, e.g., facies likely yields more robust results than segmented permeability values.
8. Adjust your partitioning strategy to the purpose of the coarsening. If you want to run multiple simulations with more or less the same flow pattern, you can use features of the specific flow patterns to aggressively coarsen parts of the reservoir with low flow. Conversely, a fully flow-adapted coarsening may not work well in workflows where well positions and flow patterns change significantly from one simulation to the next.
9. Use simple indicator functions that have an intuitive interpretation, e.g., permeability, velocity magnitude, time-of-flight/travel time, etc.
10. Adapt coarse blocks to avoid upscaling large permeability contrasts, constrain blocks to contiguous saturation regions to avoid upscaling relative permeability and capillary pressure curves, etc.
11. Check the quality of a coarse grid using single-phase flow, e.g., by computing various types of flow diagnostics to check that you preserve the volumetric communication from the fine-scale model.

The next chapter introduces you to methods you can use to compute effective petrophysical properties on a coarsened grid.

15
Upscaling Petrophysical Properties

In Chapter 2 we argued that porous rock formations are heterogeneous across a wide span of length scales, from the micrometer scale of pore channels to the kilometer scale of petroleum reservoirs or even larger scales for aquifer systems and sedimentary basins. Describing all flow processes pertinent to hydrocarbon recovery or CO_2 storage with a single model is therefore not possible. Geological characterization therefore usually involves a hierarchy of models that each covers a limited range of physical scales cover a wide range of physical scales. Section 2.3 introduced you to the concept of representative elementary volumes and mentioned various types of models used for flow studies including flow in core samples (cm scale), bed models (meter scale), sector models (tens to hundreds of meters), and field models (km scale). Geocellular models at the field/sector scale are made to represent the heterogeneity of the reservoir and possibly incorporate a measure of inherent uncertainty in reservoir simulation. Pore, core, and bed models are mainly designed to give input to the geological characterization and to derive flow parameters for simulation models. All model types must be calibrated against static and dynamic data of very different spatial (and temporal) resolution: thin sections, core samples, well logs, geological outcrops, seismic surveys, well tests, production data, core flooding and other laboratory experiments, etc. To learn more about the process of building reservoir models, you should consult the excellent textbook by Ringrose and Bentley [265].

To systematically propagate the effects of small-scale geological variations observed in core samples up to the reservoir scale, we need mathematical methods that can *homogenize* a detailed fine-scale model and replace it with an equivalent model on a coarser scale that contains much fewer parameters that represent the behavior of the fine-scale model in an averaged sense; see Figure 15.1.

Such techniques are not only used to incorporate the effect of small-scale models into macroscale models. High-resolution geocellular models used for reservoir characterization on field or sector scale tend to contain more volumetric cells than contemporary simulators can handle without having to resort to massively parallel high-performance computing. As a result, flow simulations are usually performed with models containing less details than those used for characterization. Even if you had all the computational power you needed, there are several arguments why you should still perform your simulations on coarser models. First of all, one can argue that high-resolution models tend to contain more

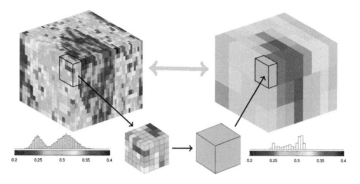

Figure 15.1 Upscaling of petrophysical properties, here represented by porosity.

details than what is justified by the information in the available data. Second, in many modeling workflows the available computational power should be spent running multiple model realizations to span the range of plausible outcomes rather than on obtaining high numerical resolution for a few highly uncertain predictions. Third, because coarser models contain fewer parameters, they are simpler to calibrate to observed reservoir responses like pressure tests and production data. Finally, a coarser model may be sufficient to predict flow patterns and reservoir responses with the accuracy needed to make a certain business decision. It is perhaps tempting to believe that with future increases in computing power, one will soon be able to close the gap in resolution between models used for reservoir characterization and models used for flow simulation. The development so far indicates that such an idea is wrong. The trend is rather that increases in computational power enable geologists and reservoir engineers to build larger and more complex geocellular models at a pace that outperforms the improvement in simulation capabilities.

For all these reasons, there is a strong need for mathematical and numerical techniques that can be used to communicate parameters and properties between models of different spatial resolution. This need will persist, and maybe even grow stronger, in the foreseeable future. This chapter discusses techniques for upscaling static rock properties.

15.1 Upscaling for Reservoir Simulation

Upscaling (or homogenization) refers to the process of propagating properties and parameters from a model of a given resolution to a model of *lower spatial resolution*. In this process, *heterogeneous regions* in a reservoir model are replaced by *homogeneous regions* to make up a coarser model of the same reservoir. Or, in the words of Chapter 14 on grid coarsening, upscaling is the process in which petrophysical properties in the cells that make up a coarse block are averaged into effective values for each coarse block; see Figure 15.1. The effective properties of the new homogeneous regions are defined so that they preserve the effects of small-scale variations in an averaged sense. How this averaging should be performed depends on the type of property to be upscaled. We distinguish between *additive properties* that can be upscaled using simple volumetric averaging and *nonaddititive properties*, for which correct averaging methods only exist in special cases and the best you can hope for in the general case is to compute accurate approximations.

Downscaling refers to the process of propagating properties from a model with a given spatial resolution to a model having *higher spatial resolution*. We can, for instance, be interested in refining coarse-scale modifications in petrophysical properties obtained during a history match on a simulation model to update the underlying geomodel. In the downscaling process, the aim is to preserve both the coarse-scale trends and the fine-scale heterogeneity structures.

The process of upscaling petrophysical parameters leads to many fundamental questions. For instance, do the partial differential equations that make up the flow model on the coarse scale take the same form as the equations modeling flow at the subgrid scale? And if so, how do we honor the fine-scale heterogeneities at the subgrid level? Even though upscaling has been a standard procedure in reservoir simulation for more than four decades, nobody has fully answered these questions rigorously, except for cases with special heterogeneous formations such as periodic or stratified media.

Homogenization is a rigorous mathematical theory for asymptotic analysis of periodic structures; see [41, 149, 135]. A relevant result states that for a periodic medium with permeability $\mathbf{K}(\frac{x}{\varepsilon})$, there exists a constant symmetric and positive-definite tensor $\mathbf{K}_0$ such that the solutions p_ε and $v_\varepsilon = -\mathbf{K}(\frac{x}{\varepsilon})\nabla p_\varepsilon$ to the elliptic problem

$$-\nabla \cdot \mathbf{K}(\frac{x}{\varepsilon})\nabla p_\varepsilon = q \tag{15.1}$$

converge uniformly as $\varepsilon \to 0$ to the solution of the homogenized equation

$$-\nabla \cdot \mathbf{K}_0 \nabla p_0 = q. \tag{15.2}$$

Homogenization theory can be used to derive homogenized tensors for the region, if we can safely assume that the region we seek to compute an effective property for is part of an infinite periodic medium. The main advantage of homogenization is that it provides mathematical methods to prove existence and uniqueness of the coarse-scale solution and also verifies that the governing equation at the macroscopic level takes the same form as the elliptic equation that governs porous media flow at the level of the representative elementary volumes (REVs). However, it is more debatable whether mathematical homogenization can be used for practical simulations, since natural rocks are rarely periodic.

As you will see shortly, the most basic upscaling techniques rely on local averaging procedures that calculate effective properties in each grid block solely from properties within the grid block. Because these averaging procedures do not consider coupling beyond the local domain, they fail to account for the effect of long-range correlations and large-scale flow patterns in the reservoir, unless their effect can be represented correctly by the forces that drive flow inside the local domain. Different flow patterns may call for different upscaling procedures, and it is generally acknowledged that global effects must also be taken into consideration to obtain robust coarse-scale simulation models.

Upscaling must also be seen in close connection with griding and coarsening methods, as discussed in Chapters 3 and 14. To make a fine-scale model, you must decide what kind of grid you want to represent the porous medium, what (local) resolution you need, and how you should orient your grid cells. Obviously, you should consider the same questions for the upscaled model. To make an upscaled simulation model as accurate and robust as possible, the grid should be designed so that grid blocks capture heterogeneities on the scale of the block. This often implies that you may need significantly more blocks if you use a regular grid than if you use an unstructured polyhedral grid. You may also want to use different coarsening factors in zones of high and low flow, near and far away from wells and fluid contacts, as discussed in Chapter 14. The importance of designing a good coarse grid should not be underestimated. Henceforth, however, we simply assume that a suitable coarse grid is available and focus on upscaling techniques.

The literature on upscaling techniques is extensive, ranging from simple averaging techniques, e.g., [151], via flow-based methods that rely on local flow problems [37, 90] to more comprehensive global [225, 133] and local–global [63, 64, 117] methods. Some attempts have been made to analyze the upscaling process, e.g., [31, 318], but so far there is generally no theory or framework for assessing the quality of an upscaling technique. In fact, upscaling techniques are rarely rigorously quantified with mathematical error estimates. Instead, the quality of upscaling techniques is usually assessed by comparing upscaled production characteristics with those obtained from a reference solution computed on an underlying fine grid. This deficiency is one of the motivations behind the development of modern multiscale methods [95, 195]. Such multiscale methods are more well developed in MRST than in any comparable software, but a detailed discussion is outside the scope of this book.

In the following, we only discuss the basic principles and try to show you how you can implement relatively simple upscaling methods; we also attempt to explain why some methods may work well for certain flow scenarios and not for others. If you are interested in a comprehensive overview, you should consult one of the many review papers devoted to this topic, e.g., [70, 311, 31, 263, 108, 91, 92]. We start by a brief discussion of how to upscale porosity and other additive properties, before we move on to discuss upscaling permeability, which is the primary example of a nonadditive property. Because of the way Darcy's law has been extended from single-phase to multiphase flow, it is common to distinguish the upscaling of absolute permeability **K** from the upscaling of relative permeability $k_{r\alpha}$. Upscaling of absolute permeability is often called *single-phase upscaling*, whereas upscaling of relative permeability is referred to as *multiphase upscaling*. The main parts of this chapter are devoted to permeability upscaling and to upscaling of the corresponding transmissibilities that account for permeability effects in finite-volume discretizations. As in the rest of the book, our discussion will to a large extent be driven by examples, for which you can find complete codes in the `upscaling` directory of the `book` module.

15.2 Upscaling Additive Properties

Porosity is the simplest example of an *additive property* and can be upscaled through a simple volumetric average. If Ω denotes the region we want to average over, the averaged porosity value is given as

$$\phi^* = \frac{1}{\Omega} \int_\Omega \phi(\vec{x}) \, d\vec{x}. \tag{15.3}$$

Implementing the computation of this volumetric average can be a bit tricky if the coarse blocks and the fine cells are not matching. In MRST, however, we always assume that the coarse grid is given as a partition of the fine grid, as explained in Chapter 14. If q denotes the vector of integers describing the coarse partition, upscaling porosity amounts to a single statement

```
crock.poro = accumarray(q,rock.poro.*G.cells.volumes)./ ...
             max(accumarray(q,G.cells.volumes),eps);
```

We use max(..,eps) to safeguard against division by zero in case your grid contains blocks with zero volume or your partition vector is not contiguous. Weighting by volume is not necessary for grids with uniform cell sizes. Similar statements were used to go from the fine-scale model on the left side of Figure 15.1 to the coarse-scale model shown on the right. Complete source code is found in the script illustrateUpscaling.m.

Other additive (or volumetric) properties like net-to-gross, (residual) saturations, and concentrations can be upscaled almost in the same way, except that you should replace the bulk average in (15.3) by a *weighted* average. If n denotes net-to-gross, the correct average would be to weight with porosity, so that

$$n^* = \left[\int_\Omega \phi(\vec{x}) \, d\vec{x} \right]^{-1} \int_\Omega \phi(\vec{x}) n(\vec{x}) \, d\vec{x}. \tag{15.4}$$

and likewise for saturations. In MRST, we can compute this upscaling as follows:

```
pv = rock.poro.*G.cells.volumes;
N  = accumarray(q,pv.*n)./ max(accumarray(q,pv),eps);
```

To verify that this is the correct average, we simply compute

$$\phi^* n^* = \left[\frac{1}{\Omega} \int_\Omega \phi(\vec{x}) \, d\vec{x} \right] \left[\int_\Omega \phi(\vec{x}) \, d\vec{x} \right]^{-1} \int_\Omega \phi(\vec{x}) n(\vec{x}) \, d\vec{x}$$
$$= \frac{1}{\Omega} \int_\Omega \phi(\vec{x}) \, n(\vec{x}) \, d\vec{x} = (\phi n)^*.$$

Using the same argument, you can easily propose that concentrations should be weighted with saturations, and so on. Rock type (or flow unit), on the other hand, is not an additive property, even though it is sometimes treated almost as if it was by applying a majority vote to identify the rock type that occupies the largest volume fraction of a block. Such a simple approach is not robust [324] and should generally be avoided.

To get more acquainted with upscaling of additive quantities, I recommend that you try to do the following computer exercises.

Computer exercises

15.2.1 Construct a $8 \times 24 \times 5$ coarse grid `CG` of the SAIGUP model and upscale the additive rock properties. Verify your upscaling by computing

```
pv = accumarray(CG.partition, poreVolume(G,rock));
max(abs(pv - poreVolume(CG, crock)))
```

Plot histograms of the fine and coarse porosities and compare. How would you make a coarse grid that preserves the span in porosity values better? To measure the quality of a given upscaling, you can for instance use the following error measure:

```
err = sum(abs(rock.poro - crock3.poro(CG.partition)))/...
      sum(rock.poro)*100;
```

It may also be helpful to plot the cell-wise discrepancy between the upscaled and the original porosity over the fine grid.

15.2.2 The file `mortarTestModel` from the `BedModels1` data set describes a sedimentary bed consisting of three different facies. Construct a coarse model that has 5×5 blocks in the lateral direction and as many blocks in the vertical direction as required to preserve distinct facies layers, as shown in the plot to the right. Upscale the porosity of the model. Do you think the resulting coarse grid is suitable for flow simulations? (Hint: Check the guidelines from Section 14.7.)

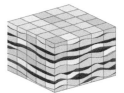

15.3 Upscaling Absolute Permeability

To study upscaling of absolute permeability, it is sufficient to consider single-phase flow in the form of a variable-coefficient Poisson equation

$$\nabla \cdot \mathbf{K} \nabla p = 0. \tag{15.5}$$

Even with such a simple equation, the choice of what is the best method to average absolute permeability generally depends on a complex interplay between the local permeability distribution and the characteristic flow directions. In certain special cases, one can develop simple methods that average permeability correctly, but in the general case, all you can do is develop computational methods that approximate the true effective permeability of the upscaled region. How accurate a given approximation is will depend on the coarse grid, the specific upscaling method, the purpose for which the upscaled values are to be used, and the complexity of the fine-scale permeability distribution.

Most techniques for upscaling absolute permeability seek an averaged tensor $\mathbf{K}^*$ that reproduces the same total flow through each homogeneous region you would obtain by solving the single-pressure equation (15.5) with the full fine-scale heterogeneity. In other words, if Ω is the homogeneous region to which we wish to assign an effective property $\mathbf{K}^*$, this property should fulfill

$$\int_\Omega \mathbf{K}(\vec{x}) \nabla p \, d\vec{x} = \mathbf{K}^* \int_\Omega \nabla p \, d\vec{x}. \tag{15.6}$$

This equation states that the net flow rate $\vec{v}_\Omega$ through Ω is related to the average pressure gradient $\nabla_\Omega p$ in Ω through the upscaled Darcy law

$$\vec{v}_\Omega = -\mathbf{K}^* \nabla_\Omega p. \tag{15.7}$$

The upscaled permeability tensor $\mathbf{K}^*$ is not uniquely defined by (15.6) for a given pressure field p, and conversely, there does not exist a unique $\mathbf{K}^*$ so that (15.6) holds for any pressure field. This reflects that $\mathbf{K}^*$ depends on the flow through Ω, which in turn is determined by the boundary conditions specified on $\partial\Omega$. The better you know the boundary conditions that a homogenized region will be subject to in subsequent simulations, the more accurate estimates you can compute for the upscaled tensor $\mathbf{K}^*$. In fact, if you know these boundary conditions exactly, you can compute the true effective permeability. In general, you will not know these boundary conditions in advance unless you have already solved your problem, and the best you can do is to make an educated and representative guess that aims to give reasonably accurate results for a specific or a wide range of flow scenarios. Another problem is that even though the permeability tensor of a physical system must be symmetric and positive definite (i.e., $\vec{z} \cdot \mathbf{K}\vec{z} > 0$ for all nonzero $\vec{z}$), there is generally no guarantee that the effective permeability tensor constructed by an upscaling algorithm fulfills the same properties. The possible absence of symmetry and positive definiteness shows that the single-phase upscaling problem is fundamentally ill-posed.

15.3.1 Averaging Methods

The simplest way to upscale permeability is to use an explicit averaging formula. *power averages* constitute a general class of such formulas,

$$\mathbf{K}^* = A_p(\mathbf{K}) = \left(\frac{1}{|\Omega|} \int_\Omega \mathbf{K}(\vec{x})^p \, d\vec{x} \right)^{1/p}. \tag{15.8}$$

Here, $p = 1$ and $p = -1$ correspond to the arithmetic and harmonic average, respectively, whereas the geometric mean is obtained in the limit $p \to 0$ as

$$\mathbf{K}^* = A_0(\mathbf{K}) = \exp\left(\frac{1}{|\Omega|} \int_\Omega \log(\mathbf{K}(\vec{x})) \, d\vec{x} \right). \tag{15.9}$$

The use of power averaging can be motivated by the so-called Wiener-bounds [317], which state that for a statistically homogeneous medium, the correct upscaled permeability will be bounded above and below by the arithmetic and harmonic mean, respectively.

To motivate (15.8), we can look at the relatively simple problem of upscaling permeability within a one-dimensional domain $[0, L]$. From the flow equation (15.5) and Darcy's law, we have that

15.3 Upscaling Absolute Permeability

$$-\big(K(x)p'(x)\big)' = 0 \quad \Longrightarrow \quad v(x) = K(x)p'(x) \equiv \text{constant}.$$

The upscaled permeability K^* must satisfy Darcy's law on the coarse scale, hence $v = -K^*[p_L - p_0]/L$. Alternatively, this formula can be derived from (15.6)

$$K^* \int_0^L p'(x)dx = K^*(p_L - p_0) \stackrel{(15.6)}{=} \int_0^L K(x)p'(x)dx = -\int_0^L v\,dx = -Lv.$$

We can now use (15.6) and Darcy's law to find the expression for K^*

$$\int_0^L p'(x)dx = -\int_0^L \frac{v}{K(x)}dx$$

$$= K^* \frac{p_L - p_0}{L} \int_0^L \frac{1}{K(x)}dx = K^* \left(\int_0^L p'(x)dx\right)\left(\frac{1}{L}\int_0^L \frac{1}{K(x)}dx\right),$$

from which it follows that the correct way to upscale K is to use the harmonic average

$$K^* = \left(\frac{1}{L}\int_0^L \frac{1}{K(x)}dx\right)^{-1}. \tag{15.10}$$

This result is universally valid only in one dimension, but also applies to the special case of a perfectly stratified isotropic medium with layers perpendicular to the direction of pressure drop as illustrated in Figure 15.2. It is a straightforward exercise to extend the analysis just presented to prove that harmonic averaging is the correct upscaling in this case. (If you are familiar with Ohm's law in electricity, you probably realize that this setup is similar to that of resistors set in parallel.)

Computing the harmonic average is straightforward in MRST:

```
vol  = G.cells.volumes;
for i=1:size(rock.perm,2)
    crock.perm(:,i) = accumarray(q,vol) ./ ...
                      accumarray(q,vol./rock.perm(:,i))
end
```

Here, we have assumed as in the previous section that q is a partition vector describing the coarse grid.

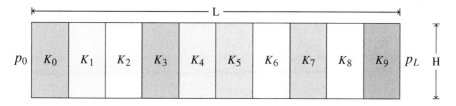

Figure 15.2 Example of a perfectly stratified isotropic medium with layers perpendicular to the direction of the pressure drop, for which harmonic averaging is the correct way to upscale absolute permeability.

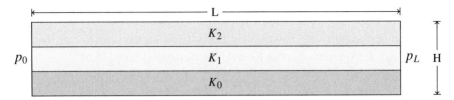

Figure 15.3 Example of a perfectly stratified isotropic medium with layers parallel to the direction of the pressure drop, for which arithmetic averaging is the correct way to upscale absolute permeability.

Another special case is when the layers of a perfectly stratified isotropic medium are parallel to the pressure drop as shown in Figure 15.3. Because the permeability is constant along the direction of pressure drop, the pressure will be linear in the x-direction with $p(x, y) = p_0 + x(p_L - p_0)/L$. Using (15.6), we can compute

$$K^* \int_0^H \int_0^L \partial_x p(x,y)\, dx\, dy = K^* H(p_L - p_0)$$

$$= \int_0^H \int_0^L K(x,y) \partial_x p(x,y)\, dx\, dy = \frac{p_L - p_0}{L} \int_0^H \int_0^L K(x,y)\, dx\, dy,$$

from which it follows that the correct upscaling is to use the arithmetic average

$$K^* = \frac{1}{LH} \int_0^H \int_0^L K(x,y)\, dx\, dy. \tag{15.11}$$

These examples show that averaging techniques can give correct upscaling in special cases, also in three dimensional space. If we now combine these two examples, we see that we can define the following upscaled permeability tensors for the isotropic media shown in Figures 15.2 and 15.3,

$$\mathbf{K}^* = \begin{bmatrix} A^x_{-1}(K) & 0 \\ 0 & A^x_1(K) \end{bmatrix} \quad \text{and} \quad \mathbf{K}^* = \begin{bmatrix} A^y_1(K) & 0 \\ 0 & A^y_{-1}(K) \end{bmatrix},$$

where the superscripts x and y on the averaging operator A from (15.8) signify that the operator is only applied in the corresponding spatial direction. These averaged permeabilities would produce the correct net flow across the domain when these models are subject to a pressure differential between the left and right boundaries or between the top and bottom boundaries. For other boundary conditions, however, the upscaled permeabilities generally give incorrect flow rates.

To represent flow in more than one direction, also for cases with less idealized heterogeneous structures modeled by a diagonal fine-scale tensor, you can compute a permeability tensor with the following diagonal components:

$$\mathbf{K}^* = \begin{bmatrix} A^{yz}_1(A^x_{-1}(\mathbf{K})) & 0 & 0 \\ 0 & A^{xz}_1(A^y_{-1}(\mathbf{K})) & 0 \\ 0 & 0 & A^{xy}_1(A^z_{-1}(\mathbf{K})) \end{bmatrix}.$$

15.3 Upscaling Absolute Permeability

In other words, we start by computing the harmonic average of each of the diagonal permeability components in the corresponding longitudinal direction, i.e., compute the harmonic average of K_{xx} along the x direction, and so on. Then, we compute the arithmetic average in the transverse directions, i.e., in the y and z directions for K_{xx}, and so on. This average is sometimes called the *harmonic-arithmetic average* and may give reasonable upscaling for layered reservoirs when the primary direction of flow is along the layers. More importantly, the harmonic-arithmetic average provides a tight lower bound on the effective permeability, whereas the opposite method, the arithmetic-harmonic average, provides a tight upper bound.

It is not obvious how to compute the harmonic-arithmetic average for a general unstructured grid, but it is almost straightforward to do it for rectilinear and corner-point grids in MRST. For brevity, we only show the details for the case when the grid has been partitioned uniformly in index space:

```
q = partitionUI(G, coarse);
vol = G.cells.volumes;
for i=1:size(rock.perm,2)
   dims = G.cartDims; dims(i)=coarse(i);
   qq = partitionUI(G, dims);
   K = accumarray(qq,vol)./accumarray(qq,vol./rock.perm(:,i));
   crock.perm(:,i) = accumarray(q,K(qq).*vol)./accumarray(q,vol);
end
```

The key idea in the implementation above is that to compute the harmonic average in one axial direction, we introduce a temporary partition qq that coincides with the coarse grid along the given axial direction and with the original fine grid in the other axial directions. This way, the call to accumarray has the effect that the harmonic average is computed for one longitudinal stack of cells at the time inside each coarse block. To compute the arithmetic average, we simply map the averaged values back onto the fine grid and use accumarray over the original partition q.

Example 15.3.1 *The script* averagingExample1 *in the* book *module shows an example of the averaging methods just discussed. The permeability field is a 40×60 sub-sample of Layer 46 from the fluvial Upper Ness formation in the SPE 10 data set. Figure 15.4 compares the effective permeabilities on a 15×15 coarse grid computed by arithmetic, harmonic, and harmonic-arithmetic averaging. We see that arithmetic averaging has a tendency to preserve high permeability values, harmonic averaging tends to preserve small permeabilities, whereas harmonic-arithmetic averaging is somewhere in between.*

Although simple averaging techniques can give correct upscaling in special cases, they tend to perform poorly in practice because the averages do not properly reflect the heterogeneity structures. Likewise, it is also generally difficult to determine which averaging to use, since the best averaging method depends both on the heterogeneity of the reservoir and the prevailing flow directions. Let us illustrate this by an example.

568 *Upscaling Petrophysical Properties*

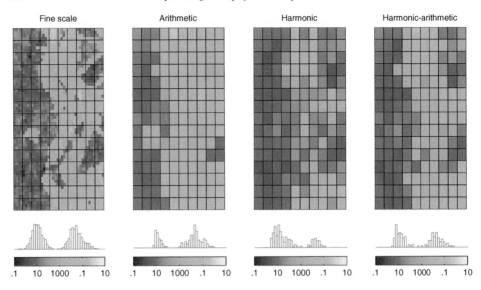

Figure 15.4 Arithmetic, harmonic, and harmonic-arithmetic averaging applied to a 40×60 subset of Layer 46 from the SPE 10 data set.

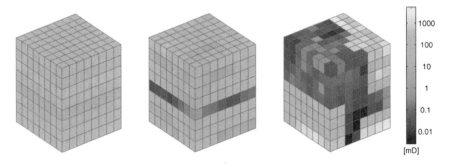

Figure 15.5 Three models used to test the accuracy of averaging techniques. From left to right: a layered model and two subsets from the Tarbert and the Upper Ness formation, respectively, from the SPE 10 data set.

Example 15.3.2 *Consider an $8 \times 8 \times 8$ reservoir with three different permeability realizations shown in Figure 15.5, which we upscale to a single coarse block using arithmetic, harmonic, and harmonic-arithmetic averaging. To assess the quality of the upscaling, we compare fine-scale and coarse-scale prediction of net flux across the outflow boundary for three different flow patterns: from west to east, from south to north, and from bottom to top. Complete source code for this simulation setup is given in the script* `averagingExample2`.

Table 15.1 reports the ratio between the outflow computed by the coarse models and the outflow computed on the original fine grid. For the layered model, arithmetic and harmonic-arithmetic averaging correctly reproduces flow in the lateral directions, whereas

Table 15.1 *Ratio between flow rate predicted by upscaled/fine-scale models for three permeability fields, flow scenarios, and upscaling methods.*

Model	Flow pattern	Arithmetic	Harmonic	Harm-arith
Layered	East→West	1.0000	0.2662	1.0000
	North→South	1.0000	0.2662	1.0000
	Top→Bottom	3.7573	1.0000	1.0000
Tarbert	East→West	1.6591	0.0246	0.8520
	North→South	1.6337	0.0243	0.8525
	Top→Bottom	47428.0684	0.3239	0.8588
Upper Ness	East→West	3.4060	0.0009	0.8303
	North→South	1.9537	0.0005	0.7128
	Top→Bottom	6776.8493	0.0020	0.3400

flow normal to the layers is correctly reproduced by harmonic and harmonic-arithmetic averaging. For the two anisotropic models from SPE 10, on the other hand, the flow rates predicted by the arithmetic and harmonic methods are generally far off. The combined harmonic-arithmetic method is more accurate, with less than 15% discrepancy for the Tarbert model and 17–76% discrepancy for the Upper Ness model. Whether this can be considered an acceptable result will depend on what purpose the simulation is to be used for.

COMPUTER EXERCISES

15.3.1 Set up a set of flow simulations with different boundary conditions and/or well patterns to test the accuracy of the three effective permeability fields computed in `averagingExample1`.

15.3.2 Implement harmonic-arithmetic averaging for the SAIGUP model. (Hint: the script `cpGridHarmonic` in the `upscaling` module computes harmonic averaging for the SAIGUP model.) Can you also apply it to the coarse grids generated for CaseB4 in the previous chapter?

15.3.3 Redo the upscaling in Figure 15.4 using a coarse grid that adapts to high-permeable or high-flow zones. Does this help preserve the distribution of the permeabilities?

15.3.4 Implement the operator A_p in (15.8) as a new utility function.

15.3.5 How would you upscale the well indices in a Peaceman well model?

15.3.2 Flow-Based Upscaling

Thus far in this section, we repeatedly used simple flow problems to argue whether a particular averaging method was good or not. Taking this idea one step further, we could impose representative boundary conditions along the perimeter of each coarse block and solve

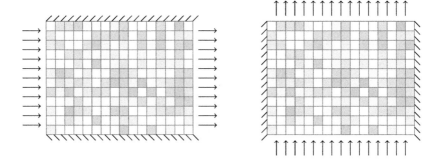

Figure 15.6 Illustration of a simple flow-based upscaling method, solving $-\nabla \cdot (\mathbf{K}\nabla p) = 0$, with $p = 1$ and $p = 0$ prescribed along the inflow and outflow boundaries respectively, and no-flow boundary conditions elsewhere.

the flow problem (15.5) numerically to determine fine-scale pressures and flow rates and corresponding net flow rates and average pressure gradients. We could then invert the flow equations discretized on the coarse block to determine effective coarse-scale permeabilities from Darcy's law. That is, once p and v have been computed, we use the same type of argument as we used to determine the formulas for the harmonic and arithmetic averages, except that we now work with discrete pressures and fluxes. This raises the immediate question of what kind of boundary conditions we should use.

Different boundary conditions have been suggested over the years. One obvious approach is to use three different boundary conditions for each block to create a pressure drop in each axial directions in the same way as for the computations reported in Table 15.1; see Figure 15.6. By specifying sealing boundary conditions along the other boundaries, we ensure that the effective flow over the block follows the same axial direction as the pressure drop. This setup emulates how permeability is measured on core samples in the laboratory and provides us with three pairs of flow rates in 3D that we can use to compute an upscaled permeability tensor with diagonal elements

$$K_{xx} = -\frac{v_x L_x}{\Delta p_x}, \qquad K_{yy} = -\frac{v_y L_y}{\Delta p_y}, \qquad K_{zz} = -\frac{v_z L_z}{\Delta p_z}.$$

Here, v_x is net flux, L_x the characteristic length of the block, and Δp_x the pressure drop inside the block in the x-direction. With this method, the off-diagonal elements are zero by construction. Strictly speaking, the assumption of sealing boundaries is only valid in the idealized case when the permeability field is symmetric with respect to the faces of the coarse grid, i.e., if the coarse block is surrounded by mirror images of itself. It can be shown that this method tends to introduce an upscaling bias towards low permeability values by thickening shale barriers and narrowing sand channels [163, 161].

Unless the grid block is located next to a sealing fault or an impermeable layer, it therefore more natural to assume that the block has open boundaries so that a pressure differential applied along one of the axial directions will induce a net flow also in transverse directions. This means that the effective permeability is a full tensor. One method to emulate open boundaries is to prescribe a constant pressure on faces perpendicular to each flow direction and linear pressure drop along the sides that are parallel to the flow direction [120, 266] so that the flow can leave or enter these sides. As for the sealing boundary, a unit pressure drop is applied in each of the axial directions. Use of linear boundary conditions is only strictly valid if the heterogeneous coarse block is embedded inside a homogeneous medium and tends to produce permeabilities with a bias towards high values [90].

Another popular option is to prescribe *periodic boundary conditions* [90], assuming that the grid block is sampled from a periodic medium so that the fluxes in and out of opposite boundaries are equal. In other words, to compute the x-component, we impose the following conditions:

$$\begin{aligned}
p(L_x, y, z) &= p(0, y, z) - \Delta p, & v(L_x, y, z) &= v(0, y, z), \\
p(x, L_y, z) &= p(x, 0, z), & v(x, L_y, z) &= v(x, 0, z), \\
p(x, y, L_z) &= p(x, y, 0), & v(x, y, L_z) &= v(x, y, 0),
\end{aligned} \quad (15.12)$$

and similarly for the other axial directions. This approach gives a symmetric and positive-definite tensor and is usually more robust than specifying sealing boundaries. Periodic boundaries tend to give permeabilities that lie in between the lower and upper bounds computed using sealing and linear boundaries, respectively.

Let us see how two of these methods can be implemented in MRST for a rectilinear or corner-point grid. For simplicity, we assume that the d-dimensional domain to be upscaled is rectangular and represented by grid G and rock structure rock. We start by setting up structures representing boundary conditions,

```
bcsides = {'XMin', 'XMax'; 'YMin', 'YMax'; 'ZMin', 'ZMax'};
for j = 1:d;
   bcl{j} = pside([], G, bcsides{j, 1}, 0);
   bcr{j} = pside([], G, bcsides{j, 2}, 0);
end
Dp = {4*barsa, 0};
L  = max(G.faces.centroids)-min(G.faces.centroids);
```

The first structure, bcsides, just contains name tags to locale the correct pair of boundary faces for each flow problem. The bcl and bcr structures are used to store a template of all the boundary conditions we must use. This is not needed to implement the pressure-drop method, but these structures are handy when setting up periodic boundary conditions. Finally, Dp and L contain the prescribed pressure drop and the characteristic length in each axial direction. With the data structures in place, the loop that does the upscaling for the pressure-drop method is quite simple:

```
for i=1:d
  bc = addBC([], bcl{i}.face, 'pressure', Dp{1});
  bc = addBC(bc, bcr{i}.face, 'pressure', Dp{2});
  xr = incompTPFA(initResSol(G, 100*barsa, 1),< G, hT, fluid, 'bc', bc);
  v(i) = sum(xr.flux(bcr{i}.face)) / sum(G.faces.areas(bcr{i}.face));
  dp(i) = Dp{1}/L(i);
end
K = convertTo(v./dp, milli*darcy);
```

That is, we loop over the axial directions and for each direction we: (i) specify a pressure from left to right, (ii) compute pressures and fluxes by solving the resulting Poisson problem, and (iii) compute average velocity v across the outflow boundary and average pressure drop d. Finally, we can compute the effective permeability by inverting Darcy's law. The implementation just described is a key algorithmic component in steady-state upscaling of relative permeabilities for multiphase flow and is therefore offered as a separate utility function upscalePermeabilityFixed in the upscaling module.

Periodic boundary conditions are slightly more involved. We start by calling a routine that modifies the grid structure so that it represents a periodic domain:

```
[Gp, bcp] = makePeriodicGridMulti3d(G, bcl, bcr, Dp);
for j=1:d, ofaces{j} = bcp.face(bcp.tags==j); end
```

Technically, the grid is extended with a set of additional faces that connect cells on opposite boundaries of the domain so that the topology becomes toroidal. The routine also sets up an appropriate structure for representing periodic boundary conditions, which we use to extract the faces across which we are to compute outflow. You can find details of how the periodic grid is constructed in the source code. With the modified grid in place, the loop for computing local flow solutions reads:

```
dp = Dp{1}*eye(d); nbcp = bcp;
for i=1:d
    for j=1:d,  nbcp.value(bcp.tags==j)=dp(j,i); end
    xr = incompTPFA(initResSol(Gp, 100*barsa, 1), Gp, hT, fluid, 'bcp', nbcp);
    for j=1:d,
        v(j,i) = sum(xr.flux(ofaces{j})) / sum(Gp.faces.areas(ofaces{j}));
    end
end
```

Inside the loop over all axial directions, the first for-loop extracts the correct pressure drop to be included in the periodic boundary conditions from the diagonal matrix dp; that is, a pressure drop along the current axial direction and zero pressure drop in the other directions. Then, we solve the flow problem and compute the average velocity across the outflow boundaries. Because of the periodic conditions, we may have outflow also across boundaries that have no associated pressure drop, which is why we generally get a full-tensor permeability. Outside the loop, we compute the average pressure drop in each axial direction and invert Darcy's law to compute the permeability tensor:

```
dp = bsxfun(@rdivide, dp, L);
K  = convertTo(v/dp, milli*darcy)
```

This implementation is offered as a utility `upscalePermeabilityPeriodic` in the `upscaling` module. Linear boundaries are not supported in MRST, but can be implemented by combining elements from the two cases just discussed: using the function `pside` to set a pressure distribution on the boundaries parallel to the direction of the applied pressure drop and solving a small linear system to compute the components of the effective permeability.

Setting up the necessary flow problems is relatively simple when the grid corresponds to just one coarse block. It is a bit more involved to do this upscaling efficiently for many blocks at a time. In the `upscaling` module, we therefore offer a utility function `upscalePerm` that computes permeability upscaling using the pressure-drop method.

Example 15.3.3 *Figure 15.7 shows two different permeability fields upscaled using flow-based upscaling with sealing and periodic boundary conditions. For the case to the right, the correlation in the permeability field is along the y-direction, and both methods compute the same diagonal upscaled tensor. Using pressure drop along the axial directions with sealing boundaries still gives a diagonal tensor if we rotate the grid so that the correlation direction is along the diagonal, whereas the periodic boundary conditions give a full tensor. By computing the eigenvalue decomposition of this tensor, we find that the upscaled tensor is diagonal with values 15 and 19.5 if we rotate the axial directions 45 degrees clockwise. Full source code for this example is given in the* `permeabilityExample1` *script.*

Example 15.3.4 *Both flow-based methods will by design correctly reproduce flow along or orthogonal to layered media for the experiment in Table 15.1. We therefore replace the layered permeability field with one having dipping layers. Moreover, we replace the top-to-bottom pressure drop by a pressure drop between diagonally opposite corners and sample from a different part of the SPE 10 model. Figure 15.8 shows the new permeability fields*

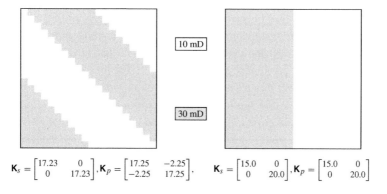

Figure 15.7 Upscaling two isotropic permeability fields using flow-based upscaling with sealing boundaries (s) or periodic boundary conditions (p).

Table 15.2 *Ratio between flow rate predicted by upscaled/fine-scale models for three different permeability fields, flow scenarios, and upscaling methods.*

Model	Flow pattern	Harm-arith	Sealing	Periodic
Layered	East→West	0.78407	1.00000	1.02594
	North→South	0.49974	1.00000	1.00000
	Corner→Corner	1.04273	1.30565	1.34228
Tarbert	East→West	0.86756	1.00000	0.56111
	North→South	0.89298	1.00000	0.53880
	Corner→Corner	0.00003	0.00027	39.11923
Upper Ness	East→West	0.83026	1.00000	0.40197
	North→South	0.71283	1.00000	0.28081
	Corner→Corner	0.09546	0.62377	2183.26643

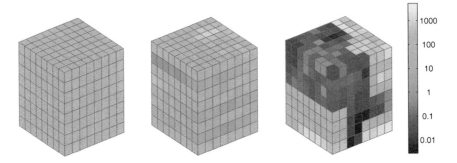

Figure 15.8 Three models used to test the accuracy of averaging techniques. From left to right: an isotropic model with dipping layers and two subsets from the Tarbert and the Upper Ness formation, respectively, from the SPE 10 data set.

and Table 15.2 reports the results of the experiment. The method with sealing boundary conditions is exact by design for the first two flow fields, but gives incorrect flow rates for the diagonal flow, in particular for the Tarbert subsample. With periodic boundary conditions, we get the exact flux when the flow direction follows regions of constant permeability from the north to the south side for the isotropic case. For the other two flow patterns, the flow directions cross dipping layers and we thus get minor deviations in the lateral direction and somewhat larger deviation for the diagonal flow. For the anisotropic subsamples, the periodic conditions generally give less accurate results than the other two methods, except for the diagonal flow on the Tarbert case. Altogether, this example illustrates quite well how ill-posed the upscaling problem is. Source code is given in the `permeabilityExample2` *script.*

COMPUTER EXERCISES

15.3.1 The data sets BedModels1 and BedModel2 are examples of fine-scale rock models developed for use in a workflow that propagates petrophysical properties from the core scale to the reservoir scale. Use the methods presented in this subsection to upscale the absolute permeability in these models.

15.3.2 Implement a method for upscaling with linear boundary conditions.

15.3.3 Estimates computed by the averaging and flow-based methods have a certain ordering, from low to high values: harmonic, harmonic-arithmetic, sealing boundaries, periodic boundaries, linear boundaries, arithmetic-harmonic, and arithmetic. Make a small test suite of heterogeneity permeability fields and verify this claim.

15.3.4 For grids that are not K-orthogonal, we know that the TPFA method will compute incorrect fine-scale flow solutions and that it generally is better to use a consistent discretization. Go back to Example 15.3.3, perturb the grid (e.g., with `twister`) and compare upscaled solutions computed by TPFA and the mimetic or MPFA method. (At the time of writing, the consistent methods do not yet implement periodic boundary conditions.)

15.4 Upscaling Transmissibility

In the discrete case, the choice of an appropriate upscaling method depends on the numerical stencil to be used for the spatial discretization of the upscaled model. In previous chapters we have seen that the two-point finite-volume method is the method of choice in reservoir simulation. When using this method, we only need grid-block permeabilities to compute transmissibilities between neighboring coarse blocks. It would therefore be more convenient if, instead of computing an effective permeability tensor, we could compute the coarse transmissibilities associated with the interface between pairs of neighboring coarse blocks directly. These transmissibilities should be defined so that they reproduce fine-scale flow fields in an averaged sense. That is, instead of upscaled block-homogenized tensors $\mathbf{K}^*$, we seek block transmissibilities T_{ij}^* satisfying

$$v_{ij} = T_{ij}^* \left(\frac{1}{|\Omega_i|} \int_{\Omega_i} p \, d\vec{x} - \frac{1}{|\Omega_j|} \int_{\Omega_j} p \, d\vec{x} \right), \tag{15.13}$$

where $v_{ij} = -\int_{\Gamma_{ij}} (\mathbf{K} \nabla p) \cdot \vec{n} \, dv$ is the total Darcy flux across Γ_{ij}.

We can compute the upscaled transmissibilities T_{ij}^* much the same way we computed upscaled permeabilities in Figure 15.9. Here, we use a pressure drop to drive a flow across the interface Γ_{ij} between two coarse blocks Ω_i and Ω_j. Thus, by solving (15.5) in the two-block domain $\Omega_i \cup \Omega_j$ subject to suitable boundary conditions, we can compute the average pressures P_i and P_j in Ω_i and Ω_j and then obtain T_{ij}^* directly from the formula

$$v_{ij} = T_{ij}^* (P_i - P_j). \tag{15.14}$$

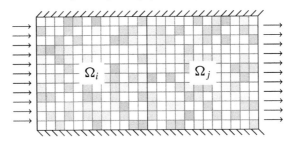

Figure 15.9 Illustration of flow-based upscaling of transmissibility.

Let us see how this upscaling can be implemented in MRST. That is, we discuss how to upscale the transmissibility associated with a single interface between two grid blocks. To this end, we assume a fine grid G and a coarse partition vector q that subdivides the grid into two coarse blocks in the x-direction. We start by constructing a coarse-grid structure CG, as introduced in Section 14.2. From the data members in this structure we can find all faces in the fine grid that lie on the interface between the coarse blocks and determine the correct sign to apply to the coarse-scale flux

```
i     = find(~any(CG.faces.neighbors==0,2));
faces = CG.faces.fconn(CG.faces.connPos(i):CG.faces.connPos(i+1)-1);
sgn   = 2*(CG.faces.neighbors(i, 1) == 1) - 1;
```

As we saw in Section 15.3.2, there are different ways we can set up a localized flow problem that gives a flux across the interface between the two blocks. Assuming that the coarse interface is more or less orthogonal to the x-axis, we use a pressure drop in this direction and no-flow boundary conditions on the other sides, as shown in Figure 15.9:

```
bc   = pside([], G, 'XMin', Dp(1));
bc   = pside(bc, G, 'XMax', Dp(2));
xr   = incompTPFA(initResSol(G, 100*barsa, 1), G, hT, fluid, 'bc', bc);
flux = sgn * sum(xr.flux(faces));
mu   = fluid.properties();
```

All that now remains is to compute pressure values P_i and P_j associated with each coarse block before we use (15.14) to compute the effective transmissibility. Following (15.13), P_i and P_j are defined as the average pressure inside each block

```
P = accumarray(q,xr.pressure)./accumarray(q,1);
T = mu*flux/(P(1) - P(2));
```

Here, we have assumed that all grid cells have the same size. If not, we need to weight the pressure average by cell volumes. As an alternative, we could also have used the pressure value at the block centroids, which would give a (slightly) different transmissibility:

```
cells = findEnclosingCell(G,CG.cells.centroids);
P     = xr.pressure(cells);
```

15.4 Upscaling Transmissibility 577

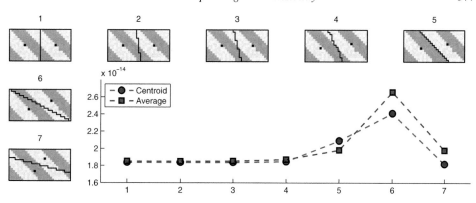

Figure 15.10 Upscaled transmissibility computed on a series of two-block configurations using either centroid values or averaged values for the block pressures. Black dots show the location of the block centroids from which pressure values are sampled.

Example 15.4.1 *In Figure 15.10 we have used the code just discussed to compute the transmissibility between two blocks covering a rectangular domain $[0,200] \times [0,100]$ m^2. We gradually rotate the interface between the two coarse blocks so that it goes from vertical towards being horizontal. Complete source code is given in the script* transmissibilityExample1 *in the* book *module. For the first four pairs of coarse blocks shown in Figure 15.10, there is only a slight difference between the transmissibilities computed using pressure values defined as block averages and pressures defined by sampling at the block centroids. The transmissibilities start to deviate as the block interface approaches the diagonal.*

A pressure drop along the x-axis is a reasonable drive mechanism for the first block pairs, but it is highly questionable for the last two block pairs. Let us, for instance, consider block pair number seven. If we instead apply the pressure drop along the y-axis to create a flow that is more perpendicular to the coarse interface, the transmissibility computed with block-averaged pressures changes from 2.0e-14 to 8.1e-14, whereas the value is 8.9e-14 with a pressure drop along both axial directions. (The latter setup is generally not well-posed, as it will have singularities at the southeast and northwest corners.) Depending upon what type of flow conditions the block pair is subject to in subsequent simulations, it might be better to set a pressure drop orthogonal to the coarse-block interface or use one of the oversampling methods that will be introduced in the next section.

An important property of the upscaled transmissibilities T_{ij}^* is that they all should be positive across all interfaces that can transmit fluids, since this will ensure that the TPFA scheme defined by $\sum_j T_{ij}^*(p_i - p_j) = q_i$ reproduces the net grid block pressures $p_\ell = \frac{1}{|\Omega_\ell|} \int_{\Omega_\ell} p \, d\vec{x}$, and hence also the coarse fluxes v_{ij}. Unfortunately, there is no guarantee that the transmissibilities defined by (15.13) are positive, for instance, if the blocks are such that the chord between the cell centroids does not cross the coarse interface. Negative transmissibilities may also occur for regular block shapes if the permeability is sufficiently heterogeneous (e.g., as in the SPE 10 model). Even worse, unique transmissibility values may not even exist, since the upscaling problem generally is not well-posed; see [318] for a more thorough discussion of existence and uniqueness. To guarantee that the resulting coarse-scale discretization is stable, you should ensure that the T_{ij}^* values are positive. A typical trick of the trade is to set $T_{ij} = \max(T_{ij}, 0)$, or in other words, just ignore ill-formed connections between neighboring grid blocks. This is generally not a satisfactory solution, and you should therefore either try to change the grid, use different pressure points to define the transmissibility, or apply some kind of fallback strategy that changes the upscaling method locally.

The `upscaling` module does not contain any simple implementation of transmissibility upscaling like the one outlined in this section. Instead, the module offers a routine, `upscaleTrans`, that is designed for the general case where both the coarse and the fine grids can be fully unstructured and have faces that do not necessarily align with the axial direction. To provide robust upscaling for a wide range of geological models, the routine relies on a more comprehensive approach that will be outlined briefly in the next section, and also has fallback strategies to reduce the number of negative permeabilities.

15.5 Global and Local–Global Upscaling

The methods described so far in this chapter have all be local in nature. Averaging methods derive upscaled quantities solely from the local heterogeneous structures, whereas flow-based methods try to account for flow responses by solving local flow problems with prescribed boundary conditions. These boundary conditions are the main factor that limits the accuracy of flow-based methods. In Table 15.2, we saw how neither of the local methods were able to accurately capture the correct net flow for the anisotropic Tarbert and Upper Ness samples with boundary conditions giving diagonal flow. The main problem for flow-based techniques is, of course, that we do not know a priori the precise flow that will occur in a given region of the reservoir during a subsequent simulation. Thus, it is generally not possible to specify the appropriate boundary conditions for the local flow problems in a unique manner unless we already have solved the flow problem.

To improve the accuracy of the upscaling, you can use a so-called *oversampling technique*, which is sometimes referred to as an overlapping method. In oversampling methods, the domain of the local flow problem is enlarged with a border region that surrounds each (pair of) grid block(s) and boundary conditions, or other mechanisms for driving flow, like wells or source terms, are specified in the region outside the domain you wish to upscale. Local flow problems are then solved in the whole enlarged region, but the effective permeability tensor is only computed inside the original coarse block. This way, you lessen

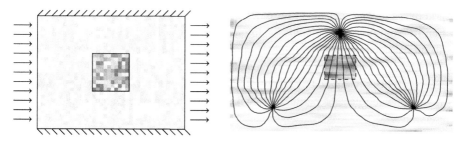

Figure 15.11 Illustration of oversampling techniques for upscaling. The size of the grid block is exaggerated in the right plot.

the impact of the localization assumption and the particular mechanism used to drive flow, and the flow across the boundaries of the coarse block is to a larger extent determined by the heterogeneity in the local region surrounding the block. The motivation for using an oversampling technique is to better account for permeability trends that are not aligned with the grid directions and possible large-scale connectivity in the permeability fields.

To also account for global flow patterns, you can solve the pressure equation once, or a couple of times, on the full geological model and impose fluxes sampled from the global flow solution as boundary conditions on the local flow problem used to upscale permeabilities or transmissibilities. In these so-called *global upscaling methods*, the pressure solution can either be computed using representative and *specific* drive mechanisms. Alternatively, you can compute the pressure solution using a set of *generic conditions* that, e.g., specify a flow through the reservoir from east to west, from north to south, and so on. From a computational point of view, this approach may seem to contradict the purpose of upscaling, but can be justified for numerical simulation of compressible and/or multiphase flow. Indeed, for such transient flows, the pressure equation must be solved multiple times throughout the simulation, and the cost of solving the global pressure equation once (or even a few times) on a fine grid will in most cases be small compared with the cost of solving the full multiphase flow problem.

Computing a global pressure solution on the fine grid can be avoided if we use a so-called *local–global* method [64, 63, 117]. In these methods, pressure solutions obtained by solving global flow problems on a coarse grid are used to set boundary conditions for local flow problems that determine the upscaled transmissibilities. Since the initial coarse-grid computation may give quite poor boundary conditions for the local flow problems, these methods use a iterative bootstrapping procedure to gradually ensure consistency between the local and the global calculations.

Global and local–global upscaling methods may suffer from negative or anomalously large transmissibility values. To avoid these, Holden and Nielsen [133], who were among the first to study global upscaling methods, proposed an iterative process involving the solution of an optimization problem that gradually perturbs the transmissibilities until they fall within a prescribed interval. A similar technique that avoids the solution of the fine-grid problem was presented in [225]. Chen and Durlofsky [63] observed that unphysical transmissibilities occur mainly in low-flow regions. Therefore, instead of using an optimization procedure that alters all transmissibilities in the reservoir, they proposed to use a thresholding procedure in which negative and very large transmissibilities are replaced by transmissibilities computed by a local method. Because the transmissibilities are altered only in low-flow regions, the perturbation will have limited impact on the total flow through the reservoir.

The script `upscaleTrans` offers various implementations of global transmissibility upscaling using either specific or generic mechanisms to drive flow in the full reservoir model. This specific approach enables you to tailor the coarse-scale model and achieve very good upscaling accuracy for a particular simulation setup. The generic approach gives you a more robust upscaling that is less accurate for a specific simulation setup, but need not be recomputed if you later wish to simulate setups with significant changes in the well pattern, aquifer support, and other factors that affect the global flow patterns in the reservoir. Local–global methods are not yet supported in MRST. Instead, the software offers various multiscale methods [95], including multiscale mixed finite elements [6, 243], multiscale finite volumes [211, 214], and the multiscale restriction-smoothed basis method [216, 215]. These methods offer an alternative approach, in which the impact of fine-scale heterogeneity is included more directly into coarse-scale flow equations as a set of basis functions that not only represent the average effect but also incorporate small-scale variation. The methods also give a natural way to reconstruct fine-scale pressure and mass-conservative flux approximations.

COMPUTER EXERCISES

15.5.1 Extend the computations reported in Table 15.2 to include oversampling methods. Does this improve the accuracy of the two flow-based methods?

15.5.2 Implement a function that performs generic global upscaling for uniform partitions of rectilinear and curvilinear grids. To drive the global flow pattern, you can use a pressure drop in each of the axial directions. How would you extend your method to more general grids and partitions?

15.5.3 Extend the `upscaling` module to include local–global upscaling methods.

15.6 Upscaling Examples

In this section we go through three examples that apply the methods discussed earlier in the chapter to upscale reservoir models. In doing so, we also introduce you to a simple form of flow diagnostics [274, 218] you can use to assess the accuracy of single-phase upscaling. The last example represents a complete workflow, from geological surfaces to grid, via fine-scale simulation to upscaled model. We end the section with a set of general advice and simple guidelines.

15.6 Upscaling Examples

15.6.1 Flow Diagnostics Quality Measure

A challenge with upscaling is to predict upfront whether the upscaling will be accurate or not. To partially answer this question and give an indication of the quality of the upscaling, we suggest to compare the cumulative well-allocation factors computed by the fine-scale and upscaled models. As shown in Section 13.1.3, a well-allocation factor is the percentage of the flux in or out of a completion that can be attributed to a pair of injection and production wells. This information is similar to what is obtained by a production logging tool.

To compute these factors, we need to first compute a global, single-phase pressure solution for the specific well pattern we want to investigate. From the resulting flow field we compute influence regions that subdivide the reservoir into subvolumes associated with pairs of injection and production wells, as discussed in Section 13.1. Global methods tend to compute at least one fine-scale pressure solution as part of the upscaling procedure, but if a global solution is not available for the specific well pattern, it is generally not very expensive to compute compared with the cost of the flow-based upscaling procedure. The same goes for the computation of a coarse-scale flow field and partition functions on the fine and the upscaled model.

To improve the predictive power of well-allocation factors, you may have to subdivide each well into two or more segments that each consists of a connected set of completed cells. This way, you can measure the communication between different parts of the wells, which is particularly important if the model to be upscaled includes layers or geological objects with significantly different permeabilities. You can obviously also investigate to what extent dynamic heterogeneity measures, sweep/drainage regions, and time-of-flights are preserved, or compare the fine and coarse fluxes to assess the accuracy of the upscaling.

15.6.2 A Model with Two Facies

We start by considering a small rectangular reservoir containing two facies with contrasting petrophysical properties. The reservoir is produced by an injector and a producer placed diagonally opposite each other and completed mainly in the high-permeability parts of the reservoir. Both wells operate at a fixed rate that amounts to the injection/production of 0.2 pore volumes per year. We neglect gravity forces to avoid potential complications from cross-flow. The fine grid consists of $40 \times 20 \times 15$ cells, which we seek to upscale to a coarse $5 \times 5 \times 15$ model. That is, we upscale by a factor 8×4 in the lateral direction and leave the vertical layers in an attempt to preserve the vertical communication in the model as accurately as possible. Figure 15.12 shows the porosities of the fine and the coarse model. To also upscale permeability, we use `upscalePerm`, which implements the flow-based method from Section 15.3.2 with sealing boundary conditions. The script `upscalingExample1` in the `book` module contains complete source code.

To get an indication of the accuracy of this upscaling, we compare the volumetric connections between the injector and producer computed for the original and the upscaled permeability. (The latter is prolongated back onto the fine grid.) With a single

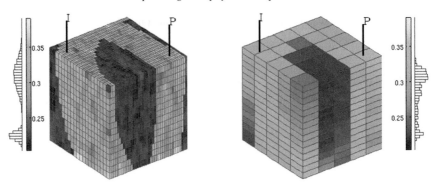

Figure 15.12 Porosity distribution in a model with two different rock types before and after upscaling.

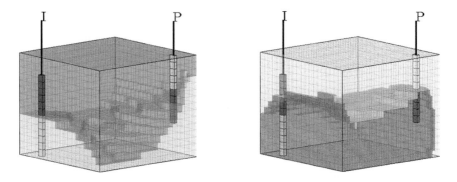

Figure 15.13 Volumetric subdivision of flooded/drainage regions by majority vote. The left plot shows the regions flooded by the upper and lower parts of the injector in blue and cyan, respectively. The right plot shows the drainage regions of the upper and lower half or the producer in yellow and red, respectively. In both plots, each cell is assigned to the completion that has the highest concentration value.

injector–producer pair that operates at fixed rates, the only difference we can expect to see is variation in the distribution of flow rates within each well. To better measure the communication within the reservoir, we therefore subdivide each well into an upper and a lower segments, so that we altogether have four well-pairs, whose well-allocation factors can be used to verify the quality of the upscaling. Figure 15.13 shows the influence regions of the fine-scale model for each of the four segments, defined as regions with "color concentration" larger than 0.5. A good upscaling method should preserve these communication volumes as accurately as possible in the coarse model.

Volumes are not always easy to compare in complex 3D models. Instead, we compare cumulative well-allocation factors. The plots in Figure 15.14 show bar charts of the cumulative flux in/out of the completions that make up each segment, from bottom to top. Here, each segment is assigned the same unique color as in Figure 15.13 and each bar is subdivided into the fraction of the total influx/outflux that belongs to the different well-pairs the segment is part of; see Section 13.1.3. To compare the fine-scale and coarse-scale

15.6 Upscaling Examples

permeability, **K** and **K***, we use colors for well allocation computed with the *coarse* model, whereas solid lines show the same quantities computed for the fine-scale model. The closer the color bars and the solid lines are, the better the upscaling.

To explain how this type of flow diagnostics should be interpreted, let us look at the lower-left bar chart, which reports the allocation factors for the upper segment of the producer (P:1, yellow color). Here, blue color signifies inflow that can be attributed to injection from the upper segment I:1 of the injector, whereas cyan color signifies inflow attributed to the lower segment I:2. Because the colored bars are mostly blue, the inflow into the upper producer segment is predominantly associated with the I:1–P:1 pair. In other words, P:1 is mainly supported by the upper segment of the injector and is hardly connected to the lower segment. You can confirm this by looking at the upper-right bar chart, which shows that the flow from the lower injector segment mainly contributes to inflow in the lower segment of the producer. The flooded volumes shown to the left in Figure 15.13 show that fluid injected from I:2 will mainly sweep the volume surrounding the lower three cells of P:2. Similarly, the right plot shows that the drainage region of P:1 only engulfs the upper parts of I:1. The flooded regions of I:1, on the other hand, engulfs both P:1 and the upper parts of P:2, and hence the upper injector segment will contribute flux to both segments of the producer, as shown in the upper-left chart of Figure 15.14. Likewise, the drainage region from P:2 engulfs both I:2 and the lower part of I:1 and hence is supported by flux from both the injector segments, as shown in the lower-right chart of Figure 15.14.

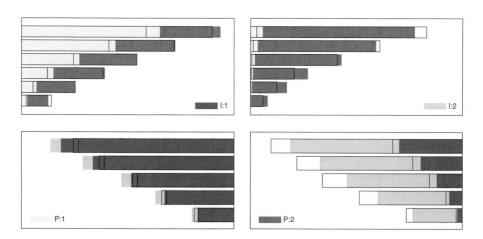

Figure 15.14 Normalized cumulative well-allocation factors for the four well completions in the two-facies model. The color bars show well-allocation factors computed with **K*** (upscaled by `upscalePerm`) on the fine grid, while the black lines show the same factors computed with the original permeability **K**. The colored bars and the black lines should coincide if the upscaled permeability reproduces the flux allocation correctly.

So, what does this flow diagnostics tell us about our upscaled permeability? First of all, $\mathbf{K}^*$ does not fully reproduce the volumetric connections of the original permeability $\mathbf{K}$. Looking at well-allocation factors for the producers in the bottom row of Figure 15.14, we see that $\mathbf{K}^*$ predicts too much flux in the upper segment (P:1, yellow color, left chart) since the colorbars exceed the solid lines at the top. Likewise, in the lower segment of the producer (P:2, red color, right chart), we get too low influx. Altogether, the allocation factors seem to suggest that the main problems are that with $\mathbf{K}^*$ we miss some of the vertical communication and exaggerate the connection between I:1 and P:1. We could have used a single number to measure the overall discrepancy in flux allocation, but in my opinion you will benefit more from a graphical presentation like this for problems with only a few wells, because it may help you identify the cause of the mismatch between the upscaled and the original permeability. (Lorenz coefficients for $\mathbf{K}$ and $\mathbf{K}^*$ are 0.2329 and 0.2039, respectively.)

To develop a better upscaling strategy, we can upscale transmissibilities and well indices using a global method. For a given flow scenario, like the one with fixed well positions and constant well controls, you can determine a set of transmissibilities and well indices that reproduce a single pressure step exactly. This may seem desirable, but has the disadvantage that the linear system becomes negative definite, which in turn may lead to unphysical solutions away from the scenario used for upscaling. To span a reasonable set of global flow directions, we instead use generic boundary conditions that pressurize the reservoir from east to west, from north to south, and from top to bottom. We can then combine these three global flow fields in several different ways to compute a transmissibility value for each coarse face; see e.g., [152]. Here, we use a simple strategy suggested by [172]: We pick the one among the three flow field that gives the largest flux across a given coarse face and insert this into (15.14) to compute the transmissibility associated with the face. A rationale for this choice is that the accuracy of the coarse model is most sensitive to the choice of transmissibilities in high-flow regions.

We upscale *well indices* using a coarse-scale version of the inflow-performance relationship (see (4.27) on page 122)

$$J_i^* = Q_i/(P_i - p_w), \tag{15.15}$$

where Q_i is the sum of the rates of all the perforations inside block i, P_i is the average block pressure, and p_w is the pressure inside the wellbore. To obtain representative flow fields for n wells, upscaleTrans solves $n-1$ pressure equations. Each solution is computed by setting a positive pressure in one well and zero pressure in all the other. We can then combine the flow fields using the same strategy as for the transmissibilities by picking the solution that gives the largest well rate, or as we will do here, add all well rates and divide by the sum of the corresponding pressure drops. (For models with many wells, we would rather upscale well indices using local or specific approaches.)

Unfortunately, there is no guarantee that the upscaled transmissibilities are nonnegative. To avoid creating an ill-conditioned discretization matrix, we therefore set all negative transmissibilities to zero, thereby blocking the corresponding coarse face for flow. This

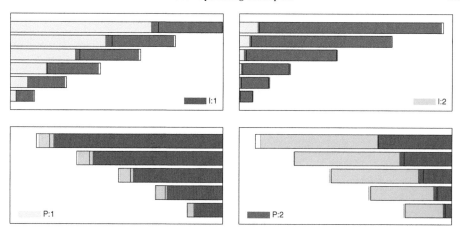

Figure 15.15 Normalized cumulative well-allocation factors for the four well completions in the two-facies model. The color bars show well-allocation factors computed on a coarse model upscaled using `upscaleTrans`, while the black lines show the same factors computed on the original model.

is in our experience an acceptable solution as long as the upscaled transmissibilities are only intended for a specific scenario, like herein, since negative transmissibilities tend to appear because flow is tangential to the coarse interface or as a result of numerical noise in regions with very low flow. If your upscaling, on the other hand, aims to serve a more general purpose with varying well patterns and/or boundary conditions, you would need to implement a more sophisticated *fallback strategy* that recomputes negative transmissibilities by an alternative method like a local flow-based method, an averaging method, or a combination of these.

Figure 15.15 shows the resulting match in well-allocation factors, which is significantly better than when using a simple permeability upscaling. In part, this is a result of improved transmissibilities, and in part because we now have upscaled the well indices.

15.6.3 SPE 10 with Six Wells

You have already encountered Model 2 from the 10th SPE Comparative Solution Project multiple times throughout the book. Because of its simple grid and strong heterogeneity and the fact that it is freely available online, this data set has become a community benchmark that is used for many different purposes. The original aim of the project was to "*compare upgridding and upscaling approaches and the ability to predict performance of a waterflood through a million cell geological model*," and in [71] you can read about the relative merits of the various methods used in the upscaling studies that were submitted to the project by August 2000. In short, the study showed that the data set is generally very difficult to upscale accurately with single-phase methods and that the best results are

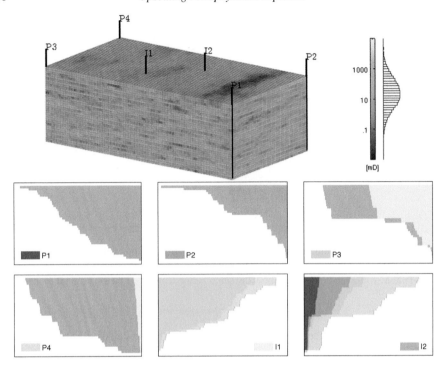

Figure 15.16 Model of the Tarbert formation with six wells, two central injectors and four peripheral producers. The 3D plot shows lateral permeability distribution, whereas the bar charts show normalized cumulative well-allocation factors.

generally obtained by methods that also account for multiphase effects. Later, however, several authors have obtained excellent results with local–global and multiscale methods.

In this section, we compare how accurate four different upscaling methods can predict single-phase flow. That is, we consider two averaging methods (harmonic and harmonic-arithmetic) and a local flow-based upscaling with sealing boundary conditions for upscaling permeability, and the global upscaling method discussed at the end of the previous example for upscaling transmissibilities and well indices. Unlike in the previous example, we use maximum flow rate to determine both transmissibilities and well indices. The script `upscalingExample2` contains full details.

We start by considering the Tarbert formation, which is found in the upper 35 layers of the model. We choose a coarsening factor of $10 \times 10 \times 3$, and to get an even number of cell layers inside each grid block, we extend the model slightly by repeating the top layer so that the fine-scale model altogether has $220 \times 60 \times 36 = 475\,200$ cells. We also replace the original five-spot pattern with a pattern consisting of producers in the four corners of the model and two injectors located near the center of the model; see Figure 15.16. All wells are controlled by bottom-hole pressure; the producers operate at 200 bar and the producers at 500 bar.

15.6 Upscaling Examples

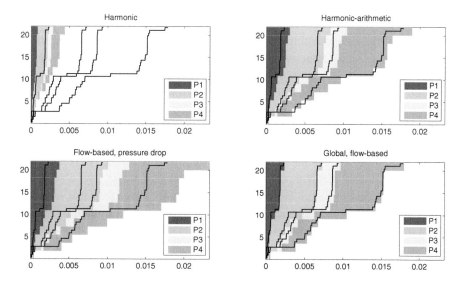

Figure 15.17 Well-allocation factors for injector I2 in the Tarbert model computed on four different coarse models.

Figure 15.17 compares well-allocation factors predict by the four upscaling methods for injector I2. This well has significant communication with all four producers and is hence more difficult to match than injector I1, which mainly communicates with P3 and P4. Harmonic averaging not surprisingly gives well rates that are far from the fine-scale targets. As explained in Section 15.3.1, the harmonic mean has a bias against smaller values and tends to underestimate the effective permeability. Because all wells are controlled by bottom-hole pressure, too low permeability means too low well rates. If the wells were controlled by total rate, as in the previous example, the well rates predicted by the harmonic average would have been more correct, but the predicted pressure buildup around the injector and the pressure drawdown in the producers would both be too high.

Comparing the local flow-based method to the harmonic-arithmetic averaging method, we see that the latter is significantly more accurate. By design, the flow-based method should give accurate prediction of flow along the axial directions of the grid. Here, however, we have flow that is strongly affected by heterogeneity and mostly goes in the lateral, diagonal direction and hence the harmonic-arithmetic method seems to be more suited, in particular if we consider the computational time of the two methods. In the current implementation, the flow-based method extracts a subgrid and assembles and inverts three local matrices for each coarse block. The computational overhead of this procedure is significant since each new local solve incurs full start-up cost. The result is that the function `upscalePerm` has orders of magnitude higher computational cost than the harmonic-arithmetic averaging, which is implemented using a few highly efficient calls to `accumarray`.

The global upscaling method reproduces the fine-scale well-allocation factors almost exactly. The plots may be a bit deceiving because of the areas where the colorbars extend beyond the solid lines representing the fine-scale well-allocation. However, bear in mind that the bars and lines represent *cumulative* factors, and thus we should only look at the discrepancy at the top of each colorbar. If you look closely, you will see that here the match is excellent. There are two reasons for this: First of all, with global boundary conditions, the boundary conditions used to localize the computation of each transmissibility account better for correlations that extend beyond the block pair and thus set up a flow that is more representative of the flow the interface will experience in the subsequent flow simulation. Secondly, the global method includes upscaling of well indices. This can have a significant impact on the well rates that determine the flux allocation. In passing, we also note that the computational cost of the global method is significantly lower than that of the local method, primarily since our implementation utilizes a highly efficient multigrid solver for the global flow problems and avoids solving a long sequence of small problems that each have a large start-up cost.

For completeness, we also include the results of a similar study for the whole SPE 10 model; see Figure 15.18. This includes the fluvial Upper Ness formation, which is notoriously difficult to upscale accurately. The harmonic-arithmetic and the global method are no longer as accurate as for the Tarbert formation. For this 1.1 million grid cell model, an efficient iterative solver like AGMG is indispensable to be able to compute the fine-scale solution and perform a global upscaling in MATLAB.

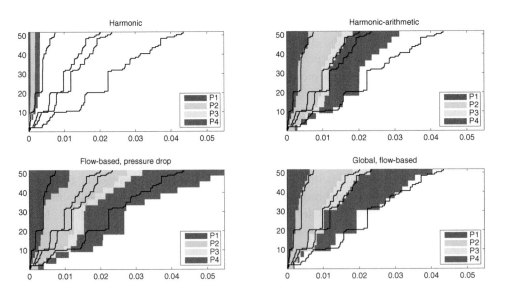

Figure 15.18 Well-allocation factors for injector I2 in the full SPE 10 model computed on four different coarse models.

15.6.4 Complete Workflow Example

Our last example shows a complete workflow for a synthetic sector model, from setting up the geocellular model, via compressible flow simulations, to coarsening/upscaling and analysis of the results using flow diagnostics. The example can thus be seen as a partial recap of the content in this book. We therefore include more code details than in the two previous examples. You can find the complete source code in upscalingExample3 in the book module.

The reservoir has a layered stratigraphy and contains three faults. The reservoir fluids form a slightly compressible two-phase oil/water system and hydrocarbons are recovered by a well pattern consisting of one injector and three producers. The model is mostly conceptual and is designed to outline and give details of the workflow rather than as a problem representing a realistic scenario in terms of well locations and fluid physics. However, unlike many other examples in the book, it is completely self-contained and does not involve any ECLIPSE files to describe reservoir geometry, wells, fluid properties, and simulation schedule.

We start by creating the main horizons that deliminate the external and internal geology of the sector. The top surface is composed of three different trends: a large-scale anticline structure in the form of an elongated arc-like shape, a set of medium-scale sinusoidal folds, and a set of small-scale random perturbations. For simplicity, we use the same surface, shifted by a constant factor, to deliminate the reservoir downward. In between, we introduce two tilted surfaces with extra small-scale random perturbations. These surfaces are clipped against the top and bottom surface to create erosional effects. The four horizons are shown in Figure 15.19. Assuming that these surfaces are given in the arrays zt, zmt, zbt, and zb, we create a basic corner-point grid as follows (see Section 3.3.1):

```
horizons = {struct('x', x, 'y', y, 'z', zt), ...
   struct('x', x, 'y', y, 'z', zmt), ...
   struct('x', x, 'y', y, 'z', zmb), ...
   struct('x', x, 'y', y, 'z', zb)};
grdecl = convertHorizonsToGrid(horizons, 'dims', [40 40], 'layers', [3 6 3]);
```

To introduce the faults, we first extract the coordinates for the corner-points and then shift the z-coordinates in selected parts of the domain:

```
[X,Y,Z]  = buildCornerPtNodes(grdecl);
i=47:80; Z(i,:,:) = Z(i,:,:) + .022*min(0,Y(i,:,:)-550);
j= 1:30; Z(:,j,:) = Z(:,j,:) + .021*min(0,X(:,j,:)-400);
j=57:80; Z(:,j,:) = Z(:,j,:) + .023*min(0,X(:,j,:)-750);
grdecl.ZCORN = Z(:);
```

The lower-left plot in Figure 15.19 shows the resulting model. The permeability consists of four layers of lognormally distributed values with average lateral permeability of 100, 400, 10, and 50 md, from top to bottom. The vertical permeability is set to one-tenth of the lateral permeability.

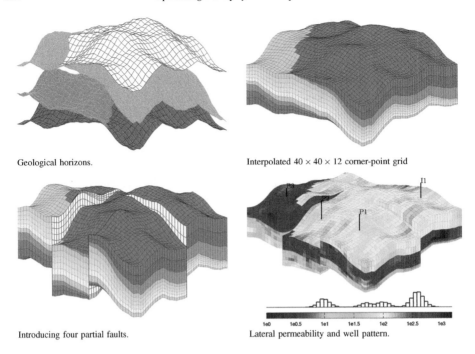

Figure 15.19 Creating a synthetic sector model whose geology is described by three volumes (two of which are eroded), three vertical faults that partially penetrate the sector, and four layers of distinctively different permeabilities. Oil is recovered by three producers supported by one injector.

The three vertical producers are completed in all layers of the model and operate at a constant bottom-hole pressure of 250 bar. The injector is also perforated in all layers and operates at a constant rate that would fill one pore volume of water over a period of ten years. As you probably recall, such wells are set up as follows:

```
W = verticalWell([], G, rock, 10, 10, [],...
            'Name', 'P1', 'comp_i', [1 0], 'Val', 250*barsa, ..
            'Type', 'bhp', 'refDepth', 50);
```

The fluid system is assumed to consist of two phases, an incompressible water phase and an oil phase with constant compressibility. If we express the oil compressibility as an expansion factor on the form, $b_o(p) = b_0 \exp[c(p - p_0)]$, we can model the fluid system as a special case of the general three-phase black-oil model from the AD-OO framework discussed in Chapter 12. The simplest fluid object in this framework models an incompressible fluid with constant densities. To get the correct behavior for the oil phase, we simply replace the function handle in the bo field by a pointer to an anonymous function that evaluates the exp function with appropriate arguments. We then call the appropriate constructor from the AD-OO framework:

```
fluid = initSimpleADIFluid('mu',   [1, 5, 0]*centi*poise, ...
                           'rho',  [1000, 700, 0]*kilogram/meter^3, ...
                           'n',    [2, 2, 0]);
fluid.bO = @(p, varargin) exp((p/barsa - 300)*0.001);

gravity reset on
model = TwoPhaseOilWaterModel(G, rock, fluid);
```

Initially, the reservoir is assumed to be in an equilibrium state with a horizontal oil–water (OW) contact separating pure oil from pure water. As explained earlier, MRST uses water, oil, and gas ordering internally, so in this case we have water in the first column and oil in the second for the saturations in the initial state

```
region = getInitializationRegionsBlackOil(model, depthOW, ...
         'datum_depth', ddepth, 'datum_pressure', p_ref);
state0 = initStateBlackOilAD(model, region);
```

The initialization is quite simple and does not perform any sub-cell integration to assign more accurate saturations in cells intersected by the OW contact; see the upper-left plot in Figure 15.20.

We define a straightforward simulation schedule consisting of five small initial control steps followed by 25 larger steps. We keep the well controls fixed throughout the simulation. If dT contains the time step, the schedule is constructed by the call simpleSchedule(dT, 'W', W). The simulation model has 16,350 grid cells and the resulting linear systems are a bit too large for MATLAB's direct solver to be efficient. We therefore set somewhat stricter tolerances and use a CPR preconditioner with an algebraic multigrid solver for the elliptic pressure system as discussed in more detail for the SPE 9 benchmark in Section 12.4.4. Figure 15.20 shows how water displaces oil over time inside the reservoir.

To reduce computational times, we partition the original $40 \times 40 \times 12$ grid uniformly in index space into a new $20 \times 20 \times 4$ coarse grid. This will generally give blocks having cells on opposite sides of faults. To ensure that as many as possible of the coarse blocks are hexahedral (except for those that are partially eroded along the top or bottom), we split any coarse block intersected by one of the faults. To do this, we introduce a temporary grid topology in which faults are set as barriers, and then perform a simple postprocessing to split any disconnected blocks:

```
cdims = [20, 20, 4];
Gf = makeInternalBoundary(G, find(G.faces.tag > 0));
p = processPartition(Gf, partitionUI(G, cdims));
```

Altogether, this produces a coarse grid with 1,437 active blocks (see Figure 15.21), which corresponds to a little more than a 22 times reduction in the number of unknowns. This should put us well within the range of models for which we safely can use MATLAB's default linear solver without the need for a CPR preconditioner.

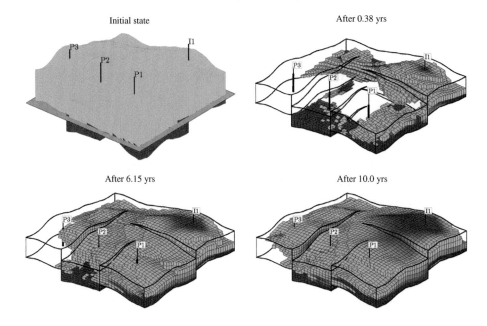

Figure 15.20 Time evolution of the water saturation for the sector model.

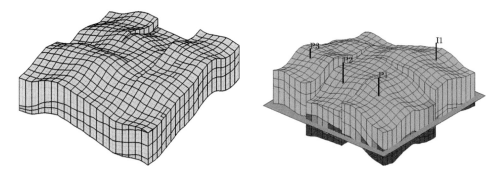

Figure 15.21 A coarse 20 × 20 × 4 partition of the sector model. New blocks arising when we split blocks penetrated by one of the faults are marked in green. The right plot shows the initial saturation as computed by `upscaleState`.

We can now directly upscale the model, schedule, and initial state. By default, the upscaling routine uses the simplest possible options, i.e., harmonic averaging of permeabilities,

```
modelC1    = upscaleModelTPFA(model, p);
scheduleC1 = upscaleSchedule(modelC1, schedule);
state0C1   = upscaleState(modelC1, model, state0);
```

15.6 Upscaling Examples

We have already seen that power-averaging of permeability is not very accurate, and thus we only consider this model as a base case. In addition, we use flow-based transmissibility upscaling with a specific flow field obtained by solving a single-phase flow problem with the given well pattern. This approach has proved to be the most accurate in the examples earlier in the section:

```
[~, TC, WC] = upscaleTrans(GC, model.operators.T_all, ...
   'Wells', W, 'bc_method', 'wells', 'fix_trans', true);

modelC2 = upscaleModelTPFA(model, p, 'transCoarse', TC);
scheduleC2 = schedule;
for i = 1:numel(scheduleC2.control)
   scheduleC2.control(i).W = WC;
end
```

To get an idea of how accurate the two upscaling methods will be, we can compare well allocation with the fine model based on the computation of a single pressure step. The plots in Figure 15.22 confirm that harmonic averaging also in this case underestimates the coarse-scale permeability and hence will predict too small flux into the three production wells. Specific transmissibility upscaling, on the other hand, reproduces the correct flux allocation very accurately and it therefore seems likely that it will produce quite accurate results. Let us now simulate the two cases. This is done as for a standard fine-scale model by calling the standard simulator interface with the coarsened model, initial data, and schedule

```
[wellSolsC1, statesC1] = simulateScheduleAD(state0C1, modelC1, scheduleC1);
```

Somewhat surprising, the production curves predicted by the specific flow-based upscaling are generally not much more accurate than those predicted by the simple harmonic

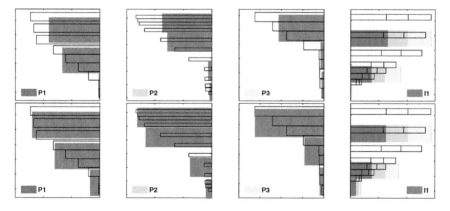

Figure 15.22 Flow diagnostics used as a quality measure for the two upscaling methods on the 20 × 20 × 4 upscaled sector model with harmonic upscaling of permeability (top) and specific flow-based transmissibility upscaling (bottom).

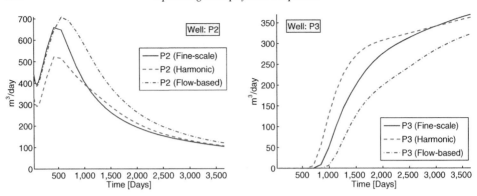

Figure 15.23 Comparison of oil rate (producer P2) and water rate (producer P3) predicted by the upscaled $20 \times 20 \times 4$ models and the original $40 \times 40 \times 12$ sector model.

permeability upscaling. Figure 15.23 shows the two production curves for which specific transmissibility upscaling compares least favorably with harmonic upscaling. This illustrates an important point: *whereas accurate reproduction of single-phase well allocation often is a necessary condition for accurate upscaling, it is not a sufficient condition.*

The question now is what has gone wrong here? The primary problem is that the grid blocks in the coarse grid do not adapt to the layering structure in the permeability field if we use a uniform subdivision of the K index. You may recall we argued that adapting to permeability layering is important when we discussed CaseB4 in Section 14.3.3. So let us rerun the upscaling exercise with a new basic partition that respects the layering

```
p0 = partitionLayers(G, cdims(1:2), L);
```

Here, L=[1,3,8,11,13] is a run-length encoding of the layer indices; this vector can be obtained as the second output of the `lognormLayers` function we used to generate the permeability field. The resulting coarse partition has exactly the same number of grid blocks, but gives significantly more accurate predictions for both upscaling methods, as you can see in Figure 15.24. We also obtain almost the same accuracy if we instead use a uniform $10 \times 10 \times 12$ partition, which supports our hypothesis that preserving layering is very important in this particular example.

Computer Exercises

15.6.1 How coarse model are you able to make and still predict the oil and water rates with reasonable accuracy?

15.6.2 Make a fine-grid model that respects the OW contact exactly. How does this affect the simulation? How would you coarsen and upscale this model?

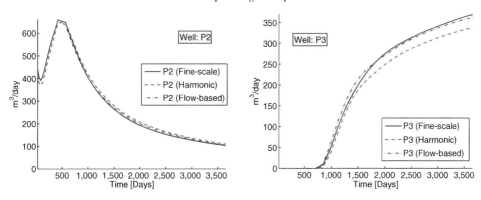

Figure 15.24 Comparison of oil rate (producer P2) and water rate (producer P3) predicted by the upscaled $20 \times 20 \times 4$ models where the coarse blocks adapt to the permeability layers of the original $40 \times 40 \times 12$ sector model.

15.6.5 General Advice and Simple Guidelines

Through the various examples in this chapter, I have tried (and hopefully succeeded) to convince you that accurate and robust upscaling is a challenging problem and that there is no single upscaling method that is unequivocally better than others. This is one motivation for the big attention so-called multiscale methods [95, 195] have received in recent years. These methods offer a promise of a more systematic and consistent way of bringing the impact of small-scale heterogeneity variations into simulations on a coarser scale.

What is the best upscaling method for a specific project will depend on many factors. The most important factor is probably what you intend to use the upscaled model for. If you are upscaling for a specific scenario, I strongly recommended that you use a method employing specific global information as this generally will enable you to perform more aggressive coarsening. If you, on the other hand, want to use the upscaled model to simulate a large variety of flow scenarios, or some yet unspecified scenario, you should pick a technique that is as robust as possible. Which technique to pick will depend on the type of heterogeneity you are facing, the flow patterns that exist in a reservoir, and so on. For relatively homogeneous reservoirs with flow patterns that mainly follow the axial directions, a simple averaging technique may be sufficient, whereas techniques that utilize some kind of global information are recommended for highly heterogeneous reservoirs with strong contrasts and large variations in correlation lengths. Other determining factors include the time and the computer resources you have available for upscaling, the number of petrophysical realizations you need to upscale, the required upscaling factor, and so on.

Regardless of your situation, I generally recommended that you try different upscaling and coarsening methods. Start by simple averaging methods that have a low computational cost, since these not only give you a first quick estimate of the upscaling uncertainty (or more generally the uncertainty in the petrophysical properties), but will also provide upper and lower bounds for more sophisticated upscaling methods. Check to see if there are features in your model that need to be resolved by adapting the coarse grid, and ensure that you do not create a coarse grid that gives problems for subsequent flow simulations; see e.g., the discussion in Section 14.7. Then you can gradually move towards more sophisticated upscaling methods. For local methods, you should try both sealing and linear pressure boundaries to provide lower and upper bounds. You may also need to check the use of oversampling methods to lessen the impact of boundary conditions used for localization.

At all stages, you should use representative single-phase simulations to validate your upscaled model against the original fine-scale model. To this end, you should compare flow fields computed both with TPFA and a consistent method to get an idea of the amount of grid-orientation effects. Likewise, I strongly recommend that you use various kinds of flow diagnostics like the well-allocation factors discussed in the examples in this chapter, partition functions that give you the volumetric connections between inflow and outflow, time-of-flight that gives time lines for displacement fronts, and so on. Herein, we have computed these quantities using finite-volume discretizations that are available in MRST, but these flow diagnostic measures can also be computed by streamline simulation [79], which is generally a very efficient tool for comparing geomodels and upscaled simulation models. Whether you choose one or the other is not that important. What is important is that you stick to using single-phase flow physics to avoid mixing in the effect of multiphase flow parameters, such as variations in relative permeabilities and capillary pressures, which need to be upscaled by other means.

Appendix

The MATLAB Reservoir Simulation Toolbox

Practical computer modeling of porous media constitutes an important part of the book and is presented through a series of examples that are intermingled with more traditional textbook material. All examples discussed in the book rely on the MATLAB Reservoir Simulation Toolbox (MRST), which is a free, open-source software that can be used for any purpose under the GNU General Public License version 3 (GPLv3).

The toolbox is primarily developed by my research group at SINTEF, which is the fourth-largest contract research organization in Europe. The software started out as a research code developed to study consistent discretization and multiscale solvers for incompressible two-phase flow on stratigraphic and unstructured polyhedral grids. Over the past 6–7 years the software has been applied to research a wide spectrum of other problems related to reservoir modeling, and, as a result, the software today has many of the same capabilities that can be found in commercial reservoir simulators. In addition, it contains a spectrum of new research ideas; some of these were discussed in Chapters 13 and 14. At least for the time being, our main motivation for continuing to maintain and develop the software is to have a versatile research platform that enables researchers at SINTEF to rapidly develop proof-of-concept implementations of new ideas and then subsequently, and rather effortlessly, turn these into software prototypes that could be verified and validated for problems of industry-standard complexity with regard to flow physics and geological description of geology. A second purpose is to support the idea of replicable/reproducible research, as well as to enable us to effectively leverage results from one research project in another.

Over the years, we have also seen that MRST is an efficient teaching tool and a good platform for disseminating new ideas. Maintaining and developing the software as a reliable community code takes a considerable effort, but by carefully documenting and releasing our research software as free, open source, we hope to contribute to simplifying the experimental programming of other researchers in the field and give a head start to students about to embark on a master's or PhD project.

A.1 Getting Started with the Software

This Appendix provides you with a brief overview of the software and the philosophy underlying its design. I show you how to obtain and install the software and explain its terms of use, as well as how we recommend that you use the software as a companion to the textbook. I also briefly discuss how you can use a scripted, numerical programming environment like MATLAB (or its open-source clone GNU Octave) to increase the productivity of your experimental programming and share examples of tricks and ways of working with MATLAB that we have found particularly useful. I end the chapter by introducing you to automatic differentiation, which is one of the key aspects that make MRST a powerful tool for rapid prototyping and enable us to write compact and quite self-explanatory codes that are well suited for pedagogical purposes. As a complement to the material presented in this chapter, you should also consult the first of two JOLTS (short online learning modules), developed in collaboration with Stanford University [188]. The JOLT gives a brief overview of the software, shows the way it looked a few years ago, tells you why and how it was created, and instructs you how to download and install it on your computer. If you are not interested in programming at all, you need not read this Appendix. However, if you choose to not work with the software alongside the textbook material, be warned, you will miss a lot of valuable insight.

A.1.1 Core Functionality and Add-on Modules

MRST is a research tool whose aim is to support research on modeling and simulation of flow in porous media. The software contains a wide variety of mathematical models, computational methods, plotting tools, and utility routines that extend MATLAB in the direction of reservoir simulation. To make the software as flexible as possible, it is organized quite similar to MATLAB and consists of a collection of core routines and a set of add-on modules, as illustrated in Figure A.1. The material presented in Part I of the book relies almost entirely on general core functionality, which includes routines and data structures for creating and manipulating grids, petrophysical data, and global drive mechanisms such as gravity, boundary conditions, source terms, and wells. The core functionality also contains an implementation of automatic differentiation – you write the formulas and specify the independent variables, the software computes the corresponding derivatives or Jacobians – based on operator overloading (see Section A.5), as well as a few routines for plotting cell and face data defined over a grid. The core functionality is considered to be stable and not expected to change significantly in future releases.

To minimize maintenance costs and increase flexibility, MRST core does not contain flow equations, discretizations, and solvers; these are implemented in various add-on modules. If you have read Chapter 1, you have already encountered the `incompTPFA` solver from the `incomp` module in Section 1.4; this module implements fluid behavior and standard solvers for incompressible, immiscible, single-phase and two-phase flow. The mathematical models, discretizations, and solution techniques underlying this module

A.1 Getting Started with the Software 599

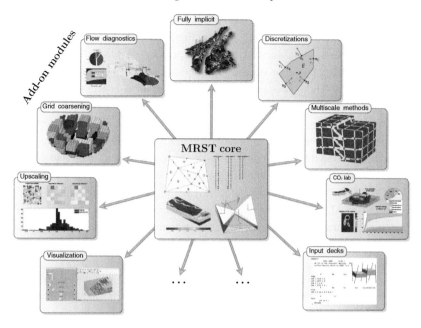

Figure A.1 MRST consists of core functionality that provides basic data structures and utility functions, and a set of add-on modules that offer discretizations and solvers, simulators for incompressible and compressible flow, and various workflow tools such as flow diagnostics, grid coarsening, upscaling, visualization of simulation output, and so on.

are extensively discussed in Parts II and III of the book. In particular, Chapters 5 and 10 outline key functionality offered in the module and discuss in detail how the solvers are implemented. The standard solvers are unfortunately not unconditionally consistent and may therefore exhibit strong grid-orientation errors and fail to converge. To get a convergent scheme, one can use one of the methods from the `mimetic` and `mpfa` modules discussed in Chapter 6, which offer consistent discretizations on general polyhedral grids [192].

Solvers for incompressible flow have been part of the software since the beginning and constitute the first family of add-on modules signified by the "Discretizations" block in Figure A.1. The `incomp`, `mimetic`, and `mpfa` modules are all implemented using a procedural (imperative) programming model from classical MATLAB, i.e., using mathematical functions that operate mainly on vectors, (sparse) matrices, structures, and a few cell arrays. These are generally robust and well documented, have remained stable over many years, and will not likely change significantly in future releases. The family of incompressible flow solvers also includes a few other discretization methods from the research front like virtual element methods and are intimately connected with two of the modules implementing multiscale methods (MsFV and MsMFE). These are not discussed herein; you can find details in the software itself, on the MRST website, or in the many papers using the software.

The second family of modules consists of simulators based on automatic differentiation and is illustrated by the "Fully implicit" block in Figure A.1. The introduction of automatic differentiation has been a big success and has not only enabled efficient development of black-oil simulators, but also opened up for a unprecedented capabilities for rapid prototyping [170, 30]. The process of writing new simulators is hugely simplified by the fact that you no longer need to compute analytic expressions for derivatives and Jacobians. This is discussed in Chapters 7 and 11. Initially, we implemented solvers based on automatic differentiation using an imperative programming model similar to the incomp family of modules. Soon it became obvious that this was a limiting factor, and a new object-oriented programming model, implementing a general framework for solvers referred to as MRST AD-OO [170, 212, 30], was introduced. The individual modules and the underlying implementations of AD-OO underwent significant changes in the period 2014–2016. From release 2016a and onward, however, the basic functionality has remained largely unchanged and has mainly been subject to bug fixes, feature enhancements, and performance improvements.

The ad-core module is the most basic part in MRST AD-OO and does not contain any complete simulators, but rather implements the common framework used for many other modules. The design of this framework deviates significantly from that of the incompressible solvers. The main motivation for introducing an object-oriented framework is to be able to simulate compressible multiphase models of industry-standard complexity. Chapters 11 and 12 only discuss how to simulate compressible multiphase models of black-oil type, but the software also offers simulators for compositional flow. All these multiphase models are significantly more complex to simulate than the basic incompressible models implemented in the incomp module. Not only are there more equations and more complex parameters and constitutive relationships, but making robust simulators also necessitates more sophisticated solution algorithms involving nonlinear solvers, preconditioners, time-step control, variable switching, etc. Moreover, industry-standard simulations generally require lots of bells and whistles to implement specific fluid behavior, well models, and group controls. A robust implementation also requires a number of subtle tricks of the trade to ensure that your simulator is able to reproduce results of leading commercial simulators. Object-orientation enables us to divide the implementation into different numerical contexts (mathematical model, nonlinear solver, time-step control, linearization, linear solver, etc.), hide unnecessary details, and only expose the details that are necessary within each specific context.

The third family of modules consists of tools that can be used as part of the reservoir modeling workflow. In Part IV, we go through the three types of tools shown in Figure A.1. "Diagnostics" signifies a family of computational methods for determining volumetric connections in the reservoir, computing well-allocation factors, measuring dynamic heterogeneity, providing simplified recovery estimates, etc. Tools from the "Grid coarsening" and "Upscaling" modules can be used to develop reduced simulation models with fewer degrees of freedom and hence lower computational costs. MRST also offers several other modules of the same type, e.g., history matching and production optimization, but these are outside the scope of the book.

The fourth family of modules consists of computational methods that have been developed to study a special problem. In Figure A.1, this is exemplified by the "CO_2 lab" module, which is a comprehensive collection of computational methods and modeling tools developed especially to study the injection and long-term migration of CO_2 in large aquifer systems. Other modules of the same type include solvers for geomechanics and various modeling frameworks and simulators for fractured media. These modules are all outside the scope of this book.

The fifth family consists of a variety of utility modules that offer graphical interfaces and advanced visualization, more comprehensive routines for reading and processing simulation models and other input data, C-acceleration of selected routines from the core module to avoid computational bottlenecks, etc. You will encounter functionality from several of these modules throughout the book, but the modules themselves will not be discussed in any detail. Last, but not least, there are also modules developed by researchers not employed by SINTEF.

A.1.2 Downloading and Installing

The main parts of MRST are hosted as a collection of software repositories on Bitbucket. Official releases are provided as self-contained archive files that can be downloaded from the website: www.mrst.no/

Assume now that you have downloaded the tarball of one of the recent releases; here, we use release 2016b as an example. Issuing the following command

```
untar mrst-2016b.tar.gz
```

in MATLAB creates a new folder `mrst-2016b` in your current working director that contains all parts of the software. Once all code has been extracted to some folder, which we henceforth refer to as the *MRST root* folder, you must navigate MATLAB there, either using the built-in file browser, or by using the `cd` command. Assuming that the files were extracted to the home folder, this would amount to the following on Linux/Mac OS:

```
cd /home/username/mrst-2016b/        % on Linux/Mac OS
cd C:\Users\username\mrst-2016b\     % on Windows
```

To activate the software, you must make sure that the MRST root folder is on MATLAB's search path. This is done by use of a startup script, which also scans your installation and determines which modules you have installed. When you are in the folder that contains the software, you the software is activated by the command:

```
startup;
```

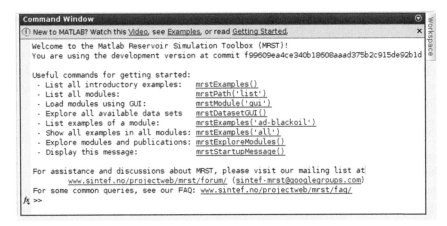

Figure A.2 The welcome message displayed when the `startup` script is run in release 2016a and later. The careful reader may notice that the user runs a development version of the software and not one of the official releases.

In MRST 2016a and newer, the startup script will display a welcome message showing that the software is initialized and ready to use; see Figure A.2. The whole procedure of downloading and installing the software, step by step, can be seen in the first MRST Jolt [188] (uses release 2014b).

At this point, a word of caution is probably in order. We generally refer to the software as a toolbox. By this we mean that it is a collection of data structures and routines that can be used alongside with MATLAB. It is, however, *not* a toolbox in the same sense as those purchased from the official vendors of MATLAB. This means, for instance, that MRST is not automatically loaded unless you start MATLAB from the MRST root folder, or make this folder your standing folder and manually issue the `startup` command. Alternatively, if you do not want to navigate to the root folder, for instance in an automated script, you can call `startup` directly

```
run /home/username/mrst-2016b/startup    % or C:\MyPath\mrst-2016b\startup
```

In versions prior to MRST 2016a, the startup script only sets up the global search path so that MATLAB is able to locate MRST's core functionality and the various modules. To verify that the software is working, you can run the simple example discussed in Section 1.4 by typing `flowSolverTutorial1`. This should produce the same plot as in Figure 1.4.

A.1.3 Exploring Functionality and Getting Help

The welcome message shown in Figure A.2 contains links to a number of functions that are useful if you want to get more acquainted with the software. Upon your first encounter,

A.1 Getting Started with the Software

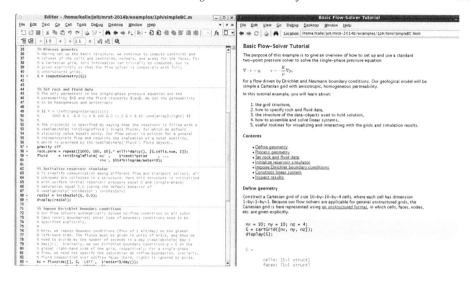

Figure A.3 Illustration of the MATLAB workbook concept. The editor window shows part of the source code for the `simpleBC` tutorial from release 2014b. (In newer versions, this tutorial has been renamed `incompIntro` and moved to the new `incomp` module.) Notice how cells are separated by horizontal lines and how each cell has a header and a text that describes what the cell does. The exception is the first cell, which summarizes the content of the whole tutorial. The right window shows the result of publishing the workbook as a webpage.

the best point to start is by typing (or clicking the corresponding blue text in the welcome message)

```
mrstExamples()
```

This will list all introductory examples found in MRST core. Some of these examples introduce you to basic functionality, whereas others highlight capabilities found in various add-on modules that implement specific discretizations, solvers, and workflow tools. These examples are designed using *cell-mode scripts*, which can be seen as a type of "MATLAB workbook" that enables you to break a script into smaller code sections (code cells) that can be run individually to perform a specific task such as creating parts of a model or making an illustrative plot; see Figure A.3 for an illustration.

In my opinion, the best way to understand tutorial examples is to go through the corresponding scripts, evaluating one section at the time. Alternatively, you can set a breakpoint on the first line and step through the script in debug mode, e.g., as shown in the fourth video of the first MRST Jolt [188]. Some of the example scripts contain quite a lot of text and are designed to be published as HTML documents, as shown to the right in Figure A.3. If you are not familiar with cell-mode scripts or debug mode, I strongly urge you to learn these useful features in MATLAB as soon as possible.

```
>> help computeTrans
 Compute transmissibilities.

  SYNOPSIS:
    T = computeTrans(G, rock)
    T = computeTrans(G, rock, 'pn', pv, ...)

  PARAMETERS:
    G    - Grid structure as described by grid_structure.

    rock - Rock data structure with valid field 'perm'.  The permeability
           is assumed to be in measured in units of metres squared (m^2).
           Use function 'darcy' to convert from darcies to m^2, e.g.,
               perm = convertFrom(perm, milli*darcy)
           if the permeability is provided in units of millidarcies.
            :
            :
  RETURNS:
    T - half-transmissibilities for each local face of each grid cell
        in the grid.  The number of half-transmissibilities equals
        the number of rows in G.cells.faces.

  COMMENTS:
    PLEASE NOTE: Face normals are assumed to have length equal to
    the corresponding face areas. ..

  SEE ALSO:
    computeGeometry, computeMimeticIP, darcy, permTensor.
```

Figure A.4 Most functions in MRST are documented in a standard format that gives a one-line summary of what the function does, specifies the synopsis (e.g., how the function should be called), explains input and output parameters, and points to related functions.

To find the syntax and understand what a specific function does, you can type

```
help computeTrans
```

which will bring up the documentation for `computeTrans` shown in Figure A.4. All core functionality is well documented in a format that follows the MATLAB standard. Functionality you are meant to utilize from any of the add-on modules should also, as a rule, be well documented. However, the format and quality of documentation may differ more, depending upon who developed the module and how mature and widely used it is. Starting with release 2017b, all documentation has been formatted so that it can be auto-extracted with Sphinx, e.g., as HTML posted on the MRST website (see Figure A.5).

As a general rule, all modules distributed as part of the MRST release are required to contain worked tutorials highlighting key functionality that most users should understand. A subset of these tutorials is also available on the MRST website. To list tutorial examples found in individual modules, you can use the `mrstExamples` command, which can also list all the tutorial examples the software offers,

```
mrstExamples('ad-blackoil')    % all examples in the black-oil module
mrstExamples('all')            % all examples across all available modules
```

To learn more about the different modules and get a full overview of all available functionality, you can type the command

A.1 Getting Started with the Software

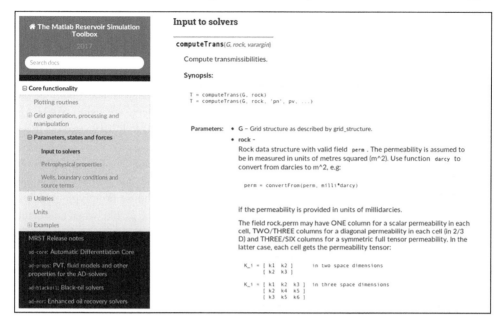

Figure A.5 Auto-generated documentation of all function, as well as all classes from the AD-OO framework, can also be found on the MRST website.

```
mrstExploreModules()
```

This brings up a graphical user interface that lists all accessible modules and outlines their purpose and functionality, including a short description of all tutorial examples found in each module, as well as a list of relevant scientific publications, with possibility to view online versions and export citations to BibTeX. Section A.3 gives a brief description of the majority of the modules.

MRST also offers simplified access to a number of public data sets. You can use the following graphical user interface to list all data sets that are known to MRST

```
mrstDatasetGUI()
```

The GUI briefly describes each data set, provides functionality for downloading and installing it in a standard location, lists the files it contains as well as the tutorial examples in which it is used. More details are given in Section A.2.

Last, but not least, a number of common queries have been listed on MRST's FAQ page: www.sintef.no/projectweb/mrst/faq/ For further assistance and discussions you may visit the user forum (www.sintef.no/projectweb/mrst/forum/) and/or subscribe to the mailing list (sintef-mrst@googlegroups.com). As you get more experience with the software, I encourage you to help others out by answering questions and contributing to discussions.

A.1.4 Release Policy and Version Numbers

Over the last few years, key parts of the software have become relatively mature and well tested. This has enabled a stable biannual release policy with one release in the spring and one in the fall. The version number of MRST refers to the biannual release schedule and does not imply a direct compatibility with the same release number for MATLAB. That is, you do not need to use MRST 2016b if you are using MATLAB R2016b, or vice versa, you do not need to upgrade your MATLAB to R2019a to use MRST 2019a.

Throughout the releases, basic functionality like grid structures has remained largely unchanged, except for occasional and inevitable bug fixes, and our primary focus has been on expanding functionality by maturing and releasing in-house prototype modules. Fundamental changes will nevertheless occur from time to time, e.g., like when automatic differentiation was introduced in 2012 and when its successor, the AD-OO framework, was introduced in 2014b. Likewise, parts of the software may sometimes be reorganized, like when the basic incompressible solvers were taken out of the core functionality and put in a separate module in 2015a. In writing this, I (regretfully) acknowledge the fact that specific code details and examples in books describing evolving software tend to become somewhat outdated. To countermand this, complete codes for almost all examples presented in the book are contained in a separate `book` module that accompanies 2015a and later releases. These are part of the prerelease test suite and should thus always be up to date.

A.1.5 Software Requirements and Backward Compatibility

MRST was originally implemented so that the minimum requirement should be MATLAB version 7.4 (R2007a). However, certain parts of the software use features that were not present in R2007a:

- The AD libraries use new-style classes (`classdef`) introduced in R2008a.
- Various scripts use new syntax for random numbers introduced in R2007b.
- Use of the tilde operator to ignore return values (e.g., `[~,i]=max(X,1)`) was introduced in R2009b.
- Some routines, like the fully implicit simulators for black-oil models, rely on accessing sub-blocks of large sparse matrices. Although these routines will run on any version from R2007a and onward, they may not be efficient on versions older than R2011b.

When it comes to visualization, things are a bit more complicated, since MATLAB 3D graphics does not behave exactly the same on all platforms. Moreover, MATLAB introduced new handle graphics in R2014b, which has been criticized by many because it is slow and because it breaks backward compatibility. We have tried to revise plotting in newer versions of MRST so that it works well both with the old and the new handle graphics, but every now and then you may stumble across certain tricks (e.g., setting grid lines semitransparent to make them thinner) that may not work well for your particular MATLAB version.

MRST can also be used to a certain extent with recent versions of GNU Octave, which is an open-source scientific programming language that is largely compatible with MATLAB. The procedural parts should work more or less out of the box, but you may encounter some problems with some of the (3D) plotting, which works a bit differently in GNU Octave. The 2017b release adds preliminary support also for the object-oriented AD-OO framework through the `octave` module. Unfortunately, the performance seems to be orders of magnitude behind recent versions of MATLAB, but this will hopefully improve in the future. The only parts of MRST you cannot expect to work are graphical user interfaces; their implementation is fundamentally different between MATLAB and GNU Octave.

MRST is designed to only use standard MATLAB. Nevertheless, we have found a few third-party packages and libraries to be quite useful:

- `MATLAB-BGL`: MATLAB does not yet have extensive support for graph algorithms. The MATLAB Boost Graph Library contains binaries for useful graph-traversal algorithms such as depth-first search, computation of connected components, etc. The library is freely available under the BSD License from the Mathwork File Exchange[1]. MRST has a particular module (see Section A.3) that downloads and installs this library.
- `METIS` is a widely used library for partitioning graphs, partitioning finite element meshes, and producing fill reducing orderings for sparse matrices [154]. The library is released[2] under a permissive Apache License 2.0.
- `AGMG`: For large problems, the linear solvers available in MATLAB are not always sufficient, and it may be necessary to use an iterative algebraic multigrid method. AGMG [238] has MATLAB bindings included and was originally published as free open-source. The latest releases have unfortunately only offered free licenses for academic research and teaching[3].
- `AMGCL` is a header-only library for preconditioning with and without algebraic multigrid[4].

`MATLAB-BGL` is required by several of the more advanced solvers that are not part of the basic functionality in MRST. Installing the other packages is recommended but not required. When installing extra libraries or third-party toolboxes you want to integrate with MRST, you must make the software aware of them. To this end, you should add a new script called `startup_user.m` and use the built-in command `mrstPath` to make sure that the routines you want to use are on the search path used by MRST and MATLAB to find functions and scripts.

---

[1] http://www.mathworks.com/matlabcentral/fileexchange/10922
[2] http://glaros.dtc.umn.edu/gkhome/metis/metis/overview
[3] http://homepages.ulb.ac.be/~ynotay/AGMG/
[4] http://github.com/ddemidov/amgcl

A.1.6 Terms of Usage

MRST is distributed as free, open-source software under the GNU Public License (GPLv3).[5] This license is a widely used example of a so-called copyleft license that offers the right to distribute copies and modified versions of copyrighted creative work, provided the same rights are preserved in modified and extended versions of the work. For MRST, this means that you can use the software for any purpose, share it with anybody, modify it to suit your needs, and share the changes you make. However, if you share any version of the software, modified or unmodified, you must grant others the same rights to distribute and modify it as in the original version. By distributing it as free software under the GPLv3 license, the developers of MRST have made sure that the software will stay free, no matter who changes or distributes it.

The development of the toolbox has to a large extent been funded by strategic research grants awarded from the Research Council of Norway. Dissemination of research results is an important evaluation criterion for these types of research grants. To provide the developers with an overview of some usage statistics for the software, we ask you to kindly register your affiliation/country upon download. This information is only used when reporting impact of the creative work to agencies co-funding its development. If you also leave an email address, we will notify you when a new release or critical bug fixes are available. Your email address will under no circumstances be shared with any third party.

If you use MRST in any scholar work, we require that the creative work of the MRST developers is courteously and properly acknowledged by referring to the MRST website and by citing this book or one of the overview papers describing the software [192, 170, 30]. Last, but not least, if you have modified parts of the software or added new functionality, I strongly encourage you to release your code and thus pay back to the community that has developed it; if not the whole code, then at least generic parts that could be of significant interest to others.

COMPUTER EXERCISES

A.1.1 Download and install the software

A.1.2 Run `flowSolverTutorial1` from the command line to verify that your installation is working.

A.1.3 Load `flowSolverTutorial1` in the editor (`editflowSolverTutorial1.m`) and run it in cell mode to evaluate one cell at the time. Use `help` or `doc` to inspect the documentation for the various functions used in the script.

A.1.4 Run `flowSolverTutorial1` line by line: Set a breakpoint on the first executable line by clicking on the small symbol next to line 27, push the "play" button, and then use the "step" button to advance a single line at the time. Change the grid size to $10 \times 10 \times 25$ and rerun the example.

---

[5] See www.gnu.org/licenses/gpl.html for more details.

A.1.5 Use `mrstExploreModules()` to locate and load `incompIntro` from the `incomp` module. Run the tutorial line by line or cell by cell, and then publish the workbook to reproduce the contents of Figure A.3.

A.1.6 Replace the constant permeability in the `incompIntro` tutorial by a random permeability field

```
rock.perm = logNormLayers(G.cartDims,[100 10 100])*milli*darcy;
```

Can you explain the changes in the pressure field?

A.1.7 Run all of the examples listed by `mrstExamples()` that have the word "tutorial" in their names. In particular,

- `gridTutorialIntro` introduces you quickly to the most fundamental parts of MRST, the grid structure, which is discussed in more detail in Chapter 3;
- `tutorialPlotting` introduces you to various basic routines and techniques for plotting grids and data defined over these;
- `tutorialBasicObjects` will give you a quick overview of a lot of the functionality that can be found in the toolbox.

A.2 Public Data Sets and Test Cases

Good data sets are in our experience essential to enable tests of new computational methods in a realistic and relevant setting. Such data sets are hard to come by, and when making MRST we have made an effort to provide simple access to a number of public data sets that can be freely downloaded. With the exception of a few illustrations in Chapter 3, which are based on data that cannot be publicly disclosed, all examples discussed in the book either use MRST to create their input data or rely on public data sets that can be downloaded freely from the internet.

To simplify dataset management, MRST offers a graphical user interface,

```
mrstDatasetGUI()
```

that lists all public data sets known to the software, gives a short description of each, and can download and unpack most data sets to the correct subfolder of the standard path. A few data sets require you to register your email address or fill in a license form, and in these cases we provide a link to the correct webpage. Figure A.6 shows some of the data sets that are available via the graphical interface. Several of these are used throughout the book.

Herein, I use the convention that data sets are stored in subfolders of a standard path, which you can retrieve by issuing the query

```
mrstDataDirectory()
```

I recommend that you adhere to this convention when using the software as a supplement to the book. If you insist on placing standard data sets elsewhere, I suggest that you use `mrstDataDirectory(<path>)` to reset the default path.

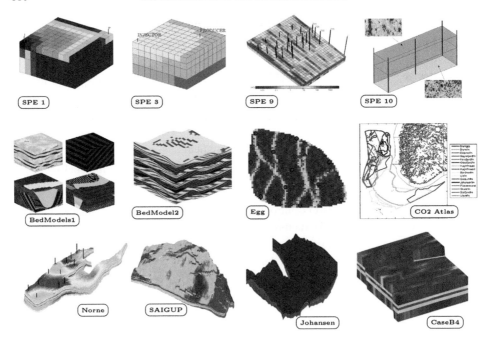

Figure A.6 Examples of the free data sets that are distributed along with MRST.

MRST also contains a number of grid factory routines and simplified geostatistical algorithms that you can use to make your own test cases. These will be discussed in more detail in Chapters 3 and 2, respectively. Here, a word of caution about exact reproducibility is in order. The grid-factory routines are mostly deterministic and should enable you to create the exact same grids each time your run them (and hence reproduce the test cases discussed in the book). Routines for generating petrophysical data rely on random numbers and will not give the same results in repeated runs. Hence, you can only expect to reproduce plots and numbers that are qualitatively similar whenever these are used, unless you make sure to store and reset the seed used by MATLAB's random number generator.

A.3 More About Modules and Advanced Functionality

As you will recall from Section A.1.1, MRST consists of core functionality and a set of add-on modules that extend, complement, and override basic features, typically in the form of specialized or more advanced solvers and workflow tools like upscaling, grid coarsening, etc. This section explains how to load and manage different modules, and tries to explain the basic characteristics of modules, or in other words the design criteria you can apply if you intend to develop your own module. For completeness, we also provide a brief overview of the comprehensive set of modules that currently are in the official release. Most of these modules are not discussed at all in the book, which for brevity needs to focus on basic flow models and modules offering standard functionality of a general interest to a wide audience.

A.3.1 Operating the Module System

The module system is a simple facility for extending and modifying the feature set. Specifically, the module system enables on-demand activation and deactivation of advanced features that are not part of the core functionality. It consists of two parts: one that handles mapping of module names to system paths and one that uses this mapping to manipulate MATLAB's search path. The search path is a list of folders used by MATLAB to locate files. A module in MRST is strictly speaking a collection of functions, object declarations, and example scripts located in a folder. Each module needs to have a unique name and reside in a different folder than other modules. When you activate a module, the corresponding folder is added to the search path, and when you deactivate, the folder is removed from the search path.

The mapping between module names and paths on your computer is maintained by the function `mrstPath`. The paths are expected to be full paths to existing folders on your computer. To determine which modules are part of your current installation, you use the function as a command

```
mrstPath
```

This will list all modules the software is aware of and can activate. To activate a particular module, you use the function `mrstModule`. As an example, calling

```
mrstModule add mimetic mpfa
```

will load the two modules for consistent discretization methods discussed in Chapter 6. The `mrstModule` function has the following additional modes of operation:

```
mrstModule <list|clear|reset|gui> [module list]
```

which will list all active modules, deactivate all modules, deactivate all modules except for those in the list, or bring up a graphical user interface with check-boxes that enable you to activate and deactivate individual modules. For the latter, you can also use the command `moduleGUI`.

All modules that come bundled with the official release will be placed in a predefined folder structure, so that they can be added automatically to the module mapping by the `startup` function. However, module folders can in principle be placed anywhere on your computer as long as they are readable. To make MRST aware of these additional modules, you must use `mrstPath` to register them in the mapping. Assuming, for instance, that you want to make the AGMG multigrid solver [238] available. This can be done as follows,

```
mrstPath register AGMG S:\mrst\modules\3rdparty\agmg
```

Once the mapping is established, the module can be activated

```
mrstModule add AGMG
```

In my experience, the best way to register modules that are not part of the default mapping that comes with the official versions of the software is to add appropriate lines to your `startup_user` file. The following is an excerpt from mine

```
mrstPath reregister distmesh ...
   /home/kalie/matlab/mrst-bitbucket/mrst-core/utils/3rdparty/distmesh
```

which adds the mesh generator `DistMesh` [248] as a module in MRST.

A.3.2 What Characterizes a Module?

MRST does not have a strict policy for what becomes a module and what does not, and if you look at the modules that are part of the official release, you will see that they differ quite a lot. Some modules are just small pieces of code that add specialized capabilities like the `mpfa` module, whereas others add comprehensive new functionality such as the `incomp`, `ad-core`/`ad-blackoil`, and `co2lab` modules. Some of these modules are robust, well-documented, and contain features that will likely not change in future releases. As such, they could have been included as part of the core functionality if we had not decided to keep this as small as possible to simplify maintenance and reduce the potential for feature conflicts. Other modules are constantly changing to support ongoing research. Until recently, all modules in the AD-OO family were of this type.

Semi-independent modules is a simple way to organize software development that promotes software reuse. By organizing a new development as a separate module you will probably be more careful to distinguish code of generic value from code that is case specific or of temporary value only. The fact that others, or your future self, may want to reuse your code can motivate the extra effort needed to document and make examples and tutorials, whose presence often is decisive when people consider to use or continue to develop the functionality you have implemented. The module concept is particularly convenient if you have specific functionality you want to activate or deactivate as you like. In-house, our team has many modules that are in varying degrees of development and/or decay. Some are accessible to the whole team, whereas others reside on one person's computer only.

You may say that any module that is part of the official release is just some code that has been organized in a certain way and released to others. However, common for all such modules is that they attempt to adhere to the following recommended design rules:

- A module should offer new functionality that is distinctly different from what is already available in the core functionality and/or in other modules.

- A module should distinguish between functionality exposed to the users of the module and functionality that is only used internally. The latter can be put in a folder called

`private` (so that it is not accessible to functions other than those in the parent folder) or some other folder that signals that this functionality is for internal use.
- A module should contain a set of tutorials/examples that explain and highlight basic functionality of the module. The examples should be as self-explanatory as possible and at most take a few minutes to run through. Examples are easier to comprehend if you section your scripts using cell mode and accompany each code section by an informative text that explains the computations to take place or discusses visual output. If special data sets are required, these should be published along with the module.
- All main routines in a module should be documented, preferably following the format used elsewhere in the software, describing input and output data, the underlying method, and assumptions and limitations.
- Modules should, as a general rule, not use functionality from the official toolboxes that are sold with MATLAB since many users do not have access to these.
- Name conflicts should be minimized to avoid messing up the search path in MATLAB. When two files with the same name appear on the search path, MATLAB will pick the one found nearest the top. To avoid potential unintended side effects, it is therefore important that files have unique names across different modules.
- To the extent possible, the implementation should try to stick to the same naming conventions as used in MRST for readability. This means using `camelCaseNames` for functions and `CamelCase` starting with a capital letter for class objects. Likewise, widely used data objects can preferably use easily recognizable names like `G` for grid, `rock` for petrophysics, `fluid` for fluid models, `W` for wells, etc.
- Preferably, the source code should be kept in a public (or private) repository on a centralized service like Bitbucket or GitHub so that you can use a version control system to keep track of its development.

I have recently written a paper [186] that discusses development of open-source software and good practices for experimental scientific programming in more detail.

A.3.3 List of Modules

This section outlines the various modules that make up the official MRST 2018a release and explains briefly the purpose and key features of each individual module.

Grid Generation and Partitioning

The core functionality of MRST includes a general data structure for fully unstructured grids, as well as a number of grid factory and utility routines, including the possibility to read and construct corner-point grids from ECLIPSE input. In addition, there are a number of add-on modules for generating and processing grids:

`agglom`: Offers a number of elementary routines that can be combined in various ways to create coarse partitions that adapt to geology or flow fields [125, 124, 194, 187] based on cell-based indicator values. More details are given in Chapter 14 and in the additional tutorial examples.

coarsegrid: Extends the unstructured grid format in MRST to also include coarse grids formed as a partition of an underlying fine grid. Such grids form a key part in upscaling and multiscale methods, but act almost like any standard MRST grid and can hence be passed to many solvers in other modules. Coarse grid generated from a partition are also useful for visualization purposes. More details are given in Chapter 14.

libgeometry: Geometric quantities like cell volumes and centroids and areas, centroids, and normals of faces must be computed by the `computeGeometry` routine. This module offers a C-accelerated version, `mcomputeGeometry`, that reduces the computational time substantially for large models. Since 2016a, the need for this C-acceleration has diminished.

opm_gridprocessing: Corner-point grids can be constructed from ECLIPSE input by the `processGRDECL` function, which is part of the core functionality in MRST. This module offers a C-accelerated version, `mprocessGRDECL` or `processgrid`, of the same processing routines. The implementation is from the Open Porous Media (OPM) initiative, which generally can be seen as a C/C++ sibling of MRST.

triangle: Provides a mex interface to `Triangle`, a two-dimensional quality mesh generator and Delaunay triangulator developed by Jonathan Richard Shewchuk, to generate grids for MRST.

upr: Contains functionality for generating unstructured polyhedral grids that align to prescribed geometric objects. Control-point alignment of cell centroids is introduced to accurately represent horizontal and multilateral wells, but can also be used to create volumetric representations of fracture networks. Boundary alignment of cell faces is introduced to accurately preserve geological features such as layers, fractures, faults, and/or pinchouts. This third-party module was originally developed as part of a master thesis by Berge [42]; see also [169].

Incompressible Discretizations and Solvers

This family of modules consist primarily of functionality for solving Poisson-type pressure equations:

incomp: Implements fluid objects for incompressible, two-phase flow. The module implements the two-point flux-approximation (TPFA) method and explicit and implicit transport solvers with single-point upstream mobility weighting, as described in detail in Parts II and III of the book. Originally part of the core module, but was moved to a separate module once the software started offering solvers for more advanced fluid models.

mimetic: The standard TPFA method in `incomp` is not consistent and may give significant grid-orientation errors for grids that are not K-orthogonal. The module implements a family of consistent, mimetic finite difference methods for incompressible (Poisson-type) pressure equations; see Section 6.4.

mpfa: Implements the MPFA-O scheme (see Section 6.4 on page 188), which is an example of a scheme that employs more degrees of freedom when constructing the discrete fluxes across cell interfaces to ensure a consistent discretization with reduced grid-orientation effects.

vem: Virtual element methods (VEM) [38, 39] constitute a unified framework for higher-order methods on general polygonal and polyhedral grids. The module solves general incompressible flow problems by first- and second-order VEM, with the possibility to choose different inner products. Originally developed by Klemetsdal as part of his master thesis [168].

adjoint: Implements strategies for production optimization based on adjoint formulations for incompressible, two-phase flow. Example: optimization of the net present value constrained by the bottom-hole pressure in wells. For optimization problems with more complex fluid physics, the newer optimization module from the AD-OO framework is recommended.

Implicit Solvers Based on Automatic Differentiation

The object-oriented AD-OO framework is based on automatic differentiation and offers a rich set of functionality for solving a wide class of model equations:

ad-core: does not contain any complete simulators by itself, but rather implements the common framework used for many other modules. This includes abstract classes for reservoir models, for time stepping, nonlinear solvers, linear solvers, functionality for variable updating/switching, routines for plotting well responses, etc. See the discussion in Chapters 11 and 12 and [170, 30] for more details.

deckformat: Contains functionality for handling complete simulation decks in the ECLIPSE format, including input reading, conversion to SI units, and construction of MRST objects for grids, fluids, rock properties, and wells. Functionality from this module is essential for the fully implicit simulators in the AD-OO framework; see Chapters 11 and 12.

ad-props: Functionality related to property calculations for the AD-OO framework. Specifically, the module implements a variety of test fluids and functions that are used to create fluids from external data sets. This module is used as part of almost all simulators in MRST studying compressible equations of black-oil or compositional type. More details are in Section 11.3.

ad-blackoil: Models and examples that extend the MRST AD-OO framework found in the ad-core module to black-oil problems. Solvers and functionality from this module are discussed at length in Chapters 11 and 12.

blackoil-sequential: Implements sequential solvers for the same set of equations as in ad-blackoil based on a fractional flow formulation wherein pressure and transport are solved as separate steps; see [215] for more details. These solvers can be significantly faster than those found in `ad-blackoil` for many problems, especially problems where the total velocity changes slowly during the simulation.

`ad-eor`: Builds on the `ad-core` and `ad-blackoil` modules and defines model equations and provides simulation tools for water-based enhanced oil recovery techniques (polymer and surfactant injection); see [30] for more details about the polymer case.

`compositional`: Introduced in release 2017b and offers solvers for compositional flow problems with cubic equations of state or tabulated K-values. Both the natural variables and overall composition formulations are included. More details about the underlying models and equations can be found in the PhD thesis by Møyner [212]. The tutorials include validation against commercial and external research simulators.

`solvent`: Introduced in release 2017b and offers an extension of the Todd–Longstaff model for miscible flooding [289] that makes it possible to simulate miscible and partially miscible displacement without using an expensive fully compositional formulation.

`ad-mechanics`: Introduced in release 2017b and includes a mechanical model for linear elasticity that can be coupled to standard reservoir flow models, such as oil-water or black-oil models. The global system is either solved simultaneously or by fixed-stress splitting [22]. The solver for linear elasticity uses the virtual element method and can handle general grids.

`optimization`: Routines for solving optimal control problems with (forward and adjoint) solvers based on the AD-OO framework. The module contains a quasi-Newton optimization routine using BFGS-updated Hessians, but can easily be set up to use any (non-MRST) optimization code.

`ad-fi`: Deprecated module containing our first implementation of AD-based solvers for black-oil models.

Workflow Tools

The next category of modules contains functionality that can be used either before or after a reservoir simulation as part of a general modeling workflow.

`diagnostics`: Flow diagnostics tools are run to establish volumetric connections and communication patterns in the reservoir and measure the heterogeneity of dynamic flow paths. Details are discussed in Chapter 13 and in [218, 261, 171].

`upscaling`: Implements methods and tutorials for averaging and flow-based upscaling of permeabilities and transmissibilities as discussed in Chapter 15.

`steady-state`: Functionality for upscaling relative permeabilities based on a steady-state assumption. This includes general steady state, as well as capillary and viscous-dominated limits. The functionality is demonstrated through a few tutorial examples. For more details, see [132, 131].

Multiscale Methods

Multiscale methods can either be used as a robust upscaling method to produce upscaled flow velocities and pressures on a coarse grid or as an approximate, iterative fine-scale solver. Instead of placing these modules into one of the categories discussed so far, I have given them a separate category, in part because of the prominent place they have played as a driving force for the development of MRST:

msmfe: Implements the multiscale mixed finite-element (MsMFE) method on stratigraphic and unstructured grids in 3D [68, 1, 4, 5, 6, 165, 222, 17, 194, 243]. Basis functions (prolongation operators) associated with the faces of a coarse grid are computed numerically by solving localized flow problems on pairs of coarse blocks. With these, you can define a reduced flow problem on a coarse grid and prolongate the resulting coarse solution back to the fine grid to get a mass-conservative flux field. Gives a good approximate solver and a robust and accurate alternative to upscaling for incompressible flow problems. MRST was originally developed solely to support research on MsMFE methods, but the module has nevertheless not been actively developed since 2012.

msfv: Implements the operator formulation [200] of the multiscale finite-volume (MsFV) method [146] for incompressible flow on structured and unstructured grids in 3D, under certain restrictions on the grid geometry and topology [211, 214]. Prolongation operators are defined by solving flow problems localized to dual coarse blocks and then used to define a reduced flow problem on the coarse primal grid and map unknowns computed on this grid back to the underlying fine grid. MsFV can either be used as an approximate solver or as a global preconditioner in an iterative solver. The module is not actively developed and is offered mostly for historic reasons. You should use the msrsb module instead.

msrsb: The multiscale restricted-smoothing basis (MsRSB) method [216, 215] is the current state-of-the-art within multiscale methods [195]. MsRSB is very robust and versatile and can either be used as an approximate coarse-scale solver that has mass-conservative subscale resolution, or as an iterative fine-scale solver that will provide mass-conservative solutions for any given tolerance. Performs well on incompressible 2-phase flow [216], compressible 2- and 3-phase black-oil type models [215, 130], as well as compositional models [217, 212]. MsRSB can utilize combinations of specialized and adapted prolongation operators to accelerate convergence [191]. In active research, and a more comprehensive version exists in-house.

Specialized Simulation Tools

This category consists of modules that implement solvers and simulators for other mathematical models than those discussed in this book.

co2lab: Combines results of more than one decade of academic research and development on CO_2 storage modeling into a unified toolchain that is easy and intuitive to use. The module is geared towards the study of long-term trapping in large-scale saline aquifers and offers computational methods and graphical user interfaces to visualize migration paths and compute upper theoretical bounds on structural, residual, and solubility trapping. It also offers efficient simulators based on a vertical-equilibrium formulation to analyze pressure build-up and plume migration and compute detailed trapping inventories for realistic storage scenarios. Last but not least, the module provides simplified access to publicly available data sets, e.g., from the Norwegian CO_2 Storage Atlas. For more details, see [19] or specific references documenting the general toolchain of methods [232, 193, 21], methods for identifying structural traps [230], and the various vertical-equilibrium formulations [20, 227, 228].

dfm: Contains two-point and multipoint solvers for discrete fracture-matrix systems [268, 297], including multiscale methods. This third-party module is developed by Sandve from the University of Bergen, with minor modifications by Keilegavlen.

dual-porosity: A module for geologic well-testing in naturally fractured reservoirs has been developed and is maintained by the Carbonate Reservoir Group at Heriot Watt University. The module implements tools to generate synthetic transient pressure responses for idealized and realistic fracture networks.

fvbiot: Developed and maintained by the University of Bergen and implements cell-centered discretizations of three different equations: (i) scalar elliptic equations (Darcy flow), using multipoint flux approximations; this is more or less equivalent to the MPFA implementation in the mpfa module, although the implementation and data structures are slightly different; (ii) Linear elasticity, using the multipoint stress-approximation (MPSA) method [237, 233, 156]; (iii) Poromechanics, e.g., coupling terms for the combined system of the first two models.

geochemistry: Implements solvers for the following models: aqueous speciation, surface chemistry, redox chemistry, equilibrium with gas and solid phases. Each of these models can be readily coupled with a flow solver. The module is developed as a collaboration between University Texas Austin and SINTEF.

hfm: Implements the embedded discrete fracture method (EDFM) on stratigraphic and unstructured grids in 2D and 3D. The module also implements a multiscale restriction-smoothed basis (MsRSB) solver; see [273] for more details. The module was originally developed as a collaboration between TU Delft and SINTEF. Recently, researchers from Heriot Watt University have contributed several new functions and examples. Most notably, this includes greatly improved support for fracture shapes and orientations in 3D.

vemmech: Offers functionality to set up solvers for linear elasticity problems on irregular grid, using the virtual element method [39, 114], which is a generalization of finite-element methods that takes inspiration from modern mimetic finite-difference methods; see [23, 229]

A.3 More About Modules and Advanced Functionality

Miscellaneous

The last category consists of modules that do not offer any computational or modeling tools but rather provide general utility functions employed by the other modules in MRST:

book: Contains all the scripts used for the examples, figures, and some of the exercises in this book.

linearsolvers: Offers bindings to external linear solvers. The initial release includes tentative support for the AMGCL header-only library for preconditioning with and without algebraic multigrid methods.

matlab_bgl: Routine for download the MATLAB Boost Graph Library.

mrst-gui: Graphical interfaces for interactive visualization of reservoir states and petrophysical data. The module includes additional routines for improved visualization (histograms, well data, etc.) as well as a few utility functions that enable you to override some of MATLAB's settings to enable faster 3D visualization (rotation, etc.).

octave: A number of patches to provide compatibility with GNU Octave.

spe10: Contains tools for downloading, converting, and loading the data into MRST; see Section 2.5.3. The module also features utility routines for extracting parts of the model, as well as a script that sets up a (crude) simulation of the full model (using the AGMG multigrid solver).

streamlines: Implements Pollock's method [253] for tracing of streamlines on Cartesian and curvilinear grids based on a set of input fluxes computed by the incompressible flow solvers in MRST.

wellpaths: Functionality for defining wells following curvilinear trajectories.

This list includes all public modules in release 2018a. By the time you read this book, more modules that are currently in the making will likely have been added to the official list.

Modules Not Part of the Official Release

The following modules are publicly available, but *not* part of the official release:

enkf: Ensemble Kalman filter (EnKF) module developed by researchers at TNO [182, 181] that contains EnKF and EnRML schemes, localization, inflation, asynchronous data, production and seismic data, updating of conventional and structural parameters.

remso: An optimization module based on multiple shooting, developed by Codas [87], which allows for great flexibility in the handling of nonlinear constraints in reservoir management optimization problems.

mrst-cap Implements a compositional flow model with capillary pressure; see https://github.com/sogoogos/mrst-cap.

You can also find other MRST codes online that have not been structured modules or have not been maintained for many years.

COMPUTER EXERCISES

A.3.1 Try to run the following tutorials and examples from various modules

- `simpleBCmimetic` from the `mimetic` module.
- `simpleUpscaleExample` from the `upscaling` module
- `gravityColumnMS` from the `msmfem` module
- `example2` from the `diagnostics` module
- `firstTrappingExample` from the `co2lab` module (notice that this example does not work unless you have MATLAB-BGL installed).

A.4 Rapid Prototyping Using MATLAB and MRST

How can you reduce the time span from the moment you get a new idea to when you have demonstrated that it works well for realistic reservoir engineering problems? In our experience, prototyping and validating new numerical methods is painstakingly slow. There are many reasons for this. First of all, there is often a strong disconnect between the mathematical abstractions and equations used to express models and numerical algorithms and the syntax of the computer language you use to implement your algorithms. This is particularly true for compiled languages, where you typically end up spending most of your time writing and tweaking loops that perform operations on individual members of arrays or matrices. Object-oriented languages like C++ offer powerful functionality that can be used to make abstractions that are both flexible and computationally efficient and enable you to design your algorithms using high-level mathematical constructs. However, these advanced features are usually alien and unintuitive to those who do not have extensive training in computer sciences. If you are familiar with such concepts and are in the possession of a flexible framework, you still face the never-ending frustration caused by different versions of compilers and (third-party) libraries that seems to be an integral part of working with compiled languages.

Experimental Programming is Efficient in a Scripting Language

Based on working with many different students and researchers over the past twenty years, I claim that using a numerical computing environment based on a scripting language like MATLAB/GNU Octave to prototype, test, and verify new models and computational algorithms is significantly more efficient than using a compiled language like Fortran, C, and C++. Not only is the syntax intuitive and simple, but there are many mechanisms that help you boost your productivity and you avoid some of the frustrations that come with compiled languages: there is no complicated build process or handling of external libraries, and your implementation is inherently cross-platform compatible.

MATLAB, for instance, provides mathematical abstractions for vectors and matrices and built-in functions for numerical computations, data analysis, and visualization that enable you to quickly write codes that are not only compact and readable, but also efficient and robust. On top of this, MRST provides additional functionality that has been developed especially for computational modeling of flow in porous media:

- an unstructured grid format that enables you to implement algorithms without knowing the specifics of the grid;
- discrete operators, mappings, and forms that are not tied to specific flow equations, and hence can be precomputed independently and used to write discretized flow equations in a very compact form that is close to the mathematical formulations of the same;
- automatic differentiation enables you to compute the values of gradients, Jacobians, and sensitivities of any programmed function without having to compute the necessary partial derivatives analytically; this can, in particular, be used to automate the formulation of fully implicit discretizations of time-dependent and coupled systems of equations;
- data structures providing unified access enable you to hide specific details of constitutive laws, fluid and rock properties, drive mechanisms, etc.

This functionality is gradually introduced and described in detail in the main parts of the book.

Interactive Development of New Programs

An equally important aspect of using a numerical environment like MATLAB is that you can develop your program differently than what you would do in a compiled language. Using the interactive environment, you can interactively analyze, change and extend data objects, try out each operation, include new functionality, and build your program as you go. This feature is essential in a debugging process, when you try to understand why a given numerical method fails to produce the results you except it to give. In fact, you can easily continue to change and extend your program during a test run: the debugger enables you to run the code line by line and inspect and change variables at any point. You can also easily step back and rerun parts of the code with changed parameters that may possibly change the program flow. Since MATLAB uses dynamic type-checking, you can also add new behavior and data members while executing a program. However, how to do this in practice is difficult to teach in a textbook. Instead, you should run and modify the various examples that come with MRST and the book. We also recommend that you try to solve the computer exercises that are suggested at the end of several of the chapters and sections in the book.

Ensuring Efficiency of Your MATLAB Code

Unfortunately, all this flexibility and productivity comes at a price: it is very easy to develop programs that are not very efficient. In the book, I therefore try to implicitly teach programming concepts you can use to ensure flexibility and high efficiency of your programs. These include, in particular, efficient mechanisms for traversing data structures like vectorization, indirection maps, and logical indexing, as well as use of advanced MATLAB functions like `accumarray`, `bsxfun`, etc. Although these will be presented in the context of reservoir simulation, I believe the techniques are of interest for readers working with lower-order finite-volume discretizations on general polyhedral grids.

As an illustration of the type of MATLAB programming that will be used, we can consider the following code, which generates five million random points in 3D and counts the number of points that lie inside each of the eight octants:

```
n = 5000000;
pt = randn(n,3);
I = sum(bsxfun(@times, pt>0, [1 2 4]),2)+1;
num = accumarray(I,1);
```

The third line computes the sign of the x, y, and z coordinates and maps the triplets of logical values to an integer number between 1 and 8. To count the number of points inside each octant, we use the function `accumarray` that groups elements from a data set and applies a function to each group. The default function is summation, and by setting a unit value in each element, we count the entries.

Next, let us compute the mean point inside each octant. A simple loop-based solution could be something like:

```
avg = zeros(8,3);
for j=1:size(pt,1)
   quad = sum((pt(j,:)>0).*[1 2 4])+1;
   avg(quad,:) = avg(quad,:)+pt(j,:);
end
avg = bsxfun(@rdivide, avg, num);
```

Generally, you should try to avoid explicit loops since they tend to be slow in MATLAB. On my old laptop with MATLAB R2014a, it took 0.47 seconds to count the number of points within each octant, but 24.9 seconds to compute the mean points. So let us try to be more clever and utilize vectorization. We cannot use `accumarray` since it only works for scalar values. Instead, we can build a sparse matrix that we multiply with the `pt` array to sum the coordinates of the points. The matrix has one row per octant and one column per point. Lastly, we use the indicator `I` to assign a unit value in the correct row for each column, and use `bsxfun` to divide the sum of the coordinates with the number of points inside each octant:

```
avg = bsxfun(@rdivide, sparse(I,1:n,1)*pt, accumarray(I,1));
```

On my computer this operation took 0.53 seconds, which is 50 times faster than the loop-based solution. An alternative solution is to expand each coordinate to a quadruple $(x, y, z, 1)$, multiply by the same sparse matrix, and use `bsxfun` to divide the first three columns by the fourth column to compute the average:

```
avg = sparse(I,1:n,1)*[pt, ones(n,1)];
avg = bsxfun(@rdivide, avg(:,1:end-1), avg(:,end));
```

These operations ran in 0.64 seconds. Which solution do you think is most elegant?

Hopefully, this simple example has inspired you to learn a bit more about efficient programming tricks if you do not already speak MATLAB fluently. MRST is generally full of tricks like this, and in the book we will occasionally show a few of them. However, if you really want to learn the tricks of the trade, the best way is to dig deep into the actual codes.

A.5 Automatic Differentiation in MRST

Automatic differentiation (AD) is a technique that exploits the fact that any function evaluation, regardless of complexity, can be broken down to a limited set of arithmetic operations ($+$, $-$, $*$, $/$, etc.) and evaluation of elementary functions like exp, sin, cos, log, and so on, that all have known derivatives. In AD, the key idea is to keep track of quantities and their derivatives simultaneously; every time an elementary operation is applied to a numerical quantity, the corresponding differential rule is applied to its derivative.

Implementing Automatic Differentiation

Consider a scalar primary variable x and a function $f = f(x)$. Their AD representations would then be the pairs $\langle x, 1 \rangle$ and $\langle f, df \rangle$, where 1 is the derivative dx/dx and df is the numerical value of the derivative $f'(x)$. Accordingly, the action of the elementary operations and functions must be defined for such pairs,

$$\langle f, df \rangle + \langle g, dg \rangle = \langle f + g, df + dg \rangle,$$
$$\langle f, df \rangle * \langle g, dg \rangle = \langle fg, f\,dg + df\,g \rangle,$$
$$\exp(\langle f, df \rangle) = \langle \exp(f), \exp(f) df \rangle.$$

In addition to this, we need to use the chain rule to accumulate derivatives; that is, if $f(x) = g(h(x))$, then $f_x(x) = g'(h(x))h'(x)$. If we have a function $f(x, y)$ of two primary variables x and y, their AD representations are $\langle x, 1, 0 \rangle$, $\langle y, 0, 1 \rangle$ and $\langle f, f_x, f_y \rangle$. This more or less summarizes the key idea behind AD; the remaining and difficult part is how to implement the idea as efficient computer code that has a low user-threshold and minimal computational overhead.

It is not very difficult to implement elementary rules needed to differentiate the function evaluations you find in a typical numerical PDE solver. To be useful, however, these rules should not be implemented as new functions, so that you need to write something like myPlus(a, myTimes(b,c)) when you want to evaluate $a + bc$. You can find many different AD libraries for MATLAB that offer both forward and reverse model for accumulating derivatives, e.g., ADiMat [286, 43], ADMAT [60, 300], MAD [291, 275, 112], or from MATLAB Central [110, 208]. An elegant solution is to use classes and operator overloading. When MATLAB encounters an expression a+b, the software will choose one out of several different addition functions depending on the data types of a and b. All we now have to do is introduce new addition functions for the various classes of data types that a and b may belong to. Neidinger [223] gives a nice introduction to how to implement this in MATLAB. MRST has its own implementation that is specially targeted at solving flow equations, i.e., working with long vectors and large sparse matrices.

The ADI Class in MRST

The ADI class in MRST relies on operator overloading as suggested in [223] and uses a relatively simple forward accumulation of derivatives. The ADI class differs from many other libraries in a subtle, but important way. Instead of working with a single Jacobian of the full discrete system as one matrix, *MRST uses a list of matrices that represent the derivatives with respect to different variables that will constitute sub-blocks in the Jacobian of the full system.* The reason for this choice is computational performance and user utility. In a typical simulation, we need to compute the Jacobian of systems of equations that depend on primary variables that each is a long vector. Oftentimes, we want to manipulate parts of the full Jacobian that represents specific subequations. This is not practical if the Jacobian of the system is represented as a single matrix; manipulating subsets of large sparse matrices is currently not very efficient in MATLAB, and keeping track of the necessary index sets can also be quite cumbersome from a user's point-of-view. Accordingly, our current choice is to let the MRST ADI class represent the derivatives of different primary variable as a list of matrices.

The following example illustrates how the ADI class works:

Example A.5.1 *We want to compute the expression $z = 3e^{-xy}$ and its partial derivatives $\partial z/\partial x$ and $\partial z/\partial y$ for the values $x = 1$ and $y = 2$. Using our previous notation, the AD-representation of z should be an object of the following form*

$$z = \langle 3e^{-xy}, -3ye^{-xy}, -3xe^{-xy} \rangle \approx \langle 0.4060, -0.8120, -0.4060 \rangle.$$

With the `ADI` *class, computing z and its partial derivatives is done as follows:*

```
[x,y] = initVariablesADI(1,2);
z = 3*exp(-x*y)
```

The first line tells MRST that x *and* y *are independent variables and instantiates two class objects, initialized with correct values. The second line is what you normally would write in MATLAB to evaluate the given expression. After the second line is executed, you have three ADI variables (pairs of values and derivatives):*

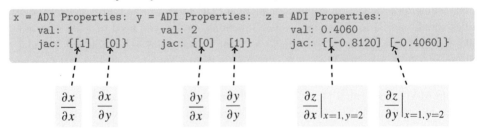

If we continue computing with these variables, each new computation will give a result that contains the value of the computation as well as the derivatives with respect to x and y.

A.5 Automatic Differentiation in MRST

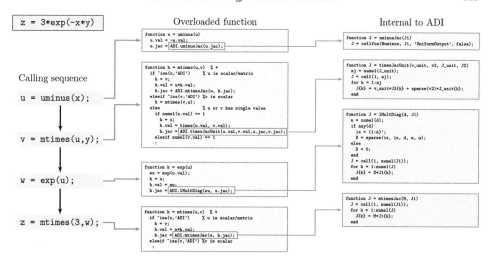

Figure A.7 Complete set of functions invoked to evaluate `3*exp(-x*y)` when x and y are ADI variables. For brevity, we have abbreviated parts of the called functions.

Let us look a bit in detail on what happens behind the curtain. We start by observing that the function evaluation 3*exp(-x*y) in reality consists of a sequence of elementary operations: −, *, exp, and *, executed in that order. In MATLAB, this corresponds to the following sequence of call to elementary functions

```
u = uminus(x);
v = mtimes(u,y);
w = exp(u);
z = mtimes(3,w);
```

To see this, you can enter the command into a file, set a breakpoint in front of the assignment to z, and use the "Step in" button to step through all details. The ADI class overloads these three functions (uminus, mtimes, and exp) by new functions that have the same names, but operate on an ADI class object for uminus and exp, and on two ADI class objects or a combination of a double and an ADI class object for mtimes. Figure A.7 gives a breakdown of the sequence of calls to overloaded and internal functions that are invoked within the ADI class library to evaluate 3*exp(-x*y) when x and y are ADI variables. The same principles apply for other functions and if x and/or y are vectors, except that the specific sequence of overloaded and internal functions from ADI will be different.

Computational Efficiency

As you can see from the example, use of AD will give rise to a whole new set of function calls that are not executed if you only evaluate a mathematical expression and do not compute its derivatives. Apart from the cost of the extra code lines that are executed, user-defined classes are fairly new in MATLAB and there is still some overhead in using class objects and accessing their properties (e.g., val and jac) compared to the built-in struct-class. The reason why using the ADI class library still pays off in most examples, is that the cost of generating derivatives is typically much smaller than the cost of the solution algorithms they will be used in, in particular when working with equations systems consisting of large sparse matrices with more than one row per cell in the computational grid. You should nevertheless still seek to limit the number of calls involving ADI class functions (including the constructor). We let the following example be a reminder that vectorization is of particular importance when using ADI classes in MRST:

Example A.5.2 *To investigate the efficiency of vectorization versus serial execution of the ADI objects in MRST, we consider the inner product of two vectors*

```
z = x.*y;
```

and compare the cost of computing z and its two partial derivatives using four different approaches:

1. *explicit expressions for the derivative, $z_x = y$ and $z_y = x$, evaluated using standard MATLAB vectors of doubles;*
2. *the overloaded vector multiply (.*) with ADI vectors for x and y;*
3. *a loop over all vector elements with matrix multiply (*=mtimes) and x and y represented as scalar ADI variables; and*
4. *same as 3, but with vector multiply (.*=times).*

MATLAB offers a stopwatch timer, which we start by the command tic *and whose elapsed time we read by the command* toc:

```
[n,t1,t2,t3,t4] = deal(zeros(m,1));
for i = 1:m
    n(i) = 2^(i-1);
    xv = rand(n(i),1); yv=rand(n(i),1);
    [x,y] = initVariablesADI(xv,yv);
    tic, z = xv.*yv; zx=yv; zy = xv;           t1(i)=toc;
    tic, z = x.*y;                              t2(i)=toc;
    tic, for k =1:n(i), z(k)=x(k)*y(k);  end;  t3(i)=toc;
    tic, for k =1:n(i), z(k)=x(k).*y(k); end;  t4(i)=toc;
end
```

Figure A.8 reports the corresponding runtimes as function of the number elements in the vector. For this simple function, using ADI is a factor 20–40 times more expensive than using direct evaluation of z and the exact expressions for z_x and z_y. Using a loop will on average be more than three orders more expensive than vectorization. Since the inner iterations multiply scalars, many programmers would implement it using matrix multiply (*) without a second thought. Replacing (*) by vector multiply (.*) reduces the cost significantly for short vectors, but the reduction diminishes as the length of the vectors increases.

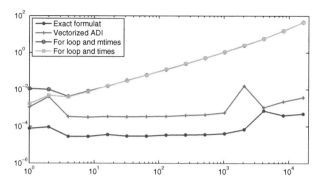

Figure A.8 Comparison of the time required for computing z=x.*y and derivatives as function of the number of elements in the vectors x and y.

Vectorization is usually significantly more efficient than loops when built-in or user-defined functions can be applied to complete vectors. The benefit of vectorization can be significantly diminished and completely disappear if the vectorized operations must allocate a lot of temporary memory and does not match the processor cache. Still, I would recommend that you try to vectorize as much of your code as possible, as this typically will lead to more readable code. As a general rule, you should always try to avoid loops that call functions that have non-negligible overhead. The ADI class in MRST has been designed to exploit vectorization on long vectors and lists of (sparse) Jacobian matrices and has not been optimized for scalar variables. As a result, there is considerable overhead when working with small ADI objects.

Using ADI to Assemble Linear(ized) Systems

The main use of ADI objects in MRST is to linearize and assemble (large) systems of discrete equations. To use ADI to assemble and solve a linear system $\mathbf{A}x = b$, we must first write the system on residual form,

$$r(x) = \mathbf{A}x - b = \mathbf{0}. \tag{A.1}$$

Since x is unknown, we assume that we have an initial guess called y. Inserting this into (A.1), we obtain

$$r(y) = \mathbf{A}y - b = \mathbf{A}(y - x),$$

which we can solve for x to obtain $x = y - \mathbf{A}^{-1}r(y)$. It follows from (A.1) that $r'(x) = \mathbf{A}$, which means that if we write a code that evaluates equations on residual form, we can use AD to assemble the system.

Example A.5.3 *As a very simple illustration of how AD can be used to assemble a linear system, let us consider the following linear 3×3 system*

$$\begin{bmatrix} 3 & 2 & -4 \\ 1 & -1 & 2 \\ -2 & -2 & 4 \end{bmatrix} \begin{bmatrix} x_1 \\ x_2 \\ x_3 \end{bmatrix} = \begin{bmatrix} -5 \\ -1 \\ 6 \end{bmatrix},$$

whose solution $x = [1\ 2\ 3]^T$ can be computed by a single line in MATLAB:

```
x = [3, 2, -4; 1, -4, 2; -2,- 2, 4]\ [-5; -1; 6]
```

To solve the same system using AD,

```
x   = initVariablesADI(zeros(3,1));
A   = [3,  2, -4; 1, -4,  2; -2,- 2.  4];
eq  = A*x + [5; 1; -6];
u   = -eq.jac{1}\eq.val
```

or we can specify the residual equations line by line and then concatenate them to assemble the whole matrix

```
x   = initVariablesADI(zeros(3,1));
eq1 = [ 3,  2, -4]*x + 5;
eq2 = [ 1, -4,  2]*x + 1;
eq3 = [-2, -2,  4]*x - 6;
eq  = cat(eq1,eq2,eq3);
u   = -eq.jac{1}\eq.val
```

Here, the first line sets up the AD variable x and initializes it to zero. The next three lines evaluate the equations that make up our linear system. Evaluating each equation results in a scalar residual value `eq1.val` and a 1×3 Jacobian matrix `eq1.jac`. In the fifth line, the residuals are concatenated to form a vector and the Jacobians are assembled into the full Jacobian matrix of the overall system.

At this point, you may argue that what I have shown you is a convoluted and expensive way of setting up and solving a simple system. However, now comes the interesting part. When solving partial differential equations on complex grids, it is often much simpler to evaluate the residual equations in each grid cell than assembling the same local equations into a global system matrix. Using AD, you can focus on the former and avoid the latter. In Section 4.4.2, we introduce discrete divergence and gradient operators, `div` and `grad`. With these, discretization of the flow equation $\nabla \cdot \mathbf{K}\mu^{-1}(\nabla p - g\rho \nabla z) = 0$ introduced in Section 1.4 can be written in a form that strongly resembles its continuous form

```
eq = div((T/mu)*.(grad(p) - g*rho*grad(z)));
```

where T is the transmissibilities (which can be precomputed for a given grid), mu and rho are constant fluid viscosity and density, g is the gravity constant, and z is the vector of cell centroids. That single line is all we need to evaluate and assemble the corresponding linear system of discrete flow equations.

A.5 Automatic Differentiation in MRST

Our primary use of AD is for compressible flow models, which typically give large systems of nonlinear discrete equations that need to be linearized and solved using a Newton–Raphson method. As a precursor to the discussion in Chapter 7, we consider a last example to outline how you can use ADI to solve a system of nonlinear equations.

Example A.5.4 *Minimizing the Rosenbrock equation*

$$f(x,y) = (a-x)^2 + b(y-x^2)^2 \qquad \text{(A.2)}$$

is a classical test problem from optimization. This problem is often called Rosenbrock's valley or banana function, since the global minimum (a, a^2) is located inside a long, narrow and relatively flat valley of parabolic shape. Finding this valley is straightforward, but it is more challenging to converge to the global minimum. A necessary condition for a global minimum is that $\nabla f(x,y) = 0$, which translates to the following two equations for (A.2)

$$g(x) = \begin{bmatrix} \partial_x f(x,y) \\ \partial_y f(x,y) \end{bmatrix} = \begin{bmatrix} -2(a-x) - 4bx(y-x^2) \\ 2b(y-x^2) \end{bmatrix} = \begin{bmatrix} 0 \\ 0 \end{bmatrix}. \qquad \text{(A.3)}$$

In the Newton–Raphson method, we assume an initial guess x and seek a better approximation $x + \Delta x$ by solving for Δx from the linearized equation

$$0 = g(x + \Delta x) \approx g(x) + \nabla g(x) \Delta x. \qquad \text{(A.4)}$$

This is quite simple using ADI in MRST:

```
[a, b, tol] = deal(1, 100, 1e-6);
[x0, incr]  = deal([-.5; 4]);
while norm(incr)>tol
   x = initVariablesADI(x0);
   eq = cat( 2*(a-x(1)) - 4*b.*x(1).*(x(2)-x(1).^2), ...
             2*b.*(x(2)-x(1).^2));
   incr = - eq.jac{1}\eq.val;
   x0 = x0 + incr;
end
```

This is just a backbone version of a Newton solver that does not contain safeguards of any kind like checking that the increments are finite, ensuring that the loop terminates after a finite number of iterations, etc. Figure A.9 illustrates how the Newton solver converges to the global minimum.

Beyond the examples and the discussion in this Appendix, we will not go more into details about the technical considerations that lie behind the implementation of AD in MRST. If you want a deeper understanding of how the ADI class works, the source code is fully open, so you are free to dissect the details to the level of your own choice.

COMPUTER EXERCISES

A.5.1 As an alternative to using AD, you can use finite differences, $f'(x) \approx [f(x+h) - f(x)]/h$, or a complex extension to compute $f'(x) \approx \text{Im}(f(x+ih))/h$. Use AD and the function

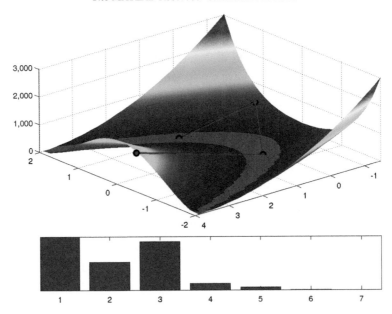

Figure A.9 Convergence of the AD-based Newton solver for the Rosenbrock problem. The upper plot shows the path taken by the nonlinear solver superimposed over the function $f(x, y)$. We have used a modified colormap to show the valley in which the function f attains its 1% lowest values. The lower plot depicts $f(x, y)^{1/10}$ for the initial guess and the six iterations needed to reduce the norm of the increment below 10^{-6}.

```
f = @(x) exp((x-.05).*(x-.4).*(x-.5).*(x-.7).*(x-.95));
```

to assess how accurate the two methods approximate $f'(x)$ at n equidistant points in the interval $x \in [0, 1]$ for different values of h.

A.5.2 The ADI class can compute log and exp, but does not yet support log2, log10, and logm. Study ADI.m and see if you can implement the missing functions. What about trigonometric and hyperbolic functions?

A.5.3 Can AD be used to compute higher-order derivatives? How or why not?

References

[1] Aarnes, J. E. 2004. On the use of a mixed multiscale finite element method for greater flexibility and increased speed or improved accuracy in reservoir simulation. *Multiscale Model. Simul.*, **2**(3), 421–439. doi:10.1137/030600655.

[2] Aarnes, J. E., Gimse, T., and Lie, K.-A. 2007b. An introduction to the numerics of flow in porous media using Matlab. In Hasle, G., Lie, K.-A., and Quak, E. (eds), *Geometrical Modeling, Numerical Simulation and Optimisation: Industrial Mathematics at SINTEF*. Berlin, Heidelberg, New York: Springer-Verlag, pp. 265–306.

[3] Aarnes, J. E., Hauge, V. L., and Efendiev, Y. 2007a. Coarsening of three-dimensional structured and unstructured grids for subsurface flow. *Adv. Water Resour.*, **30**(11), 2177–2193. doi:10.1016/j.advwatres.2007.04.007.

[4] Aarnes, J. E., Kippe, V., and Lie, K.-A. 2005. Mixed multiscale finite elements and streamline methods for reservoir simulation of large geomodels. *Adv. Water Resour.*, **28**(3), 257–271. doi:10.1016/j.advwatres.2004.10.007.

[5] Aarnes, J. E., Krogstad, S., and Lie, K.-A. 2006. A hierarchical multiscale method for two-phase flow based upon mixed finite elements and nonuniform coarse grids. *Multiscale Model. Simul.*, **5**(2), 337–363. doi:10.1137/050634566.

[6] Aarnes, J. E., Krogstad, S., and Lie, K.-A. 2008. Multiscale mixed/mimetic methods on corner-point grids. *Comput. Geosci.*, **12**(3), 297–315. doi:10.1007/s10596-007-9072-8.

[7] Aavatsmark, I. 2002. An introduction to multipoint flux approximations for quadrilateral grids. *Comput. Geosci.*, **6**, 405–432. doi:10.1023/A:1021291114475.

[8] Aavatsmark, I. 2007. Interpretation of a two-point flux stencil for skew parallelogram grids. *Comput. Geosci.*, **11**(3), 199–206. doi:10.1007/s10596-007-9042-1.

[9] Aavatsmark, I., Barkve, T., Bøe, Ø., and Mannseth, T. 1994. Discretization on non-orthogonal, curvilinear grids for multi-phase flow. In: *ECMOR IV – 4th European Conference on the Mathematics of Oil Recovery*. doi:0.3997/2214-4609.201411179.

[10] Aavatsmark, I., Eigestad, G. T., and Klausen, R. A. 2006. Numerical convergence of the MPFA O-method for general quadrilateral grids in two and three dimensions. In: Arnold, D. N., Bochev, P. B., Lehoucq, R. B., Nicolaides, R. A., and Shashkov, M. (eds), *Compatible Spatial Discretizations*. New York: Springer.

[11] Aavatsmark, I., Eigestad, G. T., Klausen, R. A., Wheeler, M. F., and Yotov, I. 2007. Convergence of a symmetric MPFA method on quadrilateral grids. *Comput. Geosci.*, **11**(4), 333–345. doi:10.1007/s10596-007-9056-8.

[12] Aavatsmark, I., Eigestad, G. T., Mallison, B. T., and Nordbotten, J. M. 2008. A compact multipoint flux approximation method with improved robustness. *Numer. Meth. Partial Diff. Eqs.*, **24**(5), 1329–1360. doi:10.1002/num.20320.

[13] Aavatsmark, I., and Klausen, R. 2003. Well index in reservoir simulation for slanted and slightly curved wells in 3D grids. *SPE J.*, **8**(01), 41–48. doi:10.2118/75275-PA.

[14] Abou-Kassem, J. H., Farouq-Ali, S. M., and Islam, M. R. 2006. *Petroleum Reservoir Simulations: A Basic Approach*. Houston: Gulf Publishing Company.

[15] AGMG. 2012. *Iterative solution with AGgregation-based algebraic MultiGrid*. http://agmg.eu. [Online; accessed July 11, 2018].

[16] Allen III, M. B., Behie, G. A., and Trangenstein, J. A. 1988. *Multiphase Flow in Porous Media: Mechanics, Mathematics, and Numerics*. New York: Springer-Verlag.

[17] Alpak, F. O., Pal, M., and Lie, K.-A. 2012. A multiscale method for modeling flow in stratigraphically complex reservoirs. *SPE J.*, **17**(4), 1056–1070. doi:10.2118/140403-PA.

[18] Alvestad, J., Holing, K., Christoffersen, K., and Stava, O. 1994. Interactive modelling of multiphase inflow performance of horizontal and highly deviated wells. In: *European Petroleum Computer Conference*. Society of Petroleum Engineers. doi:10.2118/27577-MS.

[19] Andersen, O. 2017. *Simplified models for numerical simulation of geological CO_2 storage*. PhD. thesis, University of Bergen. url: http://hdl.handle.net/1956/15477.

[20] Andersen, O., Gasda, S. E., and Nilsen, H. M. 2015. Vertically averaged equations with variable density for CO_2 flow in porous media. *Transp. Porous Media*, **107**(1), 95–127. doi:10.1007/s11242-014-0427-z.

[21] Andersen, O., Lie, K.-A., and Nilsen, H. M. 2016. An open-source toolchain for simulation and optimization of aquifer-wide CO2 storage. *Energy Procedia*, **86**(Jan.), 324–333. doi:10.1016/j.egypro.2016.01.033.

[22] Andersen, O., Nilsen, H. M., and Raynaud, X., 2017a. Coupled geomechanics and flow simulation on corner-point and polyhedral grids. In: *SPE Reservoir Simulation Conference*. doi:10.2118/182690-MS.

[23] Andersen, O., Nilsen, H. M., and Raynaud, X. 2017b. Virtual element method for geomechanical simulations of reservoir models. *Comput. Geosci.*, **21**(5–6), 877–893. doi:10.1007/s10596-017-9636-1.

[24] Appleyard, J. R., and Cheshire, I. M. 1983. Nested factorization. In: *SPE Reservoir Simulation Symposium*. doi:10.2118/12264-MS.

[25] Arbogast, T., Cowsar, L. C., Wheeler, M. F., and Yotov, I. 2000. Mixed finite element methods on nonmatching multiblock grids. *SIAM J. Num. Anal.*, **37**(4), 1295–1315. doi:10.1137/S0036142996308447.

[26] Arbogast, T., Dawson, C. N., Keenan, P. T., Wheeler, M. F., and Yotov, I. 1998. Enhanced cell-centered finite differences for elliptic equations on general geometry. *SIAM J. Sci. Comp.*, **19**(2), 404–425. doi:10.1137/S1064827594264545.

[27] Ates, H., Bahar, A., El-Abd, S., et al. 2005. Ranking and upscaling of geostatistical reservoir models using streamline simulation: A field case study. *SPE Res. Eval. Eng.*, **8**(1), 22–32. doi:10.2118/81497-PA.

[28] Aziz, K., and Settari, A. 1979. *Petroleum Reservoir Simulation*. London, New York: Elsevier Applied Science Publishers.

[29] Baker, L. E. 1988. Three-phase relative permeability correlations. In: *SPE Enhanced Oil Recovery Symposium*. doi:10.2118/17369-MS.

[30] Bao, K., Lie, K.-A., Møyner, O., and Liu, M. 2017. Fully implicit simulation of polymer flooding with MRST. *Comput. Geosci.*, **21**(5–6), 1219–1244. doi:10.1007/s10596-017-9624-5.

[31] Barker, J., and Thibeau, S. 1997. A critical review of the use of pseudorelative permeabilities for upscaling. *SPE Reservoir Engineering*, **12**(2), 138–143. doi:10.2118/35491-PA.

[32] Batycky, R. P., Thieles, M. R., Baker, R. O., and Chugh, S. H. 2008. Revisiting reservoir flood-surveillance methods using streamlines. *SPE Res. Eval. Eng.*, **11**(2), 387–394. doi:10.2118/95402-PA.

[33] Baxendale, D., Rasmussen, A., Rustad, A. B., et al. 2018. *Open porous media: Flow documentation manual*. 2018-10 Rev-2 edn. http://opm-project.org.

[34] Bear, J. 1988. *Dynamics of Fluids in Porous Media*. Mineola, NY: Dover.

[35] Bear, J. 2007. *Hydraulics of Groundwater*. Mineola, NY: Dover.

[36] Bear, J., and Bachmat, Y. 1990. *Introduction to Modeling of Transport Phenomena in Porous Media*. Dordrecht: Springer.

[37] Begg, S. H., Carter, R. R., and Dranfield, P. 1989. Assigning effective values to simulator gridblock parameters for heterogeneous reservoirs. *SPE Res. Eng.*, **4**(4), 455–463. doi:10.2118/16754-PA.

[38] Beirão da Veiga, L., Brezzi, F., Cangiani, A., et al. 2013. Basic principles of virtual element methods. *Math. Mod. Meth. Appl. Sci.*, **23**(01), 199–214. doi:10.1142/S0218202512500492.

[39] Beirão da Veiga, L., Brezzi, F., Marini, L. D., and Russo, A. 2014. The hitchhiker's guide to the virtual element method. *Math. Mod. Meth. Appl. Sci.*, **24**(08), 1541–1573. doi:10.1142/S021820251440003X.

[40] Beirao da Veiga, L., Lipnikov, K., and Manzini, G. 2014. *The Mimetic Finite Difference Method for Elliptic Problems*. New York: Springer.

[41] Benesoussan, A., Lions, J.-L., and Papanicolaou, G. 1978. *Asymptotic Analysis for Periodic Structures*. Amsterdam: Elsevier Science Publishers.

[42] Berge, R. L. 2016. *Unstructured PEBI grids adapting to geological feautres in subsurface reservoirs*. M.Sc. thesis, Norwegian University of Science and Technology. http://hdl.handle.net/11250/2411565.

[43] Bischof, C. H., Bücker, H. M., Lang, B., Rasch, A., and Vehreschild, A. 2002. Combining source transformation and operator overloading techniques to compute derivatives for MATLAB programs. *Proceedings of the Second IEEE International Workshop on Source Code Analysis and Manipulation (SCAM 2002)*. Los Alamitos, CA, pp. 65–72 doi:10.1109/SCAM.2002.1134106.

[44] Blunt, M. J. 2017. *Multiphase Flow in Permeable Media: A Pore-Scale Perspective*. Cambridge: Cambridge University Press.

[45] Braess, D. 1997. *Finite Elements: Theory, Fast Solvers, and Applications in Solid Mechanics*. Cambridge: Cambridge University Press.

[46] Branets, L., Ghai, S. S., Lyons, S. L., and Wu, X.-H. 2009a. Efficient and accurate reservoir modeling using adaptive gridding with global scale up. In: *SPE Reservoir Simulation Symposium*. doi:10.2118/118946-MS.

[47] Branets, L. V., Ghai, S. S., Lyons, S. L., and Wu, X.-H. 2009b. Challenges and technologies in reservoir modeling. *Commun. Comput. Phys.*, **6**(1), 1–23.

[48] Branets, L., Kubyak, V., Kartasheva, E., Shmyrov, L., and Kandybor, D. 2015. Capturing geologic complexity in simulation grid. In: *SPE Reservoir Simulation Symposium*. doi:10.2118/173270-MS.

[49] Brenier, Y., and Jaffré, J. 1991. Upstream differencing for multiphase flow in reservoir simulation. *SIAM J. Numer. Anal.*, **28**(3), 685–696. doi:10.1137/0728036.

[50] Brenner, S. C., and Scott, L. R. 2007. *The Mathematical Theory of Finite Element Methods*. 3rd edn. New York: Springer-Verlag.

[51] Brewer, M., Camilleri, D., Ward, S., and Wong, T. 2015. Generation of hybrid grids for simulation of complex, unstructured reservoirs by a simulator with MPFA. In: *SPE Reservoir Simulation Symposium*. doi:10.2118/173191-MS.

[52] Brezzi, F., Lipnikov, K., and Simoncini, V. 2005. A family of mimetic finite difference methods on polygonial and polyhedral meshes. *Math. Models Methods Appl. Sci.*, **15**, 1533–1553. doi:10.1142/S0218202505000832.

[53] Brezzi, F., and Fortin, M. 1991. *Mixed and Hybrid Finite Element Methods*. New York: Springer-Verlag.

[54] Brooks, R. H., and Corey, A. T. 1966. Properties of porous media affecting fluid flow. *J. Irrigation Drainage Div.*, **92**(2), 61–90.

[55] Buckingham, E. 1907. *Studies on the Movement of Soil Moisture*. Bulletin, no. 38. United States. Bureau of Soils. https://archive.org/details/studiesonmovemen38buck.

[56] Buckley, S. E., and Leverett, M. C. 1942. Mechanism of fluid displacement in sands. *Trans. AIME*, **146**(01), 107–116. doi:10.2118/942107-G.

[57] Caers, J. 2005. *Petroleum Geostatistics*. Richardson, TX: Society of Petroleum Engineers.

[58] Cao, H. 2002. *Development of techniques for general purpose simulators*. PhD. thesis, Stanford University.

[59] Castellini, A., Edwards, M. G., and Durlofsky, L. J. 2000. Flow based modules for grid generation in two and three dimensions. In: *ECMOR VII – 7th European Conference on the Mathematics of Oil Recovery*. doi:10.3997/2214-4609.201406120.

[60] Cayuga Research. *ADMAT*. www.cayugaresearch.com/admat.html. [Online; accessed Jul 11, 2018].

[61] Chavent, G., and Jaffré, J. 1982. *Mathematical Models and Finite Elements for Reservoir Simulation*. Amsterdam: North Holland.

[62] Chavent, G., and Jaffré, J. 1986. *Mathematical Models and Finite Elements for Reservoir Simulation: Single Phase, Multiphase and Multicomponent Flows through Porous Media*. Amsterdam: Elsevier.

[63] Chen, Y., and Durlofsky, L. J. 2006. Adaptive local-global upscaling for general flow scenarios in heterogeneous formations. *Transport Porous Media*, **62**, 157–182. doi:10.1007/s11242-005-0619-7.

[64] Chen, Y., Durlofsky, L. J., Gerritsen, M., and Wen, X. H. 2003. A coupled local-global upscaling approach for simulating flow in highly heterogeneous formations. *Adv. Water Resour.*, **26**(10), 1041–1060. doi:10.1016/S0309-1708(03)00101-5.

[65] Chen, Z. 2000. Formulations and numerical methods of the black oil model in porous media. *SIAM J. Numer. Anal.*, **38**(2), 489–514. doi:10.1137/S0036142999304263.

[66] Chen, Z. 2007. *Reservoir Simulation: Mathematical Techniques in Oil Recovery*. Philadelphia: Society for Industrial and Applied Mathematics.

[67] Chen, Z., and Ewing, R. E. 1997. Comparison of various formulations of three-phase flow in porous media. *J. Comput. Phys.*, **132**(2), 362–373. doi:10.1006/jcph.1996.5641.

[68] Chen, Z., and Hou, T. Y. 2003. A mixed multiscale finite element method for elliptic problems with oscillating coefficients. *Math. Comp.*, **72**, 541–576. doi:10.1090/S0025-5718-02-01441-2.

[69] Chen, Z., Huan, G., and Ma, Y. 2006. *Computational Methods for Multiphase Flows in Porous Media*. Philadelphia: Society of Industrial and Applied Mathematics. doi:10.1137/1.9780898718942.

[70] Christie, M. A. 1996. Upscaling for reservoir simulation. *J. Pet. Tech.*, **48**(11), 1004–1010. doi:10.2118/37324-MS.

[71] Christie, M. A., and Blunt, M. J. 2001. Tenth SPE comparative solution project: A comparison of upscaling techniques. *SPE Reservoir Eval. Eng.*, **4**, 308–317. doi:10.2118/72469-PA.

[72] Cirpka, O. A., Frind, E. O., and Helmig, R. 1999. Streamline-oriented grid generation for transport modelling in two-dimensional domains including wells. *Adv. Water Resour.*, **22**(7), 697–710. doi:10.1016/S0309-1708(98)00050-5.

[73] Coats, K. H. 2000. A note on IMPES and some IMPES-based simulation models. *SPE J.*, **05**(03), 245–251. doi:10.2118/65092-PA.

[74] Cordes, C., and Kinzelbach, W. 1992. Continous groundwater velocity fields and path lines in linear, bilinear, and trilinear finite elements. *Water Resour. Res.*, **28**(11), 2903–2911. doi:10.1029/92WR01686.

[75] Courant, R., Friedrichs, K., and Lewy, H. 1928. Über die partiellen Differenzengleichungen der mathematischen Physik. *Math. Ann.*, **100**(1), 32–74. doi:10.1007/BF01448839.

[76] Dafermos, C. M. 2010. *Hyperbolic Conservation Laws in Continuum Physics*. Berlin, Heidelberg: Springer.

[77] Darcy, H. P. G. 1856. *Les Fontaines Publiques de la Ville de Dijon*. Paris: Dalmont.

[78] Datta-Gupta, A., and King, M. J. 1995. A semianalytic approach to tracer flow modeling in heterogeneous permeable media. *Adv. Water Resour.*, **18**, 9–24. doi:10.1016/0309-1708(94)00021-V.

[79] Datta-Gupta, A., and King, M. J. 2007. *Streamline Simulation: Theory and Practice*. Richardson, TX: Society of Petroleum Engineers.

[80] DeBaun, D., et al. 2005. An extensible architecture for next generation scalable parallel reservoir simulation. In: *SPE Reservoir Simulation Symposium*. doi:10.2118/93274-MS.

[81] Demidov, D. 2017 (Oct.). amgcl-sdd-scaling. https://zenodo.org/record/1002948#.XE_CAFxKiUk.

[82] Deutsch, C. V., and Journel, A. G. 1998. *GSLIB: Geostatistical Software Library and User's Guide*. 2nd edn. New York: Oxford University Press.

[83] Ding, X. Y., and Fung, L. S. K. 2015. An unstructured gridding method for simulating faulted reservoirs populated with complex wells. In: *SPE Reservoir Simulation Symposium*. doi:10.2118/173243-MS.

[84] Dogru, A. H., Fung, L. S.-K., Middya, U., Al-Shaalan, T., and Pita, J. A. 2009. A next-generation parallel reservoir simulator for giant reservoirs. In: *SPE/EAGE Reservoir Characterization & Simulation Conference*. doi:10.2118/119272-MS.

[85] Dogru, A. H., Fung, L. S. K., Middya, U., et al. 2011. New frontiers in large scale reservoir simulation. In: *SPE Reservoir Simulation Symposium*. doi:10.2118/142297-MS.

[86] Douglas Jr., J., Peaceman, D. W., and Rachford Jr., H. H. 1959. A method for calculating multi-dimensional immiscible displacement. *Petrol. Trans. AIME*, **216**, 297–308.

[87] Duarte, A. C. 2016. *Contributions to production optimization of oil reservoirs*. PhD. thesis, Norwegian University of Science and Technology. http://hdl.handle.net/11250/2383090.

[88] Duff, I. S. 2004. MA57–A code for the solution of sparse symmetric definite and indefinite systems. *ACM Trans. Math. Softw.*, **30**(2), 118–144. doi:10.1145/992200.992202.

[89] Duff, I. S., and Reid, J. K. 1983. The multifrontal solution of indefinite sparse symmetric linear. *ACM Trans. Math. Software*, **9**(3), 302–325. doi:10.1145/356044.356047.

[90] Durlofsky, L. J. 1991. Numerical calculations of equivalent gridblock permeability tensors for heterogeneous porous media. *Water Resour. Res.*, **27**(5), 699–708. doi:10.1029/91WR00107.

[91] Durlofsky, L. J. 2003. *Upscaling of Geocellular Models for Reservoir Flow Simulation: A Review of Recent Progress*. Presented at 7th International Forum on Reservoir Simulation Bühl/Baden-Baden, Germany, June 23–27, 2003.

[92] Durlofsky, L. J. 2005. *Upscaling and Gridding of Fine Scale Geological Models for Flow Simulation*. Presented at 8th International Forum on Reservoir Simulation Iles Borromees, Stresa, Italy, June 20–24, 2005.

[93] Durlofsky, L. J., Jones, R. C., and Milliken, W. J. 1997. A nonuniform coarsening approach for the scale-up of displacement processes in heterogeneous porous media. *Adv. Water Resour.*, **20**(5-6), 335–347. doi:10.1016/S0309-1708(96)00053-X.

[94] Edwards, M. G., and Rogers, C. F. 1994. A flux continuous scheme for the full tensor pressure equation. In: *ECMOR IV – 4th European Conference on the Mathematics of Oil Recovery*. doi:10.3997/2214-4609.201411178.

[95] Efendiev, Y., and Hou, T. Y. 2009. *Multiscale Finite Element Methods: Theory and Applications*. New York: Springer-Verlag.

[96] Eigestad, G. T., and Klausen, R. A. 2005. On the convergence of the multi-point flux approximation o-method: Numerical experiments for discontinuous permeability. *Num. Meth. Partial Diff. Eqs.*, **21**(6), 1079–1098. doi:10.1002/num.20079.

[97] Eigestad, G., Dahle, H., Hellevang, B., Riis, F., Johansen, W., and Øian, E. 2009. Geological modeling and simulation of CO2 injection in the Johansen formation. *Comput. Geosci.*, **13**(4), 435–450. doi:10.1007/s10596-009-9153-y.

[98] Eikemo, B., Lie, K.-A., Dahle, H. K., and Eigestad, G. T. 2009. Discontinuous Galerkin methods for transport in advective transport in single-continuum models of fractured media. *Adv. Water Resour.*, **32**(4), 493–506. doi:10.1016/j.advwatres.2008.12.010.

[99] Ertekin, T., Abou-Kassem, J. H., and King, G. R. 2001. *Basic Applied Reservoir Simulation*. Richardson, TX: Society of Petroleum Engineers.

[100] Evazi, M., and Mahani, H. 2010a. Generation of Voronoi grid based on vorticity for coarse-scale modeling of flow in heterogeneous formations. *Transp. Porous Media*, **83**(3). doi:10.1007/s11242-009-9458-2.

[101] Evazi, M., and Mahani, H. 2010b. Unstructured-coarse-grid generation using background-grid approach. *SPE J.*, **15**(2), 326–340. doi:10.2118/120170-PA.

[102] Ewing, R. E., Lazarov, R. D., Lyons, S. L., Papavassiliou, D. V., Pasciak, J., and Qin, G. 1999. Numerical well model for non-Darcy flow through isotropic porous media. *Comput. Geosci.*, **3**(3-4), 185–204. doi:10.1023/A:1011543412675.

[103] Eymard, R., Gallouët, T., and Herbin, R. 1999. Convergence of finite volume schemes for semilinear convection diffusion equations. *Numer. Math.*, **82**(1), 91–116. doi:10.1007/s002110050412.

[104] Eymard, R., Gallouët, T., and Herbin, R. 2001. Finite volume approximation of elliptic problems and convergence of an approximate gradient. *App. Numer. Math.*, **37**(1-2), 31–53. doi:10.1016/S0168-9274(00)00024-6.

[105] Eymard, R., Guichard, C., and Herbin, R. 2012a. Small-stencil 3D schemes for diffusive flows in porous media. *ESAIM: Math. Model. Numer. Anal.*, **46**(2), 265–290. doi:10.1051/m2an/2011040.

[106] Eymard, R., Guichard, C., Herbin, R., and Masson, R. 2012b. Vertex-centred discretization of multiphase compositional Darcy flows on general meshes. *Comput. Geosci.*, **16**(4), 987–1005. doi:10.1007/s10596-012-9299-x.

[107] Fanchi, J. R. 2005. *Principles of Applied Reservoir Simulation*. Houston, TX: Gulf Professional Publishing.

[108] Farmer, C. L. 2002. Upscaling: A review. *Int. J. Numer. Meth. Fluids*, **40**(1–2), 63–78. doi:10.1002/fld.267.

[109] Fayers, F. J., and Matthews, J. D. 1984. Evaluation of normalized Stone's methods for estimating three-phase relative permeabilities. *SPE J.*, **24**(2), 224–232. doi:10.2118/11277-PA.

[110] Fink, M. 2007. *Automatic Differentiation for Matlab*. MATLAB Central. https://tinyurl.com/ycvp6n8a. [Online; accessed July 11, 2018].

[111] Floris, F. J. T., Bush, M. D., Cuypers, M., Roggero, F., and Syversveen, A. R. 2001. Methods for quantifying the uncertainty of production forecasts: a comparative study. *Petroleum Geoscience*, **7**(S), S87–S96. doi:10.1144/petgeo.7.S.S87.

[112] Forth, S. A. 2006. An efficient overloaded implementation of forward mode automatic differentiation in MATLAB. *ACM Trans. Math. Software*, **32**(2), 195–222. doi:10.1145/1141885.1141888.

[113] Fung, L. S. K., Ding, X. Y., and Dogru, A. H. 2014. Unconstrained Voronoi grids for densely spaced complex wells in full-field reservoir simulation. *SPE J.*, **19**(5), 803–815. doi:10.2118/163648-PA.

[114] Gain, A. L., Talischi, C., and Paulino, G. H. 2014. On the virtual element method for three-dimensional linear elasticity problems on arbitrary polyhedral meshes. *Comput. Meth. App. Mech. Engng.*, **282**, 132–160. doi:10.1016/j.cma.2014.05.005.

[115] Gao, M. 2014. *Reservoir and Surface Facilities Coupled through Partially and Fully Implicit Approaches*. M.Sc. thesis, Texas A & M University. http://hdl.handle.net/1969.1/154076.

[116] Garcia, M. H., Journel, A. G., and Aziz, K. 1992. Automatic grid generation for modeling reservoir heterogeneities. *SPE Reservoir Eng.*, **7**(6), 278–284. doi:10.2118/21471-PA.

[117] Gerritsen, M., and Lambers, J. V. 2008. Integration of local-global upscaling and grid adaptivity for simulation of subsurface flow in heterogeneous formations. *Comput. Geosci.*, **12**(2), 193–208. doi:10.1007/s10596-007-9078-2.

[118] Godunov, S. K. 1959. A difference method for numerical calculation of discontinuous solutions of the equations of hydrodynamics. *Mat. Sb. (N.S.)*, **47 (89)**, 271–306.

[119] Gries, S., Stüben, K., Brown, G. L., Chen, D., and Collins, D. A. 2014. Preconditioning for efficiently applying algebraic multigrid in fully implicit reservoir simulations. *SPE J.*, **19**(04), 726–736. doi:10.2118/163608-PA.

[120] Guérillot, D., Rudkiewicz, J. L., Ravenne, C., Renard, D., and Galli, A. 1990. An integrated model for computer aided reservoir description: From outcrop study to fluid flow simulations. *Oil Gas Sci. Technol.*, **45**(1), 71–77. doi:10.2516/ogst:1990005.

[121] Gunasekera, D., Cox, J., and Lindsey, P. 1997. The generation and application of K-orthogonal grid systems. In: *SPE Reservoir Simulation Symposium*. doi:10.2118/37998-MS.

[122] Hægland, H., Dahle, H. K., Lie, K.-A., and Eigestad, G. T. 2006. Adaptive streamline tracing for streamline simulation on irregular grids. In: Binning, P. J., Engesgaard, P. K., Dahle, H. K., Pinder, G. F., and Gray, W. G. (eds), *XVI International Conference on Computational Methods in Water Resources*. http://proceedings.cmwr-xvi.org/.

[123] Hales, H. B. 1996. A method for creating 2-d orthogonal grids which conform to irregular shapes. *SPE J*, **1**(2), 115–124. doi:10.2118/35273-PA.

[124] Hauge, V. L. 2010. *Multiscale methods and flow-based gridding for flow and transport in porous media*. PhD. thesis, Norwegian University of Science and Technology. http://hdl.handle.net/11250/258800.

[125] Hauge, V. L., Lie, K.-A., and Natvig, J. R. 2012. Flow-based coarsening for multiscale simulation of transport in porous media. *Comput. Geosci.*, **16**(2), 391–408. doi:10.1007/s10596-011-9230-x.

[126] He, C., and Durlofsky, L. J. 2006. Structured flow-based gridding and upscaling for modeling subsurface flow. *Adv. Water Resour.*, **29**(12), 1876–1892. doi:10.1016/j.advwatres.2005.12.012.

[127] He, Z., Parikh, H., Datta-Gupta, A., Perez, J., and Pham, T. 2004. Identifying reservoir compartmentalization and flow barriers from primary production using streamline diffusive time of flight. *SPE J.*, **7**(3), 238–247. doi:10.2118/88802-PA.

[128] Heinemann, Z. E., Brand, C. W., and Munka, M. 1991. Modeling reservoir geometry with irregular grids. *SPE Res. Eng.*, **6**(2), 225–232. doi:10.2118/18412-PA.

[129] Helmig, R. 1997. *Multiphase Flow and Transport Processes in the Subsurface: A Contribution to the Modeling of Hydrosystems*. Berlin, Heidelberg: Springer.

[130] Hilden, S. T., Møyner, O., Lie, K.-A., and Bao, K. 2016. Multiscale simulation of polymer flooding with shear effects. *Transp. Porous Media*, **113**(1), 111–135. doi:10.1007/s11242-016-0682-2.

[131] Hilden, S. T. 2016. *Upscaling of water-flooding scenarios and modeling of polymer flow*. PhD. thesis, Norwegian University of Science and Technology. http://hdl.handle.net/11250/2388331.

[132] Hilden, S. T., Lie, K.-A., and Raynaud, X. 2014. Steady state upscaling of polymer flooding. In: *ECMOR XIV – 14th European Conference on the Mathematics of Oil Recovery*. doi:10.3997/2214-4609.20141802.

[133] Holden, L., and Nielsen, B. F. 2000. Global upscaling of permeability in heterogeneous reservoirs; the output least squares (OLS) method. *Trans. Porous Media*, **40**(2), 115–143. doi:10.1023/A:1006657515753.

[134] Holden, H., and Risebro, N. H. 2002. *Front Tracking for Hyperbolic Conservation Laws*. New York: Springer-Verlag.

[135] Hornung, U. 1997. *Homogenization and Porous Media*. New York: Springer-Verlag.

[136] Hoteit, H., and Firoozabadi, A. 2008. Numerical modeling of two-phase flow in heterogeneous permeable media with different capillarity pressures. *Adv. Water Resour.*, **31**(1), 56–73. doi:10.1016/j.advwatres.2007.06.006.

[137] Hubbert, M. K. 1956. Darcy's law and the field equations of the flow of underground fluids. *Petrol. Trans., AIME*, **207**, 22–239.

[138] Hui, M.-H. R., Karimi-Fard, M., Mallison, B., and Durlofsky, L. J. 2018. A general modeling framework for simulating complex recovery processes in fractured reservoirs at different resolutions. *SPE J.*, **22**(1), 20–29. doi:10.2118/182621-MS.

[139] Huseby, O., Sagen, J., and Dugstad, Ø. 2012. Single well chemical tracer tests – Fast and accurate simulations. In: *SPE EOR Conference at Oil and Gas West Asia*. doi:10.2118/155608-MS.

[140] Idrobo, E. A., Choudhary, M. K., and Datta-Gupta, A. 2000. Swept volume calculations and ranking of geostatistical reservoir models using streamline simulation. In: *SPE/AAPG Western Regional Meeting*. doi:10.2118/62557-MS.

[141] Iemcholvilert, S. 2013. *A Research on production optimization of coupled surface and subsurface model*. M.Sc. thesis, Texas A & M University. http://hdl.handle.net/1969.1/151189.

[142] Islam, M. R., Hossain, M. E., Moussavizadegan, S. H., Mustafiz, S., and Abou-Kassem, J. H. 2016. *Advanced Petroleum Reservoir Simulation: Towards Developing Reservoir Emulators*. John Wiley & Sons, Inc. doi:10.1002/9781119038573.

[143] Islam, M. R., Mousavizadegan, S. H., Mustafiz, S., and Abou-Kassem, J. H. 2010. *Advanced Petroleum Reservoir Simulations*. Hoboken, NJ: John Wiley & Sons, Inc.

[144] Izgec, O., Sayarpour, M., and Shook, G. M. 2011. Maximizing volumetric sweep efficiency in waterfloods with hydrocarbon f-ϕ curves. *J. Petrol. Sci. Eng.*, **78**(1), 54–64. doi:10.1016/j.petrol.2011.05.003.

[145] Jansen, J. D. 2017. *Nodal Analysis of Oil and Gas Wells-System Modeling and Numerical Implementation*. Richardson, TX: Society of Petroleum Engineers.

[146] Jenny, P., Lee, S. H., and Tchelepi, H. A. 2003. Multi-scale finite-volume method for elliptic problems in subsurface flow simulation. *J. Comput. Phys.*, **187**, 47–67. doi:10.1016/S0021-9991(03)00075-5.

[147] Jenny, P., Tchelepi, H. A., and Lee, S. H. 2009. Unconditionally convergent nonlinear solver for hyperbolic conservation laws with S-shaped flux functions. *J. Comput. Phys.*, **228**(20), 7497–7512. doi:10.1016/j.jcp.2009.06.032.

[148] Jenny, P., Wolfsteiner, C., Lee, S. H., and Durlofsky, L. J. 2002. Modeling flow in geometrically complex reservoirs using hexahedral multiblock grids. *SPE J.*, **7**(2). doi:10.2118/78673-PA.

[149] Jikov, V. V., Kozlov, S. M., and Oleinik, O. A. 1994. *Homogenization of Differential Operators and Integral Functionals*. New York: Springer-Verlag.

[150] Jimenez, E., Sabir, K., Datta-Gupta, A., and King, M. J. 2007. Spatial error and convergence in streamline simulation. *SPE J.*, **10**(3), 221–232. doi:10.2118/92873-MS.

[151] Journel, A. G., Deutsch, C. V., and Desbarats, A. J. 1986. Power averaging for block effective permeability. In: *SPE California Regional Meeting*. doi:10.2118/15128-MS.

[152] Karimi-Fard, M., and Durlofsky, L. J. 2012. Accurate resolution of near-well effects in upscaled models using flow-based unstructured local grid refinement. *SPE J.*, **17**(4), 1084–1095. doi:10.2118/141675-PA.

[153] Karimi-Fard, M., and Durlofsky, L. J. 2016. A general gridding, discretization, and coarsening methodology for modeling flow in porous formations with discrete geological features. *Adv. Water Resour.*, **96**(Supplement C), 354–372. doi:10.1016/j.advwatres.2016.07.019.

[154] Karypis, G., and Kumar, V. 1998. A fast and high quality multilevel scheme for partitioning irregular graphs. *SIAM J. Sci. Comp.*, **20**(1), 359–392. doi:10.1137/S1064827595287997.

[155] Keilegavlen, E., Kozdon, J. E., and Mallison, B. T. 2012. Multidimensional upstream weighting for multiphase transport on general grids. *Comput. Geosci.*, **16**, 1021–1042. doi:10.1007/s10596-012-9301-7.

[156] Keilegavlen, E., and Nordbotten, J. M. 2017. Finite volume methods for elasticity with weak symmetry. *Int. J. Numer. Meth. Eng.*, **112**(8), 939–962. doi:10.1002/nme.5538.

[157] Keilegavlen, E., Nordbotten, J. M., and Aavatsmark, I. 2009. Sufficient criteria are necessary for monotone control volume methods. *Appl. Math. Letters*, **22**(8), 1178–1180. doi:10.1016/j.aml.2009.01.048.

[158] Kenyon, D. 1987. Third SPE comparative solution project: Gas cycling of retrograde condensate reservoirs. *J. Petrol. Tech.*, **39**(08), 981–997. doi:10.2118/12278-PA.

[159] Killough, J. E. 1995. Ninth SPE comparative solution project: A reexamination of black-oil simulation. In: *SPE Reservoir Simulation Symposium*. doi:10.2118/29110-MS.

[160] King, M. J. 2007. Recent advances in upgridding. *Oil Gas Sci. Technol. – Rev. IFP*, **62**(2), 195–205. doi:10.2516/ogst:2007017.

[161] King, M. J., and Mansfield, M. 1999. Flow simulation of geologic models. *SPE Res. Eval. Eng.*, **2**(4), 351–367. doi:10.2118/57469-PA.

[162] King, M. J., Burn, K. S., Muralidharan, P. W. V., et al. 2006. Optimal coarsening of 3D reservoir models for flow simulation. *SPE Reserv. Eval. Eng.*, **9**(4), 317–334. doi:10.2118/95759-PA.

[163] King, M. J., MacDonald, D. G., Todd, S. P., and Leung, H. 1998. Application of novel upscaling approaches to the Magnus and Andrew reservoirs. In: *European Petroleum Conference*. doi:10.2118/50643-MS.

[164] King, M. J., and Datta-Gupta, A. 1998. Streamline simulation: A current perspective. *In Situ*, **22**(1), 91–140.

[165] Kippe, V., Aarnes, J. E., and Lie, K.-A. 2008. A comparison of multiscale methods for elliptic problems in porous media flow. *Comput. Geosci.*, **12**(3), 377–398. doi:10.1007/s10596-007-9074-6.

[166] Klausen, R. A., Rasmussen, A. F., and Stephansen, A. 2012. Velocity interpolation and streamline tracing on irregular geometries. *Comput. Geosci.*, **16**, 261–276. doi:10.1007/s10596-011-9256-0.

[167] Klausen, R. A., and Winther, R. 2006. Robust convergence of multi point flux approximation on rough grids. *Numer. Math.*, **104**(3), 317–337. doi:10.1007/s00211-006-0023-4.

[168] Klemetsdal, Ø. S. 2016. *The virtual element method as a common framework for finite element and finite difference methods – numerical and theoretical analysis*. M.Sc. thesis, Norwegian University of Science and Technology. http://hdl.handle.net/11250/2405996.

[169] Klemetsdal, Ø. S., Berge, R. L., Lie, K.-A., Nilsen, H. M., and Møyner, O. 2017. Unstructured gridding and consistent discretizations for reservoirs with faults and complex wells. In: *SPE Reservoir Simulation Conference*. doi:10.2118/182679-MS.

[170] Krogstad, S., Lie, K.-A., Møyner, O., Nilsen, H. M., Raynaud, X., and Skaflestad, B. 2015. MRST-AD – An open-source framework for rapid prototyping and evaluation of reservoir simulation problems. In: *SPE Reservoir Simulation Symposium*. doi:10.2118/173317-MS.

[171] Krogstad, S., Lie, K.-A., Nilsen, H. M., Berg, C. F., and Kippe, V. 2017. Efficient flow diagnostics proxies for polymer flooding. *Comput. Geosci.*, **21**(5-6), 1203–1218. doi:10.1007/s10596-017-9681-9.

[172] Krogstad, S., Raynaud, X., and Nilsen, H. M. 2016. Reservoir management optimization using well-specific upscaling and control switching. *Comput. Geosci.*, **20**(3), 695–706. doi:10.1007/s10596-015-9497-4.

[173] Kružkov, S. N. 1970. First order quasilinear equations in several independent variables. *Mathematics of the USSR-Sbornik*, **10**(2), 217. doi:10.1070/SM1970v010n02ABEH002156.

[174] Kurganov, A., Noelle, S., and Petrova, G. 2001. Semidiscrete central-upwind schemes for hyperbolic conservation laws and Hamilton–Jacobi equations. *SIAM J. Sci. Comp.*, **23**(3), 707–740. doi:10.1137/S1064827500373413.

[175] Kwok, F., and Tchelepi, H. 2007. Potential-based reduced Newton algorithm for nonlinear multiphase flow in porous media. *J. Comput. Phys.*, **227**(1), 706–727. doi:10.1016/j.jcp.2007.08.012.

[176] Lake, L. W. 1989. *Enhanced Oil Recovery*. Upper Saddle River, NJ: Prentice-Hall.

[177] Lake, L. W. (ed). 2007. *Petroleum Engineering Handbook*. Richardson, TX: Society of Petroleum Engineers.

[178] Lax, P., and Wendroff, B. 1960. Systems of conservation laws. *Comm. Pure Appl. Math.*, **13**(2), 217–237. doi:10.1002/cpa.3160130205.

[179] Le Potier, C. 2009. A nonlinear finite volume scheme satisfying maximum and minimum principles for diffusion operators. *Int. J. Finite Vol.*, **6**(2), 1–20.

[180] Lee, S. H., Jenny, P., and Tchelepi, H. A. 2002. A finite-volume method with hexahedral multiblock grids for modeling flow in porous media. *Comput. Geosci.*, **6**(3-4), 353–379. doi:10.1023/A:1021287013566.

[181] Leeuwenburgh, O., and Arts, R. 2014. Distance parameterization for efficient seismic history matching with the ensemble Kalman Filter. *Comput. Geosci.*, **18**(3-4), 535–548. doi:10.1007/s10596-014-9434-y.

[182] Leeuwenburgh, O., Peters, E., and Wilschut, F. 2011. Towards an integrated workflow for structural reservoir model updating and history matching. In: *SPE EUROPEC/EAGE Annual Conference and Exhibition*. doi:10.2118/143576-MS.

[183] LeVeque, R. J. 2002. *Finite Volume Methods for Hyperbolic Problems*. Cambridge: Cambridge University Press.

[184] Leverett, M. C. 1941. Capillary behavior in porous solids. *Trans. AIME*, **142**, 159–172. doi:10.2118/941152-G.

[185] Li, X., and Zhang, D. 2014. A backward automatic differentiation framework for reservoir simulation. *Comput. Geosci.*, **18**(6), 1009–1022. doi:10.1007/s10596-014-9441-z.

[186] Lie, K.-A. 2018. On Holden's seven guidelines for scientific computing and development of open-source community software. In: Gesztesy, F., et al. (eds), *Non-Linear Partial Differential Equations, Mathematical Physics, and Stochastic Analysis: The Helge Holden Anniversary Volume*. European Mathematical Society Publishing House. pp. 389–422.

[187] Lie, K.-A., Kedia, K., Skaflestad, B., et al. 2017a. A general non-uniform coarsening and upscaling framework for reduced-order modeling. In: *SPE Reservoir Simulation Conference*. doi:10.2118/182681-MS.

[188] Lie, K.-A. 2015a. *JOLT 1: Introduction to MRST*. SINTEF ICT / ICME, Stanford University. www.sintef.no/mrst-jolts.

[189] Lie, K.-A. 2015b. *JOLT 2: Grids and petrophysical data*. SINTEF ICT / ICME, Stanford University. www.sintef.no/mrst-jolts.

[190] Lie, K.-A., Mykkeltvedt, T. S., and Møyner, O. 2018. Fully implicit WENO schemes on stratigraphic and fully unstructured grids. In: *ECMOR XVI – 16th European Conference on the Mathematics of Oil Recovery*.

[191] Lie, K.-A., Møyner, O., and Natvig, J. R. 2017a. Use of multiple multiscale operators to accelerate simulation of complex geomodels. *SPE J.*, **22**(6), 1929–1945. doi:10.2118/182701-PA.

[192] Lie, K.-A., Krogstad, S., Ligaarden, I. S., et al. 2012b. Open-source MATLAB implementation of consistent discretisations on complex grids. *Comput. Geosci.*, **16**, 297–322. doi:10.1007/s10596-011-9244-4.

[193] Lie, K.-A., Nilsen, H. M., Andersen, O., and Møyner, O. 2016. A simulation workflow for large-scale CO_2 storage in the Norwegian North Sea. *Comput. Geosci.*, **20**(3), 607–622. doi:10.1007/s10596-015-9487-6.

[194] Lie, K.-A., Natvig, J. R., Krogstad, S., Yang, Y., and Wu, X.-H. 2014. Grid adaptation for the Dirichlet–Neumann representation method and the multiscale mixed finite-element method. *Comput. Geosci.*, **18**(3), 357–372. doi:10.1007/s10596-013-9397-4.

[195] Lie, K.-A., Møyner, O., Natvig, J. R., et al. 2017b. Successful application of multiscale methods in a real reservoir simulator environment. *Comput. Geosci.*, **21**(5-6), 981–998. doi:10.1007/s10596-017-9627-2.

[196] Lie, K.-A., Natvig, J. R., and Nilsen, H. M. 2012a. Discussion of dynamics and operator splitting techniques for two-phase flow with gravity. *Int. J. Numer. Anal. Mod.*, **9**(3), 684–700.

[197] Ligaarden, I. S. 2008. *Well models for mimetic finite difference methods and improved representation of wells in multiscale methods*. M.Sc. thesis, University of Oslo. http://urn.nb.no/URN:NBN:no-19435.

[198] Lipnikov, K., Shashkov, M., Svyatskiy, D., and Vassilevski, Y. 2007. Monotone finite volume schemes for diffusion equations on unstructured triangular and shape-regular polygonal meshes. *J. Comput. Phys.*, **227**(1), 492–512. doi:10.1016/j.jcp.2007.08.008.

[199] Lipnikov, K., Shashkov, M., and Yotov, I. 2009. Local flux mimetic finite difference methods. *Numer. Math.*, **112**(1), 115–152. doi:10.1007/s00211-008-0203-5.

[200] Lunati, I., and Lee, S. H. 2009. An operator formulation of the multiscale finite-volume method with correction function. *Multiscale Model. Simul.*, **8**(1), 96–109. doi:10.1137/080742117.

[201] Mahani, H., Muggeridge, A. H., and Ashjari, M. A. 2009. Vorticity as a measure of heterogeneity for improving coarse grid generation. *Petrol. Geosci.*, **15**(1), 91–102. doi:10.1144/1354-079309-802.

[202] Mallison, B., Sword, C., Viard, T., Milliken, W., and Cheng, A. 2014. Unstructured cut-cell grids for modeling complex reservoirs. *SPE J.*, **19**(2), 340–352. doi:10.2118/163642-PA.

[203] Manzocchi, T., et al. 2008. Sensitivity of the impact of geological uncertainty on production from faulted and unfaulted shallow-marine oil reservoirs: Objectives and methods. *Petrol. Geosci.*, **14**(1), 3–15. doi:10.1144/1354-079307-790.

[204] Matringe, S. F., and Gerritsen, M. G. 2004. On accurate tracing of streamlines. In: *SPE Annual Technical Conference and Exhibition*. doi:10.2118/89920-MS.

[205] Matringe, S. F., Juanes, R., and Tchelepi, H. A. 2007. Streamline tracing on general triangular or quadrilateral grids. *SPE J.*, **12**(2), 217–233. doi:10.2118/96411-MS.

[206] Mattax, C. C., and Dalton, R. L. (eds). 1990. *Reservoir Simulation*. Society of Petroleum Engineers.

[207] McCain, Jr., W. D. 1990. *The Properties of Petroleum Fluids*. 2nd edn. Tulsa, OK: PennWell Books.

[208] McIlhagga, W. 2010. *Automatic Differentiation with Matlab Objects*. MATLAB Central. https://tinyurl.com/yavlcra4. [Online; accessed July 11, 2018].

[209] Merland, R., Caumon, G., Lévy, B., and Collon-Drouaillet, P. 2014. Voronoi grids conforming to 3d structural features. *Comput. Geosci.*, **18**(3-4), 373–383. doi:10.1007/s10596-014-9408-0.

[210] Mlacnik, M. J., Durlofsky, L. J., and Heinemann, Z. E. 2006. Sequentially adapted flow-based PEBI grids for reservoir simulation. *SPE J.*, **11**(3), 317–327. doi:10.2118/90009-PA.

[211] Møyner, O. 2012. *Multiscale finite-volume methods on unstructured grids*. M.Sc. thesis, Norwegian University of Science and Technology. http://hdl.handle.net/11250/259015.

[212] Møyner, O. 2016. *Next generation multiscale methods for reservoir simulation*. PhD. thesis, Norwegian University of Science and Technology. http://hdl.handle.net/11250/2431831.

[213] Møyner, O. 2017. Nonlinear solver for three-phase transport problems based on approximate trust regions. *Comput. Geosci.*, **21**(5-6), 999–1021. doi:10.1007/s10596-017-9660-1.

[214] Møyner, O., and Lie, K.-A. 2014. The multiscale finite-volume method on stratigraphic grids. *SPE J.*, **19**(5), 816–831. doi:10.2118/163649-PA.

[215] Møyner, O., and Lie, K.-A. 2016a. A multiscale restriction-smoothed basis method for compressible black-oil models. *SPE J.*, **21**(06), 2079–2096. doi:10.2118/173265-PA.

[216] Møyner, O., and Lie, K.-A. 2016b. A multiscale restriction-smoothed basis method for high contrast porous media represented on unstructured grids. *J. Comput. Phys.*, **304**, 46–71. doi:10.1016/j.jcp.2015.10.010.

[217] Møyner, O., and Tchelepi, H. A. 2017. A multiscale restriction-smoothed basis method for compositional models. In: *SPE Reservoir Simulation Conference*. doi:10.2118/182679-MS.

[218] Møyner, O., Krogstad, S., and Lie, K.-A. 2014. The application of flow diagnostics for reservoir management. *SPE J.*, **20**(2), 306–323. doi:10.2118/171557-PA.

[219] Muskat, M., and Wyckoff, R. D. 1937. *The Flow of Homogeneous Fluids through Porous Media*. Vol. 12. New York: McGraw-Hill.

[220] Natvig, J. R., and Lie, K.-A. 2008. Fast computation of multiphase flow in porous media by implicit discontinuous Galerkin schemes with optimal ordering of elements. *J. Comput. Phys.*, **227**(24), 10108–10124. doi:10.1016/j.jcp.2008.08.024.

[221] Natvig, J. R., Lie, K.-A., Eikemo, B., and Berre, I. 2007. An efficient discontinuous Galerkin method for advective transport in porous media. *Adv. Water Resour.*, **30**(12), 2424–2438. doi:10.1016/j.advwatres.2007.05.015.

[222] Natvig, J. R., Skaflestad, B., Bratvedt, F., et al. 2011. Multiscale mimetic solvers for efficient streamline simulation of fractured reservoirs. *SPE J.*, **16**(4), 880–888. doi:10.2018/119132-PA.

[223] Neidinger, R. 2010. Introduction to automatic differentiation and MATLAB object-oriented programming. *SIAM Review*, **52**(3), 545–563. doi:10.1137/080743627.

[224] Nessyahu, H., and Tadmor, E. 1990. Nonoscillatory central differencing for hyperbolic conservation laws. *J. Comput. Phys.*, **87**(2), 408–463. doi:10.1016/0021-9991(90)90260-8.

[225] Nielsen, B. F., and Tveito, A. 1998. An upscaling method for one-phase flow in heterogeneous reservoirs; A Weighted Output Least Squares (WOLS) approach. *Comput. Geosci.*, **2**, 92–123. doi:0.1023/A:1011541917701.

[226] Nikitin, K., Terekhov, K., and Vassilevski, Y. 2014. A monotone nonlinear finite volume method for diffusion equations and multiphase flows. *Comput. Geos.*, **18**(3-4), 311–324. doi:10.1007/s10596-013-9387-6.

[227] Nilsen, H. M., Lie, K.-A., and Andersen, O. 2016b. Robust simulation of sharp-interface models for fast estimation of CO_2 trapping capacity. *Comput. Geosci.*, **20**(1), 93–113. doi:10.1007/s10596-015-9549-9.

[228] Nilsen, H. M., Lie, K.-A., and Andersen, O. 2016a. Fully implicit simulation of vertical-equilibrium models with hysteresis and capillary fringe. *Comput. Geosci.*, **20**(1), 49–67. doi:10.1007/s10596-015-9547-y.

[229] Nilsen, H. M., Nordbotten, J. M., and Raynaud, X. 2018. Comparison between cell-centered and nodal based discretization schemes for linear elasticity. *Comput. Geosci.*, **22**(1), 233–260. doi:10.1007/s10596-017-9687-3.

[230] Nilsen, H. M., Lie, K.-A., Møyner, O., and Andersen, O. 2015b. Spill-point analysis and structural trapping capacity in saline aquifers using MRST-co2lab. *Comput. Geosci.*, **75**, 33–43. doi:10.1016/j.cageo.2014.11.002.

[231] Nilsen, H. M., Lie, K.-A., and Natvig, J. R. 2012. Accurate modelling of faults by multipoint, mimetic, and mixed methods. *SPE J.*, **17**(2), 568–579. doi:10.2118/149690-PA.

[232] Nilsen, H. M., Lie, K.-A., and Andersen, O. 2015a. Analysis of CO_2 trapping capacities and long-term migration for geological formations in the Norwegian North Sea using MRST-co2lab. *Comput. Geosci.*, **79**, 15–26. doi:10.1016/j.cageo.2015.03.001.

[233] Nordbotten, J. M. 2016. Stable cell-centered finite volume discretization for biot equations. *SIAM J. Numer. Anal.*, **54**(2), 942–968. doi:10.1137/15M1014280.

[234] Nordbotten, J. M., and Eigestad, G. T. 2005. Discretization on quadrilateral grids with improved monotonicity properties. *J. Comput. Phys.*, **203**(2), 744–760. doi:10.1016/j.jcp.2004.10.002.

[235] Nordbotten, J. M., Aavatsmark, I., and Eigestad, G. T. 2007b. Monotonicity of control volume methods. *Numer. Math.*, **106**(2), 255–288. doi:10.1007/s00211-006-0060-z.

[236] Nordbotten, J. M., and Aavatsmark, I. 2005. Monotonicity conditions for control volume methods on uniform parallelogram grids in homogeneous media. *Comput. Geosci.*, **9**(1), 61–72. doi:10.1007/s10596-005-5665-2.

[237] Nordbotten, J. M. 2015. Convergence of a cell-centered finite volume discretization for linear elasticity. *SIAM J. Numer. Anal.*, **53**(6), 2605–2625. doi:10.1137/140972792.

[238] Notay, Y. 2010. An aggregation-based algebraic multigrid method. *Electron. Trans. Numer. Anal.*, **37**, 123–140.

[239] Nutting, P. G. 1930. Physical analysis of oil sands. *AAPG Bulletin*, **14**(10), 1337–1349.

[240] Obi, E., Eberle, N., Fil, A., and Cao, H. 2014. Giga cell compositional simulation. In: *IPTC 2014: International Petroleum Technology Conference*. doi:10.2523/IPTC-17648-MS.

[241] Odeh, A. S. 1981. Comparison of solutions to a three-dimensional black-oil reservoir simulation problem. *J. Petrol. Techn.*, **33**(1), 13–25. doi:10.2118/9723-PA.

[242] Øren, P.-E., Bakke, S., and Arntzen, O. J. 1998. Extending predictive capabilities to network models. *SPE J.*, **3**(4), 324–336. doi:10.2118/52052-PA.

[243] Pal, M., Lamine, S., Lie, K.-A., and Krogstad, S. 2015. Validation of the multiscale mixed finite-element method. *Int. J. Numer. Meth. Fluids*, **77**(4), 206–223. doi:10.1002/fld.3978.

[244] Park, H.-Y., and Datta-Gupta, A. 2011. Reservoir management using streamline-based flood efficiency maps and application to rate optimization. In: *SPE Western North American Region Meeting*. doi:10.2118/144580-MS.

[245] Peaceman, D. W. 1983. Interpretation of well-block pressures in numerical reservoir simulation with nonsquare grid blocks and anisotropic permeability. *Soc. Petrol. Eng. J.*, **23**(3), 531–543. doi:10.2118/10528-PA.

[246] Peaceman, D. W. 1991. *Fundamentals of Numerical Reservoir Simulation*. New York, NY, USA: Elsevier Science Inc.

[247] Peaceman, D. W. 1978. Interpretation of well-block pressures in numerical reservoir simulation. *Soc. Petrol. Eng. J.*, **18**(3), 183—194. doi:10.2118/6893-PA.

[248] Persson, P.-O., and Strang, G. 2004. A simple mesh generator in MATLAB. *SIAM Review*, **46**(2), 329–345. doi:10.1137/S0036144503429121.

[249] Peters, L., Arts, R., Brouwer, G., et al. 2010. Results of the Brugge benchmark study for flooding optimization and history matching. *SPE Reser. Eval. Eng.*, **13**(03), 391–405. doi:10.2118/119094-PA.

[250] Pettersen, Ø. 2006. *Basics of Reservoir Simulation with the Eclipse Reservoir Simulator*. Lecture Notes. Department of Mathematics, University of Bergen. http://folk.uib.no/fciop/index_htm_files/ResSimNotes.pdf.

[251] Pettersen, Ø. 2012. Horizontal simulation grids as alternative to structure-based grids for thin oil-zone problems: A comparison study on a Troll segment. *Comput. Geosci.*, **16**(2), 211–230. doi:10.1007/s10596-011-9240-8.

[252] Pinder, G. F., and Gray, W. G. 2008. *Essentials of Multiphase Flow in Porous Media*. Hoboken, NJ: John Wiley & Sons.

[253] Pollock, D. W. 1988. Semi-analytical computation of path lines for finite-difference models. *Ground Water*, **26**(6), 743–750. doi:10.1111/j.1745-6584.1988.tb00425.x.

[254] Ponting, D. K. 1989. Corner point geometry in reservoir simulation. In: King, P. R. (ed), *ECMOR I – 1st European Conference on the Mathematics of Oil Recovery*. Oxford: Clarendon Press, pp. 45–65. doi:10.3997/2214-4609.201411305.

[255] Potempa, T. C. 1982. *Finite element methods for convection dominated transport problems*. PhD. thesis, Rice University. http://hdl.handle.net/1911/15714.

[256] Prevost, M., Edwards, M. G., and Blunt, M. J. 2002. Streamline tracing on curvilinear structured and unstructured grids. *SPE J.*, **7**(2), 139–148. doi:10.2118/78663-PA.

[257] Prevost, M., Lepage, F., Durlofsky, L. J., and Mallet, J.-L. 2005. Unstructured 3D gridding and upscaling for coarse modelling of geometrically complex reservoirs. *Petrol. Geosci.*, **11**(4), 339–345. doi:10.1144/1354-079304-657.

[258] Pyrcz, M. J., and Deutsch, C. V. 2014. *Geostatistical Reservoir Modeling*. Oxford: Oxford University Press.

[259] Rashid, B., Muggeridge, A., Bal, A.-L., and Williams, G. J. J. 2012a. Quantifying the impact of permeability heterogeneity on secondary-recovery performance. *SPE J.*, **17**(2), 455–468. doi:10.2118/135125-PA.

[260] Rashid, B., Bal, A.-L., Williams, G. J. J., and Muggeridge, A. H. 2012b. Using vorticity to quantify the relative importance of heterogeneity, viscosity ratio, gravity and diffusion on oil recovery. *Comput. Geosci.*, **16**(2), 409–422. doi:10.1007/s10596-012-9280-8.

[261] Rasmussen, A. F., and Lie, K.-A. 2014. Discretization of flow diagnostics on stratigraphic and unstructured grids. In: *ECMOR XIV – 14th European Conference on the Mathematics of Oil Recovery*. doi:10.3997/2214-4609.20141844.

[262] Raviart, P. A., and Thomas, J. M. 1977. A mixed finite element method for 2nd order elliptic equations. In: Galligani, I., and Magenes, E. (eds), *Mathematical Aspects of Finite Element Methods*. Berlin, Heidelberg, New York: Springer-Verlag.

[263] Renard, P., and De Marsily, G. 1997. Calculating equivalent permeability: A review. *Adv. Water Resour.*, **20**(5), 253–278. doi:10.1016/S0309-1708(96)00050-4.

[264] Richards, L. A. 1931. Capillary conduction of liquids through porous mediums. *J. App. Phys.*, **1**(5), 318–333. doi:10.1063/1.1745010.

[265] Ringrose, P., and Bentley, M. 2015. *Reservoir Model Design: A Practitioner's Guide*. New York: Springer.

[266] Samier, P. 1990. A finite element method for calculation transmissibilities in n-point difference equations using a non-diagonal permeability tensor. In: Guérillot, D. (ed), *ECMOR II – 2nd European Conference on the Mathematics of Oil Recovery*. Editions TECHNIP, pp. 121–130. doi:10.3997/2214-4609.201411106.

[267] Samier, P., and Masson, R., 2017. Implementation of a vertex-centered method inside an industrial reservoir simulator: Practical issues and comprehensive comparison with corner-point grids and perpendicular-bisector-grid models on a field case. *SPE J.*, **22**(02), 660–678. doi:10.2118/173309-PA.

[268] Sandve, T. H., Berre, I., and Nordbotten, J. M. 2012. An efficient multi-point flux approximation method for discrete fracture matrix simulations. *J. Comput. Phys.*, **231**(9), 3784–3800. doi:10.1016/j.jcp.2012.01.023.

[269] Schlumberger. 1999. *ECLIPSE 100 User Course*. Schlumberger GeoQuest.

[270] Schlumberger. 2014a. *ECLIPSE: Reference Manual*. 2014.1 edn. Schlumberger.

[271] Schlumberger. 2014b. *ECLIPSE Reservoir Simulation Software: Technical Description*. 2014.1 edn. Schlumberger.

[272] Schneider, M., Flemisch, B., and Helmig, R. 2017. Monotone nonlinear finite-volume method for nonisothermal two-phase two-component flow in porous media. *Int. J. Numer. Meth. Fluids*, **84**(6), 352–381. doi:10.1002/fld.4352.

[273] Shah, S., Møyner, O., Tene, M., Lie, K.-A., and Hajibeygi, H. 2016. The multiscale restriction smoothed basis method for fractured porous media. *J. Comput. Phys.*, **318**, 36–57. doi:10.1016/j.jcp.2016.05.001.

[274] Shahvali, M., Mallison, B., Wei, K., and Gross, H. 2012. An alternative to streamlines for flow diagnostics on structured and unstructured grids. *SPE J.*, **17**(3), 768–778. doi:10.2118/146446-PA.

[275] Shampine, L. F., Ketzscher, R., and Forth, S. A. 2005. Using AD to solve BVPs in MATLAB. *ACM Trans. Math. Software*, **31**(1), 79–94. doi:10.1145/1055531.1055535.

[276] Sheldon, J. W., Harris, C. D., and Bavly, D. 1960. A method for general reservoir behavior simulation on digital computers. In: *Fall Meeting of the Society of Petroleum Engineers of AIME*. doi:10.2118/1521-G.

[277] Shook, G. M., and Forsmann, J. H. 2005. *Tracer Interpretation Using Temporal Moments on a Spreadsheet*. Tech. rept. INL report 05-00400. Idaho National Laboratory.

[278] Shook, G., and Mitchell, K. 2009. A robust measure of heterogeneity for ranking earth models: The F-Phi curve and dynamic Lorenz coefficient. In: *SPE Annual Technical Conference and Exhibition*. doi:10.2118/124625-MS.

[279] Shubin, G. R., and Bell, J. B. 1984. An analysis of the grid orientation effect in numerical simulation of miscible displacement. *Comput. Methods Appl. Mech. Eng.*, **47**(1), 47–71. doi:10.1016/0045-7825(84)90047-1.

[280] Spillette, A. G., Hillestad, J. G., and Stone, H. L. 1973. A high-stability sequential solution approach to reservoir simulation. In: *Fall Meeting of the Society of Petroleum Engineers of AIME*. doi:542-MS.

[281] Stephansen, A. F., and Klausen, R. A. 2008. Mimetic MPFA. In: *ECMOR XI – 11th European Conference on the Mathematics of Oil Recovery*. doi:10.3997/2214-4609.20146365.

[282] Stone, H. L. 1970. Probability model for estimating three-phase relative permeability. *J. Petrol. Tech.*, **22**(02), 214–218. doi:10.2118/2116-PA.

[283] Stone, H. L. 1973. Estimation of three-phase relative permeability and residual oil data. *J. Pet. Technol.*, **12**(4). doi:10.2118/73-04-06.

[284] Stone, H. L., and Garder Jr., A. O. 1961. Analysis of gas-cap or dissolved-gas drive reservoirs. *SPE J.*, **1**(02), 92–104. doi:10.2118/1518-G.

[285] Stüben, K. 2001. A review of algebraic multigrid. *J. Comput. Appl. Math.*, **128**(1), 281–309. doi:10.1016/S0377-0427(00)00516-1.

[286] Technische Universität Darmstadt. *Automatic Differentiation for Matlab (ADiMat)*. http://www.adimat.de/. [Online; accessed July 11, 2018].

[287] Thiele, M. R., and Batycky, R. P. 2003. Water injection optimization using a streamline-based workflow. In: *SPE Annual Technical Conference and Exhibition*. doi:10.2118/84080-MS.

[288] Thomas, G. W. 1981. *Principles of Hydrocarbon Reservoir Simulation*. Boston: IHRDC.

[289] Todd, M. R. and Longstaff, W. J. 1972a. The development, testing, and application of a numerical simulator for predicting miscible flood performance. *J. Petrol. Tech.*, **24**(07), 874–882. doi:10.2118/3484-PA.

[290] Todd, M. R., O'Dell, P. M., and Hirasaki, G. J. 1972b. Methods for increased accuracy in numerical reservoir simulators. *Soc. Petrol. Eng. J.*, **12**(06), 515–530. doi:10.2118/3516-PA.

[291] Tomlab Optimization Inc. *Matlab Automatic Differentiation (MAD)*. http://matlabad.com/. [Online; accessed July 11, 2018].

[292] Toor, S. M., Edwards, M. G., Dogru, A. H., and Shaalan, T. M. 2015. Boundary aligned grid generation in three dimensions and CVD-MPFA discretization. In: *SPE Reservoir Simulation Symposium*. doi:10.2118/173313-MS.

[293] Toro, E. F. 2009. *Riemann Solvers and Numerical Methods for Fluid Dynamics: A Practical Introduction*. 3rd edn. Berlin: Springer-Verlag.

[294] Trangenstein, J. A. 2009. *Numerical solution of hyperbolic partial differential equations*. Cambridge: Cambridge University Press.

[295] Trangenstein, J. A., and Bell, J. B. 1989. Mathematical structure of the black-oil model for petroleum reservoir simulation. *SIAM J. Appl. Math.*, **49**(3), 749–783. doi:10.1137/0149044.

[296] Trottenberg, U., Oosterlee, C. W., and Schüller, A. 2000. *Multigrid*. Academic press.

[297] Ucar, E., Berre, I., and Keilegavlen, E. 2015. Simulation of slip-induced permeability enhancement accounting for multiscale fractures. In: *Fourtieth Workshop on Geothermal Reservoir Engineering*.

[298] van Genuchten, M. T. 1980. Closed-form equation for predicting the hydraulic conductivity of unsaturated soils. *Soil Science Soc. America J.*, **44**(5), 892–898. doi:10.2136/sssaj1980.03615995004400050002x.

[299] Varela, J. 2018. *Implementation of an MPFA/MPSA-FV solver for the unsaturated flow in deformable porous media.* M.Sc. thesis, University of Bergen. http://hdl.handle.net/1956/17905.

[300] Verma, A. 1999. ADMAT: Automatic differentiation in MATLAB using object oriented methods. In: Henderson, M. E., Anderson, C. R., and Lyons, S. L. (eds), *Object Oriented Methods for Interoperable Scientific and Engineering Computing: Proceedings of the 1998 SIAM Workshop.* Philadelphia: SIAM, pp. 174–183.

[301] Verma, S., and Aziz, K. 1997. A control volume scheme for flexible grids in reservoir simulation. In: *SPE Reservoir Simulation Symposium.* doi:10.2118/37999-MS.

[302] Voskov, D. V., and Tchelepi, H. A. 2012. Comparison of nonlinear formulations for two-phase multi-component eos based simulation. *J. Petrol. Sci. Engrg.*, **82–83**(0), 101–111. doi:10.1016/j.petrol.2011.10.012.

[303] Voskov, D. V., Tchelepi, H. A., and Younis, R. 2009. General nonlinear solution strategies for multiphase multicomponent EoS based simulation. In: *SPE Reservoir Simulation Symposium.* doi:10.2118/118996-MS.

[304] Wallis, J. R. 1983. Incomplete gaussian elimination as a preconditioning for generalized conjugate gradient acceleration. In: *SPE Reservoir Simulation Symposium.* doi:10.2118/12265-MS.

[305] Wallis, J. R., Kendall, R. P., and Little, T. E. 1985. Constrained residual acceleration of conjugate residual methods. In: *SPE Reservoir Simulation Symposium.* doi:10.2118/13536-MS.

[306] Wang, X., and Tchelepi, H. A. 2013. Trust-region based solver for nonlinear transport in heterogeneous porous media. *J. Comp. Phys.*, **253**, 114–137. doi:10.1016/j.jcp.2013.06.041.

[307] Watts, J. W. 1986. A compositional formulation of the pressure and saturation equations. *SPE Res. Eng.*, **1**(3), 243–252. doi:10.2118/12244-PA.

[308] Weiser, A., and Wheeler, M. F. 1988. On convergence of block-centered finite differences for elliptic problems. *SIAM J. Numer. Anal.*, **25**(2), 351–375. doi:10.1137/0725025.

[309] Welge, H. J. 1952. A simplified method for computing oil recovery by gas or water drive. *J. Petrol. Tech.*, **4**(04), 91–98. doi:10.2118/124-G.

[310] Wen, X. H., Durlofsky, L. J., and Edwards, M. G. 2003. Upscaling of channel systems in two dimensions using flow-based grids. *Transp. Porous Media*, **51**(3), 343–366. doi:10.1023/A:1022318926559.

[311] Wen, X.-H., and Gómez-Hernández, J. J. 1996. Upscaling hydraulic conductivities in heterogeneous media: An overview. *J. Hydrol.*, **183**, ix–xxxii. doi:10.1016/S0022-1694(96)80030-8.

[312] Wheeler, J. A., Wheeler, M. F., and Yotov, I. 2002. Enhanced velocity mixed finite element methods for flow in multiblock domains. *Comput. Geosci.*, **6**(3-4), 315–332. doi:10.1023/A:1021270509932.

[313] Wheeler, M. F., Arbogast, T., Bryant, S., et al. 1999. A parallel multiblock/multidomain approach for reservoir simulation. In: *SPE Reservoir Simulation Symposium*, pp. 51–61.

[314] Wheeler, M. F., and Yotov, I. 2006. A cell-centered finite difference method on quadrilaterals. In: Arnold, D. N., Bochev, P. B., Lehoucq, R. B., Nicolaides, R. A., and Shashkov, M. (eds), *Compatible Spatial Discretizations.* New York: Springer, pp. 189–207.

[315] Whitaker, S. 1986. Flow in porous media I: A theoretical derivation of Darcy's law. *Transp. Porous Media*, **1**(1), 3–25. doi:10.1007/BF01036523.

[316] Whitson, C. H., and Brulé, M. R. 2000. *Phase Behavior*. Richardson, TX: Society of Petroleum Engineers.

[317] Wiener, O. 1912. *Abhandlungen der Matematisch*. PhD. thesis, Physischen Klasse der Königlichen Sächsischen Gesellscaft der Wissenschaften.

[318] Wu, X.-H., Efendiev, Y., and Hou, T. Y. 2002. Analysis of upscaling absolute permeability. *Discrete Contin. Dyn. Syst. Ser. B*, **2**(2), 185–204. doi:10.3934/dcdsb.2002.2.185.

[319] Wu, X.-H., and Parashkevov, R. 2009. Effect of grid deviation on flow solutions. *SPE J.*, **14**(01), 67–77. doi:10.218/92868-PA.

[320] Wyckoff, R. D., Botset, H. G., Muskat, M., and Reed, D. W. 1933. The measurement of the permeability of porous media for homogeneous fluids. *Rev. Sci. Instrum.*, **4**(7), 394–405. doi:10.1063/1.1749155.

[321] Yanosik, J. L., and McCracken, T. A. 1979. A nine-point, finite-difference reservoir simulator for realistic prediction of adverse mobility ratio displacements. *Soc. Petrol. Eng. J.*, **19**(04), 253–262. doi:10.2118/5734-PA.

[322] Younis, R. 2009. *Advances in modern computational methods for nonlinear problems: A generic efficient automatic differentiation framework, and nonlinear solvers that converge all the time*. Ph.D. thesis, Stanford University.

[323] Younis, R., and Aziz, K. 2007. Parallel automatically differentiable data-types for next-generation simulator development. In: *SPE Reservoir Simulation Symposium*. doi:10.2118/106493-MS.

[324] Zhang, P., Pickup, G. E., and Christie, M. A. 2005. A new upscaling approach for highly heterogenous reservoirs. In: *SPE Reservoir Simulation Symposium*. doi:10.2118/93339-MS.

[325] Zhou, Y., Tchelepi, H. A., and Mallison, B. T. 2011. Automatic differentiation framework for compositional simulation on unstructured grids with multi-point discretization schemes. In: *SPE Reservoir Simulation Symposium*. doi:10.2118/141592-MS.

Index

AD-OO framework, 342, 415, 599
Add-on modules, 598
Additive property, 561
Adjacency matrix, 521, 538
Adjoint equations, 425
Agglomeration, 541
   a priori flow indicator, 551
   feasible directions, 541
   flow indicator, 547
   global flow indicator, 551
   greedy algorithms, 548
   local flow indicator, 551
   merging blocks, 546
   permissible directions, 541
   refining blocks, 547
   volume indicator, 546
Annular space, 395
Annulus, 396
Anonymous function, 205
Anticline, 25, 27
API gravity, 372
Aqueous phase, 233
Aquifer, 256
Areal sweep, 308
Assembly, 435
Automatic differentiation, 204, 623–629

Basis functions, 178, 180
Bi-stream function, 130
Biconnected components, 543
Black-oil equations
   gas component, 392
Black-oil models, 246
   equation of state, 372
   formation-volume factor, 371
   oil component, 390
   primary unknowns, 389
   primary variables, 392
   shrinkage factor, 371
   solution gas–oil ratio, 371
   vaporized oil–gas ratio, 371
   water component, 389
Bottom-hole pressure, 122
Boundary conditions
   data structure, 146
   Dirichlet, 120, 147
   hydrostatic, 161
   linear pressure, 161
   Neumann, 120, 147
   no flow, 120, 146
Bubble point, 362
Bubble-point curve, 362
Bubble-point pressure, 375
Buckley–Leverett, 259, 281, 284, 298, 417
Bundle of tubes, 236

Capillary fringe, 320, 330
Capillary pressure, 235
   Brooks–Corey, 239
   hysteresis, 237
   Leverett J, 238
   van Genuchten, 239
Carbonate rock, 24
Carman–Kozeny, **37**
Cartesian grid, 57
Cell-mode script, 603
CFL condition, 275, 278
Christmas tree, 365
Circumcenter, 63
Circumcircle, 63
Class object, 401
Clastic rock, 23
CO_2 storage, 301, 311
Coarse grid, 519
Cocurrent flow, 251
Compatible discretization, 188
Component conservation, 245
Components, 233

Compositional flow, 245
Compositional simulation, 338
Compressibility
    constant, 117
    fluid, **117**
    gas (isothermal), 373
    oil, 376
    oil (isothermal), 378
    rock, **36**
    slightly, 118
    total, 117, 252
    water, 379
Compressible flow, 117
Computational overhead, 418
Computational steering, 419
Condensate, 360
Connate water saturation, 233
Connected components, 521
Conservative method, 275
Consistent, 174
Constitutive equations, 116
Constrained pressure residual (CPR), 444
Contacted volumes, 482
Control step, 403, 416, 438, 471
Convergence
    nonlinear, 403, 438
    rate, 326
Convergent, 174
Coordinate system, 91
Corner-point grid, 45, 48, 73
Countercurrent flow, 251
Critical point, 361
Critical temperature, 361
Crossflow, 395
Curvilinear grid, 58
Cylinder coordinates, 122, 124

Darcy
    flow experiment, 113
    Henry, 36, 113
    unit, 36, 40
    velocity, 115
Darcy's law, 14, 36, **115**
Darcy–Buckingham equation, 257
Data set
    CaseB4, 79, 536, 550
    Johansen, 44, 529
    Norne, 79, 323
    SAIGUP, 46, 97, 171, 514, 532
    SPE 1, 455
    SPE 10, 41, 307, 501, 551, 585
    SPE 10, Tarbert, 495
    SPE 3, 355, 409
    SPE 9, 355, 383, 460
    SPE 1, 350, 355, 383, 399, 409
Data structure
    boundary conditions, 146

coarse grid, 519, 525
fluid properties, 144
grid, 88
reservoir states, 145
source terms, 145
wells, 147
Dead oil, 247, 342, 380
Delaunay, 63
Dew point, 362
Dew-point curve, 362
Disconnected blocks, 521
Discrete operators, 135, 203
    average faces to cells, 218
    averaging, 207
    divergence, 136, 206
    divergence (wells), 397
    gradient, 136, 206
    gradient (wells), 397
    harmonic average, 215
    upwind, 141, 222
Displacement
    favorable, 262, 308
    unfavorable, 262, 308
Dissolution matrix, 452
DistMesh, 69
Downscaling, 560
Drainage, 237
Drainage region, 480
Dry gas, 382, 383
Dulmage–Mendelsohn, 169, 309, 522

ECLIPSE input, 46, 97, 339
    ACTNUM, 48, 80
    COMPDAT, 402
    COORD, 48, 75
    DENSITY, 383
    EDIT, 340
    GRAVITY, 383
    GRID, 340
    INCLUDE, 341
    METRIC, 49
    MULTX, 48
    MULTZ, 50
    NTG, 48
    PERMX, 48
    PORO, 48
    PROPS, 340, 350, 379
    PVCDO, 380
    PVCO, 381
    PVDG, 382, 383
    PVDO, 380
    PVTG, 382
    PVTO, 381, 383
    PVTW, 382
    PVZG, 382
    REGIONS, 340
    ROCK, 383

ECLIPSE input (cont.)
  ROCKTAB, 383
  RUNSPEC, 340, 380
  SATNUM, 48, 52
  SCHEDULE, 340, 402
  SGOF, 351
  SOLUTION, 340, 400
  SUMMARY, 340
  SWOF, 351
  TABDIMS, 380
  TSTEP, 403
  TUNING, 403
  WCONINJE, 402
  WCONPROD, 402
  WELSPECS, 402
  ZCORN, 48, 75
ECLIPSE output, 350
Energy equation, 220
Energy minimization, 179
Enhanced oil recovery, 5
Enthalpy, 221
Entropy condition, 261
  Lax, 261
  Oleinik, 261
Entry pressure, 236, 320
Equation of state
  compressibility, 373
  cubic, 119
  gas deviation factor, 373
  Peng–Robinson, 119
  Redlich–Kwong, 119
  Redlich–Kwong–Soave, 119
Error
  capillary dominated flow, 330
  grid-orientation, 176, 187, 330
  mass balance, 439, 464
  normalized residual, 439, 464
  splitting, 293, 326
  temporal, 326
  truncation, 174, 276, 285
Exercises
  advanced grids, 108
  agglomeration, 553
  arithmetic vs. harmonic, 215
  automatic differentiation, 629
  averaging, 569
  boundary conditions, 164
  buoyant migration, 312
  consistent schemes, 187
  discretization operators, 142
  efficiency of prototype code, 228
  flow diagnostics optimization, 504
  flow diagnostics, SAIGUP, 516
  flow-based upscaling, 574
  fluid object, 292
  global upscaling, 580
  grid coarsening, 539
  grid structure, 96

heterogeneity measures, 489
heterogeneous q5-spot, 310
homogeneous q5-spot, 306
hyperbolic schemes, 282
incompressible solvers, 325
intro to MRST, 608
modules in MRST, 619
multisegment wells, 459
non-Newtonian fluids, 220
numerical errors, 334
quarter five-spot, 159
rock properties, 52
sector model, 453
single-phase with wells, 173
single-phase, compressible, 212
single-phase, incompressible, 169
SPE 9 benchmark, 469
structured grids, 61
three-phase intro, 359
unstructured grids, 71
unstructured stencils, 169
upscaling sector model, 594
upstream methods, 286
volume averaging, 563
wells, incompressible, 173
Expansion factor, 371
Explicit flux representation, 184

Face mobility, 215
Facies, 238
Facility model, 433
Fanning friction factor, 399
Fault trap, 27
Faults, **26**, 39, 49
Fickian diffusion, 245
Fictitious domain, 60
Finite-volume method, 132
Five-spot, 126
Flow capacity, 483
Flow diagnostics, 477
Flow equation, 250
Flow resistance, 458
Fluid object, 16, 290
  black-oil, 355
  ECLIPSE input, 350
  relative permeability, 144
  saturation, 144
  single phase, 144
Fluvial, 28, 42
Formation-volume factor, 123, 124, 247
  gas, 371, 373
  oil, 371, 375
  total, 376
  water, 379
Fractional-flow formulation
  global pressure, 254
  phase potential, 255
  phase pressure, 250–254

Fractional-flow function, 250
Fracture, 25
Fully implicit, 244
Function spaces
 H_0^{div}, 178
 $H^{\frac{1}{2}}$, 183
 L^2, 178

Gas compressibility factor, 373
Gas liberation
 differential, 365
 flash, 365
Gas lift, 396
Gas reservoir, 365
 condensate, 366
 dew-point, 366
 dry, 366
 retrograde condensate, 366
 rich, 366
 wet, 366
Gaseous phase, 233
Generating points, 63
Geochemistry, 338
Geological model, 10, **30**
Geostatistics, 40
Gibbs' phase rule, 360
Global numbering, 447
Global pressure, 254
Global-pressure formulation, 254
Godunov scheme, 278
Gravel pack, 396
Gravity
 API, 372
 specific, 372
Gravity column, 14, 267, 301
Gravity override, 315
Grid
 Cartesian, 57
 coarse, 519
 composite, 101
 conformal, 103
 corner-point, 73, 97
 curvilinear, 58
 data structure, 88
 Delaunay, 63
 disconnected blocks, 521
 DistMesh, 69
 fictitious domain, 60, 523
 geometry, 93
 multiblock, 104
 partition, 519
 PEBI (2.5D), 85
 PEBI (2D), 67
 PEBI (3D), 104
 rectilinear, 57
 stair stepped, 79
 stratigraphic, 72
 tessellation, 63, 68
 Triangle, 69
 Voronoi, 66, 104
Grid-orientation errors, 176, 330

Heat conduction, 221
Hele–Shaw cell, 307
Heterogeneity, 483
Heterogeneity measure
 dynamic, 483
 F-Φ diagram, 484
 Lorenz coefficient, 486
 static, 483
 sweep efficiency, 487
High-pressure cell, 362
Homogenization, 560
Hybrid, mixed, 182
Hydraulic conductivity, 2, 114, 257
Hydraulic head, 113
Hydrocarbon recovery
 primary production, 3
 secondary production, 4
 tertiary production, 5
Hysteresis, 237

Ideal gas, 118
Imbibition, 237
Immiscible, 232
Immobile oil, 237
Implicit discretization, 203
Incompressible flow, 117
Indefinite system, 184
Indirection map, 89
Inflow control device, 396
Inflow-performance relation, 122, 394
Influence region, 479, 497
Injectivity, 310
 index, 122
Inner product
 continuous, 178
 discrete, 180
 local, 185
 mixed hybrid, 183
Input deck, 48
Irreducible saturation, 233

Jacobian matrix, 204, 295, 296, 309
Johansen data set, **44**
JOLTS, 598, 602, 603

K-orthogonal, 176, 333
Klinkenberg effect, 239

Laplace operator, 117
Lax–Friedrichs, 275
Lax–Wendroff, 276
Leverett J-function, 238
Limestone, 22

Line-search method, 296
Linear solver
   AGMG, 153, 445, 463
   algebraic multigrid, 445
   condition number, 347
   CPR, 444, 463
   direct, MATLAB default, 152
   GMRES, 442
   ILU preconditioner, 443
   iterative methods, 341
   Krylov subspace method, 442
   `mldivide`, 152, 303, 309
   multigrid, 153
   nested factorization, 447
   orthomin, 447
   preconditioner, 341
   Thomas algorithm, 303
   triangular system, 156
   tridiagonal system, 303
Linearization, 203
Linearized problem, 434
Linearly degenerate wave, 261
Lithography, 72
Live oil, 247, 367, 381, 383
Local numbering, 447
Local-flux mimetic, 196
Logical indexing, 99
Lorenz coefficient, 486, 501

M-matrix, 197
Marine rock, 23
Mass conservation, 116
Mass fraction, 233
Mass-balance error, 439
MATLAB
  `accumarray`, 621
  `arrayfun`, 346
  `bsxfun`, 621
  `cellfun`, 305, 353
  `classdef`, 401, 423
  handle classes, 437
  indirection maps, 89
  logical indexing, 99
  run-length encoding, 89
Mesh generator, 69
Method of characteristics, 259
Metis, 538
Mimetic finite differences, 188
   consistency, 189
Mimetic inner product, 189–196
   general family, 190
   parametric family, 191
   quasi RT0, 195
   quasi two-point, 193
   simple (default), 195
   two-point, 193
Min-max angle criterion, 63

Ministep, 438
Miscible displacement, 246
Mixed finite elements, 177
Mixed formulation, 178
Mixed hybrid, 182
Mixture velocity, 397
Model
   black-oil, 430
   facility, 433
   physical, 422
   reservoir, 426
   wells (multisegment), 433
   wells (Peaceman), 433
Model object, 414
Modules, 598, 610
  `ad-blackoil`, 430, 615
  `ad-core`, 416, 600
  `ad-eor`, 616
  `ad-fi`, 616
  `ad-mechanics`, 616
  `ad-props`, 351, 615
  `adjoint`, 615
  `agglom`, 541, 613
  `blackoil-sequential`, 420, 615
  `book`, 10, 157, 619
  `co2lab`, 617
  `coarsegrid`, 614
  `compositional`, 616
  `deckformat`, 48, 350, 405, 615
  `design rules`, 612
  `dfm`, 618
  `diagnostics`, 153, 480, 616
  `dual-porosity`, 618
  `enkf`, 619
  `fvbiot`, 618
  `geochemistry`, 618
  `hfm`, 618
  `incomp`, 16, 149, 614
  `libgeometry`, 614
  `linearsolver`, 446
  `linearsolvers`, 619
  `matlab_bgl`, 619
  `mimetic`, 186, 599, 614
  `mpfa`, 196, 599, 615
  `mrst-cap`, 619
  `mrst-gui`, 53, 619
  `msmfe`, 617
  `msrsb`, 617
  `octave`, 607, 619
  `opm_processing`, 614
  `optimization`, 616
  `remso`, 619
  `solvent`, 616
  `spe10`, 42, 619
  `steady-state`, 616
  `streamlines`, 159, 619
  `triangle`, 69, 614

upr, 104, 614
upscaling, 572, 616
vem, 615
vemmech, 618
wellpaths, 619
Monotone, 197
Mudrock, 22
Multicomponent flow, 245
Multilateral well, 396
Multiplier, 39, 50, 323
Multipoint flux approximation, 196
Multisegment well, 339, 395, 455

Nested factorization, 447
Net-to-gross, **38**, 50
Newton update, 204, 296
Newton–Raphson, 203, 295
Non-neighboring connection, 74
Non-Newton fluid, 215
Norne data set, 79
Numerical context, 415

Octave, 607
Oil reservoir, 365
  black-oil, 367
  bubble-point, 367
  solution gas, 367
  volatile, 367
Oil-water contact, 237
Oleic phase, 233
Operator splitting, 292

Partition
  agglomeration, 541
  confined blocks, 543
  disconnected, 521, 542
  flow indicator, 544
  graph algorithms, 538
  grid faces, 527
  hierarchical, 554
  load balanced, 519
  METIS, 538
  near well refinement, 536
  segmentation, 542
  volume indicator, 544
Partition of unity, 479
Partition vector, 519
Pathline, 129
Peaceman, **126**, 169
PEBI grid, 67, 85
Periodic medium, 560
Permeability, 2, **36**
  absolute, 239
  anisotropic, 37
  Carman–Kozeny relation, 38
  effective, 239
  isotropic, 37

Johansen, 45
  modeling in MRST, 39
  relative, 240
  SAIGUP, 51
  SPE 10, 44
Perpendicular bisector, 66
Phase (dis)appearance, 393
Phase diagram, 360
  binary substances, 362
  bubble-point reservoir, 368
  cricondenbar, 364
  cricondentherm, 364
  critical point, 363
  gas reservoirs, 366
  single-component system, 360
Phase mobility, 244
Physical model, 416
Pillar grid, 73
Poisson's equation, 117
Polymer flooding, 216
Pore volumes injected (PVI), 326
Porosity, **35**
  Carman–Kozeny relation, 38
  Johanesen, 45
  SAIGUP, 51
  SPE 10, 43
Potential ordering, 309
Preconditioner, 442
Pressure decline, 4
Pressure equation
  global pressure, 254
  phase-potential formulation, 256
  two-phase compressible, 253
  two-phase incompressible, 250
Primary production, 3
  combination drive, 4
  gas cap drive, 4
  gravity drive, 4
  solution gas drive, 4
  water drive, 4
Production-logging tool, 482
Productivity index, 122, 394
Properties
  extrapolation, 408
  interpolation, 408
Pseudo components, 233
  gas, 246
  oil, 246

Quarter five-spot, 126, 157
  heterogeneous, 307, 327
  homogeneous, 303, 326
  rotated, 330

Radial flow, 122, 167
Rankine–Hugoniot, 260
Rarefaction wave, 261

Raviart–Thomas, 180
Rectilinear grid, 57
Relative permeability, 240
    Brooks–Corey, 241
    Corey, 241
    default in ECLIPSE 100, 349
    in ad-props, 353
    SPE 1,3,9, 356
    Stone I, 242
    three-phase, 348
    two-phase, 241, 242
    van Genuchten, 242
Repeated five-spot, 126
Representative elementary volume, **31**, 232, 560
Reservoir simulation, 6
Residence time, 478
Residence-time distribution, 479, 490
Residual equations, 294
Residual saturation, 233
Result handler, 419
Retrograde condensate gas, 387
Retrograde gas condensate, 366, 377
Reynolds number, 399
Rheology
    Newtonian fluid, 215
    non-Newtonian fluid, 215
    shear thickening, 216
    shear thinning, 216
Rich gas, 382, 387
Richards' equation, 256
Riemann fan, 262
Riemann problem, 262
Rock types, 51
Root folder, 601
Rosebrock, 629
Rotated gravity, 311
R_s-factor, 371, 372, 375, 381, 383, 385
Run-length encoding, 89
R_v-factor, 371, 372, 375

S-shape, 251
SAIGUP data set, **46**, 97
Sand screen, 396
Sandstone, 22
Saturated, 368, 375
Saturation, 232
    connate water, 233
    effective, 238, 241
    irreducible, 233
    residual, 233
Saturation equation
    phase-potential formulation, 256
    two-phase compressible, 254
    two-phase incompressible, 250
Schedule, 340, 402, 418, 457, 462
Schur-complement, 184
Secondary production, 4
    waterflooding, 4

Sedimentary environment, 23
Sedimentary rocks, 21
Self-sharpening wave, 261
Sequential fully implicit, 420
Sequential solution, 292
Shale, 22, 239
Shallow marine, **24**, 28, 42
Shear factor, 217
Shock wave, 261
Shrinkage factor, 247, 371
Simulation model, 7, 401
Skin factor, 125, 394
Solution gas-oil ratio, 247, 371, 375, 381
Source terms, 145
Sparsity pattern, 167
SPE 10 data set, 41
Specific gas, 373
Specific gravity
    gas, 373
    liquid, 372
Stability analysis, 284
Stable, 174
State object, 145
    flux, 145
    pressure, 145
    saturation, 145
    well solution, 145
Stock-tank, 364
Storage capacity, 483
Storeage cofficient, 257
Stratigraphic grid, 72
Stratigraphic trap, 27
Streakline, 129
Streamline coordinates, 304
Streamlines, **128**, 158
Structural model, 31, 47, 49
Subset extraction, 99
Summary file, 405
Supercritical state, 361
Surface area, specific, 38
Surface tension, 234
Sweep efficiency, 487
Sweep region, 480, 496
Syncline, 25

Tarbert, 42
Tessellation, 63
Test function, 178
Thermal expansion
    adiabatic, 224
    free, 223
    Joule–Thomson, 224
    reversible, 224
Thermal: single-phase, 220
Thin-tube experiment, 235
Time loop, 210
Time-of-flight, 129, 153
    backward, 478

forward, 478
per influence region, 482
Time-step
chopping, 297, 393
control, 297, 415, 463
selection, 416, 439
Topological sort, 309
Tortuosity, 38
Tracer, 129, 156
Tracer distribution, 130
Transmissibility, 16, 133
Transport equation, 250
Trial function, 180
Triple point, 360
Truncation error, 174, 285
Two-point scheme, 16
conditionally consistent, 175
derivation, 132–133
half-transmissibility, 133
implementation, 150
linear system, 133
nonlinear TPFA, 200
transmissibility, 133

Unconditionally stable, 283
Unconventional reservoirs, 239
Undersaturated, 368, 375
Unsaturated flow, 256
Upper Ness, 42
Upscaling, 559
arithmetic average, 564
fallback strategy, 585
flow diagnostics, 581
flow-based, 570
generic conditions, 579
geometric average, 564
global, 579
harmonic average, 564
harmonic-arithmetic, 567
ill-posed problem, 564
laboratory boundary conditions, 570
local-global, 579
oversampling, 578
parallel layers, 566
periodic, 571
permeability definition, 564
perpendicular layers, 565
positive transmissibilities, 577
power average, 564
specific conditions, 579
transmissibility, 575
volume average, 561
well indices, 584
Wiener bound, 564
Upstream mobility weighting, 279
Upwind scheme, 278

Vadose zone, 256
Vanishing viscosity, 273
Vaporization curve, 362
Vaporized oil–gas ratio, 247, 371, 375
Variable elimination, 436
Variable switching, 393
Velocity reconstruction, 218
Vertical equilibrium, 450
Viscosibility, 380
Viscosity
gas, 374
oil, 378
water, 379
Viscous fingering, 307
Volatile oil, 360, 367, 377
Volumetric partition, 479
Voronoi, 66, 104

Water coning, 313
Water content, 257
Waterflooding, 4
Well allocation factor, 482, 581
Well head, 365
Well model
crossflow, 395
flow rate, 208
frictional pressure drop, 398, 457
hydrostatic pressure, 208
inflow-performance relation, 122
injectivity index, 122
mobility calculation, 395
multisegment, 397
Peaceman, 126–128, 169
productivity index, 122
scaling, 443
simple valve, 398, 457
skin, 125
surface rate, 209
well control, 209, 428, 451
Well pair, 480
Well placement, 502
Well-pair region, 480
Wells, 121, 338
control, 338, 402, 423, 462
control switching, 453, 464
data structure, 147
flow rate, 147
horizontal, 455
multisegment, 339, 455
simple, 457
Wet gas, 382
Wettability, 234
contact angle, 234
non-wetting, 234
oil wet, 235
water wet, 235
wetting, 234
Workbook, 603

Usage of MRST Functions

addBC, **146**
addSource, **145**
addWell, **148**, 170
AGMGSolverAD, 445
applySuccessivePart, 555
assembleSystem, 435
assignPVTO, 384
assignRelPerm, 353
assignROCK, 352
assignSWOF, 352

BackslashSolverAD, 403, 442, 463
blackoilSectorModelExample, 451
blackoilTutorialOnePhase, 449
blackoilTutorialSPE1, 399
blackoilTutorialSPE9, 460

cartGrid, 15, **57**
checkProperty, 425
coarsenGeometry, 525
combineEquations, 436
combineMSwithRegularWells, 457
compressPartition, 498, 523
computeFandPhi, 485
computeGeometry, 15, **93**
computeLorenz, 487
computeMimeticIP, 187, 192
computeMultiPointTrans, 196
computeSweep, 487
computeTimeOfFlight, 153, 159
computeTimestep, 440
computeTOFandTracer, 480
computeTranMult, 324
computeTrans, **150**
computeWellPairs, 481, 498
convertDeckSchedule, 456

convertDeckScheduleToMRST, 402
convertHorizonsToGrid, 77
convertInputUnits, 49
convertTo, 17
covert2MSWell, 457
CPRSolverAD, 444, 463
cutGrdecl, 81

Driving forces, 428

eliminateVariable, 436
equationsBlackOil, 390, 431
evaluateRelPerm, 428
explicitTransport, 294
explosionView, 524
extrudedTriangleGrid, 96

faceAvg, 389
faceFlux2cellFlux, 171
faceUpstr, 390
FacilityModel, 433, 457
findConfinedBlocks, 543
fluxside, **147**, 161

gaussianField, 40
generateCoarseGrid, 527
getAdjointEquations, 425
getDrivingForces, 423
getEquations, 414, 423, 431
getLinearSystem, 435
getPlotAfterStep, 418
getProp, 424
getProps, 414
getUnitSystem, 49
getVariableField, 424, 427
getWellOutput, 468
glue2DGrid, 105
GMRES_ILUSolverAD, 443
gravity, 311, 450

grdecl2Rock, 49
griddedInterpolant, 353

implicitTransport, 295
incompMimetic, 187
incompMPFA, 196
incompTPFA, 16, **150**, 151
initCoreyFluid, 291
initDeckADIFluid, 351
initResSol, **145**
initSimpleADIFluid, 417, 449
initSimpleFluid, 290
initSimpleFluidPc, 291
initSingleFluid, 16, **144**, 144
initState, **145**
initVariablesADI, 209, **624**
initWellSolAD, 417, 429
insertWellEquations, 432
interactiveDiagnostics, 505
interpReg, 353, 386
isPointInsideGrid, 166

LinearizedProblem, 434
LinearSolverAD, 441
logNormLayers, 41

makeLayeredGrid, 86
makeModel3, 96
makeRock, 15, **40**
moduleGUI, 611
mrstDataDirectory, 609
mrstDatasetGUI, 355, 605, **609**
mrstExamples, 603, 604
mrstExploreModules, 605
mrstModule, **611**
mrstPath, 69, 607, **611**
MultisegmentWell, 433, 458
multisegmentWellExample, 455

NonLinearSolver, 403, 417, 436
nozzleValve, 457

outlineCoarseGrid, 520

partitionCartGrid, 522
partitionMETIS, 539
partitionUI, 520
pebi, 68, 102
PhysicalModel, 422
pickTimestep, 440
plotCellData, 15, 46, 53, 99
plotFaces, 17
plotGrid, 15, 46
plotToolbar, 53, 404, 419
plotTracerBlend, 497

plotWellAllocationPanel, 499
plotWellSols, 306, 404
pollock, 159
processFacePartition, 528
processGRDECL, 41, 45, 49, 81, 98
pside, 16, **147**, 161
psideh, 162

rampupTimesteps, 452
readEclipseDeck, 350
readEclipseSummaryFmt, 406
readEclipseSummaryUnFmt, 406, 467
readGRDECL, 48, 80
reduceToSingleVariableType, 436
refineBlocks, 548
refineGreedy, 548
refineGreedy2, 548
refineGreedy3, 548
refineGreedy4, 548
refineNearWell, 536
refineRecursiveCart, 549
refineUniform, 548
relPermWO, 354
relPermWOG, 354
removeCells, 60
removeConfinedBlocks, 544
ReservoirModel, 426
ResultHandler, 419

sampleFromBox, 549
segmentIndicator, 542
selectModelFromDeck, 401, 456
setProp, 424
setupOperatorsTPFA, 422
simpleGrdecl, 74
simpleSchedule, 418
SimpleTimeStepSelector, 403, 440
SimpleWell, 433, 458
simulateScheduleAD, 404, 418
solveAdjoint, 425
solveLinearProblem, 441
solveTimeStep, 418
solveTimestep, 438
startup, 601, 611
startup_user, 69, 607, 612
stepFunction, 424, 438

tensorGrid, **58**
tessellationGrid, 69
tetrahedralGrid, 66
ThreePhaseBlackOilModel, 401, 430, 451
toleranceCNV, 439
toleranceMB, 439

triangleGrid, 64
twister, 59
twophaseJacobian, 295
TwoPhaseOilWaterModel, 417, 431

updateForChangedControls, 429
updateState, 429, 431
updateStateFromIncrement, 429

upscalePerm, 573
upscaleTrans, 578, 580

validateModel, 425, 430
validateState, 425, 430
verticalWell, **149**, 170, 172

WaterModel, 449
wellBoreFriction, 457